Monografie Matematyczne

Instytut Matematyczny Polskiej Akademii Nauk (IMPAN)

Volume 71
(New Series)

Founded in 1932 by
S. Banach, B. Knaster, K. Kuratowski,
S. Mazurkiewicz, W. Sierpinski, H. Steinhaus

Volumes 31–62 of the series
Monografie Matematyczne were published by
PWN – Polish Scientific Publishers, Warsaw

Lev Bukovský

The Structure of the Real Line

 Birkhäuser

Lev Bukovský
Institute of Mathematics
University of P.J. Šafárik
Jesenná 5
040 01 Košice
Slovakia
lev.bukovsky@upjs.sk

2010 Mathematics Subject Classification: 03E15, 03E17, 03E25, 03E35, 03E50, 03E60, 03E65, 26A21, 28A05, 28A99, 54D99, 54G15, 54H05

ISBN 978-3-0348-0005-1   e-ISBN 978-3-0348-0006-8
DOI 10.1007/978-3-0348-0006-8

Library of Congress Control Number: 2011923518

Cover design: deblik, Berlin

Printed on acid-free paper

Springer Basel AG is part of Springer Science+Business Media

www.birkhauser-science.com

To Zuzana

Contents

Preface

V názvu té knihy musí být slovo *struktura*.

Petr Vopěnka, 1975.

The title of this book must contain the word *structure*.

We must distinguish at least four periods – sometimes overlapping – in the history of investigation of the real numbers.

The first period, devoted to understanding the real numbers system, began as early as about 2 000 B.C. in Mesopotamia, then in Egypt, India, China, Greece (Pythagorians showed that $\sqrt{2}$ is an irrational), through Eudoxus, Euclid, Archimedes, Fibonacci, Euler, Bolzano and many others, and continued till the second half of the nineteenth century. Moris Kline [1972], p. 979 says that "one of the most surprising facts in the history of mathematics is that the logical foundation of the real numbers system was not erected until the late nineteenth century".

The rigorization of analysis forced the beginning of the second period, concentrated in the second half of the nineteenth century, and primarily devoted to an exact definition of the real numbers system. The investigations showed that mathematicians needed, as a framework for such a definition, a rigorous theory of infinity. The Zermelo – Fraenkel set theory **ZFC** was accepted as the best solution. Several independent definitions of reals in the framework of set theory turned out to be equivalent and, as usual in mathematics, that was a strong argument showing that the exact definition of an intuitive notion was established correctly.

Establishing an exact notion of the real numbers system, mathematicians began in the third period to intensively study the system. Essentially there were three possibilities: to study the algebraic structure of the field of reals, to study the subsets of the reals related to the topological and measure theoretic properties of reals induced by the order – in this case I speak about the real line instead of the system of real numbers – and finally, taking into account both the algebraic and topological (or measure theoretical) properties.

The fourth period started by arousing many open unanswered questions of the later one. It turned out that they are closely connected to set theoretical questions which were unanswered in set theory. By the invention of forcing as a method for showing the non-provability of a statement in **ZFC**, mathematicians obtained a strong tool to show that many questions concerning the topological and the measure theoretical structure of the real line are undecidable, or at least non-provable, in the considered set theory. The forcing method concerns essentially questions connected with infinite sets. Since the algebraic structure (to be an algebraic number, to be linearly independent over rationals, to be a prime number, etc.) is usually connected with the properties of finite sets, forcing hardly can contribute to a solution of an algebraic problem.

Focusing on this historical point of view, in 1975, I began to work on the Slovak version of a book collecting the main results of the above mentioned second, third and mainly the fourth period. I had previously presented the design of the book to participants in the Prague Set Theory Seminar. After my presentation Petr Vopěnka spontaneously stated the sentence quoted above. And I immediately accepted this proposal that provided an appropriate emphasis in the title on the book's content.

When the book appeared, in 1979, I was pleasantly surprised by the interest it generated in countries where students understood Slovak (which included the former Czechoslovakia, Poland, and some other particular locations, e.g., Hungary). At the end of 1987 I was further surprised to receive permission[1] to publish an English edition abroad. Immediately I began to work on an English version of my book. After the political events in Czechoslovakia in 1989, I could not refuse the (at least moral) responsibility to contribute to academic politics and my work on the English language version of the book was interrupted. When the monograph by T. Bartoszyński and H. Judah [1995] appeared, I thought my rôle in writing about the structure of the real line was finished. However, at the beginning of the 21st Century, several colleagues, mainly Polish, urged me to prepare a new edition of the Slovak version of "The Structure of the Real Line". In March 2003, I was invited to participate in the Boise Extravaganza in Set Theory at Boise State University in Idaho. After the conference I visited Tomek Bartoszyński in his office and was surprised to find my Slovak book open on his desk. This started my reflection, finishing with the conclusion that a book with the intention of my Slovak edition is not a rival book to Bartoszyński and Judah's book but rather a complementary, maybe, useful monograph. This presented a final – convincing – reason for my decision to prepare a new – significantly revised – edition of "The Structure of the Real Line".

I tried to follow the spirit of the Slovak edition. I shall not discuss the history of the real system. Since I present the main consequences of the Axiom of Determinacy **AD**, that contradicts the Axiom of Choice **AC**, in the basic parts of the book I try to avoid any use of **AC**, if possible, or, at least to replace it by the Weak Axiom of Choice **wAC**, that is a consequence of **AD**. Moreover, for some readers this may be interesting. In Chapter 1, I briefly describe the framework

---

[1]Let me remind the reader, that I lived in a country where everything was strongly controlled by a political party.

for the mathematical treatment of infinity – the Zermelo-Fraenkel set theory **ZF**. Then I sketch the main topological results in the above mentioned way. A precise definition of the real line is given in Chapter 2. It is shown that the definition leads to a unique object (up to a mathematical identification) and the set theoretical framework implies its existence. Chapters 3 to 8 contain mainly the rather classical theory of the structure of the real line – the study of properties of subsets of reals, which is important from the point of view of topology and measure theory. Chapter 7 is devoted to the phenomena of the measure – category duality. Finally, with overlapping topics, Chapters 7 to 10 deal with reductions of many important problems of the structure of the real line to the sentences of set theory, which the forcing method already proved are undecidable, or at least non-provable in set theory. To make the book self-contained, I wrote an Appendix containing miscellaneous, in my opinion, necessary material. I here recall all notions and basic facts of set theory and algebra that I use in the book. Then I present a short introduction to the metamathematics of set theory. Finally, I present the main results obtained (mainly) by the forcing method in the last fifty years, which I needed to answer important questions about the structure of the real line. I tried to attribute each result to its author. Any new notion or denotation can be found in the very large Index or Index of notations.

In the introduction of each chapter I try to explain its content and mainly, whether a reader should read a particular section immediately or before reading another sections. Especially, I suppose that a reader should start with reading Section 1.1 and then, she/he can begin to read any chapter and then go to the results needed for understanding of the presented material with the help of the indexes. Each section is supplemented by a series of exercises that contains a great deal of supplementary information concerning the topics of the section.

I am deeply satisfied that the book is being published in the series Monografie Matematyczne. For many years in Czechoslovakia we had found it difficult to obtain scientific information. The main source were the (often illegal) translations of English language books and articles in Russian and, in the topics of interest to me, Polish books that were mainly published as Monografie Matematyczne. So I consider it a great honor to contribute to this most prestigious series.

I would like to thank those who contributed to the preparation of the manuscript. Mirko Repický advised me as a TeX specialist. Peter Eliaš helped me with presentation of results related to the thin sets of harmonic analysis. My postgraduate students, former or current, Jozef Haleš, Michal Staš and Jaroslav Šupina, read the manuscript, discovering and correcting many errors, both typographical and factual. They contributed significantly to the correctness of the exercises.

I wish to express my gratitude to the Institute of Mathematics of Pavol Jozef Šafárik University at Košice that created good conditions for me while I was writing the book. My work was supported by grants 1/3002/06 and 1/0032/09 of the Slovak Grant Agency VEGA.

My intention for the book was to survey progress in the study of the real line over the period encompassing the highly productive end of one century and the beginning of another. I hope that this effort will prove to be a source of useful and stimulating information for a wide variety of mathematicians.

Košice, October 1, 2010   Lev Bukovský

Chapter 1

Introduction

Również problematem o dużej doniosłości jest tego rodzaju ujęcie teorii mnogości, które – odpowiednio zacieśniając pojęcie zbioru – eliminowałoby istnienie zbiorów patologicznych (które jest konsekwencją – jak wspomnieliśmy – przede wszystkim aksjomatu wyboru), a nie uszczupliłoby przy tym istotnie wartościowych osiągnięć teorii mnogości.

Kazimierz Kuratowski and Andrzej Mostowski [1952].

Equally, there is a problem with strong consequences in establishing a set theory of the kind that – making adequately precise the notion of a set – should eliminate the existence of pathological sets (which is a consequence – as we have already said – mainly of the axiom of choice) and does not weaken the surely worthy results of set theory.

In spite of the belief that the word around us is essentially finite, mathematics, physics and some other natural sciences cannot exist without a concept of infinity. Investigating the notion of a number, mathematicians almost always met an infinity. Starting with the Pythagoreans, continuing with Newton and Leibniz, then Bolzano and Cauchy, finishing with Dedekind and Cantor. As a consequence a necessity to build an appropriate mathematical theory of infinity and to investigate the numbers in a framework of such a theory arose. It was B. Bolzano who began the study of infinity. Then G. Cantor developed an adequate theory, fruitful with important results. All known attempts to build a theory of infinity convenient for a study of numbers essentially converge to the set theory as initiated by G. Cantor. So we have chosen as a framework for investigating numbers the most common set theory called the Zermelo-Fraenkel axiomatic set theory.

In this chapter we summarize necessary terminology and facts of set theory and of topology that we shall need later. In presenting our set theory we emphasize its axiomatization showing the rôle of some of its axioms in our investigations. The axiom of choice and its weak forms will play an important rôle in our next investigations. For this reason we present some well-known results of the set topology. In particular, we must identify the results which depend on some form of the axiom of choice.

1.1   Set Theory

We assume that the reader is familiar with elementary set theory, say to the extent of a basic graduate course. Usually a set theory is developed in the framework of the Zermelo-Fraenkel axiom system, including the axiom of choice. Working mathematicians often do not notice when they have used the axiom of choice, even in an essential way. We shall always make it clear. Moreover, we shall try to indicate that a weaker form of an axiom of choice is sufficient for proving a statement. "**Theorem**" is a statement provable in **ZF**. "**Theorem** [φ]" is a statement provable in **ZF** $+ \varphi$, where φ is an additional axiom to **ZF**. Especially, " **Theorem** [**AC**]" is a statement provable in **ZFC** and "**Theorem** [**wAC**]" is a statement provable in **ZFW**. Similarly for Corollaries, Lemmas, and Exercises.

In this section we present necessary facts of set theory. We shall present only proofs of those which we consider as not standard and/or which are not usually included in basic courses.

The basic notions of set theory are that of a **set**, denoted usually by letters $a, b, \ldots, x, y, z, A, B, \ldots, X, Y, Z$ and others, and that of the **membership relation** $\in$. $x \in A$ is read as x is an **element** of A, or x **belongs** to A, or x is a **member** of A. Actually we assume that any object we shall deal with is a set. When all members of a considered set are subsets of a given set, we use the word **family** instead of set. Another notion is that of the inclusion relation $X \subseteq Y$, which is a short denotation for the formula $(\forall x)\,(x \in X \rightarrow x \in Y)$ meaning that X is a **subset** of Y. Similarly, $\emptyset$ is a constant which denotes the unique set satisfying $(\forall x)\,x \notin \emptyset$.

An **atomic formula** of set theory is a formula of the form $x = y$ or $x \in y$, where x, y are variables (or constants, generally terms). The formulas of set theory are built from atomic formulas in an obvious way by logical connectives $\neg$, $\wedge$, $\vee$, $\rightarrow$, $\equiv$, and quantifiers $\forall$ and $\exists$.

Zermelo-Fraenkel set theory ZF consists of the following axioms.

1. Axiom of Extensionality. If X and Y have the same elements, then $X = Y$.

2. Axiom of Pairing. For any x, y there exists a set X that contains exactly the elements x and y.

3. Axiom of Union. For any set X there exists a set Y such that $x \in Y$ if and only if $x \in u$ for some $u \in X$.

4. Axiom of Power Set. For any set X there exists a set Y that contains all subsets of X.

5. Axiom Scheme of Separation. For any formula $\varphi(x, x_1, \ldots, x_k)$ of set theory the following statement is an axiom: for given sets X and $x_1, \ldots, x_k$ there exists a set Y such that Y contains exactly those elements $x \in X$ that have the property $\varphi(x, x_1, \ldots, x_k)$.

6. Axiom Scheme of Replacement. For any formula $\varphi(x, y, x_1, \ldots, x_k)$ of set theory such that

$$(\forall x, y, z, x_1, \ldots, x_k)\,((\varphi(x, y, x_1, \ldots, x_k) \wedge \varphi(x, z, x_1, \ldots, x_k)) \to y = z)$$

holds true, the following statement is an axiom: for given sets X and $x_1, \ldots, x_k$ there exists a set Y such that Y contains all those elements y for which there exists an $x \in X$ such that $\varphi(x, y, x_1, \ldots, x_k)$ holds true.

7. Axiom of Infinity. There exists an infinite set.

8. Axiom of Regularity. Every non-empty set X has an $\in$-minimal element, i.e., there exists an $x \in X$ such that x and X have no common element.

One of the main technical consequences of the Axiom of Regularity is that there exists no set x such that $x \in x$.

According to the Axiom of Extensionality the sets whose existence is guaranteed by axioms 2.–6. are unique. So we can introduce a notation for them. The set X of the Pairing Axiom will be denoted by $\{x, y\}$. The set Y of the Axiom of Union will be denoted by $\bigcup X$. Especially, if $X = \{x, y\}$, then we write $x \cup y = \bigcup X$. The set Y of the Axiom of Power Set will be denoted by $\mathcal{P}(X)$. The set Y of the Axiom Scheme of Separation will be denoted by

$$\{x \in X : \varphi(x, x_1, \ldots, x_k)\}.$$

Finally, the set Y of the Axiom Scheme of Replacement will be denoted by

$$\{y : x \in X \wedge \varphi(x, y, x_1, \ldots, x_k)\}$$

We need to make precise the meaning of the notion of "an infinite set". The simplest way is to find a property of a set which implies that it is "an infinite set". We can consider a set X as infinite if X is non-empty and for each of its elements contains a new element different in some sense from all "previous" ones. It turns out that the following property of a set X is enough[1] for being infinite:

$$(\exists x)\,(x \in X) \wedge (\forall z \in X)\,(z \cup \{z\} \in X).$$

If there exists at least one set, then there exists the empty set $\emptyset$. We can specify that an infinite set contains the empty set, i.e.,

$$\emptyset \in X \wedge (\forall z \in X)\,(z \cup \{z\} \in X). \tag{1.1}$$

It is well known that using the Axiom of Replacement the existence of an infinite set implies the existence of a set with property (1.1). Moreover, the existence of a

---

[1] By the Axiom of Regularity $z \notin z$ and therefore $z \neq z \cup \{z\}$.

set with property (1.1) implies the existence of a "minimal" set with this property, i.e., the existence of a set with both properties (1.1) and

$$(\forall Y)((\emptyset \in Y \wedge (\forall z \in Y)(z \cup \{z\} \in Y)) \to X \subseteq Y). \qquad (1.2)$$

The set with properties (1.1) and (1.2) is uniquely determined and will be denoted by ω. By definition, $\emptyset \in \omega$. We shall write $0 = \emptyset$. If $n \in \omega$ we write $n + 1 = n \cup \{n\}$. An element of ω is called a **natural number**. Note that $1 = \{0\}, 2 = \{0, 1\}, \ldots, n + 1 = \{0, \ldots, n\}$. For our convenience we formulate the Axiom of Infinity as

There exists the set ω.

However if we replace the former Axiom of Infinity by the later formulation, then we need to assume that there exists the empty set.

The definition of the set ω immediately yields a useful method of proof of sentences about elements of ω – mathematical induction. Actually we can show a metatheorem, or for any particular formula φ of set theory we have a particular theorem. So, let $\varphi(x, x_1, \ldots, x_k)$ be a formula of set theory. Then

Theorem 1.1 (Theorem on Mathematical Induction). *Let $x_1, \ldots, x_k$ be given. If*

(IS1) $\varphi(0, x_1, \ldots, x_k)$,

(IS2) $(\forall n \in \omega)(\varphi(n, x_1, \ldots, x_k) \to \varphi(n + 1, x_1, \ldots, x_k))$, *then $\varphi(x, x_1, \ldots, x_k)$ holds true for any $x \in \omega$.*

Proof. Set

$$Y = \{x \in \omega : \varphi(x, x_1, \ldots, x_k)\}.$$

By (IS1) and (IS2) the set $Y \subseteq \omega$ satisfies the premise of the implication (1.2). Thus $Y = \omega$. $\square$

We assume that the reader is familiar with the theory of ordinals. Let us recall that X is a **transitive set** if $(\forall x)(x \in X \to x \subseteq X)$. An **ordinal** is a transitive set well-ordered by the relation $\eta \in \zeta \vee \eta = \zeta$. Thus an ordinal is the set of all smaller ordinals. If ξ, η are different ordinals, then either $\xi \in \eta$ or $\eta \in \xi$. If ξ is an ordinal, then $\xi + 1 = \xi \cup \{\xi\}$ is the least ordinal greater than ξ – the immediate successor of ξ. If an ordinal $\xi \neq \emptyset$ is not an immediate successor of any ordinal, then $\xi = \sup\{\eta : \eta < \xi\} = \bigcup \xi$ is called a **limit ordinal**. ω is the least limit ordinal. The ordinal sum $\xi + \eta$ is defined by transfinite induction. Any ordinal α can be expressed as $\alpha = \lambda + n$, where λ is 0 or a limit ordinal and $n \in \omega$. The fundamental property of ordinals is expressed by

Theorem 1.2. *Every well-ordered set $\langle X, \leq \rangle$ is isomorphic to a unique ordinal.*

The unique ordinal ξ is called the **order type** of the well-ordered set X and we write $\xi = \text{ot}(X) = \text{ot}(X, \leq)$.

The method of mathematical induction can be extended for well-ordered sets. So, let $\varphi(x, x_1, \ldots, x_k)$ be a formula of set theory. Then

Theorem 1.3 (Theorem on Transfinite Induction). *Assume that $\langle X, \leq \rangle$ is a well-ordered set. Let $x_1, \ldots, x_k$ be given. If for any $x \in X$,*

$$(\forall y < x)\, \varphi(y, x_1, \ldots, x_k) \rightarrow \varphi(x, x_1, \ldots, x_k)$$

holds true, then $\varphi(x, x_1, \ldots, x_k)$ holds true for any $x \in X$.

We shall use without any commentary the following result or its adequate modification:

Theorem 1.4 (Definition by Transfinite Induction). *Let ξ be an ordinal, X being a non-empty set. Let $a \in X$, $f : X \longrightarrow X$ and $g : \bigcup_{\eta < \xi} {}^{\eta}X \longrightarrow X$. Then there exists unique function $F : \xi \longrightarrow X$ such that*

a) $F(0) = a$,
b) $F(\eta + 1) = f(F(\eta))$ *for any $\eta < \xi$,*
c) $F(\eta) = g(F|\eta)$ *for any limit $\eta < \xi$.*

Sometimes we shall speak about a **class of sets**. By a class of sets we understand "a collection" of sets satisfying a given formula, for which we do not have any argument for being a set. E.g., by **V** we denote the **class of all sets**. It is easy to see that **V** is not a set. The expression $x \in$ **V** simply means "x is a set". Similarly, we can define the class **On** of all ordinals. The class **On** is not a set and the formula $x \in$ **On** simply means that x is an ordinal. Actually, a class is an object of metamathematics (a formula of the language of **ZF**). We shall often speak about the class of all topological spaces or about the class of all Polish spaces, see Section 5.2. However, we must deal with the notion of a class very carefully, e.g., saying "for all classes ... holds true" is not a formula of set theory.

We say that two sets A, B have the **same cardinality**, written $|A| = |B|$, if there exists a one-to-one mapping of A onto B. The relation $|A| = |B|$ is reflexive, symmetric and transitive. We customarily say that $|A|$ is the **cardinality** of the set A. However, we do not know what it is. A cardinality has sense only in an interrelation with some other cardinality. The set A has cardinality **not greater** than the set B, written $|A| \leq |B|$, if there exists a one-to-one mapping of A into B. Finally, the set A has cardinality **smaller** than the set B, written $|A| < |B|$, if $|A| \leq |B|$ and not $|A| = |B|$. The relation $|A| \leq |B|$ is reflexive and transitive. In **ZF** one can prove that it is antisymmetric:

Theorem 1.5 (G. Cantor – F. Bernstein). *For any sets A, B, if $|A| \leq |B|$ and $|B| \leq |A|$, then $|A| = |B|$.*

We define another relation between cardinalities of sets as

$$|X| \ll |Y| \equiv (\exists f)\,(f : Y \xrightarrow{\text{onto}} X). \tag{1.3}$$

The relation $\ll$ is reflexive and transitive. Evidently $|X| \leq |Y|$ implies $|X| \ll |Y|$, provided $X \neq \emptyset$. As we shall see in Section 9.4 we have no chance to prove in **ZF** that the relation $\ll$ is antisymmetric. One can easily see that

$$|A| \ll |B| \rightarrow |\mathcal{P}(A)| \leq |\mathcal{P}(B)|. \tag{1.4}$$

Generally the relation $|A| \le |B|$ is not dichotomous, i.e., one cannot prove in **ZF** that

$$|A| \le |B| \lor |B| \le |A|$$

for any sets A, B.

The arithmetic operations on cardinalities are defined as follows:

$$|A| + |B| = |A \cup B| \text{ provided that } A \cap B = \emptyset,$$
$$|A| \cdot |B| = |A \times B| \text{ for any sets } A, B,$$
$$|A|^{|B|} = |{}^{B}A| \text{ for any sets } A, B,$$

where ${}^{B}A$ denotes the set of all mappings from B into A. The operations satisfy the obvious laws of arithmetics.

Theorem 1.6 (G. Cantor). *There exists no mapping of X onto $\mathcal{P}(X)$. Therefore $|X| < |\mathcal{P}(X)|$. Moreover, if A, X are sets such that $|A| \ll |X|$, then $\neg(2^{|X|} \ll |A|)$ and $|A| < 2^{|X|}$.*

Proof. The former statement follows from the later one.

So assume that $f : X \xrightarrow{\text{onto}} A$. Since the mapping $F(x) = f^{-1}(\{x\})$ is an injection $F : A \xrightarrow{1-1} \mathcal{P}(X)$, we obtain $|A| \le 2^{|X|}$.

To obtain a contradiction, we shall suppose that there exists a mapping $h : A \xrightarrow{\text{onto}} \mathcal{P}(X)$. Set $g = f \circ h : X \xrightarrow{\text{onto}} \mathcal{P}(X)$. Let $E = \{x \in X : x \notin g(x)\}$. Then there exists an $e \in X$ such that $g(e) = E$. Thus

$$e \in E \equiv e \notin g(e) = E,$$

which is a contradiction. $\square$

A set A is called **finite** if $|A| < |\omega|$. A set A is called **countable** if $|A| \le |\omega|$. A set that is not finite is **infinite** and a set that is not countable is **uncountable**. If $n \in \omega$ and $|A| = |n|$ we write $|A| = n$. We shall use without reference the following result.

Theorem 1.7. *A set A is finite if and only if $|A| = n$ for some $n \in \omega$. Thus ω is the set of all finite ordinals.*

An ordinal α is called a **cardinal number** or simply **cardinal** if $|\alpha| \ne |\xi|$ for every $\xi < \alpha$. Thus every finite ordinal is a cardinal, ω is a cardinal. The infinite cardinal numbers can be enumerated by ordinals

$$\omega = \aleph_0 < \aleph_1 < \cdots < \aleph_\xi < \cdots .$$

That is why an infinite cardinal number is also called an **aleph**. Sometimes we write ω_ξ instead $\aleph_\xi$, i.e., $\omega_\xi = \aleph_\xi$. A cardinal $\aleph_\xi$ is a **limit cardinal** or a **successor cardinal** if ξ is a limit or successor ordinal, respectively.

$\aleph_1$ is the least uncountable cardinal and also the least uncountable ordinal. Instead of $|A| = |\aleph_\xi|$ we write simply $|A| = \aleph_\xi$ and in such a case $\aleph_\xi$ is called the cardinality of the set A. Similarly for inequalities. Generally we shall denote a cardinality by Fraktur letters $\mathfrak{a}$, $\mathfrak{b}$, $\mathfrak{m}$, $\mathfrak{n}$ etc.[2] When we want to emphasize that the considered cardinality is the cardinality of a well-ordered set, we shall use Greek letters κ, λ, μ etc. The smallest cardinality greater than $\mathfrak{m}$ (if it does exist!) we denote by $\mathfrak{m}^+$.

A set A **can be well ordered** if there exists a well-ordering with the field A. If an infinite set A can be well ordered, then its cardinality is an aleph. Thus instead of saying that A can be well ordered we shall say also that $|A|$ **is an aleph.**

For addition and multiplication of alephs the simple **Hessenberg Theorem** holds true:

$$\aleph_\xi + \aleph_\eta = \aleph_\xi \cdot \aleph_\eta = \aleph_{\max\{\xi,\eta\}}. \tag{1.5}$$

Let $\eta < \xi$ be two ordinals. The ordinal ξ is said to be **cofinal** with η if there exists an increasing function $f : \eta \longrightarrow \xi$ such that $\sup\{f(\zeta) : \zeta < \eta\} = \xi$, i.e., if the set $\{f(\zeta) : \zeta < \eta\}$ is cofinal in ξ. By $\mathrm{cf}(\xi)$ we denote the least ordinal η such that ξ is cofinal with η. An infinite ordinal ξ is called **regular** if $\mathrm{cf}(\xi) = \xi$. Otherwise ξ is **singular**. The ordinal $\mathrm{cf}(\xi)$ is always regular. A regular limit ordinal is always a cardinal. Not vice versa. There exist singular cardinals, e.g., $\aleph_\omega$ for which $\mathrm{cf}(\aleph_\omega) = \omega$.

We shall use the following result.

Theorem 1.8 (A. Tarski). *If $|X| \geq \aleph_0$, then $2^{|X|} + |X| = 2^{|X|}$ and for any set Y such that $|Y| + |X| = 2^{|X|}$ we have $|Y| = 2^{|X|}$.*

The statement of this theorem may be expressed as follows: if $\mathfrak{m} \geq \aleph_0$ is a cardinality, then $2^{\mathfrak{m}} + \mathfrak{m} = 2^{\mathfrak{m}}$ and for any cardinality $\mathfrak{n}$ such that $\mathfrak{n} + \mathfrak{m} = 2^{\mathfrak{m}}$ we have $\mathfrak{n} = 2^{\mathfrak{m}}$. A proof may be found in Exercise 1.6.

Let us recall that if an axiom of **ZF** assures the existence of a set, then the set is uniquely determined (and we have usually denoted it by some symbol). We shall sometimes need such an axiom, which assures the existence of a set with a property that does not determine the set uniquely.

If $\mathcal{F}$ is a family of non-empty sets, then a function $f : \mathcal{F} \longrightarrow \bigcup \mathcal{F}$ is called **a choice function** or a **selector** for $\mathcal{F}$ if $f(A) \in A$ for every $A \in \mathcal{F}$. The **Axiom of Choice AC** says that, for every family of non-empty sets, there exists a choice function. Evidently **AC** is equivalent to the statement: a Cartesian product of a family of non-empty sets is a non-empty set. The theory **ZF** + **AC** will be denoted by **ZFC**. By results of K. Gödel (11.11) and P.J. Cohen (11.17), the axiom **AC** is undecidable in **ZF**.

Using Theorem 1.7 one can easily prove by mathematical induction that, for every finite family of non-empty sets, there exists a selector. As a consequence we obtain

---

[2]Note that in Sections 5.3, 5.4 and later on, the letters $\mathfrak{a}$, $\mathfrak{b}$, $\mathfrak{d}$, $\mathfrak{m}$, $\mathfrak{p}$ denote particular cardinalities.

Theorem 1.9 (Dirichlet Pigeonhole Principle). *Let* $\mathfrak{m}$ *be a cardinality (i.e.,* $\mathfrak{m} = |B|$ *for some set* B*). Let* $f : X \longrightarrow Y$*,* Y *being finite. If* $|Y| \cdot \mathfrak{m} < |X|$*, then there exists a* $y \in Y$ *such that the cardinality of the inverse image* $f^{-1}(\{y\})$ *is not smaller than or equal to* $\mathfrak{m}$*. Equivalently: if* $\mathcal{A}$ *is a finite partition of the set* X *and* $\mathfrak{m} \cdot |\mathcal{A}| < |X|$*, then there exists a set* $A \in \mathcal{A}$ *such that* $\neg |A| \leq \mathfrak{m}$*.*

The following theorem is a basic result.

Theorem 1.10 (E. Zermelo – M. Zorn). *The following are equivalent:*

a) Axiom of Choice **AC**.

b) **Zermelo's Theorem***: Every set can be well ordered.*

c) **Zorn's Lemma***: If every chain in a poset* $\langle X, \leq \rangle$ *is bounded from above, then for any* $x \in X$ *there exists a maximal element* $a \geq x$*.*[3]

Corollary 1.11. *If* **AC** *holds true, then any set* A *is either finite or there exists an ordinal* ξ *such that* $|A| = \aleph_\xi$*. In other words,* **AC** *implies that any cardinality is a cardinal number.*

Thus if **AC** holds true, then the class of cardinalities is equal to the class of all cardinal numbers and therefore is well ordered, i.e., there exists the smallest cardinality (=cardinal number) with given property (if there exists any such). In what follows we shall use this fact without any commentary.

Theorem 1.12. *If* **AC** *holds true, then any* $\aleph_{\xi+1}$ *is a regular cardinal.*

However, as we shall see later, the Axiom of Choice must be essentially used in a proof of Theorem 1.12.

By Theorem 1.6 and Corollary 1.11, assuming **AC**, for any infinite set X the cardinality of the power set $\mathcal{P}(X)$ is an aleph greater than $|X|$. The assumption that this cardinality is the smallest possible is called the **Generalized Continuum Hypothesis** and is denoted as **GCH**. Thus **GCH** says that $(\forall \xi) \, 2^{\aleph_\xi} = \aleph_{\xi+1}$. The **Continuum Hypothesis CH** says that $2^{\aleph_0} = \aleph_1$. Thus **CH** follows from **GCH**. By results of K. Gödel (11.11) and P.J. Cohen (11.17), both **CH** and **GCH** are undecidable in **ZFC**.

A limit regular cardinal κ is called a **weakly inaccessible cardinal**. If moreover, for any $\lambda < \kappa$ we have $2^\lambda < \kappa$, then κ is called **strongly inaccessible**. Note that if $\aleph_\xi$ is weakly inaccessible, then $\aleph_\xi = \xi$. By Metatheorem 11.3 the existence of a strongly inaccessible cardinal cannot be proved in **ZF**. Neither is the existence of a weakly inaccessible cardinal provable in **ZFC**.

For sake of brevity we denote by **IC** the statement "there exists a strongly inaccessible cardinal".

In our reasoning we do not always need the full **AC**. We formulate some weak forms of the axiom of choice. The **Countable Axiom of Choice AC**$^\omega$ says that for every countable family of non-empty sets there exists a choice function. In several investigations we shall need even weaker forms. The **Weak Axiom of Choice wAC** says that for any countable family of non-empty subsets of a given set

---

[3]For the notions used in the formulation of Zorn's Lemma, see Section 11.1.

of cardinality $2^{\aleph_0}$ there exists a choice function. Finally, the **Axiom of Dependent Choice DC** says that for any binary relation R on a non-empty set A such that for every $a \in A$ there exists a $b \in A$ such that aRb, for every $a \in A$ there exists a function $f : \omega \longrightarrow A$ satisfying $f(n)Rf(n+1)$ for any $n \in \omega$ and $f(0) = a$. One can easily show that

$$\text{AC} \rightarrow \text{DC}, \quad \text{DC} \rightarrow \text{AC}^\omega, \quad \text{AC}^\omega \rightarrow \text{wAC}.$$

It is well known that no implication can be reversed. We denote by **ZFW** the theory **ZF+wAC**.

We shall often need the following simple result.

Theorem 1.13. *The following are equivalent:*

a) *The Weak Axiom of Choice* **wAC**.
b) *For any countable family of non-empty subsets of $^\omega 2$ there exists a choice function.*
c) *For every X such that $|X| \ll 2^{\aleph_0}$ and every sequence $\{A_n\}_{n=0}^\infty$ of non-empty subsets of X there exists a selector for $\{A_n\}_{n=0}^\infty$.*
d) *For every sequence $\langle A_n : n \in \omega \rangle$ of non-empty subsets $^\omega 2$ there exists an infinite $E \subseteq \omega$ and a function $f : E \longrightarrow {}^\omega 2$ such that $f(n) \in A_n$ for every $n \in E$.*

The proof is easy. The implication a) $\rightarrow$ b), b) $\rightarrow$ d), and c) $\rightarrow$ a) are trivial.

Assume b). We show that c) holds true. If $|X| \ll 2^{\aleph_0}$, then there exists a surjection $f : {}^\omega 2 \xrightarrow{\text{onto}} X$. Hence $\{f^{-1}(A_n)\}_{n=0}^\infty$ is a sequence of non-empty subsets of $^\omega 2$. By b), there exists a selector $\langle a_n \in f^{-1}(A_n) : n \in \omega \rangle$ for this sequence. Then $\{f(a_n)\}_{n=0}^\infty$ is a selector for $\{A_n\}_{n=0}^\infty$.

Assume d). To show b), consider the sequence $\langle B_n = \Pi_{k \leq n} A_k : n \in \omega \rangle$. If $E \subseteq \omega$ is infinite and $g : E \longrightarrow \bigcup_n {}^{n+1}(^\omega 2)$ is such that $g(n) \in B_n$ for every $n \in E$, we define

$$f(n) = g(m)(n), \quad \text{where } m = \min\{k \in E : n \leq k\}. \qquad \square$$

The Axiom of Choice implies that the relation $|A| \leq |B|$ is dichotomous, i.e., for any sets A, B, we have $|A| \leq |B| \vee |B| \leq |A|$. As we have already remarked, one cannot prove this statement if **ZF**. See, e.g., Theorem 9.28. However, **wAC** implies a similar statement at least in the most important case.

Theorem [wAC] 1.14. *If $A \subset X$, $|X| \ll 2^{\aleph_0}$, then either A is countable or $|A| > \aleph_0$.*

Proof. If A is not finite, then using Theorem 1.7 one can easily show by mathematical induction that the sets $\langle \Psi_n = \{f \in {}^n A : f \text{ is an injection}\} : n \in \omega \rangle$ are non-empty. Let $\langle f_n : n \in \omega \rangle$ be a choice function. We define

$$f(n) = \begin{cases} f_1(0) & \text{if } n = 0, \\ f_{n+1}(k) & \text{where } k = \min\{l \in \omega : (\forall i < n)\, f_{n+1}(l) \neq f(i)\} \text{ otherwise.} \end{cases}$$

Evidently $f : \omega \xrightarrow{1-1} A$ and therefore $\aleph_0 \leq |A|$. Thus if A is not countable, then $\aleph_0 < |A|$. $\hfill\square$

A set $\mathcal{I} \subseteq \mathcal{P}(X)$ is called an **ideal** on X if

1) $\emptyset \in \mathcal{I}$, $X \notin \mathcal{I}$,
2) if $A \in \mathcal{I}$, $B \subseteq A$, then $B \in \mathcal{I}$,
3) if $A, B \in \mathcal{I}$, then $A \cup B \in \mathcal{I}$.

For simplicity we usually assume that

4) $\{x\} \in \mathcal{I}$ for every $x \in X$.

Let κ be an uncountable regular cardinal. An ideal $\mathcal{I}$ is said to be κ-**additive** if $\bigcup \mathcal{A} \in \mathcal{I}$ for any set $\mathcal{A} \subseteq \mathcal{I}$, $|\mathcal{A}| < \kappa$. An $\aleph_1$-additive ideal is simply called σ-**additive**. A set $\mathcal{I}_0 \subseteq \mathcal{I}$ is a **base of the ideal** $\mathcal{I}$ if every element A of $\mathcal{I}$ is a subset of some $B \in \mathcal{I}_0$.

The dual notion to the notion of an ideal is the notion of a filter. A set $\mathcal{F} \subseteq \mathcal{P}(X)$ is called a **filter** on X if

1) $\emptyset \notin \mathcal{F}$, $X \in \mathcal{F}$,
2) if $A \in \mathcal{F}$, $A \subseteq B$, then $B \in \mathcal{F}$,
3) if $A, B \in \mathcal{F}$, then $A \cap B \in \mathcal{F}$.

Similarly as above, we usually assume that

4) $X \setminus \{x\} \in \mathcal{F}$ for every $x \in X$.

$\mathcal{F}$ is a filter if and only if the family $\{X \setminus A : A \in \mathcal{F}\}$ is an ideal. A filter $\mathcal{F}$ is called an **ultrafilter** if for every $A \subseteq X$, either $A \in \mathcal{F}$ or $X \setminus A \in \mathcal{F}$. It is easy to see that a filter $\mathcal{F}$ is an ultrafilter if and only if $\mathcal{F}$ is maximal with respect to ordering by inclusion. A set $\mathcal{F}_0 \subseteq \mathcal{F}$ is a **base of the filter** $\mathcal{F}$ if for every $A \in \mathcal{F}$ there exists a $B \in \mathcal{F}_0$ such that $B \subseteq A$.

For any $x \in X$ the set $\{A \subseteq X : x \in A\}$ is an ultrafilter on X. An ultrafilter of this form is called **trivial**. A filter $\mathcal{F}$ on X is called a **free filter** if $\mathcal{F}$ does not contain any finite set. Thus, an ultrafilter is free if and only if it is not a trivial ultrafilter. The **Boolean Prime Ideal Theorem BPI** says that every filter on any set can be extended to an ultrafilter. Equivalently, any ideal is contained in a maximal ideal. Similarly as in the case of an ultrafilter, one can show that a maximal ideal $\mathcal{I}$ on a set X is a **prime ideal**, i.e., for any subset $A \subseteq X$ we have either $A \in \mathcal{I}$ or $X \setminus A \in \mathcal{I}$.

Theorem 1.15. AC *implies* BPI.

The proof is based on an application of the Zermelo Theorem. It is known that the converse implication is not true, see J.D. Halpern and A. Lévy [1971].

An ultrafilter $\mathcal{G}$ on an infinite set X is called **uniform** provided that every element of $\mathcal{G}$ has cardinality $|X|$.

Assume **AC**. Then for any infinite set X, the set

$$\mathcal{F} = \{A \subseteq X : |X \setminus A| < |X|\}$$

is a filter[4]. If $\mathcal{G}$ is an ultrafilter extending $\mathcal{F}$, then every element of $\mathcal{G}$ has cardinality $|X|$. Thus, (assuming **AC**) there exists a uniform ultrafilter on any infinite set[5].

If $A \subseteq \omega$ is infinite, then there is a unique increasing **enumeration** e_A of A. The nth element of A is $e_A(n)$. Thus the enumeration e_A is uniquely determined by

$$e_A(0) < e_A(1) < \cdots < e_A(n) < e_A(n+1) < \cdots \tag{1.6}$$

and

$$A = \{e_A(n) : n \in \omega\}. \tag{1.7}$$

A bijection π from $\omega \times \omega$ onto ω is called a **pairing function**. It is well known that there exists a pairing function, e.g.,

$$\pi(n, m) = \frac{1}{2}(n + m)(n + m + 1) + m. \tag{1.8}$$

Note that $\pi(n, m) \geq n, m$ for any n, m. We denote by λ and ρ the **left inverse** and **right inverse** functions of π, respectively, i.e.,

$$\pi(\lambda(n), \rho(n)) = n, \quad \lambda(\pi(n, m)) = n, \quad \rho(\pi(n, m)) = m \tag{1.9}$$

for any $n, m \in \omega$. Note also that $\lambda(n) \leq n$ and $\rho(n) < n$ for any $n > 0$.

Using a pairing function we can "identify" the sets $^\omega X$ and $^\omega(^\omega X)$. Actually, to a $\varphi \in {}^\omega(^\omega X)$ we assign a $\psi \in {}^\omega X$ by setting $\psi(n) = \varphi(\lambda(n))(\rho(n))$. A projection of $^\omega X$ onto $^\omega X$ considered as the nth factor of the product $^\omega(^\omega X)$ is defined by

$$\text{Proj}_n(\varphi) = \psi, \text{ where } \psi(m) = \varphi(\pi(n, m)). \tag{1.10}$$

Similarly we can identify $^\omega X$ with $^\omega X \times {}^\omega X$ using the mapping Π_X defined as $\Pi_X(\alpha, \beta) = \gamma$, where

$$\gamma(n) = \begin{cases} \alpha(n/2) & \text{if } n \text{ is even,} \\ \beta((n-1)/2) & \text{if } n \text{ is odd.} \end{cases} \tag{1.11}$$

The left inverse Λ_X and right inverse R_X of Π_X from $^\omega X$ into $^\omega X$ are defined as

$$\Lambda_X(\alpha) = \{\alpha(2n)\}_{n=0}^\infty, \qquad R_X(\alpha) = \{\alpha(2n+1)\}_{n=0}^\infty. \tag{1.12}$$

Thus, for any $\alpha, \beta, \gamma \in {}^\omega X$ we have

$$\Lambda_X(\Pi_X(\alpha, \beta)) = \alpha, \quad R_X(\Pi_X(\alpha, \beta)) = \beta, \quad \Pi_X(\Lambda_X(\gamma), R_X(\gamma)) = \gamma. \tag{1.13}$$

If the set X is understood we simply write Π, Λ, R.

We shall use many natural modifications of pairing functions.

Theorem 1.16. *There exists a function* $f : \mathcal{P}(\omega) \xrightarrow{\text{onto}} \omega_1$, *i.e.,* $\aleph_1 \ll 2^{\aleph_0}$.

---

[4] Actually, this statement is equivalent to **AC**. See Exercise 1.4.
[5] Again, this assertion is equivalent to **AC**, see Exercise 1.4.

Proof. Let π be a pairing function. If $A \subseteq \omega$, then $\pi^{-1}(A) \subseteq \omega \times \omega$, i.e., $\pi^{-1}(A)$ is a binary relation. We define a function f as

$$f(A) = \begin{cases} \xi & \text{if } \pi^{-1}(A) \text{ is a well-ordering of } \omega \text{ in the ordinal type } \xi, \\ n & \text{if } |\pi^{-1}(A)| = n, \\ 0 & \text{otherwise.} \end{cases}$$

Since for every infinite ordinal $\xi < \omega_1$ there exists a well-ordering of ω in the ordinal type ξ, the function f is a surjection of $\mathcal{P}(\omega)$ onto ω_1. $\square$

Corollary 1.17. *There exists a decomposition of $\mathcal{P}(\omega)$ in ω_1 non-empty pairwise disjoint sets*

$$\mathcal{P}(\omega) = \bigcup_{\xi < \omega_1} f^{-1}(\{\xi\}), \tag{1.14}$$

where $|f^{-1}(\{\xi\})| = \mathfrak{c}$ for every $\xi \geq \omega$.

Proof. If $\langle \omega, R \rangle$ is a well-ordered set of an infinite type ξ and $g : \omega \xrightarrow[\text{onto}]{1-1} \omega$, then $\langle \omega, \{\langle n, m \rangle : \langle g(n), g(m) \rangle \in R\} \rangle$ is a well-ordered set of the type ξ as well. For different g's the sets $\{\langle n, m \rangle : \langle g(n), g(m) \rangle \in R\}$ are different. Hence for any $\xi \geq \omega$ we have $|f^{-1}(\{\xi\})| = \mathfrak{c}$. $\square$

The decomposition of (1.14) is called the **Lebesgue decomposition**. In Section 6.4 we describe a related decomposition of the set of dyadic numbers.

The famous fact, that a union of countably many countable sets is a countable set, is usually proved by using some weak form of the axiom of choice.

Theorem 1.18. *Assuming $\mathbf{AC}^\omega$, a countable union of countable sets is a countable set. Assuming $\mathbf{wAC}$, a countable union of countable subsets of a given set of cardinality $\ll 2^{\aleph_0}$ is a countable set.*

The theorem cannot be proved in $\mathbf{ZF}$, since S. Feferman and A. Lévy (11.19) have constructed a model of $\mathbf{ZF}$, in which the following holds true:

$$\mathcal{P}(\omega) \text{ is a countable union of countable sets.} \tag{1.15}$$

We show that in a proof of Theorem 1.12 we need a form of the axiom of choice even in the case of ω_1.

Theorem 1.19. *If (1.15) holds true, then $\mathrm{cf}(\omega_1) = \omega$.*

Proof. Assume that

$$\mathcal{P}(\omega) = \bigcup_{n \in \omega} \mathcal{A}_n, \quad \mathcal{A}_n \text{ is countable for any } n. \tag{1.16}$$

Let $f : \mathcal{P}(\omega) \xrightarrow{\text{onto}} \omega_1$ be the function constructed in the proof of Theorem 1.16. Set $\eta_n = \sup\{f(A) : A \in \mathcal{A}_n\}$. If some $\eta_n = \omega_1$, then we are ready. If not, then every η_n is a countable ordinal. If $\xi < \omega_1$, then there exists a set $A \in \mathcal{P}(\omega)$ such that $f(A) = \xi$. By (1.16) there exists an n such that $A \in \mathcal{A}_n$. Then $\xi \leq \eta_n$. Thus $\sup\{\eta_n : n \in \omega\} = \omega_1$. $\qquad\square$

Theorem 1.20. *The Weak Axiom of Choice* **wAC** *implies that ω_1 is a regular cardinal.*

Proof. Let $A \subseteq \omega_1$ be countable, i.e., $A = \{\xi_n : n \in \omega\}$. We can assume that $\xi_n > \omega$ for each n. We have to show that A is bounded in ω_1.

Let f be the function of Theorem 1.16. We denote $\Phi_n = \{A \subseteq \omega : f(A) = \xi_n\}$. Then by **wAC** there exists a choice function $\langle B_n \in \Phi_n : n \in \omega \rangle$. Thus $\langle \omega, \pi^{-1}(B_n) \rangle$ is a well-ordered set of order type ξ_n. We define a well-ordering R on $\omega \times \omega$ of order type equal to $\xi_0 + \cdots + \xi_n + \cdots$ as follows:

$$\langle n_1, m_1 \rangle R \langle n_2, m_2 \rangle \equiv (n_1 < n_2 \vee (n_1 = n_2 \wedge \langle m_1, m_2 \rangle \in \pi^{-1}(B_{n_1}))).$$

Let η be the order type of $\langle \omega \times \omega, R \rangle$. Since $|\omega \times \omega| = \aleph_0$, we obtain $\eta < \omega_1$. On the other hand every $\langle \omega, \pi^{-1}(B_n) \rangle$ can be naturally embedded into $\langle \omega \times \omega, R \rangle$, therefore $\xi_n \leq \eta$. Thus η is an upper bound of A. $\qquad\square$

We close with a technical result that will be useful in some investigation. Let us consider the following property $\mathrm{COF}(\xi)$ of an ordinal $\xi \leq \omega_1$:

there exists a function $F : \xi \longrightarrow {}^\omega \omega_1$ such that for any limit $\eta < \xi$, $F(\eta)$ is an increasing sequence of ordinals $\{\eta_n\}_{n=0}^\infty$ and $\eta = \sup\{\eta_n : n \in \omega\}$.

Of course, the Axiom of Choice implies that $\mathrm{COF}(\xi)$ holds true for any $\xi \leq \omega_1$. However, we want to avoid a use of **AC**.

Theorem 1.21. $\mathrm{COF}(\xi)$ *holds true for any $\xi < \omega_1$.*

Proof. If $\xi < \omega_1$, then there is a well-ordering R (we suppose that R is antireflexive) on ω such that $\mathrm{ot}(\omega, R) = \xi$. If $\eta < \xi$ is a limit ordinal, then there exists a natural number k such that η is the order type of the set $\{n \in \omega : nRk\}$. We set by induction

$$\eta_n = \max\{\eta_0, \ldots, \eta_{n-1}, \mathrm{ot}(\{m \in \omega : mRn \wedge mRk\})\} + 1.$$

If $\zeta < \eta$, then there exists an $l \in \omega$ such that lRk and ζ is the order type of the set $\{n \in \omega : nRl\}$. Then $\eta_l > \zeta$. Set $F(\eta) = \{\eta_n\}_{n=0}^\infty$. $\qquad\square$

Exercises

1.1 The Cumulative Hierarchy

The smallest transitive set containing a given set x as a subset is called the **transitive closure** of x.

a) Set $x_0 = x$, $x_{n+1} = \bigcup x_n$ for any n. Show that $\mathbf{TC}(x) = \bigcup_n x_n$ is the transitive closure of x.
  Hint: If $y \in x_n$, then $y \subseteq x_{n+1}$.

b) We define the **Cumulative Hierarchy** $\langle V_\xi, \xi \in \mathbf{On} \rangle$ by transfinite induction:

$$V_0 = \emptyset,$$

$$V_\xi = \bigcup_{\eta < \xi} V_\eta, \text{ if } \xi \text{ is a limit ordinal,}$$

$$V_{\xi+1} = \mathcal{P}(V_\xi).$$

  Show that every V_ξ is a transitive set.

c) Show that $x \in V_\xi$ if and only if $\mathbf{TC}(x) \in V_\xi$.

d) For every set x there exists an ordinal ξ such that $x \in V_\xi$, i.e., $\mathbf{V} = \bigcup_{\xi \in \mathbf{On}} V_\xi$.
  Hint: Assume that there exists a transitive set x which does not belong to any V_ξ. By the Axiom of Regularity there exists $\in$-minimal element $y \in x$ which does not belong to any V_ξ. Thus every $z \in y$ belongs to some V_η. By suitable use of an instance of the Scheme of Replacement we obtain $y \subseteq V_\zeta$ for some ζ, a contradiction.

e) The **rank** of a set x is $\mathrm{rank}(x) = \min\{\xi : x \in V_{\xi+1}\}$. Show that $x \in y \to \mathrm{rank}(x) < \mathrm{rank}(y)$.

f) $\mathrm{rank}(x) = \sup\{\mathrm{rank}(y) : y \in x\} + 1$.

g) An uncountable regular cardinal κ is a strongly inaccessible cardinal if and only if $|V_\kappa| = \kappa$.

1.2 Cardinal Arithmetics without AC

a) $\aleph_0 \le |X|$ if and only if there exists a set $Y \subseteq X$, $|X| = |Y|$, $Y \ne X$.
  Hint: If $f : X \xrightarrow[\mathrm{onto}]{1-1} Y$, $a \in X \setminus Y$, then $|\{f^{-n}(a) : n \in \omega\}| = \aleph_0$.

b) If a set X is non-empty and $|X| + |X| = |X|$, then $\aleph_0 \le |X|$.

c) If $|X| > 1$ and $|X| \cdot |X| = |X|$, then $\aleph_0 \le |X|$.

d) If the cardinalities $|X|$ and $|Y|$ are incomparable (i.e., neither $|X| \le |Y|$ nor $|Y| \le |X|$), then $|X| < |X| + |Y|$ and $|X| < |X| \cdot |Y|$.

e) If $\aleph_1$ and $2^{\aleph_0}$ are incomparable, then $2^{\aleph_0} < 2^{\aleph_1}$.

f) If $\aleph_0 \le |X|$, then $\aleph_0 + |X| = |X|$.

g) Show that for any set X the following are equivalent:

  1) $|X| \ge \aleph_0$;

  2) $|X| + 1 = |X|$;

  3) $|X| + \aleph_0 = |X|$.

1.3 Hartogs' Function

Hartogs' function $\aleph$ is defined as follows: for any set X the value $\aleph(X)$ is the first ordinal ξ such that there is no injection $f : \xi \xrightarrow{1-1} X$.

a) For every set X there exists an ordinal ξ such that $|\xi| \le |\mathcal{P}(X \times X)|$ and $|\xi| \not\le |X|$.
  Hint: Consider the set W of all well-orderings of subsets of X. To each element $R \in W$ assign its ordinal type $h(R)$. Show that $\xi = \{h(R) : R \in W\}$ is the desired ordinal.

b) Hartogs' function $\aleph(X)$ is well defined.

c) $\aleph(X) \not\leq |X|$ for any infinite set X.

d) If the set X is infinite, then $\aleph(X)$ is an aleph.

e) A set X can be well ordered if and only if $\aleph(X) > |X|$.

f) $\aleph_1$ and $2^{\aleph_0}$ are incomparable if and only if $\aleph(\mathcal{P}(\omega)) = \aleph_1$.

   Hint: $\aleph(\mathcal{P}(\omega)) \geq \aleph(\omega) = \aleph_1$, since every infinite $\xi < \omega_1$ is the order type of a well-ordering of ω.

1.4 Addition of cardinals and AC

a) If $|X| + \aleph(X) = |X| \cdot \aleph(X)$, then X can be well ordered.

   Hint: Let $|Y| = \aleph(X)$, $X \cap Y = \emptyset$. Assume that $X \times Y = A \cup B$, where $|A| = |X|$ and $|B| = \aleph(X)$. Since $\aleph(X) \not\leq |X|$, for every $a \in X$ there exists a $b \in Y$ such that $\langle a, b \rangle \in B$. Since Y can be well ordered, one define an injection of X into Y.

b) $|X| + |Y| \leq |X| \cdot |Y|$ for any infinite X, Y.

c) The following are equivalent:

   (1) $(\forall \mathfrak{m}, \mathfrak{n} \text{ infinite}) (\mathfrak{m} + \mathfrak{n} = \mathfrak{m} \vee \mathfrak{m} + \mathfrak{n} = \mathfrak{n})$.

   (2) $(\forall \mathfrak{m}, \mathfrak{n} \text{ infinite}) (\mathfrak{m} \leq \mathfrak{n} \vee \mathfrak{n} \leq \mathfrak{m})$.

   (3) $(\forall \mathfrak{m}, \mathfrak{n} \text{ infinite}) (\mathfrak{m} \cdot \mathfrak{n} = \mathfrak{m} \vee \mathfrak{m} \cdot \mathfrak{n} = \mathfrak{n})$.

   (4) $(\forall \mathfrak{m} \text{ infinite}) (\mathfrak{m}^2 = \mathfrak{m})$.

   (5) $(\forall \mathfrak{m}, \mathfrak{n} \text{ infinite}) (\mathfrak{m}^2 = \mathfrak{n}^2 \rightarrow \mathfrak{m} = \mathfrak{n})$.

   (6) **AC**.

   Hint: $\mathbf{AC} \rightarrow (1) \rightarrow (2)$, $\mathbf{AC} \rightarrow (3) \rightarrow (4) \rightarrow (5)$. (2) implies $\mathbf{AC}$, since $|X| \leq \aleph(X)$ for any infinite X.

   Assume (5). If X is infinite, set $\mathfrak{p} = |X|^{\aleph_0}$, $\mathfrak{m} = \mathfrak{p} + \aleph(\mathfrak{p})$, $\mathfrak{n} = \mathfrak{p} \cdot \aleph(\mathfrak{p})$. Evidently $\mathfrak{p} = \mathfrak{p} + 1 = 2 \cdot \mathfrak{p} = \mathfrak{p}^2$. Similarly one has $\aleph(\mathfrak{p}) = \aleph(\mathfrak{p}) + 1 = 2 \cdot \aleph(\mathfrak{p}) = (\aleph(\mathfrak{p}))^2$. Then by simple calculation one obtains $\mathfrak{m}^2 = \mathfrak{n}^2$. Thus $\mathfrak{m} = \mathfrak{n}$ and by part a) any set of cardinality $\mathfrak{p}$ can be well ordered. Note that $|X| \leq \mathfrak{p}$.

d) If for any infinite X the family $\{A \subseteq X : |A| < |X|\}$ is an ideal, then **AC** holds true.

   Hint: If $\mathbf{AC}$ fails, then there exist disjoint infinite sets X, Y such that $|X| < |X| + |Y|$ and $|Y| < |X| + |Y|$.

e) If on every infinite set there exists a uniform ultrafilter, then **AC** holds true.

   Hint: Let X, Y be as in d). If $\mathcal{F}$ were a uniform ultrafilter on $X \cup Y$, then either $X \in \mathcal{F}$ or $Y \in \mathcal{F}$.

1.5 Tarski's Lemma

Assume that M, P, Q are pairwise disjoint sets, $A = M \cup P$, $B = M \cup Q$ and $f : A \xrightarrow[\text{onto}]{1-1} B$.

We set $P_1 = \{x \in P : (\forall n > 0)\, f^n(x) \in M\}$, $Q_1 = \{x \in Q : (\forall n > 0)\, f^{-n}(x) \in M\}$, $P_2 = P \setminus P_1$, $Q_2 = Q \setminus Q_1$.

a) Show that $|P_2| = |Q_2|$.

   Hint: Set

   $$C_n = \{x \in P : (\forall k < n, k > 0)\, f^k(x) \in M \wedge f^n(x) \notin M\},$$
   $$D_n = \{x \in Q : (\forall k < n, k > 0)\, f^{-k}(x) \in M \wedge f^{-n}(x) \notin M\}.$$

Show that $f^n : C_n \xrightarrow[\text{onto}]{1-1} D_n$, $P_2 = \bigcup_{n>0} C_n$ and $Q_2 = \bigcup_{n>0} D_n$.

b) Show that $|P_1| + |M| = |M|$.

   Hint: For $x \in P_1 \cup M$ set $g(x) = x$ if $x \in M \setminus \bigcup_{n>0} f^n(P_1)$ and $g(x) = f(x)$ otherwise.

c) Show that $|Q_1| + |M| = |M|$.

d) Prove **Tarski's Lemma**: If $\mathfrak{m}, \mathfrak{p}, \mathfrak{q}$ are cardinalities such that $\mathfrak{m} + \mathfrak{p} = \mathfrak{m} + \mathfrak{q}$, then there are cardinalities $\mathfrak{n}, \mathfrak{p}_1, \mathfrak{q}_1$ such that

$$\mathfrak{p} = \mathfrak{p}_1 + \mathfrak{n}, \; \mathfrak{q} = \mathfrak{q}_1 + \mathfrak{n}, \; \mathfrak{m} + \mathfrak{p}_1 = \mathfrak{m} = \mathfrak{m} + \mathfrak{q}_1.$$

   Hint: Set $\mathfrak{n} = |P_2| = |Q_2|$.

e) Prove **Bernstein's Theorem**: If $\mathfrak{m} + \mathfrak{m} = \mathfrak{m} + \mathfrak{q}$, then $\mathfrak{m} \geq \mathfrak{q}$.

   Hint: Use Tarski's Lemma.

1.6 Tarski's Theorem

Let $\mathfrak{m}$ be a cardinality such that $\aleph_0 \leq \mathfrak{m}$.

a) $2^{\mathfrak{m}} + \mathfrak{m} = 2^{\mathfrak{m}}$.

   Hint: If $\aleph_0 \leq \mathfrak{m}$, then $\mathfrak{m} + 1 = \mathfrak{m}$.

b) If $2^{\mathfrak{m}} = \mathfrak{m} + \mathfrak{p}$ and $\mathfrak{m} = \mathfrak{m} + \mathfrak{s}$, then $\mathfrak{p} \geq 2^{\mathfrak{s}}$.

   Hint: Let $|M| = \mathfrak{m}$, $\mathcal{P}(M) = M_1 \cup P$, $M_1 \cap P = \emptyset$, $|M_1| = \mathfrak{m}$, $|P| = \mathfrak{p}$, $M = M_2 \cup S$, $M_2 \cap S = \emptyset$, $|M_2| = \mathfrak{m}$, $|S| = \mathfrak{s}$, $f : M \xrightarrow[\text{onto}]{1-1} M_1$, $g : M_2 \xrightarrow[\text{onto}]{1-1} M$. For $A \subseteq S$ set

$$h(A) = A \cup \{x \in M_2 : x \notin f(g(x))\}.$$

   If $h(A) \in M_1$, then $h(A) = f(g(x))$ for some $x \in M_2$ and we obtain a contradiction

$$x \in h(A) \equiv x \notin f(g(x)).$$

   Thus $h : \mathcal{P}(S) \xrightarrow{1-1} P$.

c) Assume that $\mathfrak{m} + \mathfrak{p} = 2^{\mathfrak{m}}$. Then $\mathfrak{p} \geq 2^{\mathfrak{m}}$.

   Hint: If $\mathfrak{m} + \mathfrak{p} = \mathfrak{m} + 2^{\mathfrak{m}}$, then by Tarski's Lemma 1.5 d) there are $\mathfrak{n}, \mathfrak{p}_1, \mathfrak{q}_1$ such that $\mathfrak{p} = \mathfrak{n} + \mathfrak{p}_1$, $2^{\mathfrak{m}} = \mathfrak{n} + \mathfrak{q}_1$ and $\mathfrak{m} + \mathfrak{p}_1 = \mathfrak{m} + \mathfrak{q}_1 = \mathfrak{m}$. By b) we obtain $\mathfrak{p} \geq 2^{\mathfrak{q}_1} > \mathfrak{q}_1$. Hence $\mathfrak{p} + \mathfrak{p} \geq \mathfrak{n} + \mathfrak{q}_1 = 2^{\mathfrak{m}} = 2^{\mathfrak{m}} + \mathfrak{p} \geq \mathfrak{p} + \mathfrak{p}$.

d) Prove Tarski's Theorem 1.8.

1.7 [AC] Cardinal Arithmetics in ZFC

We assume **AC**. κ, λ denotes infinite cardinals. κ^+ denotes the smallest cardinal greater than κ, i.e., if $\kappa = \aleph_\xi$, then $\kappa^+ = \aleph_{\xi+1}$. The **gimel function** is defined as $\beth(\kappa) = \kappa^{\mathrm{cf}(\kappa)}$. If κ_s is a cardinal for each $s \in S$, we can define the sum $\sum_{s \in S} \kappa_s$ and the product $\Pi_{s \in S} \kappa_s$ in an obvious way, e.g., $\Pi_{s \in S} \kappa_s = |\Pi_{s \in S} A_s|$, where $|A_s| = \kappa_s$. Actually, you can set $A_s = \kappa_s$.

a) Prove **König's Inequality**: if $\kappa_s < \lambda_s$ for any $s \in S$, then $\sum_{s \in S} \kappa_s < \Pi_{s \in S} \lambda_s$.

   Hint: Let $|A_s| = \kappa_s$, $|B_s| = \lambda_s$, $A_s \subseteq B_s$, A_s, $s \in S$ being pairwise disjoint. Choose a $g \in \Pi_{s \in S}(B_s \setminus A_s)$ and for $a \in A_s$ set $\psi(a) = f$, where $f(t) = g(t)$ for $t \neq s$ and $f(s) = a$. Then $\psi : \bigcup_{s \in S} A_s \xrightarrow{1-1} \Pi_{s \in S} B_s$.

Vice versa $\varphi : \bigcup_{s \in S} A_s \overset{1-1}{\longrightarrow} \Pi_{s \in S} B_s$ *is an injection, then* $|\varphi(A_s)| = \kappa_s$. *Set* $C_s = \{f(s) : f \in \varphi(A_s)\}$. *Then* $B_s \setminus C_s \neq \emptyset$ *and* $\Pi_{s \in S}(B_s \setminus C_s) \cap \mathrm{rng}(\varphi) = \emptyset$. *Thus* φ *is not a surjection.*

b) Show that $\mathrm{cf}(\kappa) = \lambda$ if and only if there exists an increasing sequence of cardinals $\langle \kappa_\xi : \xi < \lambda \rangle$ such that $\kappa = \sum_{\xi < \lambda} \kappa_\xi$.

c) Show that $\mathrm{cf}(\kappa^\lambda) > \lambda$.
   Hint: If $\kappa^\lambda = \sum_{\xi < \mu} \nu_\xi$, *then by König's Inequality we obtain*

$$\sum_{\xi < \mu} \nu_\xi < \Pi_{\xi < \mu} \nu_\xi \leq (\kappa^\lambda)^\mu.$$

d) $\beth(\kappa) \geq \mathrm{cf}(\beth(\kappa)) > \mathrm{cf}(\kappa)$ for any infinite κ.

e) If $\kappa \leq \lambda$, then $\kappa^\lambda = 2^\lambda$.

f) Prove the **Hausdorff formula**: $(\kappa^+)^\lambda = \kappa^\lambda \cdot \kappa^+$.
   Hint: $^\lambda(\kappa^+) = \bigcup_{\xi < \kappa^+} {}^\lambda\xi$.

g) Prove the **Bukovský-Hechler** formula: if $\mathrm{cf}(\kappa) < \kappa$, $\lambda < \kappa$ and $2^\mu = 2^\lambda$ for every $\lambda \leq \mu < \kappa$, then $2^\kappa = 2^\lambda$.
   Hint: If $\kappa = \sum_{\xi < \mathrm{cf}(\kappa)} \kappa_\xi$, $\lambda \leq \kappa_\xi < \kappa$, *then* $2^{\sum_{\xi < \mathrm{cf}(\kappa)} \kappa_\xi} = \Pi_{\xi < \mathrm{cf}(\kappa)} 2^\lambda = 2^\lambda$.

h) Assume $\mathrm{cf}(\kappa) < \kappa$. If for each $\lambda < \kappa$ there exists a $\mu < \kappa$ such that $2^\mu > 2^\lambda$, then $2^\kappa = \beth(\sum_{\lambda < \kappa} 2^\lambda)$.
   Hint: Note that $\mathrm{cf}(\sum_{\lambda < \kappa} 2^\lambda) = \mathrm{cf}(\kappa)$ *and*

$$\beth\left(\sum_{\lambda < \kappa} 2^\lambda\right) \leq 2^\kappa = \Pi_{\xi < \mathrm{cf}(\kappa)} \kappa_\xi \leq \left(\sum_{\lambda < \kappa} 2^\lambda\right)^{\mathrm{cf}(\sum_{\lambda < \kappa} 2^\lambda)} = \beth\left(\sum_{\lambda < \kappa} 2^\lambda\right).$$

i) Prove the **Tarski formula**: $\kappa^\lambda = \sum_{\gamma < \kappa} \gamma^\lambda$ provided that κ is limit and $\lambda < \mathrm{cf}(\kappa)$.
   Hint: Note that $^\lambda\kappa = \bigcup_{\gamma < \kappa} {}^\lambda\gamma$.

j) Show that the value $\aleph_\xi^{\aleph_\eta}$ is determined by the function $\beth$.
   Hint: By transfinite induction over ξ *using the results of* d)–i).

1.8 Axiom of Choice

For a non-empty set X we introduce the following statements:

$\mathbf{AC}(X) \equiv$ for every family of non-empty subsets of X there exists a selector,

$\mathbf{AC}_n(X) \equiv$ for every family of n elements subsets of X there exists a selector,

$\mathbf{AC}_{<\omega}(X) \equiv$ for every family of finite non-empty subsets of X there exists a selector,

$\mathbf{AC}^\omega(X) \equiv$ for every countable family of non-empty subsets of X there
exists a selector.

a) Show that:
   1) **AC** is equivalent to "**AC**(X) holds true for any set X".
   2) $\mathbf{AC}^\omega$ is equivalent to "$\mathbf{AC}^\omega(X)$ holds true for any set X".
   3) **wAC** is equivalent to "$\mathbf{AC}_\omega(\mathcal{P}(\omega))$ holds true".

b) If the set X can be well ordered, then $\mathbf{AC}(X)$ holds true.

c) If the set X can be linearly ordered, then $\mathbf{AC}_n(X)$, $1 < n < \omega$ and $\mathbf{AC}_{<\omega}(X)$ hold true.

d) If $\mathbf{AC}_{<\omega}(X)$ holds true for any set, then a countable union of finite sets is a countable set.

   Hint: If $A = \bigcup_n A_n$, A_n finite, then there exists a selector $\{f_n\}_{n=0}^{\infty}$ of the family $\{^{A_n}|A_n|\}_{n=0}^{\infty}$. Using the selector one can easily construct an injection of A into $\omega \times \omega$.

1.9 Binomial Coefficients

If $\mathfrak{m}$ is a cardinality and X is a set, we write

$$[X]^{\mathfrak{m}} = \{A \subseteq X : |A| = \mathfrak{m}\}, \quad [X]^{<\mathfrak{m}} = \{A \subseteq X : |A| < \mathfrak{m}\}.$$

Moreover we set $n! = 1 \cdot \dots \cdot n$.

a) If $|X| = |Y|$, then $|[X]^{\mathfrak{m}}| = |[Y]^{\mathfrak{m}}|$ and $|[X]^{<\mathfrak{m}}| = |[Y]^{<\mathfrak{m}}|$.

If $|X| = n$, $k, n \in \omega$, we define the **binomial coefficient** as

$$\binom{n}{k} = |[X]^k|.$$

b) Show that $\binom{n+1}{k+1} = \binom{n}{k} + \binom{n}{k+1}$ for any $k, n \in \omega$.

   Hint: Fix an $a \in X$. Then

   $$[X]^{k+1} = [X \setminus \{a\}]^{k+1} \cup \{A \subseteq X : |A| = k+1 \wedge a \in A\}.$$

c) Using b) show that $\binom{n}{k} = \frac{n!}{k!(n-k)!}$.

d) If $|X| = n$, then $2^n = \sum_{i=0}^{n} \binom{n}{i}$.

   Hint: $2^n = |\mathcal{P}(X)|$.

e) For $n > 0$ show that

   $$\sum_{i=0}^{n} (-1)^i \binom{n}{i} = 0.$$

   Hint: If $|X| = n$, $a \in X$, then $F : \mathcal{P}(X) \longrightarrow \mathcal{P}(X)$ defined as $F(A) = A \cup \{a\}$ if $a \notin A$ and $F(A) = A \setminus \{a\}$ if $a \in A$, is a bijection.

f) Show the Binomial Theorem: If R is a ring, then

   $$(x+y)^n = \sum_{i=0}^{n} \binom{n}{i} x^i \cdot y^{n-i}$$

   for any $x, y \in R$.

1.10 Pigeonhole Principle

a) If $\mathcal{A}$ is a finite partition of a set X, and $|A| \leq |B|$ for each $A \in \mathcal{A}$, then $|X| \leq |\mathcal{A}| \cdot |B|$.

   Hint: By induction over $|\mathcal{A}|$.

b) Prove Theorem 1.9.

c) Assuming the Axiom of Choice prove the Pigeonhole Principle for infinite sets: Let $\mathfrak{m}$ be a cardinality, i.e., $\mathfrak{m} = |B|$ for some set B. Let $f : X \longrightarrow Y$. If $|Y| \cdot \mathfrak{m} < |X|$, then there exists a $y \in Y$ such that the cardinality of the inverse image $f^{-1}(\{y\})$ is greater than $\mathfrak{m}$.

1.11 Choice from Finite Sets

We denote by $\mathbf{AC}_n$ the statement "$\mathbf{AC}_n(X)$ holds true for any set X".

a) Show (in **ZF**) that $\mathbf{AC}_{kn} \to \mathbf{AC}_n$ for any $n, k > 1$.

b) Assume $p|n$, p is a prime and $\mathbf{AC}_p$ holds true. Then for any set X there exists a function $g : [X]^n \longrightarrow \mathcal{P}(X)$ such that $g(A)$ is a non-empty proper subset of A for any $A \in [X]^n$.

Hint: Note that $\binom{n}{p}$ is not divisible by n. Let $h : [X]^p \longrightarrow X$ be a selector. If $A \in [X]^n$, $a \in A$, denote by $n(a)$ the number of subsets $B \subseteq A$ of cardinality p for which $h(B) = a$. Since $\sum_{a \in A} n(a) = \binom{n}{p}$, the numbers $r(a)$, $a \in A$ cannot be equal. Take $g(A)$ to be the set of those $a \in A$ for which $n(a)$ is maximal.

c) Show that $\mathbf{AC}_2 \equiv \mathbf{AC}_4$.

Hint: Let g be the function of b) *for $n = 4$, f being a selector for $[X]^2$. For $A \in [X]^4$ set*

$$h(A) = \begin{cases} \text{the only element of } g(A), & \text{if } |g(A)| = 1, \\ \text{the only element of } A \setminus g(A), & \text{if } |g(A)| = 3, \\ f(g(A)), & \text{if } |g(A)| = 2. \end{cases}$$

d) Show that

$$\mathbf{AC}_6 \equiv (\mathbf{AC}_2 \wedge \mathbf{AC}_3), \qquad (\mathbf{AC}_2 \wedge \mathbf{AC}_5) \to \mathbf{AC}_8, \qquad \mathbf{AC}_{10} \to \mathbf{AC}_8,$$
$$(\mathbf{AC}_2 \wedge \mathbf{AC}_3) \to \mathbf{AC}_8, \qquad (\mathbf{AC}_2 \wedge \mathbf{AC}_3) \to \mathbf{AC}_9, \qquad \mathbf{AC}_6 \to \mathbf{AC}_8.$$

Hint: Similarly as in c) *and use results of* a) *and* b).

1.12 Definitions of Finiteness

Define:

$\mathbf{F}_1(X) \equiv$ every non-empty subset of $\langle \mathcal{P}(X), \subseteq \rangle$ has a maximal element,

$\mathbf{F}_2(X) \equiv$ every proper subset of $\mathcal{P}(X)$ has smaller cardinality than $\mathcal{P}(X)$,

$\mathbf{F}_3(X) \equiv$ every proper subset of X has smaller cardinality than X,

$\mathbf{F}_4(X) \equiv X$ is empty or $|X| < |X| + |X|$,

$\mathbf{F}_5(X) \equiv |X| \leq 1$ or $|X| < |X| \cdot |X|$,

$\mathbf{F}_6(X) \equiv |X| = \aleph_\xi$ for no ξ.

Show that

a) X is finite if and only if $\mathbf{F}_1(X)$ holds true.

Hint: Consider the set $\{A \subseteq X : (\exists n)\, |A| = n\}$.

b) $\mathbf{F}_i(X) \to \mathbf{F}_{i+1}(X)$ for $i = 1, \ldots, 5$.

c) If $\mathbf{AC}$ holds true, then $\mathbf{F}_6(X) \to \mathbf{F}_1(X)$ for any set X.

d) $\neg \mathbf{F}_3(X) \equiv |X| \geq \aleph_0$.

1.13 Alternatives

Alonzo Church considered three mutually exclusive **Church alternatives**:

CHA(A)   There exists a selector for the Lebesgue decomposition (1.14).

CHA(B)   There exists no selector for the Lebesgue decomposition (1.14) and ω_1 is regular.

CHA(C) ω_1 is cofinal with ω.

a) If the alternative CHA(A) holds true, then ω_1 is regular.

b) The alternative CHA(A) holds true in **ZFC**.

c) The alternative CHA(C) is consistent with **ZF**.
   Hint: See Theorem 1.19.

d) The following are equivalent in **ZF**:

   (i) COF(ω_1) holds true.

   (ii) Church's alternative CHA(A) holds true.

   (iii) There exists a function $G : \omega_1 \longrightarrow {}^\omega\omega_1$ such that $G(\xi) : \omega \xrightarrow[\text{onto}]{1-1} \xi$ for each $\xi < \omega_1$, $\xi \geq \omega$.

   Hint: To show (ii) $\rightarrow$ (i) *construct the desired function* $F : \omega_1 \longrightarrow {}^\omega\omega_1$ *from a choice function for the Lebesgue decomposition similarly as in the proof of Theorem 1.21.* (iii) $\rightarrow$ (ii) *is trivial.* (i) $\rightarrow$ (iii) *can be proved by a transfinite inductive construction.*

1.14 Partition Relation

Let κ, λ denote cardinalities, n, k being natural numbers. The **partition relation** $\kappa \rightarrow (\lambda)^k_n$ is defined as follows: for any set X of cardinality κ and for any mapping $F : [X]^k \longrightarrow n$ there exists a set $Y \subseteq X$ and an $i < n$ such that $|Y| = \lambda$ and $F(x) = i$ for any $x \in [Y]^k$. The function F is called a **coloring**, $i < n$ are colors. The set Y is called a **homogeneous set**. The negation of the partition relation will be denoted by $\kappa \nrightarrow (\lambda)^k_n$.

a) Show that $6 \rightarrow (3)^2_2$, $17 \rightarrow (3)^2_3$, $5 \nrightarrow (3)^2_2$.
   Hint: Let $F : [6]^2 \longrightarrow 2$. *Fix* $i \in 6$. *Then there exists a set* $A \subseteq 6$, $|A| = 3$ *and* $k \in 2$ *such that* $F(\{i,j\}) = k$ *for* $j \in A$. *If* $F(\{x,y\}) = 1 - k$ *for all* $x, y \in A$ *we are ready. Otherwise there are* $x, y \in A$ *such that* $F(\{x,y\}) = k$. *Then* $\{i, x, y\}$ *is the desired homogeneous set.*

b) If $\kappa \rightarrow (\lambda)^k_n$ holds true and $\kappa' \geq \kappa$, $\lambda' \leq \lambda$, $k' \leq k$ and $n' \leq n$, then also $\kappa' \rightarrow (\lambda')^{k'}_{n'}$.

c) Show that $2^{2m-1} \rightarrow (m)^2_2$.
   Hint: Let $|X| = 2^{2m-1}$. *Let* $F : [X]^2 \longrightarrow 2$ *be a coloring. For* $i = 0, \ldots, 2m - 1$ *choose a set* $A_i \subseteq X$, *a point* $x_i \in A_i$ *and for* $i < 2m - 1$ *also a color* $k_i \in 2$ *such that:* $A_0 = X$, $|A_i| \geq 2^{2m-i-1}$ *for any* $i \leq 2m - 1$, $F(\{x_i, a\}) = k_i$ *for any* $a \in A_{i+1}$, $i < 2m - 1$. *There exists a set* $C \subseteq 2m - 1$ *such that* $|C| = m$ *and* $k_i = k_j$ *for any* $i, j \in C$. *The set* $\{x_i : i \in C\}$ *is homogeneous.*

d) If $m \rightarrow (n)^2_k$ and $n \rightarrow (p)^2_l$, then $m \rightarrow (p)^2_{k \cdot l}$.
   Hint: Join colors to obtain k *colors each of* l *elements.*

e) Prove the **Ramsey Theorem**: For any natural numbers m, k there exists a natural number p such that $p \rightarrow (m)^2_k$.
   Hint: By induction over k – *join two colors.*

f) Show that in e) you can replace number 2 by any natural number $n > 1$.

1.2 Topological Preliminaries

We assume that the reader is familiar with basic notions of topology and their properties. However, since, as in other branches of infinite mathematics, mathematicians freely use the axiom of choice, we must carefully check classical results of topology to determine if they really need **AC** for their proof.

Let X be a non-empty set. A family $\mathcal{O} \subseteq \mathcal{P}(X)$ is called a **topology** on X if

1) $\emptyset, X \in \mathcal{O}$,

2) if $A, B \in \mathcal{O}$, then $A \cap B \in \mathcal{O}$,

3) if $\mathcal{A} \subseteq \mathcal{O}$, then $\bigcup \mathcal{A} \in \mathcal{O}$.

The couple $\langle X, \mathcal{O} \rangle$ is called a **topological space**. A subset of X belonging to $\mathcal{O}$ is called an **open set**. An element of a topological space is usually called a **point**.

If $\mathcal{O}$ is a topology on X and $Y \subseteq X$ is a non-empty subset, then the family $\mathcal{O}|Y = \{A \cap Y : Y \in \mathcal{O}\}$ is a topology on Y. The topological space $\langle Y, \mathcal{O}|Y \rangle$ is called a **topological subspace** of $\langle X, \mathcal{O} \rangle$ and $\mathcal{O}|Y$ is the **subspace topology** on Y.

For any set $A \subseteq X$ there exists the largest open subset of A,

$$\text{Int}(A) = \bigcup \{U \in \mathcal{O} : U \subseteq A\}.$$

The set $\text{Int}(A)$ is called the **interior** of A. A set A is open if and only if $A = \text{Int}(A)$. A set $A \subseteq X$ is called **closed** if $X \setminus A$ is open. The **closure** $\overline{A}$ of a set A is the smallest closed set containing A as a subset, i.e., $\overline{A} = X \setminus \text{Int}(X \setminus A)$. The **boundary** of a set A is the set $\text{Bd}(A) = \overline{A} \cap \overline{X \setminus A}$. A set A is called **clopen** if A is both open and closed. Thus A is clopen if and only if $\text{Bd}(A) = \emptyset$. Evidently

$$\text{Int}(\text{Int}(A)) = \text{Int}(A), \quad \text{Int}(A \cap B) = \text{Int}(A) \cap \text{Int}(B), \quad \overline{\overline{A}} = \overline{A}, \quad \overline{A \cup B} = \overline{A} \cup \overline{B}.$$

A set $A \subseteq B \subseteq X$ is **topologically dense** or simply **dense in** B if $B \subseteq \overline{A}$. A set dense in X is called **dense**. A set A is called **regular open** if $A = \text{Int}(\overline{A})$. For any set A, the set $\text{Int}(\overline{A})$ is regular open. If moreover A is open, then $A \subseteq \text{Int}(\overline{A})$.

A topological space $\langle X, \mathcal{O} \rangle$ is called **Hausdorff** if for any pair of distinct points $x, y \in X$ there exist open sets U, V such that $x \in U$, $y \in V$ and $U \cap V = \emptyset$.

A Hausdorff space is called **regular** if for any closed set $A \subseteq X$ and any $x \in X \setminus A$ there exist open sets U, V such that $A \subseteq U$, $x \in V$ and $U \cap V = \emptyset$. Finally, a Hausdorff space is called **normal** if for any disjoint closed sets A, B there exist open sets U, V such that $A \subseteq U$, $B \subseteq V$ and $U \cap V = \emptyset$.

Theorem 1.22. *Let $\langle X, \mathcal{O} \rangle$ be a topological space, $Y \subseteq X$ being endowed with the subspace topology.*

a) *If $\langle X, \mathcal{O} \rangle$ is Hausdorff, then $\langle Y, \mathcal{O}|Y \rangle$ is Hausdorff as well.*

b) *If $\langle X, \mathcal{O} \rangle$ is regular, then $\langle Y, \mathcal{O}|Y \rangle$ is regular as well.*

A similar assertion is not true for normal spaces. There exists a normal space such that by omitting a point one obtains a subspace that is not normal – see Exercise 1.22.

A family $\mathcal{B} \subseteq \mathcal{O}$ is called a **base of the topology** $\mathcal{O}$ if for any $A \in \mathcal{O}$ there exists a set $\mathcal{A} \subseteq \mathcal{B}$ such that $A = \bigcup \mathcal{A}$. We shall assume that the empty set is not a member of a base. A topological space X is **separable** if there exists a countable subset dense in X.

Theorem 1.23. *If $\mathcal{B}$ is a base of the topology $\mathcal{O}$ of a Hausdorff space X, then $|\mathcal{O}| \leq 2^{|\mathcal{B}|}$ and $|X| \leq 2^{|\mathcal{B}|}$ as well. Consequently, if X is a Hausdorff space with a countable base, then $|X| \leq \mathfrak{c}$. Hence, **wAC** implies that a Hausdorff space with a countable base is separable.*

Proof. Set $F(x) = \{U \in \mathcal{B} : x \in U\} \in \mathcal{P}(\mathcal{B})$. If X is Hausdorff, then F is an injection.

A selector of a countable base $\mathcal{B}$ of X is a countable dense subset of X.   □

A set $U \subseteq X$ is called a **neighborhood** of a point $x \in X$ if $x \in \mathrm{Int}(U)$, i.e., if $x \in V \subseteq U$ for some open set V. One can easily show that

$$x \in \overline{A} \equiv A \cap U \neq \emptyset \text{ for any neighborhood } U \text{ of } x. \qquad (1.17)$$

Let $\mathcal{V}$ be a family of non-empty subsets of X, $x \in X$. $\mathcal{V}$ is called a **neighborhood base** at the point x if a set V is a neighborhood of x if and only if there exists a set $U \in \mathcal{V}$ such that $U \subseteq V$.

A point $x \in X$ is called an **accumulation point** of A if for every neighborhood U of x the intersection $U \cap A$ has at least two points. If $x \in A$ and x is not an accumulation point of A, then the point x is called an **isolated point** of the set A. Thus x is an isolated point of A if there exists a neighborhood U of x such that $U \cap A = \{x\}$. A non-empty closed set $A \subseteq X$ is called **perfect** if A does not contain any isolated point.

Theorem [wAC] 1.24 (G. Cantor – I. Bendixson). *In any Hausdorff topological space $\langle X, \mathcal{O} \rangle$ with a countable base, there exist a perfect set P and a countable set R such that $X = P \cup R$.*

Proof. Let $\mathcal{B}$ be a countable base of topology $\mathcal{O}$. We set

$$P = \{x \in X : (\forall U)\,(U \text{ neighborhood of } x \rightarrow |U| > \aleph_0)\}.$$

One can easily check that P is closed and without isolated points. Using Theorem 1.14 we obtain

$$X \setminus P \subseteq \bigcup \{U \in \mathcal{B} : U \text{ countable}\}.$$

By Theorem 1.18 we have that $R = X \setminus P$ is countable.   □

We define a similar notion for a sequence of points. If $\{a_n\}_{n=0}^{\infty}$ is a sequence of points of X, then a point a is called a **cluster point of the sequence** $\{a_n\}_{n=0}^{\infty}$ if for every neighborhood U of a and every natural number n there exists an $m > n$ such that $a_m \in U$. Evidently, every accumulation point of the set $\{a_n : n \in \omega\}$ is a cluster point of the sequence $\{a_n\}_{n=0}^{\infty}$. If $a_n = a$ for all n, then a is a cluster point of the sequence but not an accumulation point of the set. A point x is a **limit** of the sequence $\{a_n\}_{n=0}^{\infty}$, written $x = \lim_{n\to\infty} a_n$, if for any neighborhood U of x there exists an n_0 such that $a_n \in U$ for any $n \geq n_0$. A sequence $\{a_n\}_{n=0}^{\infty}$ is said to be **convergent**, if there exists an $x = \lim_{n\to\infty} a_n$.

Note the following:

$$\text{if } f : \omega \xrightarrow[\text{onto}]{1-1} \omega, \quad \lim_{n\to\infty} x_n = x, \text{ then } \lim_{n\to\infty} x_{f(n)} = x. \qquad (1.18)$$

Thus for any infinite countable $A \subseteq X$ we can write $\lim A$, since the result does not depend on the enumeration of A.

If a point $x \in X$ has a countable neighborhood base $\mathcal{V} = \{U_n : n \in \omega\}$, then $\{\bigcap_{i=0}^{n} U_i : n \in \omega\}$ is a decreasing neighborhood base. So in such a case we can always assume that the neighborhood base is decreasing

A set G is called a G_δ **set** if $G = \bigcap_n G_n$, where G_n are open. A complement of a G_δ set is called an F_σ **set**. Thus F is an F_σ set if $F = \bigcup_n F_n$, where F_n are closed. We can assume that $G_n \supseteq G_{n+1}$ and $F_n \subseteq F_{n+1}$ for each n. The family of all G_δ subsets of X and all F_σ subsets of X will be denoted as $G_\delta(X)$ and $F_\sigma(X)$, respectively.

Note the following important fact. If $\langle X, \mathcal{O} \rangle$ has a countable base, then by Theorem 1.23 we have

$$|G_\delta(X)| \ll \mathfrak{c}, \qquad |F_\sigma(X)| \ll \mathfrak{c}. \qquad (1.19)$$

Hence, we can apply **wAC** to sequences of families of G_δ and F_σ sets.

Let $\langle X_1, \mathcal{O}_1 \rangle$ and $\langle X_2, \mathcal{O}_2 \rangle$ be topological spaces. A mapping $f : X_1 \longrightarrow X_2$ is called **continuous** if $f^{-1}(U) \in \mathcal{O}_1$ (is open) for any open set $U \in \mathcal{O}_2$. A mapping $f : X_1 \xrightarrow[\text{onto}]{1-1} X_2$ is called a **homeomorphism** if both f and f^{-1} are continuous. If f is a homeomorphism, then also the inverse mapping f^{-1} is a homeomorphism. Topological spaces $\langle X_1, \mathcal{O}_1 \rangle$ and $\langle X_2, \mathcal{O}_2 \rangle$ are said to be **homeomorphic** if there exists a homeomorphism $f : X_1 \longrightarrow X_2$. Two homeomorphic spaces possess equal topological properties and therefore from the topological point of view they are usually identified. A continuous injection $f : X_1 \xrightarrow{1-1} X_2$ is called an **embedding** if $f^{-1} : f(X_1) \longrightarrow X_1$ is continuous, i.e., if $f : X_1 \longrightarrow f(X_1)$ is a homeomorphism. Thus if $f : X_1 \xrightarrow{1-1} X_2$ is an embedding, then X_1 and the subset $f(X_1)$ of the space X_2 are homeomorphic. Of course we suppose that $f(X_1)$ is endowed with the subspace topology $\mathcal{O}_2|f(X_1)$.

A function $f : X_1 \longrightarrow X_2$ is called **continuous at a point** $a \in X_1$ if $f^{-1}(V)$ is a neighborhood of a for any neighborhood $V \subseteq X_2$ of $f(a)$.

Theorem 1.25.

a) $f : X_1 \longrightarrow X_2$ *is continuous if and only if f is continuous at each point $a \in X_1$.*

b) *Let $\mathcal{V}_1$, $\mathcal{V}_2$ be neighborhood bases of $\langle X_1, \mathcal{O}_1 \rangle$ at point a and that of $\langle X_2, \mathcal{O}_2 \rangle$ at point $f(a)$, respectively. Then f is continuous at a if and only if for any $V \in \mathcal{V}_2$ there exists a $U \in \mathcal{V}_1$ such that $f(U) \subseteq V$.*

A subset $A \subseteq X$ of a topological space X is called a **retract** of X if there exists a continuous mapping $r : X \longrightarrow A$ such that $r(x) = x$ for any $x \in A$. We have a simple

Theorem 1.26. *If A is a retract of X and $f : A \longrightarrow Y$ is continuous, then there exists a continuous $F : X \longrightarrow Y$ such that $f = F|A$.*

Let A be a subset of a topological space $\langle X, \mathcal{O} \rangle$. $\mathcal{A} \subseteq \mathcal{P}(X)$ is a **cover** of A if $A \subseteq \bigcup \mathcal{A}$. We usually ask that $\emptyset \notin \mathcal{A}$ and $X \notin \mathcal{A}$. A cover $\mathcal{A}$ of a set A is called an **open cover** of A if $\mathcal{A} \subseteq \mathcal{O}$, i.e., if every member of $\mathcal{A}$ is an open set. An open cover of the set X is simply called an open cover. Similarly for **closed covers, clopen covers** etc.

A Hausdorff space $\langle X, \mathcal{O} \rangle$ is **compact** if for every open cover $\mathcal{A}$ of X there exists a finite subcover $\mathcal{A}_0 \subseteq \mathcal{A}$ of X. A non-empty subset $A \subseteq X$ is called **compact** if the topological space $\langle A, \mathcal{O}|A \rangle$ is compact. It is easy to see that a subset A of a Hausdorff topological space X is compact if and only if for every open cover $\mathcal{A}$ of A there exists a finite subcover $\mathcal{A}_0 \subseteq \mathcal{A}$ of A. In the standard proof you must use the axiom of choice for finite families. However, this is provable in **ZF** – compare the proof of Theorem 1.30. Now one can easily see that a closed subset of a compact topological space is a compact set. The family of all compact subsets of a topological space $\langle X, \mathcal{O} \rangle$ will be denoted by $\mathcal{K}(X, \mathcal{O})$ or simply $\mathcal{K}(X)$.

Theorem 1.27.

a) *If $\langle X, \mathcal{O} \rangle$ is a Hausdorff space and $A \subseteq X$ is a compact subset, then A is closed.*

b) *A compact space is normal.*

Proof. Define a function $\mathfrak{r} : \mathcal{O} \longrightarrow \mathcal{O}$ as $\mathfrak{r}(U) = \mathrm{Int}(X \setminus U)$. Then $U \cap \mathfrak{r}(U) = \emptyset$ for any $U \in \mathcal{O}$. Moreover, $\mathfrak{r}(U_1 \cup U_2) = \mathfrak{r}(U_1) \cap \mathfrak{r}(U_2)$.

Let $A \subseteq X$ be compact. We show that $\overline{A} \subseteq A$, i.e., that A is closed. So, let $x \notin A$. We want to show that $x \notin \overline{A}$. Set

$$\mathcal{W} = \{U \in \mathcal{O} : x \in \mathfrak{r}(U)\}.$$

If $y \in A$, then $x \neq y$ and therefore there are open sets U, V such that $y \in U$, $x \in V$ and $U \cap V = \emptyset$. Then $V \subseteq \mathfrak{r}(U)$ and therefore $U \in \mathcal{W}$. Thus $\mathcal{W}$ is an open cover of A. Since A is a compact set, there exist finitely many $U_0, \ldots, U_n \in \mathcal{W}$ such that $A \subseteq U_0 \cup \cdots \cup U_n$. Then $\mathfrak{r}(U_0) \cap \cdots \cap \mathfrak{r}(U_n)$ is a neighborhood of x disjoint with the set A, consequently $x \notin \overline{A}$.

Now assume that X is a compact space and A is a closed subset of X, $x \notin A$. Then A is compact and as above we can find open sets $U_0, \ldots, U_n$ such that

$$A \subseteq U = U_0 \cup \cdots \cup U_n \text{ and } x \in \mathfrak{r}(U_0) \cap \cdots \cap \mathfrak{r}(U_n) = \mathfrak{r}(U).$$

Since U, $\mathfrak{r}(U)$ are disjoint, we have shown that the space X is regular.

Assume now that A, B are closed disjoint sets. If $\mathcal{W} = \{U \in \mathcal{O} : B \subseteq \mathfrak{r}(U)\}$, then by regularity $\mathcal{W}$ is an open cover of A. Since A is compact, there exist open sets $U_0, \ldots, U_n \in \mathcal{W}$ such that

$$A \subseteq U = U_0 \cup \cdots \cup U_n \text{ and } B \subseteq \mathfrak{r}(U_0) \cap \cdots \cap \mathfrak{r}(U_n) = \mathfrak{r}(U). \qquad \square$$

Theorem 1.28.

a) *Let $\langle X, \mathcal{O} \rangle$ be a Hausdorff space. Then $\langle X, \mathcal{O} \rangle$ is compact if and only if every family $\mathcal{F}$ of closed subsets of X such that for any finite $\mathcal{F}_0 \subseteq \mathcal{F}$ the intersection $\bigcap \mathcal{F}_0$ is non-empty, has non-empty intersection $\bigcap \mathcal{F}$.*

b) *If $\langle X, \mathcal{O} \rangle$ is a compact space and $\{A_n\}_{n=0}^{\infty}$ is a decreasing sequence of non-empty closed subsets of X, then $\bigcap_n A_n \neq \emptyset$.*

The theorem follows by the de Morgan laws.

Corollary 1.29. *Every sequence of points of a compact space has a cluster point.*

Proof. The set $\bigcap_n \overline{\{a_k : k \geq n\}}$ is non-empty and any of its elements is a cluster point of $\{a_n\}_{n=0}^{\infty}$. $\qquad \square$

Theorem 1.30. *Assume that $\langle X_1, \mathcal{O}_1 \rangle$ and $\langle X_2, \mathcal{O}_2 \rangle$ are topological spaces, $\langle X_2, \mathcal{O}_2 \rangle$ is Hausdorff and $f : X_1 \longrightarrow X_2$ is continuous. If A is a compact subset of X_1, then the subset $f(A)$ of X_2 is compact as well.*

Proof. Let $\mathcal{A}$ be an open cover of $f(A)$. Then $\mathcal{B} = \{f^{-1}(U) : U \in \mathcal{A}\}$ is an open cover of the set A and therefore there exists a finite subcover $\mathcal{B}_0 \subseteq \mathcal{B}$. Then

$$\{\{U \in \mathcal{A} : f^{-1}(U) = V\} : V \in \mathcal{B}_0\}$$

is a finite family of non-empty sets. Thus there exists a choice function for this family. The range of this choice function is a finite subcover of $\mathcal{A}$. $\qquad \square$

Corollary 1.31. *If f is a one-to-one continuous mapping of a compact topological space $\langle X_1, \mathcal{O}_1 \rangle$ onto a Hausdorff topological space $\langle X_2, \mathcal{O}_2 \rangle$, then f is a homeomorphism.*

A Hausdorff topological space $\langle X, \mathcal{O} \rangle$ is **locally compact** if for every point $x \in X$ and every neighborhood U of x there exists a compact neighborhood V of x such that $V \subseteq U$. It is easy to see that a Hausdorff space is locally compact if and only if for any $x \in X$ there exists a compact neighborhood of x. A Hausdorff topological space $\langle X, \mathcal{O} \rangle$ is σ-**compact** if there are compact sets $\langle X_n \subseteq X : n \in \omega \rangle$ such that $X = \bigcup_n X_n$. Similarly, a set $A \subseteq X$ is σ-**compact** if there are compact sets $\langle A_n \subseteq A : n \in \omega \rangle$ such that $A = \bigcup_n A_n$.

Theorem 1.32. *A locally compact space with a countable base of topology is σ-compact.*

The proof is easy. Let $\mathcal{B} = \{U_n : n \in \omega\}$ be a countable base of topology. We set

$$X_n = \bigcup\{\overline{U}_k : k < n \wedge \overline{U}_k \text{ is compact}\}.$$

Evidently every X_n is compact and $\bigcup_n X_n = X$. $\qquad\square$

A set $A \subseteq X$ is called **disconnected** if there exist open sets U, V such that $U \cap A \neq \emptyset$, $V \cap A \neq \emptyset$, $U \cap V \cap A = \emptyset$ and $A \subseteq U \cup V$. If A is not disconnected, then A is called **connected**.

Theorem 1.33. *If $f : X_1 \longrightarrow X_2$ is continuous, $A \subseteq X_1$ is a connected set, then also the set $f(A)$ is connected.*

A topological space X is **zero-dimensional**, written $\mathrm{ind}(X) = 0$, if there exists a base of topology consisting of clopen sets, i.e., if for any $x \in X$ and any neighborhood U of x there exists a clopen set V such that $x \in V \subseteq U$. A subspace of a zero-dimensional space is zero-dimensional as well. A topological space X has the **large inductive dimension zero**, written $\mathrm{Ind}(X) = 0$, if for any disjoint closed sets $A, B \subseteq X$ there exists a clopen set U such that $A \subseteq U$ and $U \cap B = \emptyset$. Note that such a space is normal.

Theorem 1.34. *A compact zero-dimensional topological space has large inductive dimension zero.*

Proof. Let A, B be closed disjoint subsets of X. We set

$$\mathcal{U} = \{U \subseteq X : U \text{ clopen} \wedge U \cap A \neq \emptyset \wedge U \cap B = \emptyset\}.$$

Since X is zero-dimensional, $\mathcal{U}$ is a clopen cover of A. If $\{U_0, \ldots, U_n\} \subseteq \mathcal{U}$ is a finite subcover of A, then $U = U_0 \cup \cdots \cup U_n$ is the desired set containing the set A and disjoint with B. $\qquad\square$

A set $A \subseteq X$ is **nowhere dense** if $\mathrm{Int}(\overline{A}) = \emptyset$. Note that the closure of a nowhere dense set is nowhere dense. A set A is **meager** if there exists a sequence $\{A_n\}_{n=0}^{\infty}$ of nowhere dense sets such that $A \subseteq \bigcup_n A_n$. Evidently a meager set is a subset of a meager F_σ set. A meager set is also called a set of the **first Baire category**. A set $A \subseteq X$ is called **comeager** if $X \setminus A$ is meager.

Theorem 1.35. *For any topological space $\langle X, \mathcal{O} \rangle$ the following are equivalent:*

  a) *No non-empty open set is meager.*
  b) *$\bigcap_n A_n$ is dense in X for any sequence $\{A_n\}_{n=0}^{\infty}$ of open dense subsets.*
  c) *$X \setminus A$ is dense in X for every meager $A \subseteq X$.*

A topological space $\langle X, \mathcal{O} \rangle$ is said to **satisfy the Baire Category Theorem** if any of the conditions a)–c) of Theorem 1.35 holds true, e.g., if no non-empty open subset is of the first Baire category.

Note the following simple fact: if $\langle X, \mathcal{O} \rangle$ satisfies the Baire Category Theorem, then an F_σ-subset $A \subseteq X$ is meager if and only if $\mathrm{Int}(A) = \emptyset$.

Theorem 1.36 (E. Čech). *A locally compact space with a countable base of topology satisfies the Baire Category Theorem.*

Proof. Let $\mathcal{B}$ be a countable base of a locally compact space $\langle X, \mathcal{O} \rangle$. Let $\{A_n\}_{n=0}^{\infty}$ be a sequence of open sets dense in X. We show that $U \cap \bigcap_n A_n \neq \emptyset$ for any non-empty open set U.

Fix an enumeration of the base $\mathcal{B}$. Let U_0 be the first element of $\mathcal{B}$ such that $\overline{U}_0 \subseteq U \cap A_0$ and $\overline{U}_0$ is compact. By induction, if U_n is defined, let U_{n+1} be the first element of $\mathcal{B}$ such that $\overline{U}_{n+1} \subseteq U_n \cap A_{n+1}$. Then

$$\emptyset \neq \bigcap_n \overline{U}_n \subseteq U \cap \bigcap_n A_n.$$

Hence $\bigcap_n A_n$ is dense. $\qquad\square$

Assuming the Axiom of Choice, the theorem holds true for any locally compact space, see Exercise 1.23.

The family of all topologies on a non-empty set is ordered by the inclusion. A topology $\mathcal{O}_1$ on X is called **weaker** than the topology $\mathcal{O}_2$ on X if $\mathcal{O}_1 \subseteq \mathcal{O}_2$. We say also that $\mathcal{O}_2$ is **coarser** than $\mathcal{O}_1$. The weakest topology on a set X is $\{\emptyset, X\}$. The coarsest topology on X is the **discrete topology** $\mathcal{P}(X)$. Very often if there exists a topology with a given property, then there exists the weakest topology with this property. Especially, if $\langle \langle X_s, \mathcal{O}_s \rangle : s \in S \rangle$ are topological spaces and $\langle f_s : X \longrightarrow X_s : s \in S \rangle$ are mappings, then there exists the weakest topology on X such that all $\langle f_s : s \in S \rangle$ are continuous.

Let $\langle \langle X_s, \mathcal{O}_s \rangle : s \in S \rangle$ be topological spaces. The weakest topology on the Cartesian product $\Pi_{s \in S} X_s$ such that all projections $\langle \mathrm{proj}_s : s \in S \rangle$ are continuous is called the **product topology** and denoted by $\Pi_{s \in S} \mathcal{O}_s$. The family of all sets $\Pi_{s \in S} U_s$, where the sets $U_s \subseteq X_s$ are open and $\{s \in S : U_s \neq X_s\}$ is finite, is a base of the product topology. Thus, if $\langle X_1, \mathcal{O}_1 \rangle$ and $\langle X_2, \mathcal{O}_2 \rangle$ are topological spaces, then the family $\{A_1 \times A_2 : A_1 \in \mathcal{O}_1, A_2 \in \mathcal{O}_2\}$ is a base for the product topology $\mathcal{O}_1 \times \mathcal{O}_2$ on $X_1 \times X_2$.

Theorem [AC] 1.37 (A. Tychonoff). *Let $\langle \langle X_s, \mathcal{O}_s \rangle : s \in S \rangle$ be Hausdorff topological spaces. Then the product space $\langle \Pi_{s \in S} X_s, \Pi_{s \in S} \mathcal{O}_s \rangle$ is compact if and only if every space $\langle \langle X_s, \mathcal{O}_s \rangle : s \in S \rangle$ is compact.*

The use of **AC** in the proof is essential. Actually, **AC** is equivalent to the assertion "if $\langle X_s : s \in S \rangle$ are non-empty sets, then the product $\Pi_{s \in S} X_s$ is non-empty".

If $f : X_1 \longrightarrow X_2$, then $f \subseteq X_1 \times X_2$ can be considered as the graph of f.

Theorem 1.38. *Assume that $\langle X_1, \mathcal{O}_1 \rangle$, $\langle X_2, \mathcal{O}_2 \rangle$ are Hausdorff topological spaces and $f : X_1 \longrightarrow X_2$ is continuous. Then f is a closed subset of $X_1 \times X_2$.*

A **convergence structure** on a set X is a mapping $\lim : \mathcal{X} \longrightarrow X$ from a set $\mathcal{X} \subseteq {}^{\omega}X$. A sequence $\{x_n\}_{n=0}^{\infty}$ belonging to $\mathcal{X}$ is called **convergent** and the value $\lim(\{x_n\}_{n=0}^{\infty})$ is called the **limit** of it and denoted $\lim_{n \to \infty} x_n$. A set X endowed with a convergence structure lim is called an $\mathcal{L}^*$-**space** if the following conditions are satisfied:

(L1)   if $x_n = x$ for every n, then $\lim_{n\to\infty} x_n = x$;

(L2)   if $\lim_{n\to\infty} x_n = x$ and $\{n_k\}_{k=0}^\infty$ is increasing, then $\lim_{k\to\infty} x_{n_k} = x$;

(L3)   if $x \neq \lim_{n\to\infty} x_n$, then there exists a subsequence $\{x_{n_k}\}_{k=0}^\infty$ such that no subsequence of $\{x_{n_k}\}_{k=0}^\infty$ has limit x.

One can easily show that an $\mathcal{L}^*$-space has property (1.18).

Assume that X is an $\mathcal{L}^*$-space. For any subset $A \subseteq X$ we define the **sequential closure** of A by

$$\mathrm{scl}\,(A) = \{\, \lim_{n\to\infty} x_n : x_n \in A \text{ for every } n\}.$$

The **sequential closure** $\mathrm{scl}_\xi\,(A)$ **of order** ξ is defined by transfinite induction:

$$\mathrm{scl}_0\,(A) = A, \quad \mathrm{scl}_\xi\,(A) = \mathrm{scl}\left(\bigcup_{\eta<\xi} \mathrm{scl}_\eta\,(A)\right) \text{ for } \xi > 0. \tag{1.20}$$

If ω_1 is regular, then

$$\mathrm{scl}_{\omega_1}\,(A) = \bigcup_{\xi<\omega_1} \mathrm{scl}_\xi\,(A),$$

and therefore $\mathrm{scl}_{\omega_1}\,(A) = \mathrm{scl}_\xi\,(A)$ for any $\xi \geq \omega_1$.

Every topological space $\langle X, \mathcal{O}\rangle$ may be endowed with a convergence structure – the topological limit operation $\lim_{n\to\infty} x_n$. The set X with this convergence structure is an $\mathcal{L}^*$-space. Moreover, if X is a topological space, then $\mathrm{scl}\,(A) \subseteq \overline{A}$ for any $A \subseteq X$. The inverse inclusion need not be true. A topological space $\langle X, \mathcal{O}\rangle$ is a **Fréchet space** if $\overline{A} = \mathrm{scl}\,(A)$ for every set $A \subseteq X$, i.e., if for every $A \subseteq X$ and for every $x \in \overline{A}$ there exists a sequence $\{x_n\}_{n=0}^\infty$ of points of the set A such that $\lim_{n\to\infty} x_n = x$. Then $\mathrm{scl}_\xi\,(A) = \mathrm{scl}\,(A)$ for any $\xi > 0$.

By Theorem 1.23 we can easily show that

Theorem [wAC] 1.39. *A Hausdorff topological space with a countable base is a Fréchet space.*

We shall need the following result.

Theorem [wAC] 1.40. *Assume that X is an $\mathcal{L}^*$-space, $|X| \ll 2^{\aleph_0}$. Then the following conditions are equivalent:*

a) *If $\lim_{n\to\infty} x_n = x$, $\lim_{m\to\infty} x_{n,m} = x_n$ for every n, then there exist sequences $\{n_k\}_{k=0}^\infty$ and $\{m_k\}_{k=0}^\infty$ such that $\lim_{k\to\infty} x_{n_k,m_k} = x$.*

b) *If $\lim_{n\to\infty} x_n = x$, $\lim_{m\to\infty} x_{n,m} = x_n \neq x$ for every n, then there exist sequences $\{n_k\}_{k=0}^\infty$ and $\{m_k\}_{k=0}^\infty$ such that $\{n_k\}_{k=0}^\infty$ is increasing and $\lim_{k\to\infty} x_{n_k,m_k} = x$.*

c) *If $\lim_{n\to\infty} x_n = x$, $\lim_{m\to\infty} x_{n,m} = x_n \neq x$ for every n, then there exist increasing sequences $\{n_k\}_{k=0}^\infty$ and $\{m_k\}_{k=0}^\infty$ such that $\lim_{k\to\infty} x_{n_k,m_k} = x$.*

d) $\mathrm{scl}_1\,(A) = \mathrm{scl}_2\,(A)$ *for any $A \subseteq X$.*

The proof is easy. One can show that a) → b) → c) → d) → a). The implication c) → a) is trivial. So the equivalences a) ≡ b) ≡ c) can be proved in **ZF**. For a proof of c) → d) one needs **wAC**. □

The condition a), therefore any of its equivalent conditions b), c) and d), is called the **sequence selection property**, shortly SSP.

Corollary [wAC] 1.41. *The convergence structure of a Fréchet topological space of cardinality ≪ c possesses the sequence selection property.*

If X is a topological group, then X possesses SSP if and only if the following condition e) is satisfied:

e) If $\lim_{m\to\infty} x_{n,m} = e$ for every n, then there exist increasing sequences $\{n_k\}_{k=0}^\infty$ and $\{m_k\}_{k=0}^\infty$ such that $\lim_{k\to\infty} x_{n_k,m_k} = e$.

Indeed, if $\lim_{n\to\infty} x_n = x$, $\lim_{m\to\infty} x_{n,m} = x_n$ for every n, set $y_{n,m} = x_{n,m} \circ x_n^{-1}$. If $\lim_{k\to\infty} y_{n_k,m_k} = e$, then $\lim_{k\to\infty} x_{n_k,m_k} = x$.

In what follows we shall use this fact without any commentary.

By a **property** of a subset A of a topological space $\langle X, \mathcal{O} \rangle$ we understand a formula φ such that for a given set A the sentence $\varphi(A, X, \mathcal{O})$ is true or false. A property φ is called **topological** if for any topological spaces $\langle X_i, \mathcal{O}_i \rangle$, $i = 1, 2$ and any homeomorphism $f : X_1 \longrightarrow X_2$, the sentences $\varphi(A, X_1, \mathcal{O}_1)$ and $\varphi(f(A), X_2, \mathcal{O}_2)$ are equivalent for any $A \subseteq X_1$. Evidently properties "to be open", "to be closed", "to be nowhere dense", "to be compact" are topological. Actually, all properties introduced in this section are topological.

Similarly, a property $\psi(X, \mathcal{O})$ of a topological space $\langle X, \mathcal{O} \rangle$ is **topological** if the sentences $\psi(X_1, \mathcal{O}_1)$ and $\psi(X_2, \mathcal{O}_2)$ are equivalent for any homeomorphic spaces $\langle X_1, \mathcal{O}_1 \rangle$ and $\langle X_2, \mathcal{O}_2 \rangle$. We can also consider the property ψ of a subset A of a topological space $\langle X, \mathcal{O} \rangle$ by considering the property $\psi(A, \mathcal{O}|A)$. That was our definition of the notion "a compact set".

It may happen that a property of a subset of a topological space is actually a property of a topological space. E.g., we could define that a subset A of a Hausdorff topological space $\langle X, \mathcal{O} \rangle$ is compact if from any open cover of A one can choose a finite subcover of A. That is a property of the subset A of a topological space $\langle X, \mathcal{O} \rangle$. However, it is well known that this definition is equivalent with our definition as a property of a topological space. Also, we can define a notion of a connected topological space and then show that a subset A of a topological space $\langle X, \mathcal{O} \rangle$ is connected according to our definition if and only if the topological space $\langle A, \mathcal{O}|A \rangle$ is connected.

In the rest of this section we assume that the reader is familiar with some elementary logic, see Sections 11.4 and 11.5. Topological properties essentially fall into two disjoint classes – internal and external properties. It may happen that a set A does have a property when considered as a subset of a topological space X_1 and A (or its homeomorphic copy) does not have the property when considered as a subset of another – not homeomorphic – topological space X_2. We begin with an exact definition. Let us consider a class $\mathbf{S}$ of topological spaces. We want to say that a topological property φ is internal[6] in a class $\mathbf{S}$, if for any topological spaces $\langle X_i, \mathcal{O}_i \rangle$ in $\mathbf{S}$ and any subsets $A_i \subseteq X_i$, $i = 1, 2$ such that $\langle A_1, \mathcal{O}_1 | A_1 \rangle$ is homeomorphic to $\langle A_2, \mathcal{O}_2 | A_2 \rangle$, the set A_1 possesses the property φ if and only if A_2 possesses the property φ, i.e., if $\varphi(A_1, X_1, \mathcal{O}_1)$ is equivalent to $\varphi(A_2, X_2, \mathcal{O}_2)$. A careful reader must remark that this is actually a metamathematical notion and we should be precise in which theory the equivalence of $\varphi(A_1, X_1, \mathcal{O}_1)$ and $\varphi(A_2, X_2, \mathcal{O}_2)$ is provable. Thus the definition must be as follows. Let $\mathbf{T}$ be an extension of the set theory $\mathbf{ZF}$. A property φ is **internal for a class $\mathbf{S}$ in a theory $\mathbf{T}$** if[7]

$$\mathbf{T} \vdash (\forall \langle X_i, \mathcal{O}_i \rangle \in \mathbf{S}, i = 1, 2)(\forall A_i \subseteq X_i, i = 1, 2)\,((A_1 \text{ homeomorphic to } A_2)$$
$$\rightarrow (\varphi(A_1, X_1, \mathcal{O}_1) \equiv \varphi(A_2, X_2, \mathcal{O}_2))). \tag{1.21}$$

A topological property that is not internal should be external. However that is confusing, since we do not know how to understand it: the equivalence in (1.21) is not provable or the negation is provable? We prefer to define that a property φ is **external for a class $\mathbf{S}$ in a theory $\mathbf{T}$** if

$$\mathbf{T} \vdash \neg(\forall \langle X_i, \mathcal{O}_i \rangle \in \mathbf{S}, i = 1, 2)(\forall A_i \subseteq X_i, i = 1, 2)\,((A_1 \text{ homeomorphic to } A_2)$$
$$\rightarrow (\varphi(A_1, X_1, \mathcal{O}_1) \equiv \varphi(A_2, X_2, \mathcal{O}_2))). \tag{1.22}$$

It is easy to see that properties "to be open", "to be closed", "to be nowhere dense" are external topological properties. They were intended as properties of sets.

Assume that $\mathbf{S}_2 \subseteq \mathbf{S}_1$ and the theory $\mathbf{T}_2$ is an extension of $\mathbf{T}_1$. Then any property internal for the class $\mathbf{S}_1$ in the theory $\mathbf{T}_1$ is an internal property for the class $\mathbf{S}_2$ in the theory $\mathbf{T}_2$ as well. Similarly, any property external for the class $\mathbf{S}_2$ in the theory $\mathbf{T}_1$ is an external property for the class $\mathbf{S}_1$ in the theory $\mathbf{T}_2$.

One can easily find a property that is neither internal nor external for a suitable class and the set theory $\mathbf{ZFC}$. E.g., let $\varphi(A, X, \mathcal{O})$ denote the formula – see Section 11.5

$$(A \text{ is an open subset of } X) \vee \mathbf{V} = \mathbf{L}.$$

Evidently φ is a topological property. If $\mathbf{S}$ denotes the class of all topological spaces, then the sentence of (1.21) is undecidable in $\mathbf{ZFC}$. Thus φ is neither internal nor external for the class of all topological spaces in $\mathbf{ZFC}$.

Assume now that for any topological space $\langle X, \mathcal{O} \rangle$ from a class $\mathbf{S}$ and any $A \subseteq X$ the topological property φ satisfies (in some extension $\mathbf{T}$ of $\mathbf{ZF}$) the following condition:

$$\varphi(A, X, \mathcal{O}) \text{ holds true if and only if } \varphi(A, A, \mathcal{O} | A) \text{ holds true.} \tag{1.23}$$

---

[6] An internal property is different from the notion of an internal definition considered by topologists, see, e.g., R. Engelking [1977].
[7] Note that all objects of the definition φ, $\mathbf{S}$ and T are objects of metamathematics. Thus the notion of an internal property is a notion of metamathematics!

In other words, the property φ is formulated in the terms of the subspace topology on the set A as a property of a topological space. Evidently in such a case φ is an internal property for the class $\mathbf{S}$ in the theory $\mathbf{T}$.

Vice versa, if φ is an internal property for the class $\mathbf{S}$ of all topological spaces in $\mathbf{T}$, then (1.23) holds true. However the condition that $\mathbf{S}$ is the class of all topological spaces is essential. We show it. Let $\mathbf{S}$ denote the class of all topological spaces, φ being the property

$$(A \subseteq X \text{ is infinite}) \wedge (\langle X, \mathcal{O} \rangle \text{ is compact}).$$

Then φ is internal for $\mathbf{S}$ in any extension $\mathbf{T}$ of $\mathbf{ZF}$, however (1.23) fails. Even so, sometimes we prefer to define an internal property as a property of a set and then prove that (1.23) holds true. In such a case we emphasize the fact that an introduced property of a set is internal for a class $\mathbf{S}$ in a theory $\mathbf{T}$.

The topologists usually distinguish between internal and external property by speaking about a property of a topological space and a property of a set, respectively[8]. Sometimes trying to stress that a property is external, we say that a set possesses a property relative to a space. We present a typical example: a set $A \subseteq X$ is said to be relatively compact if the closure $\overline{A}$ is compact. This property is external, since it essentially depends on the space X in any extension of $\mathbf{ZF}$.

Important internal and external properties will be investigated in Chapter 8.

Exercises

1.15 Closure Operator

Assume that, for any set $A \subseteq X$, there is defined a closure $\overline{A}$ such that for any $A, B \subseteq X$ the following hold true:

$$\overline{\emptyset} = \emptyset, \qquad A \subseteq \overline{A} = \overline{\overline{A}}, \qquad \overline{A \cup B} = \overline{A} \cup \overline{B}.$$

a) Show that $\overline{A} \subseteq \overline{B}$ for any $A \subseteq B \subseteq X$.

b) The family $\mathcal{O} = \{A \subseteq X : \overline{X \setminus A} = X \setminus A\}$ is a topology on X.

c) For any $A \subseteq X$, the set $\overline{A}$ is the closure of A in the sense of the topology $\mathcal{O}$.

1.16 Limit of a Filter

Let $\langle X, \mathcal{O} \rangle$ be a topological space. For a point $a \in X$ we denote by $\mathcal{N}(a)$ the filter of all neighborhoods of a. Let $\mathcal{F}$ be a filter of subsets of X. A point $a \in X$ is a **limit of the filter** $\mathcal{F}$, written $a = \lim \mathcal{F}$, if $\mathcal{N}(a) \subseteq \mathcal{F}$. The point a is a **cluster point of the filter** $\mathcal{F}$ if $a \in \overline{A}$ for any $A \in \mathcal{F}$.

a) $\langle X, \mathcal{O} \rangle$ is Hausdorff if and only if any filter on X has at most one limit.

b) A limit of a filter is a cluster point of it.

c) A cluster point of an ultrafilter is a limit of it.

---

[8]Check the careful distinctions made by K. Kuratowski [1958a].

d) Let $A \subseteq X$. Then $a \in \overline{A}$ if and only if there exists a filter $\mathcal{F}$ such that $A \in \mathcal{F}$ and $\lim \mathcal{F} = a$.

   Hint: Take the filter generated by $\mathcal{N}(a)$ and the set A.

e) Assuming **BPI**, a point a is a cluster point of a filter $\mathcal{F}$ if and only if there exists an ultrafilter $\mathcal{F}' \supseteq \mathcal{F}$ such that a is a limit of $\mathcal{F}'$.

f) A Hausdorff space $\langle X, \mathcal{O} \rangle$ is compact if and only if every filter on X has a cluster point.

g) Assume **BPI**. Then a Hausdorff space $\langle X, \mathcal{O} \rangle$ is compact if and only if every ultra-filter on X has a limit.

1.17 Net

Let $\langle X, \mathcal{O} \rangle$ be a topological space, $A \subseteq X$. A triple $\langle E, \leq, f \rangle$ is called a **net** in A if $\langle E, \leq \rangle$ is a directed poset and $f : E \longrightarrow A$. If $F \subseteq E$ is a cofinal subset of E, then $\langle F, \leq, f|F \rangle$ is a **subnet** of the net $\langle E, \leq, f \rangle$. If $E = \omega$, then a net is simply a sequence. A point $a \in X$ is a **limit of the net** $\langle E, \leq, f \rangle$, written $\lim_{x \in E, \leq} f(x) = a$, (simply $\lim_{x \in E} f(x) = a$ when the ordering $\leq$ is understood) if for any neighborhood U of a there exists an $x_0 \in E$ such that $f(x) \in U$ for any $x \geq x_0$. The point a is a **cluster point of the net** $\langle E, \leq, f \rangle$ if for any neighborhood U of a and for any $x_0 \in E$ there exists an $x \geq x_0$ such that $f(x) \in U$.

The net $\langle E_1, \leq_1, f_1 \rangle$ is **finer** than the net $\langle E_2, \leq_2, f_2 \rangle$ if there exists a function $\varphi : E_1 \longrightarrow E_2$ such that $f_1 = \varphi \circ f_2$ and for every $x_0 \in E_2$ there exists a $y_0 \in E_1$ such that $\varphi(y) \geq_2 x_0$ whenever $y \geq_1 y_0$.

a) Let $g : X \longrightarrow Y$. If $\langle E, \leq, f \rangle$ is a net in X, then $\langle E, \leq, f \circ g \rangle$ is a net in Y.

b) If the mapping $g : X \longrightarrow Y$ is continuous at a point $a \in X$ and $\lim_{x \in E} f(x) = a$, then $\lim_{x \in E} g(f(x)) = g(a)$.

c) If $F \subseteq E$ is cofinal in E, then the subnet $\langle F, \leq, f|F \rangle$ is finer than the net $\langle E, \leq, f \rangle$.

d) Assume that the net $\langle E_1, \leq_1, f_1 \rangle$ is finer than a net $\langle E_2, \leq_2, f_2 \rangle$. Show that:

   (i) if $a = \lim_{x \in E_1} f_1(x)$, then a is a cluster point of the net $\langle E_2, \leq_2, f_2 \rangle$;

   (ii) if a is a cluster point of the net $\langle E_1, \leq_1, f_1 \rangle$, then a is a cluster point of the net $\langle E_2, \leq_2, f_2 \rangle$;

   (iii) if $b = \lim_{x \in E_2} f_2(x)$, then $a = \lim_{x \in E_1} f_1(x)$.

e) Assuming **AC** show that if a is a cluster point of a net $\langle E_2, \leq_2, f_2 \rangle$, then there is a finer net $\langle E_1, \leq_1, f_1 \rangle$ such that $a = \lim_{x \in E_1} f_1(x)$.

   Hint: Set

   $$E_1 = \{ \langle x, U \rangle : x \in E_2 \wedge f_2(x) \in U \wedge U \text{ neighborhood of } a \}$$

   ordered as

   $$\langle x_1, U_1 \rangle \leq_1 \langle x_2, U_2 \rangle \equiv x_1 \leq_2 x_2 \wedge U_2 \subseteq U_1.$$

   Let $\varphi(\langle x, U \rangle) = x$.

f) Assuming **AC** show that $a \in \overline{A}$ if and only if there exists a net $\langle E, \leq, f \rangle$ in A such that $a = \lim_{x \in E} f(x)$.

   Hint: Take $E = \mathcal{N}(a)$.

g) Assuming **AC** prove the **Heine Criterion**: a mapping $g : X \longrightarrow Y$ is continuous at a point $a \in X$ if and only if, for any net $\langle E, \leq, f \rangle$ in X such that $\lim_{x \in E, \leq} f(x) = a$, also $\lim_{x \in E, \leq} g(f(x)) = g(a)$.

If $\mathcal{F}$ is a filter on a set X and $f : X \longrightarrow Y$, then we write $f[\mathcal{F}] = \{A \subseteq Y : f^{-1}(A) \in \mathcal{F}\}$.

h) A point $a \in X$ is a limit of the net $\langle E, \leq, f \rangle$ if and only if a is the limit of the filter $f[\mathcal{F}(E)]$ (Fréchet filter $\mathcal{F}(E)$ is defined in Exercise 11.2).

i) Assuming **AC** show that a Hausdorff space $\langle X, \mathcal{O} \rangle$ is compact if and only if every net in X has a cluster point.
Hint: Use Exercise 1.16*, f).*

1.18 Tychonoff's Theorem

a) Assuming **AC** prove Tychonoff's Theorem 1.37.
Hint: If $\mathcal{F}$ is an ultrafilter on $\Pi_{s \in S} X_s$, x_s is a limit of $\mathrm{proj}_s[\mathcal{F}]$ for $s \in S$, then $\lim \mathcal{F} = x \in \Pi_{s \in S} X_s$, where $x(s) = x_s$ for $s \in S$.

Let $\langle X_i, \mathcal{O}_i \rangle$, $i = 1, 2$ be compact topological spaces. Let $X = X_1 \times X_2$, $\mathcal{O} = \mathcal{O}_1 \times \mathcal{O}_2$. Let $\mathcal{U}$ be an open cover of X consisting of sets of the form $U_1 \times U_2$, where $U_i \in \mathcal{O}_i$, $i = 1, 2$.

b) For any $x \in X_1$ there exist an open set $V \ni x$ and a finite set $\mathcal{U}_0 \subseteq \mathcal{U}$ such that $V \times X_2 \subseteq \bigcup \mathcal{U}_0$.

c) Prove that there exists a finite subcover of $\mathcal{U}$ covering X.
Hint: For any non-empty open $V \subseteq X_1$ set

$$\mathcal{A}(V) = \{\mathcal{U}_0 \subseteq \mathcal{U} : V \times X_2 \subseteq \bigcup \mathcal{U}_0 \wedge \mathcal{U}_0 \text{ is finite}\}.$$

Show that $\{V \in \mathcal{O}_1 : \mathcal{A}(V) \neq \emptyset\}$ is a cover of X_1. If $\{V_0, \ldots, V_n\}$ is a finite subcover, one can choose from the finite set $\{\mathcal{A}(V_0), \ldots, \mathcal{A}(V_n)\}$ a finite subcover of $\mathcal{U}$.

d) Prove in **ZF** that a topological product of finitely many compact spaces is a compact space.

1.19 Countability Axioms

If for any $x \in X$, the neighborhood filter $\mathcal{N}(x)$ has a countable base, then we say that $\langle X, \mathcal{O} \rangle$ satisfies the **first axiom of countability** or is **first countable**. If there exists a countable base of the topology $\mathcal{O}$, then we say that the space $\langle X, \mathcal{O} \rangle$ satisfies the **second axiom of countability** or is **second countable**. X is **Lindelöf** if any open cover of X contains a countable subcover.

a) The first axiom of countability follows from the second one and not vice versa.
Hint: Take a suitable discrete space.

b) Let X satisfy the first axiom of countability and assume $\mathbf{AC}^\omega$. Then $\overline{A} = \mathrm{scl}(A)$ for any $A \subseteq X$, i.e., X is Fréchet.

c) If $|X| \ll 2^{\aleph_0}$, then in b), $\mathbf{AC}^\omega$ can be replaced by **wAC**.

d) Assume $\mathbf{AC}^\omega$. If a Hausdorff space $\langle X, \mathcal{O} \rangle$ satisfies the second axiom of countability, then X is compact if and only if every sequence in X has an accumulation point.
Hint: Use Exercise 1.16*, g).*

e) The second axiom of countability cannot be replaced by the first one in d).
Hint: Consider ω_1 with interval topology.

f) If **AC** holds true, then a second countable topological space is Lindelöf.

g) A zero-dimensional Lindelöf space X has large inductive dimension zero. If moreover X is Hausdorff, then X is normal.

Hint: If $A, B \subseteq X$ are disjoint closed, take for any $x \in X$ a clopen set $W_x \ni x$ such that $W_x \cap A = \emptyset$ or $W_x \cap B = \emptyset$. Take $\{x_n\}_{n=0}^{\infty}$ such that $\bigcup_n W_{x_n} = X$. Set $U_n = W_{x_n} \setminus \bigcup_{i<n} W_{x_i}$. Then the set $\bigcup\{U_n : U_n \cap A \neq \emptyset\}$ is clopen containing A and disjoint with B.

1.20 Compactification

Let $\langle X, \mathcal{O}\rangle$ and $\langle Y, \mathcal{Q}\rangle$ be topological spaces, $f : X \xrightarrow{1-1} Y$ be a homeomorphism of $\langle X, \mathcal{O}\rangle$ onto $\langle f(X), \mathcal{Q}|f(X)\rangle$. The triple $\langle Y, \mathcal{Q}, f\rangle$ is called a **compactification of** $\langle X, \mathcal{O}\rangle$ if Y is compact and $\overline{f(X)} = Y$. A compactification $\langle Y_1, \mathcal{Q}_1, f_1\rangle$ is **weaker** than the compactification $\langle Y_2, \mathcal{Q}_2, f_2\rangle$ if there exists a continuous mapping $g : Y_2 \xrightarrow{onto} Y_1$ such that $f_1 = f_2 \circ g$. Then compactification $\langle Y_2, \mathcal{Q}_2, f_2\rangle$ is **stronger** than $\langle Y_1, \mathcal{Q}_1, f_1\rangle$.

a) Let $\langle X, \mathcal{O}\rangle$ be locally compact and not compact. Set $Y = X \cup \{\infty\}$ (we assume that $\infty \notin X$). A set $A \subseteq Y$ belongs to $\mathcal{Q}$ if either $A \in \mathcal{O}$ or $\infty \in A$ and $X \setminus A$ is compact. Show that $\langle Y, \mathcal{Q}, \mathrm{id}_X\rangle$ is a compactification of $\langle X, \mathcal{O}\rangle$.

b) Show that the compactification constructed in part a) is the weakest one.

c) If $\langle Y, \mathcal{Q}\rangle$ is compact, $X \subseteq Y$, then $\langle \overline{X}, \mathcal{Q}|\overline{X}, \mathrm{id}_X\rangle$ is a compactification of $\langle X, \mathcal{Q}|X\rangle$.

d) If $\langle Y_i, \mathcal{Q}_i, f_i\rangle$, $i = 1, 2$ are compactifications of a topological space $\langle X, \mathcal{O}\rangle$, then there exists a compactification $\langle Y, \mathcal{Q}\rangle$ stronger than both $\langle Y_i, \mathcal{Q}_i, f_i\rangle$, $i = 1, 2$.
Hint: Define $h : X \longrightarrow Y_1 \times Y_2$ by $h(x) = \langle f_1(x), f_2(x)\rangle$. Take $Y = \overline{h(X)}$.

1.21 [AC] The Strongest Compactification

a) Show that for any system $\langle\langle Y_s, \mathcal{Q}_s, f_s\rangle, s \in S\rangle$ of compactifications of $\langle X, \mathcal{O}\rangle$ there exists a compactification stronger than any of them.
Hint: Embed X in the Tychonoff product of $\langle\langle Y_s, \mathcal{Q}_s, f_s\rangle, s \in S\rangle$ and take the closure of the image by the embedding.

b) The class of all compactifications of a topological space X is a set.
Hint: If Y is a compactification of X, then by Exercise (1.16) every point of Y is a limit of a filter on X. Hence $|Y| \leq 2^{2^{|X|}}$. Therefore the class of all compactifications has cardinality at most $2^{2^{2^{|X|}}}$.

c) Conclude that if there exists a compactification of a topological space $\langle X, \mathcal{O}\rangle$, then there exists the strongest one.
Hint: By b) The class of all compactifications of a topological space X is a set. Apply the result of a).

1.22 Normal Space with a Non-normal Subspace

If X is an infinite set endowed with the discrete topology, $\infty \notin X$, we denote by $\mathcal{O}_X$ the topology on $X \cup \{\infty\}$ defined in Exercise 1.20, a).

a) If X, Y are infinite sets, then $\langle(X \cup \{\infty\}) \times (Y \cup \{\infty\}), \mathcal{O}_X \times \mathcal{O}_Y\rangle$ is normal.

b) $X \times \{\infty\}$ and $\{\infty\} \times Y$ are disjoint closed subsets of the space

$$Z = (X \cup \{\infty\}) \times (Y \cup \{\infty\}) \setminus \{\langle\infty, \infty\rangle\}$$

endowed with the subspace topology.

c) If $|X| = \aleph_0$, $|Y| = \mathfrak{c}$ and $E \subseteq X \times Y$ is such that $|\{y \in Y : \langle x, y \rangle \in E\}| < \aleph_0$ for every $x \in X$, then $|\{y \in Y : (\exists x \in X) \langle x, y \rangle \in E\}| \leq \aleph_0 < \mathfrak{c}$.

   Hint: We can assume that $X = \omega$ and $Y = \mathbb{R}$. Let $E_n = \{y \in \mathbb{R} : \langle n, y \rangle \in E\}$ be the vertical section. All E_n being finite are well ordered by the usual ordering $\leq$ of $\mathbb{R}$. Therefore there exists a unique order preserving mapping f_n from E_n onto a natural number. Use a pairing function to show that the set $\{y \in Y : (\exists x \in X) \langle x, y \rangle \in E\}$ is countable.

d) Assume $\aleph_0 \leq |X| < |Y| = \mathfrak{c}$. If $U \subseteq Z$ is open and such that $X \times \{\infty\} \subseteq U$, then there exists a $y_0 \in Y$ such that $\langle x, y_0 \rangle \in U$ for every $x \in X$.

   Hint: For every $x \in X$ the set $\{y \in Y : \langle x, y \rangle \notin U\}$ is finite.

e) If $|X| = \aleph_0$ and $|Y| = \mathfrak{c}$, then the subspace Z of $(X \cup \{\infty\}) \times (Y \cup \{\infty\})$ is not normal.

   Hint: Use the result of d).

f) Assume **AC**. If $\aleph_0 \leq |X| < \mathrm{cf}(|Y|)$, then the subspace Z of $(X \cup \{\infty\}) \times (Y \cup \{\infty\})$ is not normal.

   Hint: There exist no open disjoint subsets of the space Z separating the sets $X \times \{\infty\}$ and $\{\infty\} \times Y$.

1.23 [AC] Topologically Complete Space

A topological space $\langle X, \mathcal{O} \rangle$ is said to be **topologically complete** if there exists a compactification $\langle Y, \mathcal{Q}, f \rangle$ such that $f(X)$ is a G_δ subset of Y.

a) If $\langle X, \mathcal{O} \rangle$ is locally compact, then X is topologically complete.

   Hint: Consider the compactification of 1.20, a).

b) Any G_δ-subset of a topologically complete space endowed with the subspace topology is topologically complete.

c) Prove the Čech Theorem: If $\langle X, \mathcal{O} \rangle$ is topologically complete, then $\langle X, \mathcal{O} \rangle$ satisfies the Baire Category Theorem.

   Hint: Follow the proof of Theorem 1.36.

d) A locally compact topological space satisfies the Baire Category Theorem.

1.24 [AC] Baire Category Theorem

A topological space $\langle X, \mathcal{O} \rangle$ is called κ-**Baire** if no non-empty open subset of X is a union of κ meager sets. A family of non-empty sets $\mathcal{A}$ is called κ-**closed** if for any decreasing chain $\langle A_\xi : \xi \in \kappa \rangle$ of elements of $\mathcal{A}$ there exists an element $A \in \mathcal{A}$ such that $A \subseteq A_\xi$ for every $\xi \in \kappa$.

a) A topological space is κ-Baire if and only if the intersection of κ many open dense sets is a dense set.

b) If a topological space $\langle X, \mathcal{O} \rangle$ has a κ-closed base (not containing the empty set), then X is κ-Baire.

c) Let $X = {}^{\omega_1}2$. For $g \in {}^A 2$, $A \subseteq \omega_1$ we set $[g] = \{f \in X : g \subseteq f\}$. Let $\mathcal{O}$ be the weakest topology on X containing all sets $[g]$, where $g \in {}^A 2$, A countable. $\langle X, \mathcal{O} \rangle$ does not satisfy the first axiom of countability.

d) Show that $\langle X, \mathcal{O} \rangle$ defined in c) has an ω-closed base, therefore satisfies the Baire Category Theorem, and is not locally compact.

1.25 [AC] Convergence Structure

Let $\langle X, \lim \rangle$ be an $\mathcal{L}^*$-space.

a) $\overline{A} = \mathrm{scl}_{\omega_1}(A)$ is a closure operator on X.

b) $\lim_{n \to \infty} x_n$ in the sense of $\mathcal{L}^*$-space is the limit in the sense of the topology defined from the closure operator $\mathrm{scl}_{\omega_1}(A)$ as well.

c) If $\langle X, \lim \rangle$ satisfies any of the conditions of Theorem 1.40, then the family

$$\mathcal{O} = \{A \subseteq X : X \setminus A = \mathrm{scl}_1(X \setminus A)\}$$

is a topology on X and such that $\lim$, in the sense of this topology, is the limit in the sense of $\mathcal{L}^*$-space as well.

Historical and Bibliographical Notes

There are many excellent textbooks on set theory. We mention just some of them. One of the first monographs on set theory was F. Hausdorff [1904] published in several editions. However, some set-theoretic results were already presented by E. Borel [1898]. The monograph K. Kuratowski and A. Mostowski [1952] is a very classical source and still relevant. The book by T. Jech [2006] contains also a systematic introduction to forcing and models of set theory with related applications. A. Lévy [1979] is really a basic set theory containing all important basic results (including an exposition on how to deal with classes). We can recommend – according to the authors' preface – "the text that takes on the form of a dialogue between the authors and the reader" W. Just and M. Weese books [1996] and [1997]. B. Balcar and P. Štěpánek [2000], written in Czech, contains many results concerning infinite combinatorics. An exposition of cardinal arithmetic at an advanced level is presented by S. Shelah [1994].

The notion of a set – die Menge – was introduced by Bernard Bolzano [1950]. George Cantor [1870] investigated uniqueness of a trigonometric series. Trying to generalize his result, he needed to describe an infinite collection of points and therefore in [1872] introduced again a notion of a set (he speaks about a "Punktmenge") and that of a countable set. G. Cantor [1874] shows that the set of algebraic reals is countable and the set of all reals is uncountable. Then he started to study systematically sets of reals and mainly, cardinalities of such sets, see [1878]. Note that G. Cantor sometimes preferred the word "Mannigfaltigkeit" instead of "Menge". The story of paradoxes that appeared in the development of naive set theory is presented in many sources, see, e.g., S.C. Kleene [1952] or A.A. Fraenkel and Y. Bar-Hillel [1958]. The book [2000] by D.A. Aczel is devoted to G. Cantor's life from the point of view of his contribution to mathematics.

The first system of axioms of set theory was introduced by E. Zermelo [1908]. The axiom scheme of replacement was invented and added to the Zermelo system by A. Fraenkel [1921]. For history see A.A. Fraenkel and Y. Bar-Hillel [1958]. The proof by transfinite induction – Theorem 1.3 – was implicitly used by G. Cantor [1895] and explicitly formulated by G. Hessenberg [1906]. The definition by transfinite induction – Theorem 1.4 – was formulated and proved by J. von Neumann [1923] and [1928]. The notion of sets of the same cardinality was introduced by B. Bolzano [1950] and G. Cantor [1878]. Theorem 1.5 was conjectured and used by G. Cantor [1895], proved by R. Dedekind in 1887, however the first published proof given by F. Bernstein has appeared in E. Borel [1898]. Cantor's Theorem 1.6 was proved in [1892]. The equalities (1.5) were proved by G. Hessenberg [1906]. The equality for multiplication was proved also independently by P.E.B. Jourdain [1908]. Theorem 1.8 was announced in A. Lindenbaum and A. Tarski [1926] as a result of A. Tarski without a proof. A proof was given by W. Sierpiński [1947], see also W. Sierpiński [1965]. Exercises 1.5 and 1.6 follow W. Sierpiński [1965]. The Pigeonhole Principle Theorem 1.9 (for finite sets) was formulated as "Schubfachprinzip" by P.G. Lejeune Dirichlet in 1834.

The implication a) → b) of Theorem 1.10 was proved by E. Zermelo [1904]. The statement c) was formulated and proved by M. Zorn [1935]. A related statement was formulated and proved earlier by K. Kuratowski [1922].

G. Cantor [1878] raised the question whether is there any set of reals whose size is strictly between that of the set of all integers and that of the set of all reals. He tried to prove a negative answer to this question and actually in [1878], p. 257 he announced that he can do it by some induction.

The axiom of dependent choice was formulated by P. Bernays [1942]. The Weak Axiom of Choice $\mathbf{wAC}$ probably appeared for the first time in J. Mycielski [1964b]. Theorem 1.15 was essentially proved by A. Tarski [1930] and by M.H. Stone [1936]. The notions of a filter and an ultrafilter as a tool for a study of convergence was explicitly introduced by H. Cartan [1937a] and investigated in [1937b]. Theorem 1.16 is essentially due to H. Lebesgue [1905]. The consistency of (1.15) was shown by S. Feferman and A. Lévy [1963].

Hartogs' function of Exercise 1.3 is defined in F. Hartogs [1915]. König's Theorem, Exercise 1.7, a) was published by J. König [1905] and independently by E. Zermelo [1908]. F. Hausdorff [1904] proved the formula of Exercise 1.7 f). The formula g) was proved by L. Bukovský [1965] and later independently by S.H. Hechler [1973]. The paper L. Bukovský [1965] is devoted to a proof of the result of Exercise 1.7, j). Tarski's recurrence formula is proved in A. Tarski [1925].

A systematic study of the various forms of the axiom of choice introduced in Exercise 1.8 is presented in T. Jech [1967]. The axiom $\mathbf{AC}_n$ was introduced and studied by A. Mostowski [1945]. W. Sierpiński [1965] contains a systematic presentation of the topics. A. Lévy [1902] investigated the various definitions of finiteness. $\mathbf{F}_3(X)$ means that X is Dedekind finite as introduced by R. Dedekind [1888]. $\mathbf{F}_1(X)$ is the definition of finiteness by A. Tarski [1924]. The Church alternatives were investigated by A. Church [1927]. P. Hájek [1965] showed that all three of the Church alternatives considered in Exercise 1.13 may occur.

Exercises 1.14 are devoted to the finite version of F.P. Ramsey's results [1930]. For a systematic explanation we recommend the monograph by P. Erdős, A Hajnal, A. Máté and R. Rado [1984].

The basic source of information on topological spaces may be R. Engelking [1977]. However the common proofs of many topological results usually exploit the axiom of choice in spite of the fact that one can prove them in **ZF** or in **ZFW**. That is the reason why in Section 1.2 we have presented some (maybe not obvious) proofs. One of the first textbooks on topology (containing also the first definition of a topology using a neighborhood system) is F. Hausdorff [1914]. The classical monograph by K. Kuratowski [1958a] is a great source of information, both mathematical and historical. The Bourbaki exposition of general topology [1940] is based on filters.

A proof of part c) of Theorem 1.22, presented in Exercise 1.22, can be found, e.g., in R. Engelking [1977]. Theorem 1.24 was proved independently by G. Cantor [1883] and I. Bendixson [1883]. Theorem 1.36 and results of Exercise 1.23 are due to E. Čech [1937]. Theorem 1.37 was proved by A. Tychonoff [1930].

A limit of a filter was defined and studied by H. Cartan [1937a], [1937b] and then systematically developed in N. Bourbaki [1940]. The notion of a net was introduced by E.H. Moore and H.L. Smith [1922] and then systematically presented by J.L. Kelley [1955]. Problems of compactification were studied by A.N. Tychonoff [1930], E. Čech [1937] and M.H. Stone [1937].

M. Fréchet [1906] introduced a notion of an $\mathcal{L}^*$-space. For more information see M. Fréchet [1928], K. Kuratowski [1958a] and R. Engelking [1977].

Chapter 2

The Real Line

Lehrsatz. Wenn eine Eigenschaft M nicht allen Werthen einer veränder—lichen Größe x, wohl aber allen, die kleiner sind, als ein jewisser u, zukommt: so gibt es allemahl eine Größe U, welche die größte derjenigen ist, von denen behauptet werden kann, daß alle kleineren x die Eigenschaft M besitzen.

Bernard Bolzano [1817], p. 41.

Theorem. *If a property* M *does not apply to all values of a variable quantity* x, *but to all those that are smaller than a certain* u, *there is always a quantity* U *which is the largest of those of which it can be asserted that all smaller* x *possess the property* M.

Translation M. Kline [1972], p. 953.

We describe the real line and its fundamental properties We try to avoid any form of the axiom of choice when it is possible. When we use any form of the axiom of choice we shall stay it explicitly.

2.1   The Definition

A **real line** is a linearly ordered field $\mathcal{R} = \langle \mathbb{R}, =, \leq, +, \cdot, 0, 1 \rangle$ satisfying the **Bolzano Principle**:

every non-empty subset of $\mathbb{R}$ bounded from above has a supremum.   (2.1)

More precisely, if a set $M \subseteq \mathbb{R}$ is such that $M \neq \emptyset$ and M is bounded from above, then there exists a real s such that $s = \sup M$, i.e., s is an upper bound of M and for any other upper bound t of M one has $s \leq t$. An element of $\mathbb{R}$ is called a **real number** or simply a **real**. Note that by the Bolzano Principle every non-empty set of reals bounded from below has an infimum.

In the next section we show that in **ZF** one can prove that there exists a real line and, up to isomorphism, the real line is unique. Now, we assume that there exists a real line $\mathcal{R} = \langle \mathbb{R}, =, \leq, +, \cdot, 0, 1 \rangle$ and we start to investigate its properties.

In spite of the fact that we have already defined a set of natural numbers we present now another definition. Then we show that these notions are in a certain sense isomorphic. We define the set $\mathbb{N}$ of **natural numbers** as the smallest subset of $\mathbb{R}$ containing 0 and with any element x containing also $x + 1$. Thus the set $\mathbb{N}$ of natural numbers is defined as the set satisfying the following conditions:

1) $0 \in \mathbb{N}$,
2) $(\forall n)\,(n \in \mathbb{N} \to n + 1 \in \mathbb{N})$,
3) if $X \subseteq \mathbb{R}$ is any set such that $0 \in X$ and $(\forall x)\,(x \in X \to x + 1 \in X)$, then $\mathbb{N} \subseteq X$.

The definition can be equivalently formulated as

$$\mathbb{N} = \{x \in \mathbb{R} : (\forall X \subseteq \mathbb{R})\,((0 \in X \wedge (\forall y)\,(y \in X \to y + 1 \in X)) \to x \in X)\}.$$

Usually we denote natural numbers by letters i, j, k, l, m, n or with indexes k_1, n_1 etc.

The definition of $\mathbb{N}$ is similar to that of ω. Therefore we can obtain a similar theorem on mathematical induction. For any particular formula φ of set theory we have a particular theorem. So, let $\varphi(x, x_1, \ldots, x_k)$ be a formula of set theory. Then

Theorem 2.1 (The First Theorem on Mathematical Induction).
Let $x_1, \ldots, x_k$ be given. If

(IS1) $\varphi(0, x_1, \ldots, x_k)$,
(IS2) $(\forall n \in \mathbb{N})\,(\varphi(n, x_1, \ldots, x_k) \to \varphi(n + 1, x_1, \ldots, x_k))$,

then $\varphi(x, x_1, \ldots, x_k)$ holds true for any $x \in \mathbb{N}$.

The proof is almost the same as that of Theorem 1.1.   □

Using the First Theorem on Mathematical Induction one can prove basic properties of natural numbers.

Theorem 2.2.

a) *0 is the smallest natural number, i.e., $0 \leq n$ for any $n \in \mathbb{N}$.*
b) *If n is a natural number, $n \neq 0$, then there exists a natural number m such that $n = m + 1$.*
c) *If $m < n$, then $m + 1 \leq n$.*
d) *If n, m are natural numbers such that $|n - m| < 1$, then $n = m$.*
e) *If $m \leq n$, then there exists a natural number k such that $n = m + k$.*
f) *If n, m are natural numbers, then also $n + m$, $n \cdot m$ are natural numbers.*

Proof. To prove a) set $X = \{n \in \mathbb{N} : n \geq 0\}$ and show by mathematical induction that $X = \mathbb{N}$.

Similarly, to prove b), set $X = \{n \in \mathbb{N} : n = 0 \lor (\exists n)\, m + 1 = n\}$ and show by induction that $X = \mathbb{N}$.

Set $X_n = \{m \in \mathbb{N} : m \geq n \lor m + 1 \leq n\}$. By a) we have $X_0 = \mathbb{N}$. If $X_n = \mathbb{N}$, then it is easy to show (by induction) that $X_{n+1} = \mathbb{N}$. Thus, by induction we obtain that $X_n = \mathbb{N}$ for any n and therefore c) holds true.

d) follows easily from c) (by contradiction).

To obtain e) set $X_n = \{m \in \mathbb{N} : m < n \lor (\exists k)\, m = n + k\}$ and show by induction that $X_n = \mathbb{N}$ for every n (using b)).

Since $\mathcal{R}$ is a field, from the definition you obtain immediately that the following four **Peano axioms**[1] hold true:

$$n + 0 = n, \tag{2.2}$$
$$n + (m + 1) = (n + m) + 1, \tag{2.3}$$
$$n \cdot 0 = 0, \tag{2.4}$$
$$n \cdot (m + 1) = (n \cdot m) + n \tag{2.5}$$

for any $n, m \in \mathbb{N}$.

Using those axioms one can easily prove by induction the statement f). $\qquad \square$

The similarity of definitions of $\mathbb{N}$ and ω (note that a finite ordinal $n \in \omega$ is the set $\{k : k < n\}$) leads to

Theorem 2.3. *There exists a unique bijection $F : \mathbb{N} \xrightarrow[\text{onto}]{1-1} \omega$ such that $F(0) = \emptyset$ and for any $n \in \mathbb{N}$, $F(n + 1) = F(n) \cup \{F(n)\}$.*

Moreover, for any $n, m \in \mathbb{N}$ we have

$$|F(n + m)| = |F(n)| + |F(m)|, \quad |F(n \cdot m)| = |F(n)| \cdot |F(m)|,$$

where the signs $+, \cdot$ on the left-hand side denote the operations on the field $\mathcal{R}$ and on the right-hand side they denote the operations on cardinals as defined in Section 1.1.

Proof. We write $\mathbb{N}_n = \{k \in \mathbb{N} : k < n\}$. Let

$$\mathcal{F} = \bigcup_{n \in \mathbb{N}} \{f \in {}^{\mathbb{N}_n}\omega : f(0) = \emptyset \land (\forall k)\,(k + 1 < n \to f(k+1) = f(k) \cup \{f(k)\})\}.$$

One can easily check that $f \subseteq g$ or $g \subseteq f$ for any $f, g \in \mathcal{F}$. Moreover, for any $n \in \mathbb{N}$ there exists an $f \in \mathcal{F}$ such that $n \in \mathrm{dom}(f)$. Set $F = \bigcup \mathcal{F}$.

Since both the operations on the field $\mathcal{R}$ and the operations on cardinals satisfy Peano axioms (2.2)–(2.5), the statement follows easily by induction. $\qquad \square$

Using this bijection we shall identify the set $\mathbb{N}$ and ω. Thus $\mathbb{N} = \omega$ and consequently $n = \{k \in \mathbb{N} : k < n\} = \mathbb{N}_n$ for any $n \in \mathbb{N}$. Especially, $0 = \emptyset$ and $n + 1 = n \cup \{n\} = \{0, 1, \ldots, n\}$ for any $n \in \mathbb{N}$.

---

[1] The list of all Peano axioms is given in Section 11.4.

Theorem 2.4. *Every non-empty set of natural numbers contains the smallest element.*

Proof. Let $A \subseteq \mathbb{N}$ be a non-empty set. Since A is bounded from below, there exists a real $s = \inf A$. Since $s + 1 > s$, there exists an $n \in A$ such that $s \leq n < s + 1$. If $s < n$, then again there exists an $m \in A$ such that $s < m < n$, a contradiction, since $|n - m| < 1$. Thus $s = n \in A$. $\square$

This theorem yields another method of proof by mathematical induction. Assume that $\varphi(x, x_1, \ldots, x_k)$ is a formula of set theory. Then

Theorem 2.5 (The Second Theorem on Mathematical Induction).
Let $x_1, \ldots, x_k$ be given. If for any $n \in \mathbb{N}$ we have

(IS) $(\forall k < n)\, \varphi(k, x_1, \ldots, x_k) \to \varphi(n, x_1, \ldots, x_k),$

then $\varphi(n, x_1, \ldots, x_k)$ holds true for any $n \in \mathbb{N}$.

Proof. We set
$$X = \{n \in \mathbb{N} : \neg\varphi(n, x_1, \ldots, x_k)\}.$$

If $X \neq \emptyset$, then, by Theorem 2.4, there exists the smallest element n of the set X. Then $\neg\varphi(n, x_1, \ldots, x_k)$ and $\varphi(k, x_1, \ldots, x_k)$ for $k < n$, a contradiction with (IS).
                                                                    $\square$

Theorem 2.6 (Archimedes Principle). *The set $\mathbb{N}$ is unbounded in $\mathbb{R}$. Hence for any positive reals x, ε there is a natural number n such that $n \cdot \varepsilon > x$.*

Proof. If $\mathbb{N}$ were bounded, then there exists a supremum $s = \sup \mathbb{N}$. Then there is an $n \in \mathbb{N}$, $n > s - 1$. Hence $n + 1 > s$, a contradiction. $\square$

An **integer number** or simply an **integer** is any element of the set

$$\mathbb{Z} = \{x \in \mathbb{R} : x \in \mathbb{N} \vee -x \in \mathbb{N}\}$$

and a **rational number** or simply a **rational** is any element of the set

$$\mathbb{Q} = \{x \in \mathbb{R} : (\exists z \in \mathbb{Z})(\exists n \in \mathbb{N})\,(n \neq 0 \wedge x = z/n)\}.$$

A real that is not rational is called an **irrational number** or simply an **irrational**.

$\mathbb{Z}$ is an integral domain and $\mathbb{Q}$ is the field of fractions of $\mathbb{Z}$. The set $\mathbb{Z}$ being a union of two countable sets is a countable set. The set $\mathbb{Q}$ is the image of the countable set $\mathbb{N} \times \mathbb{Z}$ by the mapping $f(n, z) = z/(n + 1)$, therefore

$$\mathbb{Q} \text{ is countable.} \tag{2.6}$$

Consequently the set $\mathbb{R} \setminus \mathbb{Q}$ of irrationals has cardinality $\mathfrak{c}$.

Theorem 2.7. *For any reals $a < b$ there exists a rational r such that $a < r < b$.*

Proof. We can assume that $0 \leq a < b$ (the other cases can be easily reduced to that one). By Archimedes' Principle there exists a natural number n such that

$1/n < b - a$. Using again Archimedes' Principle there exists a natural number m such that $m \cdot 1/n > a$. Taking the smallest m with this property we obtain $a < m/n < b$. $\qquad \square$

Corollary 2.8. $\langle \mathbb{Q}, \leq \rangle$ *is densely ordered.*

Thus, by Theorem 11.4, a) we obtain

Corollary 2.9. *Every linearly ordered countable set can be embedded in* $\langle \mathbb{Q}, \leq \rangle$.

About 500 B.C. the Pythagoreans already knew that

$$\text{there is no } r \in \mathbb{Q} \text{ such that } r^2 = 2. \tag{2.7}$$

Actually, let us suppose that there exists an $r = m/n \in \mathbb{Q}$ such that $r^2 = 2$. We can assume that n is the smallest possible. Then $m^2 = 2n^2$ and therefore $m = 2k$ for some $k < m$. Hence $2k^2 = n^2$ and again $n = 2l$ for some $l < n$. Then $r = k/l$, a contradiction.

One can easily show that

$$\text{there exists a positive real } s \text{ such that } s^2 = 2. \tag{2.8}$$

Let $M = \{x \in \mathbb{R} : x^2 < 2\}$. We show that $s = \sup M \in \mathbb{R}$ is such that $s^2 = 2$.

Assume $s^2 > 2$. Then $s_1 = (s^2+2)/(2s) < s$ and $s_1^2 = 2+((s^2-2)/(2s))^2 > 2$. Evidently s_1 is an upper bound of M, a contradiction.

Assume now that $s^2 < 2$. Set

$$s_2 = s \left(1 + \frac{2 - s^2}{3s^2} \right) > s.$$

Since $(2 - s^2)/3s^2 < 1$, we obtain

$$s_2^2 = s^2 \left(1 + 2\frac{2 - s^2}{3s^2} + \left(\frac{2 - s^2}{3s^2} \right)^2 \right) < s^2 \left(1 + 3\frac{2 - s^2}{3s^2} \right) = 2.$$

Thus $s_2 \in M$ – again a contradiction. Hence $s^2 = 2$ and by (2.7) $s \notin \mathbb{Q}$. Consequently

$$\mathbb{Q} \neq \mathbb{R}, \text{ i.e., there exists an irrational real.}$$

From Theorem 2.7 we immediately obtain

Corollary 2.10. *For any reals $a < b$ there exists an irrational x such that $a < x < b$.*

Generally the existence of the real $\sqrt{x}$ for $x > 0$ follows by Corollary 2.17 below.

For every real x there exists the greatest integer z such that $z \leq x$. The integer z is called the **integer part** of the real x and will be denoted by $\lfloor x \rfloor$. Hence

$$\lfloor x \rfloor \leq x < \lfloor x \rfloor + 1.$$

Thus a real x can be written as

$$x = \lfloor x \rfloor + (x - \lfloor x \rfloor), \qquad (2.9)$$

where $\{x\} = x - \lfloor x \rfloor \in \langle 0, 1)$ is the **fractional part** of the real x.

If a is a real we set

$$a^+ = \begin{cases} a & \text{if } a \geq 0, \\ 0 & \text{if } a < 0. \end{cases} \qquad a^- = \begin{cases} 0 & \text{if } a \geq 0, \\ -a & \text{if } a < 0. \end{cases}$$

Then

$$a = a^+ - a^-, \qquad |a| = a^+ + a^-. \qquad (2.10)$$

Sometimes we shall use the denotation

$$\mathbb{R}^+ = \{x \in \mathbb{R} : x \geq 0\}.$$

Exercises

2.1 Divisibility

Recall that an integer u **divides** an integer v, or u is a **divisor** of v, or v is a **multiple** of u, written $u|v$, if there exists an integer z such that $v = u \cdot z$.

a) The relation $|$ is a partial ordering on $\mathbb{N}$.

b) For any natural numbers $n, q, q \neq 0$ there are unique natural numbers k, r such that $n = kq + r$ and $0 \leq r < q$.

c) Show that for any non-zero natural numbers n and m there exists the greatest common divisor. We denote it by (n, m). Moreover, there exist integers x, y such that $(n, m) = xn + ym$.
Hint: Consider the integer $k = \min\{xn + ym > 0 : x, y \in \mathbb{Z}\}$.

d) Show that the greatest common divisor k of natural numbers n, m, p is $k = ((n, m), p)$.

e) The least common multiple of naturals n, m is $nm/(n, m)$.

f) The poset $\langle \mathbb{N}, | \rangle$ is a lattice.

g) If $(k, n) = 1$ and $k|n \cdot m$, then $k|m$.
Hint: If $m = kq + r$, $0 \leq r < k$, then $nm = nkq + nr$ and therefore $k|nr$. On the other hand $1 = xk + yn$, thus $r = xrk + yrn$ and therefore $k|r$.

2.2 Prime Numbers

A natural number $n > 1$ is called a **prime** if the only positive divisors of n are 1 and n.

a) Every natural number greater than 1 is divisible by a prime.
Hint: Use the Second Theorem on Mathematical Induction.

b) An integer is irreducible in $\mathbb{Z}$ if and only if z or $-z$ is a prime.

c) For any prime p there exists a prime q such that $p < q \leq p! + 1$.

d) Prove the **Euclid Theorem**: There exists infinitely many primes.

c) Let p be a prime. If $p|n \cdot m$, then $p|n$ or $p|m$.

d) Prove **Fundamental Theorem of Arithmetic**: Any natural number $n > 1$ can be expressed as $p_0^{k_0} \cdot \ldots \cdot p_m^{k_m}$, where $p_0, \ldots, p_m$ are mutually different primes and $k_0, \ldots, k_m$ are positive natural numbers. Moreover, this expression is unique up to order.

Hint: Existence is easy by a). *The uniqueness follows by* e).

2.3 Chinese Remainder Theorem

Let us recall that integers m, n are **coprime**, if no prime divides both of them. Assume that $m_1, \ldots, m_k$ are pairwise coprime integers greater than 1. Let $M = m_1 \cdots m_k$.

a) If $\neg x \equiv y \mod M$, then $\neg x \equiv y \mod m_i$ for some i.

b) Let $x_1, \ldots, x_k$ be the remainders of x, when divided by m_i, $i = 1, \ldots, k$, respectively. If x varies from given integer c to $c + M - 1$, then $x_1, \ldots, x_k$ take different values.

c) Prove the **Chinese Remainder Theorem**: For any integers $a_1, \ldots, a_k$ there exists arbitrarily large x such that

$$x \equiv a_1 \mod m_1, \quad \ldots \quad , x \equiv a_k \mod m_k. \tag{2.11}$$

d) If x, y are solutions of the equations (2.11), then $x \equiv y \mod M$.

e) Assuming that $a_i \equiv a_j \mod m_i m_j$ for any $i, j \leq k$, one can omit the condition that $m_1, \ldots, m_k$ are pairwise coprime.

2.4 Algebraic Numbers

A real α is **algebraic** if there are $n > 0$, integers $z_0, \ldots, z_n$, $z_n \neq 0$ such that $\sum_{i=0}^{n} z_i \alpha^i = 0$, i e., if α is a root of a non-trivial polynomial with integer coefficients. A real that is not algebraic is called **transcendental**. An irrational real α is called a **Liouville number** if for any natural n there are integers p, q, $q > 1$ such that

$$\left| \alpha - \frac{p}{q} \right| < \frac{1}{q^n}. \tag{2.12}$$

a) A real α is algebraic if and only if α is algebraic over $\mathbb{Q}$ in the sense of Exercise 11.13.

b) Every rational is algebraic.

c) The set of algebraic numbers is countable. Thus, the cardinality of the set of all transcendental reals is $\mathfrak{c}$.

Hint: The set of all non-trivial polynomials with integer coefficients is countable.

d) Let α be an algebraic irrational number. Assume that $P(x) = \sum_{i=0}^{n} z_n x^n$ is the polynomial with integer coefficients of the smallest degree such that $P(\alpha) = 0$. Then $P(x)$ has no rational root.

e) Let $P(x)$ be a polynomial with integer coefficients of degree $n > 1$. Assume that $P(\alpha) = 0$. Then there exists a positive integer k such that $|P(x)| \leq k|\alpha - x|$ for any $|x - \alpha| \leq 1$.

Hint: $|\sum_{i=0}^{n} a_i x^i - \sum_{i=0}^{n} a_i \alpha^i| \leq |x - \alpha| \sum_{i=0}^{n} |a_i| n(|\alpha| + 1)^{n-1}$.

f) Let α be an algebraic real, $P(x)$ being a polynomial with integer coefficients of the smallest degree such that $P(\alpha) = 0$, $\alpha \notin \mathbb{Q}$. Then there exists a positive integer k such that

$$\left| \alpha - \frac{p}{q} \right| \geq \frac{1}{kq^n}$$

for any integers p, q, $q > 0$.

Hint: By e) *we have* $|q^n P(p/q)| \leq kq^n |\alpha - p/q|$. *Since* $q^n P(p/q)$ *is an integer, by* d) *we obtain* $|q^n P(p/q)| \geq 1$.

g) Prove the **Liouville Theorem**: No algebraic number is a Liouville number.

Hint: Assume that α *is an irrational algebraic real. Then* $|\alpha - p/q| > 1/kq^n$. *Take* m *such that* $2^m \geq k2^n$. *If* α *is a Liouville number, then* $|\alpha - p/q| < 1/q^m$ *for some* p, q. *Then* $k > q^{m-n} \geq 2^{n-m} \geq k$.

i) The real $\sum_{n=0}^{\infty} p^{-n!}$ is transcendental for any integer $p > 1$.

Hint: It is a Liouville number.

2.5 Sierpiński's Theorem

a) Let $A \subseteq \mathbb{R}$, M a positive real. If $|\sum_{i=0}^{n} a_i| \leq M$ for any finite set $\{a_0, \ldots, a_n\} \subseteq A$, then A is countable.

Hint: For any positive integer n *the set* $\{a \in A : |a| \geq 1/n\}$ *is finite.*

b) By Corollary 2.9 for any countable ordinal ξ there exists a subset of $\mathbb{R}$ order isomorphic to ξ. Show the opposite statement: if $A \subseteq \mathbb{R}$ is well ordered by ordering of reals, then A is countable.

Hint: Use Theorem 2.7.

c) Prove **Sierpiński's Theorem**: If the set $\mathbb{R}$ can be well ordered, then $2^{\aleph_0} \nrightarrow (\aleph_1)_2^2$ (for the definition of the partition relation see Exercise 1.14).

Hint: A couple $\{a, b\}$ *is colored by the color 0 if the well-ordering on* $\{a, b\}$ *agrees with the ordering of reals. Otherwise the couple has color 1.*

2.2   Topology of the Real Line

A set $A \subseteq \mathbb{R}$ is **open** if for every $x \in A$ there exists an $\varepsilon > 0$ with $(x - \varepsilon, x + \varepsilon) \subseteq A$. One can easily see that the family $\mathcal{E}$ of so-defined open subsets of $\mathbb{R}$ is a topology on $\mathbb{R}$. Every open interval is an open set. Every closed interval is a closed set. Moreover, the closure of an open interval (a, b) is the closed interval $[a, b]$. Similarly for unbounded intervals.

Every open set is a union of a set of open intervals. Using Theorem 2.7 one can easily check that the countable set

$$\{(a, b) : a, b \in \mathbb{Q} \wedge a < b\}$$

is a base of the topology $\mathcal{E}$. The set $\mathbb{Q}$ is a topologically dense subset of $\mathbb{R}$.

We begin with proving some basic properties of the topological space $\langle \mathbb{R}, \mathcal{E} \rangle$.

Theorem 2.11. *A non-empty closed bounded set has a greatest element.*

The proof is easy. If $A \neq \emptyset$ is closed and bounded, then there exists $a = \sup A$. Since $A \cap (a - \varepsilon, a + \varepsilon) \neq \emptyset$ for every $\varepsilon > 0$, we obtain $a \in \overline{A} = A$. $\square$

Theorem 2.12. *Every closed interval $[a, b]$ is a compact set. Hence, the topological space $\langle \mathbb{R}, \mathcal{E} \rangle$ is locally compact.*

Proof. We present a proof due to H. Lebesgue, which does not use any form of the axiom of choice.

Let $\{A_s : s \in S\}$ be an open cover of the interval $[a, b]$, i.e., $A_s \in \mathcal{E}$ for each $s \in S$ and $[a, b] \subseteq \bigcup_{s \in S} A_s$. Set

$$M = \left\{ x \in (a, b] : (\exists S_0 \subseteq S) \left(S_0 \text{ finite } \wedge [a, x] \subseteq \bigcup_{s \in S_0} A_s \right) \right\}.$$

We claim that $b \in M$. We show successively that $M \neq \emptyset$, $\sup M \in M$ and $\sup M = b$.

Since $a \in \bigcup_{s \in S} A_s$, there exists an $s_0 \in S$ such that $a \in A_{s_0}$. By assumption the set A_{s_0} is open, therefore there is a positive ε such that $(a - \varepsilon, a + \varepsilon) \subseteq A_{s_0}$. Since $[a, a + \varepsilon/2] \subseteq A_{s_0}$, we have $a + \varepsilon/2 \in M$. Hence $M \neq \emptyset$.

The set M is bounded by the real b. So, there exists the supremum $c = \sup M$. Evidently $a < c \leq b$. Thus, there exists an $s_1 \in S$ such that $c \in A_{s_1}$. Again, there is a positive real ε such that $(c - \varepsilon, c + \varepsilon) \subseteq A_{s_1}$. Since $c - \varepsilon < c$, there is an $x \in M$ such that $x \geq c - \varepsilon$. By the definition of the set M there exists a finite set S_0 such that $[a, x] \subseteq \bigcup_{s \in S_0} A_s$. Set $S_1 = S_0 \cup \{s_1\}$. Then S_1 is finite and

$$[a, c] = [a, x] \cup (c - \varepsilon, c] \subseteq \bigcup_{s \in S_1} A_s.$$

Therefore $c \in M$.

If $c < b$, then

$$[a, c + \varepsilon/2] \subseteq [a, x] \cup (c - \varepsilon, c + \varepsilon/2] \subseteq \bigcup_{s \in S_1} A_s,$$

consequently $c + \varepsilon/2 \in M$ – a contradiction with $c = \sup M$. Therefore $c = b$.

If U is a neighborhood of $x \in \mathbb{R}$, then $(x - \varepsilon, x + \varepsilon) \subseteq U$ for some positive ε. The closed interval $[x - \varepsilon/2, x + \varepsilon/2] \subseteq U$ is a compact neighborhood of x. $\square$

One can easily see that any compact subset of $A \subseteq \mathbb{R}$ is bounded, i.e., there exists a real c such that $|x| < c$ for any $x \in A$. Therefore using Theorem 1.30 we obtain

Corollary 2.13 (The First Weierstrass Theorem). *Any continuous function from a closed interval into $\mathbb{R}$ is bounded.*

Corollary 2.14 (The Second Weierstrass Theorem). *If $f : [a, b] \longrightarrow \mathbb{R}$ is continuous, then there exists a real $c \in [a, b]$ such that $f(x) \leq f(c)$ for any $x \in [a, b]$.*

Proof. The non-empty set $A = \{f(x) : x \in [a, b]\}$, being compact, is closed and bounded. By Theorem 2.11 there exists a maximal element of A. $\square$

We present another fundamental property of the real line.

Theorem 2.15 (The Bolzano Theorem). *If $f : [a, b] \longrightarrow \mathbb{R}$ is a continuous function, $f(a) \neq f(b)$, and c is a real between $f(a)$ and $f(b)$, i.e., $f(a) < c < f(b)$ or $f(b) < c < f(a)$, then there exists a real $x_0 \in (a, b)$ such that $f(x_0) = c$.*

Proof. We present a slight modification of the original proof by B. Bolzano, which does not use any form of the axiom of choice.

We assume that $f(a) < c < f(b)$. Set

$$M = \{x \in [a, b] : f(x) < c\}.$$

Evidently M is a non-empty bounded set of reals. Therefore there exists the supremum $x_0 = \sup M$. We show that $f(x_0) = c$.

If $f(x_0) < c$, then there exists a real $\delta > 0$ such that $|f(x) - f(x_0)| < c - f(x_0)$ for every $x \in (x_0 - \delta, x_0 + \delta)$. Then $x_0 + \delta/2 \in M$, a contradiction.

If $f(x_0) > c$, then there exists a real $\delta > 0$ such that $|f(x) - f(x_0)| < f(x_0) - c$ for every $x \in (x_0 - \delta, x_0 + \delta)$. Since $x_0 - \delta < x_0 = \sup M$, there exists $x \in M$ such that $x > x_0 - \delta$. Then $f(x) > c$, which is a contradiction with the definition of the set M.

Hence $f(x_0) = c$.   □

Theorem 2.16. *Any interval in $\mathbb{R}$ is a connected set.*

Proof. Assume that an interval I is not connected. Then there are open sets U, V such that

$$I \subseteq U \cup V, \quad U \cap V \cap I = \emptyset, \quad U \cap I \neq \emptyset, \quad V \cap I \neq \emptyset.$$

Choose $a \in I \cap U$, $b \in I \cap V$. We can assume that $a < b$. Then $[a, b] \subseteq U \cup V$. Set

$$f(x) = \begin{cases} -1 & \text{if } x \in [a, b] \cap U, \\ 1 & \text{if } x \in [a, b] \cap V. \end{cases}$$

For any set $W \subseteq \mathbb{R}$, the set $f^{-1}(W)$ is one of the sets $[a, b]$, $\emptyset$, $[a, b] \cap U$ or $[a, b] \cap V$, i.e., always a set open in $[a, b]$. Therefore f is continuous. Since 0 is between $f(a) = -1$ and $f(b) = 1$, by the Bolzano Theorem there exists a real $x_0 \in [a, b]$ such that $f(x_0) = 0$, a contradiction with the definition of f.   □

Corollary 2.17. *Let n be a positive natural number, x a real. If n is odd, then there exists a real y such that $y^n = x$, i.e., there exists the root $\sqrt[n]{x}$. If n is even and $x \geq 0$, then there exists a non-negative real y such that $y^n = x$, i.e., there exists the root $\sqrt[n]{x}$.*

Proof. One can easily see that the function $f(x) = x^n$ is continuous. Let, e.g., n be even. If $x = 0$ take $y = 0$. If $x > 0$, then

$$f(0) = 0 < x < (1 + x)^n = f(1 + x).$$

By Bolzano's Theorem there exists a real $y \in (0, 1 + x)$ such that $f(y) = x$.   □

Any two bounded closed intervals are homeomorphic. Similarly for bounded open intervals. One can easily construct a homeomorphism of $(0,1)$ and $(0,+\infty)$ and that of $(0,1)$ and $\mathbb{R}$.

We recall a well-known notion of elementary calculus. A sequence $\{x_n\}_{n=0}^{\infty}$ of reals is said to be **Bolzano-Cauchy**, shortly **B-C**, if

$$(\forall \varepsilon > 0)(\exists n_0)(\forall n, m \geq n_0) |x_n - x_m| < \varepsilon.$$

Evidently a convergent sequence is a Bolzano-Cauchy sequence. One can easily see that a Bolzano-Cauchy sequence is bounded, i.e., there exists a real K such that $|x_n| \leq K$ for every $n \in \omega$.

By Theorems 2.12 and 1.29 we obtain

Theorem 2.18 (B. Bolzano – K. Weierstrass). *Every bounded sequence of reals has a cluster point.*

One can easily check that if a is a cluster point of a Bolzano-Cauchy sequence $\{x_n\}_{n=0}^{\infty}$, then $a = \lim_{n \to \infty} x_n$. Thus

Theorem 2.19 (B. Bolzano – A.L. Cauchy). *A Bolzano-Cauchy sequence of reals is convergent.*

As an immediate consequence of the Bolzano Principle we obtain

Theorem 2.20. *A monotone bounded sequence of reals is convergent.*

Let $\{a_n\}_{n=0}^{\infty}$ be a sequence of reals. A pair of sequences $\{a_n\}_{n=0}^{\infty}$, $\{s_n\}_{n=0}^{\infty}$ is called a **series** if $s_n = \sum_{i=0}^{n} a_i$ for every $n \in \omega$. The real a_n is a **term of the series** and s_n is a **partial sum**. If there exists a real $s = \lim_{n \to \infty} s_n$, we call it the **sum of the series** and we write

$$s = \sum_{n=0}^{\infty} a_n, \text{ i.e., } \sum_{n=0}^{\infty} a_n = \lim_{n \to \infty} \sum_{i=0}^{n} a_i.$$

For the sake of brevity we shall sometimes denote a series simply as $\sum_{n=0}^{\infty} a_n$ independently of the existence of the sum. A series is **convergent** if there exists a real that is the sum of the series, otherwise the series is **divergent**. If every term a_n is non-negative, then we express that the series $\sum_{n=0}^{\infty} a_n$ is convergent by writing $\sum_{n=0}^{\infty} a_n < \infty$. $\sum_{n=0}^{\infty} a_n = \infty$ means in this case that the series $\sum_{n=0}^{\infty} a_n$ is divergent.

As an easy consequence of Theorem 2.20, we obtain an important result.

Theorem 2.21 (The Comparison Test). *Let $\{a_n\}_{n=0}^{\infty}$, $\{b_n\}_{n=0}^{\infty}$ be sequences of reals, $0 \leq |a_n| \leq b_n$ for each $n \in \omega$. If the series $\sum_{n=0}^{\infty} b_n$ is convergent, then also the series $\sum_{n=0}^{\infty} a_n$ is convergent and*

$$\left| \sum_{n=0}^{\infty} a_n \right| \leq \sum_{n=0}^{\infty} b_n.$$

If $a_{n+1} = a_n \cdot q$, q being a real, the series $\sum_{n=0}^{\infty} a_n$ is called a **geometric series**. One can easily check that

$$a_n = a_0 q^n, \quad s_n = a_0 \frac{q^{n+1} - 1}{q - 1} \text{ for every } n \in \omega.$$

Theorem 2.22.

a) *If $|q| < 1$, then the geometric series is convergent and*

$$\sum_{n=0}^{\infty} a_0 q^n = \frac{a_0}{1 - q}.$$

b) *If $a_0 \neq 0$ and $|q| \geq 1$, then the geometric series is divergent.*

The **exponential function** $\exp : \mathbb{R} \longrightarrow (0, \infty)$ is defined as

$$\exp(x) = \sum_{n=0}^{\infty} \frac{x^n}{n!}.$$

From elementary analysis we know that $\exp$ is a continuous isomorphism of the additive group $\langle \mathbb{R}, +, 0 \rangle$ and the multiplicative group $\langle (0, \infty), \cdot, 1 \rangle$ such that

$$\exp(1) = e = \sum_{n=0}^{\infty} \frac{1}{n!}.$$

The inverse function to $\exp$ is the **natural logarithm** $\ln : (0, \infty) \longrightarrow \mathbb{R}$. By Corollary 1.31 the function $\ln$ is continuous.

For a positive real a and any real x we set

$$a^x = \exp(x \ln(a)). \tag{2.13}$$

Thus $\exp(x) = e^x$. The function $f(x) = a^x$ is a continuous isomorphism of the additive group $\langle \mathbb{R}, + \rangle$ and the multiplicative group $\langle (0, \infty), \cdot \rangle$ such that $f(1) = a$. For a positive natural number n the value a^n is the nth power of a as defined above.

We shall need the following elementary result of calculus.

Theorem 2.23. *Assume that $\{a_n\}_{n=0}^{\infty}$ is a bounded sequence of non-negative reals, $a_0 > 0$. Set $s_n = \sum_{k=0}^{n} a_k$. If the series $\sum_{n=0}^{\infty} a_n$ is divergent, then the series $\sum_{n=0}^{\infty} a_n s_n^{-1-\varepsilon}$ is divergent for $\varepsilon = 0$ and convergent for any $\varepsilon > 0$.*

A pair of sequences $\{a_n\}_{n=0}^{\infty}$, $\{p_n\}_{n=0}^{\infty}$ of reals is called an **infinite product** if $p_n = \prod_{i=0}^{n} a_i$ for every $n \in \omega$. The real a_n is a **term of the product** and p_n is a **partial product**. If there exists a real $p = \lim_{n \to \infty} p_n$, we call it the **value of the product** and we write

$$p = \prod_{n=0}^{\infty} a_n, \quad \text{i.e.,} \quad \prod_{n=0}^{\infty} a_n = \lim_{n \to \infty} \prod_{i=0}^{n} a_i.$$

For the sake of brevity we shall sometimes denote an infinite product simply as $\prod_{n=0}^{\infty} a_n$ independently of the existence of the value. An infinite product is **convergent** if there exists a non-zero real that is the value of the product. Otherwise the product is **divergent**.

A proof of the next theorem can be found in standard textbooks of analysis.

Theorem 2.24. *Let* $\langle a_n : n \in \omega \rangle$, $\langle b_n : n \in \omega \rangle$ *be positive reals.*

a) *The product* $\prod_{n=0}^{\infty} a_n$ *converges if and only if the series* $\sum_{n=0}^{\infty} \ln(a_n)$ *converges.*

b) *If* $a_n < 1$ *for each* n, *then the infinite product* $\prod_{n=0}^{\infty}(1 - a_n)^{b_n}$ *converges if and only if the series* $\sum_{n=0}^{\infty} a_n b_n$ *converges.*

We shall also consider the **extended real line** $\mathbb{R}^* = \mathbb{R} \cup \{-\infty, +\infty\}$. The set $\mathbb{R}^*$ is ordered by

$$-\infty < a < b < +\infty$$

for any $a, b \in \mathbb{R}$, $a < b$. The ordering induces a topology on $\mathbb{R}^*$ such that the subspace topology on $\mathbb{R}$ coincides with the topology of the real line. Moreover, a typical neighborhood of $-\infty$ is the set $\{x \in \mathbb{R}^* : x < a\}$, where a is a real. Similarly for $+\infty$. It is easy to see that $\mathbb{R}^*$ endowed with the above-described topology is homeomorphic to $[0, 1]$ and therefore compact.

Any monotone sequence of elements of $\mathbb{R}^*$ has a limit in $\mathbb{R}^*$. Thus, a divergent series with non-negative terms has the sum $+\infty$. Any non-empty subset of $\mathbb{R}^*$ has an infimum and a supremum. If we want to emphasize that a subset of $\mathbb{R}$ has a supremum in $\mathbb{R}$ we say that the supremum is a real. Similarly in the case of an infimum, a limit or a sum of a series.

We describe arithmetic of $\mathbb{R}^*$. Let $a, b, c \in \mathbb{R}$, $b > 0, c < 0$. We set

$$
\begin{aligned}
a \pm (+\infty) &= \pm\infty, & a \pm (-\infty) &= \mp\infty, & \pm\infty + (\pm\infty) &= \pm\infty, \\
b \cdot (\pm\infty) &= \pm\infty, & c \cdot (\pm\infty) &= \mp\infty, & (-\infty) \cdot (+\infty) &= +\infty, \\
(-\infty) \cdot (-\infty) &= +\infty, & (-\infty) \cdot (+\infty) &= -\infty, & 0 \cdot (\pm\infty) &= 0.
\end{aligned}
$$

Note that $+\infty - (+\infty)$, $+\infty + (-\infty)$ and $-\infty - (-\infty)$ are not defined.

Exercises

2.6 Fixed Point and the Bolzano Theorem

If $f : X \longrightarrow X$ is a mapping, then a point $x_0 \in X$ is called a **fixed point** of f if $f(x_0) = x_0$.

a) Let $f : [a, b] \longrightarrow [a, b]$. $x_0 \in [a, b]$ is a fixed point of f if and only if $g(x_0) = 0$, where $g(x) = f(x) - x$.

b) Every continuous function $f : [0, 1] \longrightarrow [0, 1]$ has a fixed point.

c) If $p(x) = \sum_{i=0}^{n} a_i x^i$, n odd and $a_n \neq 0$, then there exists a real x_0 such that $p(x_0) = 0$.

2.7 Derivative

Let $f : (a, b) \longrightarrow \mathbb{R}$. A real d is called the **derivative** of f at a point x if

$$d = \lim_{h \to 0} \frac{f(x + h) - f(x)}{h}.$$

If the function f has a derivative at every point of (a, b), we denote by f' the function which assigns to every $x \in (a, b)$ the derivative of f at x. We say that f has a **local maximum** (**local minimum**) at $x_0 \in (a, b)$ if there exists a $\delta > 0$ such that $f(x) \leq f(x_0)$ ($f(x) \geq f(x_0)$) for every $x \in (x_0 - \delta, x_0 + \delta) \subseteq (a, b)$.

a) If f has a derivative at a point $x \in (a, b)$, then f is continuous at x.

b) Assume that f has a derivative at each point of (a, b). If f is non-decreasing (non-increasing), then $f'(x) \geq 0$ ($f'(x) \leq 0$) for every $x \in (a, b)$.

c) If f has a local maximum or local minimum at $x_0 \in (a, b)$ and has a derivative at x_0, then $f'(x_0) = 0$.

d) Prove the **Rolle Theorem**: Let $f : [a, b] \longrightarrow \mathbb{R}$, $f(a) = f(b)$. If f has a derivative in (a, b), then there exists a point $x_0 \in (a, b)$ such that $f'(x_0) = 0$.

e) Prove the **Lagrange Theorem**: Let $f : [a, b] \longrightarrow \mathbb{R}$. If f has a derivative in (a, b), then there exists a point $x_0 \in (a, b)$ such that

$$f'(x_0) = \frac{f(b) - f(a)}{b - a}.$$

f) Assume that f has a derivative at every point of (a, b). If $f'(x) > 0$ ($f'(x) < 0$) for every $x \in (a, b)$, then f is increasing (decreasing) in (a, b).

2.8 Convex Function

Let $I \subseteq \mathbb{R}$ be an interval. A function $f : I \longrightarrow \mathbb{R}$ is called **convex** if for any $a, b \in I$, $a < b$ and any $t \in [0, 1]$ we have $f(ta + (1 - t)b) \leq tf(a) + (1 - t)f(b)$.

a) If $f''(x) > 0$ for every $x \in \text{Int}(I)$, then f is convex.

b) exp is convex on $\mathbb{R}$.

c) $f(x) = x^p$ is convex on $\langle 0, \infty \rangle$ if and only if $p \geq 1$.

d) A convex function is continuous.

e) A continuous function f is convex if and only if $f(\frac{a+b}{2}) \leq \frac{f(a)+f(b)}{2}$ for any $a, b \in I$, $a < b$.

2.9 Exponentiation

If a is a positive real and $r = m/n$ is a positive rational, then we define the exponentiation by $a^{n/m} = \sqrt[m]{x^n}$ and $a^{-r} = 1/a^r$.

a) Show that $\exp(x + y) = \exp(x) \cdot \exp(y)$ for any reals x, y.
   Hint: You will need the Binomial Theorem.

b) Show that $\exp(x) = e^x$ for x rational.
   Hint: Start with a natural number x, then a positive rational.

c) Note that $a = \exp(\ln(a))$.

d) The above-introduced exponentiation a^r and a^{-r}, r rational, coincides with that defined by (2.13).

e) Show that there exists a unique continuous function $\underset{\cdot}{^*} : \mathbb{R} \longrightarrow (0, \infty)$ such that $f(1) = a > 0$ and $f(x + y) = f(x) \cdot f(y)$ for any $x, y \in \mathbb{R}$.

f) Conclude that the definition of exponentiation (2.13) agrees with the common understanding of an exponentiation.

2.10 Elementary Inequalities

Let $a_1, \ldots, a_n, b_1, \ldots, b_n$ be reals.

a) Prove the **Cauchy Inequality**

$$\left(\sum_{k=1}^{n} a_k b_k \right)^2 \leq \left(\sum_{k=1}^{n} a_k^2 \right) \left(\sum_{k=1}^{n} b_k^2 \right).$$

Moreover the equality holds true if and only if

$$(\exists \alpha, \beta)\,(\alpha^2 + \beta^2 > 0 \wedge (\forall k = 1, \ldots, n)\,(\alpha a_k = \beta b_k)). \tag{2.14}$$

Hint: Consider the discriminant of the quadratic equation $\sum_{k=1}^{n}(a_k x + b_k)^2 = 0$.

b) If $p, q > 1$, $\frac{1}{p} + \frac{1}{q} = 1$, a, b are non-negative reals, then the **Young Inequality**

$$\frac{a^p}{p} + \frac{b^q}{q} \geq ab$$

holds true. The equality holds true if and only if $a^p = b^q$.

Hint: $ab = e^{\frac{1}{p} \ln(a^p)} e^{\frac{1}{q} \ln(b^q)}$ *and* exp *is convex.*

c) If $p, q > 1$, $\frac{1}{p} + \frac{1}{q} = 1$, then **Hölder's Inequality**

$$\sum_{k=1}^{n} |a_k b_k| \leq \left(\sum_{k=1}^{n} |a_k|^p \right)^{1/p} \left(\sum_{k=1}^{n} |b_k|^q \right)^{1/q}$$

holds true. Moreover the equality holds true if and only if

$$(\exists \alpha, \beta)\,(\alpha^2 + \beta^2 > 0 \wedge (\forall k = 1, \ldots, n)\,(\alpha |a_k|^p = \beta |b_k|^q)).$$

Hint: In the inequality of b) *take*

$$a = \frac{|a_i|}{\left(\sum_{k=1}^{n} |a_k|^p \right)^{1/p}}, \quad b = \frac{|b_i|}{\left(\sum_{k=1}^{n} |b_k|^q \right)^{1/q}}$$

and sum over $i = 1, \ldots, n$.

d) For a real $p > 1$ prove the **Minkowski Inequality**

$$\left(\sum_{k=1}^{n} (a_k + b_k)^p \right)^{1/p} \leq \left(\sum_{k=1}^{n} a_k^p \right)^{1/p} + \left(\sum_{k=1}^{n} b_k^p \right)^{1/p},$$

where a_k, b_k are non-negative reals. Moreover the equality holds true if and only if (2.14) holds true.

Hint: In the trivial inequality

$$\sum_{k=1}^{n} |a_k + b_k|^p \le \sum_{k=1}^{n} |a_k||a_k + b_k|^{p-1} + \sum_{k=1}^{n} |b_k||a_k + b_k|^{p-1}$$

apply Hölder's Inequality to both terms on the right side.

2.11 Absolute Convergence

A series $\sum_{n=0}^{\infty} a_n$ is called **absolutely convergent** if the series $\sum_{n=0}^{\infty} |a_n|$ is convergent.

a) If $0 \le a_n \le b_n$ for every $n \in \omega$ and the series $\sum_{n=0}^{\infty} b_n$ is convergent, then also the series $\sum_{n=0}^{\infty} a_n$ is convergent.

b) An absolutely convergent series is convergent.
   Hint: $0 \le a_n^+ \le |a_n|$ and $0 \le a_n^- \le |a_n|$.

c) Prove Theorem 2.21.

2.12 A Convergence Test

Let $f : (0, \infty) \longrightarrow (0, \infty)$ be a non-increasing function with $\lim_{x \to \infty} f(x) = 0$. Let $\{a_k\}_{k=0}^{\infty}$ be a sequence of positive reals bounded from above by a real M and such that $\sum_{k=0}^{\infty} a_k = \infty$. Set $s_n = \sum_{i=0}^{n} a_i$.

a) If $\sum_{n=0}^{\infty} f(n) < \infty$, then $\sum_{n=0}^{\infty} a_n f(s_n) < \infty$.
   Hint: Consider the points $[s_n, f(s_n)]$ on the graph of the function f and compare the areas expressed by the considered series. Show that $\sum_{n=0}^{k} a_n f(s_n) \le \sum_{n=0}^{m} f(n)$, provided $s_k \le m + 1$.

b) If $\sum_{n=0}^{\infty} f(n) = \infty$, then $\sum_{n=1}^{\infty} a_n f(s_{n-1}) = \infty$.
   Hint: Consider the area on the graph as above. If $a_0 \le l < k + 1 \le s_m$, then $\sum_{n=l}^{k} f(n) \le \sum_{n=1}^{m} a_n f(s_{n-1})$.

c) The series $\sum_{n=0}^{\infty} a_n (f(s_{n-1}) - f(s_n))$ is convergent.
   Hint: $\sum_{k=m}^{n} a_k (f(s_{k-1}) - f(s_k)) \le M(f(s_{m-1}) - f(s_n))$ for any $m < n$.

d) Conclude that

$$\sum_{n=0}^{\infty} f(n) < \infty \equiv \sum_{n=0}^{\infty} a_n f(s_n) < \infty.$$

e) Theorem 2.23 is a special case of d) with $f(x) = x^{-1-\varepsilon}$, $\varepsilon \ge 0$.

2.13 Partitions of an Interval

Let $a < b$ be reals. A finite subset of $[a, b]$ containing the end points a, b is a **partition of the interval** $[a, b]$. If $\{a_0, \ldots, a_n\}$ is a partition, we assume that $a = a_0 < a_1 < \cdots < a_n = b$. Let $\text{PART}(a, b)$ be the set of all partitions of $[a, b]$.

a) The partially ordered set $\langle \text{PART}(a, b), \subseteq \rangle$ is directed.

b) No countable subset of $\text{PART}(a, b)$ is cofinal in $\text{PART}(a, b)$.
   Hint: Find a real $c \in (a, b)$ such that c does not belong to any element of the countable set. Take the partition $\{a, c, b\}$.

c) Every infinite subset of $\text{PART}(a, b)$ is unbounded in $\text{PART}(a, b)$.

2.14 Variation of a Function

Let $f : [a, b] \longrightarrow \mathbb{R}$ be a function. Let us denote

$$\mathrm{VAR}(f; a, b) = \sup\left\{\sum_{i=1}^{n} |f(a_i) - f(a_{i-1})| : \{a_0, \ldots, a_r\} \in \mathrm{PART}(a, b)\right\},$$

$$\mathrm{VAR}^+(f; a, b) = \sup\left\{\sum_{i=1}^{n} (f(a_i) - f(a_{i-1}))^+ : \{a_0, \ldots, a_n\} \in \mathrm{PART}(a, b)\right\},$$

$$\mathrm{VAR}^-(f; a, b) = \sup\left\{\sum_{i=1}^{n} (f(a_i) - f(a_{i-1}))^- : \{a_0, \ldots, a_n\} \in \mathrm{PART}(a, b)\right\}.$$

$\mathrm{VAR}(f; a, b)$, $\mathrm{VAR}^+(f; a, b)$, $\mathrm{VAR}^-(f; a, b)$ are called the **total variation**, the **positive variation** and the **negative variation** of f, respectively. If $\mathrm{VAR}(f; a, b) < +\infty$ we say that f is of **bounded variation** in $[a, b]$. Finally, f is of bounded variation in $\mathbb{R}$ if $\mathrm{VAR}(f; a, b) < +\infty$ for any $a < b$.

a) Show that $\mathrm{VAR}(f; a, b) = \mathrm{VAR}^+(f; a, b) + \mathrm{VAR}^-(f; a, b)$.

b) If $a < b < c$, then $\mathrm{VAR}(f; a, c) = \mathrm{VAR}(f; a, b) + \mathrm{VAR}(f; b, c)$. Similarly for $\mathrm{VAR}^+(f; a, c)$ and $\mathrm{VAR}^-(f; a, c)$.

c) If f is monotone in $[a, b]$, then $\mathrm{VAR}(f; a, b) = |f(a) - f(b)|$. Find $\mathrm{VAR}^+(f; a, b)$ and $\mathrm{VAR}^-(f; a, b)$.

d) Functions $g(x) = \mathrm{VAR}^+(f; a, x)$ and $h(x) = \mathrm{VAR}^-(f, a, x)$ are non-decreasing in $[a, b]$.

e) If f is of bounded variation in $[a, b]$, then $f(x) = g(x) - h(x)$ for any $x \in [a, b]$.

f) f is of bounded variation in $[a, b]$ if and only if $f = g - h$, where g, h are non-decreasing on $[a, b]$.

g) f is of bounded variation in $\mathbb{R}$ if and only if $f = g - h$, where g, h are non-decreasing.

h) If f is of bounded variation in $[a, b]$, then $\lim_{x \to c^+} f(x)$ exists for any $c \in [a, b)$ and $\lim_{x \to c^-} f(x)$ exists for any $c \in (a, b]$.

2.3 Existence and Uniqueness

The real line $\mathcal{R}$ contains well-defined subsets $\mathbb{N}$, $\mathbb{Z}$, $\mathbb{Q}$. The relationships between them have algebraic character: $\mathbb{Z}$ is the set of all differences of elements of $\mathbb{N}$, $\mathbb{Q}$ is the set of all fractions of elements of $\mathbb{Z}$. We need to describe a relationship between $\mathbb{Q}$ and $\mathbb{R}$ that is not more purely algebraic.

A set $A \subseteq \mathbb{Q}$ is called a **Dedekind cut** if

1) A is non-empty and $A \neq \mathbb{Q}$,

2) for any rationals x, y, if $x \in A$ and $y < x$, then also $y \in A$,

3) the set A does not have a greatest element.

For any real $x \in \mathbb{R}$ we write $D_x = \{r \in \mathbb{Q} : r < x\}$. One can easily see that D_x is a Dedekind cut. Moreover, if $x \neq y$, say $x < y$, then by Theorem 2.7 there exists a rational $x < r < y$. Thus $r \in D_y$ and $r \notin D_x$. Hence $D_x \neq D_y$.

Lemma 2.25. *If $A \subseteq \mathbb{Q}$ is a Dedekind cut, then there exists unique real $x \in \mathbb{R}$ such that $A = D_x$.*

Proof. Since $A \neq \mathbb{Q}$, there exists a rational $s \notin A$. One can easily check that s is an upper bound of A. Thus, being non-empty, the set A has a supremum $x = \sup A$. Since x is an upper bound of A and A does not have a greatest element, we obtain $A \subseteq D_x$. Let $r \in D_x$, i.e., $r < x$. Since x is a supremum of A, there exists an $s \in A$ such that $r < s$. Then by 2) we have $r \in A$. Thus $A = D_x$.   □

We start with the uniqueness of the set of reals.

Theorem 2.26. *If $\mathcal{R}_i = \langle \mathbb{R}_i, =, \leq_i, +_i, \cdot_i, 0_i, 1_i \rangle$, $i = 1, 2$ are linearly ordered fields satisfying the Bolzano Principle, then there exists a mapping $H : \mathbb{R}_1 \xrightarrow[\text{onto}]{1-1} \mathbb{R}_2$ which is an order-preserving isomorphism of the ordered fields $\mathcal{R}_1$ and $\mathcal{R}_2$.*

Proof. Denote by $\mathbb{N}_i$, $\mathbb{Z}_i$, $\mathbb{Q}_i$ the sets of natural, integer and rational reals constructed in $\mathcal{R}_i$, $i = 1, 2$.

By Theorem 2.3 there exists a unique bijection $H : \mathbb{N}_1 \xrightarrow[\text{onto}]{1-1} \mathbb{N}_2$ such that

$$H(n +_1 m) = H(n) +_2 H(m), \; H(n \cdot_1 m) = H(n) \cdot_2 H(m) \text{ for } n, \, m \in \mathbb{N}_1.$$

One can easily extend H to an order-preserving isomorphism of $\mathbb{Q}_1$ and $\mathbb{Q}_2$ by setting $H(-_1 n) = -_2 H(n)$ for $n \in \mathbb{N}_1$ and then $H(z /_1 n) = H(z) /_2 H(n)$ for $z \in \mathbb{Z}_1$ and $n \in \mathbb{N}_1$, $n \neq 0_1$.

It is easy to see that if $x \in \mathbb{R}_1$, then $H(D_x)$ is a Dedekind cut in $\mathbb{Q}_2$. By Lemma 2.25 there is a unique $y \in \mathbb{R}_2$ such that $H(D_x) = D_y$. Set $H(x) = y$. It is rather trivial to show that H is the desired order-preserving isomorphism.   □

Theorem 2.27. *There exists a real line. More precisely: in* **ZF** *one can show that there exists a linearly ordered field satisfying the Bolzano Principle.*

Proof. We set $\mathbb{N} = \omega$. Since ω is a set of cardinals, the addition and multiplication of elements of ω is defined. Moreover, four Peano axioms (2.2)–(2.5) hold true.

Define an equivalence relation $\sim$ on $\omega \times \omega$ as

$$\langle n, m \rangle \sim \langle p, q \rangle \equiv (n + q = p + m).$$

Let $\mathbb{Z}$ be the quotient set $(\omega \times \omega)/ \sim$. A natural number n will be identified with the equivalence class $\{\langle n, 0 \rangle\}_\sim$. For $n, m, p, q \in \mathbb{N}$ we define

$$\{\langle n, m \rangle\}_\sim \leq \{\langle p, q \rangle\}_\sim \equiv (n + q \leq p + m),$$
$$\{\langle n, m \rangle\}_\sim + \{\langle p, q \rangle\}_\sim = \{\langle n + p, m + q \rangle\}_\sim,$$
$$\{\langle n, m \rangle\}_\sim \cdot \{\langle p, q \rangle\}_\sim = \{\langle np + mq, mp + nq \rangle\}_\sim.$$

One can easily check that $\mathbb{Z}$ endowed with those operations is an ordered integrity domain. Moreover, for any $z \in \mathbb{Z}$, $z \neq 0$, there exists an $n \in \mathbb{N}$ such that either $z = n = \{\langle n, 0 \rangle\}_\sim$ or $z = -n = \{\langle 0, n \rangle\}_\sim$.

By a standard algebraic construction – see Theorem 11.8 – there exists up to isomorphism a unique field $\mathbb{Q}$ that is the field of fractions of $\mathbb{Z}$, i.e., there is an isomorphism of $\mathbb{Z}$ onto a subring of $\mathbb{Q}$ and every element of $\mathbb{Q}$ can be expressed as x/y with $x, y \in \mathbb{Z}$, $y \neq 0$. The ordering on $\mathbb{Q}$ is defined by saying which element of $\mathbb{Q}$ is positive.

Denote by $\mathbb{R}$ the set of all Dedekind cuts on $\mathbb{Q}$. An element $r \in \mathbb{Q}$ will be identified with $D_r \in \mathbb{R}$. For any Dedekind cuts $A, B \in \mathbb{R}$ we define

$$A \leq B \equiv A \subseteq B.$$

Immediately we obtain the trichotomy

$$A < B \vee A = B \vee B < A.$$

It is easy to show that the Bolzano Principle holds true. Indeed, let $\mathcal{M} \subseteq \mathbb{R}$ be a non-empty bounded from above set, i.e., $\mathcal{M}$ is a non-empty set of Dedekind cuts and there exists a Dedekind cut K such that $A \subseteq K$ for any $A \in \mathcal{M}$. Let $C = \bigcup \mathcal{M}$. Evidently $C \subseteq K$. Since $K \neq \mathbb{Q}$ and $\mathcal{M} \neq \emptyset$, we have $\emptyset \neq C \neq \mathbb{Q}$. If $s < r$ and $r \in C$, then $r \in A$ for some $A \in \mathcal{M}$ and therefore also $s \in A \subseteq C$. If $r \in C$ were a maximal element, then r would be maximal in some $A \in \mathcal{M}$, which is impossible. Thus C is a Dedekind cut and C is an upper bound of $\mathcal{M}$. If D is any other upper bound of $\mathcal{M}$, then evidently $C \subseteq D$. Thus C is the supremum of $\mathcal{M}$.

We define the addition of Dedekind cuts:

$$A + B = \{r \in \mathbb{Q} : (\exists s, t)\, (s \in A \wedge t \in B \wedge r = s + t)\}.$$

For any Dedekind cut A we set

$$-A = \{r \in \mathbb{Q} : (\exists s \in \mathbb{Q})\, (r < -s \wedge s \notin A)\}.$$

Evidently $A + (-A) = 0 \ (= D_0)$.

To define the multiplication we must distinguish several cases. We start with setting $A \cdot D_0 = D_0 = 0$. If $0 = D_0 < A$ and $0 = D_0 < B$ we set

$$A \cdot B = \{r \in \mathbb{Q} : (\exists s, t > 0)\, (s \in A \wedge t \in B \wedge r \leq s \cdot t)\}.$$

Now, if $A < 0$, $B > 0$ we set $A \cdot B = -(-A \cdot B)$. Similarly, if $A < 0$, $B < 0$ we set $A \cdot B = (-A) \cdot (-B)$.

We recommend to the reader as an exercise to show that $\mathbb{R}$ endowed with that structure is a real line, i.e., a linearly ordered field satisfying the Bolzano Principle. $\qquad \Box$

Checking carefully the proof of Theorem 2.27 we observe that on the base of the first four axioms of **ZF** and the axiom scheme of separation, from the existence of the set ω we have deduced the existence of a real line. Vice versa, in Section

2.1 on the base of the same axioms we have shown that there exists the set $\mathbb{N}$. However for constructing the set ω, even using $\mathbb{N}$, we need a particular case of the axiom scheme of replacement. Anyway, on the base of the first four axioms and axiom schemes of separation and replacement, the existence of the set ω is equivalent to the existence of a real line.

Now we go back to the question of uniqueness of the real line. Actually we have shown that a linearly ordered field with properties known from an elementary course of analysis is isomorphic to the real line. Let

$$\mathcal{T} = \langle T, =, \leq_T, +_T, \cdot_T, 0_T, 1_T \rangle$$

be a linearly ordered field. The mapping defined by $e(n) = n \times 1_T$ is an embedding of $\mathbb{N}$ into T. For simplicity, we shall identify $\mathbb{N}$ with the set $\{e(n) : n \in \mathbb{N}\} \subseteq T$. We define a topology $\mathcal{O}$ on T by taking the set $\{(a, b) : a, b \in T, a <_T b\}$ of all open intervals as a base. We present several properties of a linearly ordered field that are characteristic for the real line, i.e., an ordered field with such a property is already order isomorphic to the real line.

Theorem 2.28. *Let $\mathcal{T}$ be a linearly ordered field. Then the following properties of $\mathcal{T}$ are equivalent.*

 a) *The Bolzano Principle holds true in $\mathcal{T}$, and therefore $\mathcal{T}$ is order isomorphic to the real line.*

 b) *The Bolzano Theorem holds true: if $f : T \longrightarrow T$ is continuous and 0_T is between $f(a)$ and $f(b)$, then there exists a c between a and b such that $f(c) = 0_T$.*

 c) *The topological space $\langle T, \mathcal{O} \rangle$ is connected.*

 d) *The Archimedes Principle holds true and every Bolzano-Cauchy sequence of elements of T is convergent.*

 e) *The topological space $\langle T, \mathcal{O} \rangle$ is locally compact.*

 f) *Every non-decreasing bounded sequence of elements of T is convergent.*

 g) *The Archimedes Principle and the First Weierstrass Theorem hold true: if f is a continuous function from $\langle 0_T, 1_T \rangle \subseteq T$ into T, then f is bounded.*

Proof. We already know that a) implies any of b)–g), see Theorems 2.15, 2.16, 2.6, 2.19, 2.12, 2.21, and 1.30 (a compact subset of $\mathbb{R}$ is evidently bounded).

By the proof of Theorem 2.16 we obtain that b) implies c).

Proof that c) $\rightarrow$ a): Assume that $M \subseteq T$ is a non-empty bounded set without a supremum. Let U be the set of all upper bounds of M and $V = T \setminus U$. Evidently both sets U, V are non-empty. We show that both are open, contradicting the connectedness of T. If $x \in U$, then there exists an upper bound $y <_T x$ of M. Then $(y, x +_T 1_T) \subseteq U$ is a neighborhood of x. Thus U is open. Similarly, if $x \in V$, then there exists a $z \in M$, $x <_T z$. Then $(x -_T 1_T, z) \subseteq V$ is a neighborhood of x and therefore V is open, a contradiction.

Proof[2] that d) $\rightarrow$ a): Let $M \subseteq T$ be a non-empty bounded from above set. Let $a \in M$, $b >_T a$ being an upper bound of M. For any natural number n we consider the elements $c_n^i = a +_T i \cdot 2^{-n} \times (b -_T a)$, $i = 0, \ldots, 2^n$ of T. Evidently $a = c_n^0 <_T c_n^i <_T c_n^{2^n} = b$ for $0 < i < 2^n$, therefore some of c_n^i are upper bounds of M and some are not. Let k_n be the least i such that c_n^i is an upper bound of M. Note that $k_n > 0$. We set $x_n = c_n^{k_n}$ and claim that $\{x_n\}_{n=0}^{\infty}$ is a Bolzano-Cauchy sequence. Note that $\{x_n\}_{n=0}^{\infty}$ is non-increasing and every x_n is an upper bound of M.

Let $\varepsilon \in T$, $\varepsilon >_T 0_T$. Since the Archimedes Principle holds true in T, there exists a natural number n_0 such that $2^{-n_0} \times (b -_T a) <_T \varepsilon/2$. Since

$$x_{n_0} -_T 2^{-n_0} \times (b -_T a) \leq x_n \leq x_{n_0}$$

for any $n > n_0$, we obtain that $|x_n -_T x_m| <_T \varepsilon$ for any $n, m \geq n_0$.

Therefore the sequence $\{x_n\}_{n=0}^{\infty}$ is convergent. Let x be its limit. Since

$$a \leq c_n^{k_n - 1} \leq x_n \leq c_n^{k_n} \leq b,$$

we have $a \leq_T x \leq_T b$. We show that $x = \sup M$. If $y \in M$, then $y \leq_T x_n$ for any n. Thus also $y \leq_T x = \lim_{n\to\infty} x_n$ and x is an upper bound of M. Assume now that $z \leq_T x$ is an upper bound of M. By the definition of k_n we obtain that $c_n^{k_n - 1}$ is not an upper bound of M and therefore

$$c_n^{k_n - 1} = x_n -_T 2^{-n} \times (b -_T a) < z$$

for any n. Then also $x = \lim_{n\to\infty} x_n \leq_T z$ (we use again the Archimedes Principle).

Proof that e) $\rightarrow$ f): If $\langle T, \mathcal{O} \rangle$ is locally compact, then it is easy to see that every closed interval is compact. If $\{x_n\}_{n=0}^{\infty}$ is a non-decreasing sequence bounded from above by $a \in T$, then all elements of $\{x_n\}_{n=0}^{\infty}$ lie in the compact set $\langle x_0, a \rangle$. By Corollary 1.29 there exists a cluster point x of this sequence. It is easy to see that $\lim_{n\to\infty} x_n = x$.

Proof that f) $\rightarrow$ d): Let $a, \varepsilon \in T$, $\varepsilon >_T 0_T$. Since the increasing sequence $\{n \times \varepsilon\}_{n=0}^{\infty}$ cannot be convergent (it is not B-C!), by f) it is not bounded from above. Therefore a is not an upper bound of $\{n \times \varepsilon\}_{n=0}^{\infty}$ and we obtain the Archimedes Principle.

Assume that $\{x_n\}_{n=0}^{\infty}$ is a Bolzano-Cauchy sequence. Let k_l be the least natural number k such that $|x_n -_T x_m| <_T 2^{-l}$ for $n, m \geq k$. We set

$$a_0 = x_{k_0} -_T 1_T, \qquad a_{l+1} = \max\{a_l, x_{k_{l+1}} -_T 2^{-l-1} \times 1_T\}.$$

One can easily see that $\{a_n\}_{n=0}^{\infty}$ is non-decreasing, bounded from above. Thus there exists an $a = \lim_{n\to\infty} a_n$. One can easily check that also $a = \lim_{n\to\infty} x_n$.

---

[2]We essentially follow the original proof by B. Bolzano [1817] of the statement in the quotation that introduces this chapter. B. Bolzano did not note that he had used the Archimedes Principle. Moreover, we do it for supremum and Bolzano did it for infimum.

Proof that g) → f): Assume that $\{x_n\}_{n=0}^{\infty}$ is a non-decreasing bounded sequence of elements of T and the limit $\lim_{n\to\infty} x_n$ does not exist. Then for every upper bound b of the sequence $\{x_n\}_{n=0}^{\infty}$ there exists a smaller upper bound $c < b$. Fix an upper bound a of the sequence. One can easily show that the sets

$$U = \{x \in [x_0, a] : (\forall n)\, x >_T x_n\}, \quad [x_0, a] \setminus U = \{x \in [x_0, a] : (\exists n)\, x \leq_T x_n\}$$

are disjoint and open in $[x_0, a]$. Then the piecewise linear function $f : [x_0, a] \longrightarrow T$ such that $f(x_n) = n \times 1_T$ for any n and $f(x) = 0_T$ for $x \in U$ is continuous. Since the Archimedes Principle holds true, the function f is unbounded. $\square$

Even the linear ordering of the real line can be uniquely characterized. We present a simple

Theorem 2.29. *A linearly ordered set $\langle X, \leq \rangle$ is order isomorphic to $\langle \mathbb{R}, \leq \rangle$ if and only if $\langle X, \leq \rangle$ possesses the following properties:*

$$\langle X, \leq \rangle \text{ has no least neither greatest element,} \tag{2.15}$$

$$\langle X, \leq \rangle \text{ is densely ordered,} \tag{2.16}$$

$$\text{there exists a countable dense subset of } \langle X, \leq \rangle, \tag{2.17}$$

$$\langle X, \leq \rangle \text{ satisfies the Bolzano Principle.} \tag{2.18}$$

Proof. Let $\langle X, \leq \rangle$ be a linearly ordered set satisfying (2.15)–(2.17). By (2.17) there is a countable dense subset $S \subseteq X$. Evidently S is densely ordered and has no least neither greatest element. Therefore by Theorem 11.4, b), $\langle S, \leq \rangle$ is order isomorphic to $\langle \mathbb{Q}, \leq \rangle$. The isomorphism can be extended as in the proof of Theorem 2.26. $\square$

One can also find topological properties of a topological space $\langle X, \mathcal{O} \rangle$ such that $\langle X, \mathcal{O} \rangle$ satisfies them if and only if $\langle X, \mathcal{O} \rangle$ is homeomorphic to the real line. We present such results in exercises.

Exercises

2.15 [AC] A Non-Archimedian Field
Let $j \subseteq \mathcal{P}(\omega)$ be an ultrafilter. On the set ${}^\omega\mathbb{R}$ we define relations

$$f =^* g \equiv \{n \in \omega : f(n) = g(n)\} \in j, \qquad f \leq^* g \equiv \{n \in \omega : f(n) \leq g(n)\} \in j.$$

The operations are defined "by coordinates", i.e.,

$$f +^* g = h, \text{ where } h(n) = f(n) + g(n), \qquad f \cdot^* g = h, \text{ where } h(n) = f(n) \cdot g(n).$$

Finally, $\hat{a}$ is the function $\hat{a}(n) = a$ for any $a \in \mathbb{R}$.

a) Show that $=^*$ is an equivalence relation and $\leq^*$ is a linear ordering of the quotient set ${}^\omega\mathbb{R}/=^*$.

b) Show that $\mathcal{R}^* = \langle {}^\omega\mathbb{R}/=^*, =, +^*, \cdot^*, \hat{0}, \hat{1} \rangle$ is a field.

c) $\mathcal{R}^* = \langle {}^\omega\mathbb{R}/=^*, =, \leq^*, +^*, \cdot^*, \hat{0}, \hat{1} \rangle$ is a linearly ordered field.

d) The mapping F defined as $f(a) = \hat{a}$ for any $a \in \mathbb{R}$ is an embedding of $\mathcal{R}$ into $\mathcal{R}^*$.

e) Archimedes' Principle does not hold true in $\mathcal{R}^*$.

   Hint: Take $\varepsilon = \hat{1}$ and $d(n) = n$. For any m natural $m \times \hat{1} =^* m^* <^* d$.

2.16 More Uniqueness

Let $\mathcal{T} = \langle T, =, \leq_T, +_T, \cdot_T, 0_T, 1_T \rangle$ be a linearly ordered field satisfying the Archimedes Principle. Then the following properties of $\mathcal{T}$ are equivalent.

a) $\mathcal{T}$ is order isomorphic to the real line.

b) The Rolle Theorem holds true: If $f : [0_T, 1_T] \longrightarrow T$ is continuous, $f(0_T) = f(1_T)$, and f has a derivative in every point of $(0_T, 1_T)$, then there exists an $x_0 \in (0_T, 1_T)$ such that $f'(x_0) = 0_T$.

c) $[0_T, 1_T] \subseteq T$ is sequentially compact.

d) Every non-empty countable subset of T bounded from above has a supremum.

e) Any non-decreasing sequence bounded from above is convergent.

f) The comparison test holds true.

Hint: Evidently d) $\equiv$ e) $\equiv$ f). Show b) $\rightarrow$ e): If $\{a_n\}_{n=0}^{\infty}$ is a non-decreasing sequence of positive elements bounded from above by b without a limit, we set $f(x) = x^2$ for every x such that $a_0 \leq x$ and $x \leq a_n$ for some n. For $x \leq b$ and such that $x > a_n$ for all n we set $f(x) = (x - b)^2 + a_0^2$. f satisfies conditions of the Rolle Theorem in the interval $[a_0, b]$ and $f'(x) \neq 0$ for all x. Finally from d) deduce the Archimedes Principle and the Bolzano Principle.

2.17 [AC] Continuum

A connected compact topological space X with at least two points is called a **continuum**. A point x of a continuum is called a **cut point** if $X \setminus \{x\}$ is not connected. A point which is not a cut point is called a **noncut point**. In the next we suppose that X is a continuum.

a) X is infinite.

b) If $x \in X$ is a cut point, $X \setminus \{x\} = A \cup B$, A, B are disjoint non-empty open sets, then $\overline{A} = A \cup \{x\}$ is connected and therefore a continuum.

c) For any $x \in X$ there exists $y \in X$, $y \neq x$ such that y is not a cut point.

   Hint: Let $\mathcal{C}$ be the family of all proper subcontinua of X containing the point x. Order $\mathcal{C}$ by $C_1 \subseteq \text{Int}(C_2)$. Take a maximal linearly ordered $\mathcal{C}_0 \subseteq \mathcal{C}$. Show that $\bigcap \{\overline{X \setminus C} : C \in \mathcal{C}_0\} \neq \emptyset$ and take y from the intersection. Let $X \setminus \{y\} = A \cup B$, A, B being open disjoint. Assume $x \in A$. Then $\overline{A} = A \cup \{y\} \notin \mathcal{C}_0$ is a continuum. By the maximality, $\overline{A}$ is not comparable with some $C \in \mathcal{C}_0$, i.e., $A \cap C \neq \emptyset$ and $B \cap C \neq \emptyset$, a contradiction.

d) There exist $a, b \in X$, $a \neq b$ such that a, b are noncut points.

e) Assume that $x \in X$ is a cut point and $a \neq b$ are the only noncut points. Then the sets A, B of b) are (up to order) uniquely determined.

f) Assume that $x \in X$ is a cut point and $a \neq b$ are the only noncut points. If $X \setminus \{x\} = A \cup B$ with non-empty disjoint open A, B, then either $a \in A$, $b \in B$ or $a \in B$, $b \in A$.

   Hint: Assume $a, b \in A$. Then $\overline{B}$ being a continuum has two noncut points c, d.

g) Assume that $x_i \in X$, $i = 1, 2$ are cut points and $a \neq b$ are the only noncut points. If $X \setminus \{x_i\} = A_i \cup B_i$ with non-empty disjoint A_i, B_i, $a \in A_i$, then either $A_1 \subseteq A_2$ or $A_2 \subseteq A_1$.

Hint: If $A_1 \nsubseteq A_2$, then $x_2 \in A_1$ and $X = (A_1 \cap A_2) \cup ((B_1 \cap A_2) \cup B_2) \cup \{x_2\}$. By e) we obtain $A_1 \cap A_2 = A_2$.

h) A separable continuum X is homeomorphic to the unit interval $[0, 1]$ if and only if X has exactly two noncut points.

Hint: If x_1, x_2 are cut points, order the continuum by the relation $A_1 \subseteq A_2$.

i) A separable continuum X is homeomorphic to $\mathbb{T}$ if and only if $X \setminus A$ is not connected for any $A \subseteq X$, $|A| > 1$ and connected for $|A| = 1$.

2.4   Expressing a Real by Natural Numbers

For a practical reason the decomposition (2.9) of a real into integer and fractional parts is used for non-negative reals only. If $x < 0$, it is more convenient to use the decomposition

$$x = -(\lfloor -x \rfloor + (-x - \lfloor -x \rfloor)). \tag{2.19}$$

We shall fix a given natural number $p > 1$ and describe both integer and fractional part of a non-negative real by a sequence of natural numbers smaller than p. We introduce independently such a description for a natural number and a real from the interval $[0, 1]$.

So, let p be a natural number greater than 1. A **p-adic expansion of a natural number** x is a sequence $\{x_i\}_{i=0}^k$ of natural numbers such that

$$0 \leq x_i < p \text{ for } i = 0, \ldots, k, \tag{2.20}$$

$$\text{if } k > 0, \text{ then } x_k \neq 0, \tag{2.21}$$

$$x = \sum_{i=0}^k x_i \cdot p^i. \tag{2.22}$$

We shall simply write

$$x = x_k x_{k-1} \ldots x_0|_p.$$

If $p = 2$ we speak about **binary expansion**.

If x_i are natural numbers such that (2.20) holds true, then

$$\sum_{i=0}^k x_i \cdot p^i \leq (p-1) \sum_{i=0}^k p^i = p^{k+1} - 1 < p^{k+1}. \tag{2.23}$$

Theorem 2.30. *Every natural number has exactly one p-adic expansion.*

Proof. If x is a given natural number we can obtain its p-adic expansion as follows. Take the smallest k such that $p^{k+1} > x$ – the existence follows by the Archimedes

Principle, since $p^n \geq n \cdot p$. Then set $x_k = \lfloor x/p^k \rfloor$. If you have already defined $x_k, \ldots, x_l$, $0 < l \leq k$, set

$$x_{l-1} = \left\lfloor \frac{x - \sum_{i=l}^{k} x_i \cdot p^i}{p^{l-1}} \right\rfloor.$$

Since $p^0 = 1$, we obtain (2.22).

Using the inequality (2.23) one can easily show that the obtained p-adic expansion is unique. □

Now, let $x \in [0,1]$. A sequence $\{x_i\}_{i=1}^{\infty}$ of natural numbers is called a p-**adic expansion of the real** x if

$$0 \leq x_i < p \text{ for } i = 1, \ldots, k, \ldots, \tag{2.24}$$

$$x = \sum_{i=1}^{\infty} x_i \cdot p^{-i}. \tag{2.25}$$

We shall simply write

$$x = 0, x_1 x_2 \ldots x_k \ldots |_p.$$

An expansion is **finite** if there exists an i_0 such that $x_i = 0$ for each $i \geq i_0$.

Let us note that by Theorems 2.21 and 2.22, a) the series (2.25) is convergent. Moreover, if a sequence $\{x_i\}_{i=1}^{\infty}$ satisfies (2.24), then

$$\sum_{i=k}^{\infty} x_i \cdot p^{-i} \leq p^{1-k} \text{ and} \tag{2.26}$$

$$\sum_{i=k}^{\infty} x_i \cdot p^{-i} = p^{1-k} \text{ if and only if } x_i = p - 1 \text{ for every } i \geq k. \tag{2.27}$$

Similarly as above, if $p = 2$ we speak about binary expansion.

Theorem 2.31.

i) *For every real* $x \in [0,1]$ *there exists a p-adic expansion.*

ii) *The p-adic expansion of a real* $x \in [0,1]$ *is unique up to the following exception:* $x = \sum_{i=1}^{\infty} x_i \cdot p^{-i} = \sum_{i=1}^{\infty} y_i \cdot p^{-i}$ *and there exists a $k > 0$ such that*

$$x_i = y_i \text{ for } i < k, \quad y_k = x_k + 1, \tag{2.28}$$

$$x_i = p - 1 \text{ for } i > k, \quad y_i = 0 \text{ for } i > k. \tag{2.29}$$

Proof. The existence can be proved as in Theorem 2.30. Let $x \in [0,1]$. Since

$$1 = 0, (p-1) \ldots (p-1) \ldots |_p,$$

we can assume $x < 1$. Then $px < p$ and we write $x_1 = \lfloor px \rfloor$. Evidently

$$x_1 \in \{0, \ldots, p-1\}.$$

If x_i are already defined for $i < k$ we set

$$x_k = \left\lfloor p^k \cdot x - \sum_{i=0}^{k-1} x_i p^{k-i} \right\rfloor. \tag{2.30}$$

Then $\sum_{i=0}^{k} x_i p^{-i} \le x$ for every k and therefore

$$s = \sum_{i=0}^{\infty} x_i p^{-i} \le x.$$

If $s < x$, then there is a k such that $p^{-k} < x - s \le x - \sum_{i=0}^{k} x_i p^{-i}$ and therefore

$$x_k + 1 < p^k \cdot x - \sum_{i=0}^{k-1} x_i \cdot p^{k-i},$$

which is a contradiction with the definition of x_k. Thus (2.25) holds true.

If a real x has two different p-adic expansions, then using (2.26) and (2.27) one can easily show that (2.28) and (2.29) hold true. □

Using the binary expansion we define the set $\mathbb{D}$ of **dyadic numbers** as

$$\mathbb{D} = \{r \in (0, 1) : r \text{ has a finite binary expansion}\}. \tag{2.31}$$

Thus $r \in \mathbb{D}$ if and only if there exist positive integers n, m such that $r = \frac{m}{2^n}$ and $m < 2^n$. By Theorem 11.4 the set $\mathbb{D}$ ordered by $\le$ or the opposite relation $\ge$ is order isomorphic to the set $\mathbb{Q}$ of rationals.

Using 3-adic expansion we can define a set of reals, which is a very useful tool in our investigation. The **Cantor middle-third set** $\mathbb{C}$ is the set of those reals from the interval $[0, 1]$, which have a 3-adic expansion without the number 1. One can easily see that the set of all those reals $x = 0, x_1 \ldots x_n \ldots |_3$ for which $x_n = 1$ (in both expansions eventually) is the union

$$\bigcup_{k=1}^{3n-3} \left(\frac{1 + 3(k-1)}{3^n}, \frac{2 + 3(k-1)}{3^n} \right).$$

Thus

$$\mathbb{C} = [0, 1] \setminus \bigcup_{n=1}^{\infty} \bigcup_{k=1}^{3n-3} \left(\frac{1 + 3(k-1)}{3^n}, \frac{2 + 3(k-1)}{3^n} \right) \tag{2.32}$$

is a closed set. Being a closed subset of the compact set $[0, 1]$, $\mathbb{C}$ is compact.

The cardinality of the uniquely determined real line $\mathbb{R}$ is called **cardinality of the continuum** and is denoted by $\mathfrak{c} = |\mathbb{R}|$.

Theorem 2.32. $\qquad\qquad \mathfrak{c} = |\mathbb{R}| = |\mathbb{C}| = 2^{\aleph_0}.$

Proof. Evidently distinct reals determine distinct Dedekind cuts – subsets of $\mathbb{Q}$. Thus $\mathfrak{c} \leq |\mathcal{P}(\mathbb{Q})| = 2^{\aleph_0}$.

By Theorem 2.31, the mapping $h : {}^{\omega}2 \longrightarrow \mathbb{C}$ defined by

$$h(\{x_i\}_{i=0}^{\infty}) = \sum_{i=0}^{\infty} \frac{2x_i}{3^{i+1}}$$

is one-to-one (and $\mathbb{C} = \mathrm{rng}(h)$). Therefore $2^{\aleph_0} \leq |\mathbb{C}| \leq \mathfrak{c}$. The theorem follows by the Cantor-Bernstein Theorem 1.5. $\qquad\qquad\square$

By Cantor's Theorem 1.6 we obtain

Corollary 2.33. $\qquad\qquad \aleph_0 < \mathfrak{c}.$

So we obtain again the inequality $\mathbb{Q} \neq \mathbb{R}$. Moreover, we know that there exist many more irrational numbers than rational ones.

If $\mathbb{R}$ can be well ordered, then the cardinality of $\mathbb{R}$ is an aleph. Thus there exists an ordinal $\xi > 0$ such that $\mathfrak{c} = 2^{\aleph_0} = \aleph_\xi$. The continuum hypothesis **CH** is equivalent to the statement $\mathfrak{c} = \aleph_1$.

There is another description of reals from $[0, 1]$ by a sequence of positive natural numbers – a continued fraction.

Let $\{a_n\}_{n=0}^{\infty}$ be a sequence of positive natural numbers. The expression

$$\frac{1}{a_0+}\frac{1}{a_1+}\cdots\frac{1}{a_n} = \cfrac{1}{a_0 + \cfrac{1}{a_1 + \cfrac{1}{\ddots + \cfrac{1}{a_n}}}}$$

is called a **finite continued fraction**. Its value is uniquely determined. Actually, the value of a finite continued fraction is a rational number defined by induction as

$$\frac{1}{a_0+}\frac{1}{a_1+}\cdots\frac{1}{a_n+}\frac{1}{a_{n+1}} = \frac{1}{a_0+}\frac{1}{a_1+}\cdots\frac{1}{a_n + \frac{1}{a_{n+1}}}.$$

The expression

$$\frac{1}{a_0+}\frac{1}{a_1+}\cdots\frac{1}{a_n}\cdots = \cfrac{1}{a_0 + \cfrac{1}{a_1 + \cfrac{1}{\ddots + \cfrac{1}{a_n + \cdots}}}}$$

is called a **continued fraction**. Its **value** is

$$a = \frac{1}{a_0+}\frac{1}{a_1+}\cdots\frac{1}{a_n}\cdots = \lim_{n\to\infty}\frac{1}{a_0+}\frac{1}{a_1+}\cdots\frac{1}{a_n}, \tag{2.33}$$

if the limit exists. Actually we show that the limit always exists and is an irrational number.

We define three sequences $\{P_n\}_{n=0}^{\infty}$, $\{Q_n\}_{n=0}^{\infty}$, $\{R_n\}_{n=0}^{\infty}$ by induction as follows.

$$P_0 = 1, \quad P_1 = a_1, \quad Q_0 = a_0, \quad Q_1 = a_0 \cdot a_1 + 1, \tag{2.34}$$
$$P_n = a_n \cdot P_{n-1} + P_{n-2} \text{ for } n \geq 2, \tag{2.35}$$
$$Q_n = a_n \cdot Q_{n-1} + Q_{n-2} \text{ for } n \geq 2, \tag{2.36}$$
$$R_n = P_n/Q_n. \tag{2.37}$$

Evidently, P_n and Q_n are positive integers, R_n is a positive rational. Moreover, by (2.34) and (2.36) we obtain

$$Q_n \geq n + 1. \tag{2.38}$$

By induction one can prove that

$$R_n = \frac{1}{a_0+}\frac{1}{a_1+}\cdots\frac{1}{a_n}. \tag{2.39}$$

From (2.34)–(2.36) one can easily obtain by induction

$$P_n \cdot Q_{n-1} - Q_n \cdot P_{n-1} = (-1)^n. \tag{2.40}$$

Then

$$R_n - R_{n-1} = \frac{(-1)^n}{Q_n \cdot Q_{n-1}} \text{ for } n \geq 1. \tag{2.41}$$

Thus we have

$$0 < R_1 < R_3 < \cdots < R_{2n+1} < \cdots < R_{2n} < \cdots < R_2 < R_0 \leq 1. \tag{2.42}$$

Using (2.41) we obtain

$$\text{the sequence } \{R_n\}_{n=0}^{\infty} \text{ converges.} \tag{2.43}$$

By (2.40) we have

$$\text{the only positive common divisor of } Q_n, Q_{n+1} \text{ is 1.} \tag{2.44}$$

Similarly for P_n and P_{n+1}.

Now we are ready to prove the main result concerning continued fractions.

Theorem 2.34.

  a) *If $\{a_n\}_{n=0}^{\infty}$ is a sequence of positive integers, then there exists the limit (2.33) and a is a positive irrational less than 1.*

  b) *If a is a positive irrational less than 1, then there exists a unique sequence $\{a_n\}_{n=0}^{\infty}$ of positive integers such that (2.33) holds true.*

Proof. The existence of the value a was already proved. By (2.42) for any n we have $R_{2n+1} < a < R_{2n}$, especially $0 < a < 1$.

Assume that $a = p/q$ is rational, p, q have no common divisor and $q > p$. Since $R_{2k+1} < a < R_{2k}$, by (2.41) we have $|R_n - a| < 1/Q_n \cdot Q_{n-1}$. By (2.37) there exists n_0 such that $|R_n - a| \cdot Q_n < 1/q$ for any $n \geq n_0$. Then $|P_n \cdot q - Q_n \cdot p| < 1$. Since all considered numbers are integers, we obtain $P_n \cdot q = Q_n \cdot p$ for $n \geq n_0$. Thus every such Q_n is divisible by q, contradicting (2.44).

Now let $a \in (0,1)$ be irrational. We define two sequences $\{a_n\}_{n=0}^{\infty}$, $\{r_n\}_{n=0}^{\infty}$ such that a_n is a positive integer, $r_n \in (0,1)$ is irrational. We set $a_0 = \lfloor 1/a \rfloor$, $r_0 = 1/a - a_0$. If a_n, r_n are defined we set $a_{n+1} = \lfloor 1/r_n \rfloor$, $r_{n+1} = 1/r_n - a_{n+1}$. From the definitions we immediately obtain

$$a = \cfrac{1}{a_0 + r_0} = \cfrac{1}{a_0 + \cfrac{1}{a_1 + r_1}} = \cdots = \cfrac{1}{a_0 + \cfrac{1}{a_1 +}} \cdots \cfrac{1}{a_n + r_n}.$$

By (2.35), (2.36) and the definition of r_{n+1} we have

$$\frac{P_n(a_{n+1} + r_{n+1}) + P_{n-1}}{Q_n(a_{n+1} + r_{n+1}) + Q_{n-1}} = \frac{P_{n-1}(a_n + r_n) + P_{n-2}}{Q_{n-1}(a_n + r_n) + Q_{n-2}}.$$

Since

$$a = \cfrac{1}{a_0 + \cfrac{1}{a_1 + \cfrac{1}{a_2 + r_2}}} = \frac{P_1(a_2 + r_2) + P_0}{Q_1(a_2 + r_2) + Q_0} = \frac{P_2 + r_2 \cdot P_1}{Q_2 + r_2 \cdot Q_1},$$

we obtain by induction that

$$a = \frac{P_n + P_{n-1} \cdot r_n}{Q_n + Q_{n-1} \cdot r_n}$$

for every $n \geq 2$. By a simple computation we have

$$\left| a - \frac{P_n}{Q_n} \right| = \left| \frac{P_n + P_{n-1} \cdot r_n}{Q_n + Q_{n-1} \cdot r_n} - \frac{P_n}{Q_n} \right| \leq \frac{r_n}{Q_n^2} < \frac{1}{n^2}. \tag{2.45}$$

Thus $\lim_{n \to \infty} P_n/Q_n = a$.

Assume now that

$$a = \cfrac{1}{b_0 + \cfrac{1}{b_1 +}} \cdots \cfrac{1}{b_n} \cdots,$$

where b_n are positive integers. From the above definition of naturals a_n from a given real a one obtains that $b_n = a_n$ for all n. $\qquad\square$

Exercises

2.18 Omitted Proofs

a) Show that the p-adic expansion of a natural number is unique.

Hint: Let $x = \sum_{i=0}^{k} x_i \cdot p^i = \sum_{j=0}^{l} y_j \cdot p^j$. By (2.23) and (2.21) we obtain $k = l$. If $x_j \neq y_j$ for some j, take the greatest one and use (2.21).

b) Show the part (ii) of Theorem 2.31.

Hint: By induction. Assume $x = \sum_{i=1}^{\infty} x_i \cdot p^{-i} = \sum_{i=1}^{\infty} y_i \cdot p^{-i}$ and $x_k \neq y_k$ for some k. Take the smallest one and suppose $x_k < y_k$. Use (2.27).

c) Prove (2.39).

Hint: Set $a'_k = a_{k+1}$ and consider the corresponding sequences $\{P'_n\}_{n=0}^{\infty}$, $\{Q'_n\}_{n=0}^{\infty}$, $\{R'_n\}_{n=0}^{\infty}$. Show that $P_{k+1} = Q'_k$ and $Q_{k+1} = a_0 \cdot Q'_k + P'_k$. Show by induction that $R_{n+1} = \frac{1}{a_0 + R'_n}$.

2.19 Periodic Expansion

A p-adic expansion $\{x_i\}_{i=1}^{\infty}$ of a real is called **periodic** if there are natural numbers k, l such that $x_{i+l} = x_i$ for every $i > k$. Let us notice that if a real has two distinct p-adic expansions, i.e., those with properties (2.28) and (2.29), then they both are periodic.

a) If a real $x \in [0, 1]$ has a periodic p-adic expansion, then x is rational.

Hint: The tail of a periodic p-adic expansion is a geometric series.

b) We use denotation of the proof of Theorem 2.31. Assume that $x = n/m$, $0 < n < m$. Show that there is a natural number $0 \leq r_k < m$ such that

$$x_k = p^k \frac{n}{m} - \sum_{i=0}^{k-1} x_i p^i + \frac{r_k}{m}.$$

c) Let x, r_k be as in b). Show that there are natural numbers $0 \leq k, 0 < l$ such that $r_k = r_{k+l}$.

Hint: Use the Dirichlet Pigeonhole Principle.

d) Conclude that a real $x \in [0, 1]$ has a periodic p-adic expansion if and only if x is rational.

2.20 Continued Fractions

a) Let $r = n/m$, $n < m$ being mutually prime positive natural numbers. Assume that

$$
\begin{aligned}
m &= q_0 n + r_1, & 0 &< r_1 < n, \\
n &= q_1 r_1 + r_2, & 0 &< r_2 < r_1, \\
r_1 &= q_2 r_2 + r_3, & 0 &< r_3 < r_2, \\
&\;\;\vdots & &\;\;\vdots \\
r_{k-2} &= q_{k-1} r_{k-1} + r_k & 0 &< r_k < r_{k-1}, \\
r_{k-1} &= q_k r_k.
\end{aligned}
$$

Then

$$r = \cfrac{1}{q_0 +} \cfrac{1}{q_1 +} \cdots \cfrac{1}{q_k}.$$

b) Every rational real from $(0, 1)$ can be expressed as a finite continued fraction in two ways: one with last term 1 and another with last term greater than 1.

c) **Fibonacci numbers** are defined by induction as $F_0 = F_1 = 1$ and $F_{n+1} = F_n + F_{n-1}$.
   Show that $\lim_{n \to \infty} \frac{F_{n+1}}{F_n} = \frac{1+\sqrt{5}}{2}$.
   Hint: Consider the continued fraction

   $$\frac{1}{1+}\frac{1}{1+} \cdots \frac{1}{1+} \cdots .$$

d) Show that

   $$F_n = \frac{1}{\sqrt{5}}\left(\left(\frac{1+\sqrt{5}}{2}\right)^{n+1} - \left(\frac{1-\sqrt{5}}{2}\right)^{n+1}\right), \qquad F_{n+2} = \sum_{i=0}^{n} F_i + 1$$

   for every n.

e) Assume that (2.33) holds true. If there exists a positive integer $k > 0$ such that
   $a_n = a_{n+k}$ for every n, then there are integers p, q, r, $p^2 + q^2 + r^2 \neq 0$ such that
   $pa^2 + qa + r = 0$.

f) Let a, P_n, Q_n be such as in the proof of Theorem 2.34 If

   $$\left|a - \frac{p}{q}\right| < \left|a - \frac{P_n}{Q_n}\right|,$$

   p, q being integers, then $q > Q_n$.

2.21 Cantor Expansion

Let $\{p_k\}_{k=0}^{\infty}$ be a sequence of natural numbers such that $p_0 = 1$ and $p_k > 1$ for $k > 0$.

a) For any positive natural number m there are natural numbers $k, m_0, \ldots, m_k$ such
   that $m_i < p_{i+1}$, $i = 0, \ldots, k$ and $m = \sum_{i=0}^{k} m_i \cdot (p_0 \cdots p_i)$.
   *Hint: Let k be the first for which $p_0 \cdots p_{k+1} > m$. Take m_k to be the greatest for
   which $m_k \cdot (p_0 \cdots p_k) \leq m$. Continue by induction.*

b) Show that the natural numbers in a) are uniquely determined.

c) For any real $x \in (0, 1]$ there exist natural numbers $\langle x_k : k = 1, \ldots, n, \ldots \rangle$ such that
   $x_k < p_k$, for infinitely many k also $x_k < p_k - 1$ and

   $$x = \sum_{k=1}^{\infty} \frac{x_k}{p_1 \cdots \cdots p_k}. \tag{2.46}$$

   Hint: Define by induction: $y_1 = x$, $x_k = [y_k p_k]$, $y_{k+1} = y_k p_k - x_k$.

d) Show that the expansion of (2.46) is unique.

e) Assume that for any prime p there exists infinitely many k such that $p | p_k$. Then
   the real x expressed by (2.46) is irrational if and only if $x_k \neq 0$ for infinitely many
   k.
   Hint: If $x = n/m$, then there exists a k such that $(n'm) \cdot p_1 \cdots \cdots p_k$ is integer.

f) The real $e = \sum_{k=0}^{\infty} 1/k!$ is irrational.

Historical and Bibliographical Notes

As we have already said Pythagoras (or better said Pythagoreans, since that was a group) about 500 B.C. knew that $\sqrt{2}$ is not a rational number. Actually the proof of (2.7) follows the Pythagorean's proof by reductio ad absurdum. Later on, about 340–50 B.C., Eudoxus developed a theory of proportion trying to avoid the difficulties with incommensurable ratios (essentially with irrational numbers). He also knew what we call Archimedes' Principle 2.6. Actually Archimedes (second half of the 3rd century B.C.) repeatedly used this principle in his method of exhaustion. Eudoxus' theory is well explained by Euclid in his *Elements* ($\Sigma\tau o\iota\chi\varepsilon\tilde{\iota}\alpha$) about 300 B.C. In Book X of Elements, Euclid classifies types of incommensurables, in our terminology, types of irrationals. He investigates only reals of the form $\sqrt{\sqrt{r}\pm\sqrt{s}}$, where r,s are rationals. For more details see, e.g., Morris Kline [1972].

The results of Exercises 2.1 and 2.2 are essentially stated in Euclid's Elements, Book IX. The Chinese Remainder Theorem was known already in the third century A.D. in China. The problem whether there exists a transcendental real was for a long time open. Liouville's Theorem of Exercise 2.4, g), affirmatively answering the problem, was proved by J. Liouville [1844].

At the turn of the 18th and 19th centuries the mathematicians recognized that their creations had not been formulated in a style of the deductive method of Euclid. Neither were the basic properties of reals sufficiently explained. Even the methods of infinitesimal calculus (analysis), in spite of bringing important results, became inaccurate or uncertain. A call for installation of rigor in mathematics had appeared. It was A.L. Cauchy [1821] who began to introduce such rigor in analysis, i.e., in the infinitesimal calculus. He defined a notion of B-C sequence, tried to explain the notion of a continuous function and to prove basic results of analysis. Proving Theorem 2.15 in "a pure analytic way", B. Bolzano [1817] independently started a similar job. He states the Bolzano Principle 2.1 as a theorem. As a starting point of the proof he used the evident fact (in contemporary terminology): every B-C sequence of reals has a limit. His proof contains a gap. Actually he uses implicitly the Archimedes Principle – and by Theorem 2.28 d) he must do so. Anyway his $\mathfrak{Lehrsatz}$ of 1817 was historically the first formulation of the supremum (infimum) principle and that is the reason why I propose to call it the Bolzano Principle.

Theorem 2.12 (for cover by open intervals) was originally proved by E. Heine [1870] as a tool for showing that a continuous function on a closed interval is uniformly continuous. Later on, E. Borel [1895] proved it for countable open interval covers as an important result which he used in investigations in function theory. The presented proof is essentially H. Lebesgue's modification of Borel's proof for any open interval cover.

B. Bolzano has wrote (about 1830–33?) a manuscript [1831] which was not published until 1930. The manuscript contains several important results. The presentation is based on the Bolzano Principle as proved in B. Bolzano [1817] and

Theorem 2.18, which was called the Bolzano-Weierstrass Theorem (in spite of the fact that Bolzano's paper [1831] was not published). Theorems 2.13 and 2.14 are commonly attributed to K. Weierstrass who presented them in his Berlin lectures in the 1960s century. However both theorems are presented in B. Bolzano [1831].

N.K. Bary [1961] attributes the result of Exercise 2.12 to Ch. J. de la Vallée-Poussin and A. Denjoy.

The theory of reals was developed in the 19th century by several mathematicians, ending with R. Dedekind [1888]. For details see, e.g., M. Kline[1972]. However the main principle of this theory is still the Bolzano Principle.

The positional decimal numeral system as presented in Section 2.4 was developed by Hindus in the 7th century. About 825, an Arabic scholar of Baghdad, Mohammed ibn Musa al-Khowârizmî wrote a book on calculation in Hindu style[3] that become known in Europe in several Latin translations and taught Europe on the use of decimal positional notation and calculation based on it. The Latin version of the author's name was translated as *Algoritämi*. That is the origin of the word "algorithm" expressing the main property of calculation in Hindu style: an exact description of the procedure of a calculation.

Leonardo of Pisa, called Fibonacci (son of Bonacci), in his *Practica Geometria* (1220) showed that the roots of the equation $x^3 + 2x^2 + 10x = 20$ are not included in Euclid's classification. Mathematicians of the 17th and 18th centuries used freely the notion of a rational, an irrational, even implicitly those of algebraic and transcendental numbers. They did so often without any exact definition. Nevertheless L. Euler [1737] and more systematically in [1748] laid the foundation of a theory of continued fractions. Using an expansion by a continued fraction he shows that e is irrational. We recommend the reader to a basic source, A.Ya. Kchintchin [1949].

---

[3]He wrote another important book *Al'jabr w'al muqâbala*... devoted to the solution of linear equations. The word Al'jabr was later translated as algebra.

Chapter 3

Metric Spaces and Real Functions

The general theory of topological spaces had its origin at the beginning of the 20th century. Previously topological problems had usually been investigated for only those individual spaces and their subsets for which the concepts of a limit, a cluster point, the closure of a set, etc., had a clear intuitive meaning.

Zdeněk Frolík in E. Čech [1966], p. 233

If a set is endowed with a distance measure between any two points, then one can define the weakest topology on the set in which the distance is continuous. In Section 3.1 we begin with a study of properties of such topologies. Section 3.2 is devoted to the study of an important class of topological spaces – Polish spaces.

Mathematicians tried to understand the structure of some specific subsets of the real line. As a consequence, they began – mainly at the beginning of the 20th century, when set theory was intensively developed – to study those subsets of the real line which are well definable in some way. It turned out that so well-definable sets possess good properties. We present the basic properties of the simplest well-definable sets of reals (more generally, subsets of a Polish space) – Borel sets – in Section 3.3. Since we want, if possible, to avoid the axiom of choice, our presentation is sometimes more complicated.

After the first study of phenomena connected with, in contemporary terminology, the topology of the real line and related spaces, mainly the French mathematicians at the end of the 19th century realized that similar phenomena occur in the sets of real-valued functions endowed with the structure of a convergence of a sequence. Moreover, it turned out that the structure related to the convergence of real functions is close to the structure of the simplest well-definable sets of reals. Section 3.4 investigates topics related to a convergence of a sequence of real functions, especially those related to measurability of functions. Finally, Section 3.5 is devoted to the basic study of Baire Hierarchy of real functions.

3.1   Metric and Euclidean Spaces

Let X be a non-empty set. A mapping $\rho : X \times X \longrightarrow \mathbb{R}$ is called a **metric** on X if for any $x, y, z \in X$ the following holds true:

$$\rho(x, y) \geq 0, \tag{3.1}$$

$$\rho(x, y) = 0 \equiv x = y, \tag{3.2}$$

$$\rho(x, y) = \rho(y, x), \tag{3.3}$$

$$\rho(x, z) \leq \rho(x, y) + \rho(y, z). \tag{3.4}$$

The couple $\langle X, \rho \rangle$ is called a **metric space**. When the metric is understood we shall say that X is a metric space.

We begin with some examples of metrics. Set

$$\mathbb{R}_k = \mathbb{R}^k = \{\langle x_1, \dots, x_k \rangle : x_1, \dots, x_k \in \mathbb{R}\}.$$

For $x = \langle x_1, \dots, x_k \rangle$, $y = \langle y_1, \dots, y_k \rangle \in \mathbb{R}_k$ we define the **Euclidean distance** by

$$\rho(x, y) = \sqrt{(x_1 - y_1)^2 + \cdots + (x_k - y_k)^2}. \tag{3.5}$$

The Euclidean distance is a metric on $\mathbb{R}_k$. The metric space $\langle \mathbb{R}_k, \rho \rangle$ is called the **Euclidean space**. Let us note that for $k = 1$ we obtain $\mathbb{R}_k = \mathbb{R}$ and $\rho(x, y) = |x - y|$.

Let X be a non-empty set. On the set $^\omega X$ of all infinite sequences of elements of X we define the **Baire metric** as

$$\rho(\alpha, \beta) = \begin{cases} \frac{1}{n+1} & \text{if } \alpha \neq \beta \text{ and } n = \min\{k : \alpha(k) \neq \beta(k)\}, \\ 0 & \text{otherwise.} \end{cases}$$

Note that ρ is an **ultrametric**, i.e.,

$$\rho(\alpha, \beta) \leq \max\{\rho(\alpha, \gamma), \rho(\gamma, \beta)\}$$

for any $\alpha, \beta, \gamma \in {}^\omega X$.

The metric space $\langle {}^\omega \omega, \rho \rangle$ is the **Baire space** and $\langle {}^\omega 2, \rho \rangle$ is the **Cantor space**. We can identify the sets $^\omega 2$ and $\mathcal{P}(\omega)$ in a natural way: a sequence $\{a_n\}_{n=0}^\infty \in {}^\omega 2$ is identified with the set $A = \{n \in \omega : a_n = 1\}$. We shall consider $\mathcal{P}(\omega)$ with the metric induced from a Cantor space.

Let $\langle X, \rho \rangle$ be a metric space. Let $a \in X$ and r be a positive real. The set

$$\mathrm{Ball}_\rho(a, r) = \{x \in X : \rho(a, x) < r\}$$

is an **open ball with center a and radius r**. If the metric ρ is clear from the context we shall write simply $\mathrm{Ball}(a, r)$. Moreover, we shall often omit the word "open". The **closed ball** $\overline{\mathrm{Ball}}_\rho(a, r)$ is defined as

$$\overline{\mathrm{Ball}}_\rho(a, r) = \{x \in X : \rho(a, x) \leq r\}.$$

A set $A \subseteq X$ is **bounded** if $A \subseteq \mathrm{Ball}(a, r)$ for some $a \in X$ and $r > 0$. If A is non-empty and bounded, then the **diameter** of A is the real

$$\mathrm{diam}(A) = \sup\{\rho(x, y) : x, y \in A\}.$$

A metric ρ on a set X defines a topology which we shall denote by $\mathcal{O}_\rho$. Again, the subscript ρ will be omitted when it can be understood from the context. A set A is said to be open, i.e., $A \in \mathcal{O}_\rho$, if for every $x \in A$ there exists a positive real r such that $\mathrm{Ball}_\rho(x, r) \subseteq A$. One can easily check that $\mathcal{O}_\rho$ is a topology. The topology $\mathcal{O}_\rho$ will be called the **topology induced by the metric** ρ. Since every open ball belongs to $\mathcal{O}_\rho$, the set of all open balls $\{\mathrm{Ball}_\rho(x, r) : x \in X \wedge r > 0\}$ is a base of the topology $\mathcal{O}_\rho$. By (3.2) we obtain that the topology $\mathcal{O}_\rho$ is Hausdorff. It is easy to see that a closed ball $\overline{\mathrm{Ball}}_\rho(a, r)$ is a closed set (not necessarily the topological closure of the open ball $\mathrm{Ball}_\rho(a, r)$).

If $\langle X, \rho \rangle$ is a metric space, we set $\rho_1(x, y) = \min\{1, \rho(x, y)\}$. Then ρ_1 is a metric on X and induces the same topology as the metric ρ. Thus, if we need we can assume that the values of a metric are in the interval $[0, 1]$.

The topology on $\mathbb{R}_k$ induced by the Euclidean distance (3.5) is denoted by $\mathcal{E}_k$ and is called the **Euclidean topology**. It is easy to see that $\mathcal{E}_1 = \mathcal{E}$. If we do not say otherwise, we always consider $\mathbb{R}_k$ equipped with Euclidean topology.

If $x = \langle x_1, \ldots, x_k \rangle, y = \langle y_1, \ldots, y_k \rangle \in \mathbb{R}_k$, $t \in \mathbb{R}$ we set

$$x + y = \langle x_1 + y_1, \ldots, x_k + y_k \rangle, \qquad t \cdot x = tx = \langle tx_1, \ldots, tx_k \rangle. \qquad (3.6)$$

If $A, B \subseteq \mathbb{R}_k$, then the **arithmetic sum** of A and B is the set

$$A + B = \{x + y \in \mathbb{R}_k : x \in A \wedge y \in B\}.$$

Similarly, the **arithmetic difference** of A and B is the set

$$A - B = \{x - y \in \mathbb{R}_k : x \in A \wedge y \in B\}.$$

If we work on the unit interval $[0, 1]$, the sum $x + y$ and the difference $x - y$ are always taken modulo 1. The set $x + A = \{x + y : y \in A\}$ is called a **shift** of A. If ρ denotes Euclidean distance, then

$$\rho(x, y) = \rho(x + z, y + z), \qquad \mathrm{Ball}_\rho(x + y, r) = x + \mathrm{Ball}_\rho(y, r).$$

Thus for any open set A also any shift $x + A$ is open. Similarly for other topological properties.

If $a_i, b_i \in \mathbb{R}, a_i < b_i, i = 1, \ldots, k$, then the set

$$(a_1, b_1) \times \cdots \times (a_k, b_k) \subseteq \mathbb{R}_k$$

is called an **open interval** in $\mathbb{R}_k$. Similarly, the set

$$[a_1, b_1] \times \cdots \times [a_k, b_k] \subseteq \mathbb{R}_k$$

is a **closed interval** in $\mathbb{R}_k$. Any open interval is an open set and any closed interval is a closed set in the Euclidean topology.

The quotient group $\mathbb{T} = \mathbb{R}/\mathbb{Z}$ is called a **circle**. We can identify the circle $\mathbb{T}$ with the unit interval $[0, 1]$, in which we have identified the points 0 and 1. The topology of $\mathbb{T}$ is induced by metric $\rho(x, y) = \|x - y\|$ ($\|a\|$ is the distance of the real a to the nearest integer). $\mathbb{T}$ is compact. The addition on $\mathbb{T}$ is the addition modulo 1, i.e., for $x, y, z \in \mathbb{T}$ we have $x + y = z$ if and only if $(x + y) - z \in \mathbb{Z}$. Thus z is the fractional part $\{x + y\}$ of $x + y$. A real $x \in \mathbb{R}$ is considered as the element $y \in \mathbb{T}$ such that $x - y \in \mathbb{Z}$. The shift, arithmetic sum and arithmetic difference are taken modulo 1. A real-valued function $f : \mathbb{T} \longrightarrow \mathbb{R}$ can be identified with a periodic function $f : \mathbb{R} \longrightarrow \mathbb{R}$, i.e., with a function satisfying $f(x + 1) = f(x)$ for any $x \in \mathbb{R}$.

Theorem 3.1. *The Cantor middle-third set $\mathbb{C}$ is homeomorphic with the Cantor space $^\omega 2$. Thus the Cantor space $^\omega 2$ is compact.*

The proof is easy. If $x \in \mathbb{C}$ then there exists a unique 3-adic expansion

$$x = 0, x_1 \ldots x_n \ldots |_3 \text{ with } x_n = 0, 2.$$

We set $f(x) = \{x_{n+1}/2\}_{n=0}^{\infty} \in {}^\omega 2$. If $|x - y| < 3^{-k}$, then $x_i = y_i$ for every $i \le k$ and therefore $\rho(f(x), f(y)) < 1/k$. Thus f is continuous. Since f is one-to-one and $\mathbb{C}$ is compact, by Corollary 1.31 f is a homeomorphism.   □

We shall continue with a simple result which allows us to test continuity of mappings into $\mathbb{R}_k$.

Theorem 3.2. *Let $f : X \longrightarrow \mathbb{R}_k$, $f_1 : X \longrightarrow \mathbb{R}, \ldots, f_k : X \longrightarrow \mathbb{R}$ be mappings defined on a topological space $\langle X, \mathcal{O} \rangle$ such that $f(x) = \langle f_1(x), \ldots, f_k(x) \rangle$ for any $x \in X$. Then f is continuous (at a point $a \in X$) if and only if each f_i, $i = 1, \ldots, k$ is continuous (continuous at a point a).*

Now we show that Euclidean spaces $\mathbb{R}_k$ possess some important properties.

Theorem 3.3. *Any closed interval is a compact set. Thus, $\mathbb{R}_k$ is locally compact.*

Proof. For any $k > 0$ the mapping $w_k : \mathbb{C} \longrightarrow \mathbb{R}_k$ is defined as follows. If

$$x = 0, x_1 x_2 \ldots x_n \ldots |_3$$

is the 3-adic expansion of a real $x \in \mathbb{C}$ not containing 1, we set

$$w_k(x) = \langle y_1, \ldots, y_k \rangle,$$

where

$$y_i = \sum_{j=0}^{\infty} \frac{x_{i+k \cdot j}}{2^{j+1}}, \qquad i = 1, \ldots, k. \tag{3.7}$$

Using Theorem 3.2 one can easily check that w_k is a continuous mapping of $\mathbb{C}$ onto $[0, 1]^k$. Thus $[0, 1]^k$ is compact. Since any closed interval in $\mathbb{R}_k$ is a continuous image of $[0, 1]^k$, the theorem follows.   □

Now we can prove the basic result about compact subsets of Euclidean spaces.

Theorem 3.4 (E. Borel – E. Heine). *A subset $X \subseteq \mathbb{R}_k$ is compact if and only if it is closed and bounded.*

Proof. If X is bounded, then X is a subset of a closed interval. Being closed, X is compact.

Vice versa, if X is compact, then by Theorem 1.27, b) X is closed. Take any point $a \in X$. Then $\{\mathrm{Ball}(a, n) : n > 0\}$ is an open cover of X. Since X is compact, there exists a finite subcover and therefore an $n_0 > 0$ such that $X \subseteq \mathrm{Ball}(a, n_0)$. $\qquad\square$

Theorem 3.5. *Any interval in $\mathbb{R}_k$ is connected.*

Proof. Let $I \subseteq \mathbb{R}_k$ be an interval. Assume that I is not connected. Then there are open sets U, V such that $I \cap U \neq \emptyset$, $I \cap V \neq \emptyset$, $U \cap V \cap I = \emptyset$ and $I \subseteq U \cup V$. Let $a = [a_1, \ldots, a_k] \in I \cap U$ and $b = \langle b_1, \ldots, b_k \rangle \in I \cap V$. The mapping $f : [0, 1] \longrightarrow I$ defined by

$$f(t) = \langle a_1 + t(b_1 - a_1), \ldots, a_k + t(b_k - a_k) \rangle$$

is continuous. The sets U, V witness that the set $\mathrm{rng}(f)$ is not connected, a contradiction with Theorem 1.33. $\qquad\square$

Theorem 3.6 (H. Steinhaus).

$$\mathbb{C} + \mathbb{C} = [0, 2], \quad \mathbb{C} - \mathbb{C} = [-1, 1],$$

or, when we consider $\mathbb{C}$ as a subset of $\mathbb{T}$,

$$\mathbb{C} + \mathbb{C} = \mathbb{C} - \mathbb{C} = \mathbb{T}.$$

Proof. Let

$$C_n = [0, 1] \setminus \bigcup_{k=1}^{3n-3} \left(\frac{1 + 3(k-1)}{3^n}, \frac{2 + 3(k-1)}{3^n} \right).$$

Then by (2.32) we have $\mathbb{C} = \bigcap_n C_n$ and also $\mathbb{C} \times \mathbb{C} = \bigcap_n C_n \times C_n$.

Let $t \in [0, 2]$ and $A = \{\langle x, y \rangle : x, y \in [0, 1] \land x + y = t\}$. A is a compact set. One can easily see that $A \cap (C_n \times C_n) \neq \emptyset$ for any n. Thus $A \cap (\mathbb{C} \times \mathbb{C}) \neq \emptyset$. However, if $\langle x, y \rangle \in A \cap (\mathbb{C} \times \mathbb{C})$ then $t = x + y \in \mathbb{C} + \mathbb{C}$.

Note that $\mathbb{C} = 1 - \mathbb{C}$. $\qquad\square$

Let X be a non-empty set, ρ being the Baire metric on $^\omega X$. The set $^\omega X$ may be viewed as the set of all branches of the tree $\langle ^{<\omega}X, \subseteq \rangle$. Moreover, if T is a pruned subtree of $^{<\omega}X$ then $[T] \subseteq {}^\omega X$. In accordance with the notation introduced in Section 11.1 for any $s \in {}^A X$, $A \subseteq \omega$ we write

$$[s] = \{\alpha \in {}^\omega X : s \subseteq \alpha\}.$$

Recall that if $s \in {}^{<\omega}X$, then $[s]$ is the set of all branches containing the node s. One can easily see that

$$\mathrm{Ball}_\rho(\alpha, 1/(n+1)) = \overline{\mathrm{Ball}}_\rho(\alpha, 1/n + 2) = [\alpha|n].$$

Thus $\{[s] : s \in {}^{<\omega}X\}$ is a base of the topology $\mathcal{O}_\rho$. Note that also

$$\{[s] : \mathrm{dom}(s) \in [\omega]^{<\omega} \wedge \mathrm{rng}(s) \subseteq X\}$$

is a base of the topology $\mathcal{O}_\rho$.

We show

Theorem 3.7. *A subset $C \subseteq {}^\omega X$ is closed if and only if there exists a pruned subtree $T \subseteq {}^{<\omega}X$ such that C is the set $[T]$ of all branches of the tree T.*

The proof is easy. If $C = [T]$ and $\alpha \notin C$, then there exists an n such that $\alpha|n \notin T$. Then $[\alpha|n] \cap [T] = \emptyset$. Thus $[T]$ is closed. Vice versa, if $C \subseteq {}^\omega X$ is closed we set $T = \{\alpha|n : \alpha \in C \wedge n \in \omega\}$. One can easily see that T is a pruned tree and $C \subseteq [T]$. Let $\alpha \in [T]$. Then $\alpha|n \in T$ for any n. Thus for any n there exists a $\beta \in C$ such that $\alpha|n = \beta|n$. In other words $[\alpha|n] \cap C \neq \emptyset$. Since C is closed, we obtain $\alpha \in C$. □

Corollary 3.8. *Assume that X is non-empty and can be well ordered. Then any non-empty closed subset $A \subseteq {}^\omega X$ is a retract of ${}^\omega X$.*

Proof. Let $A = [T]$, T being a pruned subtree of ${}^{<\omega}X$. We start with a construction of a "retraction" $R : {}^{<\omega}X \xrightarrow{\text{onto}} T$ of trees. Set $R(\emptyset) = \emptyset$. Assume that $R(s) \in T$ is already defined and $x \in X$. If $R(s)^\frown x \in T$ we set $R(s^\frown x) = s^\frown x$. If not, then since T is a pruned tree there is (the first in the well-ordering) element $y \in X$ such that $R(s)^\frown y \in T$. We set $R(s^\frown x) = s^\frown y$. Now we define $r(\alpha) = \bigcup_n R(\alpha|n)$ for a branch $\alpha \in {}^\omega X$. r is the desired retraction. □

The mapping $\Pi_X : {}^\omega X \times {}^\omega X \xrightarrow[\text{onto}]{1-1} {}^\omega X$ defined by (1.11) is a homeomorphism.

Also the mapping $G : {}^\omega X \xrightarrow[\text{onto}]{1-1} ({}^\omega X)^\omega$ defined by $G(\alpha) = \beta$, where

$$\beta(n)(k) = \alpha(\pi(n,k)) \text{ for any } n, k \in \omega$$

is a homeomorphism. Thus

Theorem 3.9. *${}^\omega X \times {}^\omega X$ and $({}^\omega X)^\omega$ are homeomorphic to ${}^\omega X$. Especially ${}^\omega 2$, $({}^\omega 2)^n$ and $({}^\omega 2)^\omega$ are mutually homeomorphic for any $n > 0$. Similarly for ${}^\omega \omega$, $({}^\omega \omega)^n$ and $({}^\omega \omega)^\omega$.*

We know already that the Cantor space ${}^\omega 2$ is compact. That is not the case of the Baire space ${}^\omega \omega$.

Lemma 3.10. *Let $F : \omega \longrightarrow [\omega]^{<\omega}$. If $C \subseteq {}^\omega \omega$ is a closed set such that*

$$(\forall \alpha)\, (\alpha \in C \rightarrow (\forall n)\, \alpha(n) \in F(n)), \tag{3.8}$$

then C is compact. Vice versa, if $C \subseteq {}^\omega \omega$ is a compact set, then there exists a function $F : \omega \longrightarrow [\omega]^{<\omega}$ such that (3.8) holds true.

Proof. Let $C \subseteq {}^\omega\omega$ be compact. We set $F(n) = \{\alpha(n) : x \in C\}$ for every $n \in \omega$. We have to prove that $F(n)$ is finite for each n. To do so, for a given $n \in \omega$ we set $U_k = \{\alpha \in {}^\omega\omega : \alpha(n) = k\}$. Then U_k are open, pairwise disjoint sets and $\{U_k : k \in F(n)\}$ is an open cover of C. Thus $F(n)$ must be finite.

Now assume that F satisfies (3.8). We show that the product $\Pi_n F(n)$, each $F(n)$ equipped with a discrete topology, is compact.

Let $\{n_k\}_{k=0}^\infty$ be an increasing sequence of natural numbers such that $n_0 = 0$ and $|F(k)| \leq 2^{n_{k+1}-n_k}$. Let $g_k : {}^{n_{k+1}-n_k}2 \overset{\text{onto}}{\longrightarrow} F(k)$. We define a continuous surjection $h : {}^\omega 2 \longrightarrow \Pi_n F(n)$ as follows:

$$h(\alpha) = \beta, \text{ where } \beta(k) = g_k(\alpha|(n_{k+1} - n_k)) \in F(k) \text{ for any } k \in \omega.$$

Since ${}^\omega 2$ is compact, also $\Pi_n F(n)$ is compact. C is a closed subset of $\Pi_n F(n)$, therefore compact. $\square$

Theorem 3.11. *Let C be a closed subset of ${}^\omega\omega$. Then the following are equivalent:*

a) *C is compact.*

b) *If T is a subtree of ${}^{<\omega}\omega$ such that $C = [T]$, then the degree of branching of every node of T is finite.*

c) *There exists a function $\beta \in {}^\omega\omega$ such that*

$$C \subseteq \{\alpha \in {}^\omega\omega : (\forall n)\, \alpha(n) \leq \beta(n)\}.$$

Proof. Assume that C is compact. Then, being closed, $C = [T]$ for a subtree $T \subseteq {}^{<\omega}\omega$. By Lemma 3.10 there exists a function $F : \omega \longrightarrow [\omega]^{<\omega}$ such that (3.8) holds true. Then the degree of branching of any node in $T(n)$ is at most $|F(n+1)|$.

Assume that b) holds true. Thus $C = [T]$, where every node in T has a finite degree of branching. If we set

$$\beta(n) = \begin{cases} \max\{k : \emptyset \frown k \in T\} & \text{for } n = 0, \\ \max\{k : s \frown k \in T, s \in T \cap {}^n\omega\} & \text{for } n > 0, \end{cases}$$

we obtain c).

If c) holds true, then we set $F(n) = \{k : k \leq \beta(n)\}$. Then (3.8) holds true and therefore C is compact. $\square$

If a metric space $\langle X, \rho \rangle$ is separable, then the topology $\mathcal{O}_\rho$ has a countable base. Actually, if $A \subseteq X$ is a countable dense subset then

$$\{\text{Ball}_\rho(a, r) : a \in A \wedge r > 0 \wedge r \in \mathbb{Q}\}$$

is a countable base.

As a consequence of Theorem 1.39 we have

Theorem [wAC] 3.12. *A separable metric space is Fréchet.*

By Theorem 1.23 we obtain

Theorem 3.13.

a) *If X is a metric separable space, then $|X| \leq \mathfrak{c}$.*

b) **wAC** *implies that a metric space with a countable base is separable.*

One can easily see that $\mathbb{Q}^k$ is dense in the Euclidean space $\mathbb{R}_k$. Thus every $\mathbb{R}_k$ is separable.

Now let ρ be the Baire metric on the space $^\omega X$, where X is a non-empty countable set. Then the set of eventually constant sequences

$$\{\{a_n\}_{n=0}^\infty \in {}^\omega X : (\exists n_0)(\forall n > n_0)\, a_n = a_{n_0}\}$$

is a countable dense subset of $^\omega X$. Thus both Baire and Cantor spaces are separable.

Now, let $\langle X, \rho \rangle$ be a metric space. If $A \subseteq X$ is non-empty, we define

$$\rho(x, A) = \inf\{\rho(x, y) : y \in A\}.$$

If A is closed then

$$f(x) = \rho(x, A) \text{ is a continuous function and } A = f^{-1}(\{0\}). \qquad (3.9)$$

Theorem 3.14 (P.S. Urysohn). *If $A, B \subseteq X$ are disjoint non-empty closed subsets of a metric space X, then there exists a continuous function $f : X \longrightarrow [0, 1]$ such that $A = f^{-1}(\{0\})$ and $B = f^{-1}(\{1\})$. Hence, a metric space is normal.*

Proof. We set

$$f(x) = \frac{\rho(x, A)}{\rho(x, A) + \rho(x, B)} \quad \text{for } x \in X.$$

The function f is continuous and possesses the desired properties.

The sets

$$U = \{x \in X : f(x) < 1/2\}, \qquad V = \{x \in X : f(x) > 1/2\},$$

are disjoint open and $A \subseteq U$ and $B \subseteq V$. $\qquad\qquad\qquad\qquad\qquad\qquad$ □

The established property of metric spaces can be extended for normal topological spaces.

Theorem [AC] 3.15 (P.S. Urysohn). *If A, B are disjoint non-empty closed subsets of a normal topological space X, then there exists a continuous function f from X into $[0, 1]$ such that $f(x) = 0$ for $x \in A$ and $f(x) = 1$ for $x \in B$.*

Proof. Let $A, B \subseteq X$ be closed, non-empty and disjoint. We construct a family of open sets $\{U_r : r \in \mathbb{D} \cup \{0, 1\}\}$ such that $A \subseteq U_0$, $U_1 = X \setminus B$ and $\overline{U_r} \subseteq U_s$ for any $r < s$.

Let $\{r_n : n \in \omega\}$ be an enumeration of the set $\mathbb{D} \cup \{0,1\}$. We can assume that $r_0 = 0$ and $r_1 = 1$. We shall proceed by induction. Assume that U_{r_k}, $k < n$, $n > 1$ are already constructed. Let

$$s_1 = \max\{r_k : k < n \wedge r_k < r_n\}, \quad s_2 = \min\{r_k : k < n \wedge r_n < r_k\}.$$

Since X is a normal topological space, there exist disjoint open sets $U, V \subseteq X$ such that $\overline{U}_{s_1} \subseteq U$ and $X \setminus U_{s_2} \subseteq V$. We set $U_{r_n} = U$.

Now define

$$f(x) = \begin{cases} \inf\{r \in \mathbb{D} : x \in U_r\} & \text{if } x \notin B, \\ 1 & \text{if } x \in B. \end{cases}$$

One can easily show that

$$f(x) < a \equiv (\exists r \in \mathbb{D})(r < a \wedge x \in U_r), \quad f(x) > a \equiv (\exists r \in \mathbb{D})(a < r \wedge x \notin \overline{U}_r),$$

for any $a \in (0,1)$. Thus the set

$$f^{-1}((a,b)) = \bigcup\{U_r : r < b\} \cap \bigcup\{(X \setminus \overline{U}_r) : r > a\}$$

is open and therefore f is continuous.

By construction of the function f we obtain that $f(x) = 0$ for $x \in A$ and $f(x) = 1$ for $x \in B$. □

The property "to be a normal space" is not hereditary, i.e., there exists a normal topological space X and a subset $Y \subseteq X$ that, endowed with the subspace topology, is not normal (see Theorem 1.22, c) and Exercise 1.22). There is a weaker property of a topological space that is hereditary. A Hausdorff topological space X is called **completely regular** if for any closed subset $A \subseteq X$ and a point $a \in X \setminus A$ there exists a continuous function $f : X \longrightarrow [0,1]$ such that $f(a) = 0$ and $f(x) = 1$ for $x \in A$. A subset of a completely regular topological space with the subspace topology is completely regular. By Theorems 3.14 and 3.15 every metric and every normal space is completely regular.

There is another important notion. If $f : X \longrightarrow \mathbb{R}$ we write

$$\mathsf{Z}(f) = \{x \in X : f(x) = 0\}.$$

A set $A \subseteq X$ is a **zero set** if there exists a continuous function $f : X \longrightarrow [0,1]$ such that $A = \mathsf{Z}(f)$. Evidently a zero set is a G_δ set.

A normal topological space X is called **perfectly normal** if every closed subset of X is a zero-set. By Urysohn's Theorem 3.14 we obtain

Corollary 3.16. *A metric space is perfectly normal.*

If X is a topological Hausdorff space with a countable base, then X is separable. Since every continuous function from X into the reals is uniquely determined by values in a dense subset, we obtain that there exist at most $\mathfrak{c}$ real continuous functions. Thus, we can apply **wAC** to find some sequence of functions. Especially we can prove

Theorem [wAC] 3.17. *A normal topological space with a countable base is metrizable.*

We shall not need that result. However we shall often need

Theorem [AC] 3.18. *A normal topological space is perfectly normal if and only if every closed subset is a G_δ set.*

Proof. Since every zero set is a G_δ set, in a perfectly normal space every closed set is a G_δ set.

Assume that X is normal and $A \subseteq X$ is closed. Then A is a G_δ set. Therefore $A = \bigcap_n G_n$, where every G_n is open. By Theorem 3.15 there exist continuous functions $f_n : X \longrightarrow [0,1]$ such that $A \subseteq Z(f_n)$ and $f_n(x) = 1$ for every $x \in X \backslash G_n$. For $x \in X$ we set

$$f(x) = \sum_{n=0}^{\infty} 2^{-n-1} f_n(x).$$

By elementary calculus the function f is continuous and $Z(f) = A$.   □

Corollary [AC] 3.19. *If A, B are closed disjoint subsets of a perfectly normal space X, then there exists a continuous function $f : X \longrightarrow [0,1]$ such that $A = Z(f)$ and $B = Z(1 - f)$.*

Proof. Let $g, h : X \longrightarrow [0,1]$ be continuous functions such that $A = Z(g)$ and $B = Z(h)$. The function $f(x) = g(x)/(g(x) + h(x))$ satisfies the conclusion.   □

Theorem 3.20 (H. Tietze – P.S. Urysohn). *If $A \subseteq X$ is a closed subset of a metric space X and $f : A \longrightarrow [0,1]$ is continuous, then there exists a continuous mapping $F : X \longrightarrow [0,1]$ such that $F(x) = f(x)$ for $x \in A$. Moreover, if we assume axiom of choice $\mathbf{AC}$, then the assertion holds true for any normal topological space X.*

Proof. We define

$$F(x) = \begin{cases} f(x) & \text{if } x \in A, \\ \inf\{f(a) + \rho(x,a)/\rho(x,A) - 1 : a \in A\} & \text{otherwise.} \end{cases}$$

One can easily check that the so-defined function F is a continuous extension of f.

If X is a normal topological space, you can find a proof (based on $\mathbf{AC}$) in any standard textbook of topology, e.g., R. Engelking [1977].   □

Using the theorem, the mapping w_k of the proof of Theorem 3.3 can be continuously extended to a surjection $w_k : [0,1] \longrightarrow [0,1]^k$. Such an extended mapping is called a **Peano curve**. Note that w_k for $k > 1$ is not an injection, since $[0,1]$ and $[0,1]^k$ are not homeomorphic.

Corollary [wAC] 3.21. *Assume that A is a closed subset of a separable metric space X. If $\langle f_n : A \longrightarrow [0,1] : n \in \omega \rangle$ are continuous and $f_n \to 0$ on A, then there are continuous functions $\langle F_n : X \longrightarrow [0,1] : n \in \omega \rangle$ such that $F_n \to 0$ on X and $f_n = F_n|A$ for every n. Moreover, the unit interval $[0,1]$ may be replaced by $\mathbb{R}$ or* *an open interval. If we assume* **AC**, *then the assertion holds true for any normal topological space.*

Proof. By the Tietze-Urysohn Theorem 3.20, for every $n \in \omega$ there exists a continuous function $H_n : X \longrightarrow [0,1]$ such that $H_n|A = f_n$. Since A is closed, there are open sets $\langle U_n : n \in \omega \rangle$ such that $A = \bigcap_n U_n$. We can assume that $U_{n+1} \subseteq U_n$. By Urysohn's Theorem 3.14 there are continuous functions $h_n : X \longrightarrow [0,1]$ such that $h_n(x) = 1$ for $x \in A$ and $h_n(x) = 0$ for $x \in X \setminus U_n$. It suffices to set $F_n = H_n \cdot h_n$.

If $f_n : A \longrightarrow (0,1)$ and we construct a continuous extension $F_n : X \longrightarrow [0,1]$, then we improve it as follows. The set $B_n = F_n^{-1}(\{0,1\})$ is a closed set disjoint with A. So there exists continuous $g_n : X \longrightarrow [0,1]$ such that $g_n(x) = 0$ for $x \in B_n$ and $g_n(x) = 1$ for $x \in A$. Then $F_n \cdot g_n : X \longrightarrow (0,1)$ has the desired properties. $\square$

Let $\langle X_1, \rho_1 \rangle$ and $\langle X_2, \rho_2 \rangle$ be metric spaces. Let $a \in X_1$. According to Theorem 1.25 a function $f : X_1 \longrightarrow X_2$ is continuous at the point a if and only if

$$(\forall \varepsilon > 0)(\exists \delta > 0)(\forall x)\, (x \in \mathrm{Ball}_{\rho_1}(a, \delta) \rightarrow f(x) \in \mathrm{Ball}_{\rho_2}(f(a), \varepsilon)).$$

In analysis, the so-called Heine Criterion of continuity is often used. We need a form of the axiom of choice for its proof. Actually we show that the Heine Criterion is equivalent to the weak Axiom of Choice.

Theorem [wAC] 3.22 (Heine Criterion). *Assume that $|X_1| \ll \mathfrak{c}$. The function f is continuous at a point $a \in X_1$ if and only if, for every sequence $\{a_n\}_{n=0}^{\infty}$ of points of X_1 convergent to a, also $\lim_{n \to \infty} f(a_n) = f(a)$.*

Proof. Assume that f is not continuous at the point a. Then there exists an $\varepsilon > 0$ positive such that

$$\mathrm{Ball}_{\rho_1}(a, \delta) \not\subseteq f^{-1}(\mathrm{Ball}_{\rho_2}(f(a), \varepsilon))$$

for any $\delta > 0$. Especially, the set

$$\Phi_n = \mathrm{Ball}_{\rho_1}(a, 1/(n+1)) \setminus f^{-1}(\mathrm{Ball}_{\rho_2}(f(a), \varepsilon))$$

is non-empty for any $n \in \omega$. By **wAC** there exists a selector $\langle a_n \in \Phi_n : n \in \omega \rangle$. Evidently $\lim_{n \to \infty} a_n = a$ and $\lim_{n \to \infty} f(a_n) \neq f(a)$. $\square$

Theorem 3.23. *The following statements are equivalent:*

a) *The Weak Axiom of Choice* **wAC**.

b) $\mathbb{R}$ *is Fréchet.*

c) *A function $f : \mathbb{R} \longrightarrow \mathbb{R}$ is continuous at 0 if and only if for every sequence $\{x_n\}_{n=0}^{\infty}$ with $\lim_{n \to \infty} x_n = 0$ we have $\lim_{n \to \infty} f(x_n) = f(0)$.*

Proof. We already know that a) $\rightarrow$ c). We show that c) $\rightarrow$ b) and b) $\rightarrow$ a).

Assume c) and $a \in \overline{A} \setminus A$. Set $f(x) = 1$ for $x \in A$ and $f(x) = 0$ otherwise. Then f is not continuous in a. Therefore by c) there exists a sequence $\{a_n\}_{n=0}^{\infty}$ with $\lim_{n \to \infty} a_n = a$ and $\lim_{n \to \infty} f(a_n) \neq f(a)$. Thus, for infinitely many n we have $f(a_n) = 1$ and consequently $a_n \in A$.

Assume b) and that $\langle A_n : n \in \omega \rangle$ is a sequence of non-empty subsets of $^{\omega}2$. Let f_n be a bijection of $^{\omega}2$ onto interval $(1/(n+2), 1/(n+1))$. One can easily show that $0 \in \bigcup_n f_n(A_n)$. Thus by b) there exists a sequence $\{x_n\}_{n=0}^{\infty}$ of elements of the set $\bigcup_n f_n(A_n)$ with $\lim_{n\to\infty} x_n = 0$. Set $E = \{n : (\exists m)\, x_m \in f_n(A_n)\}$ and $g(n) = \min\{m : x_m \in f_n(A_n)\}$ for $n \in E$. Then $f_n^{-1}(x_{g(n)}) \in A_n$ for $n \in E$. Now assertion b) follows by Theorem 1.13. □

Exercises

3.1 [AC] Uniform Continuity
A mapping $f : X_1 \longrightarrow X_2$ is called **uniformly continuous** if

$$(\forall \varepsilon > 0)(\exists \delta > 0)(\forall x, y \in X_1)\,(\rho_1(x,y) < \delta \to \rho_2(f(x), f(y)) < \varepsilon).$$

a) If X_1 is compact, then every continuous function is uniformly continuous.
   Hint: For $x \in X_1$ take δ_x such that

$$(\forall y)\,(\rho_1(x,y) < \delta_x \to \rho_2(f(x), f(y)) < \varepsilon/2).$$

   Then there exists a finite set $\{x_0, \ldots, x_n\}$ such that $\langle \mathrm{Ball}(x_i, \delta_{x_i}/2) : i = 0, \ldots, n \rangle$ covers X_1. Take $\delta = \min\{\delta_{x_i} : i = 0, \ldots, n\}/2$.

b) A mapping $f : X_1 \longrightarrow X_2$ is said to be **Lipschitz** if there exists a positive real C such that
$$\rho_2(f(x), f(y)) \le C \cdot \rho_1(x,y) \text{ for any } x, y \in X_1.$$

   Show that a Lipschitz mapping is uniformly continuous.

c) Find a uniformly continuous mapping defined on a compact space that is not Lipschitz.
   Hint: Consider the function $f(x) = x \sin 1/x$ on $[0,1]$ and its derivative in $(0,1)$.

3.2 [AC] Small Separable Metric Spaces
a) If $f(x) = \sum_{n=0}^{\infty} a_n x^n$ converges for every $x \in \mathbb{R}$, not every a_n is 0, then the zero set $\mathsf{Z}(f)$ is countable and has no accumulation point.
   Hint: See any elementary textbook on analytic functions, e.g., W. Rudin [1966].

b) Let $\langle X : \rho \rangle$ be a separable metric space with a bounded metric. Let $\langle a_n \in X : n \in \omega \rangle$ be a dense subset of X. For $a \in X$ we define

$$h_a(x) = \sum_{n=0}^{\infty} \frac{\rho(a, a_n)}{n!} x^n.$$

   If $a \ne b$, then $h_a(x) \ne h_b(x)$ for all but countably many reals x.

c) If $|X| < \mathfrak{c}$, then there exists a real x such that $h_a(x) \ne h_b(x)$ for any $a \ne b$.

d) If $|X| < \mathfrak{c}$, then there exists a Lipschitz injection $f : X \longrightarrow \mathbb{R}$.
   Hint: Take the x of c) and set $f(a) = h_a(x)$.

3.3 [wAC] Small Inductive Dimension

The **small inductive dimension** $\mathrm{ind}(X)$ of a topological space X is defined as follows. We set $\mathrm{ind}(\emptyset) = -1$. If $\mathrm{ind}(X) \leq n - 1$ is already defined we say that $\mathrm{ind}(X) \leq n$ if for any $x \in X$, and for any neighborhood U of x there exists an open set $V \subseteq U$, $x \in V$ and such that $\mathrm{ind}(\mathrm{Bd}(V)) \leq n - 1$. Finally, $\mathrm{ind}(X) = n$ if $\mathrm{ind}(X) \leq n$ and $\mathrm{ind}(X) \leq n - 1$ is false.

a) Show by induction that $\mathrm{ind}(\mathbb{R}_k) \leq k$.

b) Show that $\mathrm{ind}(\mathbb{R}) = 1$.

c) $\mathrm{ind}(\mathbb{Q}) = \mathrm{ind}(\mathbb{R} \setminus \mathbb{Q}) = 0$.

d) $\mathrm{ind}(^\omega 2) = \mathrm{ind}(^\omega \omega) = 0$.

e) If $X = A \cup B$, then $\mathrm{ind}(X) \leq \mathrm{ind}(A) + \mathrm{ind}(B) + 1$.

f) If a Hausdorff space X has a countable base, $\mathrm{ind}(X) = 0$, then for any disjoint closed sets A, B there exists a clopen set U such that $A \subseteq U$ and $U \cap B = \emptyset$.
   Hint: Note that there exists a countable clopen base. For each $x \in X$ choose a clopen basic set U_x with $x \in U_x$ and $U_x \cap A = \emptyset$ or $U_x \cap B = \emptyset$. Enumerate them as $\langle U_n : n \in \omega \rangle$ and set $V_n = U_n \setminus \bigcup_{k<n} V_k$. Show that $U = \bigcup \{ V_n : V_n \cap B = \emptyset \}$ is the desired clopen set.

g) If a Hausdorff space X has a countable base, $X = \bigcup_n X_n$, where each X_n is closed and $\mathrm{ind}(X_n) = 0$, then $\mathrm{ind}(X) = 0$.
   Hint: Use f) and show that $\mathrm{Ind}(X) = 0$.

h) If X_n are separable metric spaces with $\mathrm{ind}(X_n) = 0$, then $\mathrm{ind}(\Pi_n X_n) = 0$.

i) Set $A_k = \{ \langle x_1, \ldots, x_n \rangle \in \mathbb{R}_n : \text{exactly } k \text{ of } x_1, \ldots, x_n \text{ are rational} \}$. Let $r_1, \ldots, r_k$ be rational, $1 \leq i_1 < \cdots < i_k \leq n$. The set

$$\{ \langle x_1, \ldots, x_n \rangle \in \mathbb{R}_n : x_{i_j} = r_j \text{ for } j = 1, \ldots, k \text{ and } x_i \text{ irrational for the other } i \}$$

is closed in A_k and has small inductive dimension zero.

k) Show that $\mathrm{ind}(A_i) = 0$ and $\mathbb{R}_n = \bigcup_{k=0}^{n} A_k$.

3.4 Covering Dimension

The **order** of an open cover $\mathcal{A}$ of a subset A of a topological space X is the greatest natural number n such that there are mutually different sets $U_0, \ldots, U_n \in \mathcal{A}$ with $U_0 \cap \cdots \cap U_n \neq \emptyset$. If there is no such natural number, then the order of $\mathcal{A}$ is ∞. The **covering dimension** $\dim(X)$ of a topological space X is defined as follows: $\dim(X) \leq n$ if every open cover of X has a refinement of order not greater than n.

a) $\dim(X) = 0$ if and only if X has a clopen base of topology. Thus, $\dim(X) = 0$ if and only if $\mathrm{ind}(X) = 0$.

b) $\dim(\mathbb{R}) = 1$.
   Hint: Consider covers by open intervals.

c) If $A \subseteq X$ is closed, then $\dim(A) \leq \dim(X)$.

3.5 Arcwise Connected Space

A continuous mapping $\varphi : [0, 1] \longrightarrow X$ is called an **arc**. The arc φ connects the points $\varphi(0)$ and $\varphi(1)$. A topological space X is called **arcwise connected** if any two points of X can be connected by an arc.

a) Arcwise connected space is connected.

b) The set

$$\{\langle x, y \rangle \in \mathbb{R}_2 : (x = 0 \wedge -1 \le y \le 1) \vee (0 < x \le 1 \wedge y = \sin 1/x)\}$$

is connected and not arcwise connected.

c) A topological space is **locally arcwise connected** if any open neighborhood U of any point $x \in X$ contains an open set $x \in V \subseteq U$ such that V is arcwise connected. Find an arcwise connected space which is not locally arcwise connected.
Hint: Take the dense comb

$$\{\langle x, y \rangle \in [0, 1]^2 : y = 0 \vee (x \text{ rational and } y \in [0, 1])\}.$$

d) Every connected and locally arcwise connected space is arcwise connected.
Hint: Given a point $x \in X$, the set of those points which can be connected with x by an arc is clopen.

3.6 Subgroups of $\mathbb{T}$
We consider topological group $\langle \mathbb{T}, + \rangle$.

a) A subgroup $G \subseteq \mathbb{T}$ is finite if and only if G contains the least positive element.

b) An infinite subgroup of $\mathbb{T}$ is topologically dense in $\mathbb{T}$.

3.2 Polish Spaces

Let $\langle X, \rho \rangle$ be a metric space. A sequence $\{x_n\}_{n=0}^\infty$ is **Bolzano-Cauchy**, briefly **B-C**, if

$$(\forall \varepsilon > 0)(\exists n_0)(\forall n, m > n_0)\, \rho(x_n, x_m) < \varepsilon.$$

Every convergent sequence is a Bolzano-Cauchy sequence. The opposite need not be true. A metric is said to be **complete** if every Bolzano-Cauchy sequence is convergent. If there exists a complete metric that induces the topology $\mathcal{O}$ of a separable topological space $\langle X, \mathcal{O} \rangle$, then X is called a **Polish space** and the topology $\mathcal{O}$ is called **Polish topology**. A Polish space X is called **perfect** if X is a perfect set.

If X is a topological group with topology of a Polish space, then we say that X is a **Polish group**.

Let $\langle x_n = \langle x_n^1, \ldots, x_n^k \rangle : n \in \omega \rangle$ be a sequence of points of $\mathbb{R}_k$. It is easy to see that $\lim_{n \to \infty} x_n = x$, where $x = \langle x^1, \ldots, x^k \rangle \in \mathbb{R}_k$, if and only if $\lim_{n \to \infty} x_n^i = x^i$ for every $i = 1, \ldots, k$. Similarly, the sequence $\{x_n\}_{n=0}^\infty$ is Bolzano-Cauchy if and only if each sequence of reals $\{x_n^i\}_{n=0}^\infty$, $i = 1, \ldots, k$, is Bolzano-Cauchy. Thus the Euclidean space $\mathbb{R}_k$ is complete and also a Polish space. Moreover, $\mathbb{R}_k$ with the addition defined by (3.6) is a Polish commutative group. The circle $\mathbb{T}$ with the addition modulo 1 is a Polish group. The Euclidean distance on an open interval is not a complete metric. However, there exists a complete metric that induces the same topology.

We show that any space ${}^\omega X$, $X \neq \emptyset$ with the Baire metric is complete. Let $\{\alpha_n\}_{n=0}^\infty$ be a Bolzano-Cauchy sequence of points of ${}^\omega X$. We define a sequence α as follows. Let $k \in \omega$. Then there exists an n_0 such that $\rho(\alpha_n, \alpha_m) < 1/(k+1)$ for any $n, m > n_0$. Then $\alpha_n(k) = \alpha_m(k)$ for $n, m > n_0$. Let $\alpha(k)$ be this common value. One can easily see that $\lim_{n\to\infty} \alpha_n = \alpha$.

We already know that the sets of eventually constant functions are countable dense subsets of the Cantor space ${}^\omega 2$ and the Baire space ${}^\omega \omega$, respectively. Hence both ${}^\omega 2$ and ${}^\omega \omega$ are Polish spaces. The Cantor space ${}^\omega 2$ is a Polish group. Similarly we can consider ${}^\omega \omega$ as a Polish group identifying it with ${}^\omega \mathbb{Z}$.

We shall often need the following simple consequence of Theorem 3.1.

Theorem 3.24. *If $T \subseteq {}^{<\omega}\omega$ is a perfect tree, then $[T] \subseteq {}^\omega \omega$ contains a subset homeomorphic to the Cantor middle-third set.*

Proof. If T is a perfect tree, then by Theorem 11.5 T contains a subtree S similar to the tree ${}^{<\omega}2$. Then $[S] \subseteq [T]$ and by Theorem 3.1 $[S]$ is homeomorphic to the Cantor middle-third set. $\qquad\square$

Theorem 3.25. *The Baire space ${}^\omega \omega$ is homeomorphic to the set $\mathrm{Ir} = (0,1) \setminus \mathbb{Q}$ equipped with the subspace Euclidean topology.*

Proof. For $a = \{a_n\}_{n=0}^\infty \in {}^\omega \omega$ we set

$$h(a) = \frac{1}{a_0 +} \frac{1}{a_1 +} \cdots \frac{1}{a_n +} \cdots ,$$

and we denote by $R_n(a)$ the real defined by (2.37). By Theorem 2.34 we obtain that $h : {}^\omega \omega \xrightarrow[\text{onto}]{1-1} \mathrm{Ir}$. If $\rho(a,b) < 1/(n+2)$, then $a_i = b_i$ for $i = 0, \ldots, n$. Thus $R_n(a) = R_n(b)$ and by (2.45) we have

$$|h(a) - h(b)| < \frac{2}{n^2}.$$

Hence, h is continuous. We show that also h^{-1} is continuous.

We begin with an observation. Let $a \in (0,1)$ be an irrational number. Then there exists an $\varepsilon > 0$ depending on a such that $\lfloor 1/a \rfloor = \lfloor 1/x \rfloor$ for any irrational $x \in (0,1)$ such that $|a - x| < \varepsilon$. Actually, if $n = \lfloor 1/a \rfloor$, then $1/(n+1) < a < 1/n$. Take $\varepsilon = \min\{1/n - a, a - 1/(n+1)\}$.

Assume that $a, x \in (0,1)$ are irrationals and

$$a = \frac{1}{a_0 +} \frac{1}{a_1 +} \cdots \frac{1}{a_n +} \cdots , \qquad x = \frac{1}{x_0 +} \frac{1}{x_1 +} \cdots \frac{1}{x_n +} \cdots .$$

The sequence $\{a_n\}_{n=0}^\infty$ is defined similarly as in the proof of Theorem 2.34:

$$a_0 = \lfloor 1/a \rfloor, \ r_0 = 1/a - a_0, \ a_{n+1} = \lfloor 1/r_n \rfloor, \ r_{n+1} = 1/r_n - a_{n+1}.$$

Similarly are defined x_n and s_n. Moreover, we have

$$a = \frac{P_n(a) + P_{n-1}(a) \cdot r_n}{Q_n(a) + Q_{n-1}(a) \cdot r_n}, \qquad x = \frac{P_n(x) + P_{n-1}(x) \cdot s_n}{Q_n(x) + Q_{n-1}(x) \cdot s_n}.$$

To prove that h^{-1} is continuous in a it suffices to find a sequence $\{\varepsilon_n\}_{n=0}^{\infty}$ of positive reals such that

$$|a - x| < \varepsilon_n \to a_0 = x_0, \ldots, a_n = x_n. \tag{3.10}$$

Let $\varepsilon_0 > 0$ be such that $\lfloor 1/a \rfloor = \lfloor 1/x \rfloor$ for any $|x - a| < \varepsilon_0$. Now assume that ε_n satisfying (3.10) is already defined. Let $\delta > 0$ be such that $\lfloor 1/r_n \rfloor = \lfloor 1/y \rfloor$ whenever $|r_n - y| < \delta$. If $|a - x| < \varepsilon_n$ then

$$P_n = P_n(a) = P_n(x), \ Q_n = Q_n(a) = Q_n(x) \text{ and } s_n = \frac{P_n - Q_n \cdot x}{Q_{n-1} \cdot x - P_{n-1}}.$$

Thus s_n depends continuously on x. Let $\varepsilon_{n+1} < \varepsilon_n$ be such that $|r_n - s_n| < \delta$ for $|a-x| < \varepsilon_{n+1}$. Then $a_0 = x_0, \ldots, a_n = x_n, a_{n+1} = x_{n+1}$ for any $|x-a| < \varepsilon_{n+1}$. □

One can easily see that both spaces ${}^{\omega}2$ and ${}^{\omega}\omega$ are zero-dimensional. Moreover

Theorem 3.26. *For a separable metric space X the following are equivalent:*

a) *X is zero-dimensional.*
b) *X is homeomorphic to a subset of ${}^{\omega}2$.*
c) *X is homeomorphic to a subset of ${}^{\omega}\omega$.*

Proof. Since ${}^{\omega}\omega$ is zero-dimensional and ${}^{\omega}2$ is homeomorphic to a subset of ${}^{\omega}\omega$, we have only to prove that a zero-dimensional separable metric space is homeomorphic to a subset of ${}^{\omega}2$.

Let $\langle U_n : n \in \omega \rangle$ be a clopen base of X. We define a mapping $f : X \longrightarrow {}^{\omega}2$ as follows: $f(x) = \alpha$ where $\alpha(n) = 1$ if $x \in U_n$ and $\alpha(n) = 0$ otherwise. One can easily see that f is the desired embedding. □

Theorem 3.27 (R. Baire). *Every Polish space satisfies the Baire Category Theorem.*

We begin with a lemma which is interesting on its own.

Lemma 3.28 (G. Cantor). *Let X be a Polish space. If*

$$C_0 \supseteq C_1 \supseteq \cdots \supseteq C_n \supseteq \cdots$$

are non-empty closed sets such that $\lim_{n \to \infty} \operatorname{diam}(C_n) = 0$, then $\bigcap_n C_n \neq \emptyset$.

Proof. Let $\{r_n : n \in \omega\}$ be a dense subset of X. Set $\varepsilon_n = \operatorname{diam}(C_n)$. For every n let m be the smallest integer such that $\rho(r_m, C_n) < \varepsilon_n$. Set $x_n = r_m$. Note that $\rho(x_n, C_k) < \varepsilon_n$ for any $n \geq k$. Since $\lim_{n \to 0} \varepsilon_n = 0$ the sequence $\{x_n\}_{n=0}^{\infty}$ is Bolzano-Cauchy and therefore there exists an $x = \lim_{n \to \infty} x_n$. If $x \notin C_n$, then for a sufficiently large $m \geq n$ we would have $\rho(x_m, C_n) < \rho(x, C_n)/2$ and $\rho(x_m, x) < \rho(x, C_n)$, a contradiction. Thus $x \in \bigcap_n C_n$. □

Proof of the theorem. Repeat the proof of Theorem 1.36 with $U_n = \overline{\mathrm{Ball}}(a_n, \delta_n)$, where a_n is the first $r_m \in \mathrm{Ball}(a_{n-1}, \delta_{n-1}) \cap A_n$ and $\delta_n \leq 2^{-n}$ is such that $\overline{\mathrm{Ball}}(a_n, \delta_n) \subseteq U_{n-1} \cap A_n$. Then apply the lemma. $\qquad\square$

Theorem [wAC] 3.29 (K. Kuratowski). *Let X be a separable metric space, Y being a Polish space. Assume that $A \subseteq X$ and $f : A \longrightarrow Y$ is continuous. Then there exists a G_δ set $G \supseteq A$ and a continuous function $g : G \longrightarrow Y$ such that $g(x) = f(x)$ for $x \in A$.*

Proof. Let us denote

$$C = \bigcap_n \bigcup \{ U \subseteq X : U \text{ open} \wedge U \cap A \neq \emptyset \wedge \mathrm{diam}(f(A \cap U)) < 1/(n+1) \}.$$

Then $C \supseteq A$ is a G_δ set. Let $G = C \cap \overline{A}$. If $x \in G \setminus A$, then there exists a sequence $\langle x_k \in A : k \in \omega \rangle$ with $\lim_{k \to \infty} x_k = x$. By definition of the set C, the sequence $\{ f(x_k) \}_{k=0}^\infty$ is B-C. Let $g(x) = \lim_{k \to \infty} f(x_k)$. One can easily see that the definition is correct, i.e., the resulting value $g(x)$ does not depend on the choice of the sequence $\{ x_k \}_{k=0}^\infty$. Set $g(x) = f(x)$ for $x \in A$. Then g is continuous. $\qquad\square$

Theorem [wAC] 3.30. *A subset A of a Polish space is a Polish space if and only if A is a G_δ set.*

Proof. Assume that $A \subseteq X$ is a Polish space with the subspace topology. The identity mapping $\mathrm{id}_A : A \xrightarrow{1-1} A$ is continuous. By Theorem 3.29 there exists a G_δ set $A \subseteq G \subseteq \overline{A}$ and a continuous function $g : G \longrightarrow A$ extending id_A. Since $g(x) = x$ for any $x \in A$, this equality must hold true also for any $x \in \overline{A}$. Thus $A = G$.

Now, let $A = \bigcap_n G_n$ be a G_δ-subset of X, $\langle G_n : n \in \omega \rangle$ being open. Let ρ be a complete metric on X. We define a metric σ on A as follows:

$$\sigma(x,y) = \rho(x,y) + \sum_{k=0}^\infty \min \left\{ 2^{-k}, \left| \frac{1}{\rho(x, X \setminus G_k)} - \frac{1}{\rho(y, X \setminus G_k)} \right| \right\}.$$

Let $\{ x_n \}_{n=0}^\infty$ be a B-C sequence in the metric σ. By definition of σ, the sequence $\{ x_n \}_{n=0}^\infty$ is a B-C sequence in metric ρ as well. Thus there exists an $x = \lim_{n \to \infty} x_k$.

We need to show that $x \in A$. One can easily see that for every fixed n the sequence $\{ 1/\rho(x_k, X \setminus G_n) \}_{k=0}^\infty$ of reals is B-C and therefore there exists a real $a = \lim_{k \to \infty} 1/\rho(x_k, X \setminus G_n)$. Then $\rho(x, X \setminus G_n) = 1/a \neq 0$. Hence $x \in G_n$. Since n was arbitrary, we obtain $x \in A$. $\qquad\square$

Note that the second part of the proof did not use any choice.

Denote

$$\mathcal{M}(X, \mathcal{O}) = \{ A \subseteq X : A \text{ is meager} \}.$$

We shall write simply $\mathcal{M}(X)$, when the topology $\mathcal{O}$ is understood.

Theorem [wAC] 3.31. *If $\langle X, \mathcal{O} \rangle$ is a non-meager topological space with a countable base of topology, then the family $\mathcal{M}(X, \mathcal{O})$ is a σ-ideal with an F_σ base.*

Proof. Since every meager set is a subset of an F_σ meager set, the ideal $\mathcal{M}(X)$ has an F_σ base.

If $A_n \in \mathcal{M}(X)$, then the family

$$\Phi_n = \{\{F_m\}_{m=0}^\infty : (\forall m)\, (F_m \text{ is closed nowhere dense}) \wedge A_n \subseteq \bigcup_m F_m\}$$

is non-empty. By (1.19) we can apply **wAC**. Thus there exists a choice sequence $\langle \{F_{n,m}\}_{m=0}^\infty \in \Phi_n : n \in \omega \rangle$. Then

$$\bigcup_n A_n \subseteq \bigcup_n \bigcup_m F_{n,m}. \qquad \square$$

Let $\langle X, \mathcal{O} \rangle$ be a topological space. A set $A \subseteq X$ has the **Baire Property** if there exists an open set U such that the sets $A \setminus U$, $U \setminus A$ are meager. Equivalently, there are an open set U and meager sets P, Q such that $A = (U \setminus P) \cup Q$. Immediately from the definition we obtain that $A \subseteq X$ has the Baire Property if and only if there are an open set U and meager F_σ sets P, Q such that $U \setminus Q \subseteq A \subseteq U \cup P$. Thus we have

Theorem 3.32. *A subset A of a perfectly normal space X has the Baire Property if and only if there are an F_σ set F and a G_δ set G such that $G \subseteq A \subseteq F$ and $F \setminus G$ is meager.*

We denote by $\text{BAIRE}(X, \mathcal{O})$ or simply $\text{BAIRE}(X)$ the family of all subsets of X with the Baire Property. Again when the topology $\mathcal{O}$ or the space X is understood we write simply $\text{BAIRE}(X)$ or BAIRE.

Theorem [wAC] 3.33. *For any topological space $\langle X, \mathcal{O} \rangle$ with a countable base the family $\text{BAIRE}(X, \mathcal{O})$ is closed under complement and countable unions. Hence, $\text{BAIRE}(X, \mathcal{O})$ is closed under countable intersections as well. Assuming* **AC** *one can omit the condition of a countable base.*

Proof. Since every open set has the Baire Property, we have $\emptyset, X \in \text{BAIRE}(X)$. If $A = (U \setminus P) \cup Q$, U open, P, Q meager, then

$$X \setminus A = (\text{Int}(X \setminus U) \setminus Q) \cup (((X \setminus U) \setminus \text{Int}(X \setminus U)) \setminus Q) \cup (P \setminus Q)$$

and $(X \setminus U) \setminus \text{Int}(X \setminus U)$ is nowhere dense. Thus $X \setminus A \in \text{BAIRE}(X)$.

Let $\langle A_n \in \text{BAIRE}(X) : n \in \omega \rangle$. By **wAC** there exist open sets $\langle U_n : n \in \omega \rangle$ and meager F_σ sets $\langle P_n : n \in \omega \rangle$, $\langle Q_n : n \in \omega \rangle$ such that

$$A_n \setminus U_n \subseteq P_n, \qquad U_n \setminus A_n \subseteq Q_n.$$

Since

$$\bigcup_n A_n \setminus \bigcup_n U_n \subseteq \bigcup_n (A_n \setminus U_n) \subseteq \bigcup_n P_n, \quad \bigcup_n U_n \setminus \bigcup_n A_n \subseteq \bigcup_n (U_n \setminus A_n) \subseteq \bigcup_n Q_n,$$

we obtain $\bigcup_n A_n \in \text{BAIRE}(X)$.

The assertion about countable intersections follows by the de Morgan laws.

$$\square$$

We shall define below a notion of a σ-algebra. The theorem says that assuming **wAC**, $\text{BAIRE}(X, \mathcal{O})$ is a σ-algebra provided that $\langle X, \mathcal{O} \rangle$ has a countable base.

Theorem 3.34. *Let a topological space X satisfy the Baire Category Theorem. Then for every set $A \subseteq X$ with the Baire Property there exists a unique regular open set V such that $A \setminus V$ and $V \setminus A$ are meager.*

Proof. Let $A \setminus U$ and $U \setminus A$ be meager, U being open. Since $A \setminus \text{Int}(\overline{U}) \subseteq A \setminus U$, $\text{Int}(\overline{U}) \setminus A \subseteq (\overline{U} \setminus U) \cup (U \setminus A)$ and $\overline{U} \setminus U$ are nowhere dense, the regular open set $V = \text{Int}(\overline{U})$ is the desired one.

Assume that $U \neq V$ are regular open sets such that $A \setminus U$, $A \setminus V$, $U \setminus A$ and $V \setminus A$ are meager. If $U \nsubseteq V$, then since V is regular also $U \nsubseteq \overline{V}$. Then $U \setminus V \supseteq U \setminus \overline{V} \neq \emptyset$. By the Baire Category Theorem $U \setminus V$ is not meager. On the other side we have $U \setminus V \subseteq (A \setminus V) \cup (U \setminus A)$. Since both sets $A \setminus V$, $U \setminus A$ are meager, we have reached a contradiction. □

Theorem [wAC] 3.35. *Assume that X is a Hausdorff space with a countable base and satisfies the Baire Category Theorem. For any $A \subseteq X$ there exists an F_σ set F such that*

$$A \subseteq F \wedge (\forall B \in \text{BAIRE}(X)) \, (A \subseteq B \rightarrow F \setminus B \text{ is meager}). \qquad (3.11)$$

Proof. Let $\mathcal{B}$ be a countable base of the topology $\mathcal{O}$. We set

$$U = \bigcup \{V \in \mathcal{B} : V \cap A \text{ is meager}\}.$$

Evidently $A \cap U$ is meager. Let W be an F_σ meager set such that $A \cap U \subseteq W$. Set $F = (X \setminus U) \cup W$. Then F is an F_σ set and $A \subseteq F$.

Assume that $A \subseteq B$ where B possesses the Baire Property. Then there exist G open, P, Q meager such that $B = (G \setminus Q) \cup P$. Evidently $F \setminus B \subseteq (F \setminus G) \cup Q$. If $F \setminus G$ were not meager, then, being an F_σ set, there exists a non-empty open set $V \subseteq F \setminus G$. Since $A \subseteq G \cup P$, we obtain $A \cap V \subseteq P$. By definition of U we have $V \subseteq U$ and, since $V \subseteq F$, we obtain $V \subseteq W$, a contradiction. □

We introduce a property of a metric space which implies compactness of a Polish space. A metric space X is **totally bounded** if for any $\varepsilon > 0$, the space X can be covered by finitely many balls of radius ε.

Theorem [wAC] 3.36. *A Polish space X is compact if and only if X is totally bounded.*

Proof. We may suppose $\text{diam}(X) = 1$. Let $\{U_n : n \in \omega\}$ be an open cover of X. To get a contradiction we assume that $\bigcup_{k=0}^{n} U_k \neq X$ for every n. For every n we construct finitely many closed sets $C_{n,0}, \ldots, C_{n,k_n}$, $k_n \in \omega$ of diameter less than 2^{-n} such that $\bigcup_{i=0}^{k_n} C_{n,i} = X \setminus \bigcup_{k=0}^{n} U_k$ and every $C_{n+1,i}$ is a subset of some $C_{n,j}$. Set $k_0 = 0$ and $C_{0,0} = X \setminus U_0$. Assume that k_n and $C_{n,0}, \ldots, C_{n,k_n}$ are already constructed. Since X is totally bounded, there exist finitely many closed balls $B_0, \ldots, B_m$ of diameter less than $\leq 2^{-n+1}$ covering the set $C_{n,0} \cup \cdots \cup C_{n,k_n}$. Enumerate as $\langle C_{n+1,i} : i \leq k_{n+1} \rangle$ all non-empty sets $C_{n,l} \cap B_j \cap (X \setminus \bigcup_{k=0}^{n+1} U_k)$, where $l \leq k_n$ and $j \leq m$. The family $\{C_{n,i} : i \leq k_n, n \in \omega\}$ ordered by the inverse inclusion is a countable tree and each node has finite branching degree. Thus by König's Lemma 11.7 there exists an infinite branch

$$C_{0,0} \supseteq C_{1,i_1} \supseteq \cdots \supseteq C_{n,i_n} \supseteq \cdots .$$

By Cantor's Lemma 3.28 the intersection $\bigcap_n C_{n,i_n}$ is non-empty, which contradicts $\bigcup_n U_n = X$.

The opposite implication is trivial. □

Exercises

3.7 Distance of Two Sets
Let $\langle X, \rho \rangle$ be a metric space. If $A, B \subseteq X$ we define the **distance of sets** A, B as

$$\rho(A, B) = \inf\{\rho(x, y) : x \in A \wedge y \in B\}.$$

a) If $\rho(A, B) > 0$, then $A \cap B = \emptyset$.

b) Suppose that $A \cap B = \emptyset$. If X is a Polish space with a complete metric ρ, and A, B are closed, then $\rho(A, B) > 0$.

c) Show that all conditions are necessary, i.e., if X is not a Polish space or ρ is not complete, or A, B are not closed, then it may happen that $A \cap B = \emptyset$ and $\rho(A, B) = 0$.

3.8 [wAC] Compactness of Metric Spaces
We assume that $\langle X, \rho \rangle$ is a metric space. X is called **sequentially compact** if every sequence of points of X has an accumulation point.

A positive real ε is a **Lebesgue number** of an open cover $\mathcal{A}$ if for every $x \in X$ there exists a $U \in \mathcal{A}$ such that $\mathrm{Ball}(x, \varepsilon) \subseteq U$.

a) A compact separable metric space is sequentially compact. Assuming **AC** one can omit the assumption of separability, i.e., **AC** implies that a compact metric space is separable.

b) A compact separable metric space is complete. Assuming **AC** one can omit the assumption of separability.

c) A separable sequentially compact space is totally bounded.

d) If X is separable sequentially compact, then every open cover has a Lebesgue number.

e) If X is totally bounded and every open cover has a Lebesgue number, then X is compact.

f) If X is totally bounded and complete, then X is sequentially compact.

g) If every open cover has a Lebesgue number, then the space is complete.

ɔ) Let X be separable. Then the following are equivalent:

  1) X is compact.

  2) X is sequentially compact.

  3) X is totally bounded and complete.

3.9 [wAC] Product and Sum of Polish Spaces
Let $\langle X_n : n \in \omega \rangle$ be Polish spaces, ρ_n being complete metric on X_n. We assume that values of each ρ_n is in $[0, 1]$. Denote the product $X = \Pi_{n \in \omega} X_n$. We define

$$\rho(x, y) = \sum_{n=0}^{\infty} \rho_n(x(n), y(n)) \cdot 2^{-n} \tag{3.12}$$

for $x, y \in X$.

  a) ρ is a metric on X.

  b) The metric ρ induces on X the product topology, i.e., the weakest topology in which all projections on X_n, $n \in \omega$ are continuous.

  c) $\langle X, \rho \rangle$ is a complete space.

  d) $\langle X, \rho \rangle$ is a Polish space.
   Hint: If $A_n = \{a_{n,m} : m \in \omega\}$ is a dense subset of X_n then the set of all sequences $\{a_{n,k_n}\}_{n=0}^{\infty}$, where $k_n = k_m$ for sufficiently large n, m is a countable dense subset of X.

If $\langle X_1, \mathcal{O}_1 \rangle$, $\langle X_2, \mathcal{O}_2 \rangle$ are topological spaces, then the **sum** of them is the topological space $\langle X_1 \cup X_2, \mathcal{O} \rangle$, where

$$U \in \mathcal{O} \equiv (X_1 \cap U \in \mathcal{O}_1 \wedge X_2 \cap U \in \mathcal{O}_2).$$

Similarly for more spaces.

  e) Both X_1 and X_2 are clopen sets in the sum.

  f) The sum of two Polish spaces is a Polish space.

3.10 Universal Polish Space
The metric space ${}^\omega[0, 1]$ with the metric (3.12), where each ρ_n is the Euclidean distance on $[0, 1]$ is called the **Hilbert cube**. We emphasize that all results of this exercise may be proved without using the axiom of choice.

  a) A Hilbert cube is a compact Polish space.
   Hint: Construct a continuous mapping of $\mathbb{C}$ onto ${}^\omega[0, 1]$.

  b) If X is a separable metric space, then there exists a homeomorphism f of X onto a subset $f(X) \subseteq {}^\omega[0, 1]$.
   Hint: We can assume that $\operatorname{diam}(X) \leq 1$. If $\{a_n : n \in \omega\}$ is a dense subset of X define $f(x) = \{\rho(x, a_n)\}_{n=0}^{\infty}$.

  c) X is a Polish space if and only if X is homeomorphic to a G_δ subset of the Hilbert cube ${}^\omega[0, 1]$.

  d) A Polish space is a topologically complete space (see Exercise 1.23).

  e) X is a compact Polish space if and only if X is homeomorphic to a closed subset of the Hilbert cube ${}^\omega[0, 1]$.

f) Every Polish space has a compactification.

g) Every compact Polish space is a continuous image of $\mathbb{C}$.

   Hint: Use a) *and Corollary 3.8.*

3.11 $^{\omega}\mathbb{R}$

Consider the topological space $^{\omega}\mathbb{R}$ with the product topology.

a) A closed subset $A \subseteq {}^{\omega}\mathbb{R}$ is compact if and only if there exist closed intervals $\langle I_n : n \in \omega \rangle$ such that $A \subseteq \Pi_n I_n$.

   Hint: Each projection is continuous.

b) The topological space $^{\omega}\mathbb{R}$ is not locally compact.

c) $^{\omega}\mathbb{R}$ is homeomorphic to a G_δ subset of the Hilbert cube $^{\omega}[0,1]$.

3.12 Banach Space

Let B be a vector space over the field of reals $\mathbb{R}$. A mapping $p : B \longrightarrow \langle 0, \infty \rangle$ is called a **seminorm** if $p(x + y) \le p(x) + p(y)$, $p(\alpha x) = |\alpha| p(x)$ for every $x, y \in B$ and $\alpha \in \mathbb{R}$. If moreover $p(x) = 0 \equiv x = 0$, then p is called a **norm** and we prefer to write $p(x) = \|x\|$. In this case we set $\rho(x, y) = \|x - y\|$ and ρ is a metric. A vector space B with a norm $\| \ \|$ is called a **Banach space** if $\langle B, \rho \rangle$ is a complete metric space. Sometimes a Banach space is assumed to be separable, i.e., a Polish space. We do not assume the separability.

a) Show that ρ is a metric.

b) If for $[\alpha_1, \ldots, \alpha_k] \in \mathbb{R}_k$ we set $\|[\alpha_1, \ldots, \alpha_k]\| = \sqrt{\sum_{i=1}^{k} x_i^2}$, then we obtain a norm on $\mathbb{R}_k$. The corresponding metric is the Euclidean distance on $\mathbb{R}_k$ and $\mathbb{R}_k$ with the operations defined by (3.6) is a separable Banach space.

c) Let ℓ^∞ be the set of all bounded sequences of reals and for an element $\{\alpha_n\}_{n=0}^{\infty}$ of ℓ^∞ we write $\|\{\alpha_n\}_{n=0}^{\infty}\|_\infty = \sup\{|\alpha_n| : n \in \omega\}$. Show that ℓ^∞ with the norm $\| \ \|_\infty$ is a separable Banach space.

d) Let $p \ge 1$ be a real. For a sequence $\{\alpha_n\}_{n=0}^{\infty}$ of reals we set

$$\|\{\alpha_n\}_{n=0}^{\infty}\|_p = \sqrt[p]{\sum_{i=0}^{\infty} |\alpha_i|^p}$$

   ($\|\{\alpha_n\}_{n=0}^{\infty}\|_p = \infty$ if the series diverges). Let $\ell^p = \{\{\alpha_n\}_{n=0}^{\infty} : \|\{\alpha_n\}_{n=0}^{\infty}\|_p < \infty\}$. Show that ℓ^p with the norm $\| \ \|_p$ is a separable Banach space.

   Hint: Use the results of Exercise 2.10 *to show that ℓ^p is a vector space and $\| \ \|_p$ is a norm. If $\{\{\alpha_{n,k}\}_{n=0}^{\infty}\}_{k=0}^{\infty}$ is a B-C sequence, then $\{\alpha_{n,k}\}_{k=0}^{\infty}$ is a B-C sequence for each n. Let $\alpha_n = \lim_{k \to \infty} \alpha_{n,k}$. For any $\varepsilon > 0$ there exists a k_0 such that $\sum_{n=0}^{\infty} |\alpha_{n,k} - \alpha_{n,m}|^p < \varepsilon$ for any $k, m \ge k_0$. For arbitrary $l > 0$ there exists an $m_0 \ge k_0$ such that $\sum_{n=0}^{l} |\alpha_{n,m_0} - \alpha_n|^p < \varepsilon$. Thus $\sum_{n=0}^{l} |\alpha_{n,k} - \alpha_n|^p < 2\varepsilon$ for any $k \ge k_0$. Since l was arbitrary, we obtain $\sum_{n=0}^{\infty} |\alpha_{n,k} - \alpha_n|^p \le 2\varepsilon$.*

e) The Hilbert cube $^{\omega}[0,1]$ is homeomorphic to a subset of ℓ^p.

   Hint: Take $A = \{\{\alpha_n\}_{n=0}^{\infty} : (\forall n) |\alpha_n|^p \le 2^{-n}\}$.

3.13 Linear Operator

Let B_i be Banach spaces with norms $\| \ \|_i$, $i = 1, 2$. A linear mapping (see Exercise 11.10) $T : B_1 \longrightarrow B_2$ is called a **linear operator**. T is **bounded** if there exists a positive real K such that $\|T(x)\| \le K \cdot \|x\|$ for every $x \in B_1$.

a) T is bounded if and only if the set $\{T(x) : x \in \mathrm{Ball}(0, 1) \subseteq B_1\}$ is bounded.

b) T is bounded if and only if there exists a neighborhood U of 0 in B_1 such that the set $\{T(x) : x \in U\}$ is bounded.

c) If T is bounded, then we define the **norm of linear operator** $\|T\|$ as

$$\|T\| = \sup\{\|T(x)\|_2 : x \in \mathrm{Ball}(0, 1) \subseteq B_1\}.$$

Show that

$$\|T\| = \inf\{K \in \mathbb{R} : (\forall x \in B_1)\, \|T(x)\|_2 \leq K\|x\|_1\}.$$

d) If T is bounded, then T is continuous.

e) If T is continuous at 0, then T is bounded.

   Hint: If $\|T(x) - T(0)\| \leq 1$ for every $\|x\| < \delta$ then the diameter of the set $T(\mathrm{Ball}(0, 1))$ is not greater than $2/\delta$.

f) If T is continuous, then T is bounded.

g) If there exists a point $x_0 \in B_1$ such that T is continuous at x_0, then T is continuous.

   Hint: Show that T is continuous at 0 and use c).

3.14 Open Mapping Theorem

Assume that B_i, $i = 1, 2$ are Banach spaces, $T : B_1 \longrightarrow B_2$ is a continuous linear operator. We set $U_n = \mathrm{Ball}(0, 2^{-n}) \subseteq B_1$.

a) If $\overline{T(U_0)}$ has non-empty interior, then $\overline{T(U_0)}$ is a neighborhood of 0.

   Hint: $\overline{T(U_1)}$ and $\overline{T(U_0)}$ are homeomorphic. If $V \subseteq \overline{T(U_1)}$ is non-empty open, then

$$0 \in V - V \subseteq \overline{T(U_1)} - \overline{T(U_1)} \subseteq \overline{T(U_1) - T(U_1)} \subseteq \overline{T(U_0)}.$$

b) If $T(B_1)$ is not meager, then $\overline{T(U_0)}$ has non-empty interior.

   Hint: $\overline{T(U_0)}$ and $\overline{nT(U_0)}$ are homeomorphic and $T(B_1) = \bigcup_{n=1}^{\infty} nT(U_0)$.

c) Assume **AC**. If $T(B_1)$ is not meager, then $\overline{T(U_1)} \subseteq T(U_0)$.

   Hint: Consider any $y_1 \in \overline{T(U_1)}$. By induction construct $y_n \in \overline{T(U_n)}$. By a) and b), $\overline{T(U_{n+1})}$ is a neighborhood of 0, thus $(y_n - \overline{T(U_{n+1})}) \cap T(U_n) \neq \emptyset$. Take $x_n \in U_n$ such that $T(x_n) \in y_n - \overline{T(U_{n+1})}$ and set $y_{n+1} = y_n - T(x_n)$. By completeness there exists an $x = \sum_{n=1}^{\infty} x_n$. Show that $y_1 = T(x) \in T(U_0)$.

d) Assuming **AC** prove the **Open Mapping Theorem**: If $T(B_1)$ is non-meager, then $T(B_1) = B_2$ and T is an open mapping, i.e., $T(U) \subseteq B_2$ is open for every open $U \subseteq B_1$.

   Hint: If $U \subseteq B_1$ is a neighborhood of 0, then by c) and a), $T(U)$ is a neighborhood of 0 in B_2. Conclude that T is open. Since $T(B_1)$ is open we have $\mathrm{Ball}(0, 1) \subseteq T(B_1)$ and therefore $T(B_1) = B_2$.

e) If B_1, B_2 are separable, then **wAC** is sufficient for a proof of the Open Mapping Theorem.

3.15 Convex Sets

B is a Banach space. A set $A \subseteq B$ is **convex** if for any $x, y \in A$ and any real $t \in [0, 1]$ also $tx + (1 - t)y \in A$.

a) If $T : B_1 \longrightarrow B_2$ is a linear operator and $A \subseteq B_1$ is convex, then also $T(A)$ is convex.

b) If $p : B \longrightarrow \mathbb{R}$ is a seminorm, then the set $U = \{x \in B : p(x) < 1\}$ is a convex set. If moreover p is continuous, then U is a convex neighborhood of 0.

c) If U is a convex neighborhood of 0, then the **Minkowski functional of** U,

$$p(x) = \inf\{\alpha > 0 : \alpha^{-1}x \in U\},$$

is a seminorm on B and $\{x \in B : p(x) < 1\} \subseteq U \subseteq \{x \in B : p(x) \leq 1\}$.

3.16 Hahn-Banach Theorem

B is a Banach space, p is a seminorm. A linear operator $f : B \longrightarrow \mathbb{R}$ is called a **linear functional**.

a) Let A be a vector subspace of B, $a \notin A$. Assume that $f : A \longrightarrow \mathbb{R}$ is a linear functional such that $f(x) \leq p(x)$ for every $x \in A$. Let $C = \{x + ta : x \in A \wedge t \in \mathbb{R}\}$ be the vector subspace generated by A and a. Then there exists a linear functional $F : C \longrightarrow \mathbb{R}$ such that $F(x) = f(x)$ for $x \in A$ and $F(x) \leq p(x)$ for every $x \in C$.
   Hint: For any $x, y \in A$ we have $f(x) + f(y) \leq p(x - a) + p(y + a)$. Show that

$$\alpha = \sup\{f(x) - p(x - a) : x \in A\} \leq \inf\{p(x + a) - f(x) : x \in A\} = \beta$$

   and take the value $\alpha \leq F(a) \leq \beta$.

b) Assuming **AC** prove the **Hahn-Banach Theorem**: If f is a linear functional defined on a vector subspace A of a Banach space B such that $f(x) \leq p(x)$ for every $x \in A$, then there exists a linear functional $F : B \longrightarrow \mathbb{R}$ such that $F(x) = f(x)$ for every $x \in A$ and $F(x) \leq p(x)$ for every $x \in B$.
   Hint: Apply the Zorn Lemma 1.10 to the set of all couples $[A, F]$, where $A \subseteq B$ is a vector subspace, F is a linear functional extending f with $F(x) \leq p(x)$, partially ordered by

$$[A_1, F_1] \leq [A_2, F_2] \equiv (A_1 \text{ subspace of } A_2 \wedge F_1 = F_2|A_1).$$

c) Prove the following consequence of the Hahn-Banach Theorem: If $A \subseteq B$ is convex, $0 \notin \overline{A}$, then there exists a bounded linear functional F on B and an $\varepsilon > 0$ such that $F(x) > \varepsilon$ for every $x \in A$.
   Hint: Let $\text{Ball}(0, \delta) \cap A = \emptyset$. Take $a \in A$ and set $U = \text{Ball}(0, \delta) - A + a$. U is a convex neighborhood of 0 and $a \notin U$. Let p be the Minkowski functional of U. Since $p(a) \geq 1$, the linear functional $f(ta) = t$ defined on the vector subspace generated by a is bounded by p. Let F be a linear functional extending f and bounded by p. Since $F(x) \leq p(x) \leq 1$ for $x \in U$, by Exercise 3.13 b) F is continuous. $F(\text{Ball}(0, \delta))$ and $F(A)$ are convex subsets of $\mathbb{R}$, i.e., intervals. If $x \in \text{Ball}(0, \delta)$ and $y \in A$, then

$$F(x) - F(y) + 1 = F(x - y + a) \leq (x - y + a) < 1.$$

   Thus $F(x) < F(y)$. Since $0 \in F(\text{Ball}(0, 1))$ and by Exercise 3.14, d) $F(\text{Ball}(0, \delta))$ is open, we are done.

3.17 [AC] Dual Space

For a Banach space B we denote by B^* the set of all bounded linear functionals on B. It is natural to endow B^* with the structure of a vector space setting $(f+g)(x) = f(x)+g(x)$, $(\alpha f)(x) = \alpha f(x)$. B^* with norm $\| \ \|$ defined in Exercise 3.13, c) is called the **dual space** of B.

a) Show that B^* is complete, i.e., B^* is a Banach space.

b) Let $\frac{1}{p} + \frac{1}{q} = 1$, $p > 1$. Show that the dual space of ℓ^p is isomorphic to ℓ^q.
   Hint: Use the Hölder Inequality 2.10, c).

c) Show that the unit closed ball $\overline{\mathrm{Ball}}(0,1)$ in the Banach space ℓ^p, $p \geq 1$ is not compact.
   Hint: Consider the sequence $\{e_i\}_{i=0}^{\infty}$ of unit vectors $e_i = \{\delta_{ij}\}_{j=0}^{\infty}$, where $\delta_{ij} = 1$ if $i = j$ and $\delta_{ij} = 0$ otherwise.

d) For every $x \in B$ we define a linear functional F_x on B^* by $F_x(f) = f(x)$ for any $f \in B^*$. Show that $F_x \in B^{**}$ for any $x \in B$ and $\|F_x\| = \|x\|$. Moreover, the set $\{F_x : x \in B\}$ is closed in B^{**}.

e) The **weak* topology** on a dual space B^* of a Banach space B is the weakest topology in which every linear functional $\langle F_x : x \in B \rangle$ is continuous. Prove the **Banach-Alaoglu Theorem**: The closed unit ball $\overline{\mathrm{Ball}}(0,1)$ in a Banach space B^* is compact in the weak* topology.
   Hint: Consider the set $X = \Pi_{x \in B}[-\|x\|, \|x\|]$ with the product topology. By the Tychonoff Theorem 1.37 X is compact. An element $f \in \overline{\mathrm{Ball}}(0,1)$ is a function from B with value $f(x) \in [-\|x\|, \|x\|]$. Thus $\overline{\mathrm{Ball}}(0,1) \subseteq X$. It suffices to show that every continuous functional is continuous in the subspace topology of the product and that $\overline{\mathrm{Ball}}(0,1)$ is closed in X.

3.3 Borel Sets

We begin by recalling some notions. An **algebra of sets** is a family $\mathcal{S}$ of subsets of a given non-empty set X such that

1) $\emptyset$, $X \in \mathcal{S}$,

2) if A, $B \in \mathcal{S}$, then also $A \cup B$, $A \cap B$, $A \setminus B \in \mathcal{S}$.

To emphasize that an algebra contains subsets of the given set X we shall sometimes say that $\mathcal{S}$ is an algebra of subsets of X. $\mathcal{P}(X)$ and $\{\emptyset, X\}$ are trivial examples of algebras of sets.

Let κ be an uncountable cardinal. An algebra of sets $\mathcal{S}$ is κ-**additive** if for every subset $\mathcal{A} \subseteq \mathcal{S}$, $|\mathcal{A}| < \kappa$ also $\bigcup \mathcal{A} \in \mathcal{S}$. Evidently then also $\bigcap \mathcal{A} \in \mathcal{S}$. An $\aleph_1$-additive algebra is called a σ-**algebra**. Thus, by Theorem 3.33, if X is a Polish space, then **wAC** implies that the family $\mathrm{BAIRE}(X)$ is a σ-algebra.

Let us remark that if κ is a singular cardinal, then every κ-additive algebra is also κ^+-additive. So in this chapter we assume that κ is a regular cardinal.

Theorem 3.37. *Let X be a non-empty set, $\mathcal{A}$ being a family of subsets of X. Then there exists a κ-additive algebra of sets $\mathcal{S} \subseteq \mathcal{P}(X)$ such that $\mathcal{A} \subseteq \mathcal{S}$ and $\mathcal{S}$ is the least one with this property, i.e. if $\mathcal{T}$ is a κ-additive algebra of subsets of X containing $\mathcal{A}$ as a subset, then also $\mathcal{S} \subseteq \mathcal{T}$.*

Proof. It suffices to set

$$\mathcal{S} = \{A \in \mathcal{P}(X) : (\forall \mathcal{T})\, ((\mathcal{A} \subseteq \mathcal{T} \wedge \mathcal{T} \text{ is a } \kappa\text{-additive algebra }) \to A \in \mathcal{T})\}.$$

Since $\mathcal{P}(X) \supseteq \mathcal{A}$ is a κ-additive algebra, the assertion follows. □

If $\mathcal{A}$, $\mathcal{S}$ are such as in the theorem, we say that $\mathcal{A}$ κ-**generates** $\mathcal{S}$ or $\mathcal{A}$ is a set of κ-**generators** of $\mathcal{S}$.

If $\langle X, \mathcal{O} \rangle$ is a topological space, then $\mathrm{BOREL}(X, \mathcal{O})$ will denote the smallest σ-algebra of subsets of X containing all open subsets of X. The existence of such a family follows by Theorem 3.37. An element of $\mathrm{BOREL}(X, \mathcal{O})$ is called a **Borel set**. When the topology or the space is understood from the context, or when the statement holds true for any topological space, we shall write simply $\mathrm{BOREL}(X)$ or even BOREL instead of $\mathrm{BOREL}(X, \mathcal{O})$. Every open and every closed set is a Borel set. Both G_δ sets and F_σ sets are Borel sets. We could continue in definitions of simple Borel sets: a set A is called a $G_{\delta\sigma}$ **set** if $A = \bigcap_n A_n$, where $\langle A_n : n \in \omega \rangle$ are F_σ sets. Similarly we can define an $F_{\sigma\delta}$ set. Going on we define $G_{\delta\sigma\delta}$, $F_{\sigma\delta\sigma}$ sets, etc. All those sets are a part of a much larger hierarchy. For an ordinal $\xi < \omega_1$ we define by induction the families $\mathbf{\Sigma}_\xi^0(X, \mathcal{O})$ and $\mathbf{\Pi}_\xi^0(X, \mathcal{O})$ of subsets of a given topological space $\langle X, \mathcal{O} \rangle$ as follows:

$$\mathbf{\Sigma}_0^0(X, \mathcal{O}) = \mathbf{\Pi}_0^0(X, \mathcal{O}) = \text{ the set of all clopen subsets of } \langle X, \mathcal{O} \rangle,$$

$$\mathbf{\Sigma}_1^0(X, \mathcal{O}) = \mathcal{O} = \text{ the set of all open subsets of } \langle X, \mathcal{O} \rangle,$$

$$\mathbf{\Pi}_1^0(X, \mathcal{O}) = \text{ the set of all closed subsets of } \langle X, \mathcal{O} \rangle,$$

$$\mathbf{\Sigma}_\xi^0(X, \mathcal{O}) = \left\{ \bigcup_n A_n : A_n \in \bigcup_{\eta < \xi} \mathbf{\Pi}_\eta^0(X, \mathcal{O}), n \in \omega \right\},$$

$$\mathbf{\Pi}_\xi^0(X, \mathcal{O}) = \{X \setminus A : A \in \mathbf{\Sigma}_\xi^0(X, \mathcal{O})\} = \left\{ \bigcap_n A_n : A_n \in \bigcup_{\eta < \xi} \mathbf{\Sigma}_\eta^0(X, \mathcal{O}), n \in \omega \right\}$$

for $1 < \xi < \omega_1$. Thus $\mathbf{\Sigma}_2^0(X, \mathcal{O})$ is the family of F_σ subsets of X, $\mathbf{\Pi}_2^0(X, \mathcal{O})$ is the family of G_δ subsets of X, $\mathbf{\Sigma}_3^0(X, \mathcal{O})$ is the family of $F_{\sigma\delta}$ subsets of X, etc.

We introduce also the **ambiguous family**

$$\mathbf{\Delta}_\xi^0(X, \mathcal{O}) = \mathbf{\Sigma}_\xi^0(X, \mathcal{O}) \cap \mathbf{\Pi}_\xi^0(X, \mathcal{O}).$$

When the topology is understood we simply write $\mathbf{\Sigma}_\xi^0(X)$ or $\mathbf{\Pi}_\xi^0(X)$; even when no confusion can arise we shall write $\mathbf{\Sigma}_\xi^0$, $\mathbf{\Pi}_\xi^0$, $\mathbf{\Delta}_\xi^0$ instead of $\mathbf{\Sigma}_\xi^0(X, \mathcal{O})$, $\mathbf{\Pi}_\xi^0(X, \mathcal{O})$, $\mathbf{\Delta}_\xi^0(X, \mathcal{O})$, respectively.

Immediately from the definition one obtains

$$\mathbf{\Pi}_\eta^0 \subseteq \mathbf{\Sigma}_\xi^0, \quad \mathbf{\Sigma}_\eta^0 \subseteq \mathbf{\Pi}_\xi^0, \quad \mathbf{\Pi}_\eta^0 \subseteq \mathbf{\Delta}_\xi^0, \quad \mathbf{\Sigma}_\eta^0 \subseteq \mathbf{\Delta}_\xi^0 \text{ for any } \eta < \xi < \omega_1. \tag{3.13}$$

Thus, for ξ limit we have

$$\mathbf{\Sigma}_\xi^0(X, \mathcal{O}) = \left\{ \bigcup_n A_n : A_n \in \bigcup_{\eta < \xi} \mathbf{\Sigma}_\eta^0(X, \mathcal{O}) \right\}.$$

We shall use the following almost trivial results without any commentary.

Theorem 3.38. *Let X_1, X_2 be topological spaces.*

a) *Let $f : X_1 \longrightarrow X_2$ be continuous. If $A \subseteq X_2$ is Borel, then $f^{-1}(A) \subseteq X_1$ is Borel. Moreover, if $A \in \mathbf{\Sigma}_\xi^0(X_2)$, then $f^{-1}(A) \in \mathbf{\Sigma}_\xi^0(X_1)$. Similarly for $\mathbf{\Pi}_\xi^0$ and $\mathbf{\Delta}_\xi^0$.*

b) *For any $\xi < \omega_1$, the families $\mathbf{\Sigma}_\xi^0$, $\mathbf{\Pi}_\xi^0$ and $\mathbf{\Delta}_\xi^0$ are closed under finite unions and finite intersections.*

c) *For any $\xi < \omega_1$, if $B \in \mathbf{\Sigma}_\xi^0(X_1)$, then*

$$B \times X_2 \in \mathbf{\Sigma}_\xi^0(X_1 \times X_2), \quad X_2 \times B \in \mathbf{\Sigma}_\xi^0(X_2 \times X_1).$$

Similarly for $\mathbf{\Pi}_\xi^0$ and $\mathbf{\Delta}_\xi^0$.

Note the following (see the notation (11.3)):

$$\text{If } A \subseteq X, \text{ then } \text{BOREL}(A, \mathcal{O}|A) = \text{BOREL}(X, \mathcal{O})|A. \tag{3.14}$$

Similarly for the families $\mathbf{\Sigma}_\xi^0$, $\mathbf{\Pi}_\xi^0$, and $\mathbf{\Delta}_\xi^0$ for any $\xi < \omega_1$.

For a normal topological space $\langle X, \mathcal{O} \rangle$, any of the inclusions $\mathbf{\Sigma}_1^0 \subseteq \mathbf{\Sigma}_2^0$ and $\mathbf{\Pi}_1^0 \subseteq \mathbf{\Pi}_2^0$ is equivalent to the condition that $\langle X, \mathcal{O} \rangle$ is perfectly normal. By a transfinite induction we obtain

Theorem 3.39. *If $\langle X, \mathcal{O} \rangle$ is perfectly normal space, then*

$$\mathbf{\Sigma}_\eta^0(X, \mathcal{O}) \subseteq \mathbf{\Sigma}_\xi^0(X, \mathcal{O}), \quad \mathbf{\Pi}_\eta^0(X, \mathcal{O}) \subseteq \mathbf{\Pi}_\xi^0(X, \mathcal{O}) \tag{3.15}$$

for any $\eta < \xi < \omega_1$. Hence

$$\mathbf{\Pi}_\eta^0(X, \mathcal{O}) \subseteq \mathbf{\Delta}_\xi^0(X, \mathcal{O}), \quad \mathbf{\Sigma}_\eta^0(X, \mathcal{O}) \subseteq \mathbf{\Delta}_\xi^0(X, \mathcal{O}) \tag{3.16}$$

for any $\eta < \xi < \omega_1$.

Thus by Corollary 3.16 we obtain

Corollary 3.40. *If $\langle X, \rho \rangle$ is a metric space, then (3.15) and (3.16) hold true.*

From the definitions one obtains

$$\mathbf{\Sigma}_\xi^0, \mathbf{\Pi}_\xi^0, \mathbf{\Delta}_\xi^0 \subseteq \text{BOREL} \text{ for every } \xi < \omega_1.$$

Actually

Theorem 3.41. *If ω_1 is a regular cardinal then*

$$\mathrm{BOREL}(X,\mathcal{O}) = \bigcup_{\xi<\omega_1} \mathbf{\Sigma}^0_\xi(X,\mathcal{O}) = \bigcup_{\xi<\omega_1} \mathbf{\Pi}^0_\xi(X,\mathcal{O}) = \bigcup_{\xi<\omega_1} \mathbf{\Delta}^0_\xi(X,\mathcal{O}). \qquad (3.17)$$

Especially, the weak axiom of choice **wAC** *implies that* (3.17) *holds true.*

Proof. By (3.13) we have

$$\bigcup_{\xi<\omega_1} \mathbf{\Sigma}^0_\xi(X,\mathcal{O}) = \bigcup_{\xi<\omega_1} \mathbf{\Pi}^0_\xi(X,\mathcal{O}).$$

We show that this family is a σ-algebra of sets containing all open sets.

Since $\mathbf{\Sigma}^0_1 \subseteq \bigcup_{\xi<\omega_1} \mathbf{\Sigma}^0_\xi$, the family contains all open sets. If $A \in \bigcup_{\xi<\omega_1} \mathbf{\Sigma}^0_\xi$ then $A \in \mathbf{\Sigma}^0_\xi$ for some $\xi < \omega_1$ and $X \setminus A \in \mathbf{\Pi}^0_\xi \subseteq \mathbf{\Sigma}^0_{\xi+1}$.

Now let $A_n \in \bigcup_{\xi<\omega_1} \mathbf{\Pi}^0_\xi$ for $n \in \omega$. Then there exists $\xi_n < \omega_1$ such that $A_n \in \mathbf{\Pi}^0_{\xi_n}$. To avoid a use of any axiom of choice take ξ_n the least possible. By the assumption that ω_1 is regular there exists an $\eta < \omega_1$ such that $\xi_n < \eta$ for any n. Then $\bigcup_n A_n \in \mathbf{\Sigma}^0_\eta$. $\qquad\square$

By Theorem 3.41 the family of Borel sets can be stratified by ω_1 subfamilies. If $\langle X, \mathcal{O}\rangle$ is a perfectly normal space, then those subfamilies together with the inclusions (3.13), (3.15) and (3.16) form the **Borel Hierarchy**:

Later on, in Theorem 6.12, we show that for an uncountable Polish space X all inclusions are proper.

If X is a Hausdorff topological space with a countable base of open sets, then assuming **AC**, one can prove by induction that $|\mathbf{\Sigma}^0_\xi| \leq 2^{\aleph_0}$ for any $\xi < \omega_1$. Thus

Theorem [AC] 3.42. *If X is a Hausdorff topological space with a countable base, then $|\mathrm{BOREL}(X)| \leq 2^{\aleph_0}$. Thus, if $|X| = 2^{\aleph_0}$, then there are non-Borel subsets of X.*

Proof. By **AC** we have

$$|\mathrm{BOREL}(X)| \leq \left| \bigcup_{\xi<\omega_1} \mathbf{\Sigma}^0_\xi(X) \right| \leq \aleph_1 \cdot 2^{\aleph_0} = 2^{\aleph_0}.$$

Since $|\mathcal{P}(X)| > 2^{\aleph_0}$, there exist non-Borel subsets of X. $\qquad\square$

Without the axiom of choice we can prove a weaker result. We construct a mapping from $\mathcal{P}(\omega)$ onto $\bigcup_{\xi<\omega_1} \mathbf{\Sigma}^0_\xi$.

We begin with a modification of the proof of Theorem 1.16 by constructing functions

$$h : \mathcal{P}(\omega) \xrightarrow{\text{onto}} \omega_1 \times \mathcal{P}(\omega), \qquad h_\xi : \mathcal{P}(\omega) \xrightarrow{\text{onto}} \xi \times \mathcal{P}(\omega) \text{ for any } 0 < \xi < \omega_1. \quad (3.18)$$

The idea is simple: for a set $A \subseteq \omega$ we consider two sets $\{k \in \omega : 2k \in A\}$ and $\{k \in \omega : 2k + 1 \in A\}$. The former set will code a countable ordinal and the later one will be the desired subset of ω.

Let f be the function $f : \mathcal{P}(\omega) \xrightarrow{\text{onto}} \omega_1$ constructed in the proof of Theorem 1.16. For a set $A \subseteq \omega$ we let

$$h(A) = \langle f(\{k \in \omega : 2k \in A\}), \{k \in \omega : 2k + 1 \in A\}\rangle,$$

$$h_\xi(A) = \begin{cases} h(A) & \text{if } f(\{k \in \omega : 2k \in A\}) < \xi, \\ \langle 0, \{k \in \omega : 2k + 1 \in A\}\rangle & \text{otherwise.} \end{cases}$$

It is easy to see that h, h_ξ are desired mappings.

We use the function h for a coding of Borel sets by subsets of ω.

Theorem 3.43. *Let $\langle X, \mathcal{O}\rangle$ be a Hausdorff topological space with countable base $\mathcal{B}$ of open sets. Then there exist mappings*

$$H : \mathcal{P}(\omega) \xrightarrow{\text{onto}} \bigcup_{\eta<\omega_1} \mathbf{\Sigma}^0_\eta(X), \qquad H_\xi : \mathcal{P}(\omega) \xrightarrow{\text{onto}} \bigcup_{\eta<\xi} \mathbf{\Sigma}^0_\eta(X).$$

Thus, if ω_1 is a regular cardinal then

$$H : \mathcal{P}(\omega) \xrightarrow{\text{onto}} \text{BOREL}(X),$$

i.e. $|\text{BOREL}(X)| \ll \mathfrak{c}$.

Proof. We define a mapping

$$F : \omega_1 \times \mathcal{P}(\omega) \xrightarrow{\text{onto}} \bigcup_{\xi<\omega_1} \mathbf{\Sigma}^0_\xi(X)$$

such that $F|\xi \times \mathcal{P}(\omega)$ is a surjection onto $\bigcup_{\eta<\xi} \mathbf{\Sigma}^0_\eta(X)$. Then it suffices to set $H = h \circ F$ and $H_\xi = h_\xi \circ F$.

In the proof, we fix a pairing function $\pi : \omega \times \omega : \xrightarrow[\text{onto}]{1-1} \omega$.

Let $\mathcal{B} = \{U_n : n \in \omega\}$ be an enumeration of an open base. We are not interested in values $F(0, A)$, so we can set $F(0, A) = \emptyset$ for any $A \subseteq \omega$.

For a set $A \subseteq \omega$ we set

$$F(1, A) = \bigcup_{n \in A} U_n.$$

Then $\{F(1, A) : A \subseteq \omega\} = \mathcal{O} = \mathbf{\Sigma}^0_1$.

Assume that $F(\eta, A)$ is defined for any $\eta < \xi$ and any $A \subseteq \omega$. For a subset $A \subseteq \omega$ we define $F(\xi, A) = \bigcup_n X_n \in \Sigma_\xi^0$, where

$$X_n = X \setminus F(h_\xi(\{k \in \omega : \pi(n, k) \in A\})) \in \bigcup_{\eta < \xi} \Sigma_\eta^0.$$

Using **wAC**, one can easily prove by a transfinite induction that $\Sigma_\xi^0 \subseteq \mathrm{rng}(F)$ for any $\xi < \omega_1$.   □

Corollary [wAC] 3.44. *If X is a separable metric space with $|X| = 2^{\aleph_0}$, then* BOREL $\neq \mathcal{P}(X)$.

A family $\mathcal{O} \subseteq \mathcal{P}(X)$ is called a σ-**topology** on X if

1) $\emptyset, X \in \mathcal{O}$,

2) if $A, B \in \mathcal{O}$, then $A \cap B \in \mathcal{O}$,

3) if $\mathcal{A} \subseteq \mathcal{O}$ is countable, then $\bigcup \mathcal{A} \in \mathcal{O}$.

Many results concerning topology can be proved also for a σ-topology. Evidently, every topology is a σ-topology. Vice versa, a σ-topology with a countable base is a topology. However we shall deal with many important σ-topologies that are not topologies.

Corollary [wAC] 3.45. *If X is a topological space with a countable base of open sets, then every $\Sigma_\xi^0(X)$ is closed under countable unions and every $\Pi_\xi^0(X)$ is closed under countable intersections. Thus $\Sigma_\xi^0(X)$ is a σ-topology.*

Proof. Let $\langle X_n \in \Sigma_\xi(X) : n \in \omega \rangle$. We put

$$\Phi_n = \{A \in \mathcal{P}(\omega) : H_{\xi+1}(A) = X_n\}.$$

By **wAC** there exists a selector $\langle A_n \in \Phi_n : n \in \omega \rangle$. If $A = \{\pi(n, k) : k \in A_n\}$, then $F(\xi, A) = \bigcup_n X_n$.   □

If X is a topological space with a countable base and **wAC** holds true, then by Theorem 3.33 the family BAIRE$(X, \mathcal{O})$ is a σ-algebra containing all open sets. Since BOREL$(X, \mathcal{O})$ is the smallest σ-algebra containing all open sets, we obtain

Theorem [wAC] 3.46. *Every Borel subset of a topological space with a countable base has Baire property.*

We shall need a rather technical result concerning Borel sets.

Lemma [wAC] 3.47 (N. Luzin). *Let $\langle X, \mathcal{O} \rangle$ be a perfectly normal topological space with a countable base. If $A \in \Sigma_\xi^0$, $\xi > 2$, then there are pairwise disjoint sets $A_n \in \bigcup_{\eta < \xi} \Pi_\eta^0$ such that $A = \bigcup_n A_n$.*

Proof. We begin with a remark: any $C \in \Sigma_\zeta^0$, $\zeta > 1$, can be represented as $\bigcup_n B_n$, where $\langle B_n : n \in \omega \rangle$ are pairwise disjoint and $B_n \in \Delta_\zeta^0$ for every n. Actually, if $C = \bigcup_n C_n$, where $C_n \in \bigcup_{\eta < \zeta} \Pi_\eta^0$, we set $B_0 = C_0$, $B_{n+1} = C_{n+1} \setminus \bigcup_{i=0}^n C_i$.

Let $A = \bigcup_n C_n$, $C_n \in \bigcup_{\eta < \xi} \mathbf{\Pi}^0_\eta$, $\xi > 2$. Then for a given n, $X \backslash \bigcup_{i<n} C_i \in \mathbf{\Sigma}^0_{\eta_n}$ for some $1 < \eta_n < \xi$. By Theorem 3.43 we can use **wAC**, therefore by the remark we have $X \backslash \bigcup_{i<n} C_i = \bigcup_i B_{n,i}$, where the sets $\langle B_{n,i} : i \in \omega \rangle$ are pairwise disjoint and belong to $\mathbf{\Delta}^0_{\eta_n}$. Then the sets $\langle C_n \cap B_{n,i} : i, n \in \omega \rangle$ are pairwise disjoint, belong to $\bigcup_{\eta < \xi} \mathbf{\Pi}^0_\eta$, and

$$A = \bigcup_n \left(C_n \backslash \bigcup_{i<n} C_i \right) = \bigcup_n \left(C_n \cap \bigcup_i B_{n,i} \right) = \bigcup_{n,i} (C_n \cap B_{n,i}). \qquad \square$$

Theorem [wAC] 3.48. *Let $\langle X, \mathcal{O} \rangle$ be a perfectly normal topological space with a countable base . Then $\mathrm{BOREL}(X)$ is the smallest family $\mathcal{F}$ of subsets of X with the following properties:*

a) *every open subset belongs to $\mathcal{F}$;*

b) *$\mathcal{F}$ is closed under countable intersections;*

c) *$\mathcal{F}$ is closed under countable unions of pairwise disjoint sets.*

Proof. Let $\mathcal{F}$ be the smallest family satisfying conditions a)–c). By a), b) and perfect normality we have $\mathbf{\Pi}^0_2 \subseteq \mathcal{F}$. By Lemma 3.47 we obtain $\mathbf{\Sigma}^0_2 \subseteq \mathbf{\Sigma}^0_3 \subseteq \mathcal{F}$. Using Luzin's Lemma again we can continue by induction to show that $\mathbf{\Sigma}^0_\xi \subseteq \mathcal{F}$ for any $\xi > 2$.

Since $\mathrm{BOREL}(X)$ satisfies the conditions a)–c), we have $\mathcal{F} \subseteq \mathrm{BOREL}(X)$. $\square$

Exercises

3.18 Turning a Borel Set into a Clopen Set

Let $\langle X, \mathcal{O} \rangle$ be a Polish space.

a) If A is a closed or an open subset of X, then there exists a topology $\mathcal{O}'$ on X such that $\langle X, \mathcal{O}' \rangle$ is a Polish space, A is clopen in $\mathcal{O}'$, and $\mathrm{BOREL}(X, \mathcal{O}') = \mathrm{BOREL}(X, \mathcal{O})$.
   Hint: Take the sum (see Exercise 3.9) of topologies $\mathcal{O}|A$ and $\mathcal{O}|(X \backslash A)$.

b) If $\langle \mathcal{O}_n : n \in \omega \rangle$ are Polish topologies on X stronger than $\mathcal{O}$, then the weakest topology $\mathcal{O}'$ stronger than any $\langle \mathcal{O}_n : n \in \omega \rangle$ is Polish.
   Hint: Consider $X_n = X$ with topology $\mathcal{O}_n$ and induced by a metric ρ_n. Take the product $Y = \Pi_{n \in \omega} X_n$ with the product topology. The map $f : X \xrightarrow{1-1} \Pi_{n \in \omega} X_n$ defined as $f(x) = \{x\}_{n=0}^\infty$ is a homeomorphism of $\langle X, \mathcal{O}' \rangle$ onto $f(X)$. Show that $\bigcup_n \mathcal{O}_n$ is a base of the topology $\mathcal{O}'$.

c) If in b) we assume that $\mathcal{O} \subseteq \mathcal{O}_n \subseteq \mathrm{BOREL}(X, \mathcal{O})$, then

$$\mathrm{BOREL}(X, \mathcal{O}) = \mathrm{BOREL}(X, \mathcal{O}').$$

d) Show that the family

$$\{A \subseteq X : (\exists \mathcal{O}' \supseteq \mathcal{O} \text{ Polish topology})(A, X \backslash A \in \mathcal{O}' \wedge \mathrm{BOREL}(X, \mathcal{O}) = \mathrm{BOREL}(X, \mathcal{O}'))\}$$

is a σ-algebra containing $\mathcal{O}$.

e) For any Borel set $A \subseteq X$ there exists a Polish topology $\mathcal{O}'$ on X such that A is clopen in $\mathcal{O}'$ and $\text{BOREL}(X, \mathcal{O}) = \text{BOREL}(X, \mathcal{O}')$.

f) For countably many Borel sets $\langle A_n \subseteq X : n \in \omega \rangle$ there exists a Polish topology $\mathcal{O}'$ such that all A_n are clopen in $\mathcal{O}'$ and $\text{BOREL}(X, \mathcal{O}) = \text{BOREL}(X, \mathcal{O}')$.

3.19 Alexandroff-Hausdorff Theorem

a) If $A \subseteq X$ is a perfect subset of a Polish space X, then there exists a pruned tree $\langle T, \supseteq \rangle$ of open subsets of X of height ω, such that

  i) if $U \neq V \in T$, $U \subseteq V$, then $\overline{U} \subseteq V$;

  ii) $\langle T, \supseteq \rangle$ is isomorphic to $^{\omega}2$;

  iii) if $f \in [T]$, then $\bigcap_n f(n)$ is a singleton.

  Hint: Fix an enumeration of basic open sets. By induction: if $U \in T(n)$, then $U \cap A$ has more than one point. Take V_1, V_2 disjoint open sets, the first in the given enumeration with diameter less than 2^{-n}, such that $\overline{V_i} \subseteq U$.

b) If $A \subseteq X$ is an uncountable closed subset of a Polish space X, then there exists a continuous injection $f : {}^{\omega}2 \xrightarrow{1-1} A$.

c) Prove the Alexandroff-Hausdorff Theorem (see Theorem 6.7): If $B \subseteq X$ is an uncountable Borel subset of a Polish space X, then B contains a subset homeomorphic to the Cantor middle-third set $\mathbb{C}$.

  Hint: Turn B into a closed subset of a Polish space.

3.20 [wAC] σ-reduction Theorem

In the next X is a Polish space.

a) If $A \in \mathbf{\Sigma}^0_\xi(X)$, $n \in \omega$, $\xi > 0$, then there exist pairwise disjoint sets $A_n \in \mathbf{\Delta}^0_\xi(X)$ such that $A = \bigcup_n A_n$. Note that for $\xi > 2$ the result follows from Luzin's Lemma 3.47.

  Hint: If $A = \bigcup_n B_n$, $B_n \in \bigcup_{\eta < \xi} \mathbf{\Pi}^0_\eta(X)$, set $A_n = B_n \setminus \bigcup_{i < n} B_i$.

b) **σ-reduction Theorem:**[1] If $A_n \in \mathbf{\Sigma}^0_\xi(X)$, $n \in \omega$, $\xi > 1$, then there exist pairwise disjoint $B_n \in \mathbf{\Sigma}^0_\xi(X)$ such that $B_n \subseteq A_n$ for every n and $\bigcup_n A_n = \bigcup_n B_n$. If $\bigcup_n A_n = X$, then $B_n \in \mathbf{\Delta}^0_\xi(X)$.

  Hint: If $A_n = \bigcup_i A_{n,i}$ with $A_{n,i} \in \bigcup_{\eta < \xi} \mathbf{\Sigma}^0_\eta(X)$, set ($\pi$ is a pairing function)

$$B_{n,i} = A_{n,i} \setminus \bigcup_{\pi(m,j) < \pi(n,i)} A_{m,j}, \quad B_n = \bigcup_i B_{n,i}.$$

c) If $A \subseteq B$, $A \in \mathbf{\Pi}^0_\xi$, $B \in \mathbf{\Sigma}^0_\xi$, $\xi > 1$, then there exists a set $C \in \mathbf{\Delta}^0_\xi$ such that $A \subseteq C \subseteq B$.[2]

  Hint: Use the result of Exercise b) for the couple $X \setminus A$, B.

d) If $A \in \mathbf{\Delta}^0_{\xi+1}(X)$, $\xi > 1$, then there exist $A_n \in \mathbf{\Delta}^0_\xi(X)$ such that

$$A = \bigcup_n \bigcap_{i=n}^{\infty} A_i = \bigcap_n \bigcup_{i=n}^{\infty} A_i. \tag{3.19}$$

---

[1] See also Section 5.2 and Theorem 6.9.
[2] See Section 5.2.

If ξ is limit, then we can assume that there are $\xi_n < \xi$ such that $A_n \in \Delta^0_{\xi_n}(X)$.
Hint: Since $A = \bigcup_n B_n$, $A = \bigcap_n C_n$ with $B_n \subseteq B_{n+1} \in \Pi^0_\xi$, $C_n \supseteq C_{n+1} \in \Sigma^0_\xi$, by
d), there exist $A_n \in \Delta^0_\xi$ such that $B_n \subseteq A_n \subseteq C_n$.

e) If $A \in \Delta^0_{\xi+1}(X)$, $\xi > 1$, then there exist sets $B_n \in \Delta^0_\xi(X)$, $B_n \subseteq B_{n+1}$ and
$C_n \in \Delta^0_\xi(X)$, $C_n \supseteq C_{n+1}$ such that

$$A = \bigcup_n B_n = \bigcap_n C_n.$$

Hint: If (3.19) holds true, then by c) for every n there exist $B_n, C_n \in \Delta^0_\xi$ such that
$\bigcap_{i=n}^\infty A_i \subseteq B_n \subseteq A$ and $A \subseteq C_n \subseteq \bigcup_{i=n}^\infty A_i$.

3.21 Weak Reduction Property

Let $\mathcal{A}$ be a family of subsets of a set X. $\mathcal{A}$ has the **weak reduction property** if for any
sets $A_i \in \mathcal{A}$, $i = 1, 2$, $A_1 \cup A_2 = X$ there exist disjoint sets $B_i \in \mathcal{A}$, $B_1 \cup B_2 = X$ and
such that $B_i \subseteq A_i$, $i = 1, 2$.

a) If a family has the reduction property (see Section 5.2), then it has also the weak
reduction property.

b) Assume $\mathcal{A}$ has the weak reduction property and is closed under finite unions and
finite intersections. If $A, B \in \mathcal{A}$, $X \setminus A \subseteq B$, then there exists a set C such that
$C, X \setminus C \in \mathcal{A}$ and $X \setminus A \subseteq C \subseteq B$.

c) Assume $\mathcal{A}$ has the weak reduction property and is closed under finite unions and
finite intersections. If $A_1 \cup A_2 \cup A_3 = X$ with $A_i \in \mathcal{A}$, then there exist pairwise
disjoint sets $B_i \in \mathcal{A}$ such that $B_i \subseteq A_i$ and $B_1 \cup B_2 \cup B_3 = X$.
Hint: Take C_1, C_2 for couple A_1, $A_2 \cup A_3$, take D_1, D_2 for couple $C_1 \cup (C_2 \cap A_2)$,
$C_2 \cap A_3$. Set $B_1 = D_1 \cap C_1$, $B_2 = D_1 \cap C_2$, $B_3 = D_2$.

d) If $\mathcal{A}$ has the weak reduction property and is closed under finite unions and finite
intersections, then for any n and any $\langle A_i \in \mathcal{A} : i \leq n \rangle$, $A_1 \cup \cdots \cup A_n = X$ there are
pairwise disjoint sets $\langle B_i \in \mathcal{A} : i \leq n \rangle$, $B_i \subseteq A_i$ such that $B_1 \cup \cdots \cup B_n = X$.

e) Every family $\Sigma^0_\xi(X, \mathcal{O})$, $\xi > 1$ has the weak reduction property.
Hint: See Exercise 3.20, b).

3.4 Convergence of Functions

Assume that X, Y are topological spaces, $A \subseteq X$. Let us consider mappings
$f : X \longrightarrow Y$, $\langle f_n : X \longrightarrow Y : n \in \omega \rangle$. We shall say that the sequence $\{f_n\}_{n=0}^\infty$
converges pointwise to f on A if $\lim_{n\to\infty} f_n(x) = f(x)$ for any $x \in A$, written
"$f_n \to f$ on A", or simply "$f_n \to f$" when the set A is understood or $A = X$.
The limit $\lim_{n\to\infty} f_n(x)$ is of course understood in the topology of the space Y. If
$Y = \mathbb{R}$, then the notation "$f_n \nearrow f$ on A" means that $f_n \to f$ on A and moreover,
$f_n(x) \leq f_{n+1}(x)$ for any n and any $x \in A$. Similarly, "$f_n \searrow f$ on A" means that
$f_n \to f$ on A and $f_n(x) \geq f_{n+1}(x)$ for any n and any $x \in A$.

If $\langle Y, \rho \rangle$ is a metric space, we say that $\{f_n\}_{n=0}^\infty$ **converges uniformly** to f on A
if there exists a sequence of positive reals $\{\varepsilon_n\}_{n=0}^\infty$, such that $\rho(f_n(x), f(x)) \leq \varepsilon_n$

for any n and any $x \in A$, and $\lim_{n \to \infty} \varepsilon_n = 0$. We shall write "$f_n \rightrightarrows f$ on A" or "$f_n \rightrightarrows f$". Let us note that we can take $\varepsilon_n = \sup\{\rho(f_n(x), f(x)) : x \in A\}$.

Moreover, from elementary calculus we know

Theorem 3.49. *Assume that X is a topological space, Y is a metric space. Let $\{f_n\}_{n=0}^{\infty}$ be a sequence of continuous functions from X into Y. If $f_n \rightrightarrows f$ on a set $A \subseteq X$, then f is continuous on A.*

Theorem 3.50. *Let X be a compact topological space. If $f_n, f : X \longrightarrow \mathbb{R}$ are continuous and $f_n \nearrow f$ on X, then $f_n \rightrightarrows f$ on X.*

Proof. Set $U_{n,m} = \{x \in X : f(x) - f_m(x) < 2^{-n}\}$. Since every $U_{n,m}$ is open, $U_{n,m} \subseteq U_{n,m+1}$ and $\bigcup_i U_{n,i} = X$ for every n, there exists a function $\varphi \in {}^{\omega}\omega$ such that $U_{n,\varphi(n)} = X$. Thus $f(x) - f_m(x) < 2^{-n}$ for $m \geq \varphi(n)$. □

If $\langle Y, \rho \rangle$ is a metric space, we say that $\{f_n\}_{n=0}^{\infty}$ **converges quasi-normally** to f on A if there exists a sequence $\{\varepsilon_n\}_{n=0}^{\infty}$ of positive reals converging to zero and such that

$$(\forall x \in A)(\exists k)(\forall n \geq k)\rho(f_n(x), f(x)) < \varepsilon_n. \tag{3.20}$$

The sequence $\{\varepsilon_n\}_{n=0}^{\infty}$ is called a **control** or we say that the sequence $\{\varepsilon_n\}_{n=0}^{\infty}$ **witnesses** the quasi-normal convergence. We shall write "$f_n \xrightarrow{\text{QN}} f$ on A".

Let us recall that a series $\sum_{n=0}^{\infty} f_n(x)$ **converges normally** on X if there is a sequence $\{\varepsilon_n\}_{n=0}^{\infty}$ such that $\sum_{n=0}^{\infty} \varepsilon_n < +\infty$ and $|f_n(x)| \leq \varepsilon_n$ for every $x \in X$ and every n. Similarly, a series $\sum_{n=0}^{\infty} f_n(x)$ **converges pseudo-normally** on X if there is a sequence $\{\varepsilon_n\}_{n=0}^{\infty}$ such that $\sum_{n=0}^{\infty} \varepsilon_n < +\infty$ and

$$(\forall x \in X)(\exists n_0)(\forall n \geq n_0)\,|f_n(x)| \leq \varepsilon_n.$$

Note that if a series converges pseudo-normally, then the sequence of partial sums converges quasi-normally.

Note the following simple property of quasi-normal convergence.

Theorem 3.51. *Let $f_n \xrightarrow{\text{QN}} f$ on X. For any sequence $\varepsilon_n \to 0$ of positive reals there exists an increasing sequence $\{n_k\}_{k=0}^{\infty}$ such that $f_{n_k} \xrightarrow{\text{QN}} f$ on X with the control $\{\varepsilon_k\}_{k=0}^{\infty}$.*

Of course quasi-normal convergence implies pointwise convergence and uniform convergence implies quasi-normal convergence. The simple relationship to uniform convergence is described by

Theorem 3.52. *Let $\{f_n\}_{n=0}^{\infty}$ and f be functions from X into a metric space Y. Then the following conditions are equivalent:*

a) $f_n \xrightarrow{\text{QN}} f$ *on X.*

b) *There exists a sequence $\{X_k\}_{k=0}^{\infty}$ of sets such that $X = \bigcup_k X_k$ and $f_n \rightrightarrows f$ on X_k for each k.*

c) *There exists a non-decreasing sequence $\{X_k\}_{k=0}^{\infty}$ of sets such that $X = \bigcup_k X_k$ and $f_n \rightrightarrows f$ on X_k for each k.*

Moreover, if X is a topological space and $\{f_n\}_{n=0}^{\infty}$ are continuous, then conditions a)–c) are equivalent with

d) *There exists a non-decreasing sequence $\{X_k\}_{k=0}^{\infty}$ of closed subsets of X such that $X = \bigcup_k X_k$ and $f_n \rightrightarrows f$ on X_k for each k.*

Proof. Assume that $f_n \xrightarrow{\text{QN}} f$ on X with a control $\{\varepsilon_n\}_{n=0}^{\infty}$. We set

$$X_k = \{x \in X : (\forall n, m \geq k)\, \rho(f_n(x), f_m(x)) \leq \varepsilon_n + \varepsilon_m\}.$$

Then $X = \bigcup_k X_k$ holds true, $f_n \rightrightarrows f$ on X_k and $X_k \subseteq X_{k+1}$ for every k. Moreover, if $\{f_n\}_{n=0}^{\infty}$ are continuous, then X_k are closed subsets of X.

Conditions b) and c) are trivially equivalent since if a sequence converges uniformly to f on sets A, B, then it converges so also on $A \cup B$.

Assume now that $f_n \rightrightarrows f$ on every X_k, $X = \bigcup_k X_k$ and $X_k \subseteq X_{k+1}$ for every k. Let

$$\varepsilon_{n,k} = \sup\{\rho(f_i(x), f(x)) : x \in X_k \wedge i \leq n\}.$$

Then $\varepsilon_{n,k} \to 0$ for every k. Since $X_k \subseteq X_{k+1}$, we obtain $\varepsilon_{n,k} \leq \varepsilon_{n,k+1}$ for any n and k. Moreover, the sequence $\{\varepsilon_{n,k}\}_{n=0}^{\infty}$ is non-increasing for every k.

We can easily construct an increasing sequence $\{m_k\}_{k=0}^{\infty}$ of natural numbers such that

$$(\forall n)\,(n \geq m_k \to \varepsilon_{n,k} < 2^{-k}).$$

We define $\delta_n = 2^{-k}$ for $m_k \leq n < m_{k+1}$. Thus $\lim_{n \to \infty} \delta_n = 0$. If $x \in X$, then $x \in X_k$ for some k. Then $x \in X_l$ for any $l \geq k$. Hence for any $l \geq k$ and any $m_l \leq n < m_{l+1}$ we have $\rho(f_n(x), f(x)) \leq \varepsilon_{n,l} < \delta_n$, i.e., $o(f_n(x), f(x)) < \delta_n$ for every $n \geq m_k$. □

Corollary [wAC] 3.53. *If $X = \bigcup_k X_k$, $f_n \xrightarrow{\text{QN}} f$ on every X_k, then $f_n \xrightarrow{\text{QN}} f$ on X as well.*

Assume that X, Y are topological spaces. Let $\{f_n\}_{n=0}^{\infty}$ be a sequence of functions from X into Y, $f : X \longrightarrow Y$. We say that $\{f_n\}_{n=0}^{\infty}$ **converges discretely** to f on $A \subseteq X$ if

$$(\forall x \in A)(\exists k)(\forall n \geq k)\, f_n(x) = f(x).$$

We shall write $f_n \xrightarrow{\text{D}} f$ on A.

If $f_n \xrightarrow{\text{D}} f$ on X, then $X = \bigcup_k X_k$, $X_k \subseteq X_{k+1}$ and $f_k(x) = f(x)$ for any $x \in X_k$. Moreover, if $\{f_n\}_{n=0}^{\infty}$ are continuous, then X_k are closed subsets of X. Indeed, similarly as in the proof of Theorem 3.52, we set

$$X_k = \{x \in A : (\forall m, n \geq k)\, f_m(x) = f_n(x)\}.$$

Evidently, if $f_n \xrightarrow{\text{D}} f$ on X, then $f_n \xrightarrow{\text{QN}} f$ on X.

Assume that $\mathcal{A}$ is a family of subsets of X and Y is a topological space. A function $f : X \longrightarrow Y$ is $\mathcal{A}$-**measurable** if $f^{-1}(U) \in \mathcal{A}$ for every open $U \subseteq Y$. Often we shall deal with a σ-topology $\mathcal{A}$ is on X. Then $\mathcal{A}$-measurable means continuous with respect to σ-topology $\mathcal{A}$. If $\mathcal{A} = \mathrm{BOREL}(X)$ we say that function f is **Borel measurable**. If $\mathcal{A} = \mathrm{BAIRE}(X)$ we speak about **Baire measurable** functions.

Theorem 3.54. *Assume that Y is a topological space with a countable base $\mathcal{B}$ of topology. Let $f : X \longrightarrow Y$.*

    a) *If $\mathcal{A}$ is a σ-topology on X, then a function $f : X \longrightarrow Y$ is $\mathcal{A}$-measurable if and only if $f^{-1}(U) \in \mathcal{A}$ for any $U \in \mathcal{B}$.*

    b) *If $\mathcal{A}$ is a σ-algebra and f is $\mathcal{A}$-measurable, then $f^{-1}(B) \in \mathcal{A}$ for any Borel set $B \subseteq Y$.*

Corollary 3.55. *Let Y_i be a Polish space, $\mathcal{A}_i$ being a family of subsets of X_i, $i = 1, 2$. Let $\mathcal{A}$ be a σ-topology on $X_1 \times X_2$ containing all $A_1 \times A_2$, $A_i \in \mathcal{A}_i$ for $i = 1, 2$. If $f_i : X_i \longrightarrow Y_i$ is an $\mathcal{A}_i$-measurable function, $i = 1, 2$, then the function $F : X_1 \times X_2 \longrightarrow Y_1 \times Y_2$ defined by*

$$F(x_1, x_2) = \langle f_1(x_1), f_2(x_2) \rangle \text{ for } \langle x_1, x_2 \rangle \in X_1 \times X_2$$

is $\mathcal{A}$-measurable.

Proof. Assume that $\mathcal{B}_1$, $\mathcal{B}_2$ comprise countable base of the topology on Y_1, Y_2, respectively. Then $\{U_1 \times U_2 : U_i \in \mathcal{B}_i, i = 1, 2\}$ is a countable base of the product topology on $Y_1 \times Y_2$. We are ready, since

$$F^{-1}(U_1 \times U_2) = f_1^{-1}(U_1) \times f_2^{-1}(U_2) \in \mathcal{A}. \qquad \square$$

The corollary allows us to prove an important result.

Theorem 3.56. *If a function $f : X \longrightarrow Y$ is Borel measurable, then the graph $f \subseteq X \times Y$ is a Borel set.*

Proof. Assume that f is Borel measurable. By Corollary 3.55 the function

$$F(x, y) = \langle f(x), y \rangle \text{ for } \langle x, y \rangle \in X \times Y$$

is Borel measurable. The set $D = \{\langle y, y \rangle : y \in Y\} \subseteq Y \times Y$ is closed. Therefore the set $f = F^{-1}(D)$ is Borel. $\qquad \square$

    Later, in Section 6.2, we show that the implication of the theorem can be reversed.

    We shall need a simple result concerning measurability of quasi-normal limits of continuous functions.

Theorem [AC] 3.57. *Assume that X is a normal topological space, $\{f_n\}_{n=0}^{\infty}$ is a sequence of real functions on X, $f : X \longrightarrow \mathbb{R}$. If $f_n \overset{\mathrm{QN}}{\longrightarrow} f$ on X and all f_n are continuous, then there exist a sequence $\{g_n\}_{n=0}^{\infty}$ of continuous functions such that $g_n \overset{\mathrm{D}}{\longrightarrow} f$ on X. Consequently, the function f is $\mathbf{\Delta}_2^0$-measurable. Moreover, if X is a separable metric space, then we do not need any form of the axiom of choice.*

Proof. If $\{U_n : n \in \omega\}$ is a countable base of the topology on $\mathbb{R}$, then for any open set $U \subseteq \mathbb{R}$ we have

$$f(x) \in U \equiv (\exists n)(\exists k)\,(\overline{U_n} \subseteq U \wedge (\forall m \geq k)\, f_m(x) \in \overline{U_n}).$$

Thus, f is F_σ-measurable.

By Theorem 3.52 there exists a non-decreasing sequence of closed subsets $\langle X_k : k \in \omega \rangle$ such that $X = \bigcup_k X_k$ and $f_n \rightrightarrows f$ on X_k for each k. Thus $f|X_k$ is continuous. By the Tietze-Urysohn Theorem 3.20 there exist real continuous functions $\{g_k\}_{k=0}^\infty$ defined on X such that $g_k|X_k = f|X_k$ for every k. Then $g_n \xrightarrow{\mathrm{D}} f$ on X.

Note that if X is a metric space, then the function g_n is defined by a formula from f_n and we do not need any form of the axiom of choice.

Since for any open U we have

$$f(x) \in U \equiv (\forall n)(\exists m > n)\, g_m(x) \in U,$$

the function f is G_δ-measurable. $\qquad\square$

A function $f : X \longrightarrow \mathbb{R}$ is **upper $\mathcal{A}$-measurable** if $\{x \in X : f(x) < a\} \in \mathcal{A}$ for any real a. Similarly f is **lower $\mathcal{A}$-measurable** if $\{x \in X : f(x) > a\} \in \mathcal{A}$ for any real a. If $\mathcal{A}$ is a σ-topology, then f is $\mathcal{A}$-measurable if and only if it is both upper and lower $\mathcal{A}$-measurable. If $\mathcal{O}$ is the topology of X, then an upper $\mathcal{O}$-measurable function is called **upper semicontinuous**. Similarly **lower semicontinuous**. Evidently a function is continuous if and only if it is both upper and lower semicontinuous. Let us note that a function f is upper $\mathcal{A}$-measurable if and only if $-f$ is lower $\mathcal{A}$-measurable.

We shall consider the set of all real functions $^X\mathbb{R}$ defined on a set X. The set $^X\mathbb{R}$ is partially ordered by the relation

$$f \leq g \equiv f(x) \leq g(x) \text{ for every } x \in X.$$

If $f, g \in {}^X\mathbb{R}$, then we set

$$\max\{f, g\}(x) = \max\{f(x), g(x)\} \text{ for any } x \in X,$$
$$\min\{f, g\}(x) = \min\{f(x), g(x)\} \text{ for any } x \in X.$$

The set $\boldsymbol{\Phi}$ is called a **lattice of functions** if $\max\{f, g\}, \min\{f, g\} \in \boldsymbol{\Phi}$ for any $f, g \in \boldsymbol{\Phi}$. One can easily see that a lattice of functions is a lattice in the partial ordering $\leq$. The opposite is not true.

Let us remark that the set $^X\mathbb{R}$ is a vector space over $\mathbb{R}$ with pointwise operations of addition and multiplication by a real

$$(f + g)(x) = f(x) + g(x), \quad (\alpha f)(x) = \alpha f(x) \text{ for } x \in X.$$

Similarly we define the multiplication of elements $^X\mathbb{R}$ as

$$(f \cdot g)(x) = f(x) \cdot g(x) \text{ for } x \in X.$$

All defined operations are continuous in the product topology of $^X\mathbb{R}$. Moreover, if $h_n = f_n + g_n$ for every n and $\{f_n\}_{n=0}^{\infty}$ and $\{g_n\}_{n=0}^{\infty}$ converge in any of the considered ways to f and g, respectively, then $\{h_n\}_{n=0}^{\infty}$ converges to $f+g$. Similarly for multiplication, the multiplication by a real, and the operations max and min.

If $\mathcal{A}$ is a family of subsets of a set X, then in accordance with terminology introduced later in Section 5.2, we denote

$$\neg\mathcal{A} = \{A \subseteq X : X \setminus A \in \mathcal{A}\},$$

$$\Sigma^0[\mathcal{A}] = \left\{\bigcup_n A_n : A_n \in \mathcal{A}, n \in \omega\right\},$$

$$\Pi^0[\mathcal{A}] = \left\{\bigcap_n A_n : A_n \in \mathcal{A}, n \in \omega\right\}.$$

Lemma 3.58. *Let $\mathcal{A}$ be a σ-topology on X. Assume that f, g, h, d_n, f_n, g_n, h_n, $n \in \omega$ are real functions defined on X. Moreover, assume that $f_n \searrow f$, $g_n \nearrow g$, and $h_n \rightrightarrows h$ on X.*

a) *If $d_0, \ldots, d_m$ are upper (lower) $\mathcal{A}$-measurable, then also $\max\{d_0, \ldots, d_m\}$ and $\min\{d_0, \ldots, d_m\}$ are such.*

b) *If every f_n is upper $\mathcal{A}$-measurable, then f is upper $\mathcal{A}$-measurable as well. If every f_n is lower $\mathcal{A}$-measurable, then f is lower $\Sigma^0[\Pi^0[\mathcal{A}]]$-measurable.*

c) *If every g_n is lower $\mathcal{A}$-measurable, then g is lower $\mathcal{A}$-measurable as well. If every g_n is upper $\mathcal{A}$-measurable, then g is upper $\Sigma^0[\Pi^0[\mathcal{A}]]$-measurable.*

d) *If every real function h_n is upper (lower) $\mathcal{A}$-measurable, then h is upper (lower) $\mathcal{A}$-measurable as well.*

Proof. Let $d = \max\{d_0, \ldots, d_m\}$. Then

$$d(x) > a \equiv d_i(x) > a \text{ for some } i = 0, \ldots, m,$$
$$d(x) < a \equiv d_i(x) < a \text{ for all } i = 0, \ldots, m.$$

Since $\mathcal{A}$ is a σ-topology, the statement a) follows. Similarly for minimum.

The statements b) and c) follow by the equivalences

$$f(x) < a \equiv (\exists n)\, f_n(x) < a,$$
$$f(x) > a \equiv (\exists k > 0)(\forall n)\, f_n(x) > a + 1/k,$$
$$g(x) < a \equiv (\exists k > 0)(\forall n)\, g_n(x) < a - 1/k,$$
$$g(x) > a \equiv (\exists n)\, g_n(x) > a.$$

Let $\varepsilon_n = \sup\{|h_n(x) - h(x)| : x \in X\}$. Then

$$\max\{h_0 - \varepsilon_0, \ldots, h_n - \varepsilon_n\} \searrow h, \quad \min\{h_0 + \varepsilon_0, \ldots, h_n + \varepsilon_n\} \nearrow h.$$

The statement d) follows by a)–c). $\qquad\square$

If X and Y are topological spaces, we denote by $C(X, Y)$ the set of all continuous functions from X into Y. If Y is a metric space, then we denote by $C^*(X, Y)$ the set of all bounded continuous functions from X into Y. Thus $C^*(X, Y) \subseteq C(X, Y)$. Evidently $C(X, Y)$ is a subset of the set ^XY of all functions from X into Y. The set ^XY can be equipped with the product topology. In this topology a typical neighborhood of a function $f \in {}^XY$ is the set

$$\{g \in {}^XY : g(x_1) \in U_1 \wedge \cdots \wedge g(x_n) \in U_n\}$$

for some $x_1, \ldots, x_n \in X$ and neighborhoods U_i of $f(x_i)$, $i = 1, \ldots, n$ in Y. The subset $C(X, Y)$ equipped with the subspace topology of this product topology will be usually denoted by $C_p(X, Y)$.

If $Y = \mathbb{R}$ we write simply $C(X)$, $C^*(X)$ and $C_p(X)$. In what follows by a **real function** on X we shall understand any function $f : X \longrightarrow \mathbb{R}$. If $f \in C(X)$ we speak about a **continuous real function**.

Note the following. If $A \subseteq X$ is a dense subset, then $f|A \neq g|A$ for any distinct $f, g \in C(X)$. Thus $|C(X)| \leq \mathfrak{c}^{|A|}$. Hence

$$\text{if } X \text{ is a separable topological space, then } |C(X)| \leq \mathfrak{c}. \qquad (3.21)$$

Theorem 3.59. *A sequence $\{f_n\}_{n=0}^\infty$ of functions from ^XY converges to a function $f \in {}^XY$ in the product topology if and only if $f_n \to f$ on X. The same holds true for the subspace $C_p(X, Y)$.*

Proof. Assume that $f_n \to f$ on X and $U \subseteq {}^XY$ is a neighborhood of f. Then

$$\{h \in {}^XY : h(x_1) \in U_1 \wedge \cdots \wedge h(x_k) \in U_k\} \subseteq U$$

for some $x_1, \ldots, x_k \in X$ and some neighborhoods U_i of $f(x_i)$, $i = 1, \ldots, k$. Since $f_n \to f$ on X, there exist natural numbers n_i such that $f_n(x_i) \in U_i$ for every $n \geq n_i$, $i = 1, \ldots, k$. If $n_0 = \max\{n_1, \ldots, n_k\}$, then $f_n \in U$ for every $n \geq n_0$.

The opposite implication is evident. $\qquad\square$

According to the theorem, the product topology on ^XY and the subspace topology of $C_p(X, Y)$ is called the **topology of pointwise convergence**.

If $\langle Y, \rho \rangle$ is a metric space, then one can define a metric σ on $C^*(X, Y)$ by

$$\sigma(f, g) = \sup\{\rho(f(x), g(x)) : x \in X\}. \qquad (3.22)$$

Evidently a sequence of functions $\{f_n\}_{n=0}^\infty$ belonging to $C^*(X, Y)$ converges to a function $f \in C^*(X, Y)$ in the metric σ if and only if $f_n \rightrightarrows f$ on X. $C^*(X, Y)$ is usually understood with this metric.

From an elementary course of mathematical analysis the reader knows

Theorem 3.60. *If Y is a complete metric space, then $C^*(X, Y)$ with metric* (3.22) *is a complete metric space.*

According to the following simple result, we can usually replace in our consideration a real function by a function with values in $[0, 1]$.

Theorem 3.61. *Let $\langle f_n : X \longrightarrow \mathbb{R} : n \in \omega \rangle$ be a sequence of real functions. We set $g_n(x) = \min\{|f_n(x)|, 1\}$ for any $x \in X$ and any n. Then $f_n \to 0$ on X ($f_n \rightrightarrows 0$ on X, $f_n \xrightarrow{QN} 0$ on X) if and only if $g_n \to 0$ on X ($g_n \rightrightarrows 0$ on X, $g_n \xrightarrow{QN} 0$ on X). Moreover, if every f_n, $n \in \omega$ is continuous, then every g_n, $n \in \omega$ is continuous as well.*

Using the Tietze-Urysohn Theorem 3.20 we show a usefull result.

Lemma 3.62. *Let A be a closed subset of a metric space X.*

a) *If $C_p(X)$ is a Fréchet space, then $C_p(A)$ is also Fréchet.*

b) *Assume that **wAC** holds true. If the convergence structure on $C_p(X)$ possesses the sequence selection property, then so does $C_p(A)$.*

Moreover, assuming **AC**, *both statements hold true for any normal topological space X.*

Proof. Assume that $C_p(X)$ is Fréchet, $A \subseteq X$ is closed, $Y \subseteq C_p(A)$ and $f \in \overline{Y}$. We write

$$Z = \{g \in C_p(X) : g|A \in Y\}.$$

By Theorem 3.20 there exists a function $F \in C_p(X)$ such that $F|A = f$. We claim that $F \in \overline{Z}$.

Actually, any neighborhood of F contains a basic subset of the form

$$U = \{g \in C_p(X) : \quad |g(x_1) - f(x_1)| < \varepsilon \wedge \cdots \wedge |g(x_k) - f(x_k)| < \varepsilon \wedge,$$
$$|g(y_1) - F(y_1)| < \varepsilon \wedge \cdots \wedge |g(y_l) - F(y_l)| < \varepsilon\},$$

where $x_1, \ldots, x_k \in A$, $y_1, \ldots, y_l \in X \setminus A$, $\varepsilon > 0$. Since $f \in \overline{Y}$, there exists a function $h \in Y$ such that

$$|h(x_1) - f(x_1)| < \varepsilon \wedge \cdots \wedge |h(x_k) - f(x_k)| < \varepsilon.$$

The finite set $\{y_1, \ldots, y_l\}$ is closed and disjoint with A, thus, using again Theorem 3.20, there exists a function $H \in Z$ such that $H(y_i) = F(y_i)$ for $i = 1, \ldots, l$ and $H|A = h$. Hence $H \in U \cap Z$. Therefore $F \in \overline{Z}$. Since $C_p(X)$ is Fréchet, there exists a sequence $\langle F_n \in Z : n \in \omega \rangle$ such that $F_n \to F$ on X. Then $F_n|A \to f$ on Y and by definition of the set Z we obtain $F_n|A \in Y$.

A proof of assertion b) goes in a similar way by using Corollary 3.21. $\square$

Theorem 3.63. *There exists a sequence* $\{h_n\}_{n=0}^{\infty}$ *of continuous real functions defined on the Cantor middle-third set* $\mathbb{C}$ *such that* $h_n \to 0$ *on* $\mathbb{C}$ *and for any increasing sequence* $\{n_k\}_{k=0}^{\infty}$ *of natural numbers there exists a* $z \in \mathbb{C}$ *such that*

$$\sum_{k=0}^{\infty} h_{n_k}(z) = +\infty. \tag{3.23}$$

Proof. If $x = \sum_{k=0}^{\infty} x_k 3^{-k-1} \in \mathbb{C}$, $x_k \in \{0,2\}$, we define

$$h_n(x) = \begin{cases} 0 & \text{if } x_n = 0, \\ 1/m_n & \text{where } m_n = |\{k < n : x_k = 2\}| + 1, \text{ otherwise.} \end{cases}$$

Every function h_n is continuous and $h_n \to 0$ on $\mathbb{C}$.

If $\{n_k\}_{k=0}^{\infty}$ is an increasing sequence, we set $z = \sum_{k=0}^{\infty} z_k 3^{-k}$, where $z_{n_k} = 2$ and $z_n = 0$ otherwise. Then $h_{n_k}(z) = 1/(k+1)$ and therefore the equality (3.23) holds true. $\qquad\square$

Corollary 3.64. *The topological space* $C_p(\mathbb{C})$ *does not possess the sequence selection property, therefore it is not a Fréchet space. Moreover, if* $\mathbb{C}$ *can be embedded in a topological space* X, *then* $C_p(X)$ *neither possesses the sequence selection property nor is a Fréchet space.*

Proof. Let $\{h_n\}_{n=0}^{\infty}$ be the sequence of the theorem. We write $f_{n,m} = 2^n \cdot h_m$. Then $f_{n,m} \to 0$ for every fixed n. Assume that $C_p(\mathbb{C})$ possesses the sequence selection property. Then there exist increasing sequences $\{n_k\}_{k=0}^{\infty}$ and $\{m_k\}_{k=0}^{\infty}$ such that $f_{n_k,m_k} \to 0$ on $\mathbb{C}$. Evidently, for any $x \in \mathbb{C}$ there exists a k_0 such that $f_{n_k,m_k}(x) < 1$ for every $k \geq k_0$. Thus $h_{m_k}(x) < 2^{-n_k}$ for every $k \geq k_0$. Hence $\sum_{k=0}^{\infty} h_{m_k}(x) < +\infty$. Since x was arbitrary, we have a contradiction.

For X containing $\mathbb{C}$, the statement follows by Lemma 3.62. $\qquad\square$

Exercises

3.22 Discrete Convergence

a) If $f_n \xrightarrow{D} f$ on A, then also $f_n \xrightarrow{QN} f$ on A.

b) $f_n \xrightarrow{D} f$ on A if and only if there exists an increasing sequence of sets $\{A_n\}_{n=0}^{\infty}$ such that $A = \bigcup_n A_n$ and $(\forall k \geq n)(\forall x \in A_n) f_k(x) = f(x)$. Moreover, if all f_n are continuous, then we can assume that A_n are closed.

c) Find continuous functions $\langle f_n : [0,1] \longrightarrow [0,1] : n \in \omega \rangle$ such that $f_n \xrightarrow{D} f$ on $[0,1]$ and f is not continuous.

d) Find a sequence of functions $\langle f_n : [0,1] \longrightarrow [0,1] : n \in \omega \rangle$ such that $f_n \xrightarrow{QN} f$ on $[0,1]$ and not $f_n \xrightarrow{D} f$ on $[0,1]$.

3.23 Algebra of Functions

A vector subspace $\Phi \subseteq {}^X\mathbb{R}$ is called an **algebra** (of functions) if Φ is closed under the product of functions.

a) An algebra of functions is a ring.

b) If X is a topological space, then $C(X)$ and $C^*(X)$ are algebras of functions.

c) If X is a completely regular topological space, $|X| > 1$, then neither $C(X)$ nor $C^*(X)$ is an integrity domain.

d) The set of all polynomials defined on an interval $[a, b]$ is a subalgebra of $C([a, b])$.

e) If $\Phi_i \subseteq {}^{X_i}\mathbb{R}$, $i = 1, 2$ are algebras of functions, then the set of all functions $\sum_{k=0}^n f_k \cdot g_k$, where $f_k \in \Phi_1$, $g_k \in \Phi_2$, $k \leq n$, $n \in \omega$ is an algebra of functions on $X_1 \times X_2$.

3.24 Algebra of Functions and Uniform Convergence

Let $\Phi \subseteq C^*(X)$. We say that $\Phi \subseteq {}^X\mathbb{R}$ **separates points** in X if for any $x, y \in X$, $x \neq y$ there is a function $f \in \Phi$ such that $f(x) \neq f(y)$. $\overline{\Phi}$ denotes the closure of Φ in the topology of uniform convergence, i.e., in the topology induced by the metric (3.22). In the next exercises, let Φ be a subalgebra of $C^*(X)$.

a) Show that $p_n \rightrightarrows \sqrt{x}$ on $[0, 1]$, where

$$p_0(x) = 0, \quad p_{n+1}(x) = p_n(x) + \frac{1}{2}(x - p_n^2(x)) \text{ for } x \in [0, 1].$$

   Hint: Show that $p_n(x) \leq p_{n+1}(x) \leq \sqrt{x}$. If $p_n(x) \leq \sqrt{x} - \varepsilon$, then $p_{n+1}(x) \geq p_n(x) + \varepsilon^2/2$.

b) If $f \in \Phi$, then $|f| \in \overline{\Phi}$.
   Hint: If f is bounded by a real $c > 0$ we set $f_n(x) = c \cdot p_n(\frac{1}{c}f^2(x))$. Then $f_n \rightrightarrows |f|$.

c) If $f, g \in \overline{\Phi}$, then also $\max\{f, g\}, \min\{f, g\} \in \overline{\Phi}$.

d) Assume that Φ contains all constant functions and separates points. Then for any reals α, β and any points $x, y \in X$, $x \neq y$, there exists a function $f \in \Phi$ such that $f(x) = \alpha$ and $f(y) = \beta$.

3.25 [AC] Stone-Weierstrass Theorem

Assume that X is a compact topological space. Φ is a subalgebra of $C(X)$.

a) Assume that Φ contains all constant functions, separates points and is closed under max. For any $\varepsilon > 0$, any $p \in X$ and any $f \in C(X)$ there exists a function $h \in \Phi$ and a neighborhood U of p such that $h(x) < f(x) + \varepsilon$ for any $x \in X$ and $h(x) > f(x) - \varepsilon$ for any $x \in U$.
   Hint: For $q \in X$ take a function $h_q \in \Phi$ such that $h_q(p) = f(p)$ and $h_q(q) = f(q)$. By compactness there exists finitely many $q_0, \ldots, q_k$ such that for every $x \in X$, there exists an $i \leq k$ with $h_{q_i}(x) < f(x) + \varepsilon$. Set $h = \max\{h_{q_0}, \ldots, h_{q_k}\}$.

b) Assume that Φ contains all constant functions, separates points and is closed under max. For any $\varepsilon > 0$ and any $f \in C(X)$ there exists a function $g \in \Phi$ such that $\sigma(f, g) < \varepsilon$.

c) Prove the **Stone-Weierstrass Theorem**: If an algebra $\Phi \subseteq C(X)$ contains all constant functions and separates points, then $\overline{\Phi} = C(X)$.

d) The set of all polynomials is dense in $C([0, 1])$.

e) The metric space $C([0,1])$ is separable and therefore Polish.

f) If Y_1 and Y_2 are compact topological spaces, then for any $f \in C(Y_1 \times Y_2)$ and any $\varepsilon > 0$, there exist an $n \in \omega$ and functions $g_1, \ldots, g_n \in C(Y_1)$, $h_1, \ldots, h_n \in C(Y_2)$ such that $\sigma(\sum_{i=1}^{n} g_i \cdot h_i, f) < \varepsilon$ (σ is the metric defined by (3.22)).

3.26 Separability of $C^*(X)$

Assume that X is a metric separable space, A being countable and dense in X.

a) For any $a, b \in A$, $a \neq b$ construct a continuous function $f_{a,b} : X \longrightarrow [0,1]$ such that $f(x) < 1/4$ if $\rho(a,x) < 1/4\rho(a,b)$ and $f(x) > 3/4$ if $\rho(b\ x) < 1/4\rho(a,b)$.

b) The set $\{f_{a,b} : a, b \in A\}$ separates points in X.

Hint: Let $\varepsilon = \rho(x,y) > 0$. Take $a, b \in A$ such that $\rho(a,x) < 1/8\varepsilon$, $\rho(b,y) < 1/8\varepsilon$ and show that $f_{a,b}(x) \neq f_{a,b}(y)$.

c) The set $\Phi = \{\alpha \cdot f_{a,b} + \beta \cdot f_{c,d} : a, b, c, d \in A, \alpha, \beta \in \mathbb{Q}\}$ is countable. Let $\mathcal{A}$ be the smallest subring of $C^*(X)$ closed under max, min scalar multiplication $\alpha \cdot f$ for rational α, containing all rational constants and containing Φ. Show that $\mathcal{A}$ is countable.

d) Show that $\overline{\mathcal{A}} = C^*(X)$.

Hint: Let $\mathcal{D}$ be the smallest subalgebra closed under max, min, containing all constants and Φ. Show that $\overline{\mathcal{D}} = C^*(X)$ and $\overline{\mathcal{A}} \supseteq \mathcal{D}$.

e) If X is a separable metric space, then $C^*(X)$ is a Polish space.

f) Show that in a)–e) X may be a normal separable topological space.

3.27 [AC] Banach Space and $C^*(X)$

X is a topological space. Note that if X is compact, then $C^*(X) = C(X)$.

a) $C^*(X)$ is a Banach space with the norm $\|f\| = \sup\{|f(x)| : x \in X\}$.

b) Let B be a Banach space. Set $X = \overline{\mathrm{Ball}}(0,1) \subseteq B^*$ with the weak* topology. Prove the **Kuratowski Theorem**: The Banach space B is isomorphic with a closed subspace of the Banach space $C(X)$.

Hint: Follow the terminology and notation of Exercise 3.17. By the Banach-Alaoglou Theorem, the topological space X is compact. For any $x \in B$ the functional F_x is continuous on X equipped with the weak* topology. Moreover $\|x\| = \|F_x\|$.

3.28 Completion of Metric Space

Let $\langle X, \rho \rangle$ be a metric space, metric σ on $C^*(X)$ is defined by (3.22).

a) Take a fixed $p \in X$. For any $q \in X$ we define $F_q \in {}^X\mathbb{R}$ as $F_q(x) = \rho(x,q) - \rho(x,p)$. Show that F_q is continuous and bounded.

b) $\sigma(F_x, F_y) = \rho(x,y)$ for any $x, y \in X$.

c) For any metric space $\langle X, \rho \rangle$ there exists a complete metric space $\langle \tilde{X}, \tilde{\rho} \rangle$ such that $X \subseteq \tilde{X}$, X is dense in $\tilde{X}$ and ρ and $\tilde{\rho}$ coincide on X.

Hint: Take $\tilde{X} = \overline{\{F_q : q \in X\}} \subseteq C^*(X)$.

3.5   Baire Hierarchy

We suppose that X is a separable metric space and Y is a perfect Polish space.

By Corollary 3.64 the topological space $^{[0,1]}\mathbb{R}$ does not possess the sequence selection property and the sequential closure $\mathrm{scl}_1\,(\mathrm{C}_p([0,1]))$ of the subset $\mathrm{C}_p([0,1])$ of $^{[0,1]}\mathbb{R}$ is not equal to its topological closure $\overline{\mathrm{C}_p([0,1])}$. Actually we obtain a hierarchy ($\xi < \omega_1$):

$$\mathrm{scl}_0\,(\mathrm{C}_p([0,1])) \subseteq \mathrm{scl}_1\,(\mathrm{C}_p([0,1])) \subseteq \cdots \subseteq \mathrm{scl}_\xi\,(\mathrm{C}_p([0,1])) \subseteq \cdots \subseteq \overline{\mathrm{C}_p([0,1])}.$$

We show later that this hierarchy is proper. We start our study from another side.

It is obvious that any $\mathcal{A}_1$-measurable function is $\mathcal{A}_2$-measurable provided that $\mathcal{A}_1 \subseteq \mathcal{A}_2$. Thus the Borel hierarchy ($1 < \eta < \xi < \omega_1$)

$$\mathbf{\Sigma}_1^0 \subseteq \cdots \subseteq \mathbf{\Sigma}_\eta^0 \subseteq \cdots \subseteq \mathbf{\Sigma}_\xi^0 \subseteq \mathbf{\Sigma}_{\xi+1}^0 \subseteq \cdots \subseteq \mathrm{BOREL}$$

induces a corresponding hierarchy of measurable functions

$$\mathbf{\Sigma}_1^0\text{-measurable} \subseteq \cdots \subseteq \mathbf{\Sigma}_\eta^0\text{-measurable} \subseteq \cdots \subseteq \mathbf{\Sigma}_\xi^0\text{-measurable} \subseteq$$

$$\mathbf{\Sigma}_{\xi+1}^0\text{-measurable} \subseteq \cdots \subseteq \text{Borel measurable.} \qquad (3.24)$$

Since a characteristic function χ_A of a subset $A \subseteq X$ is $\mathcal{A}$-measurable if and only if both sets $A, X \setminus A$ belongs to $\mathcal{A}$, by Theorem 6.12, the hierarchy is proper.

Theorem 3.65. *If ω_1 is regular cardinal, then a function $f : X \longrightarrow Y$ is Borel measurable if and only if f is $\mathbf{\Sigma}_\xi^0$-measurable for some $\xi < \omega_1$.*

Proof. The implication from right to left is trivial. Therefore we assume that $f : X \longrightarrow Y$ is Borel measurable. Let $\mathcal{B}$ be a countable base of the topology of Y. For every $U \in \mathcal{B}$ the set $f^{-1}(U)$ is Borel. Since ω_1 is regular, by Theorem 3.41 we have $\mathrm{BOREL}(X) = \bigcup_{\xi<\omega_1}\mathbf{\Sigma}_\xi^0(X)$. Let η_U be the smallest ordinal $\eta < \omega_1$ such that $f^{-1}(U) \in \mathbf{\Sigma}_\eta^0(X)$. Since $\mathcal{B}$ is countable and ω_1 is regular, there exists an ordinal $\xi < \omega_1$ such that $\eta_U < \xi$ for any $U \in \mathcal{B}$. However $\mathbf{\Sigma}_\xi^0(X)$ is closed under countable unions, so we have that $f^{-1}(V) \in \mathbf{\Sigma}_\xi^0(X)$ for any open set V. Thus the function f is $\mathbf{\Sigma}_\xi^0(X)$-measurable. $\qquad\square$

Let $\mathbf{BP}(X, Y)$ be the smallest family of functions from X into Y containing all continuous functions and closed under pointwise limits, i.e., if $f_n \to f$ on X, $f_n \in \mathbf{BP}(X, Y)$, then also $f \in \mathbf{BP}(X, Y)$. A function belonging to $\mathbf{BP}(X, Y)$ is called a **Baire function** or an **analytically representable function**.

We define a hierarchy of Baire functions. A continuous function $f : X \longrightarrow Y$ is said to be of **Baire class** 0. A function f is said to be of **Baire class** ξ if there exists a sequence $\{f_n\}_{n=0}^\infty$ of functions of Baire classes smaller than ξ such that $f_n \to f$ on X. We define

$\mathbf{BP}_0(X, Y) = $ the set of all continuous functions from X into Y,

$$\mathbf{BP}_\xi(X, Y) = \{f \in {}^X Y : (\exists f_n \in \bigcup_{\eta<\xi} \mathbf{BP}_\eta(X, Y), n \in \omega)\,(f_n \to f \text{ on } X)\} \text{ for } \xi > 0.$$

Using notation introduced by (1.20), for any $0 < \xi < \omega_1$ we have

$$\mathbf{BP}_\xi(X, Y) = \bigcup_{\eta < \xi+1} \mathbf{BP}_\eta(X, Y) = \mathrm{scl}\left(\bigcup_{\eta < \xi} \mathbf{BP}_\eta(X, Y)\right) = \mathrm{scl}_\xi\left(C_p(X, Y)\right).$$

Thus, a function $f \in {}^X Y$ is of Baire class ξ if and only if $f \in \mathbf{BP}_\xi(X, Y)$.

If $Y = \mathbb{R}$ we shall simply write $\mathbf{BP}(X)$, $\mathbf{BP}_\xi(X)$.

Theorem 3.66. *Assume that ω_1 is regular. Then*

$$\mathbf{BP}(X, Y) = \bigcup_{\xi < \omega_1} \mathbf{BP}_\xi(X, Y).$$

The proof is easy. Evidently $\mathbf{BP}_\xi(X, Y) \subseteq \mathbf{BP}(X, Y)$ for any $\xi < \omega_1$. On the other hand, since ω_1 is regular, $\bigcup_{\xi < \omega_1} \mathbf{BP}_\xi(X, Y)$ contains all continuous functions and is closed under pointwise limits. Thus

$$\mathbf{BP}(X, Y) \subseteq \bigcup_{\xi < \omega_1} \mathbf{BP}_\xi(X, Y). \qquad \square$$

Hence we obtain the **Baire Hierarchy** $(\eta < \xi < \omega_1)$:

$$\mathbf{BP}_0(X, Y) \subseteq \cdots \subseteq \mathbf{BP}_\eta(X, Y) \subseteq \cdots \subseteq \mathbf{BP}_\xi(X, Y) \subseteq$$
$$\mathbf{BP}_{\xi+1}(X, Y) \subseteq \cdots \subseteq \mathbf{BP}(X, Y). \qquad (3.25)$$

We shall prove that hierarchies (3.24) and (3.25) are identical up to a shift of indexes.

Note that by definition we have

$$\mathbf{\Sigma}^0[\neg \mathbf{\Sigma}^0_\xi(X)] = \mathbf{\Sigma}^0_{\xi+1}(X) \qquad (3.26)$$

for any $\xi < \omega_1$ and

$$\mathbf{\Sigma}^0[\neg \mathbf{\Sigma}^0[\bigcup_{\eta < \xi} \mathbf{\Sigma}^0_\eta(X)]] = \mathbf{\Sigma}^0_{\xi+1}(X) \qquad (3.27)$$

for any $\xi < \omega_1$ limit.

Lemma 3.67. *Assume that $\mathcal{A}$ is a family of subsets of X. Let $f_n \to f$ on X. If every $f_n : X \longrightarrow Y$, $n \in \omega$ is $\mathcal{A}$-measurable, then f is $\mathbf{\Sigma}^0[\neg \mathbf{\Sigma}^0[\mathcal{A}]]$-measurable. Moreover, if $\mathcal{A}$ is a σ-topology, then f is $\mathbf{\Sigma}^0[\neg \mathcal{A}]$-measurable.*

Proof. Let $\mathcal{B}$ be a countable base of the topology on Y. For any open subset $U \subseteq Y$ we have

$$f(x) \in U \equiv (\exists V \in \mathcal{B}, \overline{V} \subseteq U)(\exists k)(\forall n \geq k)\, f_n(x) \in \overline{V}.$$

Thus

$$f^{-1}(U) = \bigcup_{V \in \mathcal{B}, \overline{V} \subseteq U} \bigcup_k \bigcap_{n \geq k}(X \setminus f_n^{-1}(Y \setminus \overline{V})) = \bigcup_{V \in \mathcal{B}, \overline{V} \subseteq U} \bigcup_k (X \setminus \bigcup_{n \geq k} f_n^{-1}(Y \setminus \overline{V})).$$

So by definition $f^{-1}(U) \in \mathbf{\Sigma}^0[\neg \mathbf{\Sigma}^0[\mathcal{A}]]$. If $\mathcal{A}$ is a σ-topology, then $\mathbf{\Sigma}^0[\mathcal{A}] = \mathcal{A}$. $\square$

Corollary 3.68. *If $\mathcal{S}$ is a σ-algebra, $f_n \to f$, f_n are $\mathcal{S}$-measurable functions, then f is $\mathcal{S}$-measurable as well.*

We can obtain a finer result.

Theorem 3.69. *Every $f \in \mathbf{BP}_\xi(X,Y)$ is $\Sigma^0_{\xi+1}(X)$-measurable for any $\xi < \omega_1$. Thus, every analytically representable function is Borel measurable.*

Proof. For $\xi = 0$ the theorem follows from definition. For $\xi > 0$ the theorem follows by Lemma 3.67, (3.26) and (3.27) by transfinite induction. □

The shift of indexes in the theorem is due to historically accepted terminology. In the literature it is more than common that a pointwise limit of a continuous function is a function of the first Baire class and the family of F_σ sets is Σ^0_2.

There is a fundamental result on the first Baire class functions.

Theorem [wAC] 3.70 (R. Baire). *If $f : X \longrightarrow Y$ is F_σ-measurable, then the set*

$$\mathrm{disc}(f) = \{x \in X : f \text{ is not continuous at } x\}$$

is an F_σ meager set.

Proof. Let $\mathcal{B}$ be a countable base of the topology on Y. We show that

$$\mathrm{disc}(f) = \bigcup_{U \in \mathcal{B}} (f^{-1}(U) \setminus \mathrm{Int}(f^{-1}(U))). \qquad (3.28)$$

Actually, assume that f is not continuous at a point x. Thus there exists a set $U \in \mathcal{B}$ such that $x \in f^{-1}(U)$ and $f^{-1}(U)$ is not a neighborhood of x. Hence $x \notin \mathrm{Int}(f^{-1}(U))$.

Vice versa, if $x \in f^{-1}(U) \setminus \mathrm{Int}(f^{-1}(U))$ for some $U \in \mathcal{B}$, then $f(x) \in U$ and $f^{-1}(U)$ is not a neighborhood of x. Thus, f is not continuous at x.

By assumption the set $f^{-1}(U)$ is an F_σ set. Thus also $f^{-1}(U) \setminus \mathrm{Int}(f^{-1}(U))$ is a meager F_σ set. Hence the set $\mathrm{disc}(f)$ is F_σ and meager as well. □

If $\mathbf{\Phi} \subseteq {}^A\mathbb{R}$ is a set of real functions on A we write

$$
\begin{aligned}
\mathbf{\Phi}^\uparrow &= \{f \in {}^A\mathbb{R} : (\exists f_n \in \mathbf{\Phi}, n \in \omega)\, f_n \nearrow f \text{ on } A\}, \\
\mathbf{\Phi}^\downarrow &= \{f \in {}^A\mathbb{R} : (\exists f_n \in \mathbf{\Phi}, n \in \omega)\, f_n \searrow f \text{ on } A\}.
\end{aligned}
$$

Lemma [wAC] 3.71. *Assume that $|\mathbf{\Phi}| \ll \mathfrak{c}$. Then*

  a) *if $\mathbf{\Phi}$ is a lattice of functions, then $\mathbf{\Phi}^\uparrow, \mathbf{\Phi}^\downarrow, \mathrm{scl}\,(\mathbf{\Phi})$ are lattices of functions;*

  b) *if $\mathbf{\Phi}$ is a vector subspace of ${}^A\mathbb{R}$, then $\mathbf{\Phi}^\uparrow, \mathbf{\Phi}^\downarrow, \mathrm{scl}\,(\mathbf{\Phi})$ are vector subspaces;*

  c) *if $\mathbf{\Phi}$ is a lattice of functions, then we have $(\mathbf{\Phi}^\uparrow)^\uparrow = \mathbf{\Phi}^\uparrow$, $(\mathbf{\Phi}^\downarrow)^\downarrow = \mathbf{\Phi}^\downarrow$ and $\mathrm{scl}\,(\mathbf{\Phi}) = (\mathbf{\Phi}^\downarrow)^\uparrow \cap (\mathbf{\Phi}^\uparrow)^\downarrow$.*

Proof. Let us note that $|\mathbf{\Phi}^\uparrow| \ll \mathfrak{c}$, $|\mathbf{\Phi}^\downarrow| \ll \mathfrak{c}$, $|\mathrm{scl}\,(\mathbf{\Phi})| \ll \mathfrak{c}$ provided that $|\mathbf{\Phi}| \ll \mathfrak{c}$. Thus we can use the weak axiom of choice **wAC**. Proofs of parts a) and b) are immediate.

If $f \in (\mathbf{\Phi}^\uparrow)^\uparrow$, then $f = \lim_{n \to \infty} f_n$ for some $\langle f_n \in \mathbf{\Phi}^\uparrow : n \in \omega \rangle$, and $f_n \le f_{n+1}$ for any n. Every $f_n = \lim_{m \to \infty} g_{n,m}$, $g_{n,m} \in \mathbf{\Phi}$, $g_{n,m} \le g_{n,m+1}$ for any m. We set

$$h_n = \max\{g_{0,n}, \ldots, g_{n,n}\}.$$

Then $h_n \le h_{n+1}$ and $\lim_{n \to \infty} h_n = f$. Since $\mathbf{\Phi}$ is a lattice of functions, we obtain $h_n \in \mathbf{\Phi}$ and therefore $f \in \mathbf{\Phi}^\uparrow$. The opposite inclusion is trivial. We can proceed similarly in the downward case.

If $f \in \mathrm{scl}\,(\mathbf{\Phi})$, then there are $f_n \in \mathbf{\Phi}$ such that $f = \lim_{n \to \infty} f_n$. We denote

$$g_{n,m} = \max\{f_n, f_{n+1}, \ldots, f_{n+m}\}, \qquad g_n = \lim_{m \to \infty} g_{n,m}.$$

Since $g_{n,m} \le g_{n,m+1}$ for every n, m, the functions g_n are well defined (the limits do exist!) and $g_n \in \mathbf{\Phi}^\uparrow$. Since $g_n \ge g_{n+1}$ and $f = \lim_{n \to \infty} g_n$, we obtain $f \in (\mathbf{\Phi}^\uparrow)^\downarrow$. Similarly one can show that $f \in (\mathbf{\Phi}^\downarrow)^\uparrow$.

Let $f \in (\mathbf{\Phi}^\downarrow)^\uparrow \cap (\mathbf{\Phi}^\uparrow)^\downarrow$. Then (using **wAC**) there exist $g_{n,m}, h_{n,m} \in \mathbf{\Phi}$ such that $g_{n,m} \searrow g_n$, $h_{n,m} \nearrow h_n$, $g_n \nearrow f$ and $h_n \searrow f$ on X. We set

$$f_m = \max\{\min\{g_{k,m}, h_{0,m}, \ldots, h_{k,m}\} : k = 0, \ldots, m\}.$$

Evidently $f_m \in \mathbf{\Phi}$. We show that $f = \lim_{n \to \infty} f_n$. Let $\varepsilon > 0$, $x \in X$. We find an m_0 such that $f_m(x) > f(x) - \varepsilon$ for each $m \ge m_0$. Since $g_n \nearrow f$ on X, there exists an n_0 such that $g_n(x) > f(x) - \varepsilon$ for every $n \ge n_0$. Since $h_n(x) \ge f(x)$ and $h_{n,m} \nearrow h_n$ on X, there exists an $m_0 \ge n_0$ such that $h_{i,m}(x) > f(x) - \varepsilon$ for $i = 0, \ldots, n_0$ and every $m \ge m_0$. Then for $m \ge m_0$ we obtain

$$f_m(x) \ge \min\{g_{n_0,m}(x), h_{0,m}(x), \ldots, h_{n_0,m}(x)\} > f(x) - \varepsilon.$$

Similarly one can find an m_0 such that $f_m(x) < f(x) + \varepsilon$ for each $m \ge m_0$. □

Theorem [wAC] 3.72.

$$\mathbf{BP}_\xi(X) = \mathbf{BP}_\xi(X)^\uparrow \cap \mathbf{BP}_\xi(X)^\downarrow$$

Proof. If $f \in \mathbf{BP}_0(X)^\uparrow \cap \mathbf{BP}_0(X)^\downarrow$, then by Lemma 3.58, b), c) the function f is both lower and upper semicontinuous and therefore continuous.

By Lemma 3.71, c) and d), if $\mathbf{\Phi} = \mathrm{scl}\,(\mathbf{\Psi})$, then $\mathbf{\Phi} = \mathbf{\Phi}^\uparrow \cap \mathbf{\Phi}^\downarrow$. Thus, for $\xi > 0$ the statement follows by definition of $\mathbf{BP}_\xi(X)$. □

Corollary [wAC] 3.73. *Every* $\mathbf{BP}_\xi(X)$ *is a vector subspace of* $^X\mathbb{R}$ *and a lattice of functions.*

A function $h : X \longrightarrow \mathbb{R}^+$ is called **simple** if $\mathrm{rng}(h)$ is finite. It is easy to see that a function h is simple if and only if there are mutually disjoint sets

$\langle A_k \subseteq X : k = 0, \ldots, n \rangle$ and reals $0 < \alpha_0 < \cdots < \alpha_n$ such that

$$f(x) = \sum_{k=0}^{n} \alpha_k \chi_{A_k}(x). \tag{3.29}$$

A simple function f is lower $\mathcal{A}$-measurable if and only if $X \in \mathcal{A}$ and $\bigcup_{i=k}^{n} A_i \in \mathcal{A}$ for every $k = 0, \ldots, n$. Moreover, if f is lower $\mathcal{A}$-measurable we obtain

$$f = \sum_{k=0}^{n} \beta_k \chi_{B_k},$$

with $B_k \in \mathcal{A}$. It suffices to set $B_k = \bigcup_{i=k}^{n} A_i$, $\beta_0 = \alpha_0$ and $\beta_k = \alpha_k - \alpha_{k-1}$ for $k > 0$. Thus a simple lower $\mathcal{A}$-measurable function is a linear combination of characteristic functions of sets from $\mathcal{A}$.

Lemma 3.74. *Let $\mathcal{A}$ be a σ-topology on X. If $f : X \longrightarrow \mathbb{R}^+$ is a lower $\mathcal{A}$-measurable function, then there exists a sequence $\{f_n\}_{n=0}^{\infty}$ of simple lower $\mathcal{A}$-measurable functions such that $f_n \nearrow f$.*

Proof. Let $f : X \longrightarrow \mathbb{R}^+$ be lower $\mathcal{A}$-measurable. For every n, every $i \le n2^n$ and every $x \in X$ we set

$$A_{n,i} = \left\{ x \in X : f(x) > \frac{i}{2^n} \right\}, \qquad f_n(x) = \sum_{i=1}^{n2^n} \frac{1}{2^n} \chi_{A_{n,i}}(x).$$

Evidently $A_{n,i} \supseteq A_{n,i+1}$ for any $i < n2^n$. For a real a the set $\{x \in X : f_n(x) > a\}$ is one of sets $\emptyset$, $A_{n,i}, i = 0, \ldots, n2^n$, X. Thus f_n is lower $\mathcal{A}$-measurable. One can easily check that $f_n \nearrow f$. $\qquad\square$

Lemma [wAC] 3.75. *If $f : X \longrightarrow \mathbb{R}^+$ is lower semicontinuous, then there exist continuous functions f_n such that $f_n \nearrow f$.*

Proof. By Lemma 3.74 there are simple lower semicontinuous functions h_n such that $h_n \nearrow f$. Since a simple lower semicontinuous function is a linear combination of characteristic functions of open sets, by Lemma 3.71, b) we have to show that $\chi_A \in \mathrm{C}_p(X)^{\uparrow}$ for any A open.

Let $A \subseteq X$ be open. Then there are closed sets A_n such that $A = \bigcup_n A_n$. Moreover, we can assume that $A_n \subseteq A_{n+1}$ for every n. By Theorem 3.14 there exist continuous functions f_n such that $f_n(x) = 1$ for $x \in A_n$ and $f_n(x) = 0$ for $x \in X \setminus A$. We can assume that $f_n \le f_{n+1}$. One can easily see that $f_n \nearrow \chi_A$. $\quad\square$

Lemma [wAC] 3.76. *Let $0 < \xi < \omega_1$. If f is a lower $\mathbf{\Sigma}_{\xi}^{0}(X)$-measurable function with non-negative values, then $f \in (\bigcup_{\eta < \xi} \mathbf{BP}_{\eta}(X))^{\uparrow}$.*

Proof. For $\xi = 1$ the lemma follows by Lemma 3.75. For $\xi > 1$ we prove the lemma by transfinite induction. So we suppose the statement holds true for any $\eta < \xi < \omega_1$.

By Lemmas 3.74, and 3.71, b) we have to show that $\chi_A \in (\bigcup_{\eta < \xi} \mathbf{BP}_\xi(X))^\uparrow$ for any $A \in \mathbf{\Sigma}_\xi^0(X)$. So let $A \in \mathbf{\Sigma}_\xi^0(X)$. Then there are sets $A_n \in \bigcup_{\eta < \xi} \mathbf{\Pi}_\xi^0(X)$, $n \in \omega$, such that $A = \bigcup_n A_n$. We can assume that $A_n \subseteq A_{n+1}$ for every $n \in \omega$. We show that $\chi_{A_n} \in \bigcup_{\eta < \xi} \mathbf{BP}_\xi(X)$.

Let $\eta < \xi$ be such that $A_n \in \mathbf{\Pi}_\eta^0(X)$. By the inductive assumption we obtain that $\chi_{X \setminus A_n} \in (\bigcup_{\zeta < \eta} \mathbf{BP}_\zeta(X))^\uparrow$. By definition

$$\left(\bigcup_{\zeta < \eta} \mathbf{BP}_\zeta(X) \right)^\uparrow \subseteq \mathrm{scl}\left(\bigcup_{\zeta < \eta} \mathbf{BP}_\zeta(X) \right) = \mathbf{BP}_\eta(X).$$

Thus by Corollary 3.73 we have $\chi_{A_n} = 1 - \chi_{X \setminus A_n} \in \mathbf{BP}_\eta(X)$. Since $\chi_{A_n} \nearrow \chi_A$, we are ready. □

Note the following: all considered properties of a function f are preserved by the multiplication $a \cdot f$ by a positive real a and by a shift $f + b$ by a real b. E.g., we can assume that the function is bounded from below instead of being non-negative. Replacing the function f by $-f$ we must replace the word "upper" by "lower" and vice versa, "non-decreasing sequence" by "non-increasing one".

Theorem [wAC] 3.77 (H. Lebesgue – F. Hausdorff). *Assume that $f : X \longrightarrow \mathbb{R}$. If f is $\mathbf{\Sigma}_{\xi+1}^0$-measurable, then $f \in \mathbf{BP}_\xi(X)$. If f is $\mathbf{\Sigma}_\xi^0$-measurable and ξ is a limit ordinal, then $f \in (\bigcup_{\eta < \xi} \mathbf{BP}_\eta(X))^\downarrow \cap (\bigcup_{\eta < \xi} \mathbf{BP}_\eta(X))^\uparrow$.*

Proof. Assume that $H : \mathbb{R} \longrightarrow (0, 1)$ is an order-preserving homeomorphism and $G : (0, 1) \longrightarrow \mathbb{R}$ is the inverse to H. In the proof we use essentially the fact that all considered families of functions are lattices of functions.

Assume that f is $\mathbf{\Sigma}_{\xi+1}^0$-measurable with values in $(0, 1)$. Then by Lemmas 3.75 and 3.76 we have $f \in (\mathbf{BP}_\xi(X))^\uparrow$. Similarly, applying the lemmas to the function $1 - f$ we obtain that $f \in (\mathbf{BP}_\xi(X))^\downarrow$. So, by Theorem 3.72 we obtain that $f \in \mathbf{BP}_\xi(X)$.

Assume now that $f : X \longrightarrow \mathbb{R}$ is a $\mathbf{\Sigma}_{\xi+1}^0$-measurable function. Then the function $f \circ H$ is $\mathbf{\Sigma}_{\xi+1}^0$-measurable with values in $(0, 1)$. Since $f \circ H \in \mathbf{BP}_\xi$, there are functions $g_n : X \longrightarrow \mathbb{R}$ such that $g_n \to f \circ H$ and $g_n \in \bigcup_{\eta < \xi} \mathbf{BP}_\eta(X)$. However the range of g_n need not be a subset of $(0, 1)$. Thus, we denote

$$f_n = \min\{\max\{g_n, 1/(n+1)\}, n/(n+1)\}.$$

Then $f_n \circ G \in \mathbf{BP}_\xi(X)$ and $f_n \circ G \to f$.

For ξ limit the proof is almost the same. □

Corollary [wAC] 3.78. *For any $\xi < \omega_1$, a real function f defined on a metric separable space X is $\mathbf{\Sigma}_{\xi+1}^0$-measurable if and only if $f \in \mathbf{BP}_\xi(X)$.*

As an easy consequence we obtain the main result of this section.

Theorem [wAC] 3.79. *A function from a separable metric space into $\mathbb{R}$ is Borel measurable if and only if it is a Baire function, i.e., analytically representable.*

Exercises

3.29 Examples of Baire Functions

The **Dirichlet function** $\chi : \mathbb{R} \longrightarrow \mathbb{R}$ is defined as follows: $\chi(x) = 1$ for $x \in \mathbb{Q}$ and $\chi(x) = 0$ for $x \in \mathbb{R} \setminus \mathbb{Q}$.

  a) Show that $\chi(x) = \lim_{n \to \infty} \lim_{m \to \infty} (\cos n! \pi x)^{2m}$.

  b) Find sequences $\langle f_n : n \in \omega \rangle$, $\langle g_{n,m} : n, m \in \omega \rangle$, $g_{n,m}$ continuous and such that $g_{n,m} \searrow f_n$ and $f_n \nearrow \chi$.

  c) The set $\mathrm{disc}(\chi)$ is not meager, however there exists a meager set $D \subseteq \mathbb{R}$ such that $\chi|(\mathbb{R} \setminus D)$ is continuous. Conclude that $\chi \in \mathbf{BP}_2(\mathbb{R}) \setminus \mathbf{BP}_1(\mathbb{R})$.

  d) Every monotonic function $f : \mathbb{R} \longrightarrow \mathbb{R}$ is in the first Baire class.
  Hint: Make corresponding simple functions continuous.

3.30 Non-negative and Non-positive Part of a Function

The **non-negative part** f^+ and the **non-positive part** f^- of a real function f are defined as $f^+ = \max\{0, f\}$ and $f^- = \max\{0, -f\}$.

  a) Note that
  $$f = f^+ - f^- \text{ and } |f| = f^+ + f^-.$$

  b) If $\mathcal{A}$ is a σ-topology on X, then $f : X \longrightarrow \mathbb{R}$ is $\mathcal{A}$-measurable if and only if both f^+ and f^- are $\mathcal{A}$-measurable.

  c) Find conditions for upper and lower measurability of f in terms of properties of f^+ and f^-.

  d) Show that $f \in \mathbf{BP}_\xi(X)$ if and only if $f^+, f^- \in \mathbf{BP}_\xi(X)$.

3.31 [wAC] Turning a Borel Function into a Continuous One

$\langle X, \mathcal{O} \rangle$ and $\langle Y, \mathcal{T} \rangle$ are Polish spaces.

  a) If $f : X \longrightarrow Y$ is Borel measurable, then there exists a Polish topology $\mathcal{O}'$ on X such that f is continuous in this topology and $\mathrm{BOREL}(X, \mathcal{O}') = \mathrm{BOREL}(X, \mathcal{O})$.
  Hint: Use Exercise 3.18, f).

  b) If $f : X \longrightarrow Y$ is a Borel isomorphism, then there exist Polish topologies $\mathcal{O}'$ and $\mathcal{T}'$ on X and Y, respectively, such that f is a homeomorphism in those topologies, $\mathrm{BOREL}(X, \mathcal{O}') = \mathrm{BOREL}(X, \mathcal{O})$ and $\mathrm{BOREL}(Y, \mathcal{T}') = \mathrm{BOREL}(Y, \mathcal{T})$.

3.32 Baire Classes and Uniform Convergence

If $\mathbf{\Phi} \subseteq {}^X Y$ is a family of functions we write

$$\overrightarrow{\mathbf{\Phi}} = \{ f \in {}^X Y : (\exists f_n \in \mathbf{\Phi}, n \in \omega)\, f_n \rightrightarrows f \text{ on } X \}.$$

  a) If $\mathbf{\Phi} \subseteq {}^X \mathbb{R}$ is a lattice, $f + a \in \mathbf{\Phi}$ for any $f \in \mathbf{\Phi}$ and any real a, then $\overrightarrow{\mathbf{\Phi}} \subseteq \mathbf{\Phi}^\uparrow \cap \mathbf{\Phi}^\downarrow$.
  Hint: If $f_n \rightrightarrows f$ and $\varepsilon_n = \sup\{|f_n(x) - f(x)| : x \in X\}$, set

  $$g_n = \max\{f_0 - \varepsilon_0, \ldots, f_n - \varepsilon_n\}, \quad h_n = \min\{f_0 + \varepsilon_0, \ldots, f_n + \varepsilon_n\}.$$

  Then $g_n \nearrow f$ and $h_n \searrow f$.

  b) Show that $\overrightarrow{\mathbf{BP}}_\xi(X) = \mathbf{BP}_\xi(X)$ for any $\xi < \omega_1$ and any metric separable space X.

3.33 [wAC] Separation of Sets

Let us recall that $Z(f) = \{x \in X : f(x) = 0\}$.

a) $A \in \mathbf{\Pi}^0_{\xi+1}(X)$ if and only if there exists a function $f \in \mathbf{BP}_\xi(X)$ such that $A = Z(f)$.
Hint: If $A \in \mathbf{\Pi}^0_{\xi+1}(X)$, then $1 - \chi_A$ is lower $\mathbf{\Sigma}^0_{\xi+1}$-measurable. Using Lemma 3.76 find $\mathbf{\Sigma}^0_\xi$-measurable $f_n : X \longrightarrow [0,1]$ such that $f_n \nearrow 1 - \chi_A$. Set $f = \sum_{n=0}^\infty 2^{-n} f_n$ and use the result of Example 3.32, b).

b) If $A, B \in \mathbf{\Pi}^0_{\xi+1}(X)$, $A \cap B = \emptyset$, then there exists a function $f \in \mathbf{BP}_\xi(X)$ such that $A = Z(f)$ and $B = Z(f - 1)$.
Hint: If $Z(g) = A$ and $Z(h) = B$, take $f = g \cdot (2 - h)/2 \cdot (g + h)$.

3.34 [wAC] Quasi-normal Hierarchy

Let X be a Polish space. For any subset $A \subseteq {}^X\mathbb{R}$ we define the **quasi-normal closure** of A by

$$\mathrm{Qcl}(A) = \{f \in {}^X\mathbb{R} : (\exists \{f_n\} \in {}^\omega A)\, f_n \xrightarrow{\mathrm{QN}} f \text{ on } A\}.$$

The **Quasi-normal Hierarchy** $\langle \mathbf{BQ}_\xi(X) : \xi < \omega_1 \rangle$ is defined as: $\mathbf{BQ}_0(X) = C(X)$ and

$$\mathbf{BQ}_\xi(X) = \mathrm{Qcl}(\bigcup_{\eta < \xi} \mathbf{BQ}_\eta(X)).$$

The **discrete closure** $\mathrm{Dcl}(A)$ and the **Discrete Hierarchy** $\langle \mathbf{BD}_\xi(X) : \xi < \omega_1 \rangle$ are defined similarly.

a) $A \subseteq \mathrm{Dcl}(A) \subseteq \mathrm{Qcl}(A) \subseteq \mathrm{scl}\,(A)$ for any $A \subseteq {}^X\mathbb{R}$.

b) $\mathbf{BD}_\xi(X) \subseteq \mathbf{BQ}_\xi(X) \subseteq \mathbf{BP}_\xi(X)$ for any $\xi < \omega_1$.

c) If $f_n \xrightarrow{\mathrm{QN}} f$ on X, f_n are $\mathbf{\Sigma}^0_\xi(X)$-measurable, then there exist sets $A_n \in \mathbf{\Pi}^0_\xi(X)$ such that $X = \bigcup_n A_n$ and $f_n \rightrightarrows f$ on every A_n.
Hint: If ε_n is a control of the quasi-normal convergence, set

$$A_n = \{x \in X : (\forall k, m \geq n)\, |f_m(x) - f_k(x)| \leq \varepsilon_m + \varepsilon_k\}.$$

d) If $f_n \xrightarrow{\mathrm{D}} f$ on X, f_n are $\mathbf{\Sigma}^0_\xi(X)$-measurable, then there exist sets $A_n \in \mathbf{\Pi}^0_\xi(X)$ such that $X = \bigcup_n A_n$ and $f_n|A_n = f|A_n$ for every n.
Hint: Set

$$A_n = \{x \in X : (\forall k, m \geq n)\, f_m(x) = f_k(x)\}.$$

e) If $f \in \mathbf{BD}_\xi(X)$, $\xi > 0$, then f is $\mathbf{\Delta}^0_{\xi+1}$-measurable
Hint: We proceed by transfinite induction. If $f_n \xrightarrow{\mathrm{D}} f$, $f_n \in \bigcup_{\eta < \xi} \mathbf{BQ}_\eta(X)$, then f_n are $\mathbf{\Delta}^0_\xi$-measurable. Thus f is $\mathbf{\Sigma}^0_{\xi+1}$-measurable. If $\langle A_n : n \in \omega \rangle$ are those of d), then $f^{-1}(U) = \bigcap_n (f_n^{-1}(U) \setminus A_n)$.

f) If $f \in \mathbf{BD}_\xi(X)$, then for every $n \in \omega$ there exist set $A_n \in \mathbf{\Delta}^0_{\xi+1}(X)$ and continuous function g_n such that $X = \bigcup_n A_n$ and $f|A_n = g_n|A_n$.
Hint: By transfinite induction. If $f_n \xrightarrow{\mathrm{D}} f$, $f_n \in \mathbf{BD}_\xi(X)$, then $f|X_n = f|X_n$ with $X_n \in \mathbf{\Pi}^0_{\xi+1}(X)$. By inductive assumption $f_n|A_{n,m} = g_{n,m}|A_{n,m}$, $g_{n,m}$ continuous. Then $f|(X_n \cap A_{n,m}) = g_{n,m}|(X_n \cap A_{n,m})$. For ξ limit similarly.

g) If $A \in \mathbf{\Delta}_2^0(X)$, then $\chi_A \in \mathbf{BD}_1(X)$.

   Hint: If $A = \bigcup_n F_n = \bigcap_n G_n$, F_n closed increasing, G_n open decreasing, then there exist continuous functions f_n such that $f_n(x) = 1$ for $x \in F_n$ and $f_n(x) = 0$ for $x \in X \setminus G_n$. Then $f_n \xrightarrow{\mathrm{D}} \chi_A$ on X.

h) If $A \in \mathbf{\Delta}_{\xi+1}^0(X)$, $\xi \geq 1$, then $\chi_A \in \mathbf{BD}_\xi(X)$.

   Hint: Let $\xi \geq 2$. By induction, assume that the statement is true for $\eta < \xi$. Take the sets $B_n \in \mathbf{\Delta}_\xi^0(X)$ of Exercise 3.20, e). If ξ is non-limit, then $\chi_{B_n} \in \mathbf{BD}_{\xi-1}(X)$ and $\chi_{B_n} \xrightarrow{\mathrm{D}} \chi_A$. If ξ is limit, then $B_n = \bigcup_k C_n^k$ with $C_n^k \in \mathbf{\Delta}_{\xi_k}^0(X)$, $\xi_k < \xi$ for every n.

i) Let $\mathcal{A}$ be a family of subsets of X with weak reduction property (see Exercise 3.21) and closed under finite intersections and finite unions. Every $\mathcal{A}$-measurable function with non-negative real values is a quasi-normal limit of simple $\mathcal{A} \cap \neg\mathcal{A}$-measurable functions.

   Hint: Let f be $\mathcal{A}$-measurable. Then $f_n \xrightarrow{\mathrm{D}} f$ on X and f_n are $\mathcal{A}$-measurable. For $i < 2^{2n}$ we set

$$A_{n,i} = \{x \in X : (i-1)2^{-n} < f(x) < (i+1)2^{-n}\}.$$

   Moreover, let $A_{n,2^{2n}} = \{x \in X : f(x) > 2^n - 2^{-n}\}$. Take $B_{n,i} \in \mathcal{A} \cap \neg\mathcal{A}$ pairwise disjoint such that

$$\bigcup_{i \leq 2^{2n}} B_{n,i} = X, \quad B_{n,i} \subseteq A_{n,i}.$$

   Set $g_n(x) = i2^{-n}$ for $x \in B_{n,i}$, $i \leq 2^{2n}$. Then $g_n \xrightarrow{\mathrm{QN}} f$ on X.

j) $\mathbf{BP}_\xi(X) \subseteq \mathrm{Qcl}(\mathbf{BD}_\xi(X))$.

   Hint: Use the result of i) and k).

k) $\mathbf{BP}_\xi(X) \subseteq \mathbf{BQ}_{\xi+1}(X)$.

l) $\bigcup_{\xi < \omega_1} \mathbf{BQ}_\xi(X) = \mathrm{BAIRE}(X)$.

Historical and Bibliographical Notes

The notion of a metric space was introduced by M. Fréchet in his thesis [1906]. K. Kuratowski's two volumes monograph [1958a] and [1958b] is an encyclopedia of results and historical information concerning separable metric spaces.

H. Steinhaus [1920] proved Theorem 3.6. Urysohn's Lemma, Theorem 3.15, is proved in P. Urysohn [1925]. In this paper the notion of a perfectly normal space appeared implicitly. It was defined explicitly by E. Čech [1932].

E. Borel [1895] proved the theorem about finite coverings for open intervals. The Tietze-Urysohn Theorem was proved by H. Tietze [1915] for metric spaces and by P. Urysohn [1925] for normal spaces. The small inductive dimension was independently introduced by P. Urysohn [1922] and K. Menger [1923], [1928].

The notion of a complete metric space was introduced by M. Fréchet [1906]. Metric separable spaces and especially complete metric separable spaces were intensively studied, mainly by Polish mathematicians. That was the main reason why N. Bourbaki [1961] called such spaces "espace polonais". Theorem 3.27 was proved by R. Baire in 1889 for the real line and then by F. Hausdorff [1914] for any Polish space. Theorem 3.36 was essentially proved by M. Fréchet [1910] in spite of the fact that the notion of a totally bounded metric space was introduced by F. Hausdorff [1914].

The results of A.N. Tychonoff [1930] imply the results of Exercise 3.10. The results of Exercises 3.12–3.14 and 3.16–3.17 were essentially obtained by S. Banach [1932]. For a more comprehensive presentation, including results of Exercise 3.15, see, e.g., W. Rudin [1973].

E. Borel [1898] defined the σ-algebra $\mathrm{BOREL}(X)$ of Borel subsets of a metric space X. F. Hausdorff [1914] introduced the Borel Hierarchy and declared as a fact the equalities (3.17).

Á. Császár and M. Laczkovich [1975] studied quasi-normal convergence under the name equal convergence. Then Z. Bukovská [1991] independently introduced and studied the notion of a quasi-normal convergence. The discrete convergence of Exercise 3.22 was studied by Á. Császár and M. Laczkovich [1975]. Exercises 3.23–3.25 follow ideas of the generalization of the classical Weierstrass Theorem by M.H. Stone [1947].

R. Baire [1898a] proved Theorem 3.70. Then he introduced the Baire Hierarchy (3.25). H. Lebesgue [1905] and F. Hausdorff [1914] proved Theorem 3.77 and therefore also Theorem 3.79. There is a nice Russian presentation of the Baire Hierarchy by I.P. Natanson [1957]. The hierarchies of Exercise 3.34 were introduced and studied by Á. Császár and M. Laczkovich [1975], [1979] and [1990].

Chapter 4

Measure Theory

On notera d'autre part que, dans la pratique, on ne se soucie guère de préciser quelles sont les portions d'espace que l'on considère comme "mesurables"; bien entendu, il est indispensable de fixer ce point sans ambiguïté dans toute théorie mathématique de la mesure; c'est ce qu'on fait par exemple quand, en géométrie élémentaire, on définit l'aire des polygones ou le volume des polyèdres; dans tous les cas, la famille des ensembles "mesurables" doit naturellement être telle que la réunion de deux quelconques d'entre eux sans point commun soit encore "mesurable".

Nicolas Bourbaki [1952], p. 1

We suppose that the reader is acquainted with basic measure theory. However the common presentation of measure theory uses the Axiom of Choice without any comment. To replace the use of Axiom of Choice by a Weak Axiom of Choice or to avoid it at all, we present some facts about measure on a topological space, a brief construction of Lebesgue measure and we prove some of its basic properties. In Section 4.3 we recall the definition of the Lebesgue integral and we prove the basic results related to it. Finally, Section 4.4 is devoted to a brief presentation of the results which we shall intensively use: the Fubini Theorem and the Ergodic Theorem.

4.1 Measure

Let $\mathcal{S}$ be a σ-algebra of subsets of a non-empty set X. A function μ from $\mathcal{S}$ into $[0, \infty]$ is called a **measure** if

$$\mu(\emptyset) = 0, \quad \mu(X) \neq 0, \tag{4.1}$$

$$\mu(\bigcup_n A_n) = \sum_{n=0}^{\infty} \mu(A_n) \text{ for any pairwise disjoint } \langle A_n \in \mathcal{S} : n \in \omega \rangle. \tag{4.2}$$

From the definition we immediately obtain that for any $A, B \in \mathcal{S}$ and for any sequence $\langle A_n \in \mathcal{S} : n \in \omega \rangle$ the following holds true:

$$\mu(A) \leq \mu(B), \qquad\qquad \text{if } A \subseteq B, \tag{4.3}$$

$$\mu(A) = \lim_{n \to \infty} \mu(A_n), \qquad \text{if } A = \bigcup_n A_n \text{ and } A_n \subseteq A_{n+1} \text{ for every } n, \tag{4.4}$$

$$\mu(A) = \lim_{n \to \infty} \mu(A_n), \qquad \text{if } A = \bigcap_n A_n, \ A_n \supseteq A_{n+1} \text{ for every } n,$$

$$\text{and } \mu(A_0) < \infty. \tag{4.5}$$

The triple $\langle X, \mathcal{S}, \mu \rangle$ is called a **measure space**. If the σ-algebra $\mathcal{S}$ is understood we speak about a measure on X. A measure μ is said to be **complete** if for any $A \subseteq B \subseteq X$, $B \in \mathcal{S}$, $\mu(B) = 0$ also $A \in \mathcal{S}$. If μ is a complete measure then $\langle X, \mathcal{S}, \mu \rangle$ is called a **complete measure space**. A measure μ is **finite** if $\mu(X) < \infty$. A measure μ is σ-**finite** if there are sets $\langle X_n \in \mathcal{S} : n \in \omega \rangle$ with $\mu(X_n) < \infty$ for every n and such that $\bigcup_n X_n = X$. Of course, we can suppose that $X_n \subseteq X_{n+1}$. A formula $\mathcal{V}$ is **true almost everywhere** or is **true for almost every** $x \in X$ if $\mu(\{x \in X : \neg\mathcal{V}(x)\}) = 0$. A measure μ is called **diffused** if $\{x\} \in \mathcal{S}$ and $\mu(\{x\}) = 0$ for any $x \in X$. Finally, μ is a **probabilistic measure** if $\mu(X) = 1$.

One can easily check that there is no diffused measure on a countable set. The construction of a non-trivial diffused measure on an uncountable set is not simple. We describe a method of such a construction in a particular case.

Let X be a non-empty set. A function $\mu^* : \mathcal{P}(X) \longrightarrow [0, \infty]$ is an **outer measure** on X if

$$\mu^*(\emptyset) = 0, \quad \mu^*(X) \neq 0, \tag{4.6}$$

$$\mu^*(A) \leq \mu^*(B), \qquad\qquad \text{for any } A \subseteq B \subseteq X, \tag{4.7}$$

$$\mu^*\left(\bigcup_n A_n\right) \leq \sum_{n=0}^{\infty} \mu^*(A_n), \qquad \text{for any } \langle A_n \subseteq X : n \in \omega \rangle. \tag{4.8}$$

Let μ^* be an outer measure on X. A set $A \subseteq X$ is **Carathéodory measurable**, simply μ^*-**measurable**, if

$$\mu^*(B) = \mu^*(B \cap A) + \mu^*(B \setminus A) \text{ for any } B \subseteq X. \tag{4.9}$$

We denote by $\mathcal{S}(\mu^*)$ the family of all μ^*-measurable subsets of X.

The following is a fundamental result:

Theorem 4.1. *Let μ^* be an outer measure on a set X. Then $\mathcal{S}(\mu^*)$ is a σ-algebra and $\mu^* | \mathcal{S}(\mu^*)$ is a complete measure. Moreover, if $\langle A_n \in \mathcal{S}(\mu^*) : n \in \omega \rangle$ are pairwise disjoint and $B \subseteq X$, then*

$$\mu^*\left(B \cap \bigcup_n A_n\right) = \sum_{n=0}^{\infty} \mu^*(B \cap A_n). \tag{4.10}$$

The proof is rather technical. We present all details mainly to ensure that no choice has been used.

Evidently both $\emptyset$ and X are μ^*-measurable.

Let $C, D \in \mathcal{S}(\mu^*)$ and let $B \subseteq X$. Using the measurability of sets C and D we obtain

$$\mu^*(B) = \mu^*(B \cap C \cap D) + \mu^*((B \cap C) \setminus D) + \mu^*((B \setminus C) \cap D) + \mu^*((B \setminus C) \setminus D).$$

Replacing B by $B \setminus (C \setminus D)$ we obtain

$$\mu^*(B \setminus (C \setminus D)) = \mu^*(B \cap C \cap D) + \mu^*((B \cap D) \setminus C) + \mu^*((B \setminus C) \setminus D). \quad (4.11)$$

Thus

$$\mu^*(B) = \mu^*(B \cap (C \setminus D)) + \mu^*(B \setminus (C \setminus D))$$

and therefore $C \setminus D \in \mathcal{S}(\mu^*)$.

If $C, D \in \mathcal{S}(\mu^*)$, then $C \cup D = X \setminus ((X \setminus C) \setminus D) \in \mathcal{S}(\mu^*)$.

If $C, D \in \mathcal{S}(\mu^*)$ are disjoint, then

$$\mu^*(B \cap (C \cup D)) = \mu^*((B \cap (C \cup D)) \cap C) + \mu^*((B \cap (C \cup D)) \setminus C)$$
$$= \mu^*(B \cap C) + \mu^*(B \cap D). \quad (4.12)$$

Now let $\langle A_n \in \mathcal{S}(\mu^*) : n \in \omega \rangle$ be pairwise disjoint. From (4.12) we obtain by induction

$$\mu^*(B \cap \bigcup_{i=0}^{n} A_i) = \sum_{i=0}^{n} \mu^*(B \cap A_i).$$

Hence

$$\sum_{i=0}^{n} \mu^*(B \cap A_i) \leq \mu^* \left(B \cap \bigcup_{i=0}^{\infty} A_i \right)$$

and together with (4.8) we obtain (4.10). Since $\bigcup_{i=0}^{n} A_i$ is μ^*-measurable, we have

$$\mu^*(B) = \mu^* \left(B \cap \bigcup_{i=0}^{n} A_i \right) + \mu^* \left(B \setminus \bigcup_{i=0}^{n} A_i \right) \geq \sum_{i=0}^{n} \mu^*(B \cap A_i) + \mu^* \left(B \setminus \bigcup_{i=0}^{\infty} A_i \right)$$

and therefore

$$\mu^*(B) \geq \sum_{i=0}^{\infty} \mu^*(B \cap A_i) + \mu^* \left(B \setminus \bigcup_{i=0}^{\infty} A_i \right). \quad (4.13)$$

Using (4.8) and (4.10) we obtain

$$\mu^*(B) = \mu^* \left(B \cap \bigcup_{i=0}^{\infty} A_i \right) + \mu^* \left(B \setminus \bigcup_{i=0}^{\infty} A_i \right).$$

Thus $\bigcup_{i=0}^{\infty} A_i \in \mathcal{S}(\mu^*)$. Taking $B = X$ in (4.10) we obtain (4.2) for μ^*.

If $A \subseteq C$, $\mu^*(C) = 0$, then for any B we have

$$\mu^*(B) \leq \mu^*(B \cap A) + \mu^*(B \setminus A) = \mu^*(B \setminus A) \leq \mu^*(B)$$

and therefore $A \in \mathcal{S}(\mu^*)$. Thus $\mu^*|\mathcal{S}(\mu^*)$ is a complete measure. $\qquad \square$

Assume now that $\langle X, \mathcal{O} \rangle$ is a topological space and $\langle X, \mathcal{S}, \mu \rangle$ is a measure space. The measure μ is called a **Borel measure** on X if $\mathcal{S} = \text{BOREL}(X)$ and μ is **locally finite**, i.e., every point $x \in X$ has a neighborhood U with $\mu(U) < \infty$.[1] If μ is locally finite, then $\mu(K) < +\infty$ for any compact set K. A measure μ is **moderated** if there are open sets $\langle G_n : n \in \omega \rangle$ such that $X = \bigcup_n G_n$ and $\mu(G_n) < +\infty$ for each n. A moderated measure is both locally finite and σ-finite. If X has a countable base (or X is Lindelöf – see Exercise 8.20), then a locally finite measure is moderated. Thus, every Borel measure on a metric separable space is moderated.

The **support** of a Borel measure μ on X is the set

$$\text{supp}(\mu) = \{x \in X : (\forall U)\,(U \text{ neighborhood of } x \to \mu(U) > 0)\}.$$

The support is a closed set, and $\mu(U) > 0$ for each open set $U \cap \text{supp}(\mu) \neq \emptyset$. If the topology $\mathcal{O}$ has a countable base, then $\mu(X \setminus \text{supp}(\mu)) = 0$.

If $\langle X, \mathcal{S}, \mu \rangle$ is a measure space and $\langle X, \mathcal{O} \rangle$ is a topological group, then the measure μ is **shift invariant**, if $a + A \in \mathcal{S}$, $\mu(a + A) = \mu(A)$, $A + a \in \mathcal{S}$ and $\mu(A + a) = \mu(A)$ for any $A \in \mathcal{S}$ and any $a \in X$. If X is infinite, then a finite shift invariant measure is diffused.

Let us suppose that $\langle X, \mathcal{S}, \mu \rangle$ is a measure space, where $\langle X, \mathcal{O} \rangle$ is a topological space, and $\text{BOREL}(X) \subseteq \mathcal{S}$. A set $A \in \mathcal{S}$ is said to be

a) **outer regular** if
$$\mu(A) = \inf\{\mu(U) : A \subseteq U,\ U \text{ open}\}, \tag{4.14}$$

b) **inner regular** if
$$\mu(A) = \sup\{\mu(F) : F \subseteq A,\ F \text{ closed}\}, \tag{4.15}$$

c) **Radon** if
$$\mu(A) = \sup\{\mu(K) : K \subseteq A,\ K \text{ compact}\}. \tag{4.16}$$

Note the following. Assume that X has a countable base and **wAC** holds true. Then a set $A \in \mathcal{S}$, $\mu(A) < \infty$, is outer regular if and only if there exists a nonincreasing sequence of open sets $\langle U_n \supseteq A : n \in \omega \rangle$ such that $\mu(\bigcap_n U_n \setminus A) = 0$. If moreover μ is moderated, then the assumption $\mu(A) < \infty$ may be omitted. Similarly for inner regularity and Radon property – see Theorem 4.9.

The measure μ is **outer regular**, **inner regular**, if every set $A \in \mathcal{S}$ is outer regular, inner regular, respectively. A measure, which is both inner and outer regular, is **regular**. A regular measure is **Radon**, if every set $A \in \mathcal{S}$ is Radon. Every diffused outer regular measure is locally finite. Every σ-finite outer regular measure is moderated. A finite measure is inner regular if and only if it is outer regular.

---

[1] Actually a Borel measure μ can be defined on a σ-algebra $\mathcal{S} \supset \text{BOREL}(X)$. However speaking about some properties of a Borel measure we are interested just in Borel sets, e.g., a Borel measure $\mu : \mathcal{S} \longrightarrow [0, \infty]$ is inner regular if every Borel set is inner regular.

Lemma [wAC] 4.2. *Assume that X has a countable base of topology and μ is a Borel measure on X. Let $A = \bigcup_n A_n$. If every A_n is outer regular, inner regular, Radon, then A is also outer regular, inner regular, Radon, respectively.*

Proof. We can assume that $\mu(A_n) < +\infty$ for every n, since otherwise all statements are trivial.

For given $\varepsilon > 0$ take an open $G_n \supseteq A_n$ such that $\mu(G_n \setminus A_n) < 2^{-n-1}\varepsilon$. Then $\mu(\bigcup_n G_n \setminus \bigcup_n A_n) < \varepsilon$.

If $a < \mu(A)$, then there exists an m such that $\varepsilon = \mu(\bigcup_{k=0}^m A_k) - a > 0$. Take a closed set $F_k \subseteq A_k$ such that $\mu(A_k \setminus F_k) < \varepsilon/(m+1)$, $k \leq m$. Then $F = \bigcup_{k=0}^m F_k$ is closed, $F \subseteq A$ and $\mu(F) > a$.

The proof for the Radon case is equal – just replace "closed" by "compact". $\qquad\square$

Lemma [wAC] 4.3. *If μ is a Borel measure on a separable metric space X and every open set A is inner regular, then μ is regular.*

Proof. We start with a claim:

> if $G \subseteq X$ is open, $\mu(G) < \infty$, then a Borel set $A \subseteq G$
> is inner regular if and only if $G \setminus A$ is outer regular.

Actually, assume that A is inner regular. Let $\varepsilon > 0$ be given. Since $\mu(A) < \infty$, there exists a closed set $F \subseteq A$ such that $\mu(A \setminus F) < \varepsilon$. Then $G \setminus A \subseteq G \cap (X \setminus F)$, $G \cap (X \setminus F)$ is open and

$$\mu((G \cap (X \setminus F)) \setminus (G \setminus A)) = \mu(A \setminus F) < \varepsilon.$$

If A is outer regular, then there exists an open set $U \supseteq A$ such that $\mu(U \setminus A) < \varepsilon/2$. We can assume that $U \subseteq G$. Then $G \setminus A = (G \setminus U) \cup (U \setminus A)$. The set $G \setminus U$ is an F_σ set, therefore there exists an increasing sequence of closed sets $\langle F_n : n \in \omega \rangle$ such that $\bigcup_n F_n = G \setminus U$. Since $G \setminus U$ has finite measure, there exists an m such that $\mu((G \setminus U) \setminus F_m) < \varepsilon/2$. Then $\mu((G \setminus A) \setminus F_m) < \varepsilon$.

If $\{V_n : n \in \omega\}$ is a countable base of topology, we set

$$G_n = \bigcup \{V_k : k \leq n \wedge \mu(V_k) < \infty\}.$$

Since μ is locally finite, we obtain that $\bigcup_n G_n = X$ and $\mu(G_n) < \infty$, i.e., μ is moderated. Let

$$\mathcal{T} = \{A \in \text{BOREL}(X) : (\forall n) (G_n \cap A \text{ is outer and inner regular})\}.$$

By the claim, $\mathcal{T}$ is closed under complement.

By Lemma 4.2, $\mathcal{T}$ is closed under countable unions.

By the assumption, $\mathcal{T}$ contains every open set. Thus $\text{BOREL}(X) = \mathcal{T}$. $\qquad\square$

Lemma [wAC] 4.4. *Every Borel measure on a separable metric space X is regular. If moreover X is σ-compact, then every Borel measure on X is Radon.*

Proof. Assume that X is a separable metric space, $X = \bigcup_n G_n$, $\mu(G_n) < +\infty$, $G_n \subseteq G_{n+1}$, G_n are open for every n.

If $U \subseteq X$ is open, then there exist closed sets $F_{0,n} \subseteq \cdots \subseteq F_{k,n} \ldots$ such that $U \cap G_n = \bigcup_k F_{k,n}$. Then $U = \bigcup_n (\bigcup_{i=0}^n \bigcup_{k=0}^n F_{k,i})$ and (4.15) follows by (4.4).

Thus, by Lemma 4.3 the measure μ is regular.

Assume now that X is σ-compact. To show that a regular measure μ is Radon, it suffices to show that every closed set is Radon. So, let $A \subseteq X$ be closed. Since X is σ-compact, there exist compact sets $\langle K_n : n \in \omega \rangle$ such that $X = \bigcup_n K_n$. Then each K_n is covered by some G_m and therefore $\mu(K_n) < +\infty$. Then $A = \bigcup_n \bigcup_{k \leq n} (A \cap K_k)$ and (4.16) follows by (4.4). □

Lemma [wAC] 4.5. *A finite Borel measure μ on a Polish space X is Radon.*

Proof. By Lemma 4.4 the measure μ is inner regular. Therefore we have to show that (4.16) holds true for any closed set A. Since a closed set is again a Polish space, it suffices to show that (4.16) holds true for $A = X$.

Let $\langle B_{n,k} : k \in \omega \rangle$ be a sequence of closed sets such that $X = \bigcup_k B_{n,k}$ and $\text{diam}(B_{n,k}) < 2^{-n}$ for every n. Let $\varepsilon > 0$ be given. Since $\mu(X) < \infty$, for each m there exists a k_m such that $\mu(X \backslash \bigcup_{i=0}^{k_m} B_{m,i}) < \varepsilon \cdot 2^{-m-1}$. Then $K = \bigcap_m \bigcup_{i=0}^{k_m} B_{m,i}$ is a closed totally bounded set. By Theorem 3.36, the set K is compact. Since $\mu(X \backslash K) < \varepsilon$, we are done. □

Lemma [wAC] 4.6. *Let μ be a Borel measure on a topological space X with countable base. Then there exists a probabilistic Borel measure μ_P such that:*

  a) *$\mu(A) = 0 \equiv \mu_P(A) = 0$ for any Borel set $A \subseteq X$.*
  b) *μ_P is Radon if and only if μ is Radon.*

Proof. If μ is a Borel measure on X, then there exist pairwise disjoint Borel sets $\langle X_n : n \in \omega \rangle$ such that $X = \bigcup_n X_n$ and $0 < \mu(X_n) < \infty$. For any Borel set A we put

$$\mu_P(A) = \sum_{n=0}^{\infty} \frac{\mu(A \cap X_n)}{\mu(X_n)} 2^{-n-1}.$$

It is easy to verify that μ_P is a probabilistic Borel measure.

The statement a) follows immediately from the definition.

Assume that μ_P is Radon. If A is Borel, then there exists a non-decreasing sequence of compact sets $\langle K_n \subseteq A : n \in \omega \rangle$ with $\mu_P(A) = \sup\{\mu_P(K_n) : n \in \omega\}$. Let $K = \bigcup_n K_n$. Then $\mu_P(A \backslash K) = 0$. By a) also $\mu(A \backslash K) = 0$. Hence μ is Radon.

Assume now that μ is Radon and $a < \mu_P(A)$. Then there exists an m such that

$$\varepsilon = \sum_{n=0}^{m} \frac{\mu(A \cap X_n)}{\mu(X_n)} 2^{-n-1} - a > 0.$$

Take compact sets $K_n \subseteq A \cap X_n$ such that $\mu(K_n) > \mu(A \cap X_n) - \varepsilon \cdot \mu(X_n)$. Then $\bigcup_{n=0}^m K_n \subseteq A$ and $\mu_P(\bigcup_{n=0}^m K_n) > a$. $\qquad\square$

Theorem [wAC] 4.7. *Any Borel measure on a Polish space is Radon.*

Proof. The theorem follows immediately by Lemmas 4.6 and 4.5. $\qquad\square$

Assume that **wAC** holds true. We show that for any diffused Borel measure μ on a perfect Polish space X, there exist sets $A, B \subseteq X$ such that B is Borel, $\mu(B) = 0$, $A \subseteq B$ and A is not Borel. Actually, it is easy to see that for any countable set $Y \subseteq X$ and any $\varepsilon > 0$ there exists an open set $U \supseteq Y$ such that $\mu(U) < \varepsilon$. Thus, if $S \subseteq X$ is a countable dense subset of X, then there exists a G_δ set $B \supseteq S$ such that $\mu(B) = 0$. Evidently B is uncountable. Since B with the subspace topology is a Polish space, by Corollary 3.44 there exists a non-Borel set $A \subseteq B$. Hence the measure μ is not complete.

We can easily improve this "defect" of a Borel measure. Assume that μ is a Borel measure on a topological space X. For any set $A \subseteq X$ we define

$$\mu^*(A) = \inf\{\mu(B) : A \subseteq B \wedge B \in \text{BOREL}(X)\}. \qquad (4.17)$$

If X has a countable base, then assuming **wAC** one can show that μ^* is an outer measure. Moreover, for any $A \subseteq X$ there exists a Borel set $B \supseteq A$ such that $\mu^*(A) = \mu(B)$. Thus, $\mu^*(B) = \mu(B)$ for a Borel set B. We obtain

Theorem [wAC] 4.8. *Let μ be a Borel measure on a separable metric space. Then μ^* is an outer measure on X and the following hold true:*

a) $\text{BOREL}(X) \subseteq \mathcal{S}(\mu^*)$ *and* $\mu^*|\mathcal{S}(\mu^*)$ *is an extension of* μ.
b) $\mu^*|\mathcal{S}(\mu^*)$ *is a complete moderated measure.*
c) *If μ is regular, then $\mu^*|\mathcal{S}(\mu^*)$ is regular as well.*
d) *If μ is Radon, then $\mu^*|\mathcal{S}(\mu^*)$ is Radon as well.*

If μ is regular, then $\mu^*|\mathcal{S}(\mu^*)$ is uniquely determined by the conditions a)–c).

In what follows we consider a Borel measure μ, we shall often deal automatically with such a complete extension $\mu^*|\mathcal{S}(\mu^*)$.

If μ is a Borel measure on X, then

$$\mathcal{N}(X, \mu) = \{A \subseteq X : \mu^*(A) = 0\} \qquad (4.18)$$

denotes the set of all subsets of μ^*-measure zero. We shall write simply $\mathcal{N}(\mu)$ or $\mathcal{N}(X)$ when the space or the measure is understood, respectively. If X is a topological space with countable base, then **wAC** implies that $\mathcal{N}(X, \mu)$ is a σ-ideal. Moreover, by Theorem 4.8 we obtain

$$\mathcal{N}(X, \mu) \subseteq \mathcal{S}(\mu^*).$$

We shall need a finer version of regularity or of being Radon for Borel measures.

Theorem [wAC] 4.9. *Assume that μ is a Borel measure on a separable metric space* $\langle X, \mathcal{O} \rangle$.

a) *If $A \subseteq X$, then there exists a G_δ set $G \supseteq A$ such that for any Borel $B \supseteq A$ we have $\mu(G \setminus B) = 0$.*

b) *If $A \subseteq X$, then there exists an F_σ set $F \subseteq A$ such that for any Borel $B \subseteq A$ we have $\mu(B \setminus F) = 0$.*

c) *If $A \in \mathcal{S}(\mu^*)$, then there are an F_σ set $F \subseteq A$ and a G_δ set $G \supseteq A$ such that $\mu(A \setminus F) = \mu(G \setminus A) = 0$.*

d) *If X is Polish or σ-compact, then for every set $A \in \mathcal{S}(\mu^*)$ there exist compact sets $\langle K_n \subseteq A : n \in \omega \rangle$ such that $\mu(A \setminus \bigcup_n K_n) = 0$.*

Proof. Let $A \subseteq X$. Since μ is moderated, $X = \bigcup_n X_n$, where the sets X_n are Borel pairwise disjoint and of finite measure. We show that for a given $\varepsilon > 0$ we shall find an open set $U \supseteq A$ such that for any Borel $B \supseteq A$ we have $\mu(U \setminus B) < \varepsilon$. Actually, since $\mu(A \cap X_n) < \infty$, there exist open sets U_n such that $A \cap X_n \subseteq U_n$ and $\mu(U_n \setminus B) < \varepsilon \cdot 2^{-n-1}$ for every Borel $B \supseteq A$ and every $n \in \omega$. The set $U = \bigcup_n U_n$ is the desired.

a) For each n take an open set G_n such that $\mu(G_n \setminus B) < 2^{-n}$ for any Borel $B \supseteq A$ and set $G = \bigcap_n G_n$. Then for any Borel $B \supseteq A$ we have $\mu(G \setminus B) = 0$.

Assertion b) follows from a).

c) If $A \in \mathcal{S}(\mu^*)$, then by Theorem 4.8, c), there are F_σ sets $F_n \subseteq A \cap X_n$ such that $\mu(F_n) = \mu(A \cap X_n)$. Then the F_σ set $F = \bigcup_n F_n \subseteq A$ is such that $\mu(A \setminus F) = 0$. If E is an F_σ set such that $\mu((X \setminus A) \setminus E) = 0$, then $\mu((X \setminus E) \setminus A) = 0$ and $X \setminus E$ is a G_δ set.

To show d), we use the fact that μ is Radon (see Theorem 4.7 and Lemma 4.4). $\square$

Sometimes we need to show that two measures are identical. We shall use the fact that a regular Borel measure is uniquely determined by values on basic open sets. More precisely

Theorem 4.10. *Let $\mathcal{B}$ be a countable base of the topology on X closed under finite intersections. If μ_i, $i = 1, 2$ are outer regular Borel measures on X such that $\mu_1(A) = \mu_2(A)$ for every $A \in \mathcal{B}$, then $\mu_1(A) = \mu_2(A)$ for every $A \in \mathrm{BOREL}(X)$.*

Proof. We can assume that both measures are finite. One can easily prove by induction that $\mu_1(A_1 \cup \cdots \cup A_n) = \mu_2(A_1 \cup \cdots \cup A_n)$ for any $A_1, \ldots, A_n \in \mathcal{B}$. Actually the inductive step follows from the equality

$$\mu_i(A_1 \cup \cdots \cup A_n) + \mu_i(A_{n+1}) = \mu_i((A_1 \cup \cdots \cup A_n) \cup A_{n+1})$$
$$+ \mu_i((A_1 \cap A_{n+1}) \cup \cdots \cup (A_n \cap A_{n+1})).$$

If $A = \bigcup_{n=0}^\infty A_n$, $A_n \in \mathcal{B}$, then $\mu_i(A) = \lim_{n \to \infty} \mu_i(A_0 \cup \cdots \cup A_n)$ for $i = 1, 2$ and hence $\mu_1(A) = \mu_2(A)$. Now the statement follows by outer regularity. $\square$

The next result shows that a measure on a subset can be extended to the whole space and a measure can be restricted to a subset

Theorem [wAC] 4.11. *Assume that $A \subseteq X$ is a subset of a separable metric space X.*

a) *If μ is a Borel measure on X and $\mu^*(A) > 0$, then $\mu^*|\mathrm{BOREL}(A)$ is a Borel measure on $\langle A, \mathcal{O}|A \rangle$ such that $\nu(E) = \mu^*(E)$ for any $E \in \mathrm{BOREL}(A)$. Moreover, ν is diffused if and only if $\mu(\{x\}) = 0$ for any $x \in A$.*

b) *If ν is a finite Borel measure on A, then there exists a Borel measure μ on X such that $\nu(E) = \mu^*(E)$ for any $E \in \mathrm{BOREL}(A)$. Moreover, μ is diffused if and only if ν is diffused.*

Proof. Note that by (3.14) we have $\mathrm{BOREL}(A) = \{A \cap E : E \in \mathrm{BOREL}(X)\}$. For simplicity let us write $\nu^*(E) = \mu^*(E)$ for any $E \subseteq A$. Evidently ν^* is an outer measure on A. We show that every $E \in \mathrm{BOREL}(A)$ is ν^*-measurable.

So assume that $E = A \cap G$, $G \in \mathrm{BOREL}(X)$ and $E \subseteq A$. Since $B \setminus E = B \setminus G$, $B \cap E = B \cap G$ and G is μ^*-measurable, we obtain

$$\nu^*(B) = \mu^*(B) = \mu^*(B \cap G) + \mu^*(B \setminus G) = \nu^*(B \cap E) + \nu^*(B \setminus E).$$

Assume now that ν is a finite Borel measure on A. We define a measure μ on X setting $\mu(B) = \nu(A \cap B)$ for any Borel set $B \subseteq X$. Evidently μ is locally finite and therefore Borel. Since X is perfectly normal, μ is regular. Let $E \in \mathrm{BOREL}(A)$. Then $E = A \cap B$, where $B \in \mathrm{BOREL}(X)$. Since the measure μ is regular, there is a non-increasing sequence of open sets $\langle G_n \supseteq B : n \in \omega \rangle$ such that $\mu(B) = \lim_{n \to \infty} \mu(G_n) = \mu(\bigcap_n G_n)$. Since

$$\mu^*(E) = \inf\{\mu(G) : E \subseteq G \wedge G \text{ is open}\} \leq \mu\left(\bigcap_n G_n\right) = \mu(B) = \nu(E)$$

and $\mu(B) \leq \mu(C)$ for any $C \in \mathrm{BOREL}(X)$, $E \subseteq C$, we are ready. $\square$

We close the section with an important result about diffused Borel measures on a Polish space. We begin with an auxiliary result interesting on its own.

Lemma [wAC] 4.12. *If X is a separable metric space, μ is a diffused Borel measure on X with $\mathrm{supp}(\mu) = X$, $U \subseteq X$ is non-empty open, then there exists a nowhere dense closed set $F \subseteq U$ such that $\mu(U)/2 < \mu(F) < \mu(U)$.*

Proof. Let $\{r_n : n \in \omega\}$ be a countable dense subset of U. Since we suppose that $\mathrm{supp}(\mu) = X$, we have $\mu(U) > 0$. Because μ is diffused, we obtain $\mu(\{r_n\}) = 0$ and by outer regularity, there are open sets $\langle U_n \subseteq U : n \in \omega \rangle$ such that $r_n \in U_n$ and $0 < \mu(U_n) < 2^{-n-2}\mu(U)$ for any $n \in \omega$. Then there exists a non-decreasing sequence $\langle F_n : n \in \omega \rangle$ of closed sets such that $\bigcup_n F_n = U \setminus \bigcup_{n \in \omega} U_n$. By the choice of sets U_n we obtain $\mu(U)/2 < \mu(U \setminus \bigcup_n U_n) < \mu(U)$. Thus there exists an m such that $\mu(F_m) > \mu(U)/2$. Then $F = F_m$ is as required. $\square$

Theorem [wAC] 4.13. *If X is a separable metric space, μ is a diffused Borel measure on X, then there exists a meager F_σ set $F \subseteq X$ such that $\mu(X \setminus F) = 0$.*

Proof. Let $C = \mathrm{supp}(\mu)$. We can assume that $\mu(C) < \infty$. Using the lemma one can easily construct by induction a sequence of mutually disjoint nowhere dense closed sets $\langle F_n \subseteq C : n \in \omega \rangle$ such that $\mu(F_0) > \mu(C)/2$, $F_{n+1} \subseteq C \setminus \bigcup_{i=0}^{n} F_i$, $\mu(C \setminus \bigcup_{i=0}^{n} F_i)/2 < \mu(F_{n+1}) < \mu(C \setminus \bigcup_{i=0}^{n} F_i)$. Then $F = \bigcup_n F_n$ is meager and

$$\mu(X \setminus F) = \mu(C \setminus F) = \mu\left(\bigcap_n (X \setminus F_n) \right) = 0. \qquad \square$$

Exercises

4.1 [AC] Completion of a Measure
Let $\mathcal{S}$ be a σ-algebra of subsets of X, $\mathcal{I} \subseteq \mathcal{S}$ being a σ-ideal of $\mathcal{S}$. We write

$$\mathcal{I}^* = \{A \subseteq X : (\exists B)\,(B \in \mathcal{I} \wedge A \subseteq B)\}, \quad \mathcal{S}^* = \{(A \setminus B) \cup C : A \in \mathcal{S} \wedge B, C \in \mathcal{I}^*\}.$$

    a) $\mathcal{I}^*$ is a σ-ideal of $\mathcal{P}(X)$.

    b) $\mathcal{S}^*$ is the smallest σ-algebra containing $\mathcal{S} \cup \mathcal{I}$.

    c) $\mathcal{I}^*$ is a σ-ideal of $\mathcal{S}^*$.

    d) If $\langle X, \mathcal{S}, \mu \rangle$ is a measure space and $\mathcal{I} = \{A \subseteq X : \mu^*(A) = 0\}$, then $\langle X, \mathcal{S}^*, \nu \rangle$ is a complete measure space, where $\nu((A \setminus B) \cup C) = \mu(A)$ for any $A \in \mathcal{S}$, $B, C \in \mathcal{I}$.

4.2 Range of a Measure
Let $\langle X, \mathcal{S}, \mu \rangle$ be a measure space. A set $A \in \mathcal{S}$ is called an **atom of the measure** μ if $\mu(A) > 0$ and $\mu(B) = 0$ or $\mu(B) = \mu(A)$ for any $B \subseteq A$, $B \in \mathcal{S}$. The measure μ is **atomless** if there is no atom of the measure μ.

    a) If μ is atomless, $A, B \in \mathcal{S}$, $B \subseteq A$, $\mu(B) < \mu(A)$, then there exists a set $C \in \mathcal{S}$ such that $B \subseteq C \subseteq A$ and $\mu(B) < \mu(C) < \mu(A)$.
    Hint: Consider the subsets of $A \setminus B$.

    b) If μ is atomless, $A, B \in \mathcal{S}$, $B \subseteq A$, $\mu(B) < \mu(A)$, then there exists a set $C \in \mathcal{S}$ such that $B \subseteq C \subseteq A$ and $\mu(B) < \mu(C) < \mu(B) + 1/2(\mu(A) - \mu(B))$.

    c) Assume **wAC** and $|\mathcal{S}| \ll \mathfrak{c}$. If μ is atomless, then $\mathrm{rng}(\mu) = [0, \mu(X)]$. Hence $\mathfrak{c} \ll |\mathcal{S}|$.

4.3 [AC] Non-Regular Measure
The notion of a Bernstein set and its existence is investigated in Section 7.2.

    a) If $\mathcal{S} \subseteq \mathcal{P}(X)$ is a σ-algebra, $A \subseteq X$, then

$$\mathcal{S}[A] = \{B \cap A : B \in \mathcal{S}\} \cup \{B \setminus A : B \in \mathcal{S}\}$$

is a σ-algebra, $\mathcal{S} \subseteq \mathcal{S}[A]$, $A \in \mathcal{S}[A]$.

    b) Show that $\mathcal{S}[A]$ is the smallest σ-algebra extending $\mathcal{S}$ and containing A.

    c) Let $\langle X, \mathcal{S}, \mu \rangle$ be a probabilistic measure space, $A \notin \mathcal{S}$. If neither A nor $X \setminus A$ is contained in a set $B \in \mathcal{S}$ of measure less than $1/2$, then there exists a measure ν defined on $\mathcal{S}[A]$, extending μ and such that $\nu(A) = 1/2$.

d) Let μ be a diffused probabilistic Borel measure on a perfect Polish space X. Let $A \subseteq X$ be a Bernstein set. Then there exists a measure ν defined on $\mathrm{BOREL}(X)[A]$ extending the measure μ and such that $\nu(A) = 1/2$.

e) The measure ν of part c) is not regular.

   Hint: Neither A nor $X \setminus A$ is inner or outer regular.

4.4 Weighted Counting Measure

Let $x_n \in X$, $a_n \geq 0$ being reals, $n \in \omega$. For any $A \subseteq X$ we set $\mu(A) = \sum_{n=0}^{\infty} \chi_A(x_n) \cdot a_n$.

a) $\langle X, \mathcal{P}(X), \mu \rangle$ is a non-diffused measure space.

b) If $\sum_{n=0}^{\infty} a_n = 1$, then μ is a probabilistic measure.

c) If X is a Hausdorff topological space, then every subset of X is Radon.

d) If X is a non-empty set we define for any $A \subseteq X$,

$$\nu(A) = \begin{cases} |A| & \text{if } A \text{ is finite,} \\ +\infty & \text{otherwise.} \end{cases}$$

   ν is a non-diffused measure defined on $\mathcal{P}(X)$.

4.5 Inner Measure

A function $\mu_* : \mathcal{P}(X) \longrightarrow [0, \infty]$ is called an **inner measure** if conditions (4.6), (4.7) and the condition

$$\mu_*\left(\bigcup_n A_n\right) \geq \sum_{n=0}^{\infty} \mu_*(A_n) \text{ for any pairwise disjoint } \langle A_n \subseteq X : n \in \omega \rangle \qquad (4.19)$$

are satisfied.

a) Let $\langle X, \mathcal{S}, \mu \rangle$ be a measure space. For any $A \subseteq X$ we define

$$\mu_*(A) = \sup\{\mu(C) : C \subseteq A \wedge C \in \mathcal{S}\}.$$

   Show that μ_* is an inner measure, provided that an adequate version of the axiom of choice holds true.

b) Assume **wAC** and $|\mathcal{S}| \ll \mathfrak{c}$. For every subset $A \subseteq X$ there exist sets $B, C \in \mathcal{S}$ such that $B \subseteq A \subseteq C$, $\mu^*(A) = \mu(C)$ and $\mu_*(A) = \mu(B)$.

c) Assume **wAC** and $|\mathcal{S}| \ll \mathfrak{c}$. If a measure μ is complete, then a set A belongs to $\mathcal{S}$ if and only if $\mu_*(A) = \mu^*(A)$.

d) Assume **wAC**. If μ is a regular Borel measure on a separable metric space, then

$$\mu_*(A) = \sup\{\mu(B) : B \subseteq A \wedge B \text{ closed}\},$$
$$\mu_*(A) = \mu(F) \text{ for some } F_\sigma \text{ set } F \subseteq A.$$

4.6 [wAC] Measure and Topology

Let $\langle X, \mathcal{O} \rangle$ be a locally compact topological space with countable base, μ being a diffused finite Borel measure on X.

a) Prove that if U is an open non-empty set of positive measure, then there exists a closed set $B \subseteq U$ such that $\mu(B) > 0$ and $\mathrm{Int}(B) = \emptyset$.

   Hint: Note that a space with countable base is separable and follow the proof of Theorem 4.13.

b) Find a closed set $A \subseteq X$ such that $\mu(A) = \mu(X)$ and for any non-empty open set U, $\mu(U) > 0$ if and only if $U \cap A \neq \emptyset$.

Hint: Take $A = X \setminus \bigcup \{U : U$ basic open and $\mu(U) = 0\}$.

c) Show that in b) we can replace "finite" by "moderated".

4.7 Signed Measure

Let $\mathcal{S}$ be a σ-algebra of subsets of a set X. A function $\mu : \mathcal{S} \longrightarrow \mathbb{R}$ is a **signed measure** if $\mu(\emptyset) = 0$ and μ is σ-additive, i.e., (4.2) holds true.

a) If μ_1 and μ_2 are finite measures, $\alpha, \beta \in \mathbb{R}$, then $\alpha \cdot \mu_1 + \beta \cdot \mu_2$ is a signed measure.

b) Let $x_n \in X$, $a_n \in \mathbb{R}$ for $n \in \omega$ and $\sum_{n=0}^{\infty} |a_n| < \infty$. Then $\mu(A) = \sum_{n=0}^{\infty} \chi_A(x_n) \cdot a_n$ is a signed measure defined on $\mathcal{P}(X)$.

c) Any family of pairwise disjoint sets with non-zero signed measure is countable.

4.8 [AC] Hahn Decomposition

Let μ be a signed measure. A set $A \in \mathcal{S}$ is called **positive** if $\mu(A) > 0$ and $\mu(B) \geq 0$ for any $B \subseteq A$, $B \in \mathcal{S}$. Similarly we define a **negative** set.

a) A subset B of a positive (negative) set with $\mu(B) \neq 0$ is a positive (negative) set.

b) If $A \in \mathcal{S}$ is not negative, $\mu(A) < 0$, then there exists a negative set $B \subseteq A$, $B \in \mathcal{S}$.

Hint: Let k_0 be the least natural number such that there exists a set $E_0 \subseteq E$ such that $\mu(E_0) \geq 1/k_0$. By induction, if $E \setminus \bigcup_{i=0}^{n} E_i$ is negative, we are ready. If not, let k_{n+1} be the least integer such that there exists a set $E_{n+1} \subseteq E \setminus \bigcup_{i=0}^{n} E_i$ with $\mu(E_{n+1}) \geq 1/k_{n+1}$. If the construction does not stop, then $E \setminus \bigcup_n E_n$ is a negative set.

c) There exists a positive set A such that for any positive set B we have $\mu(B \setminus A) = 0$.

Hint: Consider the family $\mathbf{F}$ of pairwise disjoint positive sets. By Zorn's Lemma, Theorem 1.10, c) there exists a maximal element $\mathcal{A} \in \mathbf{F}$. By Exercise 4.7, $\mathcal{A}$ is countable. Set $A = \bigcup \mathcal{A}$.

d) There exists a **Hahn decomposition**: $X = A \cup B$, $A \cap B = \emptyset$, A is positive or $\mu(A) = 0$ and B is a negative set or $\mu(B) = 0$.

Hint: If A is the set from c), then $X \setminus A$ is a negative set.

e) Every signed measure μ is of the form $\mu^+ - \mu^-$, where μ^+, μ^- are finite measures.

Hint: Set $\mu^+(E) = \mu(A \cap E)$ and $\mu^-(E) = -\mu(B \cap E)$.

f) $|\mu| = \mu^+ + \mu^-$ is a measure called **total variation** of μ. For any set $A \in \mathcal{S}$ we have $|\mu(A)| \leq |\mu|(A)$.

g) Show that everything stated in a)–f) holds true if we allow a signed measure to admit either value $+\infty$ or $-\infty$ (not both!).

h) Find the Hahn decomposition of the signed measure constructed in Exercise 4.7, b).

i) Find μ^+ and μ^- for the measure of Exercise 4.7, b).

4.2 Lebesgue Measure

We construct an outer measure λ_k^* on $\mathbb{R}_k$ such that the corresponding measure of an interval will be what we should call the volume of that interval.

For an open interval

$$I = (a_1, b_1) \times \cdots \times (a_k, b_k), \ a_i < b_i \text{ for } i = 1, \ldots, k,$$

we set

$$\mathrm{Vol}_k(I) = (b_1 - a_1) \cdots (b_k - a_k).$$

The closure of the interval I is $\overline{I} = [a_1, b_1] \times \cdots \times [a_k, b_k]$, the boundary of I is the set $\mathrm{Bd}(I) = \overline{I} \setminus I$ and the half-open interval $\vec{I}$ is $(a_1, b_1] \times \cdots \times (a_k, b_k]$. We set $\mathrm{Vol}_k(\overline{I}) = \mathrm{Vol}_k(\vec{I}) = \mathrm{Vol}_k(I)$.

If $a_i = a_{i,0} < a_{i,1} < \cdots < a_{i,m_i} = b_i$, $i = 1, \ldots, k$, then the set of half-open intervals

$$\{(a_{1,j_1}, a_{1,j_1+1}] \times \cdots \times (a_{k,j_k}, a_{k,j_k+1}] : j_i < m_i \wedge i = 1, \ldots, k\}$$

is called a **full partition** of the half-open interval $\vec{I}$. If $\{J_i : i \le m\}$ is a full partition of a half-open interval J, then

$$\mathrm{Vol}_k(J) = \sum_{i \le m} \mathrm{Vol}_k(J_i). \tag{4.20}$$

Let us note the following: if I_1 and I_2 are half-open intervals, then the intersection $I_1 \cap I_2$ (if non-empty) is a half-open interval, the union $I_1 \cup I_2$ and the difference $I_1 \setminus I_2$ (if non-empty) is a union of pairwise disjoint half-open intervals.

If $\{J_i : i \le m\}$ is a partition of a half-open interval J into disjoint half-open intervals, then there exists a full partition of J such that the set of those intervals that are subsets of J_i forms a full partition of J_i for any $i \le m$. As a corollary we obtain that the equality (4.20) holds true for any partition of a half-open interval into half-open intervals.

By a similar argument we can show that

$$\mathrm{Vol}_k(J) \le \sum_{i=0}^{m} \mathrm{Vol}_k(J_i) \text{ for any half-open intervals with } J \subseteq \bigcup_{i=0}^{m} J_i. \tag{4.21}$$

Actually, let $\{I_0, \ldots, I_n\}$ be a full partition of a half-open interval J into half-open intervals determined by all end points of intervals $J_0, \ldots, J_m$ lying in J. We can assume that every interval J_j meets the interval J. Then each of intervals $I_0, \ldots, I_n$ is a subset of J_j or is disjoint with J_j. If $D_i = \{j \le n : I_j \subseteq J_i\}$, then $\mathrm{Vol}_k(J_i) \ge \sum_{j \in D_i} \mathrm{Vol}_k(I_j)$ and therefore

$$\mathrm{Vol}_k(J) = \sum_{i=0}^{n} \mathrm{Vol}_k(I_i) = \sum_{i=0}^{m} \sum_{j \in D_i} \mathrm{Vol}_k(I_j) \le \sum_{i=0}^{m} \mathrm{Vol}_k(J_i).$$

Now, for any $A \subseteq \mathbb{R}_k$ we let

$$\lambda_k^*(A) = \inf\{\sum_{n=0}^{\infty} \mathrm{Vol}_k(I_n) : \langle I_n : n \in \omega \rangle \text{ are open intervals} \wedge A \subseteq \bigcup_n I_n\}.$$

Evidently if we replace open intervals by half-open or closed intervals, we obtain the same result.

If $k = 1$ we shall omit the index k and simply write λ^*, Vol and λ.

Let us remark that the set of all sequences $\{I_n\}_{n=0}^{\infty}$ of open intervals has cardinality $\mathfrak{c}$. Thus we can use the weak axiom of choice **wAC**.

Theorem [wAC] 4.14. λ_k^* *is an outer measure on* $\mathbb{R}_k$.

Proof. Conditions (4.6) and (4.7) are trivially satisfied. We show that

$$\lambda_k^*\left(\bigcup_n A_n\right) \leq \sum_{n=0}^{\infty} \lambda_k^*(A_n) + \varepsilon$$

for any $A_n \subseteq \mathbb{R}_k$, $n \in \omega$ and any $\varepsilon > 0$. Actually, for every $n \in \omega$ there exists a sequence of open intervals $\{I_{n,m}\}_{m=0}^{\infty}$ such that

$$\sum_{m=0}^{\infty} \mathrm{Vol}_k(I_{n,m}) < \lambda_k^*(A_n) + \varepsilon/2^{n+1}, \quad A_n \subseteq \bigcup_m I_{n,m}.$$

Hence

$$\lambda_k^*\left(\bigcup_n A_n\right) < \sum_{n=0}^{\infty}\sum_{m=0}^{\infty} \mathrm{Vol}_k(I_{n,m}) < \sum_{n=0}^{\infty}(\lambda_k^*(A_n) + \varepsilon/2^{n+1}) = \sum_{n=0}^{\infty} \lambda_k^*(A_n) + \varepsilon. \quad \square$$

One can easily show that for any open interval I we have

$$\lambda_k^*(\mathrm{Bd}(I)) = 0 \text{ and } \lambda_k^*(I) = \lambda_k^*(\vec{I}) = \lambda_k^*(\overline{I}). \tag{4.22}$$

The corresponding measure λ_k is called **Lebesgue measure**. The σ-algebra $\mathcal{S}(\lambda_k^*)$ will be denoted simply by $\mathcal{L}(\mathbb{R}_k)$. A set $A \subseteq \mathbb{R}_k$ is called **Lebesgue measurable** if $A \in \mathcal{L}(\mathbb{R}_k)$. The family $\mathcal{N}(\mathbb{R}_k, \lambda_k)$ of all subsets of $\mathbb{R}_k$ of Lebesgue measure 0 will be simply denoted as $\mathcal{N}_k$. If the space $\mathbb{R}_k$ is understood from the context we simply write $\mathcal{L}$. Similarly for λ, λ^* and $\mathcal{N}$.

The basic result, which is usually considered as trivial, is the following assertion.

Theorem [wAC] 4.15. *If* $I = (a_1, b_1) \times \cdots \times (a_k, b_k)$ *is an open interval, then all intervals* $I, \overline{I}, \vec{I}$ *are Lebesgue measurable and*

$$\lambda_k(I) = \lambda_k(\overline{I}) = \lambda_k(\vec{I}) = (b_1 - a_1) \cdots (b_k - a_k). \tag{4.23}$$

Proof. By (4.22) it suffices to prove the statements of the theorem for one of the intervals $I, \overline{I}, \vec{I}$. We show it for a closed interval $\overline{I}$. We begin by showing that

$$\lambda_k^*(\overline{I}) = \mathrm{Vol}_k(I). \tag{4.24}$$

The inequality $\lambda_k^*(\overline{I}) \leq \mathrm{Vol}_k(I)$ follows from the definition. If $\lambda_k^*(\overline{I}) < \mathrm{Vol}_k(I)$, then there exist open intervals $\langle I_n : n \in \omega \rangle$ such that $\sum_n \mathrm{Vol}_k(I_n) < \mathrm{Vol}_k(I)$ and $\overline{I} \subseteq \bigcup_n I_n$. Since $\overline{I}$ is a compact set, there exists a natural number m such that $\overline{I} \subseteq \bigcup_{n=0}^{m} I_n$. Then $\vec{I} \subseteq \bigcup_{n=0}^{m} \vec{I}_n$ and by (4.21) we obtain

$$\mathrm{Vol}_k(\overline{I}) = \mathrm{Vol}_k(\vec{I}) \leq \sum_{n=0}^{m} \mathrm{Vol}_k(\vec{I}_n) = \sum_{n=0}^{m} \mathrm{Vol}_k(I_n) < \mathrm{Vol}_k(I),$$

which is a contradiction.

We have to show that the closed interval $\overline{I}$ is λ_k^*-measurable, i.e., $\overline{I}$ satisfies condition (4.9) for any set B. Now, suppose, to get a contradiction, that $\overline{I}$ is not λ_k^*-measurable. Then there exists a set $B \subseteq \mathbb{R}_k$ such that

$$\lambda_k^*(B) < \lambda_k^*(B \cap \overline{I}) + \lambda_k^*(B \setminus \overline{I}).$$

By definition of λ_k^* there exists a sequence $\{I_n\}_{n=0}^{\infty}$ of open intervals such that

$$B \subseteq \bigcup_n I_n, \quad \sum_{n=0}^{\infty} \mathrm{Vol}_k(I_n) < \lambda_k^*(B \cap \overline{I}) + \lambda^*(B \setminus \overline{I}). \tag{4.25}$$

For every n the intersection $\vec{I}_n \cap \vec{I}$ is a half-open interval and the difference $\vec{I}_n \setminus \vec{I}$ is a finite union $\bigcup_{i=0}^{m_n} J_{n,i}$ of half-open intervals. Thus by (4.20) we have

$$\mathrm{Vol}_k(\vec{I}_n \cap \vec{I}) + \mathrm{Vol}_k\left(\bigcup_{i=0}^{m_n} J_{n,i}\right) = \mathrm{Vol}_k(I_n).$$

Since

$$B \cap \vec{I} \subseteq \bigcup_n (\vec{I}_n \cap \vec{I}), \quad B \setminus \vec{I} \subseteq \bigcup_n \bigcup_{i=0}^{m_n} J_{n,i},$$

we obtain

$$\lambda_k^*(B \cap \vec{I}) + \lambda^*(B \setminus \vec{I}) \leq \sum_n \mathrm{Vol}_k(\vec{I}_n \cap \vec{I}) + \sum_n \sum_{i=0}^{m_n} \mathrm{Vol}_k(J_{n,i}) = \sum_n \mathrm{Vol}_k(I_n),$$

which is a contradiction. $\qquad\square$

We can summarize the obtained results concerning the Lebesgue measure as

Theorem [wAC] 4.16. *Lebesgue measure is a complete moderated shift invariant Radon measure.*

Proof. By Theorem 4.15 every open interval is Lebesgue measurable. By the definition of Borel sets we obtain BOREL $\subseteq \mathcal{L}$. The outer regularity follows directly from the definition of the outer measure λ_k^*. Since the measure of a closed interval is finite, Lebesgue measure is moderated. The shift invariance is immediate. The theorem follows by Lemma 4.4 and Theorem 4.8. $\square$

We shall need the following

Theorem 4.17. *If $A \subseteq \mathbb{R}_k$, $\lambda_k^*(A) > 0$ and $1 > \varepsilon > 0$, then there exists an open interval $I \subseteq \mathbb{R}_k$ such that $\lambda_k(I) < \varepsilon$ and $\lambda_k^*(A \cap I) \geq (1 - \varepsilon)\lambda_k(I)$.*

Proof. Let $\langle I_n : n \in \omega \rangle$ be open intervals such that $\sum_n \lambda(I_n) \leq \lambda^*(A)/(1-\varepsilon)$ and $A \subseteq \bigcup_n I_n$. We can assume that $\lambda(I_n) < \varepsilon$ for each n. Since

$$(1 - \varepsilon) \sum_n \lambda(I_n) \leq \lambda^*(A) \leq \sum_n \lambda^*(I_n \cap A),$$

at least for one n we have $(1-\varepsilon)\lambda(I_n) \leq \lambda^*(I_n \cap A)$. Take $I = I_n$ for such an n. $\square$

The circle $\mathbb{T}$ is usually endowed with the Lebesgue measure λ restricted to subsets of $\mathbb{T}$. By (2.32) we have $\lambda(\mathbb{C}) = 0$ and therefore we define another measure on $\mathbb{C}$. In Section 3.1 we have constructed a continuous surjection $w_1 : \mathbb{C} \xrightarrow{\text{onto}} [0, 1]$ by (3.7). We denote by $\lambda_{\mathbb{C}}$ the Borel measure on $\mathbb{C}$ defined by

$$\lambda_{\mathbb{C}}(A) = \lambda(w_1(A)) \text{ for any } A \in \text{BOREL}(\mathbb{C}). \tag{4.26}$$

One can easily check that the definition is correct and $\lambda_{\mathbb{C}}$ is a Borel measure on the Cantor middle-third set $\mathbb{C}$. Using the natural homeomorphism $f : \mathbb{C} \longrightarrow {}^{\omega}2$ constructed in the proof of Theorem 3.1 we can consider $\lambda_{\mathbb{C}}$ as a Borel measure on ${}^{\omega}2$. One can easily see that $\lambda_{\mathbb{C}}([s]) = 2^{-n}$ for any $s \in {}^n 2$ and for any $A \subseteq {}^{\omega}2$ we have

$$\lambda_{\mathbb{C}}^*(A) = \inf \left\{ \sum_{s \in T} 2^{-\text{length}(s)} : T \subseteq {}^{<\omega}2 \wedge A \subseteq \bigcup_{s \in T} [s] \right\},$$

where $\text{length}(s) = n$ if $s \in {}^n 2$.

Exercises

4.9 [wAC] Lebesgue Density Theorem
Let X be a locally compact metric space, μ being a Borel measure on X. Assume that $A, B \subseteq X$ are measurable. If $x \in X$, then

$$\text{dens}_*(x, A) = \liminf_{r \to 0} \frac{\mu(A \cap \text{Ball}(x, r))}{\mu(\text{Ball}(x, r))}, \quad \text{dens}^*(x, A) = \limsup_{r \to 0} \frac{\mu(A \cap \text{Ball}(x, r))}{\mu(\text{Ball}(x, r))}$$

is called the **lower density** of A at x and the **upper density** of A at x, respectively.

  a) $\text{dens}_*(x, A) + \text{dens}^*(x, X \setminus A) = 1$ for any $A \subseteq X$.

  b) If $\mu(A \setminus B) = \mu(B \setminus A) = 0$, then $\text{dens}_*(x, A) = \text{dens}_*(x, B)$ for any $x \in X$.

c) If $A \subseteq B$, then $\operatorname{dens}_*(x, A) \leq \operatorname{dens}_*(x, B)$.

d) $\operatorname{dens}_*(x, A) = \operatorname{dens}_*(x, B) = 1$ if and only if $\operatorname{dens}_*(x, A \cap B) = 1$.
Hint: Show that

$$\frac{\mu(\operatorname{Ball}(x, r) \cap A)}{\mu(\operatorname{Ball}(x, r))} + \frac{\mu(\operatorname{Ball}(x, r) \cap B)}{\mu(\operatorname{Ball}(x, r))} \leq \frac{\mu(\operatorname{Ball}(x, r) \cap A \cap B)}{u(\operatorname{Ball}(x, r))} + 1$$

for any $r > 0$.

e) If $A \subseteq \mathbb{R}_k$ is a Lebesgue measurable set, then for any positive $\varepsilon < 1$ we have $\lambda_k(A_\varepsilon) = 0$, where $A_\varepsilon = \{x \in A : \operatorname{dens}_*(x, A) < \varepsilon\}$.
Hint: Assume $\lambda^*(A_\varepsilon) > 0$. Let $A \subseteq U$, U open and $\lambda(U) < \lambda^*(A_\varepsilon)/(1 - \varepsilon)$. Write

$$\mathcal{A} = \{I \subseteq U : I \ \text{closed interval} \wedge \lambda(A \cap I) \leq (1 - \varepsilon)\operatorname{Vol}_k(I)\}.$$

Note that for any sequence $\langle I_n : n \in \omega \rangle$ of pairwise disjoint intervals from $\mathcal{A}$ we have $\lambda^*(A_\varepsilon \setminus \bigcup_n I_n) > 0$. Construct a sequence $\langle I_n : n \in \omega \rangle$ of pairwise disjoint intervals from $\mathcal{A}$ such that

$$\operatorname{Vol}_k(I_{n+1}) > \sup\{\operatorname{Vol}_k(I) : I \in \mathcal{A} \wedge (\forall i \leq n) \, I \cap I_i = \emptyset\}.$$

The set $B = A_\varepsilon \setminus \bigcup_n I_n$ has positive outer measure. Take n_0 sufficiently large such that $\sum_{n=n_0}^{\infty} \operatorname{Vol}_k(I_n) < \lambda_k^*(B) \cdot 3^k$. Let J_n be an interval with the same center as that of I_n but three times larger sides. Still $\lambda_k^*(A_\varepsilon \setminus \bigcup_{n=n_0}^{\infty} J_n) > 0$. Take an $x \in A_\varepsilon \setminus \bigcup_{n=n_0}^{\infty} J_n$, an interval $x \in I \in \mathcal{A}$ and find a contradiction.

f) Prove the **Lebesgue Density Theorem**: The set

$$\Phi(A) = \{x \in \mathbb{R}_k : \operatorname{dens}_*(x, A) = 1\}$$

is measurable and $\lambda_k((A \setminus \Phi(A)) \cup (\Phi(A) \setminus A)) = 0$.
Hint: Since $\Phi(A) \setminus A \subseteq (\mathbb{R}_k \setminus A) \setminus \Phi(\mathbb{R}_k \setminus A)$, it suffices to show that $\lambda_k(A \setminus \Phi(A)) = 0$. Note that $\Phi(A) \setminus A = \bigcup_n \{x \in X : \operatorname{dens}_*(x, A) < \frac{n}{n+1}\}$.

4.10 [wAC] Metric Outer Measure
Let X be a metric separable space, μ^* being an outer measure on X. μ^* is called a **metric outer measure** on X if for any subsets $A, B \subseteq X$ with $\rho(A, B) > 0$ we have

$$\mu^*(A \cup B) = \mu^*(A) + \mu^*(B).$$

a) If μ is a Borel measure, then μ^* is a metric outer measure.

b) Let $A \cap F = \emptyset$, F being closed. Write

$$A_n = \left\{ x \in A : \rho(x, F) \geq \frac{1}{n+1} \right\}, \quad C_n = A_{n+1} \setminus A_n.$$

Then $\rho(F, A_n) > 0$ and

$$A = A_{2n} \cup \left(\bigcup_{k=n}^{\infty} C_{2k} \right) \cup \left(\bigcup_{i=n}^{\infty} C_{2k+1} \right).$$

c) If both series

$$\sum_{k=0}^{\infty} \mu^*(C_{2k}), \quad \sum_{k=0}^{\infty} \mu^*(C_{2k+1}) \tag{4.27}$$

converge, then $\mu^*(A) = \lim_{n \to \infty} \mu^*(A_n)$.

d) If one of the series (4.27) diverges, then

$$\mu^*(A) = \lim_{n \to \infty} \mu^*(A_n) = \infty.$$

e) If μ^* is a metric outer measure on X, then every closed set F is μ^*-measurable. *Hint: Let $B \subseteq X$. Take $A = B \setminus F$ and use the inequalities*

$$\mu^*(B) \geq \mu^*((B \cap F) \cup A_n), \quad \rho(B \cap F, A_n) \geq \rho(F, A_n) > 0.$$

f) If μ^* is a metric outer measure on X, then every Borel set is μ^*-measurable.

4.11 Lebesgue-Stieltjes Measure

If $f : \mathbb{R} \to \mathbb{R}$ is non-decreasing, for a set $A \subseteq \mathbb{R}$ we define

$$\lambda_f^*(A) = \inf \left\{ \sum_{i=0}^{\infty} (f(y_i) - f(x_i)) : A \subseteq \bigcup_{i=0}^{\infty} (x_i, y_i] \right\}.$$

a) Show that λ_f^* is an outer measure on $\mathbb{R}$.

b) Show that $\text{Borel}(\mathbb{R}) \subseteq \mathcal{S}(\lambda_f^*)$.

c) Let $f(x) = x$ for $x < 0$, $f(x) = x + 1$ for $x \geq 0$, $g(x) = x$ for $x \leq 0$, and $g(x) = x + 1$ for $x > 0$. What is the difference between outer measures λ_f^* and λ_g^*?

d) λ_f^* is a metric outer measure.

e) Every Borel set is λ_f^*-measurable.

f) If $f : \mathbb{R} \to \mathbb{R}$ is of bounded variation on $\mathbb{R}$ (see Exercise 2.14), then $f = g - h$ with g, h non-decreasing. We define

$$\lambda_f(A) = \lambda_g(A) - \lambda_h(A)$$

for any $A \in \text{Borel}(\mathbb{R})$. λ_f is a signed measure on $\text{Borel}(\mathbb{R})$. Show that (see Exercise 4.8, e)) $\lambda_f^+ = \lambda_g$, $\lambda_f^- = \lambda_h$ and $|\lambda_f| = \lambda_g + \lambda_h$.

g) λ_f is diffused if and only if f is continuous.

h) Let $f : \mathbb{R} \to \mathbb{R}$ be a non-decreasing function. We set $H(x) = \lim_{h \to 0^+} f(x + h)$ and $G(x) = \lim_{h \to 0^-} f(x + h)$ for $x \in \mathbb{R}$. Show that $\lambda_f - \lambda_G$ and $\lambda_H - \lambda_f$ are weighted counting measures (see Exercise 4.4).

4.12 Hausdorff Measure

In this exercise we allow that a measure is equal identically to zero.

Let $\langle X, \rho \rangle$ be a metric space, $s > 0$ a real. We construct a function $\mathcal{H}^s$ defined on $\text{Borel}(X)$ called **Hausdorff measure**. We show that this measure is interesting for at most one value of s. This value is called the **Hausdorff dimension** of X. We show that the value of the Hausdorff dimension may be an irrational real.

a) For any $\varepsilon > 0$, $A \subseteq X$ we write

$$\mathcal{H}^{s*}_{\varepsilon}(A) = \inf\left\{ \sum_{n=0}^{\infty}(\operatorname{diam}(U_n))^s : A \subseteq \bigcup_n U_n \wedge (\forall n)(U_n \text{ open } \wedge \operatorname{diam}(U_n) < \varepsilon) \right\}.$$

Show that $\mathcal{H}^{s*}_{\varepsilon_1}(A) \geq \mathcal{H}^{s*}_{\varepsilon_2}(A)$ for any $\varepsilon_1 < \varepsilon_2$.

b) Let $\mathcal{H}^{s*}(A) = \lim_{\varepsilon \to 0+} \mathcal{H}^{s*}_{\varepsilon}$. Show that $\mathcal{H}^{s*}$ is an outer measure on X.

c) $\mathcal{H}^{s*}$ is a metric outer measure. Thus, the σ-algebra of all $\mathcal{H}^{s*}$-measurable sets contains all Borel sets. The corresponding measure will be denoted by $\mathcal{H}^s$.

d) If $s_1 < s_2$ and $\mathcal{H}^{s_1}(A) < \infty$, then $\mathcal{H}^{s_2}(A) = 0$.

e) There exists a non-negative real r such that

$$\mathcal{H}^s(X) = \begin{cases} 0 & \text{for every } s > r, \\ +\infty & \text{for every } s < r . \end{cases}$$

f) Let $X = [0,1]^k$ be endowed with Euclidean distance. Show that there exist positive reals c_k such that $\mathcal{H}^{k*}(A) = c_k \cdot \lambda^*_k(A)$ for any $A \subseteq X$.

g) The real r of e) is called the Hausdorff dimension of X. Show that the Hausdorff dimension of $\mathbb{R}_k$ is k.

h) If $s = \log 2/\log 3$, then $0 < \mathcal{H}^s(\mathbb{C}) < \infty$. Thus the Hausdorff dimension of the Cantor middle-third set is $\log 2/\log 3$.

i) From equilateral closed triangle $S_0 \subseteq \mathbb{R}_2$ of side 1 with vertexes $[0,0]$, $[1,0]$, $[1/2, \sqrt{3}/2]$ we remove the open triangle with vertexes $[1/2, 0]$, $[1/4, \sqrt{3}/4]$, $[3/4, \sqrt{3}/4]$. We obtain a set S_1 that is a union of three equilateral closed triangles of side $1/2$. From each of the triangles remove the central triangle in a similar way and we will obtain the set S_2. Continue by induction in constructing S_n for each n. The intersection $S = \bigcap_n S_n$ is called the **Sierpiński gasket**. Show that the Hausdorff dimension of the Sierpiński gasket is $\log 3/\log 2$.

4.3 Elementary Integration

Let $\langle X, \mathcal{S}, \mu \rangle$ be a measure space with a σ-finite measure. An $\mathcal{S}$-measurable function $f : X \longrightarrow \mathbb{R}^*$ will be simply called measurable. If $f = f^+ - f^-$ is the decomposition in non-negative and non-positive parts, then f is measurable if and only if both f^+ and f^- are measurable. We define an integral of some measurable functions in three steps: for a simple measurable function[2], then for a non-negative measurable function and finally we extend our definition to measurable functions using the above-mentioned decomposition.

Let $h = \sum_{i=0}^{n} a_i \chi_{A_i}$, $a_i > 0$, $A_i \in \mathcal{S}$, $i = 0, \ldots, n$ be a simple measurable function. For any $E \in \mathcal{S}$ we define

$$\int_E h(x)\, d\mu(x) = \sum_{i=0}^{n} a_i \cdot \mu(E \cap A_i).$$

---

[2]Note that a simple function is non-negative by definition.

Usually we shall simply write $\int_E h \, d\mu$. One can easily see that the value $\int_E f \, d\mu$ does not depend on the particular representation of a simple function as a linear combination of characteristic functions of measurable sets: any such representation may be reduced to the unique representation with pairwise disjoint sets A_i.

By simple computation we obtain

$$\int_E (f+g) \, d\mu = \int_E f \, d\mu + \int_E g \, d\mu, \quad \int_E (a \cdot f) \, d\mu = a \cdot \int_E f \, d\mu \qquad (4.28)$$

for any simple measurable functions f, g and a non-negative real a.

Lemma 4.18. *Let h be a simple measurable function. Then the function ν defined as $\nu(E) = \int_E h(x) \, d\mu(x)$ for $E \in \mathcal{S}$ is a measure on $\mathcal{S}$.*

The proof is immediate. Let $E = \bigcup_k E_k$, $\langle E_k : k \in \omega \rangle$ being pairwise disjoint measurable sets. Then

$$\nu(E) = \sum_{i=0}^{n} a_i \cdot \mu(E \cap A_i) = \sum_{i=0}^{n} a_i \cdot \sum_{k=0}^{\infty} \mu(E_k \cap A_i)$$

$$= \sum_{k=0}^{\infty} \sum_{i=0}^{n} a_i \cdot \mu(E_k \cap A_i) = \sum_{k=0}^{\infty} \nu(E_k). \qquad \square$$

By Lemma 3.74 for any non-negative measurable function $f : X \longrightarrow \mathbb{R}^*$ there are simple measurable functions $\langle f_n : n \in \omega \rangle$ such that $f_n \nearrow f$. Thus

$$f = \sup\{h : h \le f \wedge h \text{ measurable and simple}\}. \qquad (4.29)$$

We define the **integral** $\int_E f \, d\mu = \int_E f(x) \, d\mu(x)$ of a measurable non-negative function f over a measurable set $E \in \mathcal{S}$ as

$$\int_E f \, d\mu = \sup\{\int_E h \, d\mu : h \le f \wedge h \text{ measurable and simple}\}.$$

From the definition we immediately obtain that for any measurable non-negative functions f, g, we have $\int_E f \, d\mu \le \int_E g \, d\mu$ provided that $f(x) \le g(x)$ for any $x \in E$.

Lemma 4.19. *If $\langle f_n : n \in \omega \rangle$ are measurable functions with non-negative values, $f_n \nearrow f$ on X, then*

$$\int_E f \, d\mu = \lim_{n \to \infty} \int_E f_n \, d\mu$$

for any measurable set $E \in \mathcal{S}$.

Proof. By Theorem 3.68 function f is measurable. By Lemma 4.18 we can assume that $\mu(E) < \infty$.

First let us assume that $f(x) = \infty$ for no $x \in E$.

Since $\int_E f_n \, d\mu \le \int_E f_{n+1} \, d\mu$, there exists a limit $a = \lim_{n\to\infty} \int_E f_n \, d\mu \ge 0$. As $f_n \le f$, we have $\int_E f_n \, d\mu \le \int_E f \, d\mu$ and therefore $a \le \int_E f \, d\mu$.

Assume that $a < \int_E f \, d\mu$. Then there exists a simple measurable function h such that $h \le f$ and $a < \int_E h \, d\mu \le \int_E f \, d\mu$. For any given positive real $b < 1$ we set

$$E_n = \{x \in E : b \cdot h(x) \le f_n(x)\}.$$

Evidently $E_0 \subseteq E_1 \subseteq \cdots \subseteq E_n \subseteq \cdots$ and $\bigcup_n E_n = E$. Hence

$$\int_E f_n \, d\mu \ge \int_{E_n} f_n \, d\mu \ge b \cdot \int_{E_n} h \, d\mu.$$

Since $\nu(C) = \int_C h \, d\mu$ is a measure, we obtain that $\lim_{n\to\infty} \int_{E_n} h \, d\mu = \int_E h \, d\mu$ and therefore

$$a = \lim_{n\to\infty} \int_E f_n \, d\mu \ge b \cdot \int_E h \, d\mu.$$

Since $b < 1$ was arbitrary, we obtain

$$a = \lim_{n\to\infty} \int_E f_n \, d\mu \ge \int_E h \, du,$$

which is a contradiction.

Now, let $A = \{x \in E : f(x) = \infty\}$. If $\mu(A) = 0$, then $\int_E f \, d\mu = \int_{E\setminus A} f \, d\mu$ and $\int_E f_n \, d\mu = \int_{E\setminus A} f_n \, d\mu$ for all n and hence we are done. So assume $\mu(A) > 0$. Then $\int_E f \, d\mu = \infty$. Note that $\mu(A) \le \mu(E) < \infty$. For $n, m \in \omega$ let

$$A_{n,m} = \{x \in A : f_m(x) > n\}.$$

Since $\bigcup_k \bigcup_{m \ge k} A_{n,m} = A$ for every n, there exists a sequence $\langle m_n : n \in \omega \rangle$ such that $\mu(A_{n,m_n}) > \mu(A)/2$. Then $\int_E f_{m_n} \, d\mu > n \cdot \mu(A)/2$ for every n and therefore $\lim_{n\to\infty} \int_E f_n \, d\mu = \infty$. $\qquad\square$

Lemma 4.20. *Let f, g be non-negative measurable functions, a, b non-negative reals. Then for any measurable set $E \in \mathcal{S}$ we have*

$$\int_E (a \cdot f + b \cdot g) \, d\mu = a \cdot \int_E f \, d\mu + b \cdot \int_E g \, d\mu.$$

Proof. Since f, g are measurable, by Lemma 3.74 there exist sequences of non-negative simple functions $\langle f_n : n \in \omega \rangle$ and $\langle g_n : n \in \omega \rangle$ such that $f_n \nearrow f$, $g_n \nearrow g$. By Lemma 4.19 we obtain $\lim_{n\to\infty} \int_E f_n \, d\mu = \int_E f \, d\mu$, $\lim_{n\to\infty} \int_E g_n \, d\mu = \int_E g \, d\mu$ and $\lim_{n\to\infty} \int_E (a \cdot f_n + b \cdot g_n) \, d\mu = \int_E (a \cdot f + b \cdot g) \, d\mu$. By (4.28) we have

$$\int_E (a \cdot f_n + b \cdot g_n) \, d\mu = a \cdot \int_E f_n \, d\mu + b \cdot \int_E g_n \, d\mu$$

and the assertion follows. $\qquad\square$

A measurable function f is **integrable** if $\int_X |f|\, d\mu < +\infty$. Thus a measurable function f is integrable if and only if $\int_X f^+\, d\mu < +\infty$ and $\int_X f^-\, d\mu < +\infty$. Note that then also $\int_E f^+\, d\mu < +\infty$ and $\int_E f^-\, d\mu < +\infty$ for any measurable set E. Hence we can define the **integral**

$$\int_E f\, d\mu = \int_E f^+\, d\mu - \int_E f^-\, d\mu$$

for any $E \in \mathcal{S}$.

Note that if f is integrable, then the set $\{x \in X : f(x) = +\infty \vee f(x) = -\infty\}$ has measure zero.

It is convenient to define an integral for some non-integrable functions. If $\int_E f^+\, d\mu < +\infty$ or $\int_E f^-\, d\mu < +\infty$, we set $\int_E f\, d\mu = \int_E f^+\, d\mu - \int_E f^-\, d\mu$. Thus, we allow that $\int_E f\, d\mu$ is $+\infty$ or $-\infty$. We cannot define the value $\int_E f\, d\mu$ if both integrals $\int_E f^+\, d\mu$ and $\int_E f^-\, d\mu$ are infinite. Note the following: if f, g are measurable, $f \leq g$ and f is integrable, then $\int_X g^-\, d\mu < \infty$, thus we can define $\int_E g\, d\mu$.

We prove some elementary properties of the integral.

Theorem 4.21. *Let f, g be integrable functions, a, b being reals. Then for any measurable set E we have*

a) $\int_E (a \cdot f + b \cdot g)\, d\mu = a \cdot \int_E f\, d\mu + b \cdot \int_E g\, d\mu$;

b) $\int_E f\, d\mu \leq \int_E g\, d\mu$ *provided that $f(x) \leq g(x)$ for any $x \in E$;*

c) $\left| \int_E f\, d\mu \right| \leq \int_E |f|\, d\mu$.

Proof. Since
$$c \cdot g = (c^+ \cdot g^+ + c^- \cdot g^-) - (c^+ \cdot g^- + c^- \cdot g^+)$$

for any real c and any function g, statement a) follows by Lemma 4.20.

If $f \leq g$, then $f^+ \leq g^+$ and $f^- \geq g^-$. The statement b) follows by Lemma 4.20. Since $f \leq |f|$ and $-f \leq |f|$, c) follows by b). $\qquad\square$

Theorem 4.22 (Lebesgue Monotone Convergence Theorem). *Assume that $\langle f_n : n \in \omega \rangle$ are measurable functions, $f_n \nearrow f$ and f_0 is integrable. Then*

$$\int_X f\, d\mu = \lim_{n \to \infty} \int_X f_n\, d\mu.$$

Proof. We can assume that all values of f_0 are finite. Let $g_n = f_n - f_0$. Then the functions g_n are measurable, non-negative and $g_n \nearrow (f - f_0)$. Thus by Lemma 4.19,

$$\int_X (f - f_0)\, d\mu = \lim_{n \to \infty} \int_X (f_n - f_0)\, d\mu.$$

Since f_0 is integrable, we have $\int_X (f - f_0)\, d\mu = \int_X f\, d\mu - \int_X f_0\, d\mu$ and $\int_X (f_n - f_0)\, d\mu = \int_X f_n\, d\mu - \int_X f_0\, d\mu$. The theorem follows. $\qquad\square$

Corollary 4.23. *Assume that $\langle f_n : n \in \omega \rangle$ are non-negative measurable functions. Let $f = \sum_{n=0}^{\infty} f_n$. Then*

$$\int_E f \, d\mu = \sum_{n=0}^{\infty} \int_E f_n \, d\mu.$$

Corollary 4.24 (Fatou Lemma). *If $\langle f_n : n \in \omega \rangle$ are non-negative measurable functions, then*

$$\int_X (\liminf_{n \to \infty} f_n) \, d\mu \leq \liminf_{n \to \infty} \int_X f_n \, d\mu.$$

Proof. Let $g_n(x) = \inf\{f_k(x) : k \geq n\}$. Then

$$g_n \leq f_n \text{ and } g_n \leq g_{n+1}.$$

Since

$$g_n(x) < a \quad \equiv \quad (\exists k \geq n) \, f_k(x) < a,$$
$$g_n(x) > a \quad \equiv \quad (\exists m > 0)(\forall k \geq n) \, f_k(x) \geq a + 1/m,$$

the functions g_n are measurable. One can easily check that $g_n \nearrow \liminf_{k \to \infty} f_k$ and therefore by Theorem 4.22 we obtain $\lim \int_X g_n \, d\mu = \int_X (\liminf_{n \to \infty} f_n) \, d\mu$.

If $a < \lim \int_X g_n \, d\mu$, then there exists an n_0 such that $a < \int_X g_n \, d\mu$ for every $n \geq n_0$ and therefore $\int_X f_n \, d\mu > a$ for every $n \geq n_0$. Since $a < \lim \int_X g_n \, d\mu$ was arbitrary, we obtain $\lim \int_X g_n \, d\mu \leq \liminf \int_X f_n \, d\mu$. $\square$

Theorem 4.25 (Lebesgue Dominated Convergence Theorem). *Let g, f_n be integrable functions such that $|f_n| \leq g$ for any $n \in \omega$. If f is such a function that $f_n \to f$ on X, then f is integrable and*

$$\int_X f \, d\mu = \lim_{n \to \infty} \int_X f_n \, d\mu.$$

Proof. We know that f is measurable. As $|f| \leq g$, by Lemma 4.20 the function f is integrable.

Evidently $|f_n - f| \leq 2g$. Thus $2g - |f_n - f| \geq 0$ and we can apply the Fatou Lemma, Corollary 4.24 to the sequence $\langle 2g - |f_n - f| : n \in \omega \rangle$ and obtain

$$\int_X 2g \, d\mu = \int_X \lim_{n \to \infty} (2g - |f_n - f|) \, d\mu = \liminf_{n \to \infty} \int_X (2g - |f_n - f|) \, d\mu$$
$$= \liminf_{n \to \infty} \left(\int_X 2g \, d\mu - \int_X |f_n - f| \, d\mu \right) = \int_X 2g \, d\mu - \limsup_{n \to \infty} \int_X |f_n - f| \, d\mu.$$

Since g is integrable, the integral $\int_X 2g \, d\mu$ is a real and we obtain

$$\limsup_{n \to \infty} \int_X |f_n - f| \, d\mu = 0.$$

By Theorem 4.21, a) and c) we have

$$\left| \int_X f_n \, d\mu - \int_X f \, d\mu \right| \leq \int_X |f_n - f| \, d\mu$$

and we are ready. $\square$

Exercises

4.13 Limit and Integral
Let $\langle X, \mathcal{S}, \mu \rangle$ be a measure space with a complete σ-finite measure.

a) If $f_n : X \longrightarrow [0, +\infty]$ are measurable functions, then

$$\int_X \sum_{n=0}^{\infty} f_n \, d\mu = \sum_{n=0}^{\infty} \int_X f_n \, d\mu \geq 0.$$

Hint: Use the Lebesgue Monotone Convergence Theorem 4.22.

b) Assume that $f_n : X \longrightarrow \mathbb{R}^*$ are measurable and $\sum_{n=0}^{\infty} \int_X |f_n| \, d\mu < +\infty$. Then the series $\sum_{n=0}^{\infty} f_n(x)$ converges for almost all x, $f = \sum_{n=0}^{\infty} f_n$ is integrable and

$$\int_X \sum_{n=0}^{\infty} f_n \, d\mu = \sum_{n=0}^{\infty} \int_X f_n \, d\mu.$$

Hint: The function $h(x) = \sum_{n=0}^{\infty} |f_n(x)|$ is integrable by a). Show that $\sum_{n=0}^{\infty} f_n^+(x)$ and $\sum_{n=0}^{\infty} f_n^-(x)$ converge for almost all x. Since $|\sum_{n=0}^{k} f_n(x)| \leq h(x)$, the statement follows by the Lebesgue Dominated Convergence Theorem 4.25.

c) Show that the assertion of the Lebesgue Monotone Convergence Theorem 4.22 is false if the condition "f_0 is integrable" is omitted.

d) Find a sequence $\langle f_n : n \in \omega \rangle$ of non-negative measurable functions such that

$$\int_X (\liminf_{n \to \infty} f_n) \, d\mu < \liminf_{n \to \infty} \int_X f_n \, d\mu.$$

e) Prove the **Fatou-Lebesgue Theorem**: If g is an integrable function, $\langle f_n : n \in \omega \rangle$ are measurable functions, and $|f_n(x)| \leq g(x)$ for all n and almost all x, then

$$-\infty < \int_X \liminf_{n \to \infty} f_n \, d\mu \leq \liminf_{n \to \infty} \int_X f_n \, d\mu$$

$$\leq \limsup_{n \to \infty} \int_X f_n \, d\mu \leq \int_X \limsup_{n \to \infty} f_n \, d\mu < +\infty.$$

4.14 Infinite Series and Integral
Let $\mu : \mathcal{P}(\omega) \longrightarrow [0, \infty]$ be a measure defined as $\mu(A) = |A|$ if A is finite and $\mu(A) = \infty$ otherwise.

a) A function $f : \omega \longrightarrow \mathbb{R}$ is integrable if and only if the series $\sum_{n=0}^{\infty} f(n)$ converges absolutely. Then $\int_{\omega} f \, d\mu = \sum_{n=0}^{\infty} f(n)$.

b) Formulate Theorems 4.22 and 4.25 and Corollaries 4.23 and 4.24 in the terminology of infinite series in the case of the measure space $\langle \omega, \mathcal{P}(\omega), \mu \rangle$.

4.15 Signed Measures

Suppose that $\langle X, \mathcal{S}, \mu \rangle$ is a measure space. Let f be an integrable function. We set $\nu(E) = \int_E f \, d\mu$ for any $E \in \mathcal{S}$. For terminology see Exercises 4.7 and 4.8.

a) ν is a signed measure defined on $\mathcal{S}$.

b) Show that $\{x \in X : f(x) \geq 0\}$, $\{x \in X : f(x) < 0\}$ is a Hahn decomposition of X.

c) Show that $\nu^+(E) = \int_E f^+ \, d\mu$, $\nu^-(E) = \int_E f^- \, d\mu$, $|\nu|(E) = \int_E |f| \, d\mu$ for any $E \in \mathcal{S}$.

d) Can we replace the condition of integrability of f by a weaker condition?

4.16 [AC] Absolutely Continuous Measures

Let μ be a non-negative measure, ν a signed measure, both defined on $\mathcal{S}$. We say that ν is **absolutely continuous** with respect to μ, write $\nu \ll \mu$, if $\nu(E) = 0$ for each $E \in \mathcal{S}$ such that $\mu(E) = 0$.

a) Assume that ν is defined as in Exercise 4.15. Then ν is absolutely continuous with respect to μ.

b) If ν is absolutely continuous with respect to μ, then also ν^+, ν^- and $|\nu|$ are such.

c) If ν_1 and ν_2 are absolutely continuous with respect to μ, then also $\alpha \cdot \nu_1 + \beta \cdot \nu_2$, $\alpha, \beta \in \mathbb{R}$ is such.

d) Find a proof of the **Radon-Nikodým Theorem** in a standard textbook of measure theory: If ν is absolutely continuous with respect to μ, then there exists a (unique up to measure zero set) integrable function f such that $\nu(E) = \int_E f \, d\mu$ for each $E \in \mathcal{S}$.

4.17 Riesz Theorem

Let X be a compact space. The space $C(X) = C^*(X)$ of real continuous functions is endowed with the supremum metric

$$\sigma(f, g) = \sup\{|f(x) - g(x)| : x \in X\}.$$

A linear operator (see Exercise 3.13) from $C(X)$ into $\mathbb{R}$ is called a **linear functional**. A linear functional F is **positive** if $F(f) \geq 0$ for any $f \geq 0$.

a) A positive linear functional F is continuous.
   Hint: Note that $\sigma(f, g) < \varepsilon$ if and only if $f - g + \varepsilon > 0$ and $g - f + \varepsilon > 0$.

b) If μ is a Borel measure on X, then $F(f) = \int_X f \, d\mu$ is a positive linear functional.

c) Let F be a positive linear functional. We define

$$\mu(U) = \sup\{F(f) : f \in C(X) \wedge 0 \leq f \leq \chi_U\}.$$

Show that $\mu(V) = \inf\{\mu(U) : V \subseteq U \text{ open}\}$ for any open set V.

d) For any $E \subseteq X$ we set

$$\mu^*(E) = \inf\{\mu(U) : E \subseteq U \text{ open}\}.$$

Show that μ^* is an outer measure.

e) Show that $\mu = \mu^*|\mathcal{S}(\mu^*)$ is a finite measure and $\mathrm{BOREL}(X) \subseteq \mathcal{S}(\mu^*)$.

f) If F is a positive functional and μ is defined as above then

$$F(f) = \int_X f \, d\mu$$

for any $f \in \mathrm{C}(X)$.

g) Any continuous linear functional F is of the form $F = F^+ - F^-$, where F^+, F^- are positive linear functionals.
Hint: Set $F^+(f) = \sup\{F(h) : 0 \le h \le f, h \in \mathrm{C}(X)\}$ for any $f \ge 0$.

h) Prove the **Riesz Theorem**: For any continuous linear functional F there exists a signed measure ν of bounded variation such that

$$F(f) = \int_X f \, d\nu = \int_X f \, d\nu^+ - \int_X f \, d\nu^-.$$

i) Extend the Riesz Theorem for the space $\mathrm{C}_0(X)$ of real continuous functions f on a locally compact space X such that the support $\overline{\{x \in X : f(x) \ne 0\}}$ is compact.

4.18 Riemann Integral
Let f be a bounded real function defined on $[a, b]$. If $c, d \in [a, b]$, $c < d$ we define

$$\mathrm{UB}(f; c, d) = \sup\{f(x) : x \in [c, d]\}, \qquad \mathrm{LB}(f; c, d) = \inf\{f(x) : x \in [c, d]\}.$$

If $A = \{a_0, \dots, a_n\} \in \mathrm{Part}(a, b)$ (see Exercise 2.13), $a = a_0 < a_1 < \cdots < a_{n-1} < a_n = b$, we set

$$\mathrm{US}(f; A) = \sum_{i=1}^n \mathrm{UB}(f; a_{i-1}, a_i) \cdot (a_i - a_{i-1}), \quad \mathrm{LS}(f; A) = \sum_{i=1}^n \mathrm{LB}(f; a_{i-1}, a_i) \cdot (a_i - a_{i-1}).$$

The notion of a limit of a net was defined in Exercise 1.17.

The limit of the net $\langle \mathrm{Part}(a, b), \supseteq, \mathrm{US}(f; A) \rangle$ is the **upper Riemann integral** of f on $[a, b]$. Similarly, the limit of the net $\langle \mathrm{Part}(a, b), \supseteq, \mathrm{LS}(f; A) \rangle$ is the **lower Riemann integral** of f on $[a, b]$. They are denoted as

$$\overline{\int_a^b} f(x) \, dx, \qquad \underline{\int_a^b} f(x) \, dx,$$

respectively. Finally, if $\overline{\int_a^b} f(x) \, dx = \underline{\int_a^b} f(x) \, dx$, then f is said to be **Riemann integrable** and the common value $(\mathcal{R})\int_a^b f(x) \, dx$ is called a **Riemann integral**.

a) Show that the net $\langle \mathrm{Part}(a, b), \supseteq, \mathrm{US}(f; A) \rangle$ is non-increasing, i.e.,

$$\text{if } A, B \in \mathrm{Part}(a, b), \ A \supseteq B, \text{ then } \mathrm{US}(f; A) \le \mathrm{US}(f; B).$$

Similarly, the net $\langle \mathrm{Part}(a, b), \supseteq, \mathrm{LS}(f; A) \rangle$ is non-decreasing.

b) Show that (if f is bounded) both the nets in a) have limits.
Hint: Show that

$$\lim_{A \in \mathrm{Part}(a, b)} \mathrm{US}(f; A) = \sup \left\{ \sum_{i=0}^n \mathrm{UB}(f; a_{i-1}, a_i) \cdot (a_i - a_{i-1}) : A \in \mathrm{Part}(a, b) \right\}.$$

Similarly for $\lim_{A \in \mathrm{Part}(a, b)} \mathrm{LS}(f; A)$.

c) Consider the Dirichlet function $\chi = \chi_{[0,1]\cap\mathbb{Q}}$. Show that

$$\overline{\int_0^1} \chi(x)\,dx = 1, \qquad \underline{\int_0^1} \chi(x)\,dx = 0.$$

Thus, χ is not Riemann integrable.

d) f is Riemann integrable if and only if

$$\lim_{A\in\mathrm{PART}(a,b)} (\mathrm{US}(f;A) - \mathrm{LS}(f;A)) = 0.$$

e) Assume that

$$(\forall\varepsilon > 0)(\exists\{a_0 < \cdots < a_n\} \in \mathrm{PART}(a,b))$$
$$(\forall i < n)(\forall x,y \in [a_i, a_{i+1}])\,|f(x) - f(y)| < \varepsilon.$$

Then f is Riemann integrable.

f) If f is continuous on $[a,b]$, then f is Riemann integrable.
Hint: By Exercise 3.1 f is uniformly continuous.

g) Non-decreasing and non-increasing functions are Riemann integrable.
Hint: Let f be a non-decreasing function. Consider the finite set of those points $c \in [a,b]$ for which $\lim_{x\to c^+} f(x) - \lim_{x\to c^-} f(x) \geq \varepsilon/2 \cdot (b-a)$. Find a partition A such that $\mathrm{US}(f;A) - \mathrm{LS}(f;A) < \varepsilon$.

h) A function of bounded variation is Riemann integrable (see Exercise 2.14).

4.19 Newton–Leibniz Formula
Let f be Riemann integrable on $[a,b]$, $F(x) = (\mathcal{R})\int_a^x f(t)\,dt$ for $x \in [a,b]$.

a) F is continuous on $[a,b]$.

b) If f is continuous, then $F'(x) = f(x)$ for any $x \in (a,b)$.

c) If G is such that $G'(x) = f(x)$ for any $x \in [a,b]$, then the **Newton–Leibniz formula** holds true:

$$(\mathcal{R})\int_a^b f(x)\,dx = G(b) - G(a).$$

4.20 Riemann and Lebesgue Integral
Let $f : [a,b] \longrightarrow \mathbb{R}$ be a Riemann integrable function, $\langle A_n = \{a_{n,0}, \ldots, a_{n,k_n}\} : n \in \omega\rangle$ being a sequence of partitions of $[a,b]$ such that

$$\lim_{n\to\infty} \mathrm{US}(f;A_n) = \lim_{n\to\infty} \mathrm{LS}(f;A_n) = (\mathcal{R})\int_a^b f(x)\,dx.$$

Moreover, we assume that each A_{n+1} is a refinement of A_n. Let $g_n(x) = \mathrm{LB}(f;a_{n,i-1},a_{n,i})$, $h_n(x) = \mathrm{UB}(f;a_{n,i-1},a_{n,i})$ for $x \in [a_{n,i-1},a_{n,i})$, $i = 1,\ldots,k_n$, $g_n(b) = h_n(b) = f(b)$.

a) g_n and h_n are simple Lebesgue measurable functions and $\mathrm{LS}(f;A_n) = \int_{[a,b]} g_n(x)\,dx$, $\mathrm{US}(f;A_n) = \int_{[a,b]} h_n(x)\,dx$.

b) f is Lebesgue integrable and $(\mathcal{R})\int_a^b f(x)\,dx = \int_{[a,b]} f(x)\,dx$.
Hint: $\lim_{n\to\infty} h_n \leq f \leq \lim_{n\to\infty} g_n$ and $\lim_{n\to\infty} \int_{[a,b]} h_n\,dx = \lim_{n\to\infty} \int_{[a,b]} g_n\,dx$.

4.4   Product of Measures, Ergodic Theorem

Let us fix complete measure spaces with finite measures $\langle X_i, \mathcal{S}_i, \mu_i \rangle$, $i = 1, 2$. By the standard argument we can easily enlarge the obtained results for σ-finite measures.

A complete measure space $\langle X_1 \times X_2, \mathcal{S}, \mu \rangle$ is said to be a **product of the measure spaces** $\langle X_1, \mathcal{S}_1, \mu_1 \rangle$ and $\langle X_2, \mathcal{S}_2, \mu_2 \rangle$ if $\mathcal{S}$ is the smallest σ-algebra of subsets of $X_1 \times X_2$ containing all sets $A_1 \times A_2$, $A_1 \in \mathcal{S}_1$, $A_2 \in \mathcal{S}_2$ and satisfying the completeness condition, and $\mu(A_1 \times A_2) = \mu_1(A_1) \cdot \mu_2(A_2)$ for any $A_1 \in \mathcal{S}_1$, $A_2 \in \mathcal{S}_2$. The measure μ is said to be a **product of the measures** μ_1 and μ_2. We shall write $\mathcal{S} = \mathcal{S}_1 \times \mathcal{S}_2$ and $\mu = \mu_1 \times \mu_2$.

The following basic result can be found in any textbook devoted to measure theory.

Theorem 4.26. *A product of finite measures always exists and is uniquely determined.*

Proof. We briefly sketch an idea of a proof of the theorem. Let us recall that the vertical and horizontal sections A_x and A^y were defined in Section 11.1.

Let $\mathcal{T}$ be the smallest σ-algebra of subsets of $X_1 \times X_2$ containing all sets $A_1 \times A_2$, $A_i \in \mathcal{S}_i$, $i = 1, 2$. One can easily show that $A_x \in \mathcal{S}_2$ and $A^y \in \mathcal{S}_1$ for any $A \in \mathcal{T}$ and any $x \in X_1$, $y \in X_2$.

Thus, for a set $A \in \mathcal{T}$ we can define

$$f_A(x_1) = \mu_2(A_{x_1}), \ x_1 \in X_1, \quad g_A(x_2) = \mu_1(A^{x_2}), \ x_2 \in X_2.$$

We show that f_A and g_A are non-negative integrable functions such that

$$\int_{X_1} f_A \, d\mu_1 = \int_{X_2} g_A \, d\mu_2 \tag{4.30}$$

for any $A \in \mathcal{T}$.

We write

$$\mathcal{T}' = \{A \subseteq X : f_A, g_A \text{ are integrable and (4.30) holds true}\}.$$

We show that $\mathcal{T}'$ is a σ-algebra containing all sets $A_1 \times A_2$, $A_1 \in \mathcal{S}_1$, $A_2 \in \mathcal{S}_2$.

If $A = A_1 \times A_2$, $A_1 \in \mathcal{S}_1$, $A_2 \in \mathcal{S}_2$, then $f_A(x_1) = \mu_2(A_2)$ for $x_1 \in A_1$ and 0 otherwise. Similarly for g_A. Thus f_A and g_A are integrable and (4.30) holds true. Consequently $A \in \mathcal{T}'$.

If $A \in \mathcal{T}'$, then $f_{X \setminus A}(x_1) = \mu_2(X_2) - f_A(x_1)$ and $g_{X \setminus A}(x_2) = \mu_1(X_1) - g_A(x_2)$. Thus also $X \setminus A \in \mathcal{T}'$.

Similarly we can see that $\mathcal{T}'$ is closed under countable disjoint unions, since, e.g., if $A = \bigcup_n A_n$, then $f_A = \sum_n f_{A_n}$, $g_A = \sum_n g_{A_n}$ and we can use Corollary 4.23.

Since $\mathcal{T}$ was the smallest σ-algebra containing all sets $A_1 \times A_2$, $A_i \in \mathcal{S}_i$, $i = 1, 2$, we have $\mathcal{T} \subseteq \mathcal{T}'$.

Now we define

$$\mu(A) = \int_{X_1} f_A \, d\mu_1 = \int_{X_2} g_A \, d\mu_2$$

for $A \in \mathcal{T}$. The completion of the measure space $\langle X, \mathcal{T}, \mu \rangle$ is the unique product of measure spaces $\langle X_1, \mathcal{S}_1, \mu_1 \rangle$ and $\langle X_2, \mathcal{S}_2, \mu_2 \rangle$.   □

We are interested mainly in Lebesgue measures on Euclidean spaces. We can identify $\mathbb{R}_{n+m} = \mathbb{R}_n \times \mathbb{R}_m$. Then it makes sense

Theorem [wAC] 4.27. λ_{n+m} *is the product of measures* λ_n *and* λ_m *and*

$$\lambda_{n+m}(A) = \int_{\mathbb{R}_n} f_A \, d\lambda_n = \int_{\mathbb{R}_m} g_A \, d\lambda_m \tag{4.31}$$

for any $A \in \mathcal{L}_{n+m}$, $A \subseteq \mathbb{R}_n \times \mathbb{R}_m$, *where* $f_A(x) = \lambda_m(A_x)$ *and* $g(y) = \lambda_n(A^y)$.

Proof. $\text{BOREL}(\mathbb{R}_{n+m})$ is the smallest σ-algebra containing sets $U \times V$, where $U \subseteq \mathbb{R}_n$, $V \subseteq \mathbb{R}_m$ are open. Since a set $A \subseteq \mathbb{R}_{n+m}$ is measurable if and only if A is equal to a Borel set up to a set of measure zero and λ_{n+m} is a complete measure, we obtain $\mathcal{L}_{n+m} = \mathcal{L}_n \times \mathcal{L}_m$.

We have to show

$$\lambda_{n+m}(U \times V) = \lambda_n(U) \cdot \lambda_m(V) \tag{4.32}$$

for any measurable sets $U \subseteq \mathbb{R}_n$ and $V \subseteq \mathbb{R}_m$. Without loss of generality we can assume that the sets U, V are bounded, i.e., subsets of bounded intervals $J_n \subseteq \mathbb{R}_n$ and $J_m \subseteq \mathbb{R}_m$.

Let $I \subseteq J_n$ be a given interval. We show that the family

$$\mathcal{T}_I = \{V \in \mathcal{L}_m : V \subseteq J_m \wedge \lambda_{n+m}(I \times V) = \lambda_n(I) \cdot \lambda_m(V)\}$$

is equal to $\mathcal{L}_m$. Actually, by (4.23) any interval $J \subseteq J_m$ belongs to $\mathcal{T}_I$. Evidently $\mathcal{T}_I$ is closed under countable disjoint unions. Since $I \times (J_m \setminus V) = I \times J_m \setminus (I \times V)$, $\mathcal{T}_I$ is closed under complements. Since all considered measures are complete, we obtain the statement.

Similarly, we set

$$\mathcal{T} = \{U \in \mathcal{L}_n : U \subseteq J_n \wedge (\forall V \in \mathcal{L}_m) \, \lambda_{n+m}(U \times V) = \lambda_n(U) \cdot \lambda_m(V)\}$$

and show that $\mathcal{T} = \mathcal{L}_n$.

The equality (4.31) follows by the proof of Theorem 4.26.   □

Corollary [wAC] 4.28 (Fubini Theorem). *A set* $A \subseteq \mathbb{R}_{n+m}$ *has Lebesgue measure zero if and only if the set* $\{x \in \mathbb{R}_n : \lambda_m^*(A_x) \neq 0\}$ *has Lebesgue measure zero.*

Theorem [AC] 4.29 (Fubini-Tonelli Theorem). *Assume that $\langle X, \mathcal{S}, \mu \rangle$ and $\langle Y, \mathcal{T}, \nu \rangle$ are measure spaces with complete finite measures. If f is an $\mathcal{S} \times \mathcal{T}$-measurable function on $X \times Y$, then:*

a) *If f is integrable and*

$$g(x) = \int_Y f(x,y) \, d\mu(y), \quad h(y) = \int_X f(x,y) \, d\nu(x) \tag{4.33}$$

for $x \in X$ and $y \in Y$, then g is μ-integrable, h is ν-integrable and

$$\int_{X \times Y} f \, d\mu \times \nu = \int_X g \, d\mu = \int_Y h \, d\nu. \tag{4.34}$$

b) *If f is non-negative, then the functions defined by (4.33) are $\mathcal{S}$-measurable and $\mathcal{T}$-measurable, respectively, and (4.34) holds true.*

The proof is similar to that of preceding theorems and can be found in any standard textbook on measure theory. It is easy to see that the statement holds true for characteristic functions of sets of the form $A_1 \times A_2$, $A_1 \in \mathcal{S}_1$, $A_2 \in \mathcal{S}_2$. Then one shows that the family of those $A \subseteq X_1 \times X_2$ for which the characteristic function satisfies (4.34) is a σ-algebra. Using Lemma 4.19 one obtains the statement for non-negative integrable functions. The rest is standard. □

For $k > 0$ we define a bijection $f_k : {}^\omega 2 \xrightarrow[\text{onto}]{1-1} ({}^\omega 2)^k$ setting $f_k(\alpha) = [\beta_1, \ldots, \beta_k]$, where $\beta_i(n) = \alpha(k \cdot n + i - 1)$. If $s_1, \ldots, s_k \in {}^n 2$, then $f_k^{-1}([s_1] \times \cdots \times [s_k]) = [s]$ for suitable $s \in {}^{k \cdot n} 2$. Thus, if $\lambda_{\mathbb{C}}^k$ is the product measure on $({}^\omega 2)^k$, then we obtain

$$\lambda_{\mathbb{C}}([s]) = \lambda_{\mathbb{C}}(f_k^{-1}([s_1] \times \cdots \times [s_k])) = \lambda_{\mathbb{C}}^k([s_1] \times \cdots \times [s_k]) = 2^{-k \cdot n}. \tag{4.35}$$

Let $\mathcal{U} = \{[s] : s \in {}^{k \cdot n} 2 \wedge n \in \omega\}$ and $\mathcal{U}^*$ be the family of all finite unions of elements of $\mathcal{U}$. Since every element of $\mathcal{U}^*$ is a finite union of disjoint elements of $\mathcal{U}$, by (4.35) we obtain that $\lambda_{\mathbb{C}}^k$ and $f_k^{-1} \circ \lambda_{\mathbb{C}}$ coincide on $\mathcal{U}^*$. Thus by Theorem 4.10 we obtain

Theorem [wAC] 4.30.
$$\lambda_{\mathbb{C}}(f_k^{-1}(A)) = \lambda_{\mathbb{C}}^k(A)$$

for any Borel $A \subseteq ({}^\omega 2)^k$.

One can in a similar way define a product of any system $\langle \langle X_s, \mathcal{S}_s, \mu_s \rangle : s \in S \rangle$ of measure spaces, provided that $\mu_s(X_s) \leq 1$ for every $s \in S$. Generally the Axiom of Choice is needed. If μ is the measure on the set $2 = \{0, 1\}$ defined by $\mu(\{0\}) = \mu(\{1\}) = 1/2$, then one can construct the product measure μ^ω on ${}^\omega 2$. In this case we do not need any axiom of choice. If $f : \mathbb{C} \xrightarrow[\text{onto}]{1-1} {}^\omega 2$ is the homeomorphism constructed in the proof of Theorem 3.1, then one can show that $\mu^\omega(A) = \lambda_{\mathbb{C}}(f^{-1}(A))$ for any Borel set $A \subseteq {}^\omega 2$. Therefore we shall identify the measure μ^ω with $\lambda_{\mathbb{C}}$.

We present elementary facts from the ergodic theory. For more details see, e.g., P. Billingsley [1965], H. Furstenberg [1981] or P. Halmos [1956]. Let $\langle X, \mathcal{S}, \mu \rangle$ be a measure space with a probability measure. We say that $T : X \longrightarrow X$ is a **measure preserving mapping**, if $T^{-1}(A) \in \mathcal{S}$ and $\mu(T^{-1}(A)) = \mu(A)$ for every $A \in \mathcal{S}$. A set $A \in \mathcal{S}$ is **invariant** if $T^{-1}(A) = A$. The mapping T is **ergodic** if $\mu(A) = 0$ or $\mu(X \setminus A) = 0$ for every invariant $A \in \mathcal{S}$. Similarly, a function $f : X \longrightarrow \mathbb{R}$ is **invariant** if $f(T(x)) = f(x)$ almost everywhere. T^n is defined by induction: T^0 is the identity mapping and, $T^{n+1} = T \circ T^n$.

Theorem 4.31 (Ergodic Theorem). *Let $T : X \longrightarrow X$ be a measure preserving mapping, $g : X \longrightarrow \mathbb{R}$ being integrable. Then there exists an integrable invariant function $g^* : X \longrightarrow \mathbb{R}$ such that*

$$\lim_{n \to \infty} \frac{1}{n} \sum_{i=0}^{n-1} g(T^i(x)) = g^*(x) \text{ for all } x \text{ and } \int_X g \, d\mu = \int_X g^* \, d\mu.$$

Moreover, if T is ergodic, then we can assume that g^* is a constant function and

$$\lim_{n \to \infty} \frac{1}{n} \sum_{i=0}^{n-1} g(T^i(x)) = \int_X g \, d\mu \text{ for almost all } x.$$

An idea of a proof of the theorem is sketched in Exercise 4.23 below.

Exercises

4.21 Ergodicity of a Shift

Let $T : \mathbb{T} \longrightarrow \mathbb{T}$ be continuous. For a set $A \subseteq \mathbb{T}$ we write $T^*(A) = \bigcup_n T^n(A)$. We shall consider a special case $T(x) = a + x$ for a fixed a. Everything is understood modulo 1.

a) If $a \in \mathbb{Q}$, then $T^*(\{x\})$ is finite for any $x \in \mathbb{T}$.

b) If a is a positive irrational, $0 \le c < d \le 1$, then there are a natural number n and an integer z such that $c + z < na < d + z$.
   Hint: By the Dirichlet Theorem 8.133 there exists an n such that $\|na\| < d - c$.

c) If a is irrational, then the set $T^*(\{x\})$ is dense in $\mathbb{T}$ for any x. Note that this is a simple consequence of the Kronecker Theorem 10.41.

d) If a is irrational, then T is ergodic.
   Hint: Assume $\mu(A) > 0$, $\varepsilon > 0$. Take an open interval I with properties of Theorem 4.17. Using c) find natural numbers $n_0, \ldots, n_k$ such that $\sum_{i=0}^{k} \mu(T^{n_i}(I)) > 1 - \varepsilon$ and $T^{n_i}(I)$, $i = 0, \ldots, k$ are mutually disjoint.

4.22 Shift on $^\omega 2$

The **shift** T on $^\omega 2$ is defined as

$$T(\{x_n\}_{n=0}^{\infty}) = \{x_{n+1}\}_{n=0}^{\infty}. \tag{4.36}$$

a) If $A \subseteq {}^\omega 2$ is T invariant, $A \subseteq \bigcup_n [s_n]$, where $s_n \in {}^{<\omega} 2$, then $A \subseteq \bigcup_{n,m} [s_n \frown s_m]$.

b) If $\sum_n \mu([s_n]) < \varepsilon$, then $\sum_{n,m} \mu([s_n \frown s_m]) < \varepsilon^2$.

c) If $A \subseteq {}^\omega 2$ is T invariant, $\mu^*(A) < 1$, then $\mu(A) = 0$.

d) The shift T is ergodic.

4.23 Proof of the Ergodic Theorem

Let $\langle X, \mathcal{S}, \mu \rangle$ be a measure space, $\mu(X) = 1$, $T : X \longrightarrow X$ being a measure preserving mapping. For an integrable function $f : X \longrightarrow \mathbb{R}$ we define

$$
\begin{aligned}
S_{f,n}(x) &= \sum_{i=0}^{n-1} T^i \circ f(x), \\
M_{f,n}(x) &= \max\{0, S_{f,1}(x), \ldots, S_{f,n}(x)\}, \\
\overline{f}(x) &= \limsup_{n \to \infty} \tfrac{1}{n} S_{f,n}(x), \\
\underline{f}(x) &= \liminf_{n \to \infty} \tfrac{1}{n} S_{f,n}(x).
\end{aligned}
$$

a) Show that $\int_{T^{-1}(E)} T \circ f \, d\mu = \int_E f \, d\mu$.

b) $f + T \circ M_{f,n} \geq S_{f,i+1}$ for any $i \leq n$ and therefore $f + T \circ M_{f,n} \geq M_{f,n+1} \geq M_{f,n}$.

c) Prove the **Maximal Ergodic Theorem**: If $X_n = \{x \in X : M_{f,n}(x) > 0\}$, then $\int_{X_n} f \, d\mu \geq 0$.

   Hint: By definition and a) *we have $\int_X T \circ M_{f,n} \, d\mu = \int_{X_n} M_{f,n} \, d\mu$ and by* b) *we obtain*

   $$
   \int_{X_n} (f + M_{f,n}) \, d\mu = \int_{X_n} f \, d\mu + \int_X M_{f,n} \, d\mu \geq \int_{X_n} M_{f,n} \, d\mu = \int_X M_{f,n} \, d\mu.
   $$

d) Show that $\overline{f}$ and $\underline{f}$ are invariant.

   Hint: Note that $\limsup_{n \to \infty} \tfrac{1}{n} S_{f,n}(x) = \limsup_{n \to \infty} \tfrac{1}{n}(S_{f,n+1}(x) - f(x))$ and that $T \circ S_{f,n} = S_{f,n+1} - f$.

e) For $a < b \in \mathbb{R}$ we write $B_{a,b} = \{x \in X : \underline{f}(x) < a < b < \overline{f}(x)\}$. Show that $B_{a,b}$ is invariant.

f) Show that $\mu(B_{a,b}) = 0$ for each $a < b$.

   Hint: Set $C = B_{a,b}$ and $C_n = \{x \in C : M_{(f-b),n}(x) > 0\}$. As

   $$
   \limsup_{n \to \infty} 1/n S_{(f-b),n}(x) = \overline{f}(x) - b,
   $$

   we have $C = \bigcup_n C_n$ and therefore by c) *we obtain $\int_C (f(x) - b) \, d\mu \geq 0$. Similarly $\int_C (a - f(x)) \, d\mu \geq 0$. Since $\int_{B_{a,b}} (a - b) \, d\mu = \int_{B_{a,b}} (a - f) \, d\mu + \int_{B_{a,b}} (f - b) \, d\mu \geq 0$, we obtain $\mu(B_{a,b}) = 0$ for any $a < b$.*

g) $\overline{f}(x) = \underline{f}(x)$ almost μ-everywhere.

h) $\overline{f}$ is integrable.

   Hint: Let $f_n(x) = \inf\{\tfrac{1}{k}|S_{f,k}(x)| : k \geq n\}$. Then $f_n \nearrow \overline{f}$ almost everywhere and $|\int_X f_n \, d\mu| \leq \int |f| \, d\mu$. Use the Lebesgue Monotone Convergence Theorem 4.22.

i) For any $z \in \mathbb{Z}$, $n > 0$, write

   $$
   D_{n,z} = \left\{ x \in X : \frac{z}{n} \leq \overline{f}(x) < \frac{z+1}{n} \right\}.
   $$

   Show that for any $\varepsilon > 0$,

   $$
   \int_{D_{n,z}} \left(\frac{z+1}{n} - f \right) d\mu \geq 0 \quad \text{and} \quad \int_{D_{n,z}} \left(f - \frac{z}{n} + \varepsilon \right) d\mu \geq 0.
   $$

   Hint: Apply the Maximal Ergodic Theorem to function $\frac{z+1}{n} - f$.

j) $\int_X \overline{f}(x)\,d\mu = \int_X f(x)\,d\mu$.

   Hint: Using i) *show that* $|\int_X (f - \overline{f})\,d\mu| \le 1/n$.

k) If T is ergodic, then $\overline{f}(x) = \int_X f(y)\,d\mu(y)$ almost everywhere.

   Hint: The set $D = \{x \in X : \overline{f}(x) = \int_X f(y)\,d\mu(y)\}$ *is invariant. If* $\mu(D) = 0$, *then either the invariant set* $\{x \in X : \overline{f}(x) < \int_X f(y)\,d\mu(y) - 1/n\}$ *or the invariant set* $\{x \in X : \overline{f}(x) > \int_X f(y)\,d\mu(y) + 1/n\}$ *has measure* 1 *for some positive integer* n.

l) Prove the Ergodic Theorem 4.31.

Historical and Bibliographical Notes

E. Borel [1898] introduced a measure of a Borel subset of $\mathbb{R}$ starting from open sets represented as a disjoint union of intervals. In his dissertation H. Lebesgue [1902] worked out the idea of Borel and actually constructed the complete Lebesgue measure on $\mathbb{R}$. C. Carathéodory [1918] presents a systematic theory of measure introducing the notion of a measurable set.

The basic source of information concerning measure theory is P. Halmos [1950]. We follow the terminology connected with measures on topological spaces as presented by R.J. Gardner and W.F. Pfeffer [1984].

Ergodic Theorem 4.31 is a result by G.D. Birkhoff [1931]. The proof sketched in Exercise 4.23 follows essentially P. Halmos [1956], who attributes it to F. Riesz.

Chapter 5

Useful Tools and Technologies

> A strategy is something such that you've got that you never have to think.
>
> Robert M. Solovay [1967].

This chapter contains material that will be needed in the next investigations. A reader need not study all sections immediately. The topic is heterogeneous and we indicate when the knowledge of a particular section is supposed.

In the first two sections of this chapter we present some tools and technologies that will be useful in studying descriptive set theoretical properties of subsets of a Polish space. In Section 5.1 we present the technology introduced by F. Hausdorff (under the name of an A-operation) and then developed by N.N. Luzin and M.J. Souslin, which contributes to the understanding of the fine structure of so-called projective sets. The results will be used essentially in Chapter 6. In Section 5.2 we introduce the common framework for families of sets investigated in the descriptive set theory that was mainly developed by Y.N. Moschovakis. The technology enables us, on the one hand, to formulate in a common language results for different families of sets and, on the other hand, it often unifies proofs which were historically found independently in distinct special cases.

In the third section we introduce basic facts concerning Boolean algebras, which are essential for a study of topics related to the Martin Axiom. We shall use them in Sections 7.1, 9.1 and 10.3.

In the fourth section we investigate cardinal invariants related to combinatorial properties of sets of natural number or functions from ω into ω. The notions we introduce and their properties are important results of investigation in the set theory of the real line over the last forty years. We shall need these results in Chapter 7 and in all following chapters.

Finally, the fifth section is devoted to the basic facts about infinite games, which were studied by S. Banach and S. Mazur, and later investigated by Morton Davis, J. Mycielski and others. These facts are necessary for the understanding of

Section 9.4, where we present the basic results of investigations into the descriptive set theory under the so-called Axiom of Determinacy. Also Section 10.1 supposes that the reader is familiar with this topic.

5.1   Souslin Schemes and Sieves

The terminology and notations concerning trees are summarized in Section 11.1. Assume that T is a pruned subtree of the tree $\langle {}^{<\omega}\omega, \supseteq \rangle$ and $\langle X, \mathcal{O} \rangle$ is a topological space. If $\varphi : T \longrightarrow \mathcal{P}(X)$, then the couple $\langle T, \varphi \rangle$ is called a **Souslin scheme** on X. The Souslin scheme $\langle T, \varphi \rangle$ is said to be **closed** (**open, Borel** etc.), if every value $\varphi(x)$, $x \in T$ is a closed (open, Borel etc.) set. Let us recall that we may identify any branch of T with a function $\alpha : \omega \longrightarrow \omega$. We write

$$L(T, \varphi) = \bigcup_{\alpha \in [T]} \bigcap_n \varphi(\alpha|n).$$

We say that the set $L(T, \varphi)$ is **sifted** by the Souslin scheme $\langle T, \varphi \rangle$.

A Souslin scheme $\langle T, \varphi \rangle$ is **monotone** if

$$t < s \rightarrow \varphi(s) \subseteq \varphi(t) \tag{5.1}$$

for any $t, s \in T$.

If $\langle T, \varphi \rangle$ is a Souslin scheme on X, one can define a new Souslin scheme setting $\psi(s) = \bigcap_{t \leq s} \varphi(t)$ for $s \in T$. It is easy to see that $L(T, \varphi) = L(T, \psi)$ and the Souslin scheme $\langle T, \psi \rangle$ is monotone. Hence in what follows, we always assume that a Souslin scheme is monotone.

If S is a pruned subtree of T and $\langle T, \varphi \rangle$ is a (monotone) Souslin scheme, then the Souslin scheme $\langle S, \varphi | S \rangle$ will be simply denoted as $\langle S, \varphi \rangle$. Similarly, if $s \in T$ then the Souslin scheme $\langle T^s, \varphi | T^s \rangle$ will be simply denoted as $\langle T^s, \varphi \rangle$. One can easily see that $L(S, \varphi) \subseteq L(T, \varphi)$ and

$$L(T, \varphi) = L(S, \varphi) \cup \bigcup_{v \in T \setminus S} L(T^v, \varphi). \tag{5.2}$$

Moreover, we have

$$L(T^s, \varphi) = \bigcup_{t \in \mathrm{IS}(s,T)} L(T^t, \varphi). \tag{5.3}$$

A Souslin scheme $\langle T, \varphi \rangle$ is **regular** if $\varphi(s) \cap \varphi(t) = \emptyset$ for any incomparable $s, t \in T$. If $\langle T, \varphi \rangle$ is regular then

$$L(T, \varphi) = \bigcap_n \bigcup_{t \in T(n)} \varphi(t). \tag{5.4}$$

The **kernel** of a Souslin scheme is the set

$$\mathrm{Ker}\,(T,\varphi) = \{\alpha \in [T] : \bigcap_n \varphi(\alpha|n) \neq \emptyset\}.$$

Assume that X is a metric space. We say that the Souslin scheme $\langle T, \varphi \rangle$ has **vanishing diameter** if $\lim_{n\to\infty} \mathrm{diam}(\varphi(\alpha|n)) = 0$ for any branch $\alpha \in [T]$. Then the intersection $\bigcap_n \varphi(\alpha|n)$ has at most one element. Moreover, in this case we can define the **associated map** $F : \mathrm{Ker}\,(T,\varphi) \longrightarrow X$ by setting

$$F(\alpha) = \text{the only element of } \bigcap_n \varphi(\alpha|n) \text{ for } \alpha \in \mathrm{Ker}\,(T,\varphi).$$

Assume that X is a Polish space and $\langle T, \varphi \rangle$ is a closed Souslin scheme with vanishing diameter. By Cantor's Lemma 3.28 we obtain

$$\alpha \in \mathrm{Ker}\,(T,\varphi) \equiv (\forall n)\,\varphi(\alpha|n) \neq \emptyset. \tag{5.5}$$

Especially, if $\varphi(v) \neq \emptyset$ for each $v \in T$, then $\mathrm{Ker}\,(T,\varphi) = [T]$.

If $\langle T, \varphi \rangle$ is a Souslin scheme, we can extend φ onto $^{<\omega}\omega$ by setting $\varphi(v) = \emptyset$ for $v \in {}^{<\omega}\omega \setminus T$. Evidently

$$L(T,\varphi) = L(^{<\omega}\omega, \varphi).$$

The following rather technical result will be useful later.

Theorem 5.1. *If a subset $A \subseteq X$ of a Polish space X is sifted by a (monotone) closed Souslin scheme with vanishing diameter, then A is sifted by a (monotone) open Souslin scheme. If X has a clopen base of topology, then A is sifted by a (monotone) clopen Souslin scheme.*

Proof. Let $\langle X, \rho \rangle$ be a Polish space. Assume that $A = L(T,\varphi)$, where $\langle T, \varphi \rangle$ is a monotone and closed Souslin scheme with vanishing diameter. We define $\psi(s) = \{x \in X : \rho(x, \varphi(s)) < 1/(n+1)\}$ for $s \in T(n)$. One can easily check that $L(T, \psi) = L(T, \varphi)$ and $\langle T, \psi \rangle$ is monotone and open. □

We shall study the relationship of Souslin schemes and continuous mappings from the Baire space $^{\omega}\omega$.

Lemma 5.2. *Let $\langle X, \rho \rangle$ be a metric space. If $F : {}^{\omega}\omega \longrightarrow X$ is continuous, then there exists a closed Souslin scheme $\langle {}^{<\omega}\omega, \varphi \rangle$ with vanishing diameter and with non-empty values such that $\mathrm{rng}(F) = L(^{<\omega}\omega, \varphi)$.*

Proof. For $v \in {}^{<\omega}\omega$ we set

$$\varphi(v) = \overline{F([v])}.$$

Evidently $\langle {}^{<\omega}\omega, \varphi \rangle$ is a closed monotone Souslin scheme. We need to show that it has vanishing diameter.

Let $\alpha \in {}^{\omega}\omega$, $\varepsilon > 0$. Since F is continuous there exists a $\delta > 0$ such that

$$(\forall \beta \in {}^{\omega}\omega) \left(\rho(\alpha, \beta) < \delta \to \rho(F(\alpha), F(\beta)) < \varepsilon/2 \right).$$

Note that on the left-hand side ρ denotes the Baire metric on ${}^{\omega}\omega$ and on the right-hand side ρ is the metric on X. If $n_0 \geq 1/\delta$ then $\mathrm{diam}(\varphi(\alpha|n)) < \varepsilon$ for any $n \geq n_0$.

Since $\bigcap_n \varphi(\alpha|n) = \{F(\alpha)\}$ we obtain $L({}^{<\omega}\omega, \varphi) = \mathrm{rng}(F)$. $\qquad \square$

Theorem 5.3. *Let $\langle T, \varphi \rangle$ be a Souslin scheme with vanishing diameter on a Polish space X. Then:*

  a) *The associated map $F : \mathrm{Ker}\,(T, \varphi) \longrightarrow X$ is continuous.*
  b) *If the Souslin scheme $\langle T, \varphi \rangle$ is regular, then F is one-to-one.*
  c) *If the Souslin scheme $\langle T, \varphi \rangle$ is closed, then $\mathrm{Ker}\,(T, \varphi)$ is a closed subset of the Baire space ${}^{\omega}\omega$.*

Proof. Assume that $\alpha \in \mathrm{Ker}\,(T, \varphi)$ and $\varepsilon > 0$. Then there is an $n \in \omega$ such that $\mathrm{diam}(\varphi(\alpha|n)) < \varepsilon$. If $\beta \in [\alpha|n] \cap \mathrm{Ker}\,(T, \varphi)$, then $F(\beta) \in \varphi(\alpha|n)$ and therefore F is continuous.

If $\langle T, \varphi \rangle$ is regular and $\alpha, \beta \in [T]$, $\alpha \neq \beta$, then $\alpha(n) \neq \beta(n)$ for some n. Since $F(\alpha) \in \varphi(\alpha|(n+1))$, $F(\beta) \in \varphi(\beta|(n+1))$, $\varphi(\alpha|(n+1)) \cap \varphi(\beta|(n+1)) = \emptyset$ we obtain $F(\alpha) \neq F(\beta)$.

Assume that $\alpha \in {}^{\omega}\omega \setminus \mathrm{Ker}\,(T, \varphi)$. Then by (5.5) there exists an $n \in \omega$ such that $\varphi(\alpha|n) = \emptyset$. Then $[\alpha|n] \cap \mathrm{Ker}\,(T, \varphi) = \emptyset$ and therefore $\mathrm{Ker}\,(T, \varphi)$ is closed. $\qquad \square$

Theorem [wAC] 5.4. *Assume that X is a Polish space and $\langle T, \varphi \rangle$ is a closed Souslin scheme on X with vanishing diameter. Then $L(T, \varphi)$ is either countable or contains a subset homeomorphic to the Cantor middle-third set $\mathbb{C}$.*

Proof. We can assume that $\varphi(v) \neq \emptyset$ for any $v \in T$.

Assume that $L(T, \varphi)$ is uncountable. We set

$$E = \{v \in T : L(T^v, \varphi) \text{ is uncountable}\}.$$

Evidently E is a pruned subtree of T. Using the equality (5.3) for T^v, $v \in E$, by **wAC** we obtain

$$L(E^v, \varphi) \text{ is uncountable for any } v \in E.$$

We show that

$$(\forall v \in E)(\exists u, w \in E)\,(u, w > v \wedge \varphi(u) \cap \varphi(w) = \emptyset). \tag{5.6}$$

Actually, if $v \in E$, then the set $L(E^v, \varphi)$ is uncountable. Let $x, y \in L(E^v, \varphi)$, $x \neq y$. Set $\varepsilon = \rho(x, y) > 0$. By definition there exist $\alpha, \beta \in [E^v]$ such that $x \in \bigcap_n \varphi(\alpha|n)$ and $y \in \bigcap_n \varphi(\beta|n)$. Since the Souslin scheme $\langle T, \varphi \rangle$ has vanishing diameter there exists an n such that $\mathrm{diam}(\varphi(\alpha|n)) < \varepsilon/2$ and $\mathrm{diam}(\varphi(\beta|n)) < \varepsilon/2$. Take $u = \alpha|n$ and $w = \beta|n$.

Now construct the subtree $S \subseteq E$ similarly as in the proof of Theorem 11.5, taking always two incomparable successors with disjoint values of φ. Then $\langle S, \varphi \rangle$ is a regular Souslin scheme and the theorem follows by Theorem 5.3. $\qquad \square$

Theorem 5.5. *Every Polish space is sifted by a closed Souslin scheme $\langle {}^{<\omega}\omega, \varphi \rangle$ with vanishing diameter and non-empty values.*

Proof. Let $\{U_n : n \in \omega\}$ be a countable open base of a Polish space X. We define a function φ such that $\langle {}^{<\omega}\omega, \varphi \rangle$ will be a closed Souslin scheme with vanishing diameter and $L({}^{<\omega}\omega, \varphi) = X$.

We define by induction a function ψ from ${}^{<\omega}\omega$ into non-empty open subsets of X. We set $\psi(\emptyset) = X$. Assume that the values $\psi(s)$ are already defined for any $s \in {}^n\omega$. Enumerate with eventual repetition the set of all non-empty intersections $\psi(s) \cap U_m$, $\mathrm{diam}(U_m) < 2^{-n}$ as $\{B_i : i \in \omega\}$. Set $\psi(s \frown i) = B_i$.

Let $\varphi(s)$ be the closure of $\psi(s)$. Evidently $\langle {}^{<\omega}\omega, \varphi \rangle$ is a closed Souslin scheme with vanishing diameter. We need to show that $L({}^{<\omega}\omega, \varphi) = X$.

We assume that $x \in X$ and we construct by induction a branch $\alpha \in {}^\omega\omega$ such that $x \in \bigcap_n \psi(\alpha|n) \subseteq \bigcap_n \varphi(\alpha|n)$. Set $\alpha(0) = \emptyset$. Suppose that $\alpha|k$ is already defined and $x \in \psi(\alpha|k)$. Then there exists an m such that $x \in U_m \cap \psi(\alpha|k)$ and $\mathrm{diam}(U_m) < 2^{-k}$. Thus

$$\psi(\alpha|k \frown i) = \psi(\alpha|k) \cap U_m$$

for some i. Set $\alpha(k) = i$. Evidently $x \in \bigcap_n \psi(\alpha|n) \subseteq \bigcap_n \varphi(\alpha|n)$. $\qquad \square$

Corollary [wAC] 5.6. *An uncountable Polish space contains a subset homeomorphic to the Cantor middle-third set.*

Corollary 5.7. *Any Polish space X is a continuous image of ${}^\omega\omega$.*

Proof. The assertion follows by Theorems 5.3, 5.5 and Corollary 3.8. $\qquad \square$

Corollary 5.8. *If a non-empty subset A of a Polish space X is sifted by a closed Souslin scheme with vanishing diameter, then A is a continuous image of ${}^\omega\omega$. Vice versa, if a set $A \subseteq X$ is a continuous image of ${}^\omega\omega$, then A is sifted by a closed Souslin scheme with vanishing diameter.*

Proof. Assume that $\langle T, \varphi \rangle$ is a closed Souslin scheme with vanishing diameter and such that $A = L(T, \varphi)$. By Theorem 5.3 there exists a continuous surjection $F : \mathrm{Ker}\,(T, \varphi) \xrightarrow{\mathrm{onto}} A$. However, $\mathrm{Ker}\,(T, \varphi)$ is a non-empty closed subset of ${}^\omega\omega$ and therefore a Polish space. Now, the assertion follows by Corollary 5.7.

The opposite implication follows by Lemma 5.2. $\qquad \square$

Corollary 5.9. *If a subset A of a Polish space X is sifted by a closed Souslin scheme with vanishing diameter, then A is also sifted by a closed Souslin scheme of the form $\langle {}^{<\omega}\omega, \varphi \rangle$ with vanishing diameter and non-empty values.*

Proof. The assertion follows by Corollary 5.8 and Lemma 5.2. $\qquad \square$

Theorem 5.10. *Every Polish space X with a clopen base is homeomorphic to a closed subset of the Baire space $^\omega\omega$.*

Proof. We slightly modify the proof of Theorem 5.5. Let $\{U_n : n \in \omega\}$ be a clopen base of X. We construct a regular Souslin scheme $\langle^{<\omega}\omega, \psi\rangle$ with vanishing diameter such that $X = L(^{<\omega}\omega, \psi)$. We set $\psi(\emptyset) = X$. Assume that $\psi(s)$ is defined for all $s \in {}^n\omega$. Consider the family $\{\psi(s) \cap (U_i \setminus \bigcup_{j<i} U_j) : \mathrm{diam}(U_i) < 2^{-n} \wedge i \in \omega\}$. If the family is infinite, we let $\langle\psi(s\frown i) : i \in \omega\rangle$ enumerate this family in a one-to-one manner. Otherwise enumerate the finite family in a one-to-one manner and let the other values be the empty set.

By Theorem 5.3, $\mathrm{Ker}\,(^{<\omega}\omega, \psi)$ is a closed subset of $^\omega\omega$ and one can check that the associated map $F : \mathrm{Ker}\,(^{<\omega}\omega, \psi) \longrightarrow X$ is a homeomorphism. $\qquad\square$

There is another tool equivalent to the Souslin scheme which is sometimes more convenient for investigation. For some technical reasons we order the set $\mathbb{D}$ of dyadic numbers (for a definition see Section 2.4) opposite to values of its members, i.e.,

$$x \preceq y \equiv y \leq x. \tag{5.7}$$

We set

$$\mathbb{D}_k = \left\{ x = \sum_{i=0}^{k} 2^{-n_i} : 0 < n_0 < n_1 < \cdots < n_k \right\}.$$

Then $\mathbb{D} = \bigcup_k \mathbb{D}_k$. For every k the set $\mathbb{D}_k$ is well ordered. Actually, if $A \subseteq \mathbb{D}_k$ is non-empty we choose step by step the smallest positive integers $n_0, \ldots, n_k$ such that $a = \sum_{i=0}^{k} 2^{-n_i} \in A$. Then a is the smallest element of A according to the order $\preceq$.

If $r = \sum_{i=0}^{l} 2^{-n_i}, s = \sum_{i=0}^{k} 2^{-m_i}$ are elements of $\mathbb{D}$, we define

$$r \sqsubseteq s \equiv ((l \leq k) \wedge (\forall i \leq l)\, n_i = m_i).$$

Then $\langle\mathbb{D}, \sqsubseteq\rangle$ is a poset. Evidently

$$(r \in \mathbb{D}_n \wedge s \in \mathbb{D}_m \wedge r \sqsubseteq s) \to (n \leq m \wedge r \leq s).$$

For any $s \in \mathbb{D}$ the set $\{r \in \mathbb{D} : r \sqsubseteq s\}$ is finite.

A **sieve** on X is a mapping $\Phi : \mathbb{D} \longrightarrow \mathcal{P}(X)$. A set $A \subseteq X$ is said to be **sifted** by the sieve Φ if

$$x \in A \equiv \text{ the poset } \langle\{r \in \mathbb{D} : x \in \Phi(r)\}, \preceq\rangle \text{ is not well ordered.}$$

We shall write $A = \mathrm{S}(\Phi)$. Since the poset $\langle\mathbb{D}, \preceq\rangle$ is linearly ordered, a set A is sifted by the sieve Φ if and only if

$$x \in A \equiv \text{there exists a strictly } <\text{-increasing sequence } \{r_n\}_{n=0}^{\infty}$$
$$\text{of elements of } \mathbb{D} \text{ such that } x \in \Phi(r_n) \text{ for each } n \in \omega. \tag{5.8}$$

A sieve Φ is said to be **closed** if the values $\Phi(r)$, $r \in \mathbb{D}$ are closed subsets of the topological space X. Similarly we shall speak about a **Borel sieve**, etc. A sieve Φ is called **monotone** if $\Phi(x) \subseteq \Phi(y)$ for any $y \sqsubseteq x$.

Lemma 5.11. *Let Φ be a monotone sieve. Then $x \in \mathrm{S}(\Phi)$ if and only if there exists a strictly $\sqsubseteq$-increasing sequence $s_0 \sqsubseteq s_1 \sqsubseteq \cdots \sqsubseteq s_n \sqsubseteq \cdots$ of elements of $\mathbb{D}$ such that $x \in \Phi(s_n)$ for every $n \in \omega$.*

Proof. Since $r \sqsubseteq s \to r < s$ the implication from right to left is trivial.

Assume that there exists a $<$-increasing sequence $\{r_n\}_{n=0}^{\infty}$ of elements of $\mathbb{D}$ such that $x \in \Phi(r_n)$ for every $n \in \omega$. Set $r = \sup\{r_n : n \in \omega\}$ and consider the infinite dyadic expansion[1]

$$r = \sum_{i=0}^{\infty} 2^{-m_i}, \qquad 0 < m_0 < m_1 < \cdots < m_n < \cdots .$$

For every $n \in \omega$ we let $s_n = \sum_{i=0}^{n} 2^{-m_i}$. We show that $s_n \sqsubseteq r_k$ for some k. Then $x \in \Phi(s_n)$.

Since $s_n < r$ there is a k such that $s_n < r_k < r$. Let $r_k = \sum_{i=0}^{l} 2^{-j_i}$, $j_0 < \cdots < j_l$. We show that $j_i = m_i$ for any $i \leq m = \min\{n, l\}$. Aiming to obtain a contradiction, assume that there exists a $p \leq m$ such that $j_p \neq m_p$. We assume that p is the smallest with this property. If $j_p < m_p$, then

$$r = \sum_{i<p} 2^{-m_i} + \sum_{i=p}^{\infty} 2^{-m_i} \leq \sum_{i<p} 2^{-m_i} + 2^{-m_p+1} \leq \sum_{i<p} 2^{-m_i} + 2^{-j_p} \leq r_k,$$

which is impossible. If $j_p > m_p$, then similarly

$$r_k = \sum_{i<p} 2^{-m_i} + \sum_{i=p}^{l} 2^{-j_i} \leq \sum_{i<p} 2^{-m_i} + 2^{-j_p+1} \leq \sum_{i<p} 2^{-m_i} + 2^{-m_p} = s_p \leq s_n,$$

again a contradiction. Thus $j_i = m_i$ for all $i \leq m$. Since $s_n < r_k$ we obtain $m = n$ and $l > n$. $\qquad\square$

If $r = \sum_{i=0}^{k} 2^{-n_i}$, $0 < n_0 < \cdots < n_k$, we let $\mathbf{b}(r) = \{n_i - n_{i-1} - 1\}_{i=0}^{k}$, where $n_{-1} = 0$. One can easily see that $\mathbf{b}$ is an order isomorphism of $\langle \mathbb{D}, \sqsubseteq \rangle$ onto $\langle {}^{<\omega}\omega \setminus \{\emptyset\}, \subseteq \rangle$. Especially,

$$r \sqsubseteq s \equiv \mathbf{b}(r) \subseteq \mathbf{b}(s). \tag{5.9}$$

Theorem 5.12. *A subset $A \subseteq X$ of a Polish space X is sifted by a monotone closed (open, clopen, Borel) Souslin scheme if and only if A is sifted by a monotone closed (open, clopen, Borel) sieve.*

---

[1] If r is a dyadic rational then $m_{i+1} = m_i + 1$ for all but finitely many i's.

Proof. Let a set $A \subseteq X$ be sifted by a monotone Souslin scheme $\langle {}^{<\omega}\omega, \varphi \rangle$. We set

$$\Phi(r) = \varphi(\mathbf{b}(r)) \tag{5.10}$$

and show that

$$L({}^{<\omega}\omega, \varphi) = \mathrm{S}(\Phi).$$

Assume that $x \in L({}^{<\omega}\omega, \varphi)$. Then there exists a branch $\{n_i\}_{i=0}^\infty \in [{}^{<\omega}\omega]$ such that $x \in \varphi(\{n_i\}_{i=0}^k)$ for every k. We let

$$r_k = \frac{1}{2^{n_0+1}} + \frac{1}{2^{n_0+n_1+2}} + \cdots + \frac{1}{2^{n_0+\cdots+n_k+k+1}}.$$

Then $\mathbf{b}(r_k) = \{n_i\}_{i=0}^k$, $\Phi(r_k) = \varphi(\{n_i\}_{i=0}^k)$. Since the sequence $\{r_k\}_{k=0}^\infty$ is strictly $<$-increasing, we obtain $x \in \mathrm{S}(\Phi)$.

Assume now that $x \in \mathrm{S}(\Phi)$. Then by Lemma 5.11, there exists a $\sqsubset$-increasing sequence $\{s_k\}_{k=0}^\infty$ such that $x \in \Phi(s_k)$ for every $k \in \omega$.

By (5.9), $\{\mathbf{b}(s_k)\}_{k=0}^\infty$ is a branch of ${}^\omega\omega$. Since $x \in \Phi(s_k) = \varphi(\mathbf{b}(s_k))$ for every k we obtain $x \in L({}^{<\omega}\omega, \varphi)$.

If Φ is a sieve we set $\varphi(v) = \Phi(\mathbf{b}^{-1}(v))$ for any $v \in {}^{<\omega}\omega, v \neq \emptyset$, $\varphi(\emptyset) = X$. Then (5.10) holds true and therefore $\mathrm{S}(\Phi) = L({}^{<\omega}\omega, \varphi)$. $\square$

Exercises

5.1 Topological Characterization of Cantor and Baire Space

a) ${}^\omega 2$ is up to homeomorphism the unique perfect compact zero-dimensional Polish space.
Hint: Construct a clopen regular Souslin scheme $\langle {}^{<\omega}2, \varphi \rangle$ with vanishing diameter such that $\bigcup_{s \in {}^n 2} \varphi(s) = X$ for each n.

b) If X is a zero-dimensional Polish space in which every compact subset has empty interior, then for any open set $U \subseteq X$ and any $\varepsilon > 0$ there are pairwise disjoint clopen sets $\langle U_n : n \in \omega \rangle$ of diameter less than ε such that $U = \bigcup_n U_n$.
Hint: The closure of U is not totally bounded.

c) ${}^\omega\omega$ is up to homeomorphism the unique zero-dimensional Polish space in which all compact subsets have empty interior.
Hint: Construct a clopen regular Souslin scheme $\langle {}^{<\omega}\omega, \varphi \rangle$ with vanishing diameter.

d) The set ${}^\omega\omega\!\uparrow \subseteq {}^\omega\omega$ of non-decreasing sequences of natural numbers is homeomorphic to the Baire space ${}^\omega\omega$.

5.2 ($\mathcal{A}$)-operation

The Souslin scheme was originally introduced as ($\mathcal{A}$)-**operation**. For any finite sequence $s = \langle n_0, \ldots, n_k \rangle$ of natural numbers a set F_s is given. The result of ($\mathcal{A}$)-operation is the set

$$\mathcal{A}_s F_s = \bigcup_{\alpha \in {}^\omega\omega} \bigcap_{k \in \omega} F_{\langle \alpha(0), \ldots, \alpha(k) \rangle}.$$

a) For a given sequence $\langle B_n : n \in \omega \rangle$ of subsets of X find sets $\langle F_s : s \in {}^{<\omega}\omega \rangle$ such that $\mathcal{A}_s F_s = \bigcup_n B_n$.

b) For a given sequence $\langle B_n : n \in \omega \rangle$ of subsets of X find sets $\langle F_s : s \in {}^{<\omega}\omega \rangle$ such that $\mathcal{A}_s F_s = \bigcap_n B_n$.

c) If $\varphi(s) = F_s$ for $s \in {}^{<\omega}\omega$ then $\mathcal{A}_s F_s = L({}^{<\omega}\omega, \varphi)$.

d) If $\mathcal{F} \subseteq \mathcal{P}(X)$ is a family of subsets of a set X we write

$$\mathcal{A}(\mathcal{F}) = \{\mathcal{A}_s F_s : F_s \in \mathcal{F} \text{ for every } \varepsilon \in {}^{<\omega}\omega\}.$$

Show that $\mathcal{A}(\mathcal{A}(\mathcal{F})) = \mathcal{A}(\mathcal{F})$.

e) In terminology and notations introduced in Section 7.1, Theorem 7.10 says: If a Polish ideal space $\langle X, \mathcal{I} \rangle$ possesses the hull property, then $\mathcal{A}(\text{BOREL}(X)) \subseteq \text{BOREL}(\mathcal{I})$.

5.3 κ-Souslin Scheme

Let T be a pruned subtree of ${}^{<\omega}\kappa$, where κ is a given regular cardinal. If $\varphi : T \longrightarrow \mathcal{P}(X)$ then the couple $\langle T, \varphi \rangle$ is called a κ-**Souslin scheme**. Similarly we can define related notions such as monotone, regular, vanishing diameter, $L(T, \varphi)$, sifted, etc. The topology on ${}^{\omega}\kappa$ is induced by the Baire metric.

a) If $f : {}^{\omega}\kappa \xrightarrow{\text{onto}} X$ is continuous, then there exists a closed monotone κ-Souslin scheme $\langle {}^{<\omega}\kappa, \varphi \rangle$ such that $X = L({}^{<\omega}\kappa, \varphi)$.

b) If $\langle T, \varphi \rangle$ is a κ-Souslin scheme with vanishing diameter, then there exists a continuous mapping $F : \text{Ker}\,(T, \varphi) \xrightarrow{\text{onto}} L(T, \varphi)$.

c) If in b) φ is closed, then $\text{Ker}\,(T, \varphi)$ is a closed subset of ${}^{\omega}\kappa$.

5.4 κ-Souslin Sets

A set $A \subseteq X$ is κ-**Souslin** if there exists a closed κ-Souslin scheme $\langle T, \varphi \rangle$ with vanishing diameter such that $A = L(T, \varphi)$.

a) If A is a κ-Souslin subset of a Polish space, $|A| > \kappa$ then A contains a perfect subset.

b) A subset A of a Polish space X is κ-Souslin if and only if it is a continuous image of ${}^{\omega}\kappa$.
   Hint: Use results of Exercises 5.3 and Corollary 3.8.

c) A continuous image of a κ-Souslin set is a κ-Souslin set.

d) A subset A of a Polish space X is κ-Souslin if and only if there exists a closed set $C \subseteq {}^{\omega}\kappa \times X$ such that $x \in A \equiv (\exists y)\,\langle y, x \rangle \in C$.
   Hint: Let $f : {}^{\omega}\kappa \longrightarrow X$ be continuous and such that $A = \text{rng}(f)$. Take C to be the graph of f.

e) Let B be a non-empty set. A subset $A \subseteq {}^{\omega}B$ is κ-Souslin if and only if there exists a pruned tree $T \subseteq {}^{<\omega}(\kappa \times B)$ such that $x \in A \equiv (\exists y)\,\langle y, x \rangle \in [T]$.
   Hint: Take C of d) and a tree T such that $C = [T]$.

5.5 Kleene-Brouwer ordering

The **Kleene-Brouwer ordering** $\leq_{KB}$ of ${}^{<\omega}\omega$ is defined as

$$x \leq_{KB} y \equiv (y \subseteq x \vee (\exists k)(\forall i < k)\,(x(i) = y(i) \wedge x(k) < y(k))).$$

See also Exercise 11.4. Let $\langle {}^{<\omega}\omega, \varphi \rangle$ be a Souslin scheme on a set X. Show that:

a) The bijection $\mathbf{b}$ is an isomorphism of $\langle \mathbb{D}, \preceq \rangle$ onto $\langle {}^{<\omega}\omega, \leq_{KB} \rangle$.

b) If $A = L({}^{<\omega}\omega, \varphi)$ then

$$x \in X \setminus A \equiv \langle \{ v \in {}^{<\omega}\omega : x \in \varphi(v) \}, \leq_{KB} \rangle \text{ is well ordered.}$$

c) Let $\langle {}^{<\omega}\omega, \varphi \rangle$ be a closed (or Borel) Souslin scheme on X. Set $A = L({}^{<\omega}\omega, \varphi)$. Using the Kleene-Brouwer ordering, construct a decomposition of $X \setminus A$ into ω_1 Borel sets.

d) Show that for suitable sieve Φ you obtain the same decomposition as in Theorem 6.46.

5.6 [wAC] Well-Founded Relation

Let $\prec \subseteq X \times X$ be an antireflexive transitive relation on a Polish space X. We define a set $T \subseteq {}^{<\omega}X$ as follows:

$$\langle x_0, \ldots, x_n \rangle \in T \equiv x_n \prec x_{n-1} \prec \cdots \prec x_0 \text{ for } n > 0.$$

We assume that $\emptyset \in T$ and $\langle x \rangle \in T$ for any $x \in X$.

a) Relation $\prec$ is well founded if and only if relation $\supset$ on T is well founded (see the definition of a well-founded relation in Exercise 11.4).
   Hint: Use results of Exercise 11.4.

b) Assume that $\prec$ is well founded. Let ρ be the rank function of $\prec$ and σ be the rank function of $\supset$ on T. Show that $\rho(x_n) = \sigma(\langle x_0, \ldots, x_n \rangle)$ for any $x_n \prec \cdots \prec x_0$.

c) The rank of $\prec$ is the σ-rank of $\emptyset \in T$.

Let κ be a cardinal. Assume that a well-founded relation $\prec$ on ${}^{\omega}\kappa$ is a κ-Souslin subset of ${}^{\omega}\kappa \times {}^{\omega}\kappa$. Let $S \subseteq {}^{<\omega}(\kappa \times \kappa \times \kappa)$ be a pruned tree such that

$$\langle x, y \rangle \in \prec \equiv (\exists z) \langle x, y, z \rangle \in [S]. \tag{5.11}$$

We define a subset $W \subseteq {}^{<\omega}S$ and sets $S(x, y) \subseteq {}^{<\omega}\kappa$ for $x, y \in {}^{\omega}\kappa$ as follows:

$$\langle \langle s_0, t_0, u_0 \rangle, \ldots, \langle s_n, t_n, u_n \rangle \rangle \in W \equiv (\langle s_i, t_i, u_i \rangle \in S \wedge s_i = t_{i+1} \text{ for } i < n),$$

$$u \in S(x, y) \equiv \langle x|n, y|n, u \rangle \in S \text{ for } u \in {}^{n}\kappa.$$

d) $S(x, y)$ is a subtree of ${}^{<\omega}\kappa$.

e) $x \prec y$ if and only if $[S(x, y)] \neq \emptyset$, i.e., if there exists an infinite branch of $S(x, y)$.

f) We define a relation $\sqsubset$ on W. Let $w^i = \langle \langle s_0^i, t_0^i, u_0^i \rangle, \ldots, \langle s_{n_i}^i, t_{n_i}^i, u_{n_i}^i \rangle \rangle \in W$ for $i = 1, 2$. We set $w^1 \sqsubset w^2$ if $n_1 < n_2$ and $\langle s_j^1, t_j^1, u_j^1 \rangle \subset \langle s_j^2, t_j^2, u_j^2 \rangle$ for every $j \leq n_1$. Show that $\sqsubset$ is well founded.

5.7 [wAC] Kunen-Martin Theorem

Let V be a subtree of ${}^{<\omega}\kappa$. A branch $\alpha \in [V]$ is said to be the **leftmost** branch if $\alpha(n) = \min\{\beta(n) : \beta \in V \wedge \alpha|n \subseteq \beta\}$ for each n.

Assume that $\prec$ is a well-founded relation on ${}^{\omega}\omega$ and (5.11) holds true.

a) For $x \prec y$ denote by $z_{x,y}$ the leftmost branch of the tree $S(x, y)$. For $x \in X$ set $h([x]) = \emptyset$ and for $\langle x_0, \ldots, x_n \rangle \in T$, $n > 0$ we denote by $h(\langle x_0, \ldots, x_n \rangle)$ the n-tuple

$$\langle \langle x_1|(n+1), x_0|(n+1), z_{x_1,x_0}|(n+1) \rangle, \ldots, \langle x_n|(n+1), x_{n-1}|(n+1), z_{x_n,x_{n-1}}|(n+1) \rangle \rangle.$$

Show that h is an increasing mapping from $T \setminus \{\emptyset\}$ into W.

b) Prove the **Kunen-Martin Theorem**: If $\prec$ is a κ-Souslin well-founded relation on ${}^{\omega}\kappa$ then the rank of $\prec$ is smaller than κ^+.
   Hint: Use the result of Exercise 5.6, c) and the fact that $|W| \leq \kappa$.

5.2   Pointclasses

We introduce the notion of a pointclass in a rather non-formal way. Thus the definition is not sufficiently exact, however we hope that it is sufficient for all applications. In this section by a space we shall always understand a Polish space.[2] Recall that by Theorem 3.30, a subset $A \subseteq X$ of a Polish space X with the subspace topology is a Polish space if and only if A is a G_δ subset of X.

The product $X \times Y$ is homeomorphic to $Y \times X$, thus in what follows we shall often not distinguish them without any commentary. However, sometimes the order is important, namely, we recall that if $A \subseteq X \times Y$ and $x \in X$, $y \in Y$ then a horizontal section of A and a vertical section of A are the sets

$$A^y = \{u \in X : \langle u, y \rangle \in A\}, \qquad A_x = \{v \in Y : \langle x, v \rangle \in A\},$$

respectively.

Consider a topological property $\varphi(A, X)$ of a subset A of a Polish space X. Set

$$\Gamma(X) = \{A \subseteq X : \varphi(A, X)\}.$$

A **pointclass** Γ is a class of all sets from $\Gamma(X)$, where X is a Polish space. We assume that a pointclass satisfies the following conditions:

1) if $f : X_1 \longrightarrow X_2$ is continuous and $A \in \Gamma(X_2)$, then $f^{-1}(A) \in \Gamma(X_1)$, for any Polish spaces X_1 and X_2;

2) the family $\Gamma(X)$ is closed under finite intersections and finite unions, i.e., for any $A_0, \ldots, A_k \in \Gamma(X)$ also $A_0 \cap \cdots \cap A_k \in \Gamma(X)$ and $A_0 \cup \cdots \cup A_k \in \Gamma(X)$;

3) $X \in \Gamma(X)$.

Those conditions immediately imply the basic properties of a pointclass.

Lemma 5.13. *Let Γ be a pointclass.*

a) *If $f : X_1 \xrightarrow[\text{onto}]{1-1} X_2$ is a homeomorphism, then*

$$A \in \Gamma(X_2) \equiv f^{-1}(A) \in \Gamma(X_1) \text{ for any } A \subseteq X_2.$$

b) *If a set $A \subseteq X$ with the subspace topology is a Polish space (i.e., if A is a G_δ subset of X), then*

$$\{A \cap B : B \in \Gamma(X)\} \subseteq \Gamma(A).$$

c) *If $A \in \Gamma(X \times Y)$, then $A^y \in \Gamma(X)$ and $A_x \in \Gamma(Y)$ for any $x \in X$ and any $y \in Y$.*

---

[2]Actually we can consider any reasonable class of topological space. However Polish spaces are more than enough for our purpose.

A pointclass Γ is called **hereditary** if for any Polish space X and for any G_δ subset $A \subseteq X$ we have

$$\Gamma(A) = \{A \cap B : B \in \Gamma(X)\}.$$

Let us suppose that $G_\delta(X) \subseteq \Gamma(X)$ for any X. Then Γ is hereditary if and only if $\Gamma(A) \subseteq \Gamma(X)$ for any Polish X and any G_δ subset $A \subseteq X$.

We begin with some examples. If $\varphi(A, X)$ means that "A is a Borel subset of X" then $\Gamma = \mathrm{BOREL}$ is the pointclass of all Borel subsets of Polish spaces. By Theorem 3.38 $\boldsymbol{\Sigma}^0_\xi$, $\boldsymbol{\Pi}^0_\xi$ and $\boldsymbol{\Delta}^0_\xi$ are pointclasses. They are hereditary for any $\xi > 0$. The class $\mathcal{K}$ of all compact subsets (or the class of all connected subsets) of Polish spaces is not a pointclass.

We define basic operations on pointclasses.

If Γ is a pointclass then the **dual pointclass** $\neg\Gamma$ consists of those subsets A of a Polish space X for which $X \setminus A \in \Gamma(X)$. Thus $\neg\Gamma(X) = \{X \setminus A : A \in \Gamma(X)\}$.

The **ambiguous part of** Γ is the pointclass $\Delta[\Gamma] = \Gamma \cap \neg\Gamma$. A pointclass Γ is called **self-dual** if $\Gamma = \neg\Gamma$. Thus Γ is self-dual if and only if $\Gamma = \Delta[\Gamma]$.

The pointclasses $\boldsymbol{\Sigma}^0[\Gamma]$ and $\boldsymbol{\Pi}^0[\Gamma]$ are defined as follows:

$$A \in \boldsymbol{\Sigma}^0[\Gamma](X) \equiv (\exists\{A_n\}_{n=0}^\infty \in {}^\omega\Gamma(X)) \left(A = \bigcup_n A_n \right),$$

$$A \in \boldsymbol{\Pi}^0[\Gamma](X) \equiv (\exists\{A_n\}_{n=0}^\infty \in {}^\omega\Gamma(X)) \left(A = \bigcap_n A_n \right).$$

We say that a pointclass Γ is **closed under countable unions** or **closed under countable intersections**, if $\boldsymbol{\Sigma}^0[\Gamma] = \Gamma$ or $\boldsymbol{\Pi}^0[\Gamma] = \Gamma$, respectively.

The pointclass BOREL is self-dual. The pointclass of all open subsets of Polish spaces is not self-dual. The ambiguous part of $\boldsymbol{\Sigma}^0_\xi$ is the pointclass $\boldsymbol{\Delta}^0_\xi$. One can easily see that

$$\boldsymbol{\Pi}^0_\xi = \neg\boldsymbol{\Sigma}^0_\xi, \quad \boldsymbol{\Sigma}^0_{\xi+1} = \boldsymbol{\Sigma}^0[\boldsymbol{\Pi}^0_\xi], \quad \boldsymbol{\Sigma}^0_\xi = \boldsymbol{\Sigma}^0[\bigcup_{\eta<\xi} \boldsymbol{\Sigma}^0_\eta] = \boldsymbol{\Sigma}^0[\bigcup_{\eta<\xi} \boldsymbol{\Pi}^0_\eta] \text{ for } \xi \text{ limit.}$$

Similarly

$$\boldsymbol{\Sigma}^0_\xi = \neg\boldsymbol{\Pi}^0_\xi, \quad \boldsymbol{\Pi}^0_{\xi+1} = \boldsymbol{\Pi}^0[\boldsymbol{\Sigma}^0_\xi], \quad \boldsymbol{\Pi}^0_\xi = \boldsymbol{\Pi}^0[\bigcup_{\eta<\xi} \boldsymbol{\Pi}^0_\eta] = \boldsymbol{\Pi}^0[\bigcup_{\eta<\xi} \boldsymbol{\Sigma}^0_\eta] \text{ for } \xi \text{ limit.}$$

Let X, Y be Polish spaces. A set $A \subseteq X$ is a **projection along** Y of a set $B \subseteq Y \times X$, denoted by $A = \exists^Y B$, if

$$A = \{x \in X : (\exists y \in Y) \langle y, x \rangle \in B\} = \bigcup_{y \in Y} B_y = \mathrm{proj}_2(B).$$

We denote by $\exists^Y[\Gamma](X)$ the set of all projections along Y of sets from $\Gamma(Y \times X)$. Similarly, we define the **coprojection along** Y of a set $B \subseteq Y \times X$ as

$$\forall^Y B = \{x \in X : (\forall y \in Y) \langle y, x \rangle \in B\} = \bigcap_{y \in Y} B_y.$$

Thus, $\forall^Y[\Gamma](X)$ is the set of all coprojections along Y of sets from $\Gamma(Y \times X)$. Since we do not distinguish the homeomorphic products $X \times Y$ and $Y \times X$, we shall often denote, e.g., $\exists^Y B = \{x \in X : (\exists y \in Y) \langle x, y \rangle \in B\}$ for a set $B \subseteq X \times Y$.

Note that neither $\exists^Y[\Gamma]$ nor $\forall^Y[\Gamma]$ must be a pointclass, since they need not be closed under finite intersections or finite unions, respectively.

Using the de Morgan laws, one can easily see that

$$\forall^Y[\Gamma] = \neg \exists^Y[\neg\Gamma].$$

Let $\mathcal{F}$ be a class of mappings from a Polish space to another one. A pointclass Γ is said to be **closed under inverse images of $\mathcal{F}$** if for any Polish spaces X, Y and any mapping $f : X \longrightarrow Y \in \mathcal{F}$, $f^{-1}(A) \in \Gamma(X)$ for every $A \in \Gamma(Y)$. Thus the condition 1) of the definition of a pointclass just says that a pointclass is closed under continuous inverse images. If $\mathcal{F}$ is the class of all Borel measurable mappings then Γ is said to be **closed under Borel inverse images**. E.g., the pointclass BOREL is closed under Borel inverse images. The pointclass of all open sets in a Polish space is not closed under Borel inverse images. Similarly, we say that a pointclass Γ is **closed under images of $\mathcal{F}$** if for any Polish spaces X, Y and any mapping $f : X \longrightarrow Y \in \mathcal{F}$, we have $f(A) \in \Gamma(Y)$ for any $A \in \Gamma(X)$. By definition, the pointclass of open sets is closed under inverse images of continuous mappings and closed under images of open mappings.

A set $U \subseteq Y \times X$ is said to be a **Y-universal set for $\Gamma(X)$** if $U \in \Gamma(Y \times X)$ and

$$A \in \Gamma(X) \equiv (\exists y \in Y) A = U_y$$

for any subset $A \subseteq X$. A pointclass Γ is **Y-parametrized** if for any Polish space X there exists a Y-universal set for $\Gamma(X)$.

Lemma 5.14. *Let Z be a Polish space such that $\Gamma(Z) = \neg\Gamma(Z)$. Then there exists no Z-universal set for $\Gamma(Z)$.*

Proof. Assume that $U \in \Gamma(Z \times Z)$ is a Z-universal set for $\Gamma(Z)$. Set

$$A = \{x \in Z : \langle x, x \rangle \notin U\}.$$

Since $\Gamma(Z) = \neg\Gamma(Z)$ and by the definition a pointclass is closed under inverse images of continuous mapping, we obtain $A \in \Gamma(Z)$. Thus there exists a $z \in Z$ such that $A = U_z$. Then

$$\langle z, y \rangle \in U \equiv y \in A \equiv \langle y, y \rangle \notin U,$$

for any $y \in Z$. For $y = z$ we obtain a contradiction. $\qquad\square$

Theorem 5.15. *A self-dual pointclass cannot be parametrized by any Polish space.*

Proof. Assume that Γ is self-dual and parametrized by a Polish space Z. Then there exists a Z-universal set for $\Gamma(Z)$, a contradiction with the lemma. $\qquad\square$

Theorem 5.16. *Let Y, Z be Polish spaces.*

a) *Assume that there exists a continuous $f : Z \xrightarrow{\text{onto}} Y$. If Γ is Y-parametrized, then Γ is Z-parametrized as well.*

b) *Assume that Γ is a hereditary pointclass and that there is an embedding $f : Z \xrightarrow{1-1} Y$ (f is a homeomorphism of Z onto $f(Z)$). If the pointclass Γ is Z-parametrized, then Γ is Y-parametrized as well.*

Proof. Let X be a Polish space.

a) Let $U \in \Gamma(Y \times X)$ be a Y-universal set for $\Gamma(X)$. We set

$$V = \{\langle z, x \rangle \in Z \times X : \langle f(z), x \rangle \in Y \times X\}.$$

Since the function F defined as $F(z, x) = \langle f(z), x \rangle$ is continuous, $V = F^{-1}(U)$, we obtain $V \in \Gamma(Z \times X)$. It is easy to see that $V_z = U_{f(z)}$ for any $z \in Z$. Hence, V is a Z-universal set for $\Gamma(X)$.

b) Let $U \in \Gamma(Z \times X)$ be a Z-universal set for $\Gamma(X)$. The function F defined by $F(z, x) = \langle f(z), x \rangle$ is an embedding of $Z \times X$ into $Y \times X$. It is easy to see that the set $F(U) \in \Gamma(f(Z) \times X)$ is an $f(Z)$-universal set for $\Gamma(X)$. Since Γ is hereditary, there is a set $V \in \Gamma(Y \times X)$ such that $F(U) = V \cap (f(Z) \times X)$. By Lemma 5.13, c), V is Y-universal for $\Gamma(X)$. $\square$

Theorem [wAC] 5.17. *Assume that Γ is a hereditary pointclass and X and Y are Polish spaces, Y uncountable. Then the following are equivalent:*

a) *$\Gamma(X)$ is $^\omega 2$-parametrized.*

b) *$\Gamma(X)$ is Y-parametrized.*

c) *$\Gamma(X)$ is $^\omega\omega$-parametrized.*

d) *$\Gamma(X)$ is $[0,1]$-parametrized.*

Proof. There exist continuous surjections $^\omega 2 \xrightarrow{\text{onto}} [0,1]$ and $^\omega\omega \xrightarrow{\text{onto}} Y$ (see proof of Theorem 3.3 and Corollary 5.7). Thus by Theorem 5.16 a) we have d)→a) and b)→c). Since there are embeddings $^\omega 2 \xrightarrow{1-1} Y$ and $^\omega\omega \xrightarrow{1-1} [0,1]$ by Theorem 5.16 b) we obtain a)→b) and c)→d). $\square$

A starting point for a positive result is the basic result.

Theorem 5.18. *The pointclass of all open subsets of Polish spaces is $^\omega 2$-parametrized.*

Proof. Let $\langle X, \mathcal{O} \rangle$ be a Polish space. Fix an enumeration $\{U_n : n \in \omega\}$ of a countable base of the topology $\mathcal{O}$. We define set $U \subseteq {}^\omega 2 \times X$ as

$$\langle \alpha, x \rangle \in U \equiv (\exists n)\,(\alpha(n) = 1 \wedge x \in U_n).$$

One can easily see that the set U is open and the set $\{U_\alpha : \alpha \in {}^\omega 2\}$ is exactly the set $\mathcal{O}$ of all open subsets of X. $\square$

Now we can go up in the hierarchies of pointclasses.

Theorem 5.19. *Assume that Z is a Polish space.*

a) *If Γ is Z-parametrized then $\neg\Gamma$ is also such.*

b) *Assume that the pointclass Γ is Z-parametrized. If Y is a Polish space then $\exists^Y[\Gamma]$ is also Z-parametrized.*

c) *Let $\mathbf{wAC}$ hold true. Assume that Γ is $^\omega A$-parametrized, where $A = \{0,1\}$ or $A = \omega$. If $|\Gamma(X)| \ll \mathfrak{c}$ for each Polish space X, then $\boldsymbol{\Sigma}^0[\Gamma]$ and $\boldsymbol{\Pi}^0[\Gamma]$ are $^\omega A$-parametrized as well.*

Proof. a) If $U \in \Gamma(Z \times X)$ is a Z-universal set for $\Gamma(X)$, then $(Z \times X) \setminus U$ is a Z-universal set for $\neg\Gamma(X)$.

b) Assume that $U \in \Gamma(Z \times (Y \times X))$ is a Z-universal set for $\Gamma(Y \times X)$. Then

$$V = \{\langle z, x \rangle \in Z \times X : (\exists y \in Y)\, \langle z, \langle y, x \rangle \rangle \in U\} \in \exists^Y[\Gamma](Z \times X)$$

is Z-universal for $\exists^Y[\Gamma](X)$.

c) Assume that A is $\{0,1\}$ or ω (or any non-empty countable set). The projection $\mathrm{Proj}_n : {}^\omega A \longrightarrow {}^\omega A$ was defined in (1.10) as $\mathrm{Proj}_n(\alpha)(k) = \alpha(\pi(n,k))$. We define $P_{n,X} : {}^\omega A \times X \longrightarrow {}^\omega A \times X$ by $P_{n,X}(\alpha, x) = \langle \mathrm{Proj}_n(\alpha), x \rangle$. Evidently $P_{n,X}$ is continuous and $(P_{n,X}^{-1}(U))_\alpha = U_{\mathrm{Proj}_n(\alpha)}$ for $\alpha \in {}^\omega A$.

Assume that the set U is an $^\omega A$-universal for $\Gamma(X)$. Then the set

$$V = \bigcup_n P_{n,X}^{-1}(U)$$

is $^\omega A$-universal for $\boldsymbol{\Sigma}^0[\Gamma](X)$. Actually, if $\alpha \in {}^\omega A$ then $V_\alpha = \bigcup_n U_{\mathrm{Proj}_n(\alpha)}$ belongs to $\boldsymbol{\Sigma}^0[\Gamma](X)$. Vice versa, if $B = \bigcup_n B_n$ and $B_n = U_{\alpha_n}$ then $B = V_\alpha$, where α is such that $\alpha_n = \mathrm{Proj}_n(\alpha)$ (we have used $\mathbf{wAC}$!). Then $V \in \boldsymbol{\Sigma}^0[\Gamma]({}^\omega A \times X)$.

The assertion for $\boldsymbol{\Pi}^0[\Gamma]$ follows by a). $\qquad\square$

A pointclass Γ has the **separation property** if for any Polish space X and for any sets $A_1, A_2 \in \Gamma(X)$, $A_1 \cap A_2 = \emptyset$ there exists a set $B \in \Delta[\Gamma](X)$ such that $A_1 \subseteq B$, and $A_2 \cap B = \emptyset$. A pointclass Γ has the **σ-separation property** if for any Polish space X and for any sequence $\{A_n\}_{n=0}^\infty$ of pairwise disjoint sets from $\Gamma(X)$ there exists a sequence $\{B_n\}_{n=0}^\infty$ of pairwise disjoint sets from $\Delta[\Gamma](X)$ such that $A_n \subseteq B_n$ for each n.

A pointclass Γ has the **reduction property** if for any Polish space X and for any sets $A_1, A_2 \in \Gamma(X)$, there exist sets $B_1, B_2 \in \Gamma(X)$ such that $B_1 \cap B_2 = \emptyset$, $B_1 \subseteq A_1$, $B_2 \subseteq A_2$ and $A_1 \cup A_2 = B_1 \cup B_2$. A pointclass Γ has the **σ-reduction property** if for any Polish space X and for any sequence $\{A_n\}_{n=0}^\infty$ of sets from $\Gamma(X)$ there exists a sequence $\{B_n\}_{n=0}^\infty$ of pairwise disjoint sets from $\Gamma(X)$ such that $B_n \subseteq A_n$ for each n and $\bigcup_n A_n = \bigcup_n B_n$.

Theorem 5.20.

a) *If a pointclass Γ has the reduction property, then the dual pointclass $\neg\Gamma$ has the separation property.*

b) *Let A be $\{0,1\}$ or ω. Assume that there exists an $^\omega A$-universal set for $\Gamma(^\omega A)$. Then Γ cannot have both the separation and reduction properties.*

Assume $\mathbf{wAC}$ *and* $|\Gamma(X)| \ll \mathfrak{c}$ *for any Polish space X.*

c) *If the pointclass Γ is closed under countable unions and has the separation property, then Γ has the σ-separation property.*

d) *If $\neg\Gamma \subseteq \mathbf{\Sigma}^0[\Gamma]$, then $\mathbf{\Sigma}^0[\Gamma]$ has the σ-reduction property.*

Proof. a) Assume that $A_1, A_2 \in \neg\Gamma(X)$, $A_1 \cap A_2 = \emptyset$. Then $X \setminus A_1, X \setminus A_2 \in \Gamma(X)$ and $(X \setminus A_1) \cup (X \setminus A_2) = X$. Thus there are $B_1, B_2 \in \Gamma(X)$ such that $B_1 \subseteq X \setminus A_1$, $B_2 \subseteq X \setminus A_2$, $B_1 \cap B_2 = \emptyset$ and $B_1 \cup B_2 = (X \setminus A_1) \cup (X \setminus A_2) = X$. Hence $B_2 \in \Delta[\Gamma](X)$, $A_1 \subseteq B_2$ and $A_2 \cap B_2 = \emptyset$.

b) Let U be an $^\omega A$-universal set for $\Gamma(^\omega A)$. We define two continuous functions $P_i : {}^\omega A \longrightarrow {}^\omega A$, $i = 1, 2$ as follows: if $a \in {}^\omega A$ then $P_i(a) = b_i$, where $b_1(n) = a(2n)$, $b_2(n) = a(2n+1)$. Then the function $P(a) = \langle P_1(a), P_2(a) \rangle$ is a homeomorphism from $^\omega A$ onto $^\omega A \times {}^\omega A$. Set $U_i = \{\langle a, x \rangle : \langle P_i(a), x \rangle \in U\}$, $i = 1, 2$. Evidently $U_1, U_2 \in \Gamma(^\omega A \times {}^\omega A)$.

Assume that Γ has both the reduction and separation properties. Then there are disjoint sets $B_1, B_2 \in \Gamma(^\omega A \times {}^\omega A)$ such that

$$B_1 \subseteq U_1, \ \ B_2 \subseteq U_2, \ \ U_1 \cup U_2 = B_1 \cup B_2. \tag{5.12}$$

Let $C \in \Delta[\Gamma](^\omega A \times {}^\omega A)$ be such that

$$B_1 \subseteq C \text{ and } B_2 \cap C = \emptyset. \tag{5.13}$$

We show that C is $^\omega A$-universal for $\Delta[\Gamma](^\omega A)$, contradicting Lemma 5.14. Actually, let $D \in \Delta[\Gamma](^\omega A)$. Then there are $a_1, a_2 \in {}^\omega A$ such that $D = U_{a_1}$ and $^\omega A \setminus D = U_{a_2}$. Let a be such that $P_1(a) = a_1$, $P_2(a) = a_2$. Then $D = (U_1)_a$ and $^\omega A \setminus D = (U_2)_a$. Since $(U_1)_a \cap (U_2)_a = \emptyset$, by (5.12) we obtain $D = (B_1)_a$ and $^\omega A \setminus D = (B_2)_a$. By (5.13) we have $D = C_a$.

c) Let $\{A_n\}_{n=0}^\infty$ be a sequence of pairwise disjoint sets belonging to $\Gamma(X)$. We set $F_n = \bigcup_{i \neq n} A_i$. Then $F_n \in \Gamma(X)$. By the separation property for any n there exists a set $E_n \in \Delta[\Gamma](X)$ such that $A_n \subseteq E_n$ and $E_n \cap F_n = \emptyset$. We define by induction $B_0 = E_0$ and $B_n = E_n \setminus \bigcap_{i < n} B_i$ for $n > 0$. Evidently $B_n \in \Delta[\Gamma](X)$, $n \in \omega$ are pairwise disjoint and $A_n \subseteq B_n$ for any n.

d) Let $A_n = \bigcup_k A_{n,k}$, where $A_{n,k} \in \Gamma(X)$. We set (π is a pairing function):

$$B_{n,k} = A_{n,k} \setminus \bigcup_{\pi(m,l) < \pi(n,k)} A_{m,l}.$$

The sets $B_{n,k} \in \mathbf{\Sigma}^0[\Gamma](X)$ are pairwise disjoint. Then the sets $B_n = \bigcup_k B_{n,k}$ are pairwise disjoint as well, $B_n \in \mathbf{\Sigma}^0[\Gamma](X)$, and $\bigcup_n B_n = \bigcup_n A_n$. $\square$

A mapping $\rho : A \longrightarrow \mathbf{On}$ is called a **rank**. A rank ρ is called **regular** if $\mathrm{rng}(\rho)$ is an ordinal. Two ranks ρ_1, ρ_2 on a set A are **equivalent** if for any $x, y \in A$,

$$\rho_1(x) \leq \rho_1(y) \equiv \rho_2(x) \leq \rho_2(y).$$

If ρ is a rank on a set A, one can define a prewell-ordering $\leq_\rho$ on A by

$$x \leq_\rho y \equiv \rho(x) \leq \rho(y). \tag{5.14}$$

The strict part $<_\rho$ of this prewell-ordering is defined as

$$x <_\rho y \equiv \rho(x) < \rho(y). \tag{5.15}$$

Note that if $x \leq_\rho y$ or $x <_\rho y$ then $x, y \in A$.

By definition two equivalent ranks define the same prewell-ordering. If R is a prewell-ordering on a set A, one can define by induction a regular rank σ such that $R = \leq_\sigma$. Especially, if ρ is a rank, then there exists a regular rank σ such that $\leq_\rho = \leq_\sigma$, thus

for every rank there exists an equivalent regular rank.

Assume that $A \subseteq X$, where X is a given set, and ρ is a rank on A. We define relations $\leq_\rho^*$, $<_\rho^*$ extending the prewell-ordering $\leq_\rho$ or $<_\rho$ as follows:

$$(\forall x, y \in X)\, (x \leq_\rho^* y \equiv (x \in A \wedge (y \notin A \vee (y \in A \wedge x \leq_\rho y)))), \tag{5.16}$$

$$(\forall x, y \in X)\, (x <_\rho^* y \equiv (x \in A \wedge (y \notin A \vee (y \in A \wedge x <_\rho y)))). \tag{5.17}$$

Lemma 5.21. *Let ρ be a rank on a set $A \in \Gamma(X)$. Then the following are equivalent:*

a) *Both relations $\leq_\rho^*$ and $<_\rho^*$ are in $\Gamma(X \times X)$.*
b) *There exist relations $\leq^+ \in \Gamma(X \times X)$ and $\leq^- \in \neg\Gamma(X \times X)$ such that*

$$(\forall x \in X)(\forall y \in A)\, (x \leq_\rho y \equiv x \leq^+ y), \tag{5.18}$$

$$(\forall x \in X)(\forall y \in A)\, (x \leq_\rho y \equiv x \leq^- y). \tag{5.19}$$

Proof. Assume that ρ is a rank and $\leq_\rho^*, <_\rho^* \in \Gamma(X \times X)$. If we define

$$x \leq^+ y \equiv x \leq_\rho^* y, \qquad x \leq^- y \equiv \neg y <_\rho^* x, \tag{5.20}$$

then $\leq^+ \in \Gamma(X \times X)$, $\leq^- \in \neg\Gamma(X \times X)$, and (5.18) and (5.19) hold true.

Assume now that there are $\leq^+ \in \Gamma(X \times X), \leq^- \in \neg\Gamma(X \times X)$ satisfying (5.18) and (5.19). Then by (5.16), (5.17), (5.18), (5.19) we obtain

$$(\forall x, y \in X)\, (x \leq_\rho^* y \equiv (x \in A \wedge (\neg y \leq^- x \vee x \leq^+ y))),$$

$$(\forall x, y \in X)\, (x <_\rho^* y \equiv (x \in A \wedge \neg y \leq^- x)).$$

Thus, $\leq_\rho^*, <_\rho^* \in \Gamma(X \times X)$. $\square$

Let Γ be a pointclass, X being a Polish space. A regular rank $\rho : A \longrightarrow \mathbf{On}$ on a subset A of X is called a Γ-**rank** if the relations $\leq_\rho^*, <_\rho^*$ are in $\Gamma(X \times X)$. If there exists a Γ-rank on A we say also that A **admits** a Γ-rank. The pointclass Γ is **ranked** if every $A \in \Gamma(X)$ admits a Γ-rank for each Polish space X.

Theorem 5.22. *If Γ is a ranked pointclass, then Γ has the reduction property.*

Proof. Let $A_1, A_2 \in \Gamma(X)$. Set

$$A = A_1 \times \{1\} \cup A_2 \times \{2\} \subseteq X \times \{1, 2\}.$$

One can easily see that $A \in \Gamma(X \times \{1, 2\})$ and therefore A admits a Γ-rank ρ. Set

$$B_1 = \{x \in X : \langle x, 1\rangle \leq_\rho^* \langle x, 2\rangle\}, \quad B_2 = \{x \in X : \langle x, 2\rangle <_\rho^* \langle x, 1\rangle\}.$$

Set $f(x) = \langle\langle x, 1\rangle, \langle x, 2\rangle\rangle$ for $x \in X$. Then f is continuous and $B_1 = f^{-1}(\leq_\rho^*)$. Since $\leq_\rho^* \in \Gamma(X \times X)$ we obtain that $B_1 \in \Gamma(X)$. Similarly for B_2.

By (5.16) and (5.17) we have $B_1 \cap B_2 = \emptyset$. Evidently $B_1 \cup B_2 \subseteq A_1 \cup A_2$. If $x \in A_1 \cup A_2$ then one can easily see that either $\langle x, 1\rangle \leq_\rho^* \langle x, 2\rangle$ and $x \in B_1$ or $\langle x, 2\rangle <_\rho^* \langle x, 1\rangle$ and $x \in B_2$. $\qquad\square$

Together with Theorem 5.20 we obtain

Corollary 5.23. *If there exists a $^\omega 2$-universal set for $\Gamma(^\omega 2)$ or an $^\omega\omega$-universal set for $\Gamma(^\omega\omega)$, then pointclasses Γ and $\neg\Gamma$ cannot be both ranked.*

Theorem 5.24 (Y.N. Moschovakis). *Let Y be a Polish space. If a ranked pointclass Γ is closed under $\forall^Y$, then $\exists^Y[\Gamma]$ is ranked, provided that $\exists^Y[\Gamma]$ is a pointclass at all.*[3]

Proof. Assume that X is a Polish space and $A \in \exists^Y[\Gamma](X)$. Then there exists a set $B \in \Gamma(X \times Y)$ such that

$$A = \{x \in X : (\exists y \in Y)\, \langle x, y\rangle \in B\}.$$

By assumption, there exists a Γ-rank ρ on B. We set

$$\tau(x) = \min\{\rho(x, y) : \langle x, y\rangle \in B\}.$$

Evidently, τ is a rank on the set A. One can easily see that

$$x_1 \leq_\tau^* x_2 \equiv (\exists y \in Y)(\forall z \in Y)\, \langle x_1, y\rangle \leq_\rho^* \langle x_2, z\rangle,$$
$$x_1 <_\tau^* x_2 \equiv (\exists y \in Y)(\forall z \in Y)\, \langle x_1, y\rangle <_\rho^* \langle x_2, z\rangle.$$

Thus $\leq_\tau^*, <_\tau^* \in \exists^Y[\forall^Y[\Gamma]](X \times X)$. Since Γ is closed under $\forall^Y$ we obtain that τ is an $\exists^Y[\Gamma]$-rank. $\qquad\square$

---

[3]i.e., if it is closed under finite intersections.

A sequence $\langle \rho_n : n \in \omega \rangle$ of ranks on a subset A of a Polish space X is called a **scale** on A if

for any convergent sequence $\{x_k\}_{k=0}^{\infty}$ of elements of A such that $(\forall n)(\exists m)(\forall k_1, k_2 \geq m)\, \rho_n(x_{k_1}) = \rho_n(x_{k_2})$, the limit $\lim_{k \to \infty} x_k \in A$ and $(\forall n)(\exists m)(\forall k \geq m)\, \rho_n(\lim_{k \to \infty} x_k) \leq \rho_n(x_k)$. $\quad (5.21)$

If, moreover, every rank ρ_n is a Γ-rank then $\langle \rho_n : n \in \omega \rangle$ is said to be a Γ-**scale**. If there exists a Γ-scale on A we say also that A **admits** a Γ-scale. The pointclass Γ is **scaled** if for every Polish space X every $A \in \Gamma(X)$ admits a Γ-scale.

Let $A \subseteq X \times Y$. We say that a set $B \subseteq A$ **uniformizes** the set A if

$$(\forall x \in X)\,((\exists y \in Y)\,\langle x, y \rangle \in A \equiv (\exists y \in Y)\,\langle x, y \rangle \in B), \quad (5.22)$$
$$(\forall x \in X)(\forall y_1, y_2 \in Y)\,((\langle x, y_1 \rangle \in B \wedge \langle x, y_2 \rangle \in B) \to y_1 = y_2). \quad (5.23)$$

Thus B is a function with domain $\mathrm{dom}(B) = \mathrm{dom}(A)$ and with graph lying in A. A pointclass Γ has the **uniformization property**, if for every Polish spaces X, Y and every set $A \in \Gamma(X \times Y)$ there exists a set $B \in \Gamma(X \times Y)$ which uniformizes the set A.

There is an important relationship between the notions we have introduced.

Lemma [wAC] 5.25 (The Uniformization Lemma). *If a set $A \in \Gamma(X \times {}^{\omega}\omega)$ admits a Γ-scale, then there exists a set $B \in \forall^{\omega}\omega[\Gamma](X \times {}^{\omega}\omega)$ that uniformizes A.*

Proof. Let $A \in \Gamma(X \times {}^{\omega}\omega)$, $\{\rho_n\}_{n=0}^{\infty}$ being a Γ-scale on A. Let $\leq_n^*$, $<_n^*$ denote the relations defined from the ranks ρ_n by (5.16) and (5.17). We define Γ-ranks $\langle \sqsubseteq_n^* : n \in \omega \rangle$ on A as the lexicographic ordering of $2n$-tuples

$$\langle \rho_0(\langle x, y \rangle), y(0), \rho_1(\langle x, y \rangle), y(1), \ldots, \rho_{n-1}(\langle x, y \rangle), y(n-1) \rangle.$$

Thus for any $x, u \in X$ and $y, v \in {}^{\omega}\omega$ we have

$$\langle x, y \rangle \sqsubset_n^* \langle u, v \rangle \equiv (\exists i < n)(\forall j < i)\,(\langle x, y \rangle \leq_j^* \langle u, v \rangle \wedge \langle u, v \rangle \leq_j^* \langle x, y \rangle$$
$$\wedge\; y(j) = v(j) \wedge (\langle x, y \rangle <_i^* \langle u, v \rangle$$
$$\vee\; (\langle x, y \rangle \leq_i^* \langle u, v \rangle \wedge \langle u, v \rangle \leq_i^* \langle x, y \rangle \wedge y(i) < v(i))))$$

and

$$\langle x, y \rangle \sqsubseteq_n^* \langle u, v \rangle \equiv (\langle x, y \rangle \sqsubset_n^* \langle u, v \rangle$$
$$\vee\; (\forall i < n)\,(\langle x, y \rangle \leq_i^* \langle u, v \rangle \wedge \langle u, v \rangle \leq_i^* \langle x, y \rangle \wedge y(i) = v(i)).$$

If $\langle x, y \rangle \in A$ and $\langle u, v \rangle \notin A$, then $\langle x, y \rangle <_0^* \langle u, v \rangle$. Hence $\langle x, y \rangle \sqsubset_n^* \langle u, v \rangle$ for any n as well.

Let B_n denote the set defined as

$$\langle x, y \rangle \in B_n \equiv (\forall z \in {}^{\omega}\omega)\,(\langle x, y \rangle \sqsubseteq_n^* \langle x, z \rangle).$$

Since $\sqsubseteq_{n+1}^* \subseteq \sqsubseteq_n^*$, we have $B_{n+1} \subseteq B_n \subseteq A$. Since $B_n \in \forall^{\omega}{}^{\omega}[\Gamma](X \times {}^{\omega}\omega)$ for each n, then also $B = \bigcap_n B_n$ belongs to $\forall^{\omega}{}^{\omega}[\Gamma](X \times {}^{\omega}\omega)$. We show that B uniformizes A.

Evidently $B \subseteq A$. By definition, if $\langle x, y_1 \rangle, \langle x, y_2 \rangle \in B_n$ then $y_1(i) = y_2(i)$ for each $i \leq n$. Thus, if $\langle x, y_1 \rangle, \langle x, y_2 \rangle \in B$ then $y_1 = y_2$.

We have to show that $\mathrm{dom}(A) \subseteq \mathrm{dom}(B)$. So assume that $x \in \mathrm{dom}(A)$. Then there exists a $y \in {}^{\omega}\omega$ such that $\langle x, y \rangle \in A$. We define

$$Y_0 = \{y \in {}^{\omega}\omega : \langle x, y \rangle \in A \wedge (\forall z)\,(\langle x, z \rangle \in A \rightarrow \langle x, y \rangle \sqsubseteq_0^* \langle x, z \rangle)\},$$
$$Y_{n+1} = \{y \in Y_n : \langle x, y \rangle \in A \wedge (\forall z)\,(\langle x, z \rangle \in B_n \rightarrow \langle x, z \rangle \sqsubseteq_{n+1}^* \langle x, z \rangle)\}.$$

One can easily show by induction that $Y_n \neq \emptyset$ for every n. Moreover, if $y \in Y_n$ then $\langle x, y \rangle \in B_n$.

For any $y_1, y_2 \in Y_n$ we have $\langle x, y_1 \rangle \sqsubseteq_n^* \langle x, y_2 \rangle$ and $\langle x, y_2 \rangle \sqsubseteq_n^* \langle x, y_1 \rangle$, especially $y_1(i) = y_2(i)$ for any $i \leq n$. Let $z(n)$ be the common value $y(n)$ for $y \in Y_n$. By **wAC** there exists a sequence $\langle z_n \in Y_n : n \in \omega \rangle$ such that $z_n | n = z | n$. Then $\lim_{n \to \infty} \langle x, z_n \rangle = \langle x, z \rangle$. Moreover, for $m > n$ we have $z_m \in Y_n$ and therefore $\rho_n(\langle x, z_m \rangle) = \rho_n(\langle x, z_n \rangle)$, i.e., the assumption of (5.21) is fulfilled. Since $\langle x, z \rangle \sqsubseteq_n^* \langle x, z_n \rangle$ and $\langle x, z_n \rangle \in B_n$ we obtain $\langle x, z \rangle \in B_n$. Then also $\langle x, z \rangle \in B$. $\square$

Lemma [wAC] 5.26. *Assume that Γ is a hereditary pointclass closed under $\forall^{\omega}{}^{\omega}$ and under injective continuous images. If for every Polish space X and for any set $A \in \Gamma(X \times {}^{\omega}\omega)$ there exists a set $B \in \Gamma(X \times {}^{\omega}\omega)$ which uniformizes A, then Γ has the uniformization property.*

Proof. Let $A \in \Gamma(X \times Y)$, X, Y being Polish spaces. By Theorem 6.1 there exist a closed set $C \subseteq {}^{\omega}\omega$ and a continuous bijection $f : C \xrightarrow[\mathrm{onto}]{1-1} Y$. Then the function F defined as $F(x, y) = [x, f(y)]$ is a continuous bijection $F : X \times C \xrightarrow[\mathrm{onto}]{1-1} X \times Y$. Since Γ is hereditary we obtain $A^* = F^{-1}(A) \in \Gamma(X \times C) \subseteq \Gamma(X \times {}^{\omega}\omega)$. By the assumptions there exists a set $B^* \in \Gamma(X \times {}^{\omega}\omega)$ uniformizing the set A^*. Then $B = F(B^*)$ uniformizes the set A. Since F is an injection, we have $B \in \Gamma(X \times Y)$. $\square$

From Lemmas 5.25 and 5.26 we immediately obtain:

Theorem [wAC] 5.27 (The Uniformization Theorem). *Assume that Γ is a hereditary pointclass closed under $\forall^{\omega}{}^{\omega}$ and under injective continuous images. If Γ admits a Γ-scale, then Γ has the uniformization property.*

Exercises

5.8 Γ-rank
Let Γ be a pointclass, $\Delta = \Gamma \cap \neg\Gamma$.

a) $\rho : A \xrightarrow{\text{onto}} \lambda$ is a Γ-rank if and only if there exist relations $\leq^\circ$ and $<^\circ$ in $\neg\Gamma(X \times X)$ satisfying

$$(\forall x \in X)(\forall y \in A)\,((x \leq_\rho y \equiv x \leq^\circ y) \wedge (x <_\rho y \equiv x <^\circ y)).$$

b) If $\rho : A \xrightarrow{\text{onto}} \lambda$ is a Γ-rank, $x \in A$, then $W = \{\langle y, z \rangle \in A^2 : \rho(y) < \rho(z) < \rho(x)\} \in \Delta$ is a prewell-ordering.

c) If $\rho : A \xrightarrow{\text{onto}} \lambda$ is a Γ-rank, then $A = \bigcup_{\xi < \lambda} A_\xi$ with $A_\xi \in \Delta$.
   Hint: Let $x \in A$ be such that $\rho(x) = \xi$. Set $A_\xi = \{y \in A : \rho(y) < \rho(x)\}$.

5.9 Very Good Scale
A scale $\langle \rho_n : n \in \omega \rangle$ on a set A is a **very good scale** if

1) $\rho_n(x) \leq \rho_n(y) \rightarrow \rho_m(x) \leq \rho_m(y)$ for $m < n$,
2) if $x_n \in A$ and for any n there exists an m_0 such that $\rho_n(x_{m_1}) = \rho_n(x_{m_2})$ for any $m_1, m_2 > m_0$, then there is an $x \in A$ such that $x = \lim_{n \to \infty} x_n$.

We assume that the pointclass Γ is closed under $\forall^{\omega}{}^{\omega}$ and under Borel isomorphisms.

a) Assume that $f : X \longrightarrow Y$ is continuous, f injective on $A = f^{-1}(B)$. If $\langle \rho_n : n \in \omega \rangle$ is a very good scale on B, then $\langle \sigma_n : n \in \omega \rangle$, where $\sigma_n(x) = \rho_n(f(x))$, is a very good scale on A.

b) If $A \in \Gamma(X \times {}^\omega\omega)$ admits a Γ-scale, then A admits a very good Γ-scale.
   Hint: The scale constructed in the proof of Lemma 5.25 is a very good Γ-scale.

c) Prove directly the Uniformization Theorem.
   Hint: If $\langle \rho_n : n \in \omega \rangle$ is a very good scale on $A \subseteq X \times Y$, then

$$B = \{\langle x, y \rangle \in A : (\forall n)(\forall z)\,(\langle x, z \rangle \in A \rightarrow \langle x, y \rangle \leq^*_{\rho_n} \langle x, z \rangle)\}$$

   uniformizes A.

5.10 Scale on $\exists^{\omega}{}^{\omega}[\Gamma]$
Let $B \in \Gamma(X \times {}^\omega\omega)$, $A = \{x \in X : (\exists y \in {}^\omega\omega)\,\langle x, y \rangle \in B\}$. Let $\langle \sigma_n : B \longrightarrow \kappa : n \in \omega \rangle$ be a Γ-scale on B. Let $D \subseteq B$, $D \in \Gamma$ uniformizes the set B.

a) Define $\tau_n(x) = \sigma_n(\langle x, y \rangle)$ if $x \in A$ and $\langle x, y \rangle \in D$. Show that τ_n is a scale on A.

b) If Γ is closed under $\forall^{\omega}{}^{\omega}$, then every τ_n is a $\exists^{\omega}{}^{\omega}[\Gamma]$-rank.
   Hint: Note that

$$x_1 \leq^*_{\tau_n} x_2 \equiv (\exists y_1, y_2 \in {}^\omega\omega)\,(\langle x_1, y_1 \rangle \in D \wedge \langle x_1, y_1 \rangle \leq^*_{\sigma_n} \langle x_2, y_2 \rangle)$$
$$\equiv (\forall y_1, y_2 \in {}^\omega\omega)\,(\langle x_1, y_1 \rangle \in D \wedge \langle x_1, y_1 \rangle \leq^*_{\sigma_n} \langle x_2, y_2 \rangle).$$

c) If Γ is closed under $\forall^{\omega}{}^{\omega}$, then $\langle \tau_n : n \in \omega \rangle$ is a $\exists^{\omega}{}^{\omega}[\Gamma]$-scale.

d) If a scaled pointclass Γ is closed under $\forall^{\omega}{}^{\omega}$, then the pointclass $\exists^{\omega}{}^{\omega}[\Gamma]$ is also scaled.

5.11 Boundedness Theorem
Assume that Γ is closed under $\forall^X$, $A \subseteq X$ and $\rho : A \longrightarrow \lambda$ is a Γ-rank. If $B \subseteq A$, $B \in \neg\Gamma$, $A \notin \neg\Gamma$, then there exists $x_0 \in A$ such that $\rho(y) < \rho(x_0)$ for each $y \in B$.
Hint: If not, then

$$x \in A \equiv (\exists y)\,(y \in B \wedge x \leq_\rho^* y).$$

5.12 Moschovakis Number
Let Γ be a pointclass, $\Delta = \Gamma \cap \neg\Gamma$. The ordinal

$$\delta(\Gamma) = \sup\{|W| : W \text{ is a strict prewell-ordering of a subset of } {}^\omega\omega \wedge W \in \Delta\}$$

is called the **Moschovakis Number** of Γ.

a) $\delta(\Gamma) = \delta(\neg\Gamma) = \delta(\Delta)$.

b) If $W \in \Delta$ is a prewell-ordering, then its strict part also belongs to Δ.

c) If $W \in \Delta$ is a prewell-ordering of a subset of ${}^\omega\omega$, then there exists a prewell-ordering $R \in \Delta$ of a subset of ${}^\omega\omega$ such that $\mathrm{ot}(R) = \mathrm{ot}(W) + 1$.

d) $\delta(\Gamma)$ is a limit ordinal.

e) If X is a perfect Polish space and Γ is closed under $\exists^{\omega}{}^\omega$ or $\forall^{\omega}{}^\omega$, then

$$\delta(\Gamma) = \sup\{|W| : W \text{ is a prewell-ordering} \wedge W \in \Delta(X)\}.$$

   Hint: Use Corollary 5.7.

f) If ρ is a Γ-rank, then $\mathrm{ot}(\rho) \leq \delta(\Gamma)$.
   Hint: Use the result of Exercise 5.8, b).

g) If every set of Γ is κ-Souslin, then $\delta(\Gamma) < \kappa^+$, provided that **wAC** holds true.
   Hint: Use the Kunen-Martin Theorem of Exercise 5.7, b).

5.3   Boolean Algebras

In this section we assume the Axiom of Choice **AC**.

A **Boolean algebra** $\mathbf{B} = \langle B, \vee, \wedge, -, 0, 1 \rangle$ is a set B equipped with two binary operations $\vee, \wedge$, one unary operation $-$, and two special elements $0, 1$ such that

1) $\vee, \wedge$ are associative and commutative,

2) $\vee, \wedge$ are mutually distributive, i.e., for any $x, y, z \in B$ we have

$$x \wedge (y \vee z) = (x \wedge y) \vee (x \wedge z), \quad x \vee (y \wedge z) = (x \vee y) \wedge (x \vee z), \quad (5.24)$$

3) both operations $\vee, \wedge$ are idempotent, i.e., $x \vee x = x$ and $x \wedge x = x$,

4) $x \vee 0 = x$, $x \wedge 1 = x$, $x \vee (-x) = 1$ and $x \wedge (-x) = 0$ for any $x \in B$.

It is easy to show that $x \vee (x \wedge y) = x$ and $x \wedge (x \vee y) = x$ for any $x, y \in B$. Moreover $x \vee y = y \equiv x \wedge y = x$. For $x, y \in B$ we write $x - y = x \wedge (-y)$.
If we define

$$x \leq y \equiv x \vee y = y \equiv x \wedge y = x, \quad (5.25)$$

then $\langle B, \leq \rangle$ is a poset, 0 is the least element, 1 is the greatest element, and for every $x, y \in B$, there exist $\sup\{x, y\} = x \vee y$, $\inf\{x, y\} = x \wedge y$ and the **complement** $-x$.

A Boolean algebra can be equivalently defined as a partially ordered set $\langle E, \leq \rangle$ with the least element 0, the greatest element 1 and such that every finite subset has a supremum and an infimum, the distributive laws (5.24) hold true and every element x has a complement.

A typical example of a Boolean algebra is an algebra $\mathcal{S}$ of subsets of a given set X – see Section 3.3. Then $\vee = \cup$, $\wedge = \cap$, $0 = \emptyset$, $1 = X$ and $-B = X \setminus B$. The algebra of sets $\mathrm{CO}(X, \mathcal{O})$ consisting of all clopen subsets of a topological space $\langle X, \mathcal{O} \rangle$ is an important example of such an algebra, since

Theorem [AC] 5.28 (M.H. Stone). *For any Boolean algebra* $\mathbf{B}$ *there exists, up to a homeomorphism, a unique compact topological space* $\langle X, \mathcal{O} \rangle$ *such that* $\mathrm{CO}(X, \mathcal{O})$ *is isomorphic to* $\mathbf{B}$ *and* $\mathrm{CO}(X, \mathcal{O})$ *is a base of the topology* $\mathcal{O}$.

Let $\mathbf{B} = \langle B, \vee, \wedge, -, 0, 1 \rangle$ be a Boolean algebra. Let us recall that elements $x, y \in B$ are disjoint if $x \wedge y = 0$. Let $A \subseteq B$ be an infinite set. If there exists a supremum $a = \sup A$ (in the ordering defined by (5.25)), then we shall denote it by $\bigvee A$. Especially, if $A = \{a_s : s \in S\}$, we shall write $a = \bigvee_{s \in S} a_s$. Similarly for the infimum $\bigwedge A = \inf A$ and $a = \bigwedge_{s \in S} a_s$.

Let κ be an uncountable regular cardinal. A Boolean algebra $\mathbf{B}$ is κ-**complete** if $\sup A$ and $\inf A$ exist for any $A \subseteq B$, $|A| < \kappa$. Especially, a Boolean algebra is called σ-**complete** if it is $\aleph_1$-complete. An algebra $\mathbf{B}$ is said to be **complete** if it is κ-complete for every infinite κ. Evidently, a Boolean algebra $\mathbf{B}$ is complete if and only if it is κ^+-complete, where $\kappa = |B|$. A κ-additive algebra of sets $\mathcal{S}$ (for definition see Section 3.3) is a κ-complete Boolean algebra in which $\bigvee \mathcal{A} = \bigcup \mathcal{A}$ and $\bigwedge \mathcal{A} = \bigcap \mathcal{A}$ for any $\mathcal{A} \subseteq \mathcal{S}$, $|\mathcal{A}| < \kappa$.

If $\langle X, \mathcal{O} \rangle$ is a topological space, then the set $\mathrm{RO}(X, \mathcal{O})$ of all regular open subsets of X ordered by the inclusion is a complete Boolean algebra. $\mathrm{RO}(X, \mathcal{O})$ need not be an algebra of sets. The Boolean operations are as follows:

$$-A = \mathrm{Int}(X \setminus A), \quad \bigvee \mathcal{A} = \mathrm{Int}(\overline{\bigcup \mathcal{A}}), \quad \bigwedge \mathcal{A} = \mathrm{Int}(\overline{\bigcap \mathcal{A}}).$$

Let $\mathbf{B_1} = \langle B_1, \vee_1, \wedge_1, -_1, 0_1, 1_1 \rangle$, $\mathbf{B_2} = \langle B_2, \vee_2, \wedge_2, -_2, 0_2, 1_2 \rangle$ be Boolean algebras. $\mathbf{B_1}$ is a **subalgebra** of $\mathbf{B_2}$ if $B_1 \subseteq B_2$, $0_1 = 0_2$, $1_1 = 1_2$, $x \vee_1 y = x \vee_2 y$, $x \wedge_1 y = x \wedge_2 y$, $-_1 x = -_2 x$ for any $x, y \in B_1$. Moreover, $\mathbf{B_1}$ is said to be a **complete subalgebra** of $\mathbf{B_2}$ if for every $A \subseteq B_1$ we have $x = \bigvee_2 A$, provided that $x = \bigvee_1 A$.

A Boolean algebra $\mathbf{B}$ satisfies the κ-**chain condition**, or simply is κ-**CC**, if for every subset $A \subseteq B$ of pairwise disjoint elements we have $|A| < \kappa$. Instead of $\aleph_1$-CC we speak about a **CCC** Boolean algebra, i.e., an algebra satisfying the **countable chain condition**. Let us notice the following: if $A = \{x_\xi : \xi < \lambda\}$ is a set of pairwise disjoint elements, then the set $\{\bigvee_{\eta \leq \xi} x_\eta : \xi < \lambda\}$ is a strictly increasing chain of length λ and conversely, if $\{y_\xi : \xi < \lambda\}$ is a strictly increasing chain, then $\{y_{\xi+1} - y_\xi : \xi < \lambda\}$ is a set of λ many pairwise disjoint elements.

A subset $I \subseteq B$ is called an **ideal** of **B** if

1) $0 \in I$, $1 \notin I$,

2) if $x \in I$, $y \leq x$, then $y \in I$,

3) if $x, y \in I$, then $x \vee y \in I$.

If $\mathcal{S}$ is an algebra of subsets of a set X and $\mathcal{I}$ is an ideal on X, then $\mathcal{I} \cap \mathcal{S}$ is an ideal of the Boolean algebra $\mathcal{S}$.

If I is an ideal of the Boolean algebra **B**, we can construct the **quotient algebra B**$/I$. We define an equivalence relation $\sim_I$ on B as

$$x \sim_I y \text{ if and only if } x - y \in I \text{ and } y - x \in I.$$

The universe B/I is the set of equivalence classes $B/\sim_I$. The operations are defined modulo the ideal I, e.g. (we abbreviate $\{x\}_I = \{x\}_{\sim_I}$)

$$\{x\}_I \vee \{y\}_I = \{z\}_I \text{ if and only if } (x \vee y) \sim_I z.$$

Similarly for $\wedge$ and $-$.

If $a \in B, a \neq 0_B$, then $I = \{x \in B : x \leq -a\}$ is an ideal and **B**$/I$ is isomorphic to the Boolean algebra $\{x \in B : x \leq a\}$. We shall denote it as **B**$|a$.

A mapping $h : B_1 \longrightarrow B_2$ is called a **homomorphism** of the Boolean algebra **B**$_1$ into **B**$_2$ if $h(x \vee_1 y) = h(x) \vee_2 h(y)$ and $h(-_1 x) = -_2 h(x)$ for any $x, y \in B_1$. Note that then also $h(x \wedge_1 y) = h(x) \wedge_2 h(y)$, $h(0_1) = 0_2$ and $h(1_1) = 1_2$. We shall write $h : $ **B**$_1 \longrightarrow$ **B**$_2$.

If $h : $ **B**$_1 \longrightarrow$ **B**$_2$ is a homomorphism, then the **kernel**

$$\mathrm{Ker}\,(h) = \{x \in B_1 : h(x) = 0\}$$

is an ideal. If $q : $ **B** $\longrightarrow$ **B**$/I$ is the quotient map defined by $q(x) = \{x\}_I$, then q is a homomorphism with the kernel $\mathrm{Ker}\,(q) = I$.

A subset $F \subseteq B$ is called a **filter** of **B** if

1) $0 \notin F$, $1 \in F$,

2) if $x \in F$, $y \geq x$, then $y \in F$,

3) if $x, y \in F$, then $x \wedge y \in F$.

Evidently a set F is a filter if and only if the set $I = \{x \in B : -x \in F\}$ is an ideal. Sometimes we prefer to speak about filters instead of ideals. If $\mathcal{S}$ is an algebra of subsets of a set X and $\mathcal{F}$ is a filter on X, then $\mathcal{F} \cap \mathcal{S}$ is a filter of the Boolean algebra $\mathcal{S}$. A filter F is maximal if and only if F is an **ultrafilter**, i.e., if

4) $x \in F$ or $-x \in F$ for every $x \in B$.

The axiom of choice implies that every filter can be extended to an ultrafilter. The notions of a filter and an ultrafilter defined in Section 1.1 are special cases of introduced notions in the case of Boolean algebra $\mathcal{P}(X)$.

An ideal I of a κ-complete Boolean algebra is called κ-**complete** if for any $A \subseteq I$ of cardinality $|A| < \kappa$ also $\bigvee A \in I$. If I is a κ-complete ideal in a κ-complete Boolean algebra $\mathbf{B}$, then the quotient algebra $\mathbf{B}/I$ is κ-complete.

Again, if $\mathcal{S}$ is a κ-additive algebra of subsets of a set X and $\mathcal{I}$ is a κ-additive ideal on X then $\mathcal{I} \cap \mathcal{S}$ is a κ-complete ideal of the κ-complete Boolean algebra $\mathcal{S}$.

An ideal I of a Boolean algebra $\mathbf{B}$ is said to be κ-**saturated** if every subset $A \subseteq B \setminus I$ such that $a \wedge b \in I$ for any $a, b \in A$, $a \neq b$, has cardinality $|A| < \kappa$. Note that a κ-saturated ideal is also λ-saturated for any $\lambda > \kappa$.

If a Boolean algebra $\mathbf{B}$ is κ-complete and κ-CC, then $\mathbf{B}$ is complete. However, if an ideal I is κ-saturated, then the quotient algebra $\mathbf{B}/I$ is κ-CC. Thus

Theorem [AC] 5.29. *If* $\mathbf{B}$ *is* κ-*complete,* I *is a* κ-*complete* κ-*saturated ideal, then* $\mathbf{B}/I$ *is a complete Boolean algebra.*

A partially ordered set $\langle X, \leq \rangle$ is **separative** if for any $x, y \in X$, $y \not\leq x$ there exists an element $z \leq y$, $z \neq 0_X$ and disjoint with x. If $\mathbf{B}$ is a Boolean algebra, then $\langle B, \leq \rangle$ is separative (take $z = y - x$).

Theorem [AC] 5.30. *Let* $\langle X, \leq \rangle$ *be a separative partially ordered set. Then there exists a complete Boolean algebra* $\mathbf{B} = \langle B, \vee, \wedge, -, 0, 1 \rangle$ *such that*

(i) $X \subseteq B$ *and* $\leq$ *agrees with the partial ordering of* $\mathbf{B}$,

(ii) *if any of the least and greatest elements* $0_X, 1_X$ *of* X *does exist, then it is identical with* $0, 1$ *of* $\mathbf{B}$, *respectively,*

(iii) X *is order dense in* B,

(iv) *if* $A \subseteq X$ *and the supremum* $a = \sup A \in X$ *exists, then* $a = \bigvee A$ *in* $\mathbf{B}$,

(v) *if* $A \subseteq X$ *and the infimum* $a = \inf A \in X$ *exists, then* $a = \bigwedge A$ *in* $\mathbf{B}$.

The Boolean algebra $\mathbf{B}$ *is unique up to isomorphism.*

Thus, for any Boolean algebra there exists, up to isomorphism, a unique complete Boolean algebra containing it as a complete dense subalgebra.

The Boolean algebra $\mathbf{B}$ is called a **completion** of $\langle X, \leq \rangle$ and will be denoted by r.o.$(X, \leq)$. If $\mathbf{X} = \langle X, \leq \rangle$ is a Boolean algebra, instead of r.o.$(X, \leq)$ we write comp$(\mathbf{X})$.

Note that a set $A \subseteq B \setminus \{0\}$ is predense if $\bigvee A = 1$. A Boolean algebra $\mathbf{B}$ is said to be κ-**distributive** if the intersection of any set of cardinality κ of open dense sets is a dense set. One can easily see that $\mathbf{B}$ is κ-distributive if and only if for any system $\langle A_\xi \subseteq B : \xi \in \kappa \rangle$ of predense sets there exists a predense set $A \subseteq B$ that is a common refinement of all A_ξ, $\xi \in \kappa$.

A partially ordered set X is said to be κ-**closed** if for every $\lambda \leq \kappa$, every non-increasing sequence $\langle x_\xi : \xi < \lambda \rangle$ of elements of $X \setminus \{0_X\}$ is bounded from below by a non-zero element.

Theorem [AC] 5.31. *Assume that there exists a dense* κ-*closed subset* $D \subseteq B$. *Then* $\mathbf{B}$ *is* κ-*distributive.*

The proof is easy. Let D be a dense κ-closed subset of B. Assume that $\langle A_\xi, \xi < \kappa \rangle$ are open dense subsets of B. Let $A = \bigcap_{\xi < \kappa} A_\xi$. We show that A is dense in B.

Let $a \in B$, $a \neq 0$. We define a non-increasing sequence $\langle a_\xi : \xi \in \kappa \rangle$ of elements of D such that $a_\xi \in A_\xi$ and $a_0 \leq a$. Since A_0 is open dense there exists an $a_0 \in A_0 \cap D$ such that $a_0 \leq a$. Assume that a_η are defined for $\eta < \xi$. Then there exists an element $a_\xi \in D \cap A_\xi$ such that $a_\xi \leq a_\eta$ for $\eta < \xi$.

Let b be a non-zero lower bound of the sequence $\{a_\xi\}_{\xi < \kappa}$. Then $b \in A$ and $b \leq a$. Thus A is dense. $\qquad\square$

An element $a \in B$ is called an **atom** if $a \neq 0$ and there is no $x \in B$ such that $0 < x < a$. A Boolean algebra **B** is **atomic** if for every $b \in B \setminus \{0\}$ there exists an atom $a \leq b$. It is easy to see a complete Boolean algebra **B** is atomic if and only if **B** is isomorphic to an algebra of sets of the form $\mathcal{P}(X)$. A Boolean algebra **B** is **atomless** if there is no atom in B.

Theorem 5.32. *Let* **B** *be a complete Boolean algebra. Then the following are equivalent:*

(1) **B** *is atomic and therefore isomorphic to an algebra of the form* $\mathcal{P}(X)$.

(2) **B** *is κ-distributive for any κ.*

(3) **B** *is $|B|$-distributive.*

Proof. Evidently $(1) \to (2)$ and $(2) \to (3)$. We show that $(3) \to (1)$. Actually take the set $\{\{a, -a\} : a \in B\}$ of dense subsets of B. By $|B|$-distributivity there exists a common refinement A of all $\{a, -a\}$, $a \in B \setminus \{0, 1\}$ such that $\bigvee A = 1$. Every element of A is an atom. $\qquad\square$

Usually it is more handy to investigate the above notions on a poset. Let $\langle X, \leq \rangle$ be a separative partially ordered set. Let $\mathcal{D}$ be a family of dense subsets of X. A set $G \subseteq X$ is called a $\mathcal{D}$-**generic filter** of X if

(i) $G \neq X$;

(ii) if $x \in G$ and $x \leq y$, then $y \in G$;

(iii) if $x, y \in G$, then there exists a $z \in G$ such that $z \leq x$ and $z \leq y$;

(iv) if $D \in \mathcal{D}$, then $D \cap G \neq \emptyset$.

Thus G is a $\mathcal{D}$-generic filter of a Boolean algebra **B**, if G is a filter and $D \cap G \neq \emptyset$ for each $D \in \mathcal{D}$. There is a classical result.

Theorem [AC] 5.33 (H. Rasiowa – R. Sikorski). *Suppose that* **B** *is a Boolean algebra, $a \in B$, $a \neq 0$. If $\mathcal{D}$ is a countable set of dense subsets of B, then there exists a $\mathcal{D}$-generic filter of* **B** *containing the element a.*

Proof. Let $\mathcal{D} = \{D_n : n \in \omega\}$. Since D_0 is a dense subset, there exists an $a_0 \neq 0$, $a_0 \in D_0$ such that $a_0 \wedge a \neq 0$. By induction, if $a_n \in D_n$ is defined, then there exists an $a_{n+1} \in D_{n+1}$ such that $a_{n+1} \wedge a_n \wedge \cdots \wedge a_0 \wedge a \neq 0$. The filter F defined as

$$x \in F \equiv (\exists n)\, (a_n \wedge \cdots \wedge a_0 \wedge a) \leq x$$

is the desired $\mathcal{D}$-generic filter of B. $\qquad\square$

Note that the condition of countability in the Rasiowa-Sikorski Theorem 5.33 is essential. Actually, take the set $X = {}^{<\omega}\omega_1$ ordered by the inverse inclusion $\supseteq$. Let $\mathbf{B} = \text{r.o.}(X, \supseteq)$. For any $\xi < \omega_1$ and any $m \in \omega$ we set

$$D_\xi = \{a \in X : (\exists n \in \text{dom}(a))\, a(n) = \xi\}, \quad \mathcal{D} = \{D_\xi : \xi < \omega_1\}.$$

Then D_ξ is a dense subset of B for every $\xi < \omega_1$.

We claim that there is no $\mathcal{D}$-generic filter. Actually, if F were a $\mathcal{D}$-generic filter of $\mathbf{B}$, then the set

$$\{\langle n, \xi \rangle : (\exists a \in F \cap X)\, a(n) = \xi\}$$

would be a surjection of a subset of ω onto ω_1.

However it turned out that passing to **CCC** Boolean algebras we can replace countable sets by larger ones.

We define the **Martin number** as

$$\mathfrak{m} = \min\{|\mathcal{D}| : (\exists \mathbf{B})\, ((\mathbf{B} \text{ is a complete CCC Boolean algebra})$$
$$\wedge\, (\mathcal{D} \text{ is a set of dense subsets of } B) \wedge (\text{no filter of } \mathbf{B} \text{ is } \mathcal{D}\text{-generic}))\}.$$

By the Rasiowa-Sikorski Theorem 5.33 we have $\aleph_0 < \mathfrak{m}$.

The Martin number $\mathfrak{m}$ can be characterized by many different tools. We need just to recall some well-known notions.

A poset $\langle X, \leq \rangle$ is **CCC** if any subset of pairwise disjoint elements is countable. Similarly, a topological space $\langle X, \mathcal{O} \rangle$ is **CCC** if any family of pairwise disjoint open sets is countable, i.e., if the poset $\langle \mathcal{O}, \subseteq \rangle$ is **CCC**.

Theorem [AC] 5.34. *Let κ be an infinite cardinal. Then the following are equivalent:*

a) $\kappa < \mathfrak{m}$.

b) *If $\mathbf{B}$ is a CCC complete Boolean algebra, $a \in B$, $a \neq 0$, $\mathcal{D}$ is a set of dense subsets of B, $|\mathcal{D}| \leq \kappa$ then there exists a $\mathcal{D}$-generic filter F of $\mathbf{B}$ such that $a \in F$.*

c) *If $\langle X, \leq \rangle$ is a CCC separative partially ordered set $\mathcal{D}$ is a set of dense subsets of X, $|\mathcal{D}| \leq \kappa$ then there exists a $\mathcal{D}$-generic filter F of X.*

d) *If $\langle X, \leq \rangle$ is a CCC separative partially ordered set, $a \in X$, $a \neq 0_X$, $\mathcal{D}$ is a set of dense subsets of X, $|\mathcal{D}| \leq \kappa$ then there exists a $\mathcal{D}$-generic filter F of X such that $a \in F$.*

e) *If $\langle X, \mathcal{O} \rangle$ is a CCC compact topological space, then X is not a union of $\leq \kappa$ meager subsets.*

f) *If $\langle X, \mathcal{O} \rangle$ is a CCC compact topological space, then intersection of κ open dense sets is a dense subset of X.*

Proof. The equivalence of a) and c) follows by Theorem 5.30. Similarly for b) and d). e) and f) are trivially equivalent. Also the implications a) $\rightarrow$ b) and c) $\rightarrow$ d) are trivial.

Assume d). If $\langle X, \mathcal{O} \rangle$ is a **CCC** compact topological space then $\mathrm{RO}(X, \mathcal{O})$ is a **CCC** complete Boolean algebra. Assume that $\langle A_\xi : \xi < \kappa \rangle$ are open dense subsets of X. Let U be a non-empty open set. We show that $U \cap \bigcap_{\xi < \kappa} A_\xi \neq \emptyset$. Actually the family

$$\mathcal{A}_\xi = \{A \in \mathrm{RO}(X, \mathcal{O}) : \overline{A} \subseteq A_\xi\}$$

is dense in $\mathrm{RO}(X, \mathcal{O})$. There exists a non-empty $V \in \mathrm{RO}(X, \mathcal{O})$ such that $\overline{V} \subseteq U$. Let $\mathcal{F}$ be a $\{\mathcal{A}_\xi : \xi < \kappa\}$-generic filter containing the set V. Since any finite intersection of closed sets from $\mathcal{F}$ is non-empty, by Theorem 1.28, a) there exists a point p such that $p \in \overline{B}$ for any $B \in \mathcal{F}$. It is easy to see that $p \in U \cap \bigcap_{\xi < \kappa} A_\xi$.

Suppose that f) holds true and **B** is a **CCC** complete Boolean algebra. By Stone's Theorem 5.28 there exists a compact topological space $\langle X, \mathcal{O} \rangle$ such that $\mathrm{CO}(X, \mathcal{O})$ is isomorphic to **B** and $\mathrm{CO}(X, \mathcal{O})$ is a base of the topology $\mathcal{O}$. A dense subset of B is represented by a dense subset of X and the point of X in the intersection of corresponding dense sets determines a generic filter. $\qquad\square$

Let us consider the family $\mathcal{P}(\omega)$ of all subsets of ω. $\mathcal{P}(\omega)$ is a Boolean algebra. The set

$$\mathrm{Fin} = \{A \subseteq \omega : |A| < \aleph_0\} = [\omega]^{<\omega} \tag{5.26}$$

of all finite subsets of ω is an ideal of algebra $\mathcal{P}(\omega)$. So we can consider the quotient algebra $\mathcal{P}(\omega)/\mathrm{Fin}$. Every infinite subset of ω uniquely determines an element of $\mathcal{P}(\omega)/\mathrm{Fin}$. We describe some important properties of elements of $\mathcal{P}(\omega)/\mathrm{Fin}$ by properties of corresponding subsets of ω. Actually, we can do it for any infinite set X, i.e., we can describe some properties of elements of $\mathcal{P}(X)/[X]^{<\omega}$ by properties of corresponding infinite subsets of X.

So, let X be a fixed infinite set. For $A, B \in [X]^{\geq \omega}$, we say that A is **almost contained** in B, denoted $A \subseteq^* B$, provided that $A \setminus B$ is finite. A is **almost equal** to B, written $A =^* B$, if both A is almost contained in B and B is almost contained in A, i.e., if the set $(A \setminus B) \cup (B \setminus A)$ is finite. If $A \subseteq^* B$ and $\neg A =^* B$, we shall write $A \subset^* B$. Similarly, we say that A and B are **almost disjoint** provided that the intersection $A \cap B$ is finite. A family $\mathcal{F} \subseteq [X]^{\geq \omega}$ is said to be an **almost disjoint family** if any two different elements of $\mathcal{F}$ are almost disjoint. A family $\mathcal{F} \subseteq [X]^{\geq \omega}$ has the **finite intersection property**, shortly **f.i.p.**, if every finite subset of $\mathcal{F}$ has an infinite intersection. Finally, an infinite set $A \subseteq X$ is a **pseudointersection** of $\mathcal{F}$, if A is almost contained in each $F \in \mathcal{F}$. Note that a pseudointersection is not unique at all.

In the case of $X = \omega$ (or more generally, if X is an infinite countable) one can easily see that $A \subseteq^* B$ if and only if $A \leq B$ in the quotient Boolean algebra $\mathcal{P}(\omega)/\mathrm{Fin}$ (or in the quotient algebra $\mathcal{P}(X)/[X]^{<\omega}$). Similarly "almost disjoint" means disjoint in $\mathcal{P}(\omega)/\mathrm{Fin}$ (or in $\mathcal{P}(X)/[X]^{<\omega}$) and "finite intersection property" means that intersection of any finite subset of $\mathcal{F}$ is non-zero in $\mathcal{P}(\omega)/\mathrm{Fin}$, or equivalently, that $\mathcal{F}$ is a base of a filter of $\mathcal{P}(\omega)/\mathrm{Fin}$ (similarly for X). It is easy to see that a filter $\mathcal{F}$ on ω is induced by a filter of $\mathcal{P}(\omega)/\mathrm{Fin}$ if and only if $\mathcal{F}$ is free.

Finally, a family $\mathcal{F} \subseteq [\omega]^\omega$ has f.i.p. if and only if $\mathcal{F}$ is a base of a filter extending the filter Fin.

Theorem 5.35. *There exists a family of almost disjoint subsets of ω of cardinality $\mathfrak{c}$.*

Proof. One can easily see that it is enough to construct such a family of subsets of any infinite countable set. So, consider the set $^{<\omega}\omega = \bigcup_{n=0}^{\infty} {}^n\omega$. Evidently $|^{<\omega}\omega| = \aleph_0$.

For any $\alpha \in {}^\omega\omega$ we set

$$F(\alpha) = \{s \in {}^{<\omega}\omega : s \subseteq \alpha\}. \tag{5.27}$$

If $\alpha, \beta \in {}^\omega\omega$, $\alpha \neq \beta$, then $|F(\alpha) \cap F(\beta)| < \aleph_0$. Thus $\mathcal{A} = \{F(\alpha) : \alpha \in {}^\omega\omega\}$ is an almost disjoint family of infinite subsets of $^{<\omega}\omega$ and $|\mathcal{A}| = 2^{\aleph_0}$. □

Lemma 5.36. *If $\{A_n\}_{n=0}^{\infty}$ is a $\subseteq^*$-decreasing sequence of infinite subsets of ω, then*

   a) *there are infinite subsets $B_n \subseteq A_n$ such that $B_{n+1} \subseteq B_n$ for every n,*
   b) *there is an infinite set $A \subseteq \omega$ such that $A \subseteq^* A_n$ for every n.*

Proof. One can simply set $B_n = \bigcap_{i=0}^{n} A_i$ and obtain assertion a).

Since every B_n is infinite one can construct an increasing sequence $\{a_n\}_{n=0}^{\infty}$ of natural numbers such that $a_n \in B_n$. Evidently $A = \{a_i : i \in \omega\} \subseteq^* B_n \subseteq A_n$ for every n. □

We denote by $\mathfrak{k}$ the cardinality

$$\mathfrak{k} = |\mathcal{P}(\omega)/\text{Fin}|. \tag{5.28}$$

Theorem 5.37. *The inequalities $2^{\aleph_0} \leq \mathfrak{k}$, $\mathfrak{k} \ll 2^{\aleph_0}$ hold true. Moreover, if the set $\mathcal{P}(\omega)$ can be well ordered, then $\mathfrak{k} = 2^{\aleph_0}$.*

Proof. The former inequality follows by Theorem 5.35 and the later one follows by definitions. □

A family $\mathcal{A}$ of almost disjoint subsets of ω is called a **maximal almost disjoint family**, shortly a MAD family, if for every infinite $B \subseteq \omega$ there is a set $A \in \mathcal{A}$ such that $A \cap B$ is infinite, or equivalently, there is no larger almost disjoint family. By the Zorn Lemma, Theorem 1.10 every almost disjoint family is contained in a MAD family. A MAD family is a partition of Boolean algebra $\mathcal{P}(\omega)/\text{Fin}$. If $\mathcal{A}_1$, $\mathcal{A}_2$ are MAD families, then $\mathcal{A}_1$ is a **refinement** of $\mathcal{A}_2$ if for every $A_1 \in \mathcal{A}_1$ there is an $A_2 \in \mathcal{A}_2$ such that $A_1 \subseteq^* A_2$. Evidently, any two MAD's $\mathcal{A}_1$, $\mathcal{A}_2$ have a common refinement

$$\{A_1 \cap A_2 : A_1 \in \mathcal{A}_1 \wedge A_2 \in \mathcal{A}_2 \wedge (A_1 \cap A_2 \text{ is infinite})\}.$$

Theorem [AC] 5.38.

   a) *Any countable family of dense subsets of $\mathcal{P}(\omega)/\text{Fin}$ has a common refinement. Hence, $\mathcal{P}(\omega)/\text{Fin}$ is $\aleph_0$-distributive.*
   b) *Any countable family of MAD families has a common refinement which is a MAD family.*

Proof. Assume that $\{\mathcal{A}_n : n \in \omega\}$ is a countable family of dense subsets. We can assume that every $\mathcal{A}_{n+1}$ is a refinement of $\mathcal{A}_n$. We define

$$\mathcal{A} = \{A \in [\omega]^\omega : (\forall n)(\exists A_n \in \mathcal{A}_n)\, A \subseteq^* A_n\}.$$

It is easy to see that $\mathcal{A}$ is a common refinement of all $\mathcal{A}_n$.

If every $\mathcal{A}_n$ is a MAD family, then $\mathcal{A}$ is an almost disjoint family. We need to show that $\mathcal{A}$ is a MAD family.

Let $B \in [\omega]^\omega$. Since every $\mathcal{A}_n$ is a MAD family, there are sets $B_n \in \mathcal{A}_n$ such that $B \cap B_n$ is infinite. Moreover, $B_{n+1} \subseteq^* B_n$ for every n. By Lemma 5.36, b) there exists an infinite set A such that $A \subseteq^* B$. Then $A \in \mathcal{A}$. $\square$

Let X be an infinite set, $|X| = \kappa$. If $A \subseteq X$, then we write $(1)A = A$ and $(0)A = X \setminus A$. A family $\mathcal{F} \subseteq \mathcal{P}(X)$ is called an **independent family** on X if for any finite $\mathcal{A} \subseteq \mathcal{F}$ and any mapping $f : \mathcal{A} \longrightarrow \{0,1\}$ we have

$$\left| \bigcap_{A \in \mathcal{A}} (f(A))A \right| = \kappa.$$

Theorem [AC] 5.39. *For every infinite set X there exists an independent family on X of cardinality $2^{|X|}$.*

Proof. Let $\kappa = |X|$. We write

$$Y = \{\langle A, f \rangle : A \in [\kappa]^{<\omega} \wedge f : \mathcal{P}(A) \longrightarrow \{0,1\}\}.$$

Since $|Y| = \kappa$ it suffices to find an independent family on Y.

We set $\mathcal{F} = \{\mathrm{C}_B : B \subseteq \kappa\}$, where $\mathrm{C}_B = \{\langle A, f \rangle \in Y : f(A \cap B) = 1\}$. Since $\mathrm{C}_{B_1} \neq \mathrm{C}_{B_2}$ for $B_1 \neq B_2$, we obtain $|\mathcal{F}| = 2^\kappa = 2^{|X|}$.

We show that $\mathcal{F}$ is an independent family. Assume that $\mathrm{C}_{B_0}, \ldots, \mathrm{C}_{B_n} \in \mathcal{F}$ are mutually different and $g : \{0, \ldots, n\} \longrightarrow 2$. For a finite set $A \subseteq \kappa$ satisfying

$$A \cap B_i \neq A \cap B_j \text{ for every } i, j \leq n, i \neq j, \tag{5.29}$$

we define $h : \mathcal{P}(A) \longrightarrow \{0,1\}$ as follows. For any $D \subseteq A$ we set

$$h(D) = \begin{cases} g(i) & \text{if } D = A \cap B_i, \\ 0 & \text{otherwise.} \end{cases}$$

Then $\langle A, h \rangle \in \bigcap_{i=0}^n (g(i))\mathrm{C}_{B_i}$. Since there exists κ many sets A satisfying condition (5.29) we obtain $|\bigcap_{i=0}^n (g(i))\mathrm{C}_{B_i}| = \kappa$. $\square$

Theorem [AC] 5.40 (B. Pospíšil). *For every infinite set X there exist $2^{2^{|X|}}$ many ultrafilters on X.*

Proof. Let $\mathcal{F}$ be an independent family on X of cardinality $2^{|X|}$. For any function $f : \mathcal{F} \longrightarrow 2$ there exists an ultrafilter containing the sets $\{(f(A))A, A \in \mathcal{F}\}$. For different f's the ultrafilters are different. $\square$

Exercises

5.13 [AC] Direct Sum

$\mathbf{B} = \langle B, \vee, \wedge, -, 0, 1 \rangle$ is a Boolean algebra.

a) If $a \in B$, $a \neq 0_B$, then the poset $B|a = \{x \in B : x \leq a\}$ is a Boolean algebra $\mathbf{B}|a$ with operations $\vee, \wedge, -x = a - x, 0_{B|a} = 0_B, 1_{B|a} = a$.

b) $\mathbf{B}|a$ is isomorphic to the quotient algebra $\mathbf{B}/I$, where $I = \{x \in B : x \wedge a = 0\}$.

c) If $\mathbf{B}$ is complete, then $\mathbf{B}|a$ is complete.

A Boolean algebra $\mathbf{B}$ is a **direct sum** of Boolean algebras $\langle \mathbf{B}_s : s \in S \rangle$ if there exists a partition $\langle a_s : s \in S \rangle$ of B such that every $\mathbf{B}_s$ is isomorphic to $\mathbf{B}|a_s$ for each $s \in S$.

d) Any non-atomless complete Boolean algebra is a direct sum of a complete atomic Boolean algebra and a complete atomless Boolean algebra.

5.14 [AC] Filter and Ultrafilter

$\mathbf{B} = \langle B, \vee, \wedge, -, 0, 1 \rangle$ is a Boolean algebra.

a) An ultrafilter is a maximal filter.

b) For any filter F of B and any $a \in B$, $-a \notin F$, there exists a maximal filter extending F and containing a.

   Hint: Apply the Zorn Lemma, Theorem 1.10.

c) A filter F is maximal if and only if F is an ultrafilter.

d) If F is an ultrafilter of a Boolean algebra B, then for any $a, b \in B$ we have

$$a \vee b \in F \equiv (a \in F \vee b \in F), \ a \wedge b \in F \equiv (a \in F \wedge b \in F), \ -a \in F \equiv \neg(a \in F).$$

5.15 [AC] Stone Representation

Let $\mathbf{S}(\mathbf{B})$ the set of all ultrafilters of a Boolean algebra $\mathbf{B}$. Set $\mathfrak{s}(a) = \{F \in \mathbf{S}(\mathbf{B}) : a \in F\}$ for any $a \in B$.

a) $\{\mathfrak{s}(a) : a \in B\}$ is an algebra of sets.

b) The mapping $\mathfrak{s}$ is an isomorphism of the Boolean algebra $\mathbf{B}$ to the algebra of sets $\{\mathfrak{s}(a) : a \in B\}$.

c) If $\mathcal{O}(\mathbf{B})$ is the topology on $\mathbf{S}(\mathbf{B})$ with the base $\{\mathfrak{s}(a) : a \in B\}$, then $\langle \mathbf{S}(\mathbf{B}), \mathcal{O}(\mathbf{B}) \rangle$ is a compact zero-dimensional topological space.

d) A set $U \subseteq \mathbf{S}(\mathbf{B})$ is clopen (in topology $\mathcal{O}(\mathbf{B})$) if and only if there exists an element $a \in B$ such that $U = \mathfrak{s}(a)$.

e) Let $A \subseteq B$, $A \neq \emptyset$, $a \in B$. Show that $a = \bigvee A$ if and only if $\overline{\bigcup_{x \in A} \mathfrak{s}(x)} = \mathfrak{s}(a)$. Similarly, $a = \bigwedge A$ if and only if $\text{Int}(\bigcap_{x \in A} \mathfrak{s}(x)) = \mathfrak{s}(a)$.

f) Let $\mathbf{B}_i$, $i = 1, 2$ be Boolean algebras, $h : \mathbf{B}_1 \longrightarrow \mathbf{B}_2$ being a homomorphism. We set $H(F) = h^{-1}(F)$ for any ultrafilter F of B_2. Show that $H : \mathbf{S}(\mathbf{B}_2) \longrightarrow \mathbf{S}(\mathbf{B}_1)$ is continuous.

g) H is one-to-one if and only if h is a surjection. H is a surjection if and only if h is one-to-one.

h) Boolean algebra $\mathbf{B}$ is complete if and only if the space $\langle \mathbf{S}(\mathbf{B}), \mathcal{O}(\mathbf{B}) \rangle$ is extremally disconnected[4].

i) Show that $\text{RO}(\mathbf{S}(\mathbf{B}), \mathcal{O}(\mathbf{B}))$ is isomorphic to the completion $\text{comp}(\mathbf{B})$ of $\mathbf{B}$.

---

[4] A topological space is **extremally disconnected** if the closure of any open set is open.

5.16 [**AC**] $\beta\omega$

The Stone space $\mathbf{S}(\mathcal{P}(\omega))$ is denoted by $\beta\omega$. Thus $\beta\omega$ is the set of all ultrafilters on ω considered as a topological space with topology $\mathcal{O}(\mathcal{P}(\omega))$.

a) For $n \in \omega$ denote by $j(n)$ the ultrafilter of all subsets of ω containing the natural number n. Show that $\langle \beta\omega, \mathcal{O}(\mathcal{P}(\omega)), j \rangle$ is a compactification of ω endowed with the discrete topology.

b) We shall identify $n \in \omega$ with $j(n)$. Thus $\omega \subseteq \beta\omega$. Show that $\beta\omega \setminus \omega$ is the Stone space of the Boolean algebra $\mathcal{P}(\omega)/\mathrm{Fin}$.

c) Show that $|\beta\omega| = 2^{\mathfrak{c}}$.

   Hint: Use Pospíšil Theorem 5.40.

d) Formulate and prove similar results for any infinite cardinal κ.

5.17 [**AC**] **The Martin Number**

Consider the complete Boolean algebra $\mathrm{C} = \mathrm{r.o.}(^{<\omega}2, \supseteq)$.

a) Show that C is **CCC**.

b) For any $f \in {}^{\omega}2$ the set $D_f = \{a \in {}^{<\omega}2 : a \not\subseteq f\}$ is dense in C.

c) Let $\mathcal{F} \subseteq {}^{\omega}2$. If F is a $\{D_f : f \in \mathcal{F}\}$-generic filter of C then

$$g = \{\langle n, m \rangle : (\exists a \in F)\, a(n) = m\} \in {}^{\omega}2$$

   and $g \neq f$ for any $f \in \mathcal{F}$.

d) Conclude that $\mathfrak{m} \leq \mathfrak{c}$.

5.18 [**AC**] **Infinite Operations in a Boolean algebra**

$\mathbf{B} = \langle B, \vee, \wedge, -, 0, 1 \rangle$ is a complete Boolean algebra.

a) Let $A \subseteq B$. Prove the **de Morgan Law**:

$$a = \bigvee A \equiv -a = \bigwedge \{-x : x \in A\}.$$

b) Let $C, D \subseteq B$. Prove the **distributive law**:

$$\bigvee C \wedge \bigvee D = \bigvee \{c \wedge d : c \in C \wedge d \in D\}, \quad \bigwedge C \vee \bigwedge D = \bigwedge \{c \vee d : c \in C \wedge d \in D\}.$$

c) A homomorphism $h : \mathbf{B}_1 \longrightarrow \mathbf{B}_2$ is said to be κ-**complete** if $\bigvee h(A) = h(\bigvee A)$ for any set $|A| < \kappa$, $A \subseteq B_1$. Show that h is κ-complete if and only if the kernel $\{x \in B_1 : h(x) = 0\}$ is a κ-complete ideal.

d) A set $A \subseteq B \setminus \{0\}$ is a partition of $\mathbf{B}$ if and only if $\bigvee A = 1$ and for any two distinct elements $x, y \in A$ we have $x \wedge y = 0$.

e) Show that any two partitions have a common refinement.

f) A Boolean algebra is κ-distributive if and only if any set of partitions of cardinality κ has a common refinement.

5.19 [**AC**] **Distributive Laws**

For simplicity we suppose that all Boolean algebras we shall deal with in this exercise are complete.

Let κ, λ be cardinals. A complete Boolean algebra $\mathbf{B}$ is said to be (κ, λ)-**distributive** if for any system $\langle a_{\xi,\eta} : \xi < \kappa, \eta < \lambda \rangle$ of elements of B, if

$$\bigwedge_{\xi < \kappa} \bigvee_{\eta < \lambda} a_{\xi,\eta} = 1,$$

then also

$$\bigvee_{\varphi \in {}^\kappa \lambda} \bigwedge_{\xi < \kappa} a_{\xi,\varphi(\xi)} = 1.$$

a) If a Boolean algebra $\mathbf{B}$ is (κ, λ)-distributive and $\gamma \leq \kappa$, $\delta \leq \lambda$, then $\mathbf{B}$ is also (γ, δ)-distributive.

b) $\mathbf{B}$ is (κ, λ)-distributive if and only if any set $\{A_\xi : \xi < \kappa\}$ of partitions, $|A_\xi| \leq \lambda$, has a common refinement.

c) The following are equivalent:

    (1) $\mathbf{B}$ is κ-distributive.

    (2) $\mathbf{B}$ is (κ, λ)-distributive for any λ.

    (3) $\mathbf{B}$ is $(\kappa, |B|)$-distributive.

d) If a Boolean algebra $\mathbf{B}$ is $(\kappa, 2)$-distributive, then $\mathbf{B}$ is also $(\kappa, 2^\kappa)$-distributive.
   Hint: Let $\bigwedge_{\xi \in \kappa} \bigvee_{s \in {}^\kappa 2} a_{\xi,s} = 1$, $a_{\xi,s_1} \wedge a_{\xi,s_2} = 0$ *for* $s_1 \neq s_2$. *Set*

$$b_{\xi,\eta,i} = \bigvee \{a_{\xi,s} : s(\eta) = i\} \text{ for } \eta \in \kappa.$$

If $\varphi \in {}^\kappa ({}^\kappa 2)$, $\varphi(\xi) \neq s \in {}^\kappa 2$, *then* $\bigwedge_\eta b_{\xi,\eta,\varphi(\xi)(\eta)} \wedge a_{\xi \, s} = 0$. *Thus*

$$\bigwedge_\xi \bigwedge_\eta b_{\xi,\eta,\varphi(\xi)(\eta)} = \bigwedge_\xi a_{\xi,\varphi(\xi)}.$$

ĩ) For κ regular, construct a complete Boolean algebra that is (λ, γ)-distributive for any γ and any $\lambda < \kappa$, and is not $(\kappa, 2)$-distributive.
   Hint: Consider the complete Boolean algebra generated by the partial ordering by inclusion of the set ${}^{<\kappa} 2$.

g) Let κ be a regular cardinal. If $\mathbf{B}$ is κ-distributive, then $\langle \mathbf{S}(\mathbf{B}), \mathcal{O}(\mathbf{B}) \rangle$ is κ-Baire.
   Hint: See Exercises 1.24, a) *and* 5.15, e).

5.20 Lindenbaum algebra

We assume that the reader is familiar with basic mathematical logic as shortly summarized in the Appendix. $\mathcal{L}$ denotes a language of predicate calculus, $\mathbf{T}$ a theory in this language.

a) On the set of all formulas $\mathcal{F}(\mathcal{L})$ we define an equivalence relation $\sim_{\mathbf{T}}$ (if $\mathbf{T}$ is understood, simply $\sim$) by

$$\varphi \sim_{\mathbf{T}} \psi \text{ if and only if } \mathbf{T} \vdash (\varphi \equiv \psi).$$

In the quotient set $\mathcal{F}(\mathcal{L})/\sim_{\mathbf{T}}$ we define operations as follows:

$$\{\varphi\}_\sim \vee \{\psi\}_\sim = \{\varphi \vee \psi\}_\sim, \quad \{\varphi\}_\sim \wedge \{\psi\}_\sim = \{\varphi \wedge \psi\}_\sim, \quad -\{\varphi\}_\sim = \{\neg\varphi\}_\sim.$$

Show that if $\mathbf{T}$ is consistent, then $\langle \mathcal{F}(\mathcal{L})/\sim, \vee, \wedge, -, 0, 1 \rangle$, where 0 is the set of all refutable in $\mathbf{T}$ formulas and 1 is the set of all provable in $\mathbf{T}$ formulas, is a Boolean algebra. We call it a **Lindenbaum Algebra** of the theory $\mathbf{T}$ and denote it by $\mathfrak{L}(\mathbf{T})$.

b) Denote by $\mathcal{T}$ the set of all terms of the language $\mathcal{L}$. Then

$$\bigvee_{t\in\mathcal{T}}\{\varphi(t)\}_\sim = \{(\exists x)\varphi(x)\}_\sim,\qquad \bigwedge_{t\in\mathcal{T}}\{\varphi(t)\}_\sim = \{(\forall x)\varphi(x)\}_\sim.$$

c) The theory $\mathbf{T}$ is complete if and only if $\{\{\varphi\}_\sim \in \mathfrak{L}(\mathbf{T}) : \varphi$ is closed$\} = \{\mathbf{0}, \mathbf{1}\}$.

d) Assume that $\mathbf{T}_1$, $\mathbf{T}_2$ are consistent theories in same language. If $\mathbf{T}_2$ is stronger than $\mathbf{T}_1$, then the map $i(\{\varphi\}_{\sim_1}) = \{\varphi\}_{\sim_2}$ is a homomorphism of $\mathfrak{L}(\mathbf{T}_1)$ onto $\mathfrak{L}(\mathbf{T}_2)$.

e) If Θ is a syntactic model of $\mathbf{T}_1$ in $\mathbf{T}_2$ induced by a translation, then Θ induces a homomorphism θ of $\mathfrak{L}(\mathbf{T}_1)$ into $\mathfrak{L}(\mathbf{T}_2)$ defined by $\theta(\{\varphi\}_{\sim_1}) = \{\Theta(\varphi)\}_{\sim_2}$.

5.21 [AC] Homogenity

A Boolean algebra $\mathbf{B}$ is called **homogeneous** if $B|a$ is isomorphic to B for any $a \in B$, $a \neq 0$.

a) The Boolean algebra $\mathcal{P}(\omega)/\mathrm{Fin}$ is homogeneous.

b) Assume that $\mathbf{B}$ is a homogenuous Boolean algebra. Then the following are equivalent:

    1) $\mathbf{B}$ is not κ-distributive.

    2) There exist open dense sets $\langle A_\xi : \xi < \kappa \rangle$ such that for any $a \in B$, $a \neq 0$, there exists a $\xi < \kappa$ such that $a \notin A_\xi$.

c) If $\mathcal{P}(\omega)/\mathrm{Fin}$ is not κ-distributive, then there exists a sequence $\langle \mathcal{G}_\xi : \xi < \kappa \rangle$ of open dense subsets of $[\omega]^\omega$ such that for any infinite $A \subseteq \omega$ there exists a $\xi < \kappa$ and there exist distinct $C, D \in \mathcal{G}_\xi$ such that both intersections $A \cap C$, $A \cap D$ are infinite.

5.22 [AC] Cardinal property

Let B be a complete Boolean algebra. A function μ defined on B with values cardinal numbers is called a **cardinal property** if $\mu(a) \leq \mu(b)$ for any $a \leq b$. B is μ-**homogeneous** if μ is constant on $B \setminus \{0\}$.

a) Show that the following functions are cardinal properties:

$$\mu(a) = \min\{\kappa : \text{there is no partition of } a \in B \text{ of cardinality } \kappa\},$$
$$\mu(a) = |\,B|a\,|,\qquad \mu(a) = \min\{\kappa : B \text{ is not } \kappa\text{-distributive}\}.$$

b) If μ is a cardinal property, then any Boolean algebra $\mathbf{B}$ is a direct sum of μ-homogeneous Boolean algebras.
Hint: For every $a \neq 0$ the set $\{\mu(b) : 0 < b \leq a\}$ has the smallest element $\mu(c)$.

5.4 Infinite Combinatorics

In this section we shall work in **ZFC**, i.e., we assume that the Axiom of Choice holds true. The reader can easily check that for some particular investigations some weaker forms of the axiom of choice are sufficient.

For a partially preordered set $\langle X, \leq \rangle$ without maximal elements we define two natural **cardinal invariants**:

$$\mathfrak{b}(X, \leq) = \min\{|A| : A \subseteq X \wedge A \text{ is unbounded from above in } X\},$$
$$\mathfrak{d}(X, \leq) = \min\{|A| : A \subseteq X \wedge A \text{ is cofinal in } X\}.$$

We shall simply say "bounded" instead of "bounded from above". Similarly for "unbounded".

Theorem [AC] 5.41. *For any partially preordered set* $\langle X, \leq \rangle$ *without maximal elements, the cardinal* $\mathfrak{b}(X, \leq)$ *is regular and* $\mathfrak{b}(X, \leq) \leq \mathrm{cf}(\mathfrak{d}(X, \leq)) \leq \mathfrak{d}(X, \leq)$.

Proof. Let $A \subseteq X$ be unbounded, $|A| = \mathfrak{b}(X, \leq)$. Assume that $|A|$ is not regular, i.e., that $\kappa = \mathrm{cf}(|A|) < |A|$. Thus there are sets $\langle A_\xi : \xi \in \kappa \rangle$ such that $A = \bigcup_{\xi < \kappa} A_\xi$ and $|A_\xi| < |A|$ for every ξ. Then each A_ξ is bounded by some $x_\xi \in X$. The set $\{x_\xi : \xi < \kappa\}$ is unbounded, which is a contradiction.

Now assume that $\kappa = \mathrm{cf}(\mathfrak{d}(X, \leq)) < \mathfrak{b}(X, \leq)$, A is cofinal and $|A| = \mathfrak{d}(X, \leq)$. Then there are sets A_ξ, $\xi < \kappa$ such that $A = \bigcup_{\xi < \kappa} A_\xi$ and $|A_\xi| < \mathfrak{d}(X, \leq)$ for every ξ. Thus no set A_ξ is cofinal. Therefore, for each $\xi < \kappa$ there exists $x_\xi \in X$ such that x_ξ is not bounded by any element of A_ξ. Since $\kappa < \mathfrak{b}(X, \leq)$, the set $\{x_\xi : \xi < \kappa\}$ must be bounded. Thus there exists an upper bound $x \in X$ of this set. Since A is a cofinal set, there exists an $a \in A$ such that $x \leq a$. However there is a ξ such that $a \in A_\xi$. Then $x_\xi \leq x \leq a$, which is a contradiction. $\square$

We shall use without commentary the following simple result.

Theorem [AC] 5.42. *If* $A \subseteq X$ *is a cofinal subset, then*

$$\mathfrak{b}(X, \leq) = \mathfrak{b}(A, \leq) \text{ and } \mathfrak{d}(X, \leq) = \mathfrak{d}(A, \leq).$$

We describe a basic tool for comparing cardinal invariants of two preordered sets. Let $\langle X_1, \leq_1 \rangle$ and $\langle X_2, \leq_2 \rangle$ be partially preordered sets. A pair of mappings $\langle \varphi, \varphi^* \rangle$ is called a **Tukey connection** from $\langle X_1, \leq_1 \rangle$ into $\langle X_2, \leq_2 \rangle$ if

$$\varphi : X_1 \longrightarrow X_2, \qquad \varphi^* : X_2 \longrightarrow X_1,$$
$$\text{if } \varphi(x) \leq_2 y, \text{ then } x \leq_1 \varphi^*(y) \text{ for any } x \in X_1 \text{ and } y \in X_2.$$

We write $\langle X_1, \leq_1 \rangle \preceq_{\mathrm{T}} \langle X_2, \leq_2 \rangle$ if there exists a Tukey connection from $\langle X_1, \leq_1 \rangle$ into $\langle X_2, \leq_2 \rangle$.

Let us note the following. Assume that there exists a mapping $\varphi : X_1 \longrightarrow X_2$ such that $\varphi^{-1}(A)$ is bounded in X_1 for every bounded set $A \subseteq X_2$. Then (using the Axiom of Choice) we can define $\varphi^* : X_2 \longrightarrow X_1$ by setting

$$\varphi^*(y) = \text{an upper bound of } \varphi^{-1}(\{x \in X_2 : x \leq_2 y\}).$$

Then the couple $\langle \varphi, \varphi^* \rangle$ is a Tukey connection from $\langle X_1, \leq_1 \rangle$ into $\langle X_2, \leq_2 \rangle$.

Theorem [AC] 5.43. *If* $\langle X_1, \leq_1 \rangle \preceq_T \langle X_2, \leq_2 \rangle$, *then*

$$\mathfrak{b}(X_2, \leq_2) \leq \mathfrak{b}(X_1, \leq_1) \ and \ \mathfrak{d}(X_1, \leq_1) \leq \mathfrak{d}(X_2, \leq_2).$$

Proof. Let $A \subseteq X_1$ and $|A| < \mathfrak{b}(X_2, \leq_2)$. We show that A is a bounded subset of X_1. Actually, we set $B = \{\varphi(x) : x \in A\} \subseteq X_2$. Since $|A| < \mathfrak{b}(X_2, \leq_2)$ the set B is bounded, i.e., there exists a $y \in X_2$ such that $\varphi(x) \leq_2 y$ for every $x \in A$. Then $x \leq_1 \varphi^*(y)$ for every $x \in A$, i.e., A is bounded.

Let $B \subseteq X_2$ be a cofinal subset. We show that $A = \{\varphi^*(y) : y \in B\}$ is a cofinal subset of X_1. Actually, let $x \in X_1$. Since B is cofinal, there exists a $y \in B$ such that $\varphi(x) \leq_2 y$. Then $x \leq_1 \varphi^*(y)$, i.e., A is cofinal in X_1.   □

The set $^\omega \omega$ of all functions from ω into ω is preordered by

$$\alpha \leq^* \beta \equiv \{n \in \omega : \neg \alpha(n) \leq \beta(n)\} \text{ is a finite set.}$$

The preordering $\leq^*$ is called **eventual domination** and a cofinal subset of $\langle ^\omega \omega, \leq^* \rangle$ is called a **dominating family**. A bounded subset of $\langle ^\omega \omega, \leq^* \rangle$ is said to be **eventually bounded**. The corresponding strict preordering will be denoted by $<^*$.

The **bounding number** $\mathfrak{b}$ and the **dominating number** $\mathfrak{d}$ are defined as

$$\mathfrak{b} = \mathfrak{b}(^\omega \omega, \leq^*), \qquad \mathfrak{d} = \mathfrak{d}(^\omega \omega, \leq^*).$$

Any countable subset of $^\omega \omega$ is bounded. Actually, if $A = \{\alpha_n : n \in \omega\}$ is a countable set, then the function β defined by $\beta(n) = \max\{\alpha_0(n), \ldots, \alpha_n(n)\}$ is an upper bound of A. Thus $\mathfrak{b} > \aleph_0$. By Theorem 5.41 the cardinal $\mathfrak{b}$ is regular and we have

$$\aleph_0 < \mathfrak{b} \leq \mathrm{cf}(\mathfrak{d}) \leq \mathfrak{d} \leq 2^{\aleph_0}. \tag{5.30}$$

We shall need often a subset of $^\omega \omega$ of strictly increasing sequences

$$^\omega \omega{\uparrow} = \{\alpha \in {}^\omega \omega : \alpha \text{ is increasing}\}. \tag{5.31}$$

One can easily show that $^\omega \omega{\uparrow}$ is cofinal in $^\omega \omega$. Moreover

Theorem [AC] 5.44. *There exists an eventually increasing well-ordered unbounded sequence of increasing functions* $\{\alpha_\xi\}_{\xi < \mathfrak{b}}$.

The proof is easy. By transfinite induction one can construct the required sequence. To obtain an increasing function replace at each step a function α by $\sum_{i=0}^{n} \alpha(i)$.   □

If κ is a regular cardinal, then a sequence $S = \langle \alpha_\xi \in {}^\omega \omega : \xi \in \kappa \rangle$ is called a κ-**scale** if $\alpha_\xi <^* \alpha_\eta$ for any $\xi < \eta < \kappa$ and the set S is dominating in $^\omega \omega$. If there exists a κ-scale, then evidently $\aleph_0 < \kappa \leq 2^{\aleph_0}$.

Theorem [AC] 5.45. *Let* κ *be an uncountable cardinal not greater than* $2^{\aleph_0}$. *Then there exists a* κ-*scale if and only if* $\kappa = \mathfrak{b} = \mathfrak{d}$.

Proof. If S is a κ-scale, then one easily see that $\kappa = \mathfrak{b} = \mathfrak{d}$.

Let $\kappa = \mathfrak{b} = \mathfrak{d}$. Then there exists a dominating family $\{\alpha_\xi : \xi < \kappa\}$. By transfinite induction we define $\beta_\xi \in {}^\omega\omega$ as an upper bound of the bounded family $\{\beta_\eta : \eta < \xi\} \cup \{\alpha_\xi\}$. Evidently the sequence $\langle \beta_\xi : \xi < \kappa \rangle$ is a κ-scale. $\qquad\square$

We introduce some combinatorial cardinal invariants of the preordered set $\langle [\omega]^\omega, \subseteq^* \rangle$. We begin with some notions. If we need to emphasize that a family $\mathcal{A} \subseteq [\omega]^\omega$ is dense in the preordered set $\langle [\omega]^\omega, \subseteq^* \rangle$, or in accordance with Section 5.3, dense in the Boolean algebra $\mathcal{P}(\omega)/\mathrm{Fin}$, we say that $\mathcal{A}$ is an **almost dense** family. Similarly, if X is an infinite countable set and $X_\xi \subseteq X$, $\xi \in \kappa$, then we say that the sequence $\langle X_\xi : \xi < \kappa \rangle$ is **almost decreasing** if $X_\eta \subseteq^* X_\xi$ for any $\eta < \xi < \kappa$. An almost decreasing sequence $\mathcal{T} = \langle A_\xi : \xi < \kappa \rangle$ of subsets of ω is called a **tower** if $\mathcal{T}$ has no pseudointersection, i.e., if

$$(\forall A \in [\omega]^\omega)(\exists T \in \mathcal{T})\, \neg A \subseteq^* T.$$

The **tower number** is defined as

$$\mathfrak{t} = \min\{|\mathcal{T}| : \mathcal{T} \text{ is a tower}\}. \tag{5.32}$$

Evidently $\mathfrak{t}$ is a regular cardinal. Lemma 5.36, b) may be formulated as $\mathfrak{t} > \aleph_0$. By definition the set $[\omega]^\omega$ is a $\mathfrak{t}$-closed subset of Boolean algebra $\mathcal{P}(\omega)/\mathrm{Fin}$.

Theorem [AC] 5.46. *If $\aleph_0 \leq \kappa < \mathfrak{t}$, then $2^\kappa = \mathfrak{c}$.*

Proof. We shall construct a mapping $F : {}^{<\kappa}2 \longrightarrow [\omega]^\omega$ such that the sets $F(f)$ and $F(g)$ are almost disjoint for any distinct $f, g \in {}^\kappa 2$.

The construction is easy. Let $F(\emptyset) = \omega$. Assume that $F(f)$ is already defined for all $f \in {}^\zeta 2$ and all $\zeta < \xi$. Now, let $f \in {}^\xi 2$. If $\xi = \eta + 1$, we take two infinite disjoint subsets $A_0, A_1 \subseteq F(f|\eta)$ and set $F(f) = A_i$, where $f(\eta) = i$, $i = 0, 1$. If $\xi \leq \kappa$ is a limit ordinal, then the sequence $\langle F(f|\eta) : \eta < \xi \rangle$ is almost decreasing. Since $\xi < \mathfrak{t}$ there exists a lower bound B of this family. Set $F(f) = B$.

If $f, g \in {}^\kappa 2$ are distinct, then there is the least $\xi < \kappa$ for which $f(\xi) \neq g(\xi)$. In this case we have taken $F(f|(\xi+1)), F(g|(\xi+1))$ disjoint subsets of $F(f|\xi)$ and $F(f) \subseteq^* F(f|(\xi+1))$, $F(g) \subseteq^* F(g|(\xi+1))$. $\qquad\square$

Similarly as the tower number we define the **pseudointersection number**

$$\mathfrak{p} = \min\{|\mathcal{F}| : (\mathcal{F} \subseteq [\omega]^\omega \text{ has f.i.p. }) \wedge (\forall A \in [\omega]^\omega)(\exists F \in \mathcal{F})\, \neg A \subseteq^* F\}. \tag{5.33}$$

Thus the family $\mathcal{F}$ has no pseudointersection.

Theorem [AC] 5.47. $\qquad \mathfrak{p} \leq \mathfrak{t}$ *and* $\aleph_0 < \mathfrak{p} \leq 2^{\aleph_0}$.

Proof. Since a tower has f.i.p., we obtain the first inequality.

Let $\mathcal{F} = \{A_n : n \in \omega\}$ be a countable family with finite intersection property. Then $\bigcap_{i<n} A_i$, $n \in \omega$ is a $\subseteq^*$-decreasing sequence and by Lemma 5.36, b) there is a lower bound of it. Thus $\mathfrak{p} > \aleph_0$. $\qquad\square$

Let $\mathcal{F}$ be a free ultrafilter on ω. Then $\mathcal{F}$ has f.i.p. and $|\mathcal{F}| = 2^{\aleph_0}$. Since for any $A \subseteq \omega$ we have $A \in \mathcal{F}$ or $X \setminus A \in \mathcal{F}$, $\mathcal{F}$ has no lower bound. Thus $\mathfrak{p} \leq 2^{\aleph_0}$. □

We do not need the following important result, therefore we present it without a proof. A proof is given in Exercise 9.5.

Theorem [AC] 5.48. $\mathfrak{p}$ *is a regular cardinal.*

However we shall need the following result.

Theorem [AC] 5.49. $\mathfrak{m} \leq \mathfrak{p}$.

Proof. We show that every family $\mathcal{F} \subseteq \mathcal{P}(\omega)$ of cardinality $\kappa < \mathfrak{m}$ with f.i.p. has a lower bound in the $\subseteq^*$ ordering.

Let $\mathcal{F} \subseteq \mathcal{P}(\omega)$ have f.i.p., $|\mathcal{F}| = \kappa < \mathfrak{m}$. Without loss of generality we can assume that $\mathcal{F}$ is closed under finite intersections. We denote

$$P = \{\langle I, A \rangle : I \subseteq \omega \text{ is finite}, A \in \mathcal{F}\}.$$

We define a partial ordering on P as

$$\langle J, B \rangle \preceq \langle I, A \rangle \equiv (I \subseteq J \subseteq (I \cup A) \wedge B \subseteq A).$$

If $\langle I, A \rangle$, $\langle J, B \rangle$ are incompatible, then $I \neq J$ (otherwise $\langle I, A \cap B \rangle$ is smaller than both of them). Thus $\langle P, \preceq \rangle$ is **CCC**. For an $n \in \omega$ and a $C \in \mathcal{F}$ we set

$$D_n = \{\langle I, A \rangle \in P : (\exists m \geq n)\, m \in I\}, \qquad D_C = \{\langle I, A \rangle \in P : A \subseteq C\}.$$

It is easy to check that any D_n and D_C is a dense subset of P. Let

$$\mathcal{D} = \{D_n : n \in \omega\} \cup \{D_C : C \in \mathcal{F}\}.$$

Since $|\mathcal{D}| \leq \kappa$ there exists a $\mathcal{D}$-generic filter F of P. Set

$$E = \bigcup \{I : \langle I, A \rangle \in F\}.$$

Since F meets D_n there exists an $m \geq n$ in E. Thus E is infinite.

Let $A \in \mathcal{F}$ and $\langle I_1, B_1 \rangle \in F \cap D_A$. We claim that $E \setminus A \subseteq I_1$, i.e., $E \subseteq^* A$.

Assume that there exists an $n \in E \setminus A$ such that $n \notin I_1$. Since $B_1 \subseteq A$, we have $n \notin B_1$. Since $n \in E$ there exists $\langle I_2, B_2 \rangle \in F$ such that $n \in I_2$. F is a filter, therefore there exists $\langle I_0, B_0 \rangle \in F$ such that $\langle I_0, B_0 \rangle \preceq \langle I_1, B_1 \rangle$ and $\langle I_0, B_0 \rangle \preceq \langle I_2, B_2 \rangle$. Then $I_0 \subseteq I_1 \cup B_0$, $I_2 \subseteq I_0$ and $B_0 \subseteq B_1$. Since $n \in I_2$ and $n \notin I_1$ we obtain $n \in B_1$, a contradiction.

Thus E is a lower bound of $\mathcal{F}$. Consequently $\kappa < \mathfrak{p}$. □

Lemma 5.38 suggests that we introduce a cardinal invariant. The cardinal

$$\mathfrak{h} = \min\{\kappa : \mathcal{P}(\omega)/\mathrm{Fin} \text{ is not } \kappa\text{-distributive}\} \tag{5.34}$$

is called the **distributivity number**. By Lemma 5.38 we have $\mathfrak{h} > \aleph_0$ and one can easily show that $\mathfrak{h}$ is a regular cardinal.

If $A \in [\omega]^\omega$ we have defined the increasing enumeration e_A of A by (1.6) and (1.7). One can easily see that

$$(\forall f \in {}^\omega\omega)(\forall A \in [\omega]^\omega)(\exists B \subseteq A)\, e_B >^* f. \tag{5.35}$$

Actually, we take $n_0 \in A$, $n_0 \geq f(0)$ and by induction $n_{k+1} \in A$ such that $n_{k+1} > n_k$ and $n_{k+1} \geq f(k+1)$. Then $B = \{n_k : k \in \omega\}$ is desired.

If $A \subseteq^* B$, then $e_B \leq^* e_A + n$ for some natural number n. Therefore also $e_B <^* e_A + \mathrm{id}_\omega$. Using this fact we prove

Theorem [AC] 5.50 (B. Balcar – J. Pelant – P. Simon).

$$\aleph_0 < \mathfrak{t} \leq \mathfrak{h} \leq \mathfrak{b}.$$

Proof. We already know that $\aleph_0 < \mathfrak{t}$ (Lemma 5.36). Since $[\omega]^\omega$ is a $\mathfrak{t}$-closed dense subset of $\mathcal{P}(\omega)/\mathrm{Fin}$ the inequality $\mathfrak{t} \leq \mathfrak{h}$ follows by Theorem 5.31.

We want to show that $\kappa < \mathfrak{b}$ for any $\kappa < \mathfrak{h}$. Consider a set $\{\alpha_\xi \in {}^\omega\omega : \xi \in \kappa\}$. We define by induction dense families $\langle \mathcal{G}_\xi : \xi < \kappa \rangle$. Let $\mathcal{G}_0 = \{A \subseteq \omega : e_A >^* \alpha_0\}$. By (5.35) the family $\mathcal{G}_0$ is open dense. Assume that $\mathcal{G}_\eta$ are defined for $\eta < \xi$. Since $\xi < \mathfrak{h}$ there is a common dense refinement $\mathcal{H}_\xi$ of $\langle \mathcal{G}_\eta : \eta < \xi \rangle$. We set

$$\mathcal{G}_\xi = \{A \subseteq \omega : (\exists B \in \mathcal{H}_\xi)\, A \subseteq^* B \wedge e_A >^* \alpha_\xi\}.$$

Again by (5.35) the family $\mathcal{G}_\xi$ is an open dense common refinement of $\langle \mathcal{G}_\eta : \eta < \xi \rangle$. Since $\kappa < \mathfrak{h}$, there exists a common refinement $\mathcal{G}$ of all $\mathcal{G}_\xi$, $\xi < \kappa$. Let $B \in \mathcal{G}$. For any $\xi < \kappa$ there exists a set $A_\xi \in \mathcal{G}_\xi$ such that $B \subseteq^* A_\xi$. Then we obtain that $e_B + \mathrm{id}_\omega >^* e_{A_\xi} >^* \alpha_\xi$ for any $\xi < \kappa$. Thus $\kappa < \mathfrak{b}$. $\square$

Theorem [AC] 5.51. *If $\kappa = \mathfrak{t} = \mathfrak{b}$, then there exists a tower $\mathcal{T} = \langle A_\xi : \xi < \kappa \rangle$ such that the family of functions $\{e_{A_\xi} : \xi < \kappa\}$ is unbounded in ${}^\omega\omega$. If $\kappa = \mathfrak{t} = \mathfrak{d}$ we can assume that $\langle e_{A_\xi} : \xi < \kappa \rangle$ is a κ-scale.*

Proof. Let $\langle B_\xi : \xi < \kappa \rangle$ be a tower and $\{\alpha_\xi : \xi < \kappa\}$ be an unbounded family. By Theorem 5.44 we can assume that $\langle \alpha_\xi : \xi < \kappa \rangle$ is increasing. We construct the tower $\mathcal{T}$ by transfinite induction. We let $A_0 = B_0$. If A_η, $\eta < \xi$ are already constructed, we construct A_ξ as follows. Since $\langle A_\eta : \eta < \xi \rangle$ is not a tower, there exists a set $A \in [\omega]^\omega$ such that $A \subseteq^* A_\eta$ for every $\eta < \xi$. By (5.35) there exists a set $A_\xi \in [A]^\omega$ such that $\alpha_\xi \leq^* e_{A_\xi}$.

One can easily see that $\mathcal{T} = \langle A_\xi : \xi < \kappa \rangle$ is a tower and the family of enumerations $\{e_{A_\xi} : \xi < \kappa\}$ is unbounded.

If $\kappa = \mathfrak{t} = \mathfrak{d}$ we begin with a scale $\langle \alpha_\xi : \xi < \kappa \rangle$. $\square$

Let $A, B \subseteq \omega$. We say that the set B **splits** the set A if both sets $A \cap B$ and $A \setminus B$ are infinite. The **splitting number** $\mathfrak{s}$ is the least size of a **splitting family**, i.e., the least size of a family $\mathcal{S} \subseteq [\omega]^\omega$ such that every infinite subset $A \subseteq \omega$ is split by some set from $\mathcal{S}$. The **reaping number** $\mathfrak{r}$ is the least size of a **reaping family**, i.e., the least size of a family $\mathcal{S} \subseteq [\omega]^\omega$ such that no infinite subset of ω splits every member of $\mathcal{S}$.

Note that all those introduced notions have a simple interpretation in the Boolean algebra $\mathcal{P}(\omega)/\mathrm{Fin}$.

Theorem [AC] 5.52. $\mathfrak{h} \leq \mathfrak{s} \leq \mathfrak{d}$ and $\mathfrak{b} \leq \mathfrak{r}$.

Proof. If $\mathcal{S} \subseteq [\omega]^\omega$, $|\mathcal{S}| < \mathfrak{h}$, then $\mathcal{S}$ is not splitting, since any element of a common refinement of the system of dense sets $\{\{S, \omega\backslash S\} : S \in \mathcal{S}\}$ is not split by an element of $\mathcal{S}$. Thus $\mathfrak{h} \leq \mathfrak{s}$.

For any increasing $\alpha \in {}^\omega\omega\uparrow$ we define the function $\bar{\alpha}$ by induction

$$\bar{\alpha}(0) = \alpha(0), \quad \bar{\alpha}(n+1) = \alpha(\bar{\alpha}(n))$$

and set

$$S_\alpha = \{k \in \omega : (\exists n)\, \bar{\alpha}(2n) \leq k < \bar{\alpha}(2n+1)\}.$$

Then $\{k \in \omega : (\exists n)\, \bar{\alpha}(2n+1) \leq k < \bar{\alpha}(2n+2)\} \subseteq \omega \backslash S_\alpha$. If $X \in [\omega]^\omega$ and $e_X <^* \alpha$, then for all sufficiently large n we have

$$\bar{\alpha}(n) \leq e_X(\bar{\alpha}(n)) < \alpha(\bar{\alpha}(n)) = \bar{\alpha}(n+1).$$

Hence the set S_α splits X.

Consider now a dominating family $F \subseteq {}^\omega\omega$ consisting of increasing functions. Then $\{S_\alpha : \alpha \in F\}$ is a splitting family. Hence $\mathfrak{s} \leq \mathfrak{d}$.

Assume that $\mathcal{S} \subseteq [\omega]^\omega$ and $|\mathcal{S}| < \mathfrak{b}$. Then the family $\{e_X : X \in \mathcal{S}\}$ is bounded. Thus there exists an increasing function $\alpha \in {}^\omega\omega$ such that $e_X <^* \alpha$ for any $X \in \mathcal{S}$. Then the set S_α splits any set $X \in \mathcal{S}$. Hence $\mathcal{S}$ is not a reaping family and therefore $\mathfrak{b} \leq \mathfrak{r}$. $\square$

If $\mathcal{A}, \mathcal{B}$ are subsets $[\omega]^\omega$, then the notation $\mathcal{A} \subseteq^* \mathcal{B}$ means that $A \subseteq^* B$ for every element A, B of $\mathcal{A}, \mathcal{B}$, respectively. Instead of $\mathcal{A} \subseteq^* \{B\}$ we shall simply write $\mathcal{A} \subseteq^* B$.

Theorem [AC] 5.53 (F. Hausdorff). *There exist two sequences $\langle A_\xi : \xi < \omega_1 \rangle$ and $\langle B_\xi : \xi < \omega_1 \rangle$ of infinite subsets of ω such that*

    a) *$\langle A_\xi : \xi < \omega_1 \rangle$ is $\subset^*$-increasing,*

    b) *$\langle B_\xi : \xi < \omega_1 \rangle$ is $\subset^*$-decreasing,*

    c) *$\langle A_\xi : \xi < \omega_1 \rangle \subset^* \langle B_\xi : \xi < \omega_1 \rangle$,*

    d) *there exists no infinite $C \subseteq \omega$ such that $\langle A_\xi : \xi < \omega_1 \rangle \subseteq^* C \subseteq^* \langle B_\xi : \xi < \omega_1 \rangle$.*

Any couple of sequences with properties a)–d) is called a **Hausdorff gap**. We begin with simple auxiliary results.

Lemma 5.54. *Assume that $A_n, B_n, A \in [\omega]^\omega$ for every n.*

    a) *If $A_n \subset^* A_{n+1} \subset^* A$ for every $n \in \omega$, then there exists an infinite set $B \subseteq \omega$ such that $A_n \subset^* B \subset^* A$ for every n.*

b) If $A \subset^* A_{n+1} \subset^* A_n$ for every $n \in \omega$, then there exists an infinite set $B \subseteq \omega$ such that $A \subset^* B \subset^* A_n$ for every n.

c) If $A_n \subset^* A_{n+1} \subset^* B_{m+1} \subset^* B_m$ for every $n, m \in \omega$, then there exists an infinite set $B \subseteq \omega$ such that $A_n \subset^* B \subset^* B_n$ for every n.

Proof. The assertion a) follows from assertion b) by the equivalence

$$E \subseteq^* F \equiv (\omega \setminus F) \subseteq^* (\omega \setminus E).$$

So assume that $A \subset^* A_{n+1} \subset^* A_n$ for every n. Then also $(A_{n+1} \setminus A) \subset^* (A_n \setminus A)$ for every n and by Lemma 5.36 there exists an infinite $C \subseteq^* (A_n \setminus A)$ for every n. Set $B = A \cup C$.

To show c), we set $B = \bigcup_n (A_n \cap B_0 \cap \cdots \cap B_n)$. Since

$$B \setminus B_n \subseteq \bigcup_{k<n} (A_k \setminus B_n), \qquad A_n \setminus B \subseteq \bigcup_{k \leq n} (A_n \setminus B_k),$$

we obtain $A_n \subset^* B \subset^* B_n$ for every n. □

If $A \subseteq^* B$, then $\mathrm{sf}(A, B)$ is the least integer m such that $n \in A \to n \in B$ for every $n \geq m$. Note that if $A \subseteq^* B$, then $B \setminus A \subseteq \mathrm{sf}(A, B)$ and therefore $|B \setminus A| \leq \mathrm{sf}(A, B)$. If $\mathcal{A}$ is a subset of $[\omega]^\omega$, we say that B is **close** to $\mathcal{A}$, if $A \subseteq^* B$ and for each $n \in \omega$ there are only finitely many elements A of $\mathcal{A}$ with $\mathrm{sf}(A, B) = n$. Evidently, the set $\mathcal{A}$ must be countable. Note the following. If $\mathcal{A} \subseteq^* C \subseteq^* B$, B is close to $\mathcal{A}$, then C is close to $\mathcal{A}$ as well.

Lemma 5.55. *If $A_n \subset^* A_m \subset^* A$ for any $n < m$, then there exists an infinite set $B \subset^* A$ close to $\{A_n\}_{n \in \omega}$.*

Proof. Since

$$A_n \setminus A_{n-1} \subseteq^* A_n \setminus (A_0 \cup \cdots \cup A_{n-1})$$

and $A_n \setminus A_{n-1}$ is infinite, there exists a set $E_n \subset A_n \setminus (A_0 \cup \cdots \cup A_{n-1})$ with $|E_n| = n$. Set $B = A \setminus \bigcup_k E_k$. Then

$$A_n \setminus B = (A_n \setminus A) \cup \bigcup_k (A_n \cap \bar{E}_k).$$

Since $A_n \cap E_k = \emptyset$ for $k > n$ we obtain $A_n \subseteq^* B$ for any n. Since $A_n \subset^* A_{n+1}$ we have also $A_n \subset^* B$.

We show that B is close to $\{A_n\}_{n<\omega}$. By definition $E_n \subseteq A_n \setminus B$ and therefore $|A_n \setminus B| \geq n$. In other words, if $\mathrm{sf}(A_n, B) = k$, then $n \leq k$. □

Lemma [AC] 5.56. *Let $\gamma < \omega_1$ be a limit ordinal. If $\langle A_\xi : \xi < \gamma \rangle$ is $\subset^*$-increasing, B is such that $A_n \subset^* B$ and B is close to $\langle A_\xi : \xi < \eta \rangle$, everything for each $\eta < \gamma$, then there exists an infinite set $C \subset^* B$ close to $\langle A_\xi : \xi < \gamma \rangle$.*

Proof. Let us assume that $\langle \gamma_n : n \in \omega \rangle$ is an increasing sequence of ordinals with $\sup\{\gamma_n : n \in \omega\} = \gamma$. We consider two cases.

If B is close to $\langle A_\xi : \xi < \gamma \rangle$, then by Lemma 5.54, a) there exists a set C such that $A_{\gamma_n} \subset^* C \subset^* B$ for every n. Then the set C is close to $\langle A_\xi : \xi < \gamma \rangle$ as well.

Assume now, that B is not close to $\langle A_\xi : \xi < \gamma \rangle$. Set

$$N_k = \{\xi < \gamma : \mathrm{sf}(A_\xi, B) \le k\}.$$

Then there exists a k_0 such that N_k is infinite for every $k \ge k_0$. We can assume that $k_0 = 0$; otherwise omit finitely many $A_\xi, \xi \in \bigcup_{k < k_0} N_k$. Since for $\eta < \gamma$, the set B is close to $\langle A_\xi : \xi < \eta \rangle$, no $N_k \subseteq \eta$, i.e., every N_k is cofinal in γ. For the same reason, for every $\xi \in N_k$ the set $\xi \cap N_k$ is finite. Thus every set N_k has the order type ω. Using Lemma 5.55 we define by induction a sequence $\langle B_n : n \in \omega \rangle$ such that

$$\{A_\xi : \xi \in N_n\} \subset^* B_n \subset^* B_{n-1} \subset^* B$$

and B_n is close to $\{A_\xi : \xi \in N_n\}$ for every n. Since every N_n is cofinal in γ we have

$$\{A_\xi : \xi < \gamma\} \subset^* B_n$$

for every n. By part c) of Lemma 5.54 there exists a set C such that

$$A_{\gamma_n} \subset^* C \subset^* B_n$$

for every n.

We show that C is close to $\langle A_\xi : \xi < \gamma \rangle$. Let $n \in \omega$ be given. We set $k = \max\{n, \mathrm{sf}(C, B)\}$. Then $\{\xi < \gamma : \mathrm{sf}(A_\xi, C) \le k\} \subseteq N_k$. Since B_k is close to $\langle A_\xi : \xi \in N_k \rangle$, C is close as well. Then the set

$$\{\xi < \gamma : \mathrm{sf}(A_\xi, C) \le n\} \subseteq \{\xi < \gamma : \mathrm{sf}(A_\xi, C) \le k\}$$

is finite. $\qquad\square$

Proof of the theorem. We construct by transfinite induction a $\subseteq^*$-increasing sequence $\langle A_\xi : \xi < \omega_1 \rangle$ and a $\subseteq^*$-decreasing sequence $\langle B_\xi : \xi < \omega_1 \rangle$ of subsets of ω such that $\langle A_\xi : \xi < \omega_1 \rangle \subseteq^* \langle B_\xi : \xi < \omega_1 \rangle$ and B_ζ is close to $\langle A_\eta : \eta < \zeta \rangle$ for each $\zeta < \omega_1$.

Take any infinite $A_0 \subset^* B_0 \subseteq \omega$. If A_ξ, B_ξ are constructed for any $\xi \le \gamma$, take any $A_\gamma \subset^* A_{\gamma+1} \subset^* B_{\gamma+1} \subset^* B_\gamma$. Since B_γ is close to $\{A_\xi : \xi \le \gamma\}$, the set $B_{\gamma+1}$ will be close to $\{A_\xi : \xi \le \gamma + 1\}$.

Assume now that γ is limit and A_ξ, B_ξ are constructed for any $\xi < \gamma$. Since $\mathrm{cf}(\gamma) = \omega$, by Lemma 5.54 there exists a set C such that

$$\{A_\xi : \xi < \gamma\} \subset^* C \subset^* \{B_\xi : \xi < \gamma\}.$$

Note that C is close to $\{A_\xi : \xi < \eta\}$ for every $\eta < \gamma$. By Lemma 5.56 there exists a set $B_\gamma \subseteq^* C^*$ close to $\{A_\xi : \xi < \gamma\}$. By Lemma 5.54, a) there exists a set A_γ such that $\{A_\xi : \xi < \gamma\} \subset^* A_\gamma \subset^* B_\gamma$.

We have to prove d). To get a contradiction assume that there exists a C with $A_\xi \subseteq^* C \subseteq^* B_\xi$ for each $\xi < \omega_1$. We write $N_k = \{\xi < \omega_1 : \mathrm{sf}(A_\xi, C) = k\}$. Then there exists a k such that $|N_k| = \aleph_1$. Take a $\xi \in N_k$ such that $\xi \cap N_k$ is infinite. Since B_ξ is close to $\{A_\eta : \eta < \xi\}$ and $C \subseteq^* B_\xi$, we have a contradiction. $\qquad\square$

Since $F(A) = \chi_A$ is an order-preserving mapping from the poset $\langle [\omega]^\omega, \subset^* \rangle$ into the poset $\langle {}^\omega 2, <^* \rangle$, there exists a Hausdorff gap $\langle \langle \alpha_\xi : \xi < \omega_1 \rangle, \langle \beta_\xi : \xi < \omega_1 \rangle \rangle$ consisting of reals from ${}^\omega 2$.

The theorem has an important consequence.

Theorem [AC] 5.57. *Any perfect Polish space X is a union $X = \bigcup_{\xi < \omega_1} X_\xi$ of a strictly increasing sequence $\langle X_\xi : \xi < \omega_1 \rangle$ of G_δ subsets of X.*

Proof. We start with the Cantor space ${}^\omega 2$. Let $\langle \langle \alpha_\xi : \xi < \omega_1 \rangle, \langle \beta_\xi : \xi < \omega_1 \rangle \rangle$ be a Hausdorff gap in ${}^\omega 2$. We let

$$A_\xi = \{\gamma \in {}^\omega 2 : \alpha_\xi <^* \gamma <^* \beta_\xi\}.$$

It is easy to see that A_ξ is an F_σ set for each $\xi < \omega_1$. Moreover $\bigcap_{\xi < \omega_1} A_\xi = \emptyset$.

If X is a perfect Polish space, then by Corollary 5.6 a homeomorphic copy of ${}^\omega 2$ is a closed subset of X. Then $X_\xi = X \setminus A_\xi$ is a G_δ set and $X = \bigcup_{\xi < \omega_1} X_\xi$. $\qquad\square$

Exercises

5.23 [AC] Rothberger Theorem

If $\mathfrak{p} = \aleph_1$, then $\mathfrak{t} = \aleph_1$.

Hint: If $\mathfrak{d} = \aleph_1$ the assertion follows in a trivial way. Assume $\mathfrak{d} > \aleph_1$. Let $\langle A_\xi : \xi < \omega_1 \rangle$ have f.i.p. and let no infinite set be almost contained in all A_ξ. By induction construct a tower $\langle B_\xi : \xi < \omega_1 \rangle$: $B_{\xi+1} = B_\xi \cap A_\xi$ and for ξ limit apply the result of Exercise 5.24, e) for a cofinal sequence $\langle A_{\eta_n} : n \in \omega \rangle$.

5.24 [AC] Equivalent Definition of $\mathfrak{b}$ and $\mathfrak{d}$

An **interval partition** of ω is a partition of ω into finite intervals. We always assume that an interval partition $\mathbf{I} = \{I_n : n \in \omega\}$ is enumerated in such a way that $I_n = [i_n, i_{n+1})$, $0 = i_0 < i_1 < \cdots < i_n < \cdots$. The interval partition $\mathbf{I} = \{I_n : n \in \omega\}$ **dominates** the interval partition $\mathbf{J} = \{J_n : n \in \omega\}$ if for all but finitely many k there is an n such that $J_n \subseteq I_k$. We shall write $\mathbf{J} \subseteq^* \mathbf{I}$. Let $\mathcal{IP}$ be the set of all interval partitions.

a) If an interval partition $\mathbf{I} = \{I_n : n \in \omega\}$ does not dominate an interval partition $\mathbf{J} = \{J_n : n \in \omega\}$, then there exist infinitely many intervals $I \in \mathcal{I}$ such that there are two intervals $J_1, J_2 \in \mathcal{J}$, $I \subseteq J_1 \cup J_2$ and neither J_1 nor J_2 is a subset of I.

b) For an interval partition $\mathbf{I} = \{I_n : n \in \omega\}$ set $\varphi(\mathbf{I}) = f \in {}^\omega \omega{\uparrow}$, where $f(m) = i_{n+2} - 1$ for $m \in I_n$. For an increasing function $f \in {}^\omega \omega{\uparrow}$ set $\varphi^*(f) = \mathbf{I}$, where $i_n = \bar{f}(n)$ (see the proof of Theorem 5.52). Show that $\langle \varphi, \varphi^* \rangle$ is a Tukey connection from $\langle \mathcal{IP}, \subseteq^* \rangle$ into $\langle {}^\omega \omega{\uparrow}, <^* \rangle$.

Hint: Assume $f = \varphi(\mathbf{I}) <^* g$, $\mathbf{I} = \{[i_n, i_{n+1}) : n \in \omega\}$. Then $f(\bar{g}(n)) < \bar{g}(n+1)$ for sufficiently large n. If $\bar{g}(n) \in [i_k, i_{k+1})$, then $f(\bar{g}(n)) = i_{k+2} - 1 < \bar{g}(n+1)$. Hence $[i_{k+1}, i_{k+2}) \subseteq [\bar{g}(n), \bar{g}(n+1))$.

c) Show that $\langle \varphi^*, \varphi \rangle$ is a Tukey connection from $\langle {}^{\omega}\omega{\uparrow}, <^* \rangle$ into $\langle \mathcal{IP}, \subseteq^* \rangle$.

   Hint: Assume $g \in {}^{\omega}\omega{\uparrow}$, $\varphi^*(g) \subseteq^* \mathbf{I} = \{[i_n, i_{n+1}) : n \in \omega\}$ and $f = \varphi(\mathbf{I})$. We have to show $g \leq^* f$. If $m \in [i_k, i_{k+1})$ and k is large, then there exists an n such that $[i_k, i_{k+1}) \subseteq [\bar{g}(n), \bar{g}(n+1))$. Since $m < \bar{g}(n)$ we obtain $g(m) < \bar{g}(n_1) \leq f(m) + 1$.

d) Conclude that $\mathfrak{b} = \mathfrak{b}(\mathcal{IP}, \subseteq^*)$ and $\mathfrak{d} = \mathfrak{d}(\mathcal{IP}, \subseteq^*)$.

e) Let $\langle A_n : n \in \omega \rangle$ be a decreasing sequence of infinite subsets of ω. Let $f \in {}^{\omega}\omega$ be increasing. The set $B_f = \{m \in \omega : (\exists n)\,(m < f(n) \wedge m \in A_n)\}$ is almost contained in every A_n, $n \in \omega$.

   Hint: If $f(n) \leq m$, $m \in B_f$, then $m \in A_n$.

f) Let $A_n = (\omega \setminus n) \times \omega$ for $n \in \omega$, $\mathcal{F} \subseteq {}^{\omega}\omega$ being a dominating family. For $f \in {}^{\omega}\omega$ we set $C_f = \{\langle n, m \rangle \in \omega \times \omega : f(n) \leq m\}$. There is no infinite set D such that D is almost contained in every $\langle A_n : n \in \omega \rangle$ and $D \cap C_f$ is infinite for every $f \in \mathcal{F}$.

   Hint: Assume that there exists such a set D. The set $D_n = \{k : \langle n, k \rangle \in D \vee k = 0\}$ is finite, since $D_n \subseteq D \setminus A_n$. Let $g(n) = \max D_n$. Show that $g \leq^* f$ for no $f \in \mathcal{F}$.

g) Show that $\mathfrak{d}$ is the smallest cardinal κ with the following property: if $\langle A_n : n \in \omega \rangle$ is a decreasing sequence of infinite subsets of ω, $\mathcal{D} \subseteq [\omega]^{\omega}$ is such that every A_n intersects every element of $\mathcal{D}$ in an infinite set and $|\mathcal{D}| < \kappa$, then there is an infinite set D almost contained in every A_n and having infinite intersection with every $C \in \mathcal{D}$.

   Hint: Let $f_C(n) = n$th element of $A_n \cap C$. Let f dominate the family $\{f_C : C \in \mathcal{D}\}$. Using notation of part e), take $D = B_f$.

5.25 [AC] Base Matrix Tree

a) Let $\mathcal{G}$ be a MAD family of subsets of ω. For each $A \in \mathcal{G}$, let $\mathcal{H}_A$ be a MAD family of subsets of A. Then $\bigcup_{A \in \mathcal{G}} \mathcal{H}_A$ is a MAD family of subsets of ω.

b) If $\mathcal{G}$ is a MAD family of subsets of ω, then there exists a MAD family $\mathcal{H}$ such that $|\{A \in \mathcal{H} : A \subseteq^* B\}| = \mathfrak{c}$ for every $B \in \mathcal{G}$.

   Hint: Use a), Theorem 5.35 and Zorn's Lemma, Theorem 1.10.

c) If $\langle \mathcal{H}_\xi : \xi < \kappa \rangle$, $\kappa < \mathfrak{h}$ are open dense subfamilies of $\langle [\omega]^{\omega}, \subseteq^* \rangle$, then $\bigcap_{\xi < \kappa} \mathcal{H}_\xi$ is an open dense family as well.

d) If $\mathcal{G}$ is a MAD family of subsets of ω, then $\{A \in [\omega]^{\omega} : (\exists B)\,(B \in \mathcal{G} \wedge A \subseteq^* B)\}$ is an open dense family.

e) If $\mathcal{H}$ is an open dense family, then there exists a MAD family $\mathcal{G}$ such that

$$\{A \in [\omega]^{\omega} : (\exists B)\,(B \in \mathcal{G} \wedge A \subseteq^* B)\} \subseteq \mathcal{H}.$$

   Hint: Use Zorn's Lemma, Theorem 1.10, c).

f) Let $\mathcal{A}$ be a MAD. Then there exists a MAD $\mathcal{B}$ such that every $A \in \mathcal{A}$ contains $\mathfrak{c}$ many subsets in $\mathcal{B}$ and the following holds true: if an infinite $X \subseteq \omega$ meets $\mathfrak{c}$ many elements of $\mathcal{A}$, then there exists a $B \in \mathcal{B}$ such that $B \subset^* X$.

   Hint: Enumerate $\mathcal{A}$ and continue by transfinite induction.

g) Show that there exists a family $\mathbf{H}$ of infinite subsets of ω such that $\langle \mathbf{H}, \supset^* \rangle$ is a **base matrix tree**, i.e.,

1) $\langle \mathbf{H}, \supset^* \rangle$ is a tree of height $\mathfrak{h}$,

2) the αth level is a MAD family for every $0 < \alpha < \mathfrak{h}$,

3) the branching degree of each node is $\mathfrak{c}$,

4) $\mathbf{H}$ is dense in $\langle [\omega]^\omega, \supseteq^* \rangle$.

Hint: Let $\langle \mathcal{D}_\xi : \xi < \mathfrak{h} \rangle$ be open dense families with no common refinement. At an odd stage $2\xi + 1$ use the result of e) and take an open dense family $\mathcal{H}_{2\xi+1}$ included in all $\mathcal{H}_\eta$ for $\eta < 2\xi + 1$ and $\mathcal{D}_\xi$. At even stages use the result of f). Set $\mathbf{H} = \bigcup_{\xi < \mathfrak{c}} \mathcal{H}_\xi$. We need to show that for every infinite $X \subseteq \omega$ there exists a $\xi < \mathfrak{h}$ such that X meets $\mathfrak{c}$ many elements of $\mathcal{H}_\xi$. Construct a subtree $\mathbf{G}$ of $\mathbf{H}$ of height ω such that at the nth level X meets at least 2^n elements of $\mathbf{G}$.

5.26 [AC] Other Small Cardinals

Let $\mathfrak{a} = \min\{ |\mathcal{A}| : \mathcal{A} \subseteq [\omega]^\omega$ is a MAD family$\}$.

a) Show that
$$\mathfrak{a} = \min\{ |\mathcal{A}| : \mathcal{A} \cup \{\{n\} \times \omega : n \in \omega\} \text{ is a MAD family on } \omega \times \omega \}.$$

Hint: Take $\langle A_n \in \mathcal{A} : n \in \omega \rangle$ such that $\bigcup \mathcal{A} = \bigcup_n A_n$, make them (really) disjoint and find a surjection $h : \omega \longrightarrow \omega \times \omega$ such that $h(A_n) = \{n\} \times \omega$.

b) Show that $\mathfrak{b} \leq \mathfrak{a}$.

Hint: Let $\mathcal{A}$ be the family of a). For any $A \in \mathcal{A}$ we can find an $f_A \in {}^\omega \omega$ such that A is "under" the graph of f_A. If g were a strict upper bound of $\{f_A : A \in \mathcal{A}\}$, then the set $\{\langle n, m \rangle : m \leq g(n)\} \notin \mathcal{A}$, would be almost disjoint from any set $A \in \mathcal{A} \cup \{\{n\} \times \omega : n \in \omega\}$.

A family $\mathcal{G} \subseteq [\omega]^\omega$ is **groupwise dense** if for every $A \subseteq^* B \in \mathcal{G}$ also $A \in \mathcal{G}$ and for every interval partition $\mathbf{I} = \{I_n : n \in \omega\}$ there exists a set $K \subseteq \omega$ such that $\bigcup_{n \in K} I_n \in \mathcal{G}$. The **groupwise density number** is
$$\mathfrak{g} = \min\{ \kappa : \mathcal{G}_\xi \text{ is groupwise dense for each } \xi < \kappa \text{ and } \bigcap_{\xi < \kappa} \mathcal{G}_\xi = \emptyset \}.$$

c) A groupwise dense subset of $[\omega]^\omega$ is almost dense.

Hint: If $A \subseteq \omega$ is infinite, find an interval partition $\{I_n : n \in \omega\}$ such that $I_n \cap A \neq \emptyset$ for every n.

d) Show that $\mathfrak{h} \leq \mathfrak{g}$.

Hint: A groupwise dense family is a cover of ω.

e) Show that $\mathfrak{g} \leq \mathfrak{d}$.

Hint: For any increasing $f \in {}^\omega \omega \uparrow$ the set
$$\mathcal{G}_f = \{ X \in [\omega]^\omega : (\forall m)(\exists n > m)(\exists k \geq n) \, (k \in X \wedge k < f(n)) \}$$

is groupwise dense. If $\mathcal{D} \subseteq {}^\omega \omega \uparrow$ is a dominating family, then $\{\mathcal{G}_f : f \in \mathcal{D}\}$ has empty intersection.

The **ultrafilter number** is $\mathfrak{u} = \min\{ |\mathcal{B}| : \mathcal{B}$ is a base of a non-trivial ultrafilter on $\omega \}$.

f) Show that $\mathfrak{r} \leq \mathfrak{u}$.

   Hint: A base of a non-trivial ultrafilter is a reaping family.

In Section 5.3, we defined the notion of an independent family and, for a set $A \subseteq \omega$, we wrote $(1)A = A$ and $(0)A = \omega \setminus A$.

g) Show that there exists a maximal independent family of subsets of ω.

   Hint: Apply Zorn's Lemma, Theorem 1.10, c).

The **independence number** $\mathfrak{i}$ is the minimal size of a maximal independent family of subsets of ω.

h) Show that $\mathfrak{r} \leq \mathfrak{i}$.

   Hint: Let $\mathcal{A}$ be a maximal independent family of size $\mathfrak{i}$. Show that

$$\{ \bigcap_{A \in \mathcal{A}_0} (f(A))A : \mathcal{A}_0 \in [\mathcal{A}]^{<\omega} \wedge f : \mathcal{A}_0 \longrightarrow 2 \}$$

   is a reaping family.

In the next $\mathcal{A} \subseteq [\omega]^\omega$ is an independent family of cardinality smaller than $\mathfrak{d}$, and the sets $\langle A_n \in \mathcal{A} : n \in \omega \rangle$ are mutually distinct.

i) For every $\alpha \in {}^\omega 2$ there exists an infinite set $B_\alpha \subseteq \omega$ such that

   (1) $B_\alpha \subseteq^* \bigcap_{i<n}(\alpha(i))A_i$ for all n,

   (2) $B_\alpha \cap \bigcap_{A \in \mathcal{A}_0}(f(A))A$ is infinite for any $\mathcal{A}_0 \in [\mathcal{A} \setminus \{A_n : n \in \omega\}]^{<\omega}$ and any $f : \mathcal{A}_0 \longrightarrow 2$.

   Hint: Apply the result of Exercise 5.24, f).

j) If $Z \subseteq {}^\omega 2$ is countable, then there exist pairwise disjoint infinite sets $C_\alpha \subseteq B_\alpha$ for every $\alpha \in Z$.

k) If $Z_1, Z_2 \subseteq {}^\omega 2$ are disjoint countable dense sets, then

$$\bigcup_{\alpha \in Z_i} C_\alpha \cap \bigcap_{A \in \mathcal{A}_0} (f(A))A$$

   is infinite for any $\mathcal{A}_0 \in [\mathcal{A}]^{<\omega}$, $i = 1, 2$.

l) Show that $\mathfrak{d} \leq \mathfrak{i}$.

   Hint: If $\mathcal{A} \subseteq [\omega]^\omega$ is as above, then it cannot be a maximal independent family, since both the set $C = \bigcup_{\alpha \in Z_1} B_\alpha$ and $\omega \setminus C \supseteq \bigcup_{\alpha \in Z_2} B_\alpha$ intersect every set $\bigcap_{A \in \mathcal{A}_0}(f(A))A$, $\mathcal{A}_0 \in [\mathcal{A}]^{<\omega}$ in an infinite set.

5.27 [AC] Gaps

κ, λ are infinite cardinals. A couple of sequences $\langle \{A_\xi\}_{\xi<\kappa}, \{B_\xi\}_{\xi<\lambda} \rangle$ of infinite subsets of ω is called a (κ, λ^*)-**gap** if it satisfies the conditions a)–d) of Theorem 5.53, replacing ω_1 by κ, λ, respectively. Thus the Hausdorff gap is an (ω_1, ω_1^*)-gap.

a) If there exists an (ω, λ^*)-gap, then $\lambda \geq \mathfrak{b}$.

   Hint: Set $f_\xi(n) = \max(A_n \setminus B_\xi) + 1$ (maximum of empty set is 0).
   If $\lambda < \mathfrak{b}$ there exists $g \in {}^\omega\omega$ bounding all f_ξ. The set $\bigcup_n (A_n \setminus g(n))$ witnesses that $\langle \{A_n\}_{n<\omega}, \{B_\xi\}_{\xi<\lambda} \rangle$ is not a gap.

b) There exists an $(\omega, \mathfrak{b}^*)$-gap.

   Hint: Let $\{f_\xi\}_{\xi < \mathfrak{b}}$ be the unbounded sequence of Theorem 5.44, $\omega = \bigcup_n C_n$, C_n being infinite and pairwise disjoint. Set $A_n = \bigcup_{k \leq n} C_k$ and $B_\xi = \bigcup_k (C_k \setminus f_\xi(k))$. If $C_n \subseteq^* D \subseteq B_\xi$ for any n, ξ, consider the function

   $$f(n) = \min\{m \in C_n : (\forall k \geq m)\,(k \in C_n \to k \in D)\}.$$

5.28 [AC] Booth's Lemma

If $f_n : X \longrightarrow [0,1]$, $n \in \omega$, $r \in [0,1] \cap \mathbb{Q}$, and $x \in X$, we let $L_{r,x} = \{n \in \omega : f_n(x) \leq r\}$.

a) If an infinite set $A \subseteq \omega$ does not split any of sets $L_{r,x}$, $r \in \mathbb{Q} \cap [0,1]$ for a given $x \in X$, then there exists the limit $\lim_{n \in A} f_n(x) = \inf\{r \in \mathbb{Q} \cap [0,1] : A \subseteq^* L_{r,x}\}$.

b) Assume that $\mathcal{S} \subseteq [\omega]^\omega$ is a splitting family. For every $A \in \mathcal{S}$ we set $f_n(A) = 1$ if $n \in A$, $f_n(A) = 0$ otherwise. Show that no subsequence of the sequence $\{f_n\}_{n=0}^\infty$ is pointwise convergent.

c) Show **Booth's Lemma**: The splitting number $\mathfrak{s}$ is the least size of a set X such that there exists a sequence $\{f_n\}_{n=0}^\infty$ of functions from X into $[0,1]$ without a pointwise convergent subsequence.

5.5   Games Played by Infinitely Patient Players

Let X, $|X| \geq 2$ be a set, $P \subseteq {}^\omega X$. The topology of ${}^\omega X$ is that defined by the Baire metric. We introduce an **infinite game** $\mathrm{GAME}_X(P)$ played by two players I and II as follows. In the nth move, player I chooses an element $x_{2n} \in X$ and player II chooses an element $x_{2n+1} \in X$. We assume that each player knows all preceding moves of both players. The result is a **run** $\{x_n\}_{n=0}^\infty$. Player I wins if $\{x_n\}_{n=0}^\infty \in P$. Otherwise player II wins.

The pairing function $\Pi_X : {}^\omega X \times {}^\omega X \xrightarrow[\text{onto}]{1-1} {}^\omega X$ and its inverse functions Λ_X and R_X were defined by (1.11) and (1.13), respectively. If the sequence $\{x_n\}_{n=0}^\infty$ is a run in $\mathrm{GAME}_X(P)$, then the **play of player I** is the sequence $\{x_{2n}\}_{n=0}^\infty = \Lambda(\{x_n\}_{n=0}^\infty)$ and the **play of player II** is the sequence $\{x_{2n+1}\}_{n=0}^\infty = R(\{x_n\}_{n=0}^\infty)$. Vice versa, if the play of player I is the sequence α and the play of player II is the sequence β, then the run is $\Pi(\alpha, \beta)$.

A **strategy** in game $\mathrm{GAME}_X(P)$ is a function $f : {}^{<\omega} X \longrightarrow X$. We say that player I (player II) **follows the strategy** f if he/she plays

$$x_{2n} = f(\{x_i\}_{i < 2n}), \qquad (x_{2n+1} = f(\{x_i\}_{i < 2n+1})) \tag{5.36}$$

for every $n \in \omega$, respectively. If player II plays $\{x_{2n+1}\}_{n=0}^\infty$ and player I follows the strategy f, then the run is

$$\bar{f}_I(\{x_{2n+1}\}_{n=0}^\infty) = \Pi(\{x_{2n}\}_{n=0}^\infty, \{x_{2n+1}\}_{n=0}^\infty), \tag{5.37}$$

where x_{2n} is determined by (5.36). Similarly if player II follows the strategy f, then the run is $\bar{f}_{II}(\{x_{2n}\}_{n=0}^{\infty})$. One can easily see that $\bar{f}_I : {}^{\omega}X \xrightarrow{1-1} {}^{\omega}X$ and $\bar{f}_{II} : {}^{\omega}X \xrightarrow{1-1} {}^{\omega}X$ are continuous injections.

A strategy f is **winning** for player I (player II) if for every run of the game player I (player II) wins provided that he/she followed the strategy f. Finally, the game $\text{GAME}_X(P)$ is **determined** if one of the players has a winning strategy, i.e., if there exists a function $f : {}^{<\omega}X \longrightarrow X$ which is either a winning strategy for player I or a winning strategy for player II. The assertion that the game $\text{GAME}_X(A)$ is determined for any set $A \in \Gamma({}^{\omega}X)$ will be denoted as $\text{DET}_X(\Gamma)$.

We shall need a modification of the introduced game $\text{GAME}_X(P)$. Let X, P be as above. In the **star-game** $\text{GAME}_X^*(P)$ in the nth move, player I plays a finite sequence $a_{2n} \in {}^{<\omega}X$ (possibly empty) and player II plays an element $x_{2n+1} \in X$. As above we assume that each player knows all preceding moves. Player I wins if the resulting run belongs to P. Otherwise player II wins. A strategy for player I in game $\text{GAME}_X^*(P)$ is a function $f : {}^{<\omega}X \longrightarrow {}^{<\omega}X$ and the function $\bar{f}_I$ is defined similarly. One can easily see that $\bar{f}_I : {}^{\omega}X \xrightarrow{1-1} {}^{\omega}X$ is a continuous injection. $\text{DET}_X^*(\Gamma)$ means that the star-game $\text{GAME}_X^*(P)$ is determined for every $P \in \Gamma({}^{\omega}X)$.

Finally we introduce the **Banach-Mazur game** $\text{GAME}_X^{**}(P)$. The game is defined as above with only one difference: both players play a non-empty finite sequence of elements of X. All related notions are defined similarly and $\text{DET}_X^{**}(\Gamma)$ means that the Banach-Mazur game $\text{GAME}_X^{**}(P)$ is determined for every set $P \in \Gamma({}^{\omega}X)$.

Theorem 5.58. *Let Γ be a pointclass.*

a) *If there are maps $h : X \longrightarrow Y$ and $g : Y \longrightarrow X$ such that $g \circ h = \text{id}_Y$, then $\text{DET}_X(\Gamma)$ implies $\text{DET}_Y(\Gamma)$, $\text{DET}_X^*(\Gamma)$ implies $\text{DET}_Y^*(\Gamma)$, and $\text{DET}_X^{**}(\Gamma)$ implies $\text{DET}_Y^{**}(\Gamma)$.*

b) *$\text{DET}_{\omega}(\Gamma)$ implies both $\text{DET}_{\omega}^*(\Gamma)$ and $\text{DET}_{\omega}^{**}(\Gamma)$.*

Proof. a) The idea of the proof is simple. We shift the game from Y to X by g, we use the winning strategy in corresponding runs of this game and then we shift the run by h back to Y.

Define $H(\{x_n\}_{n=0}^{\infty}) = \{h(x_n)\}_{n=0}^{\infty}$. Then $H : {}^{\omega}X \xrightarrow{\text{onto}} {}^{\omega}Y$ is continuous. Suppose that $P \in \Gamma({}^{\omega}Y)$ and set $Q = H^{-1}(P) \in \Gamma({}^{\omega}X)$. We shall play on X. Assume that φ is a winning strategy for player I in $\text{GAME}_X(Q)$. We define a strategy ψ in $\text{GAME}_Y(P)$ by setting $\psi(\{y_i\}_{i<n}) = h(\varphi(\{g(y_i)\}_{i<n}))$ for $n \geq 0$.

Let $\{y_n\}_{n=0}^{\infty}$ be a run in $\text{GAME}_Y(P)$ in which player I follows strategy ψ. We want to show that $\{y_n\}_{n=0}^{\infty} \in P$. Set $x_n = g(y_n)$ for each n. Consider a run in the game on X, in which player II plays x_{2n+1} for any n and player I follows the strategy φ. By induction we show that $x_{2n} = \varphi(\{x_i\}_{i<2n})$ for each n and therefore

$\{x_n\}_{n=0}^{\infty} \in Q$. Actually, by definition of the strategy ψ we obtain

$$x_{2n} = g(y_{2n}) = g(\psi(\{y_i\}_{i<2n})) = g(h(\varphi(\{x_i\}_{i<2n}))) = \varphi(\{x_i\}_{i<2n}).$$

Thus $\{x_n\}_{n=0}^{\infty} \in Q = H^{-1}(P)$ and therefore $\{y_n\}_{n=0}^{\infty} = H(\{x_n\}_{n=0}^{\infty}) \in P$.

We can proceed similarly in the case when player II has a winning strategy and also for games GAME_X^* and GAME_X^{**}.

b) We fix a bijection $\theta : \omega \xrightarrow[\text{onto}]{1-1} {}^{<\omega}\omega$. Define a continuous map $\Theta : {}^{\omega}\omega \longrightarrow {}^{\omega}\omega$ by

$$\Theta(\{x_n\}_{n=0}^{\infty}) = \theta(x_0) \frown x_1 \frown \cdots \frown \theta(x_{2n}) \frown x_{2n+1} \frown \cdots .$$

Consider game $\mathrm{GAME}_{\omega}^*(P)$, $P \in \Gamma({}^{\omega}\omega)$. Let $Q = \Theta^{-1}(P)$. Assume that player II has a winning strategy φ in $\mathrm{GAME}_{\omega}(Q)$. We define a strategy ψ for player II in $\mathrm{GAME}_{\omega}^*(P)$ as follows. Let $a_i \in {}^{<\omega}\omega$ for $i = 0, 2, \ldots, 2n$ and $a_i \in \omega$ for $i = 1, 3, \ldots, 2n - 1$. We write $x_i = \theta^{-1}(a_i)$ for $i = 0, 2, \ldots, 2n$ and $x_i = a_i$ for $i = 1, 3, \ldots, 2n - 1$. Set $\psi(\{a_i\}_{i<2n+1}) = \varphi(\{x_i\}_{i<2n-1})$. Similarly as above one can easily see that ψ is a winning strategy for player II in $\mathrm{GAME}_{\omega}^*(P)$.

If player I has a winning strategy the proof runs similarly. $\square$

Especially, for any pointclass Γ we obtain

$$\mathrm{DET}_{\omega}(\Gamma) \to \mathrm{DET}_2(\Gamma), \quad \mathrm{DET}_{\omega}^*(\Gamma) \to \mathrm{DET}_2^*(\Gamma), \quad \mathrm{DET}_{\omega}^{**}(\Gamma) \to \mathrm{DET}_2^{**}(\Gamma).$$

Lemma 5.59. *If $P \subseteq {}^{\omega}X$ is countable, then player II has a winning strategy in both games $\mathrm{GAME}_X(P)$ and $\mathrm{GAME}_X^*(P)$.*

The proof is a Cantor's diagonal argument. Enumerate the set P. In the nth move player II plays in such a way that the resulting run will avoid the nth element of P. $\square$

Lemma 5.60. *If player I has a winning strategy in any of games $\mathrm{GAME}_X(P)$ or $\mathrm{GAME}_X^*(P)$, then the set P contains a perfect subset homeomorphic to the Cantor middle-third set.*

Proof. If f is a winning strategy for player I, then $\bar{f}_I({}^{\omega}X) \subseteq P$. Since the Cantor middle-third set is a compact subset of ${}^{\omega}X$, the assertion follows. $\square$

Actually we can prove much more.

Theorem 5.61 (M. Davis). *Player I has a winning strategy in $\mathrm{GAME}_2^*(P)$ if and only if the set P contains a perfect subset. Player II has a winning strategy in $\mathrm{GAME}_2^*(P)$ if and only if the set P is countable.*

Proof. Assume that $P \subseteq {}^{\omega}2$ contains a perfect subset. We describe (informally) a winning strategy for player I in game $\mathrm{GAME}_2^*(P)$. Let us recall that for a finite sequence $s \in {}^{<\omega}2$ we write $[s] = \{\alpha \in {}^{\omega}2 : s \subseteq \alpha\}$. The family $\{[s] : s \in {}^{<\omega}2\}$ is an open base of the topology on ${}^{\omega}2$.

Let $\{x_i\}_{i \leq n}$ be the result of the run after k moves and $[\{x_i\}_{i \leq n}] \cap P \neq \emptyset$. Since P is a perfect set there are two different elements $\alpha, \beta \in [\{x_i\}_{i \leq n}] \cap P$. Take the first $m > n$ such that $\alpha(m) \neq \beta(m)$. In the next step player I plays the finite sequence

$$\{x_i\}_{i=n+1}^{m-1} = \{\alpha(i)\}_{i=n+1}^{m-1} = \{\beta(i)\}_{i=n+1}^{m-1}$$

(if $m = n+1$, this sequence is empty). Player II can answer x_m equal to 0 or 1. In both cases we have $[\{x_i\}_{i \leq m}] \cap P \neq \emptyset$. The resulting run will be an element of P.

Assume that there is a winning strategy f for player II in $\mathrm{GAME}_2^*(P)$. We introduce some notation. Let R denote the set of all partial runs following the strategy f. A typical element of R has the form $s_0 \frown f(s_0) \frown \cdots \frown s_n \frown f(s_n)$.

We claim that

$$(\forall \alpha \in P)(\exists s \in R, \alpha \in [s])(\forall t \in R)\,((s \subseteq t, s \neq t) \to \alpha \notin [t]). \tag{5.38}$$

Actually, if not, then one can construct a strictly increasing sequence $s_n \in R$, $s_n \subseteq s_{n+1}$, $\alpha \in [s_n]$. Then $\alpha = \bigcup_n s_n \in P$ is the run following strategy f, contradicting the assumption that f is a winning strategy for player II.

Moreover, it is easy to see that the element s of (5.38) is unique. Since R is countable also P is countable.   □

Theorem 5.62 (S. Banach – S. Mazur). *Player* I (*player* II) *has a winning strategy in* $\mathrm{GAME}_\omega^{**}(P)$ *if and only if there exists a non-empty open set* $U \subseteq {}^\omega\omega$ *such that* $U \setminus P$ *is meager (the set* P *is meager).*

Proof. Let P be meager. Then there exist open dense sets $\langle U_n : n \in \omega \rangle$ such that $P \cap \bigcap_n U_n = \emptyset$. If the partial run is b, then in the nth move player II plays as follows: since U_n is open dense, the intersection $[b] \cap U_n$ is non-empty and player II plays a non-empty $a \in {}^{<\omega}\omega$ such that $[b \frown a] \subseteq [b] \cap U_n$. The resulting run will be in $\bigcap_n U_n$ and player II wins.

If there exists a non-empty open set U such that $U \setminus P$ is meager, then player I starts with $a \in {}^{<\omega}\omega$ such that $[a] \subseteq U$. Since $[a] \setminus P$ is meager too there are open dense subsets $\langle U_n \subseteq [a] : n \in \omega \rangle$ such that $[a] \cap \bigcap_n U_n \subseteq P$. Now, if the partial run in the nth move is $b \supseteq a$, then player I chooses such a sequence c that $[b \frown c] \subseteq U_n$. Then the resulting run will be in P.

Assume now that f is a winning strategy for player II in $\mathrm{GAME}_\omega^{**}(P)$. We denote by R the set of all partial runs in which player II follows the winning strategy f. For any $a \in R$ the set $B_a = [a] \setminus \bigcup\{[b] : a \subseteq b, a \neq b, b \in R\}$ is closed and nowhere dense. Note that B_a is the set of those runs extending a which do not follow the strategy f. For any $x \in {}^\omega\omega$, if any partial run $a \subseteq x$ can be extended to a partial run $b \neq a$, $b \subseteq x$ following the strategy f, then $x \notin P$. Thus

$$x \in P \to (\exists a \in R)\, x \in B_a.$$

Since R is countable the set P is meager.

Let f be a winning strategy for player I in $\text{GAME}_\omega^{**}(P)$. Set $a = f(\emptyset)$, i.e., a is the first move of player I. Then f is a winning strategy for player II in $\text{GAME}_\omega^{**}([a] \setminus P)$. Thus $[a] \setminus P$ is meager.   □

One can easily prove that $\text{GAME}_X(P)$ is determined for P closed or open – see Exercise 5.30. However, the basic result is

Theorem [AC] 5.63 (D.A. Martin). *Let X be a non-empty set. If $P \subseteq {}^\omega X$ is a Borel set, then $\text{GAME}_X(P)$ is determined, i.e., $\text{DET}_X(\text{BOREL})$ holds true.*

A proof can be found, e.g., in D.A. Martin [1975] or A.S. Kechris [1995]. For more complicated sets (analytic) the determinacy is not provable in **ZFC**, see T. Jech [2006].

Exercises

5.29 Non-determined Game

a) If ${}^\omega 2$ can be well ordered, then there exists a set $P \subseteq {}^\omega 2$ such that the game $\text{GAME}_2(P)$ is not determined.

   Hint: Let $\{\gamma_\xi : \xi < \mathfrak{c}\}, \{f_\xi : \xi < \mathfrak{c}\}, \{g_\xi : \xi < \mathfrak{c}\}$ be enumerations of ${}^\omega 2$, all strategies of player I and all strategies of player II, respectively. By transfinite induction find sequences $\{\alpha_\xi : \xi < \mathfrak{c}\}, \{\lambda_\xi : \xi < \mathfrak{c}\}$ such that $\bar{f}_{\xi I}(\alpha_\xi) \notin \{\lambda_\eta : \eta < \xi\}$ and $\bar{g}_{\xi II}(\lambda_\xi) \notin \{\alpha_\eta : \eta < \xi\}$. Take $P = \{\lambda_\xi : \xi < \mathfrak{c}\}$.

b) Show similar results for the star-game and the Banach-Mazur game.

5.30 Gale-Steward Theorem

Let $T \subseteq {}^{<\omega}\omega$ be a pruned tree. Elements of ${}^{<\omega}\omega$ are partial runs in $\text{GAME}_\omega([T])$.

a) Assume that II has no winning strategy in the game $\text{GAME}_\omega([T])$. A partial run $s \in {}^{2n}\omega$ is **not losing** for I if II has no winning strategy in $\text{GAME}_\omega(\{\alpha \in {}^\omega\omega : s \subseteq \alpha\})$. Show that if s is not losing for I, then there exists an m such that $s \frown m \in T$ and $s \frown m \frown k$ is not losing for I for any k.

b) Assuming that II does not have a winning strategy in $\text{GAME}_\omega([T])$ define a winning strategy for I.

   Hint: Use the result of a).

c) Prove the **Gale-Steward Theorem**: If $A \subseteq {}^\omega\omega$ is closed, then $\text{GAME}_\omega(A)$ is determined.

d) If $A \subseteq {}^\omega\omega$ is open, then $\text{GAME}_\omega(A)$ is determined.

e) If $A \subseteq {}^\omega\omega$ is closed or open, then $\text{GAME}_\omega^*(A)$ and $\text{GAME}_\omega^{**}(A)$ are determined.

5.31 Star-game

If X is a Polish space with a countable base $\mathcal{B}$, for a set $A \subseteq X$ we can define a **star-game** $\text{GAMEG}^*(X, A)$ as follows: at the nth step player I plays two basic open sets U_n^0, U_n^1 of diameter less than 2^{-n} with disjoint closure, and player II picks one of them $i_n = 0, 1$. If x is the only point of $\bigcap_n U_n^{i_n}$, then I wins if $x \in A$.

a) I has a winning strategy in $\text{GAMEG}^*(X, A)$ if and only if A contains a perfect subset.

b) II has a winning strategy in $\text{GAMEG}^*(X, A)$ if and only if A is countable.

c) If A is closed, then $\mathrm{GAMEG}^*(X, A)$ is determined.

d) Show that the star-game $\mathrm{GAMEG}_2^*(A)$ is the game $\mathrm{GAMEG}^*(^\omega 2, A)$.

5.32 Banach-Mazur Game

If X is a Polish space with a countable base $\mathcal{B}$, for a set $A \subseteq X$ we can define the **Banach-Mazur game** $\mathrm{GAMEG}^{**}(X, A)$ as follows: in the n-th move player I plays an open basic set $U_n \subseteq V_{n-1}$ with $\mathrm{diam}(U_n) < 2^{-n}$ and player II plays an open basic set $V_n \subseteq U_n$. Player II wins if $\bigcap_n U_n = \bigcap_n V_n \subseteq A$.

a) Player II has a winning strategy in $\mathrm{GAMEG}^{**}(X, A)$ if and only if A is comeager.

   Hint: If A is not comeager, then there exist a non-empty open set U_0 and open dense sets $\langle G_n : n \in \omega \rangle$ such that $U_0 \cap A \cap \bigcup_n G_n = \emptyset$. Player I starts with U_0 and for a move V_n by players II answers by a non-empty open $U_{n+1} \subseteq V_n \cap G_n$. Conversely, if A is comeager, $\bigcap_n W_n \subseteq A$, W_n open dense, at n-th move, player II plays a subset of W_n.

b) I has a winning strategy in $\mathrm{GAMEG}^{**}(X, A)$ if and only if A is meager in a non-empty open set U.

   Hint: Let f be a winning strategy for player I. Let $U = f(\emptyset)$ and

   $$G_n = \bigcup \{ f(\{V_i\}_{i=0}^n) : (\forall i \le n)\, V_i \in \mathcal{B} \}.$$

   Then G_n is open dense and $U \cap A \cap \bigcap_n G_n = \emptyset$.

c) If A is a G_δ set, then the Banach-Mazur game $\mathrm{GAMEG}^{**}(X, A)$ is determined.

5.33 [wAC] Unfolded Game

Let X be a perfect Polish space, $C \subseteq X \times {}^\omega\omega$ and

$$A = \{ x \in X : (\exists \alpha)\, [x, \alpha] \in C \}.$$

a) Let $f : X \longrightarrow Y$ be continuous, $A \subseteq Y$. If player I (player II) has a winning strategy in $\mathrm{GAMEG}^{**}(X, f^{-1}(A))$, then she/he has a winning strategy in $\mathrm{GAMEG}^{**}(Y, A)$ as well.

b) If I has a winning strategy in $\mathrm{GAMEG}^*(X \times {}^\omega\omega, C)$, then I has a winning strategy in $\mathrm{GAMEG}^*(X, A)$.

c) If II has a winning strategy in $\mathrm{GAMEG}^*(X \times {}^\omega\omega, C)$, then II has a winning strategy in $\mathrm{GAMEG}^*(X, A)$.

d) If C is closed, then $\mathrm{GAMEG}^*(X, A)$ is determined.

e) If C is closed, then either A is countable or A contains a perfect subset.

f) If I has a winning strategy in $\mathrm{GAMEG}^{**}(X \times {}^\omega\omega, C)$, then I has a winning strategy in $\mathrm{GAMEG}^{**}(X, A)$.

g) If II has a winning strategy in $\mathrm{GAMEG}^{**}(X \times {}^\omega\omega, C)$, then II has a winning strategy in $\mathrm{GAMEG}^{**}(X, A)$.

h) If C is closed, then $\mathrm{GAMEG}^{**}(X, A)$ is determined.

i) If C is closed, then A possesses the Baire Property.

Historical and Bibliographical Notes

A Souslin scheme as "an operation $(\mathcal{A})$" was introduced and investigated by F. Hausdorff as a generalization of infinite countable union and intersection of a series of sets. M. Souslin [1917] used a Souslin scheme to define and to investigate an analytic set. N.N. Luzin [1927] investigated the Lebesgue decomposition and introduces the notion of a sieve, in French "le crible". Theorem 5.12 was essentially proved by N.N. Luzin and W. Sierpiński [1923].

A notion of a pointclass was introduced by Y.N. Moschovakis [1980]. The results of Section 5.2 were presented in several papers by Y.N. Moschovakis and then in [1980]. Sometimes our presentation follows A. Kechris [1995]. However, a close analogy between properties of projective sets and sets of integers related to the recursive theory was investigated by A. Mostowski [1946].

A study of Boolean algebras started with the symbolization of language of logic by G. Boole [1847]. Then several authors investigated Boolean algebras and its properties. We mention the paper by A. Tarski [1939]. The Boolean algebra of Exercise 5.20 was introduced by A. Lindenbaum (unpublished) and later on by A. Tarski – compare footnotes in H. Rasiowa and R. Sikorski [1963], pp. 245–246. Theorem 5.28 was proved by M.H. Stone [1936] and started a new era of study of Boolean algebras. His main results are contained in Exercise 5.15. Basic literature on Boolean algebras is R. Sikorski [1964] and the handbook J.D. Monk and R. Bonnet [1989]. Theorem 5.33 was proved by H. Rasiowa and R. Sikorski [1950].

Theorem 5.39 was proved by G. Fichtenholz and L. Kantorovitch [1934] for a countable set, and then generalized for any infinite set by F. Hausdorff [1936b]. B. Pospíšil [1937] discovered it independently as a tool for proving Theorem 5.40.

F. Rothberger [1939] and [1941] started to investigate the properties of the bounding number $\mathfrak{b}$. He was courageous enough to assume that $\mathfrak{b} < \mathfrak{c}$. For basic properties of $\mathfrak{b}$ and other cardinal invariants (small uncountable cardinals) we recommend E.K. van Douven [1984], J.E. Vaughan [1990], A. Blass [1993] or A. Blass [2010]. Note that A. Blass [1993], [2010] prefers to speak about cardinal characteristic instead of cardinal invariant. A Tukey connection was implicitly used by D. Fremlin [1984a]. That was P. Vojtáš [1993], who explicitly introduced the notion of a Galois-Tukey connection and presented its basic properties. Theorem 5.47 was proved by D.A. Martin and R.M. Solovay [1970]. B. Balcar, J. Pelant and P. Simon [1980] introduced the number $\mathfrak{h}$ and proved Theorem 5.50. See also B. Balcar and P. Simon [1989]. Moreover they showed that there exists a base matrix tree of Exercise 5.25. The presentation of Exercise 5.25 follows A. Blass [2010]. A Hausdorff gap was constructed by F. Hausdorff [1909]. Our presentation partially follows A. Błaszczyk and S. Turek [2007].

Rothberger's Theorem, Exercise 5.23 was proved in F. Rothberger [1948]. The groupwise density number $\mathfrak{g}$ was introduced and investigated by A. Blass [1989]. Booth's Lemma, Exercise 5.28 was proved by D. Booth [1970].

In spite of the fact that S. Banach and S. Mazur obtained the first important results related to the infinite games in 1930s, the first published papers devoted to infinite games appeared later, e.g., D. Gale and F.M. Stewart [1953], J. Mycielski and A. Zieba [1955], M. Davis [1964], and J. Mycielski [1964a]. Actually D. Gale and F.M. Stewart showed determinacy of $\mathrm{GAME}_\omega(A)$ for A closed. We recommend R. Telgársky [1987] for a detailed history of infinite games.

The most important result is Theorem 5.63 proved by D.A. Martin [1975]. However, H. Friedmann [1971] shows that the Axiom of Replacement is essentially used in this proof. Actually, he shows that there exists a model of the Zermelo set theory $\mathbf{Z} = \mathbf{ZF} - $ "Replacement", in which Martin's Theorem 5.63 does not hold true. In 1962 J. Mycielski and H. Steinhaus [1962] proposed the Axiom of Determinacy $\mathbf{AD}$ as an alternative to the Axiom of Choice $\mathbf{AC}$. The main consequences of $\mathbf{AD}$ were published by J. Mycielski [1964b], [1966] and in a common paper with S. Świerczkowski [1964].

The idea of an unfolded game goes back to R.M. Solovay and then to D.A. Martin and A. Kechris, see, e.g., A. Kechris [1995].

Chapter 6

Descriptive Set Theory

L'étude de propriétés des ensembles projectifs est difficile; cependant pour comprendre ce que sont les ensembles projectifs et quels sont les problèmes qui s'y posent, on n'a pas besoin de connaissances spéciales.

Wacław Sierpiński [1950].

The study of properties of the projective sets is difficult; nevertheless, one needs no special knowledge to understand what the projective sets are and what problems they pose.

In 1905 Henri Lebesgue [1905] published a large paper *Sur les fonctions représentables analytiquement*, which strongly influenced the next investigations in a domain of mathematics that we call today the descriptive set theory. The paper was mainly devoted to the study of the Baire Hierarchy of real functions. Moreover, a proof of one theorem was wrong (the theorem is actually true). H. Lebesgue used the argument that a continuous image of a Borel set is a Borel set. M.J. Souslin [1917] observed the error and found a counterexample: a Borel set with a continuous image not being Borel. That was a beginning of the study of a new important class of subsets of the real line and Polish spaces.

We assume that the reader is familiar with the topic of Sections 3.3, 5.1 and 5.2. Section 6.1 is devoted to the study of basic properties of Borel subsets of a Polish space. In Section 6.2 we present fundamental properties of analytic sets. In Section 6.3 we describe the Projective Hierarchy. Moreover, we show the classical relationship of this hierarchy with logic, mainly with a form of a formula that defines a projective set. Finally Section 6.4 contains deep fundamental properties of $\mathbf{\Pi}_1^1$ and $\mathbf{\Sigma}_2^1$ sets.

6.1   Borel Hierarchy

In Section 3.3 we introduced the Borel Hierarchy. Now we present some of its properties. We apply the Souslin schemes to obtain fundamental results about Borel sets.

By slight modification of the proof of Theorem 5.5 we obtain

Theorem [wAC] 6.1. *For any Polish space X there exists a closed set $C \subseteq {}^{\omega}\omega$ and a continuous bijection $f : C \xrightarrow[\text{onto}]{1-1} X$.*

For a proof we shall need an auxiliary result[1]

Lemma [wAC] 6.2. *For every F_σ subset A of a Polish space X and for every $\varepsilon > 0$, there exist pairwise disjoint F_σ sets $\langle A_n : n \in \omega \rangle$ of diameter less than ε such that $A = \bigcup_n A_n$ and $\overline{A_n} \subseteq A$ for every n.*

Proof. Since A is an F_σ set, there exists a sequence of closed sets $\langle B_n : n \in \omega \rangle$ such that $A = \bigcup_n B_n$. We may suppose that $\operatorname{diam}(B_n) < \varepsilon$ (if not, cover the space X by countably many closed sets of diameter $< \varepsilon$ and take the intersections). We set $F_n = \bigcup_n (B_n \setminus \bigcup_{k<n} B_k)$. Then $A = \bigcup_n F_n$. The sets F_n are pairwise disjoint. Since every F_n is an F_σ set, there exist closed sets $\langle C_m^n : m \in \omega \rangle$ such that $F_n = \bigcup_m C_m^n$. We may suppose that $C_m^n \subseteq C_{m+1}^n$. Then $F_n = \bigcup_m (C_m^n \setminus C_{m-1}^n)$ (where $C_{-1}^n = \emptyset$). The F_σ sets $C_m^n \setminus C_{m-1}^n$, $n, m \in \omega$ are pairwise disjoint and

$$\overline{C_m^n \setminus C_{m-1}^n} \subseteq C_m^n \subseteq A.$$

If $\langle A_k : k \in \omega \rangle$ is an enumeration of $\langle C_m^n \setminus C_{m-1}^n : n, m \in \omega \rangle$, we are done.   □

Proof of Theorem 6.1. We construct a regular F_σ Souslin scheme $\langle {}^{<\omega}\omega, \varphi \rangle$ with vanishing diameter such that $L({}^{<\omega}\omega, \varphi) = X$ and the kernel $C = \operatorname{Ker}({}^{<\omega}\omega, \varphi)$ is closed.

We set $\varphi(\emptyset) = X$. Assume that the values $\varphi(s)$, $s \in {}^k\omega$ are defined. By Lemma 6.2 there are pairwise disjoint F_σ sets A_n, $n \in \omega$ with diameter less than 2^{-k-1} and such that $\varphi(s) = \bigcup_n A_n$ and $\overline{A_n} \subseteq \varphi(s)$. Set $\varphi(s \frown n) = A_n$.

Now we construct a closed Souslin scheme by setting $\psi(s \frown n) = \overline{\varphi(s \frown n)}$. Since $\psi(s \frown n) \subseteq \varphi(s) \subseteq \psi(s)$ we obtain $\bigcap_n \psi(\alpha|n) = \bigcap_n \varphi(\alpha|n)$ for any $\alpha \in {}^{\omega}\omega$. Hence $\operatorname{Ker}({}^{<\omega}\omega, \psi) = C$ and the continuous associated map $f : C \longrightarrow X$ for $\langle {}^{<\omega}\omega, \varphi \rangle$ is also the associated map for $\langle {}^{<\omega}\omega, \psi \rangle$. Since $\langle {}^{<\omega}\omega, \varphi \rangle$ is a regular Souslin scheme, the associated map f is injective.   □

Theorem [wAC] 6.3. *If B is a Borel subset of a Polish space X, then there exist a closed set $C \subseteq {}^{\omega}\omega$ and a continuous bijection $f : C \xrightarrow[\text{onto}]{1-1} B$.*

---

[1]Compare with Theorem 5.20, d).

Proof. We define a family $\mathcal{F}$ of subsets of X as follows

$$A \in \mathcal{F} \equiv (A \subseteq X \wedge (\exists f)(\exists C)\,(C \subseteq {}^{\omega}\omega \text{ closed } \wedge f : C \xrightarrow[\text{onto}]{1-1} A \text{ continuous})).$$

We show that the family $\mathcal{F}$ is closed under countable pairwise disjoint unions, under countable intersections, and contains every open subset of X. Then by Theorem 3.48 the family $\mathcal{F}$ contains every Borel subset of X and the theorem follows.

Since every open subset of a Polish space is a Polish space, by Theorem 6.1 the family $\mathcal{F}$ contains every open subset of X.

Assume that $A_n \in \mathcal{F}$, $f_n : C_n \xrightarrow[\text{onto}]{1-1} A_n$, f_n are continuous and $C_n \subseteq {}^{\omega}\omega$ are closed for every $n \in \omega$.

The projections $\mathrm{Proj}_n : {}^{\omega}\omega \xrightarrow{\text{onto}} {}^{\omega}\omega$ defined by (1.10) are continuous. We denote $D_n = \mathrm{Proj}_n^{-1}(C_n)$ and $g_n = \mathrm{Proj}_n \circ f_n$. Then any of the mappings g_k maps the closed set

$$D = \{x \in \bigcap_n D_n : (\forall n, m)\, g_n(x) = g_m(x)\} \tag{6.1}$$

one-to-one onto $\bigcap_n A_n$.

Assume now that $\langle A_n \in \mathcal{F} : n \in \omega \rangle$ are pairwise disjoint sets. We define mappings $S_n : {}^{\omega}\omega \longrightarrow {}^{\omega}\omega$ setting $S_n(\alpha) = \beta$, where $\beta(0) = n$ and $\beta(k+1) = \alpha(k)$ for $k \in \omega$. Then $f : \bigcup_n S_n(C_n) \xrightarrow[\text{onto}]{1-1} \bigcup_n A_n$, where $f = \bigcup_n (S_n^{-1} \circ f_n)$, is continuous. Moreover $\bigcup_n S_n(C_n)$ is a closed subset of ${}^{\omega}\omega$. $\square$

Theorem [wAC] 6.4. *Let X be a Polish space. The family of subsets of X sifted by a closed Souslin scheme $\langle {}^{<\omega}\omega, \varphi \rangle$ with vanishing diameter contains all non-empty closed subsets of X and is closed under countable intersections and countable unions.*

Proof. A non-empty closed subset A of a Polish space X with the subspace topology is a Polish space and therefore, by Theorem 5.5, is sifted by a closed Souslin scheme.

Assume that $A_n \subseteq X$, $n \in \omega$ are sifted by Souslin schemes $\langle {}^{<\omega}\omega, \varphi_n \rangle$ with vanishing diameter[2]. Define a mapping φ on ${}^{<\omega}\omega$ as follows:

$$\varphi(\emptyset) = X, \quad \varphi(n \frown s) = \varphi_n(s) \text{ for } s \in {}^{<\omega}\omega.$$

Then

$$L({}^{<\omega}\omega, \varphi) = \bigcup_n L({}^{<\omega}\omega, \varphi_n),$$

and therefore $\bigcup_n A_n$ is sifted by a closed Souslin scheme with vanishing diameter.

By Corollary 5.8 there are continuous surjections $f_n : {}^{\omega}\omega \xrightarrow{\text{onto}} A_n$ for every $n \in \omega$. Similarly as in the proof of Theorem 6.3, the maps $g_n = \mathrm{Proj}_n \circ f_n$

---

[2]We have used the axiom **wAC**.

are continuous and every g_n maps the closed set D defined by (6.1) onto the intersection $\bigcap_n A_n$. $\square$

By Theorem 3.48 we obtain

Corollary [wAC] 6.5. *Every Borel subset of a Polish space is sifted by a closed Souslin scheme with vanishing diameter.*

Corollary [wAC] 6.6. *Let X be a Polish space. The family of subsets of X which are a continuous image of the Baire space $^\omega\omega$ contains all non-empty closed subsets of X and is closed for countable intersections and unions. Thus, every Borel subset of X is a continuous image of the Baire space $^\omega\omega$.*

By Theorem 5.4 we obtain

Corollary [wAC] 6.7 (Alexandroff-Hausdorff Theorem). *Any Borel subset of a Polish space is either countable or contains a copy of the Cantor middle-third set and therefore has the cardinality $\mathfrak{c}$.*

Corollary [wAC] 6.8. *Assume that X is a perfect Polish space. If $A \subseteq X$ is a non-meager set possessing the Baire Property, then there exists a Borel meager set $B \subseteq A$ such that $|B| = \mathfrak{c}$. If μ is a diffused Borel measure on X and $A \subseteq X$ is measurable of positive measure, then there exists a Borel measure zero set $B \subseteq A$ such that $|B| = \mathfrak{c}$.*

Proof. If $A \subseteq X$ has the Baire Property, then by Theorem 3.32 there exists a G_δ set $G \subseteq A$ such that $A \setminus G$ is meager. Therefore G is uncountable and contains a copy of the Cantor middle-third set. Any meager Borel subset of the Cantor middle-third set is also a Borel meager subset of A.

Similarly, if $A \subseteq X$ is measurable then by Theorem 4.4 there is an F_σ set $F \subseteq A$ of equal measure. Since μ is diffused, F cannot be countable, therefore F contains a copy of the Cantor middle-third set $\mathbb{C}$. If $\mu(\mathbb{C}) = 0$, we are done. Otherwise apply Theorem 4.13. $\square$

Now we exploit a general result about pointclasses for Borel Hierarchy.

Theorem [wAC] 6.9. *If X is a Polish space, then $\mathbf{\Sigma}^0_\xi(X)$ has the σ-reduction property and $\mathbf{\Pi}^0_\xi(X)$ has the separation property for every $\xi > 1$.*

Proof. By definition we have $\mathbf{\Sigma}^0_\xi(X) = \mathbf{\Sigma}^0(\bigcup_{\eta<\xi} \mathbf{\Pi}^0_\eta(X))$ for any $\xi > 1$. By Theorem 3.39, $\neg\bigcup_{\eta<\xi} \mathbf{\Pi}^0_\eta(X) \subseteq \mathbf{\Sigma}^0_\xi(X)$. The theorem follows by Theorem 5.20. $\square$

Theorem [wAC] 6.10. *The pointclasses $\mathbf{\Sigma}^0_\xi$ and $\mathbf{\Pi}^0_\xi$ are $^\omega 2$-parametrized for all $\xi > 0$.*

Proof. Fix a separable metric space X and a countable ordinal $\xi > 0$. We construct by induction a family $\{U_\eta : 0 < \eta \leq \xi\}$ such that every $U_\eta \subseteq {}^\omega 2 \times X$ is $\mathbf{\Sigma}^0_\eta(X)$-universal. Then $^\omega 2 \times X \setminus U_\eta$ is $\mathbf{\Pi}^0_\eta(X)$-universal.

Let $P_{n,X}$ be the function defined in the proof of Theorem 5.19, c). Let f be a function of Theorem 1.21, a).

Take for U_1 the universal set for open sets defined by (5.18). $U_{\eta+1}$ is constructed from U_η as

$$U_{\eta+1} = \{\langle a, x\rangle : (\exists n)\,\langle \mathrm{Proj}_n(a), x\rangle \in ({}^\omega 2 \times X \setminus U_\eta)\} = \bigcup_n P_{n,X}^{-1}({}^\omega 2 \times X \setminus U_\eta).$$

If $\eta \le \xi$ is a limit ordinal then $f(\eta) = \{\eta_n : n \in \omega\}$ is a non-decreasing sequence such that $\eta = \sup\{\eta_n : n \in \omega\}$. We set

$$U_\eta = \{\langle a, x\rangle : (\exists n)\,\langle \mathrm{Proj}_n(a), x\rangle \in ({}^\omega 2 \times X \setminus U_{\eta_n})\} = \bigcup_n P_{n,X}^{-1}({}^\omega 2 \times X \setminus U_{\eta_n}).$$

Then $U_\xi \in \mathbf{\Sigma}_\xi^0({}^\omega 2 \times X)$ and it is easy to see that U_ξ is a universal set for $\mathbf{\Sigma}_\xi^0(X)$. $\qquad\square$

Corollary [wAC] 6.11. *Let Y be an uncountable Polish space. The pointclasses $\mathbf{\Sigma}_\xi^0$ and $\mathbf{\Pi}_\xi^0$ are Y-parametrized for all $\xi > 0$. However, there is no Y-universal set for the family $\mathbf{\Delta}_\xi^0(Y)$, $0 < \xi < \omega_1$.*

Proof. The former part follows from the theorem by Theorem 5.16, b) and Corollary 6.7. The latter one follows by Lemma 5.14. $\qquad\square$

Theorem [wAC] 6.12. *If X is an uncountable Polish space X, then all inclusions $\mathbf{\Delta}_\xi^0(X) \subseteq \mathbf{\Sigma}_\xi^0(X)$, $\mathbf{\Delta}_\xi^0(X) \subseteq \mathbf{\Pi}_\xi^0(X)$, $\mathbf{\Sigma}_\xi^0(X) \subseteq \mathbf{\Delta}_{\xi+1}^0(X)$, $\mathbf{\Pi}_\xi^0(X) \subseteq \mathbf{\Delta}_{\xi+1}^0(X)$ are proper for any $0 < \xi < \omega_1$. Consequently, the Borel Hierarchy is proper for any uncountable Polish space.*

Proof. Since the pointclass $\mathbf{\Delta}_\xi^0$ is self-dual, for each Polish space X and each $\xi > 0$ any of the equations $\mathbf{\Delta}_\xi^0(X) = \mathbf{\Sigma}_\xi^0(X)$, $\mathbf{\Delta}_\xi^0(X) = \mathbf{\Pi}_\xi^0(X)$, $\mathbf{\Sigma}_\xi^0(X) = \mathbf{\Delta}_{\xi+1}^0(X)$, $\mathbf{\Pi}_\xi^0(X) = \mathbf{\Delta}_{\xi+1}^0(X)$ implies the equation $\mathbf{\Sigma}_\xi^0(X) = \mathbf{\Pi}_\xi^0(X)$. Thus we have to show that $\mathbf{\Sigma}_\xi^0(X) \ne \mathbf{\Pi}_\xi^0(X)$.

Suppose, to get a contradiction, that $\mathbf{\Sigma}_\xi^0(X) = \mathbf{\Pi}_\xi^0(X)$ for a $\xi > 0$. Let U be an ${}^\omega 2$-universal set for $\mathbf{\Sigma}_\xi^0(X)$. By Corollary 5.6 there exists a set $C \subseteq X$ and a homeomorphism $h : {}^\omega 2 \xrightarrow[\text{onto}]{1-1} C$. We set

$$D = \{x \in C : \langle h^{-1}(x), x\rangle \notin U\}.$$

Then $D \in \mathbf{\Pi}_\xi^0 = \mathbf{\Sigma}_\xi^0$. Therefore there exists an $a \in {}^\omega 2$ such that $D = U_a$. Then

$$\langle a, h(a)\rangle \in U \equiv h(a) \in D \equiv \langle a, h(a)\rangle \notin U,$$

which is a contradiction.

If **wAC** holds true, then by Corollary 6.7 there exists an embedding of ${}^\omega 2$ into any uncountable Polish space. $\qquad\square$

Theorem [wAC] 6.13. *If X, Y are uncountable Polish spaces, then there is no Y-universal set for $\mathrm{BOREL}(X)$.*

Proof. The theorem follows from Theorems 5.17 and 5.15.   □

We present the simplest consequences of game theory to the structure of Borel sets. Let X, Y be topological spaces, A, B being their subsets, respectively. We say that A is **Wadge reducible** to B if there exists a continuous mapping $f : X \longrightarrow Y$ such that $A = f^{-1}(B)$. We shall write $\langle X, A \rangle \leq_W \langle Y, B \rangle$, or simply $A \leq_W B$ if the spaces X, Y are understood.

Theorem [AC] 6.14 (W. Wadge). *If $A, B \subseteq {}^{\omega}\omega$ are Borel sets, then either $A \leq_W B$ or $B \leq_W ({}^{\omega}\omega \setminus A)$.*

Proof. Let

$$P = \{\{x_n\}_{n=0}^{\infty} \in {}^{\omega}\omega : \neg(\{x_{2n}\}_{n=0}^{\infty} \in A \equiv \{x_{2n+1}\}_{n=0}^{\infty} \in B)\}.$$

Then P is a Borel subset of ${}^{\omega}\omega$ and by Theorem 5.63 the game $\mathrm{GAME}_{\omega}(P)$ is determined. If there exists a winning strategy f for player II, then to each play of player I we assign the play of player II which follows the winning strategy f, i.e., we set $g(\{x_n\}_{n=0}^{\infty}) = \{y_n\}_{n=0}^{\infty}$, where

$$y_n = f(\langle x_0, f(x_0), \ldots, x_{n-1}, f(\langle x_0, f(x_0), \ldots, x_{n-1} \rangle), x_n \rangle).$$

Evidently g is continuous and $g^{-1}(B) = A$. Thus $A \leq_W B$.

If there exists a winning strategy f for player I, we define similarly as above $h(\{x_n\}_{n=0}^{\infty}) = \{y_n\}_{n=0}^{\infty}$, where

$$y_n = f(\langle f(\emptyset), x_0, \ldots, f(\langle f(\emptyset), x_0, f(x_0) \ldots, x_{n-1} \rangle) \rangle).$$

Then $h^{-1}({}^{\omega}\omega \setminus A) = B$.   □

A subset A of a topological space X is **Wadge Γ-hard** if every $B \in \Gamma({}^{\omega}\omega)$ is Wadge reducible to A. If moreover $A \in \Gamma(X)$ we say that A is **Wadge Γ-complete**.

Theorem [AC] 6.15 (W. Wadge). *A subset $A \subseteq {}^{\omega}\omega$ is Wadge Σ_{ξ}^{0}-complete if and only if $A \in \Sigma_{\xi}^{0} \setminus \Pi_{\xi}^{0}$. Similarly, a set $A \subseteq {}^{\omega}\omega$ is Wadge Π_{ξ}^{0}-complete if and only if $A \in \Pi_{\xi}^{0} \setminus \Sigma_{\xi}^{0}$.*

Proof. If A is Wadge Σ_{ξ}^{0}-complete then $A \in \Sigma_{\xi}^{0} \setminus \Pi_{\xi}^{0}$ by Theorem 6.12.

So assume that $A \in \Sigma_{\xi}^{0} \setminus \Pi_{\xi}^{0}$. Let $B \in \Sigma_{\xi}^{0}({}^{\omega}\omega)$. Then by Theorem 6.14 either $A \leq_W ({}^{\omega}\omega \setminus B)$ or $B \leq_W A$. The first alternative is impossible.   □

By Theorem 5.10 we have

Theorem 6.16. *If $A \subseteq X$ is Wadge Γ-hard, then $\langle Y, B \rangle \leq_W \langle X, A \rangle$ for every $B \in \Gamma(Y)$ and every Polish space Y with a clopen base.*

Of course the assertion need not be true for arbitrary Polish space Y.

Exercises

6.1 [wAC] Baire Hierarchy is Proper

a) The characteristic function χ_A is Γ-measurable if and only if $A \in \Gamma \cap \neg\Gamma$.

b) If $\mathbf{BP}_\eta(X) = \mathbf{BP}_\xi(X)$ for some $\eta < \xi < \omega_1$, then $\mathbf{BP}(X) = \mathbf{BP}_\eta(X)$.

c) Assume that X is an uncountable Polish space and $\xi > 0$. Then there exists a function $f \in \mathbf{BP}_\xi(X)$ such that $f \notin \bigcup_{\eta < \xi} \mathbf{BP}_\eta(X)$.

d) If X is an uncountable Polish space, then each inclusion in the hierarchy (3.24) is proper.

e) Describe the Borel Hierarchy and the Baire Hierarchy for a countable Polish space. *Hint: Consider the set of accumulation points.*

6.2 [wAC] Generalized Separation Property

A pointclass Γ has the **generalized separation property** if for any sequence $\{A_n\}_{n=0}^\infty$ of sets from $\Gamma(X)$ with $\bigcap_n A_n = \emptyset$ there exists a sequence $\{B_n\}_{n=0}^\infty$ of sets from $\Delta[\Gamma](X)$ such that $A_n \subseteq B_n$ for each n and $\bigcap_n B_n = \emptyset$.

a) If Γ is closed under countable unions and has the σ-reduction property, then $\neg\Gamma$ has the generalized separation property. *Hint: If $\bigcap_n A_n = \emptyset$, $A_n \in \neg\Gamma$, then there exist pairwise disjoint sets $B_n \in \Gamma$ such that $\bigcup_n B_n = X$ and $B_n \subseteq X \setminus A_n$. Evidently each $B_n \in \Delta[\Gamma]$.*

b) $\mathbf{\Pi}_\xi^0$ has the generalized separation property for every $\xi > 1$.

c) If Γ is closed under countable intersections and has the generalized separation property, then Γ has the separation property.

6.3 [wAC] Decomposition of Borel Sets

a) Any set $A \in \mathbf{\Sigma}_\xi^0$, $\xi > 0$ can be expressed as a countable union of pairwise disjoint sets from $\mathbf{\Delta}_\xi^0$. *Hint: If $A = \bigcup_n A_n$, $A_n \in \bigcup_{\eta < \xi} \mathbf{\Pi}_\eta^0$, then $A_n \setminus \bigcup_{k<n} A_k \in \mathbf{\Delta}_\xi^0$, $n \in \omega$ are pairwise disjoint and $A = \bigcup_n (A_n \setminus \bigcup_{k<n} A_k)$.*

b) Any set $A \in \mathbf{\Sigma}_\xi^0$, $\xi > 1$ can be expressed as a union of pairwise disjoint sets from $\bigcup_{\eta < \xi} \mathbf{\Pi}_\eta^0$. *Hint: By a) $A = \bigcup_n A_n$ with $A_n \in \mathbf{\Delta}_\xi^0$. Then $X \setminus \bigcup_{k<n} A_k = \bigcup_i B_{n,i}$ with pairwise disjoint $B_{n,i} \in \mathbf{\Delta}_\eta^0$, $\eta < \xi$. Note that $A = \bigcup_n \bigcup_i (A_n \cap B_{n,i})$.*

c) Any set $A \in \mathbf{\Delta}_\xi^0$, $\xi > 1$ can be expressed as

$$A = \bigcup_n \bigcap_{k \geq n} A_k = \bigcap_n \bigcup_{k \geq n} A_k$$

with $A_k \in \bigcup_{\eta < \xi} \mathbf{\Delta}_\eta^0$, $k \in \omega$. *Hint: If $A = \bigcup_n B_n$ and $X \setminus A = \bigcup_n C_n$, $B_n, C_n \in \bigcup_{\eta < \xi} \mathbf{\Pi}_\eta^0$, take $A_n \in \bigcup_{\eta < \xi} \mathbf{\Delta}_\eta^0$ such that $B_n \subseteq A_n \subseteq X \setminus C_n$.*

6.4 Wadge Γ-hard Sets

a) If $A \subseteq X$ is Wadge Γ-hard (Wadge Γ-complete), then $X \setminus A$ is Wadge $\neg\Gamma$-hard (Wadge $\neg\Gamma$-complete).

b) If A is Wadge Γ-hard (Wadge Γ-complete) and $A \leq_W B$, then B is also Wadge Γ-hard (Wadge Γ-complete).

c) If a Wadge Γ-hard set is in $\neg\Gamma$, then Γ is self-dual.

6.5 Wadge Game

a) Assume that $Q \subseteq {}^\omega\omega$ is a countable dense subset and $B \subseteq {}^\omega\omega$ is an F_σ set. Find a winning strategy for player II in $\text{GAME}_\omega(P)$, where

$$P = \{\{x_n\}_{n=0}^\infty \in {}^\omega\omega : (\forall n) (\{x_{2n}\}_{n=0}^\infty \in B \equiv \{x_{2n+1}\}_{n=0}^\infty \in Q)\}.$$

Hint: Let $Q = \{\gamma_n : n \in \omega\}$, $B = \bigcup_n B_n$, $B_n \subseteq B_{n+1}$ being closed. Assume that the partial run after n steps is $x_0, \dots, x_{2n-1}$. If player II plays x_{2n}, then player I looks for the first m such that $\gamma_m|n = \{x_{2i+1}\}_{i=0}^{n-1}$. The existence of such m follows from the density of Q. If $[\{x_i\}_{i=0}^{2n}] \cap B_{n+1} = \emptyset$, player I plays $x_{2n+1} \neq \gamma_m(n)$. Otherwise player I plays $x_{2n+1} = \gamma_m(n)$. Note that if in two successive steps player II plays disjoint sets, then player I finds the same m.

b) Conclude that any countable dense subset of ${}^\omega 2$ or ${}^\omega\omega$ is Wadge F_σ-hard.

6.2   Analytic Sets

Let X be Polish space. A set $A \subseteq X$ is called **analytic** if there exist a Polish space Y, a Borel subset $B \subseteq Y$ and a continuous mapping $f : Y \longrightarrow X$ such that $A = f(B)$. The family of all analytic subsets of the space X is denoted by $\mathbf{\Sigma}_1^1(X)$. A subset $A \subseteq X$ of a Polish space X is called **co-analytic** if $X \setminus A$ is analytic. The family of all co-analytic subsets of the space X is denoted by $\mathbf{\Pi}_1^1(X)$.

Theorem [wAC] 6.17. *Let A be a subset of a Polish space X. Then the following are equivalent:*

a) *A is analytic.*

b) *A is sifted by a closed Souslin scheme with vanishing diameter.*

c) *A is sifted by a closed Souslin scheme $\langle {}^{<\omega}\omega, \varphi \rangle$ with vanishing diameter.*

d) *A is sifted by a closed sieve.*

e) *There exists a continuous $f : {}^\omega\omega \longrightarrow X$ such that $A = f({}^\omega\omega)$.*

f) *There exist a Polish space Y, a Borel subset $B \subseteq Y$ and a continuous surjection $f : B \xrightarrow{\text{onto}} A$.*

g) *There exist a Polish space Y and a Borel subset $B \subseteq Y \times X$ such that $A = \text{proj}_2(B)$.*

h) *There exists a closed set $B \subseteq {}^\omega\omega \times X$ such that $\text{proj}_2(B) = A$. Thus*

$$\mathbf{\Sigma}_1^1(X) = \exists^{{}^\omega\omega}[\mathbf{\Pi}_1^0]({}^\omega\omega \times X). \tag{6.2}$$

i) *There exists a Borel measurable $f : {}^\omega\omega \longrightarrow X$ such that $A = f({}^\omega\omega)$.*

Proof. The following implications are trivial: c) $\rightarrow$ b), a) $\rightarrow$ f), g) $\rightarrow$ a), h) $\rightarrow$ g), e) $\rightarrow$ i). By Theorem 5.12 we have c) $\equiv$ d). By Theorem 5.3 we obtain that b) $\rightarrow$ a). By Corollary 5.8 we have c) $\equiv$ e). By Corollary 6.6 we have implication f) $\rightarrow$ e). If $f : {}^\omega\omega \xrightarrow{\text{onto}} A$ is continuous, then by Theorem 1.38 the set $f \subseteq {}^\omega\omega \times X$ is closed and $\text{proj}_2(f) = A$. Thus e) $\rightarrow$ h). If $f : {}^\omega\omega \longrightarrow X$ is Borel measurable, then by Theorem 3.56 the graph $f \subseteq {}^{\omega}\omega \times X$ is a Borel set and $\mathrm{proj}_2(f) = A$. Thus i) implies g).

We have shown the following implications:

$$
\begin{array}{ccccc}
 & & \text{c)} \longrightarrow \text{b)} & & \\
 & \nearrow & & \searrow & \\
\text{f)} \longrightarrow \text{e)} & \longrightarrow \text{h)} \longrightarrow & \text{g)} \longrightarrow \text{a)} \longrightarrow \text{f)} \\
 & \searrow & \nearrow & & \\
 & & \text{i)} & &
\end{array}
$$

Hence, the theorem follows.   □

Corollary [wAC] 6.18. *The families $\mathbf{\Sigma}_1^1$ and $\mathbf{\Pi}_1^1$ are closed under countable unions and countable intersections. Hence, both $\mathbf{\Sigma}_1^1$ and $\mathbf{\Pi}_1^1$ are pointclasses.*

Proof. The corollary follows from part c) of the theorem and Theorem 6.4.   □

Corollary [wAC] 6.19 (M.J. Souslin). *An uncountable analytic subset of a Polish space contains a perfect subset homeomorphic copy to the Cantor middle-third set. Thus every analytic set is either countable or of cardinality continuum.*

Proof. The corollary follows from part b) of the theorem and Theorem 5.4.   □

One can easily check that in the proof of the theorem all implications but f) $\rightarrow$ e) (and therefore a) $\rightarrow$ d)) were proved without using **wAC**. We do not know a proof of a) $\rightarrow$ d) without **wAC**. Therefore we shall assume sometimes that a set is sifted by a closed sieve instead of being analytic.

Evidently

$$\mathrm{BOREL}(X) \subseteq \mathbf{\Sigma}_1^1(X), \qquad \mathrm{BOREL}(X) \subseteq \mathbf{\Pi}_1^1(X).$$

The **ambiguous family** $\mathbf{\Delta}_1^1$ is defined as

$$\mathbf{\Delta}_1^1 = \mathbf{\Sigma}_1^1 \cap \mathbf{\Pi}_1^1.$$

Let $0 < k < n$. Then we can consider $\mathbb{R}_n$ as the product $\mathbb{R}_k \times \mathbb{R}_{n-k}$. Thus we can define the projection $\mathrm{proj}_{n,k} : \mathbb{R}_n \longrightarrow \mathbb{R}_k$ as

$$\mathrm{proj}_{n,k}(x_1, \ldots, x_n) = \langle x_1, \ldots, x_k \rangle.$$

Corollary [wAC] 6.20. *If $A \subseteq \mathbb{R}_k$ then for any $n > k$ the following are equivalent:*

a) *A is analytic.*
b) *There exist a G_δ set $B \subseteq \mathbb{R}_n$ and a continuous mapping $f : \mathbb{R}_n \longrightarrow \mathbb{R}_k$ such that $A = f(B)$.*
c) *There exists a Borel set $B \subseteq \mathbb{R}_n$ such that $\mathrm{proj}_{n,k}(B) = A$.*
d) *There exist a Borel set $B \subseteq \mathbb{R}_n$ and a continuous mapping $f : \mathbb{R}_n \longrightarrow \mathbb{R}_k$ such that $A = f(B)$.*

Proof. The implications b) → c), c) → d), and d) → a) are trivial.

If $A \subseteq \mathbb{R}_k$ is analytic then, by h) of the theorem, there exists a closed set $B \subseteq {}^\omega\omega \times \mathbb{R}^k$ such that $\mathrm{proj}_2(B) = A$. Since ${}^\omega\omega$ is homeomorphic to a G_δ subset of $\mathbb{R}_{n-k}$, the corollary follows. $\square$

We shall call two disjoint subsets $A, B \subseteq X$ **Borel separable** if there exists a Borel set $C \subseteq X$ such that $A \subseteq C$ and $B \subseteq X \setminus C$.

Theorem [wAC] 6.21 (Luzin Separation Theorem). *Any two disjoint analytic subsets of a Polish space are Borel separable.*

Lemma [wAC] 6.22. *If $A = \bigcup_n A_n$, $B = \bigcup_m B_m$ are disjoint and such that A_n, B_m are Borel separable for any $n, m \in \omega$, then A, B are Borel separable.*

Proof. Since $|\mathrm{BOREL}| \ll \mathfrak{c}$ (by Theorem 1.13 using **wAC**) one can choose for any n, m a Borel set $C_{n,m}$ such that $A_n \subseteq C_{n,m}$, $B_m \cap C_{n,m} = \emptyset$. Let

$$D = \bigcup_n \bigcap_m C_{n,m}.$$

Then D is a Borel set and it is easy to see that $A \subseteq D$ and $B \subseteq X \setminus D$. $\square$

Proof of the theorem. Assume that $A, B \subseteq X$ are disjoint analytic sets which are not Borel separable. By Theorem 6.17, c) there exist closed Souslin schemes $\langle {}^{<\omega}\omega, \varphi \rangle$ and $\langle {}^{<\omega}\omega, \psi \rangle$ with vanishing diameters such that $A = L({}^{<\omega}\omega, \varphi)$ and $B = L({}^{<\omega}\omega, \psi)$. By (5.3) and Lemma 6.22 we can construct by induction branches $\alpha, \beta \in {}^\omega\omega$ such that $L(({}^{<\omega}\omega)^{\alpha|n}, \varphi) \subseteq \varphi(\alpha|n)$ and $L(({}^{<\omega}\omega)^{\beta|n}, \psi) \subseteq \psi(\beta|n)$ are not Borel separable for any $n \in \omega$. Let $x \in \bigcap_n \varphi(\alpha|n)$ and $y \in \bigcap_n \psi(\beta|n)$. Then $x \in A$ and $y \in B$. Hence $x \neq y$. Since both Souslin schemes have vanishing diameters there exists an $n \in \omega$ such that $\varphi(\alpha|n) \subseteq \mathrm{Ball}(x, \varepsilon/2)$ and $\psi(\beta|n) \subseteq \mathrm{Ball}(y, \varepsilon/2)$, where $\varepsilon = \rho(x, y)$. Then the open set $\mathrm{Ball}(x, \varepsilon/2)$ separates the sets $L(({}^{<\omega}\omega)^{\alpha|n}, \varphi)$, and $L(({}^{<\omega}\omega)^{\beta|n}, \psi)$, which is a contradiction. $\square$

Corollary [wAC] 6.23 (M.J. Souslin). *Let X be a Polish space. If $A \subseteq X$ is both analytic and co-analytic, then A is Borel. Thus*

$$\mathrm{BOREL} = \mathbf{\Delta}_1^1 = \mathbf{\Sigma}_1^1 \cap \mathbf{\Pi}_1^1. \tag{6.3}$$

Proof. Since A and $X \setminus A$ are disjoint analytic sets, there exists a Borel set B separating A and $X \setminus A$. However if a set B separates A and $X \setminus A$, then $A = B$. $\square$

Thus Luzin Theorem 6.21 together with Corollary 6.23 say that $\mathbf{\Sigma}_1^1$ has the separation property. By Theorem 5.20, c) and Corollary 6.18 we obtain

Corollary [wAC] 6.24. *The pointclass $\mathbf{\Sigma}_1^1$ possesses the σ-separation property.*

We present some applications of obtained results.

Lemma [wAC] 6.25. *Let X be a Polish space. If $C \subseteq {}^\omega\omega$ is a closed subset and $f : {}^\omega\omega \longrightarrow X$ is continuous and injective on C, then $f(C)$ is a Borel set.*

Proof. Let $T \subseteq {}^{<\omega}\omega$ be a tree such that $[T] = C$. We define a Souslin scheme $\langle T, \varphi \rangle$ as follows. We set $\varphi(\emptyset) = X$. For $n > 0$ the sets $f([s])$, $s \in T(n) = T \cap {}^n\omega$ are pairwise disjoint analytic sets. Therefore by Corollary 6.24 there are pairwise disjoint Borel sets $B_s \supseteq f([s])$, $s \in T(n)$. We set $\varphi(s) = B_s \cap \overline{f([s])}$. Since

$$f([s]) \subseteq \varphi(s) \subseteq \overline{f([s])}$$

and for any branch $h \in [T]$,

$$\bigcap_{n \in \omega} f([h|n]) = \bigcap_{n \in \omega} \overline{f([h|n])} = \{f(h)\},$$

we obtain $L(T, \varphi) = f([T])$. On the other hand the Souslin scheme φ is regular and therefore the set

$$L(T, \varphi) = \bigcap_n \bigcup_{s \in T(n)} \varphi(s)$$

is Borel.   □

Theorem [wAC] 6.26 (N.N. Luzin). *Assume that $A \subseteq X$ is a Borel subset of a Polish space X, $f : X \longrightarrow Y$ is continuous and injective on A. Then $f(A)$ is a Borel set.*

Proof. By Theorem 6.3 there is a closed subset $C \subseteq {}^\omega\omega$ and a continuous injection $g : C \longrightarrow X$ such that $g(C) = A$. By Lemma 6.25 the set $f(A) = f(g(C))$ is Borel.   □

Now we prove that the result of Theorem 3.56 can be reversed.

Theorem [wAC] 6.27. *If $f : X \longrightarrow Y$ has a Borel graph, then f is Borel measurable.*

Proof. If $B \subseteq Y$ is a Borel set we obtain

$$x \in f^{-1}(B) \equiv (\exists y)\,(y \in B \wedge \langle x, y \rangle \in f) \equiv (\forall y)\,(\langle x, y \rangle \in f \to y \in B).$$

Thus, by Corollary 6.23 the set $f^{-1}(B)$ is Borel.   □

Theorem [wAC] 6.28. *For any Polish space X there exists an ${}^\omega\omega$-universal set for $\Sigma^1_1(X)$.*

Proof. Let $U \subseteq {}^\omega\omega \times {}^\omega\omega \times X$ be a closed set ${}^\omega\omega$-universal for the family of all closed subsets of ${}^\omega\omega \times X$. The existence follows from Theorem 6.10. We claim that the analytic set

$$V = \{\langle a, x \rangle \in {}^\omega\omega \times X : (\exists b)\,\langle a, b\ x \rangle \in U\}$$

is ${}^\omega\omega$-universal for $\Sigma^1_1(X)$.

Evidently V is an analytic subset of $^\omega\omega \times X$. Let $A \in \mathbf{\Sigma}_1^1(X)$. Then there is a continuous function $f : {}^\omega\omega \xrightarrow{\text{onto}} A$. f is a closed subset of $^\omega\omega \times X$ therefore there exists an $a \in {}^\omega\omega$ such that $f = U_a$. Then

$$x \in A \equiv (\exists b)\, \langle b, x \rangle \in f \equiv (\exists b)\, \langle a, b, x \rangle \in U \equiv \langle a, x \rangle \in V \equiv x \in V_a. \qquad \square$$

Thus, by Theorem 5.17 we have

Corollary [wAC] 6.29. *If Y is an uncountable Polish space, then for any Polish space X there exists an analytic Y-universal set for $\mathbf{\Sigma}_1^1(X)$.*

By Lemma 5.14 we obtain another important consequence.

Corollary [wAC] 6.30. *Every uncountable Polish space X contains an analytic set $A \subseteq X$, which is not co-analytic, therefore not Borel.*

A mapping $\psi : {}^{<\omega}\omega \longrightarrow X$ is called a **Hurewicz scheme** if the following conditions are fulfilled:

(1) $(\forall v \in {}^{<\omega}\omega)(\forall n, m)\,(n \neq m \to \psi(v^\frown n) \neq \psi(v^\frown m))$,
(2) $(\forall v \in {}^{<\omega}\omega)\,\psi(v) = \lim_{n\to\infty} \psi(v^\frown n)$,
(3) $(\forall v \in {}^{<\omega}\omega)\,\lim_{n\to\infty} \operatorname{diam}(\{\psi(u) : u \geq v^\frown n\}) = 0$,
(4) $(\forall \alpha \in {}^\omega\omega)\,\lim_{n\to\infty} \operatorname{diam}(\{\psi(u) : u \geq \alpha|n\}) = 0$.

The basic property of a Hurewicz scheme is contained in the next result actually saying that we have a good control over closure of the range of a Hurewicz scheme.

Lemma 6.31. *If ψ is a Hurewicz scheme and $x \in \overline{\operatorname{rng}(\psi)} \setminus \operatorname{rng}(\psi)$, then there exists a branch $\alpha \in {}^\omega\omega$ such that $x = \lim_{n\to\infty} \psi(\alpha|n)$.*

Proof. Let $x = \lim_{n\to\infty} x_n \in \overline{\operatorname{rng}(\psi)} \setminus \operatorname{rng}(\psi)$, $x_n = \psi(u_n)$ for $n \in \omega$. We write

$$T = \{v \in {}^{<\omega}\omega : (\exists n)\, v \leq u_n\}.$$

We show that every node of T has finite branching degree. Assume not. Then there exists a node $v \in T$ and there exist sequences $\{m_k\}_{k=0}^\infty$, $\{n_k\}_{k=0}^\infty$ such that $v^\frown m_k \in T$ and $v^\frown m_k \leq u_{n_k}$. We can assume that $\{m_k\}_{k=0}^\infty$ is increasing. Since

$$\rho(\psi(v), \psi(u_{n_k})) \leq \rho(\psi(v), \psi(v^\frown m_k)) + \operatorname{diam}(\{\psi(u) : u \geq v^\frown m_k\}),$$

we obtain $x = \lim_{k\to\infty} \psi(v^\frown m_k) = \lim_{k\to\infty} \psi(u_{n_k}) = \psi(v)$, which is a contradiction.

By König's Lemma Theorem 11.7 there exists an infinite branch $\alpha \in {}^\omega\omega \cap [T]$. Again, let n_k be such that $\alpha|k \leq u_{n_k}$. By property (4) of a Hurewicz scheme we have $\lim_{k\to\infty} \rho(\psi(\alpha|k), \psi(u_{n_k})) = 0$ and therefore

$$x = \lim_{k\to\infty} \psi(u_{n_k}) = \lim_{k\to\infty} \psi(\alpha|k). \qquad \square$$

A Hurewicz scheme $\psi : {}^{<\omega}\omega \longrightarrow X$ is said to be **regular** if there exists a family $\langle V_v : v \in {}^{<\omega}\omega \rangle$ of open sets such that

$$\psi(v) \in V_v \setminus \bigcup_n \overline{V_{(v^\frown n)}}, \tag{6.4}$$

the sets $\langle V_{v^\frown n} : n \in \omega \rangle$ are pairwise disjoint subsets of V_v $\qquad$ (6.5)

for each $v \in {}^{<\omega}\omega$. One can easily see that for distinct branches $\alpha, \beta \in {}^{\omega}\omega$ the limits $\lim_{n\to\infty} \psi(\alpha|n)$ and $\lim_{n\to\infty} \psi(\beta|n)$ are distinct provided that ψ is regular.

If $A \subseteq B$ then it is natural to say that a set C **is between** sets A, B, if $A \subseteq C \subseteq B$.

Lemma [wAC] 6.32. *Let $A, B \subseteq X$, U being open. If $A \cap U \subseteq B$ and there exists no F_σ set between sets $A \cap U, B$, then there exist infinitely many points $p \in U \setminus B$ such that*

for every open $p \in V \subseteq U$, there exists no F_σ set between $A \cap V, B$. $\qquad$ (6.6)

Proof. Fix a countable open base $\{U_n : n \in \omega\}$. Assume that there is no such point p. Set

$$K = \{n \in \omega : U_n \subseteq U \land (\text{there is an } F_\sigma \text{ set between } A \cap U_n, B)\}.$$

For $n \in K$, let F_n be an F_σ set between $A \cap U_n, B$. If we set $W = \bigcup_{n \in K} U_n$ and $F = \bigcup_{n \in K} F_n$, then the F_σ set $(F \cap U) \cup (U \setminus W)$ is between $A \cap U, B$, which is a contradiction.

If there are only finitely many such points p in $U \setminus B$, then apply the lemma to the open set U with those points omitted. $\qquad \square$

Theorem [wAC] 6.33 (A.S. Kechris – A. Louveau – W.H. Woodin). *Let X be a perfect Polish space, $A \subseteq B \subseteq X$, A being analytic. If there exists no F_σ set between A, B, then there exists a countable set $L \subseteq X \setminus B$ without isolated points such that $\overline{L} \setminus L \subseteq A$ is homeomorphic to ${}^{\omega}\omega$.*

Proof. Let $\langle {}^{<\omega}\omega, \varphi \rangle$ be a closed Souslin scheme with vanishing diameter such that $A = L(T, \varphi)$. For simplicity, for any $v \in {}^{<\omega}\omega$ we write

$$A_v = L(({}^{<\omega}\omega)^v, \varphi) = \bigcup_{\alpha \supseteq v} \bigcap_n \varphi(\alpha n).$$

We construct functions $\psi : {}^{<\omega}\omega \longrightarrow X \setminus B$, $F : {}^{<\omega}\omega \longrightarrow {}^{<\omega}\omega$ and a family of open sets $\langle V_v : v \in {}^{<\omega}\omega \rangle$ satisfying (6.4) and (6.5) such that

a) ψ is a regular Hurewicz scheme,
b) if $\alpha \in {}^{\omega}\omega$, then $\bigcup_n F(\alpha|n)$ is a branch as well,
c) $\psi(v) \in \varphi(F(v)) \setminus A_{F(v)}$ for any $v \in {}^{<\omega}\omega$,
d) there exists no F_σ set between $A_{F(v)} \cap U, B$ for any open $U \ni \psi(v)$.

Since $A \subseteq \varphi(\emptyset)$ taking $U = X$ we have $p \in \varphi(\emptyset) \setminus A$ such that (6.6) holds true. Set $\psi(\emptyset) = p$, $F(\emptyset) = \emptyset$, and $V_\emptyset = X$. Assume that $\psi(v), F(v), V_v$ are already defined for a $v \in {}^k\omega$. Note that $A_{F(v)} = \bigcup_n A_{F(v)^\frown n}$. Let $n \in \omega$. Then there exists an m such that there is no F_σ set between $A_{F(v)^\frown m} \cap U_n, B$, where

$$U_n = \mathrm{Ball}\left(\psi(v), 2^{-\sum_{i<k}(v(i)+1)-n}\right) \cap V_v.$$

By Lemma 6.32 there is a point $p \in U_n \setminus B$ satisfying (6.6). We set $\psi(v^\frown n) = p$ and $F(v^\frown n) = F(v)^\frown m$. Since there exist infinitely many such points p, we can assume that $\psi(v^\frown n), n \in \omega$ are mutually different. Evidently $\lim_{n\to\infty} \psi(v^\frown n) = \psi(v)$. Since every $\psi(v^\frown n) \in V_v$ one can find by induction open sets $\langle V_{v^\frown n} : n \in \omega \rangle$ satisfying conditions (6.4) and (6.5).

By simple computation one can verify that conditions (1)–(4) are satisfied and therefore ψ is a Hurewicz scheme.

Evidently $\lim_{n\to\infty} \psi(\alpha|n) \in \bigcap_n \varphi(F(\alpha|n)) \subseteq A$ for any $\alpha \in {}^\omega\omega$.

Define $G : {}^\omega\omega \longrightarrow (\overline{L} \setminus L)$ as $G(\alpha) = \lim_{n\to\infty} \psi(\alpha|n)$. One can easily check that G is a homeomorphism. $\square$

Corollary [wAC] 6.34 (Hurewicz Theorem). *Assume that D is an analytic subset of a Polish space X such that D is not an F_σ set. Then there exists a closed subset of D homeomorphic to ${}^\omega\omega$.*

Proof. Take $A = B = X \setminus D$ in the theorem. The set $\overline{L} \setminus L$ is homeomorphic to ${}^\omega\omega$. $\square$

Corollary [wAC] 6.35 (W. Hurewicz). *A Polish space X is σ-compact if and only if X does not contain a closed subset homeomorphic to ${}^\omega\omega$.*

Proof. By Theorem 3.11 the space ${}^\omega\omega$ is not σ-compact.

Assume that X is not σ-compact. Let $Y \supseteq X$ be a compact Polish space which is a compactification of X. Then X is not an F_σ subset of Y. Moreover, being a Polish space, X is analytic. Therefore there exists a closed subset of X homeomorphic to ${}^\omega\omega$. $\square$

Exercises

6.6 The Borel Cantor-Bernstein Theorem

A map $f : X \xrightarrow[\text{onto}]{1-1} Y$ is a **Borel isomorphism** if both f and f^{-1} are Borel measurable. X and Y are **Borel isomorphic** if there exists a Borel isomorphism from X onto Y.

a) If $f : X \xrightarrow[\text{onto}]{1-1} Y$ is Borel measurable, then f is a Borel isomorphism, provided that **wAC** holds true.

Hint: Use Theorems 3.56 and 6.27 on Borel graphs.

b) If $f : X \xrightarrow{1-1} Y$ and $g : Y \xrightarrow{1-1} X$ are Borel measurable, then there are Borel sets $A \subseteq X$ and $B \subseteq Y$ such that $f(A) = Y \setminus B$ and $g(B) = X \setminus A$.

Hint: Follow any standard proof of the Cantor-Bernstein Theorem 1.5. Let $Y_0 = Y \setminus f(X)$, $X_n = g(Y_n)$, $Y_{n+1} = f(X_n)$. Denote $A = X \setminus \bigcup_n X_n$, $B = \bigcup_n Y_n$.

c) Prove the Borel Cantor-Bernstein Theorem: If $f : X \xrightarrow{1-1} Y$ and $g : Y \xrightarrow{1-1} X$ are Borel measurable, then X, Y are Borel isomorphic.

d) Assuming **wAC**, show that any two uncountable Polish spaces are Borel isomorphic.

6.7 [AC] Capacity

Let X be a locally compact topological space. A mapping $\nu : \mathcal{P}(X) \longrightarrow [0, \infty]$ is called a **capacity** on X if the following conditions are satisfied:

(C1) if $A \subseteq B$ then $\nu(A) \leq \nu(B)$,

(C2) $\nu(\bigcup_n A_n) = \sup\{\nu(A_n) : n \in \omega\}$ for any increasing sequence $\langle A_n \subseteq X : n \in \omega \rangle$,

(C3) $\nu(K) < \infty$ for any compact $K \subseteq X$,

(C4) $\nu(\bigcap_n K_n) = \inf\{\nu(K_n) : n \in \omega\}$ for any decreasing sequence $\{K_n\}_{n=0}^{\infty}$ of compact subsets of X.

a) If μ is a moderated regular Borel measure on a locally compact topological space X, then the outer measure μ^* is a capacity on X.

b) Let X, Y be locally compact, $f : Y \longrightarrow X$ be a continuous mapping, ν being a capacity on X. Set $\mu(A) = \nu(f(A))$ for any $A \subseteq Y$. Then μ is a capacity on Y

c) Let μ be a finite Borel measure on $\mathbb{T}$. If we set $\nu(E) = \mu^*(E - E)$ for any $E \subseteq \mathbb{T}$, then ν is a capacity.

6.8 [AC] Choquet's Theorem

If ν is a capacity on X then a subset $A \subseteq X$ is called **capacitable** if

$$\nu(A) = \sup\{\nu(K) : K \subseteq A \wedge K \text{ is compact}\}. \tag{6.7}$$

a) If μ is a finite Borel measure on a locally compact space X and μ^* is the corresponding outer measure, then $A \subseteq X$ is μ^*-measurable if and only if A is capacitable.

b) Let $f : X \longrightarrow Y$ be continuous, X, Y locally compact and μ, ν be capacities of Exercise 6.7, b). Then $f(A)$ is μ-capacitable for any ν-capacitable $A \subseteq X$.

c) Let ν be a capacity on a compact Polish space X. Then every G_δ-subset of a compact set is capacitable.

Hint: Let $G = \bigcap_n G_n$, $G_n = \bigcup_m F_{n,m}$, G_n open, $F_{n,n} \subseteq f_{n,m+1}$ compact. For any $a < \nu(G)$ construct by induction $K_n = K_{n-1} \cap \bigcup_{m=0}^{m_r} F_{n,m}$ such that $\nu(K_n) > a$. Then $K = \bigcap_n K_n \subseteq G$ is compact and $\nu(K) \geq a$.

d) Prove Choquet's Theorem: Any analytic subset of a compact Polish space is capacitable for any capacity.

Hint: Let $f : {}^\omega\omega \xrightarrow{\text{onto}} A$. Let $p : [0, 1] \times X \xrightarrow{\text{onto}} X$ be projection. If ν is a capacity on X use p for constructing a capacity μ on $[0, 1] \times X$. Note $f \subseteq [0, 1] \times X$ is a G_δ set.

6.9 [wAC] Unfolded Game Again

For terminology, notation and results see Exercise 5.33.

a) Find a proof of Corollary 6.19 based on an unfolded game.

Hint: If $A = \{x \in X : (\exists y)\langle x, y \rangle \in X \times {}^\omega\omega\}$, C closed, then $\text{GameG}^*(X, A)$ is determined.

b) Find a proof of the Luzin-Sierpiński Theorem 7.11, a), based on an unfolded game.

Hint: If $A = \{x \in X : (\exists y)\langle x, y \rangle \in C \times {}^\omega\omega\}$, C closed, then $\text{GameG}^{**}(X, A)$ is determined.

6.3   Projective Hierarchy

For any Polish space X, $n \in \omega$ the families of projective sets are defined as follows:

$$\mathbf{\Pi}^1_0(X) = \mathbf{\Sigma}^1_0(X) = \text{Borel}(X), \tag{6.8}$$

$$\mathbf{\Sigma}^1_{n+1}(X) = \exists^{\omega\omega}[\mathbf{\Pi}^1_n]({}^\omega\omega \times X), \tag{6.9}$$

$$\mathbf{\Pi}^1_{n+1}(X) = \neg\mathbf{\Sigma}^1_{n+1}(X), \tag{6.10}$$

$$\mathbf{\Delta}^1_{n+1}(X) = \mathbf{\Sigma}^1_{n+1}(X) \cap \mathbf{\Pi}^1_{n+1}(X), \tag{6.11}$$

for any $0 < n \in \omega$.

Thus $A \subseteq X$ is a $\mathbf{\Sigma}^1_{n+1}$ set if A is a (continuous) projection of a $\mathbf{\Pi}^1_n$-subset of ${}^\omega\omega \times X$. A $\mathbf{\Pi}^1_n$ set is a complement of a $\mathbf{\Sigma}^1_n$ set and $\mathbf{\Delta}^1_n$ sets are the ambiguous sets. A set A is called **projective** if $A \in \mathbf{\Sigma}^1_n$ or $A \in \mathbf{\Pi}^1_n$ for some $n > 0$.

One can easily show that

$$|\mathbf{\Sigma}^1_n| = |\mathbf{\Pi}^1_n| \ll \mathfrak{c}.$$

The classical notation of projective sets is as follows. **A** denotes the family of analytic sets $\mathbf{\Sigma}^1_1$. **CA** is the family of co-analytic sets $\mathbf{\Pi}^1_1$, **PCA** is the family of all projections of co-analytic sets, i.e., $\mathbf{\Sigma}^1_2$, **CPCA** is the family of complements of **PCA**, i.e., $\mathbf{\Pi}^1_2$, etc.

Lemma 6.36. *For any $n \geq 0$ the inclusions*

$$\mathbf{\Sigma}^1_n(X) \subseteq \mathbf{\Sigma}^1_{n+1}(X), \qquad \mathbf{\Pi}^1_n(X) \subseteq \mathbf{\Pi}^1_{n+1}(X), \tag{6.12}$$

$$\mathbf{\Pi}^1_n(X) \subseteq \mathbf{\Sigma}^1_{n+1}(X), \qquad \mathbf{\Sigma}^1_n(X) \subseteq \mathbf{\Pi}^1_{n+1}(X) \tag{6.13}$$

hold true. Therefore also

$$\mathbf{\Pi}^1_n(X) \subseteq \mathbf{\Delta}^1_{n+1}(X), \qquad \mathbf{\Sigma}^1_n(X) \subseteq \mathbf{\Delta}^1_{n+1}(X) \tag{6.14}$$

hold true for any $n \geq 0$.

Proof. Since BOREL $\subseteq \mathbf{\Pi}_1^1$ by (6.9) we obtain $\mathbf{\Sigma}_1^1 \subseteq \mathbf{\Sigma}_2^1$ and therefore by (6.10), we have $\mathbf{\Pi}_1^1 \subseteq \mathbf{\Pi}_2^1$.

By induction, if $\mathbf{\Pi}_n^1 \subseteq \mathbf{\Pi}_{n+1}^1$ then again by (6.9) we obtain $\mathbf{\Sigma}_{n+1}^1 \subseteq \mathbf{\Sigma}_{n+2}^1$ and by (6.10) also $\mathbf{\Pi}_{n+1}^1 \subseteq \mathbf{\Pi}_{n+2}^1$.

If $A \in \mathbf{\Pi}_n^1(X)$ then $A = \operatorname{proj}_2({}^\omega\omega \times A)$ and therefore $A \in \mathbf{\Sigma}_{n+1}^1(X)$. $\square$

The introduced families of projective sets with inclusions (6.12)–(6.14) form the **Projective Hierarchy**

Let us remember that BOREL $= \mathbf{\Delta}_1^1$.

Theorem [wAC] 6.37.

a) *The families $\mathbf{\Sigma}_n^1(X)$, $\mathbf{\Pi}_n^1(X)$ and $\mathbf{\Delta}_n^1(X)$ are closed under Borel measurable inverse images, especially under continuous inverse images.*

b) *The Borel measurable image of a $\mathbf{\Sigma}_n^1$ set is a $\mathbf{\Sigma}_n^1$ set for any $n > 0$.*

c) *The families $\mathbf{\Sigma}_n^1(X)$ and $\mathbf{\Pi}_n^1(X)$ are closed under countable unions and countable intersections. Thus the family $\mathbf{\Delta}_n^1(X)$ is a σ-field of subsets of X.*

d) *If $f : X \xrightarrow[\text{onto}]{1-1} Y$ is a continuous bijection from a Polish space X into a Polish space Y and $A \subseteq X$ is a $\mathbf{\Sigma}_n^1$, $\mathbf{\Pi}_n^1$ or $\mathbf{\Delta}_n^1(X)$ set, $n > 0$, then $f(A)$ is a $\mathbf{\Sigma}_n^1$, $\mathbf{\Pi}_n^1$ or $\mathbf{\Delta}_n^1(X)$ set, respectively, as well.*

Proof. a) The assertion follows by induction from identities

$$f^{-1}(Y \setminus A) = X \setminus f^{-1}(A), \qquad f^{-1}(\operatorname{proj}_2(B)) = \operatorname{proj}_2(F^{-1}(B))$$

for $f : X \longrightarrow Y$ and $F : {}^\omega\omega \times X \longrightarrow {}^\omega\omega \times Y$ defined by $F(a, x) = \langle a, f(x) \rangle$.

b) Let $f : X \longrightarrow Y$ be a Borel measurable mapping, X, Y being Polish spaces. Assume that $A \in \mathbf{\Sigma}_n^1(X)$. Then there exists a $\mathbf{\Pi}_{n-1}^1$ set $B \subseteq {}^\omega\omega \times X$ such that $A = \operatorname{proj}_2(B)$. Then $f(A) = \operatorname{proj}_2(C)$ where

$$C = \{\langle\langle z, x\rangle, y\rangle \in ({}^\omega\omega \times X) \times Y : \langle z, x\rangle] \in B \wedge \langle x, y\rangle \in f\} \in \mathbf{\Pi}_{n-1}^1.$$

c) It suffices to prove by induction that every family $\mathbf{\Sigma}_n^1(X)$ is closed under countable unions and countable intersections. By Corollary 6.18 it is true for $n = 1$.

Assume that it is true for n. Then also $\mathbf{\Pi}_n^1(X)$ is closed under countable unions and countable intersections. One can easily see that

$$\mathbf{\Sigma}^0[\exists^{\,\omega}\omega[\Gamma]] = \exists^{\,\omega}\omega[\mathbf{\Sigma}^0[\Gamma]]$$

and therefore $\mathbf{\Sigma}_{n+1}^1(X)$ is closed under countable unions.

Let $\langle A_k \in \mathbf{\Sigma}_{n+1}^1(X) : k \in \omega \rangle$ be given. Then there are sets $B_k \in \mathbf{\Pi}_n^1(^{\omega}\omega \times X)$ such that $A_k = \mathrm{proj}_2(B_k)$. If $P_{n,X} : {}^{\omega}\omega \times X \longrightarrow {}^{\omega}\omega \times X$ is the function defined in the proof of Theorem 5.19, c), we obtain (set $\alpha(\pi(n,k)) = \alpha_n(k)$)

$$x \in \bigcap_n A_n \equiv (\forall n)(\exists \alpha_n)\, \langle \alpha_n, x \rangle \in B_n$$

$$\equiv (\exists \alpha)(\forall n)\, P_{n,X}(\alpha, x) \in B_n \equiv x \in \mathrm{proj}_2(\bigcap_n P_{n,X}^{-1}(B_n)).$$

By inductive assumption $\bigcap_n P_{n,X}^{-1}(B_n) \in \mathbf{\Pi}_n^1(^{\omega}\omega \times X)$, hence $\bigcap_n A_n \in \mathbf{\Sigma}_{n+1}^1(X)$.

The assertion d) follows by induction from Luzin's Theorem 6.26. □

Corollary [wAC] 6.38. *Each class $\mathbf{\Sigma}_n^1$, $\mathbf{\Pi}_n^1$, $\mathbf{\Delta}_n^1$, $n \geq 0$ is a pointclass.*

Similarly as in Theorem 6.17 one can give equivalent descriptions of the families of the Projective Hierarchy. We present the most important ones.

Theorem [wAC] 6.39. *Let A be a subset of a Polish space X. Then the following are equivalent:*

a) $A \in \mathbf{\Sigma}_{n+1}^1(X)$.

b) *There exists a Polish space Y and there exists a set $B \in \mathbf{\Pi}_n^1(Y \times X)$ such that $A = \mathrm{proj}_2(B)$.*

c) *There exist a Polish space Y, a continuous map $f : Y \longrightarrow X$ and a set $B \in \mathbf{\Pi}_n^1(Y)$ such that $A = f(B)$.*

Proof. One can easily see that a) $\rightarrow$ b) $\rightarrow$ c).

Assume that c) holds true, i.e., $A = f(B)$, Y is a Polish space, $B \in \mathbf{\Pi}_n^1(Y)$ and $f : Y \longrightarrow X$ is continuous. By Corollary 5.7 there exists a continuous mapping $g : {}^{\omega}\omega \xrightarrow{\text{onto}} Y$. Using Theorem 6.37 one can easily check that the set

$$C = \{\langle a, x \rangle \in {}^{\omega}\omega \times X : g(a) \in B \wedge \langle g(a), x \rangle \in f\}$$

is a $\mathbf{\Pi}_n^1$-set. Then $A = \mathrm{proj}_2(C) \in \mathbf{\Sigma}_{n+1}^1(X)$, since for any $x \in X$ we have

$$x \in A \equiv (\exists y)\,(y \in B \wedge f(y) = x)$$
$$\equiv (\exists a)\,(a \in {}^{\omega}\omega \wedge g(a) \in B \wedge \langle g(a), x \rangle \in f) \equiv (\exists a)\,\langle a, x \rangle \in C. □$$

Theorem [wAC] 6.40. *For every Polish space X there exists a $\mathbf{\Sigma}_n^1(^{\omega}\omega \times X)$ set ${}^{\omega}\omega$-universal for $\mathbf{\Sigma}_n^1(X)$. Similarly for $\mathbf{\Pi}_n^1$.*

Proof. The theorem follows immediately from Theorems 6.28 and 5.19 by induction. □

By Theorem 5.17 we immediately obtain

Corollary [wAC] 6.41. *Let Y be an uncountable Polish space. Then for every Polish space X there exists a $\boldsymbol{\Sigma}_n^1(Y \times X)$ set Y-universal for $\boldsymbol{\Sigma}_n^1(X)$. Similarly for $\boldsymbol{\Pi}_n^1$.*

Now we can show a fundamental result concerning the Projective Hierarchy.

Theorem [wAC] 6.42. *The Projective Hierarchy is proper for any uncountable Polish space, i.e., if X is an uncountable Polish space, then each of the inclusions $\boldsymbol{\Delta}_n^1(X) \subseteq \boldsymbol{\Sigma}_n^1(X)$, $\boldsymbol{\Delta}_n^1(X) \subseteq \boldsymbol{\Pi}_n^1(X)$, $\boldsymbol{\Sigma}_n^1(X) \subseteq \boldsymbol{\Delta}_{n+1}^1(X)$, $\boldsymbol{\Pi}_n^1(X) \subseteq \boldsymbol{\Delta}_{n+1}^1(X)$ is proper for any $n > 0$.*

Proof. We can literally repeat the proof of Theorem 6.12. Since the pointclass $\boldsymbol{\Delta}_n^1$ is self-dual, for every Polish space X any of the equations $\boldsymbol{\Delta}_n^1(X) = \boldsymbol{\Sigma}_n^1(X)$, $\boldsymbol{\Delta}_n^1(X) = \boldsymbol{\Pi}_n^1(X)$, $\boldsymbol{\Sigma}_n^1(X) = \boldsymbol{\Delta}_{n+1}^1(X)$, $\boldsymbol{\Pi}_n^1(X) = \boldsymbol{\Delta}_{n+1}^1(X)$ implies the equation $\boldsymbol{\Sigma}_n^1(X) = \boldsymbol{\Pi}_n^1(X)$. Thus we have to show that $\boldsymbol{\Sigma}_n^1(X) \neq \boldsymbol{\Pi}_n^1(X)$ for any uncountable Polish space X and any $n > 0$.

However, if $\boldsymbol{\Sigma}_n^1(X) = \boldsymbol{\Pi}_n^1(X)$ for some $n > 0$ and some uncountable Polish space X, then by Lemma 5.14 there exists no X-universal set for $\boldsymbol{\Sigma}_n^1(X)$ – a contradiction with Theorem 6.40. □

The results of descriptive set theory are closely related to some results of mathematical logic, namely to those of recursion theory. The starting point of that was an observation made by K. Kuratowski and A. Tarski that the logical complexity of the formula describing a set of reals allows one to estimate the class of descriptive hierarchy to which the set belongs. We describe a variant of such method.

We shall deal with a language of logic which contains three kinds of variables:

0) **The zero-order variables** $n, m, \ldots, r_1, \ldots, r_k, \ldots$ denote elements of countable sets $\mathbb{N}$, $\mathbb{Z}$, $\mathbb{Q}$, $^{<\omega}2$ or $^{<\omega}\omega$. One can always identify which of those sets is actually dealt with.

1) **The first-order variables** $x, y, z, \ldots$, denote elements of any (uncountable) Polish space, especially reals as elements of $\mathbb{R}$, $[0, 1]$, $\mathbb{T}$, the first-order variables $\alpha, \beta, \gamma, \ldots$ denote elements $^{\omega}2$ or $^{\omega}\omega$.

2) **The second-order variables** $X, Y, Z, \ldots, A, B, C, \ldots$ denote subsets of a Polish space, especially sets of reals.

We are not very precise since we often do not distinguish between a variable and its value. However we hope that the explanation will be sufficiently clear.

A term is a numerical expression, i.e., an expression formed from variables of zero and first order and individual constants $0, 1$ (eventually others, e.g., $2, 3$) by using the operation $+, -, \cdot$ or taking a value $r(n)$ for $r \in {}^{<\omega}\omega$ and $n \in \omega$. The partial operation of division $/$ must be used with appropriate care. We allow also other operations, however we suppose that they are continuous (or at least Borel measurable). Especially we shall need the pairing functions π, Π, their inverses $\lambda, \rho, \Lambda, R$ and the projections $\mathrm{proj}_{X,Y}, \mathrm{Proj}_n$. Of course, new operations introduced by definitions are allowed too, e.g., the exponentiation x^n etc. We suppose that it is always clear whether a term denotes an object of zero order or an object of first order. The notion of an **atomic formula** can be described as follows. The formula $t = s$ is atomic for any type of terms t, s. If the terms t, s denote elements of the true real line $\mathbb{R}$, then any formula of the form $t < s, t \leq s$ is an atomic formula. If t is a term denoting a real from $^\omega 2$ or $^\omega \omega$, then $t(n) = m$ is an atomic formula. Also a formula $t \in X$, where t is a term of first order and X is a second-order variable, is an atomic formula.

Let $\varphi(r_1, \ldots, r_k, x_1, \ldots, x_n, y_1, \ldots, y_m, A_1, \ldots, A_l)$ be a first-order formula with variables $x_1, \ldots, x_n$ in Polish spaces $X_1, \ldots, X_n$, variables $y_1, \ldots, y_m$ denote elements of some Polish spaces $Y_1, \ldots, Y_l$, and $A_1, \ldots, A_l$ denote subsets of Polish spaces $Y_1, \ldots, Y_l$, respectively. We denote by

$$\mathsf{V}_\varphi(r_1, \ldots, r_k, y_1, \ldots, y_m, A_1, \ldots, A_l) \qquad (6.15)$$
$$= \{\langle x_1, \ldots, x_n \rangle \in X_1 \times \cdots \times X_n :$$
$$\varphi(r_1, \ldots, r_k, x_1, \ldots, x_n, y_1, \ldots, y_m, A_1, \ldots, A_l)\}$$

the set defined by the formula φ and parameters $r_1, \ldots, r_k, y_1, \ldots, y_m, A_1, \ldots, A_l$.[3]

Of course we allow introduction of other predicates, consequently other atomic formulas. The main criterion for introducing new operations and new predicates is the condition that, for any atomic first-order formula

$$\varphi(r_1, \ldots, r_n, x_1, \ldots, x_k, y_1, \ldots, y_k, A_1, \ldots, A_m)$$

with variables $x_1, \ldots, x_k$ in Polish spaces $X_1, \ldots, X_k$, with variables $y_1, \ldots, y_m$ in Polish spaces $Y_1, \ldots, Y_m$, for given values of zero-order variables $r_1, \ldots, r_n$ and given values of variables $A_1, \ldots, A_m$ denoting Borel subsets of Polish spaces $Y_1, \ldots, Y_m$, the set

$$\mathsf{V}_\varphi(r_1, \ldots, r_k, y_1, \ldots, y_m, A_1, \ldots, A_l) \text{ is a Borel set.} \qquad (6.16)$$

A quantifier $\forall \square, \exists \square$ is a **zero-order quantifier** or a **first-order quantifier** depending on whether $\square$ denotes a zero-order variable or a first-order variable.

A formula formed from atomic formulas by using the **logical connectives** $\neg$, $\wedge$, $\vee$, $\rightarrow$ and $\equiv$ is said to be a Σ_0^0 **formula**. A Σ_0^1 **formula** is any formula obtained from Σ_0^0 formulas by logical connectives and quantifiers of zero order. A **first order formula** is any formula obtained from Σ_0^1 formulas by logical connectives and quantifiers of zero order and/or quantifiers of first order.

Let us remark that we do not allow the quantifiers of the second order. The variables of second order will always play a rôle of parameters.

---

[3] Actually, now we mix speaking in metamathematics and mathematics. A formula φ is an object of metamathematics and parameters are mathematical objects.

We present a general result on changing the order of some zero-order and first-order quantifiers over variables $\alpha, \beta, \ldots$ denoting elements of $^\omega 2$ or $^\omega \omega$. Let us note that we can eliminate variable x denoting elements of a Polish space X by using, e.g., Corollary 5.7: we introduce a new continuous operation $f : {}^\omega\omega \xrightarrow{\text{onto}} X$ and replace each occurrence of x by term $f(\alpha)$. The reader can fill in all details.

Theorem 6.43. *For any first-order formula φ the following hold true:*

$$(\exists n)(\exists m)\, \varphi(n, m, \ldots) \equiv (\exists k),\, \varphi(\lambda(k), \rho(k), \ldots), \tag{6.17}$$

$$(\forall n)(\forall m)\, \varphi(n, m, \ldots) \equiv (\forall k)\, \varphi(\lambda(k), \rho(k), \ldots), \tag{6.18}$$

$$(\exists \alpha)(\exists \beta)\, \varphi(\alpha, \beta, \ldots) \equiv (\exists \gamma),\, \varphi(\Lambda(\gamma), R(\gamma), \ldots), \tag{6.19}$$

$$(\forall \alpha)(\forall \beta)\, \varphi(\alpha, \beta, \ldots) \equiv (\forall \gamma)\, \varphi(\Lambda(\gamma), R(\gamma), \ldots), \tag{6.20}$$

$$(\exists n)(\exists \alpha)\, \varphi(n, \alpha, \ldots) \equiv (\exists \alpha)(\exists n)\, \varphi(n, \alpha, \ldots), \tag{6.21}$$

$$(\forall n)(\forall \alpha)\, \varphi(n, \alpha, \ldots) \equiv (\forall \alpha)(\forall n)\, \varphi(n, \alpha, \ldots), \tag{6.22}$$

$$(\exists n)(\forall \alpha)\, \varphi(n, \alpha, \ldots) \equiv (\forall \alpha)(\exists n)\, \varphi(n, \mathrm{Proj}_n(\alpha), \ldots), \tag{6.23}$$

$$(\forall n)(\exists \alpha)\, \varphi(n, \alpha, \ldots) \equiv (\exists \alpha)(\forall n)\, \varphi(n, \mathrm{Proj}_n(\alpha), \ldots). \tag{6.24}$$

Proof. The proof of equivalence (6.17) is based on the fact that for any natural numbers n, m there is a natural number k such that $\lambda(k) = n$ and $\rho(k) = m$. The equivalence (6.18) follows from (6.17) by de Morgan's Law.

The proofs of equivalences (6.19) and (6.23) are based on the fact that for any sequence $\{\alpha_n\}_{n=0}^\infty$ of reals there is a real α such that $\alpha_n = \mathrm{Proj}_n(\alpha)$ for any n. (6.20) follows from (6.19) and (6.24) follows from (6.23) by de Morgan law.

The equivalences (6.21), (6.22) trivially hold true. □

Corollary 6.44. *Any first-order formula is equivalent to a formula of one of the following types:*

1) Σ_0^1 *formula,*
2) $(\forall \alpha_1)(\exists \alpha_2)(\forall \alpha_3) \cdots (Q\alpha_n)\psi$, *where ψ is a Σ_0^1 formula,*
3) $(\exists \alpha_1)(\forall \alpha_2)(\exists \alpha_3) \ldots (Q\alpha_n)\psi$, *where ψ is a Σ_0^1 formula.*

Proof. From logic it is well known that for any formula φ there exists an equivalent formula of the form $(Q_1 a_1) \ldots (Q_n a_n)\, \varphi$, where Q_i is a quantifier ($\exists$ or $\forall$) and φ does not contain any quantifier. Using (6.21)–(6.24) one can order the quantifiers in such a way that the zero-order quantifiers follow after all first-order quantifiers. If the so-obtained formula does not contain any first-order quantifier, we are ready – it is a Σ_0^1 formula. Otherwise, using (6.19) and (6.20) one units two or more neighboring equal quantifiers of first order. You obtain a formula, which begins with a block of changing first-order quantifiers followed by a Σ_0^1 formula. □

A formula equivalent to a formula of the type 2) is called a Π_n^1 **formula.** Similarly, a formula equivalent to a formula of the type 3) is called a Σ_n^1 **formula.** Finally, a formula is a Δ_n^1 **formula** if it is equivalent to both a Π_n^1 formula and a Σ^1_n formula. A formula of the type 1) is both Σ^1_0 and Π^1_0 formulae. Thus, also a Δ^1_0 formula. The introduced notation is not an accident.

Immediately from the definitions (6.15) of the set V_φ we obtain

$$V_{\varphi \vee \psi} = V_\varphi \cup V_\psi, \quad V_{\varphi \wedge \psi} = V_\varphi \cap V_\psi, \quad V_{\neg\varphi} = X_1 \times \cdots \times X_n \setminus V_\varphi, \qquad (6.25)$$

$$V_{(\exists r_1)\,\varphi} = \bigcup_{r_1} V_\varphi, \quad V_{(\forall r_1)\,\varphi} = \bigcap_{r_1} V_\varphi, \qquad (6.26)$$

$$V_{(\exists x_1)\,\varphi} = \mathrm{proj}_2(V_\varphi), \qquad V_{(\forall x_1)\,\varphi} = X_2 \times \cdots \times X_n \setminus \mathrm{proj}_2(V_{\neg\varphi}). \qquad (6.27)$$

Theorem 6.45 (K. Kuratowski – A. Tarski). *Let $X_1, \ldots, X_n, Y_1, \ldots, Y_l$ be Polish spaces, $r_1, \ldots, r_k$, $y_1 \in Y_1, \ldots, y_m \in Y_m$, $A_1, \ldots, A_l$ be given, $A_1, \ldots, A_l$ being Borel subsets of Polish spaces $Y_1, \ldots, Y_l$, respectively. Then:*

a) *If φ is a Σ^1_0 formula, then the set $V_\varphi(r_1, \ldots, r_k, y_1, \ldots, y_m, A_1, \ldots, A_l)$ is a Borel subset of $X_1 \times \cdots \times X_n$.*

b) *If φ is a Σ^1_n formula, then the set $V_\varphi(r_1, \ldots, r_k, y_1, \ldots, y_m, A_1, \ldots, A_l)$ is a $\boldsymbol{\Sigma}^1_n(X_1 \times \cdots \times X_n)$ set.*

c) *If φ is a Π^1_n formula, then the set $V_\varphi(r_1, \ldots, r_k, y_1, \ldots, y_m, A_1, \ldots, A_l)$ is a $\boldsymbol{\Pi}^1_n(X_1 \times \cdots \times X_n)$ set.*

d) *If φ is a Δ^1_n formula, then the set $V_\varphi(r_1, \ldots, r_k, y_1, \ldots, y_m, A_1, \ldots, A_l)$ is a $\boldsymbol{\Delta}^1_n(X_1 \times \cdots \times X_n)$ set.*

Proof. The assertions b), c) and d) follow from a) by (6.27). Thus we have to show a).

If φ is a Σ^1_0 formula and the parameters $A_1, \ldots, A_l$ are Borel subsets of some Polish spaces, then by (6.16), (6.25) and (6.26) the set V_φ is Borel, since all the unions and intersections in (6.26) are countable. $\qquad \square$

Exercises

6.10 Injective Images of Projective Sets

a) If $A \subseteq X$ is Borel, $f : A \xrightarrow{1-1} Y$ is Borel measurable, then $f(A)$ is a Borel set.
   Hint: Apply Theorems 3.56 and 6.26: $f(A)$ is a continuous injective projection of f and f is Borel in $X \times Y$.

b) If $f : X \xrightarrow{1-1} Y$ is Borel measurable, $A \in \boldsymbol{\Pi}^1_n(X)$, then $f(A) \in \boldsymbol{\Pi}^1_n(Y)$.
   Hint: $f(A) = f(X) \setminus f(X \setminus A)$, $f(X)$ is Borel and $f(X \setminus A) \in \boldsymbol{\Sigma}^1_n(Y)$ by Theorem 6.37.

6.11 Closure Properties of Projective Sets

a) Let X, Y be Polish spaces. For any $f : X \longrightarrow Y$ the following are equivalent:
   (1) The graph of f is a $\boldsymbol{\Sigma}^1_n$ set.
   (2) The graph of f is a $\boldsymbol{\Delta}^1_n$ set.
   (3) f is $\boldsymbol{\Sigma}^1_n$-measurable.
   (4) f is $\boldsymbol{\Delta}^1_n$-measurable.

Hint: (1) $\rightarrow$ (2) *since* $\langle x, y \rangle \notin f \equiv (\exists z)(z \neq y \wedge \langle x, z \rangle \in f)$. *Proof of* (2) $\rightarrow$ (4) *is similar to that of Theorem 6.27. Proof of* (3) $\rightarrow$ (1) *is similar to that of Theorem 3.56.*

b) Every family $\mathbf{\Sigma}_n^1$, $\mathbf{\Pi}_n^1$, $\mathbf{\Delta}_n^1$ is closed under preimages by $\mathbf{\Delta}_n^1$-measurable functions.
Hint: Note that $x \in f^{-1}(A) \equiv (\exists y)(\langle x, y \rangle \in f \wedge y \in A) \equiv (\forall y)(\langle x, y \rangle \in f \rightarrow y \in A).$

c) There exists a function with $\mathbf{\Pi}_1^1$ graph that is not a $\mathbf{\Delta}_1^1$ set.
Hint: See Theorem 6.62 below.

d) $\mathbf{\Sigma}_n^1$ is closed under a Souslin scheme operation.
Hint: Note that $\alpha \in L(T, \varphi) \equiv (\exists \beta)(\forall n)\, \alpha \in \varphi(\beta|n).$

6.12 Definition of Projective Sets by a Formula

We introduce a new predicate $\mathbf{OB}(n, x)$ with intended interpretation "x belongs to U_n", where $\langle U_n : n \in \omega \rangle$ is a countable open base of X. So $\mathbf{OB}(n, x)$ is also a new atomic formula. The condition (6.16) is satisfied.

a) Construct a Σ_0^1 formula φ such that

i) for any open set $A \subseteq X$ there exists an $\alpha \in {}^\omega 2$ such that

$$A = \{x \in X : \varphi(\alpha, x)\} \tag{6.28}$$

ii) for any $\alpha \in {}^\omega 2$ the set $\{x \in X : \varphi(\alpha, x)\}$ is open.

Hint: See the proof of Theorem 5.18.

b) Construct a Σ_1^1 formula φ such that

i) for any analytic set $A \subseteq X$ there exists an $\alpha \in {}^\omega 2$ such that (6.28) holds true;

ii) for any $\alpha \in {}^\omega 2$ the set $\{x \in X : \varphi(\alpha, x)\}$ is analytic.

Hint: Use Theorem 6.17, h).

c) Construct a Σ_1^1 formula φ and a Π_1^1 formula ψ such that for any Borel set $A \subseteq X$ there are $\alpha, \beta \in {}^\omega 2$ such that

$$A = \{x \in X : \varphi(\alpha, x)\} = \{x \in X : \psi(\beta, x)\}.$$

Hint: Take φ from b) and let ψ to be $\neg\varphi$.

d) There is no Δ_1^1 formula φ such that for any Borel set $A \subseteq {}^\omega 2$ there exists an $\alpha \in {}^\omega 2$ such that

$$A = \{x \in X : \varphi(\alpha, x)\}.$$

Hint: See the proof of Theorem 5.15.

e) Construct a Σ_n^1 formula φ such that

i) for any $\mathbf{\Sigma}_n^1$ set $A \subseteq X$ there exists an $\alpha \in {}^\omega 2$ such that $A = \{x \in X : \varphi(\alpha, x)\}$;

ii) for any $\alpha \in {}^\omega 2$ the set $\{x \in X : \varphi(\alpha, x)\}$ is a $\mathbf{\Sigma}_n^1$ set.

f) A similar assertion holds true for Π_n^1.

6.4   Co-analytic and Σ^1_2 Sets

Theorem 6.46 (N.N. Luzin – W. Sierpiński). *If a subset A of a Polish space X is sifted by a closed sieve, then there exists a sequence of pairwise disjoint Borel sets $\langle A_\xi : \xi \in \omega_1 \rangle$ such that $X \setminus A = \bigcup_{\xi \in \omega_1} A_\xi$. Moreover, we can assume that*

   a) *$\bigcup_{k<n} A_k$ is a G_δ set for $n \in \omega$,*
   b) *$\bigcup_{\eta<\xi} A_\eta$ is a Σ^0_ξ set for ξ limit,*
   c) *if $\xi = \lambda + n + 1$, where λ is limit and $n \in \omega$, then $\bigcup_{\eta<\xi} A_\eta$ is a $\Pi^0_{\lambda+1}$ set.*

Proof. Assume that A is sifted by a closed sieve Φ. For any $x \in X$ we write

$$M_x(\Phi) = \{r \in \mathbb{D} : x \in \Phi(r)\}. \tag{6.29}$$

By definition we obtain

$$x \in X \setminus A \equiv \langle M_x(\Phi), \preceq \rangle \text{ is well ordered.} \tag{6.30}$$

Let

$$A_\xi(\Phi) = \{x \in X : \langle M_x(\Phi), \preceq \rangle \text{ has order type } \xi\}. \tag{6.31}$$

Since the set $\mathbb{D}$ is countable the order type of a well-ordered set $M_x(\Phi)$ is an ordinal smaller than ω_1. Therefore $X \setminus A = \bigcup_{\xi<\omega_1} A_\xi(\Phi)$.

We show by induction over $\xi < \omega_1$ that for any closed sieve Φ all sets

$$B_\xi(\Phi) = \bigcup_{\eta<\xi} A_\eta(\Phi) = \{x \in X : \mathrm{ot}(\langle M_x(\Phi), \preceq \rangle) < \xi\}$$

satisfy conditions a)–c) of the theorem.

For $n \in \omega$ we can easily see that

$$x \notin B_n(\Phi) \equiv (\exists r_1, \ldots, r_n, \text{ mutually different}) \, x \in \Phi(r_1) \cap \cdots \cap \Phi(r_n). \tag{6.32}$$

Thus $X \setminus B_n(\Phi)$ is an F_σ set.

Assume now that $\xi < \omega_1$ is infinite and for every $\eta < \xi$ and for every closed sieve Φ the sets $B_n(\Phi)$ satisfy conditions a)–c).

If ξ is a limit ordinal, then the assertion follows by induction, since

$$B_\xi(\Phi) = \bigcup_{\eta<\xi} B_\eta(\Phi).$$

Assume that $\xi = \lambda + n + 1$, λ is limit. For any $r \in \mathbb{D}$ we define a closed sieve

$$\Phi_r(s) = \begin{cases} \Phi(s) & \text{if } s \prec r, \\ \emptyset & \text{otherwise.} \end{cases} \tag{6.33}$$

Evidently

$$M_x(\Phi_r) = \{s \in M_x(\Phi) : s \prec r\}.$$

If $M_x(\Phi)$ is a well-ordered set, then the set $M_x(\Phi_r)$ is an initial segment of $M_x(\Phi)$ for any $r \in M_x(\Phi)$. Vice versa, every initial segment of $M_x(\Phi)$ has the form $M_x(\Phi_r)$ for suitable $r \in M_x(\Phi)$. Thus

$$x \in B_\xi(\Phi) \equiv (\forall r \in M_x(\Phi)) \operatorname{ot}(M_x(\Phi_r)) < \lambda + n \equiv x \in \bigcap_{r \in M_x(\Phi)} B_{\lambda+n}(\Phi_r).$$

Since by inductive assumption every $B_{\lambda+n}(\Phi_r) \in \mathbf{\Pi}_{\lambda+1}^0$ (even to $\mathbf{\Sigma}_\lambda^0$ set if $n = 0$) we obtain $B_\xi(\Phi) \in \mathbf{\Pi}_{\lambda+1}^0(X)$.

Since $A_\xi(\Phi) = B_{\xi+1}(\Phi) \setminus B_\xi(\Phi)$, every set $A_\xi(\Phi)$ is Borel. $\quad\square$

Corollary [AC] 6.47. *Any uncountable co-analytic set has cardinality $\aleph_1$ or $\mathfrak{c}$.*

If an analytic set A is sifted by a sieve Φ, then the sets $\langle A_\xi(\Phi) : \xi < \omega_1 \rangle$ are called **constituents** of the co-analytic set $X \setminus A$.

We shall study properties of a decomposition of the power set of a countable set into $\aleph_1$ many families introduced in the proof of Theorem 1.16. To simplify our investigation, we define a decomposition of $\mathcal{P}(\mathbb{D})$ based on the properties of a subset $A \subseteq \mathbb{D}$ linearly ordered by the relation $\preceq$ defined by (5.7).

Let ξ be an ordinal. We define

$$\mathbf{WO}_\xi = \{A \subseteq \mathbb{D} : \operatorname{ot}(A, \preceq) = \xi\}, \tag{6.34}$$

$$\mathbf{WO} = \{A \subseteq \mathbb{D} : \langle A, \preceq \rangle \text{ is well ordered}\}. \tag{6.35}$$

By Theorem 11.4 we obtain that $\mathbf{WO}_\xi \neq \emptyset$ for every countable ordinal ξ and

$$\mathbf{WO} = \bigcup_{\xi < \omega_1} \mathbf{WO}_\xi.$$

However one can easily show that

$$|\mathbf{WO}_\xi| = \mathfrak{c} \text{ for any } \xi \geq \omega \text{ and } |\mathcal{P}(\mathbb{D}) \setminus \mathbf{WO}| = \mathfrak{c}. \tag{6.36}$$

Let $d : \omega \longrightarrow \mathbb{D}$ be a bijection. Any subset $A \subseteq \mathbb{D}$ can be coded by an element $t = \bar{d}(A) \in {}^\omega 2$ by setting

$$t(n) = 1 \equiv d(n) \in A.$$

If the bijection d is given we do not distinguish between a set $A \subseteq \mathbb{D}$ and its code $\bar{d}(A) \in {}^\omega 2$. Especially, we shall identify the sets $\mathbf{WO}_\xi$, $\mathbf{WO}$ with the subsets $\{\bar{d}(A) : A \in \mathbf{WO}_\xi\}$, $\{\bar{d}(A) : A \in \mathbf{WO}\}$ of ${}^\omega 2$, respectively. The family of sets

$$\{\{\bar{d}(A) : A \in \mathbf{WO}_\xi\} : \xi < \omega_1\} \cup \{{}^\omega 2 \setminus \{\bar{d}(A) : A \in \mathbf{WO}\}\} \tag{6.37}$$

is called the **Lebesgue decomposition** of ${}^\omega 2$.

We define a sieve Φ by

$$\Phi(r) = \{t \in {}^{\omega}2 : t(d^{-1}(r)) = 1\} \tag{6.38}$$

and show that

$$S(\Phi) = {}^{\omega}2 \setminus \mathbf{WO}. \tag{6.39}$$

Actually, we have

$$t \in \Phi(d(n)) \equiv t(n) = 1.$$

On the other hand

$$t = \bar{d}(\{d(n) : t(n) = 1\}).$$

Thus

$$M_t(\Phi) = \{d(n) : t \in \Phi(d(n))\} = \{d(n) : t(n) = 1\}.$$

Hence $t \notin \mathbf{WO}$ if and only if the set $\{d(n) : n \in \omega \wedge t(n) = 1\}$ is not well ordered, therefore if and only if $t \in S(\Phi)$. Moreover, the set $\mathbf{WO}_\xi \subseteq {}^{\omega}2$ is the constituent

$$A_\xi(\Phi) = \{t \in {}^{\omega}2 : \langle\{d(n) : t(n) = 1\}, \preceq\rangle \text{ has order type } \xi\}$$

of the co-analytic set $\mathbf{WO}$ for every $\xi < \omega_1$.

Since the values of Φ are clopen sets, by (6.32), the sets $\bigcup_{k<n} \mathbf{WO}_k$ are closed for any $n \in \omega$.

Hence by Theorem 6.46 we obtain

Corollary [wAC] 6.48.

a) *The sets $\mathbf{WO}_n$ are both F_σ and G_δ sets for any $n \in \omega$.*

b) *If ξ is a limit ordinal, then $\mathbf{WO}_\xi \in \mathbf{\Pi}^0_{\xi+1}({}^{\omega}2)$.*

c) *If $\xi = \lambda + n + 1$, where λ is a limit ordinal and $n \in \omega$, then $\mathbf{WO}_\xi \in \mathbf{\Delta}^0_{\lambda+2}({}^{\omega}2)$.*

d) *$\aleph_1 \leq |\mathrm{BOREL}|$.*

We generalize notions introduced in Section 6.1. A subset B of a Polish space Y is called **Borel Γ-hard** if for any Polish space X and any set $A \in \Gamma(X)$, there exists a Borel measurable mapping $f : X \longrightarrow Y$ such that $A = f^{-1}(B)$. If moreover $B \in \Gamma(Y)$ then B is called **Borel Γ-complete**. Note the difference: in case of Wadge Γ-hard or Wadge Γ-complete sets we dealt with subsets of Baire spaces only and reduction was realized by a continuous mapping.

One can easily see that if a subset A of a Polish space is Borel Γ-hard (complete) then $X \setminus A$ is Borel $\neg\Gamma$-hard (complete).

Theorem [wAC] 6.49. *If $A \in \mathbf{\Sigma}^1_1(X)$ then there exists a Borel measurable mapping $f : X \longrightarrow {}^{\omega}2$ such that $\langle f^{-1}(\mathbf{WO}_\xi) : \xi \in \omega_1\rangle$ are the constituents of the set $X \setminus A$. If A is sifted by a clopen sieve, then f is continuous. Hence, the set $\mathbf{WO}$ is Borel $\mathbf{\Pi}^1_1$-complete and therefore ${}^{\omega}2 \setminus \mathbf{WO}$ is Borel $\mathbf{\Sigma}^1_1$-complete.*

Proof. Let $A \subseteq X$ be a Σ^1_1 set sifted by a closed sieve Φ. We define a mapping $f : X \longrightarrow {}^\omega 2$ as follows: for any $x \in X$ let $f(x) = t$, where

$$t(n) = 1 \equiv x \in \Phi(d(n)).$$

We claim that f is Borel measurable. Actually, if $s \in {}^{<\omega}2$ then

$$f^{-1}([s]) = \bigcap_{k \in \mathrm{dom}(s)} (s(k))\Phi(d(k)),$$

where $(0)B = X \setminus B$ and $(1)B = B$ (compare Section 5.4).

Note the following: if $f(x) = t$ then

$$M_x(\Phi) = \{d(n) : t(n) = 1\}.$$

Thus $A_\xi(\Phi) = f^{-1}(\mathbf{WO}_\xi)$ for every $\xi < \omega_1$. □

Corollary [wAC] 6.50. *The co-analytic set* $\mathbf{WO}$ *is not analytic.*

Proof. By Theorem 6.30 there exists a co-analytic set A that is not Borel and therefore is not analytic. Since $A = f^{-1}(\mathbf{WO})$ neither $\mathbf{WO}$ is analytic. □

Since every Polish space contains a closed subset homeomorphic to ${}^\omega 2$ using (6.36) we obtain another

Corollary [wAC] 6.51. *If X is perfect, then there exists a Σ^1_1-complete set $A \subseteq X$ such that $|A| = |X \setminus A| = \mathfrak{c}$.*

The notion of a Γ-rank has been introduced in Section 5.2. We have shown that a ranked pointclass has some nice properties. We show that the pointclass $\mathbf{\Pi}^1_1$ is ranked. We start with an auxiliary result.

Lemma 6.52. *There exists a regular $\mathbf{\Pi}^1_1$-rank on $\mathbf{WO}$ with the range ω_1.*

Proof. We define a rank ρ on $\mathbf{WO}$ by $\rho(x) = \xi$ if $x \in \mathbf{WO}_\xi$. Evidently $\mathrm{rng}(\rho) = \omega_1$. By Lemma 5.21 it suffices to find relations $\leq^+ \in \mathbf{\Pi}^1_1({}^\omega 2 \times {}^\omega 2)$ and $\leq^- \in \mathbf{\Sigma}^1_1({}^\omega 2 \times {}^\omega 2)$ such that (5.18) and (5.19) hold true (for any $y \in \mathbf{WO}$ and any $x \in {}^\omega 2$).

We can code a function F from a subset of $\mathbb{D}$ into $\mathbb{D}$ by an element $\alpha \in {}^\omega 2$ setting

$$\alpha(\pi(n, m)) = 1 \equiv (d(n) \in \mathrm{dom}(F) \wedge F(d(n)) = d(m)).$$

The set C will be the set of all codes of an increasing function from its domain, a subset of $\mathbb{D}$, into $\mathbb{D}$. Thus

$$\begin{aligned} C = \{\alpha \in {}^\omega 2 : &(\forall n, m_1, m_2)\,(\alpha(\pi(n, m_1)) = \alpha(\pi(n, m_2)) = 1 \to m_1 = m_2) \\ &\wedge (\forall n_1, n_2, m_1, m_2)\,((d(n_1) < d(n_2) \wedge \alpha(\pi(n_1, m_1)) = \alpha(\pi(n_2, m_2)) = 1) \\ &\to d(m_1) < d(m_2))\}. \end{aligned}$$

By Theorem 6.45 the set C is Borel, actually a closed set.

Now, it is easy to see that the relations defined by $(x, y \in {}^{\omega}2)$

$$x \leq^{+} y \equiv (x, y \in \mathbf{WO} \wedge \neg y <_{\rho} x) \tag{6.40}$$
$$\equiv (x, y \in \mathbf{WO} \wedge \neg (\exists \alpha)(\exists k)\,(\alpha \in C \wedge y(k) = 1 \wedge (\forall n)\,(x(n) = 1$$
$$\rightarrow (\exists m)\,(d(m) < d(k) \wedge y(m) = 1 \wedge \alpha(\pi(n, m)) = 1))).$$
$$x \leq^{-} y \equiv (\exists \alpha)\,(\alpha \in C \wedge (\forall n)\,(x(n) = 1 \tag{6.41}$$
$$\rightarrow (\exists m)\,(y(m) = 1 \wedge \alpha(\pi(n, m)) = 1))),$$

are $\mathbf{\Pi}_1^1$ and $\mathbf{\Sigma}_1^1$ sets satisfying (5.18) and (5.19), respectively. $\square$

Theorem [wAC] 6.53. *The pointclass $\mathbf{\Pi}_1^1$ is ranked.*

Proof. Assume that $A \in \mathbf{\Pi}_1^1(X)$. Then there exists a closed sieve Φ on X such that $X \setminus A = \mathrm{S}(\Phi)$. By Theorem 6.49 there exists a Borel measurable mapping $f : X \longrightarrow {}^{\omega}2$ such that $A_{\xi}(\Phi) = f^{-1}(\mathbf{WO}_{\xi})$ for each $\xi < \omega_1$. Define a rank σ on A by $\sigma(x) = \xi$ for $x \in A_{\xi}(\Phi)$. We define $f \times f : X \times X \longrightarrow {}^{\omega}2 \times {}^{\omega}2$ by $f \times f(x, y) = \langle f(x), f(y) \rangle$. Let ρ be the rank of the proof of Lemma 6.52. One can easily check that

$$\leq_{\sigma}^{*} = (f \times f)^{-1}(\leq_{\rho}^{*}), \qquad <_{\sigma}^{*} = (f \times f)^{-1}(<_{\rho}^{*}).$$

Since $\leq_{\rho}^{*}$ and $<_{\sigma}^{*}$ are $\mathbf{\Pi}_1^1$ sets, $f \times f$ is continuous, the rank σ is a $\mathbf{\Pi}_1^1$-rank. $\square$

Theorem [wAC] 6.54 (Boundedness Theorem). *Let A, B be analytic subsets of a Polish space X. If $B \subseteq \bigcup_{\xi \in \omega_1} A_{\xi}$, where $\langle A_{\xi} : \xi \in \omega_1 \rangle$ are constituents of the set $X \setminus A$, then there exists a $\xi_0 < \omega_1$ such that $B \subseteq \bigcup_{\xi < \xi_0} A_{\xi}$.*

We begin with

Lemma [wAC] 6.55. *If $D \subseteq \bigcup_{\xi < \omega_1} \mathbf{WO}_{\xi}$ is an analytic set, then there exists a countable ordinal ξ_0 such that $D \subseteq \bigcup_{\xi < \xi_0} \mathbf{WO}_{\xi}$.*

Proof. Assume that D is "unbounded", i.e.,

$$(\forall \eta < \omega_1)(\exists \xi > \eta)\,(D \cap \mathbf{WO}_{\xi} \neq \emptyset).$$

Then $x \in \mathbf{WO}$ if and only if the order type of $\{d(n) : x(n) = 1\}$ is not greater than the order type of $\{d(n) : y(n) = 1\}$ for some $y \in D$. Thus

$$x \in \mathbf{WO} \equiv (\exists y)(\exists \alpha)\,(y \in D \wedge \alpha \in C \wedge (\forall n)\,(x(n) = 1$$
$$\rightarrow (\exists m)\,(y(m) = 1 \wedge \alpha(\pi(n, m)) = 1)))).$$

By Theorem 6.45 the set $\mathbf{WO}$ is analytic, a contradiction. $\square$

Proof of Theorem 6.54. Assume that A, B are analytic subsets of a Polish space X, $A = \mathrm{S}(\Phi)$ for a closed sieve Φ on X. Moreover, let $B \subseteq \bigcup_{\xi \in \omega_1} A_{\xi}(\Phi)$. By Theorem 6.49 there exists a Borel measurable mapping $f : X \longrightarrow {}^{\omega}2$ such that $A_{\xi}(\Phi) = f^{-1}(\mathbf{WO}_{\xi})$ for each $\xi < \omega_1$. By Theorem 6.37, b), the set $f(B) \subseteq \mathbf{WO}$ is analytic. Thus by Lemma 6.55, $f(B) \subseteq \bigcup_{\xi < \xi_0} \mathbf{WO}_{\xi}$ for a $\xi_0 < \omega_1$. Then we have $B \subseteq \bigcup_{\xi < \xi_0} A_{\xi}(\Phi)$. $\square$

We shall study another classical property of the pointclass $\mathbf{\Pi}_1^1$.

Theorem [wAC] 6.56 (P.S. Novikoff – M. Kondô – J.W. Addison). *The pointclass* $\mathbf{\Pi}_1^1$ *has the uniformization property.*

The theorem follows by the Uniformization Theorem 5.27 from

Theorem [wAC] 6.57. *The pointclass* $\mathbf{\Pi}_1^1$ *is scaled.*

Proof. Let $\Theta = \{\langle \xi, \eta \rangle \in \omega_1 \times \omega_1 : \eta \leq \xi\}$ be ordered lexicographically

$$\langle \xi, \eta \rangle <_{\text{lex}} \langle \zeta, \theta \rangle \equiv (\xi < \zeta \vee (\xi = \zeta \wedge \eta < \theta)).$$

Then $\text{ot}(\Theta, <_{\text{lex}}) = \omega_1$. Let Υ be the (unique) isomorphism of Θ and ω_1.

If A is a $\mathbf{\Pi}_1^1$-subset of a Polish space X, then by Theorems 5.1 and 5.12 there exists an open sieve Φ such that $X \setminus A = \text{S}(\Phi)$. By (6.30) we have

$$x \in A \equiv \langle M_x(\Phi), \preceq \rangle \text{ is well ordered,}$$

where $M_x(\Phi)$ is defined by (6.29).

The sieve Φ_r was defined by (6.33) and $d : \omega \xrightarrow[\text{onto}]{1-1} \mathbb{D}$ was fixed above. For every $n \in \omega$ define a rank ρ_n on A by

$$\rho_n(x) = \Upsilon(\beta, \gamma), \text{ where } \beta = \text{ot}(M_x(\Phi), \preceq) \text{ and } \gamma = \text{ot}(M_x(\Phi_{d(n)}), \preceq)).$$

The rank ρ_n induces a prewell-ordering on A,

$$x \leq_n y \equiv \rho_n(x) \leq \rho_n(y).$$

Similarly as in the proof of Lemma 6.52 we can construct $\leq_n^+ \in \mathbf{\Pi}_1^1$ and $\leq_n^- \in \mathbf{\Sigma}_1^1$ such that (5.18) and (5.19) hold true. Thus every ρ_n is a $\mathbf{\Pi}_1^1$-rank.

Let $\lim_{k \to \infty} x_k = x$, $x_k \in A$. Assume that for every n there exists k_n such that $\rho_n(x_k) = \alpha_n$ for every $k \geq k_n$. We have to show that $x \in A$ and $\rho_n(x) \leq \alpha_n$.

By assumptions for any n and any $k \geq k_n$ we have $\alpha_n = \Upsilon(\beta_n, \gamma_n)$, where $\beta_n = \text{ot}(M_{x_k}(\Phi), \preceq)$ and $\gamma_n = \text{ot}(M_{x_k}(\Phi_{d(n)}), \preceq)$. Thus for $k \geq \max\{k_i, k_j\}$ we have $\beta_i = \beta_j = \text{ot}(M_{x_k}(\Phi), \preceq)$. Therefore, there exists a common value $\beta = \beta_n$ for all n.

We show that

$$d(i) \prec d(j) \rightarrow \gamma_i < \gamma_j \text{ for any } d(i), d(j) \in M_x(\Phi). \tag{6.42}$$

Thus, assume that $d(i) \prec d(j)$, $d(i), d(j) \in M_x(\Phi)$. Then $x \in \Phi(d(i)) \cap \Phi(d(j))$. Since the values of the sieve Φ are open sets there exists a k such that $k \geq k_i$, $k \geq k_j$ and $x_k \in \Phi(d(i)) \cap \Phi(d(j))$. Then $d(i) \in M_{x_k}(\Phi_{d(j)})$ and therefore the set $M_{x_k}(\Phi_{d(i)})$ is a proper initial segment of $M_{x_k}(\Phi_{d(j)})$. Thus $\gamma_i = \text{ot}(M_{x_k}(\Phi_{d(i)}))$ is smaller than $\gamma_j = \text{ot}(M_{x_k}(\Phi_{d(j)}))$.

By (6.42) $\langle M_x(\Phi), \preceq \rangle$ is well ordered and therefore $x \in A$.

If we set $g(t) = \gamma_{d^{-1}(t)}$ for $t \in M_x(\Phi)$, then by (6.42) g is an increasing mapping from $\langle M_x(\Phi), \preceq \rangle$ into ordinals. Hence

$$\mathrm{ot}(M_x(\Phi_{d(n)}), \preceq) = \mathrm{ot}(\{t \in M_x(\Phi) : t < d(n)\}, \preceq) \leq \gamma_n.$$

Then

$$\mathrm{ot}(M_x(\Phi), \preceq) = \sup\{\mathrm{ot}(M_x(\Phi_{d(n)}), \preceq) : d(n) \in M_x(\Phi)\} \leq \sup\{\gamma_n : n \in \omega\}.$$

However, for any n and $k \geq k_n$ we have $\gamma_n = \mathrm{ot}(M_{x_k}(\Phi_{d(n)}), \preceq) \leq \beta$. Thus

$$\rho_n(x) = \Upsilon(\mathrm{ot}(M_x(\Phi)), \mathrm{ot}(M_x(\Phi_{d(n)}))) \leq \Upsilon(\beta, \gamma_n) = \alpha_n. \qquad \square$$

Using Theorem 6.56 we can prove a classical result.

Theorem [wAC] 6.58. *Every Σ_2^1 set is a union of $\aleph_1$ Borel sets.*

Proof. Assume that $A = \mathrm{proj}(B) \subseteq X$ is a projection of Π_1^1 set $B \subseteq X \times {}^\omega\omega$. By the Uniformization Theorem 6.56 we may assume that for every $x \in A$ there exists exactly one y such that $\langle x, y \rangle \in B$. Then the projection proj is injective on the set B. By Theorem 6.46 there are Borel sets $\langle B_\xi : \xi < \omega_1 \rangle$ such that $B = \bigcup_{\xi < \omega_1} B_\xi$. Then $A = \bigcup_{\xi < \omega_1} \mathrm{proj}(B_\xi)$ and by Luzin's Theorem 6.26 every $\mathrm{proj}(B_\xi)$ is Borel. $\qquad \square$

Corollary [AC] 6.59. *Every uncountable Σ_2^1 set has cardinality $\aleph_1$ or $\mathfrak{c}$.*

From Theorems 6.53 and 5.24 we immediately obtain

Theorem [wAC] 6.60 (Y.N. Moschovakis). *The pointclass Σ_2^1 is ranked.*

Thus, by Theorems 5.22 and 5.20, a) we have

Corollary [wAC] 6.61. *The pointclass Σ_2^1 possesses the reduction property and the pointclass Π_2^1 possesses the separation property.*

We can also extend the uniformization property to Σ_2^1.

Theorem [wAC] 6.62 (M. Kondô). *The pointclass Σ_2^1 has the uniformization property.*

Proof. Assume that $A \in \Sigma_2^1(X \times Y)$, where X, Y are Polish spaces. Then there exists a set $B \in \Pi_1^1({}^\omega\omega \times (X \times Y))$ such that $A = \mathrm{proj}_2(B)$. By Theorem 6.56 there exists a uniformization $C \in \Pi_1^1({}^\omega\omega \times (X \times Y))$ of B such that

$$(\exists a, y) \langle a, \langle x, y \rangle \rangle \in C \equiv (\exists a, y) \langle a, \langle x, y \rangle \rangle \in B,$$

$$(\forall a_1, a_2, y_1, y_2) ((\langle a_1, \langle x, y_1 \rangle \rangle \in C \wedge \langle a_2, \langle x, y_2 \rangle \rangle \in C) \rightarrow (a_1 = a_2 \wedge y_1 = y_2)).$$

One can easily see that the set $D = \mathrm{proj}_2(C) \in \Sigma_2^1$ and uniformizes the set A. $\qquad \square$

Corollary [wAC] 6.63. *There exists a Π_1^1 set of cardinality $\aleph_1$ if and only if there exists a Σ_2^1 set of cardinality $\aleph_1$.*

Proof. Assume that $A \in \Sigma_2^1(X)$ has cardinality $\aleph_1$. Then $A = \mathrm{proj}(B)$ for a set $B \in \Pi_1^1(X \times {}^\omega\omega)$. By the Uniformization Theorem 6.56 there exists a set $C \subseteq B$, $C \in \Pi_1^1$ that uniformizes the set B. Evidently $|C| = \aleph_1$. $\qquad \square$

Exercises

6.13 [wAC] Moschovakis Numbers δ_n^1
We set (compare Exercise 5.12):

$$\delta_n^1 = \delta(\mathbf{\Delta}_n^1(^\omega\omega)) = \delta(\mathbf{\Sigma}_n^1(^\omega\omega)) = \delta(\mathbf{\Pi}_n^1(^\omega\omega)).$$

a) $\delta_1^1 \leq \aleph_1$.
   Hint: By Exercise 5.8, b) every segment is Borel, hence $\aleph_0$-Souslin. By the Kunen-Martin Theorem proved in Exercise 5.7 every segment is countable.

b) Every $\mathbf{\Pi}_1^1$-rank has order type less than or equal to ω_1.
   Hint: Use the result of Exercise 5.12, f).

c) $\delta_1^1 \geq \aleph_1$
   Hint: There exists a $\mathbf{\Pi}_1^1$-rank of order type ω_1. Use the result of Exercise 5.12, e).

d) $\delta_1^1 = \aleph_1$.

6.14 [wAC] Shoenfield Theorem

a) Every $\mathbf{\Pi}_1^1$ set $A \subseteq {}^\omega\omega$ is $\aleph_1$-Souslin.
   Hint: Let $\langle \rho_n : n \in \omega \rangle$ be a very good scale on A. Set

$$\varphi(s) = \{x \in A : (\forall i < n)\, s(i) = \rho_i(x)\}$$

   for any $s \in {}^n\omega_1$.

b) Every $\mathbf{\Pi}_1^1$-subset of a Polish space X is $\aleph_1$-Souslin.
   Hint: Standard reduction using a surjection $f : {}^\omega\omega \longrightarrow X$.

c) Every $\mathbf{\Sigma}_2^1$-subset of a Polish space X is $\aleph_1$-Souslin.
   Hint: Use Exercise 5.4, c).

d) Every $\mathbf{\Sigma}_2^1$ prewell-ordering has cardinality less than $\aleph_2$.
   Hint: See Kunen-Martin Theorem of Exercise 5.7, b).

e) $\delta_2^1 \leq \aleph_2$.

6.15 [wAC] Constituents
Let $\langle A_\xi : \xi < \omega_1 \rangle$ be constituents of the $\mathbf{\Pi}_1^1(^\omega\omega)$ set ${}^\omega\omega \times {}^\omega\omega \setminus U$, where U is an ${}^\omega\omega$-universal set for $\mathbf{\Sigma}_1^1(^\omega\omega)$.

a) For every $\eta < \omega_1$ there exists a $\xi < \omega_1$ such that $A_\xi \notin \mathbf{\Sigma}_\eta^0$.
   Hint: Assume that for an η there is no ξ with the property. Let $B \subseteq {}^\omega\omega$ be a Borel set, $B \notin \mathbf{\Sigma}_\eta^0$. Then ${}^\omega\omega \setminus B = U^\alpha$ for some $\alpha \in {}^\omega\omega$. Since $\{\alpha\} \times B = \{\alpha\} \times {}^\omega\omega \setminus U$, by Boundedness Theorem 6.54 there exists an ξ_0 such that $\{\alpha\} \times B \subseteq \bigcup_{\xi < \xi_0} A_\xi$. Then $\{\alpha\} \times B = \bigcup_{\xi < \xi_0} A_\xi \cap \{\alpha\} \times {}^\omega\omega$. Since all A_ξ are in $\mathbf{\Sigma}_\eta^0$, we obtain a contradiction.

b) For every $\eta < \omega_1$ there exists a $\xi < \omega_1$ such that $\mathbf{WO}_\xi \notin \mathbf{\Sigma}_\eta^0$.
   Hint: See Theorem 6.49.

6.16 Cardinality of $\mathbf{WO}_\xi$

a) For any $x \in (0, 1)$ the poset $\langle \mathbb{D}, \preceq \rangle$ is isomorphic to $\langle \mathbb{D} \cap (x, 1), \preceq \rangle$.

b) Conclude that $|\mathbf{WO}_\xi| = \mathfrak{c}$ for every $\xi \geq \omega$.

c) Show that $|\mathcal{P}(^\omega 2) \setminus \mathbf{WO}| = \mathfrak{c}$.

6.17 [wAC] Analytic Set is a Union of $\aleph_1$ Borel Sets

Let $\langle {}^{<\omega}\omega, \varphi \rangle$ be a Souslin scheme. By induction we define $\langle \varphi^\xi(s) : s \in {}^{<\omega}\omega \rangle$ for any $\xi < \omega_1$: $\varphi^0(s) = \varphi(s)$, $\varphi^\xi(s) = \bigcap_{\eta<\xi} \varphi^\eta(s)$ for ξ limit and $\varphi^{\xi+1}(s) = \varphi^\xi(s) \cap \bigcup_n \varphi^\xi(s^\frown n)$. Set

$$S^\xi = \bigcup_n \varphi^\xi(n), \quad T^\xi = \bigcup_{s \in {}^{<\omega}\omega} (\varphi^\xi(s) \setminus \varphi^{\xi+1}(s)).$$

a) Show that $S^\xi \setminus T^\xi \subseteq L(T, \varphi)$ and $L(T, \varphi) \subseteq S^\xi$ for any $\xi < \omega_1$.

b) Show that $\bigcap_{\xi<\omega_1} T^\xi = \emptyset$.

c) Conclude that $L(T, \varphi) = \bigcap_{\xi<\omega_1} S^\xi = \bigcup_{\xi<\omega_1} (S^\xi \setminus T^\xi)$.

d) If $\langle {}^{<\omega}\omega, \varphi \rangle$ is closed, then every set S^ξ, T^ξ is Borel.

e) If A is an analytic set, then there exist Borel sets $\langle A_\xi : \xi < \omega_1 \rangle$ and $\langle B_\xi : \xi < \omega_1 \rangle$ such that $A = \bigcap_{\xi<\omega_1} A_\xi = \bigcup_{\xi<\omega_1} B_\xi$.

f) Any uncountable Polish space is a union of $\aleph_1$ non-empty pairwise disjoint Borel sets.

   Hint: Let $A \subseteq X$ be an analytic non-Borel set. Then $X \setminus A = \bigcup_{\xi<\omega_1} A_\xi$, where $\langle A_\xi : \xi < \omega_1 \rangle$ are pairwise disjoint constituents of $X \setminus A$. Uncountably many of them are non-empty. Take the differences $B_\xi \setminus \bigcup_{\eta<\xi} B_\eta$ of sets from e) to cover A.

6.18 [wAC] σ-Reduction

Let ρ be the $\mathbf{\Pi}_1^1$-rank on **WO**. If $A_n \in \mathbf{\Pi}_1^1(X)$, then we can define a $\mathbf{\Pi}_1^1$-rank ρ_n on A_n as in the proof of Theorem 6.53.

a) Show that the relation $x \in A_n \wedge (y \in A_n \to \rho_n(x) < \rho_m(y))$ is $\mathbf{\Pi}_1^1$.

   Hint: Let f_n be a Borel measurable mapping $f_n : X \longrightarrow {}^\omega 2$ of Theorem 6.49 such that $A_n = f_n^{-1}(\mathbf{WO})$. Then

$$(x \in A_n \wedge (y \in A_n \to \rho_n(x) < \rho_m(y))) \equiv f_n(x) <_\rho^* f_m(y).$$

b) The pointclass $\mathbf{\Pi}_1^1$ has the σ-reduction property.

   Hint: If ρ_n is the $\mathbf{\Pi}_1^1$-rank on A_n defined as above, $n \in \omega$, we set

$$B_n = \{x \in A_n : (\forall k \neq n)\,(x \in A_k \to \rho_k(x) > \rho_n(x))\}.$$

c) The pointclass $\mathbf{\Sigma}_2^1$ has the σ-reduction property.

Historical and Bibliographical Notes

In the introduction to this chapter we have already mentioned an error in H. Lebesgue's paper [1905]. M.J. Souslin [1917] introduced the notion of a set (A), later called analytic, and announced Theorem 6.4 and Corollary 6.23. Then he showed that every analytic set of reals is a projection of a $\mathbf{\Pi}_3^0$-subset of $\mathbb{R}_2$, i.e., the equivalences of c) and g) of Theorem 6.17. Then N.N. Luzin [1917] essentially proved Theorems 5.5, 6.3 and 6.26. N.N. Luzin also announced that M.J. Souslin can show

Corollary 6.19. N.N. Luzin [1927] analyzing Souslin's result Corollary 6.23 proved Theorem 6.21. The paper contains detailed proofs of several announced results by M.J. Souslin and N.N. Luzin: an analytic set is Lebesgue measurable and possesses the Baire Property, Corollary 6.19 and Theorem 6.26. Some of those results were already proved in a common paper by N.N. Luzin and W. Sierpiński [1918].

Theorem 6.7 was proved by P.S. Alexandroff [1916] and F. Hausdorff [1916]. Theorem 6.9 was essentially proved by K. Kuratowski [1936] (the σ-reduction property) and W. Sierpiński [1924b] (the separation property). Theorems 6.10, 6.12 and the results of Exercise 6.1 are due to H. Lebesgue [1905]. The Wadge reducibility follows A. Kechris [1995].

The notion of a Hurewicz scheme was introduced by M. Staš [2008] as a simplification of the W. Hurewicz [1928] notion "Häufungssystem". Lemma 6.31 is proved in W. Hurewicz [1928]. Simple Lemma 6.32 is due to M. Staš [2008] as well as the proof of Theorem 6.33, originally proved by A. Kechris, A. Louveau and H. Woodin [1987] as a strengthening of W. Hurewicz [1928].

The notion of a capacity was defined by G. Choquet [1953]. He also proved the Choquet Theorem of Exercise 6.8. We follow N. Bourbaki [1961].

H. Lebesgue [1905] says that Borel sets may be called analytic, since they can be defined by analytic equalities and inequalities. However, M.J. Souslin [1917], evidently influenced by his teacher N.N. Luzin, used for "an analytic set" a succinct name "a set (A)" and showed that this notion is larger than that of a Borel set. Later N.N. Luzin preferred the name "analytic set". H. Lebesgue [1918] wrote about the importance of a projection as an operation over sets. After showing in [1925a] that an analytic subset of $\mathbb{R}_k$ is a projection of a Borel subset of $\mathbb{R}_{k+1}$, N.N. Luzin [1925b] introduced the Projective Hierarchy and immediately in [1925c] showed that the hierarchy is proper. Actually Theorems 6.40 and 6.42 were implicitly announced. Complete proofs were published by N.N. Luzin [1930]. Theorem 6.37 is contained in W. Sierpiński [1929].

Theorem 6.45 (it would be better to say "the employed method") was essentially shown by K. Kuratowski and A. Tarski [1931]. See also K. Kuratowski [1931]. Actually the contemporary definitions of projective sets are usually based on the logical complexity of the used formulas.

The question concerning the cardinality of a complement of a co-analytic set was raised by N.N. Luzin [1917], later in [1925c]. Actually N.N. Luzin [1925c], p. 1818–1819 says more: *on ne sait pas et l'on ne saura jamais si la projection même d'un complémentaire analytique à deux dimensions (supposée non dénombrable) a la puissance du continu, si elle n'est pas des ensembles "qui ne sont pas Z"*[4], *ni même si elle est measurable*. Theorem 6.46 was proved by N.N. Luzin and W. Sierpiński [1918]. Corollary 6.59 gives a partial answer to Luzin's question. The result of Exercise 6.17 is due to N.N. Luzin and W. Sierpiński [1923], however the presented proof is that by W. Sierpiński [1926]. A construction of a co-analytic

---

[4]Possessing the Baire Property.

set with unbounded Borel complexity of constituents was presented in N.N. Luzin and W. Sierpiński [1929].

P. Vopěnka and L. Bukovský [1964] noticed that the result of Corollary 6.59 is the best possible. Actually, in the Cohen model for $\neg\mathbf{CH}$ there exists an uncountable $\mathbf{\Delta}_2^1$ set of cardinality $\aleph_1$.

Corollary 6.50 was proved by N.N. Luzin and W. Sierpiński [1923]. Boundedness Theorem 6.54 was proved by N.N. Luzin [1930]. Theorem 6.56 was proved in a different setting independently by P.S. Novikoff [1935] and M. Kondô [1938]. M. Kondô [1938] remarked that also Theorem 6.62 holds true. Theorems 6.53 and 6.60 concerning the existence of a rank were obtained by Y.N. Moschovakis [1980].

S.C. Kleene [1943] introduced a hierarchy of subsets of ω related to the algorithmic complexity. A. Mostowski [1946] introduced another hierarchy and realized a similarity to the Projective Hierarchy. Then J.W. Addison [1958a], p. 127 wrote that "after long discussions here in Warszawa it has been decided to propose" to denote the Borel pointclasses by $\mathbf{\Sigma}_\xi$, etc. and projective pointclasses by $\mathbf{\Sigma}_n^1$, etc. The common theory of both hierarchies has been mainly developed by Y.N. Moschovakis and systematically explained in his [1980] monograph.

Chapter 7

Decline and Fall of the Duality

In mathematics, duality has numerous meanings. Generally speaking,
duality is a metamathematical involution. Some duality concepts are
closely related and there are explicit theorems governing their relationships. Others are more intuitively related, with no precise correspondence. ... Generally speaking, a duality translates concepts, theorems
or mathematical structures into other concepts, theorems or structures,
in a one-to-one fashion, often (but not always) by means of an involution
operation: if the dual of A is B, then the dual of B is A.

[Wikipedia].

If R is a partial ordering on a set X, then the inverse relation R^{-1} is also a partial
ordering on X. Every notion of the partially ordered set $\langle X, R^{-1} \rangle$ is actually a
notion of the partially ordered set $\langle X, R \rangle$. A minimal element of a set $A \subseteq X$ in the
ordering R^{-1} is a maximal element of A in the ordering R, the greatest element in
R^{-1} is the least element in R, etc. We say that those notions are dual, i.e., minimal
is the dual notion of maximal. From every assertion about partially ordered sets
we obtain the dual assertion by replacing each notion by a corresponding dual
notion. A proof of an assertion can be translated to a "dual" proof of the dual
assertion. We speak about **duality**.

Considering properties of measure and topological properties connected with
the Baire Property and the first Baire category leads to a feeling that there exists
some kind of duality between measure and category: a notion of measure theory
has a dual notion in topology, for a statement about measure there is a dual
statement about category and vice versa. Actually, some results were obtained as
a result of looking for a dual statement. Namely, G. Vitali's result Corollary 7.19
was originally stated for Lebesgue measure and then a similar result was easily
proved for the category case as well. Similarly, Kuratowski-Ulam Theorem 7.33
was proved as a dual statement to the Fubini Theorem, Corollary 4.28. A great
deal of dual results holds true and is presented in Section 7.2, and, of course, in

Sections 7.1 and 7.3. Moreover, under some assumptions there exists a mapping transforming properties of measure to the properties of category and vice versa. However, it turns out that this phenomena is not generally true. There exists an important asymmetry between properties of measure and category. We present them in Section 7.4.

The results of S. Shelah (11.44) and (11.45) presented in Section 11.5 of Appendix should be considered as a strong failure of the measure-category duality.

The majority of results that we present needs the axiom of choice **AC**. Actually, we need **AC** even to define the considered cardinal invariants.

The reader is supposed to be familiar with the terminology and results of Sections 5.3 and 5.4.

7.1   Duality of Measure and Category

We introduce some general notions which will serve as a common framework for obtaining fundamental results on measure and category considered as duals. We begin with a presentation of proofs which are common for both cases.

Let $\mathcal{I} \subseteq \mathcal{P}(X)$ be a family of subsets of a given set X. If

a) $\mathcal{I}$ is hereditary, i.e., if $A \in \mathcal{I}$ and $B \subseteq A$, then also $B \in \mathcal{I}$,

b) $\mathcal{I}$ contains all singletons, i.e., $\{x\} \in \mathcal{I}$ for every $x \in X$,

c) $X \notin \mathcal{I}$,

then we say that $\mathcal{I}$ is a **family of thin sets**. Note that an ideal on a set X is a family of thin sets. When we consider subsets of a topological space X we automatically include in the definition the additional condition:

d) no non-empty open set belongs to $\mathcal{I}$.

The intended interpretation is that the family $\mathcal{I}$ is a family, in some sense, of small subsets of the set X. A subfamily $\mathcal{A} \subseteq \mathcal{I}$ is a **base** of the family of thin sets $\mathcal{I}$ if for every $B \in \mathcal{I}$ there exists a set $A \in \mathcal{A}$ such that $B \subseteq A$. If $\mathcal{I} \subseteq \mathcal{P}(X)$ is an ideal, then $\mathcal{I}$ is a family of thin subsets of the set X, provided that in the case when X is a topological space, $\mathcal{I}$ does not contain a non-empty open set. We define four **cardinal invariants** of a family of thin sets $\mathcal{I}$ as follows:

$$\mathrm{add}(\mathcal{I}) = \min\{|\mathcal{A}| : \mathcal{A} \subseteq \mathcal{I} \wedge \bigcup \mathcal{A} \notin \mathcal{I}\},$$
$$\mathrm{cof}(\mathcal{I}) = \min\{|\mathcal{A}| : \mathcal{A} \subseteq \mathcal{I} \wedge \mathcal{A} \text{ is a base of } \mathcal{I}\},$$
$$\mathrm{cov}(\mathcal{I}) = \min\{|\mathcal{A}| : \mathcal{A} \subseteq \mathcal{I} \wedge \bigcup \mathcal{A} = X\},$$
$$\mathrm{non}(\mathcal{I}) = \min\{|A| : A \notin \mathcal{I} \wedge A \in \mathcal{P}(X)\}.$$

If we consider the family $\mathcal{I}$ as a set ordered by inclusion, we obtain

$$\mathrm{add}(\mathcal{I}) = \mathfrak{b}(\mathcal{I}, \subseteq), \quad \mathrm{cof}(\mathcal{I}) = \mathfrak{d}(\mathcal{I}, \subseteq). \tag{7.1}$$

Sometimes we shall use also the fifth cardinal invariant

$$\mathrm{size}(\mathcal{I}) = \min\{\kappa : (\forall A \in \mathcal{I})\,|A| < \kappa\}.$$

Theorem [AC] 7.1. *Let $\mathcal{I}$ be a family of thin sets. Then*

   a) $\mathrm{add}(\mathcal{I}) \le \mathrm{cov}(\mathcal{I})$, $\mathrm{add}(\mathcal{I}) \le \mathrm{cf}(\mathrm{non}(\mathcal{I})) \le \mathrm{non}(\mathcal{I}) \le |X|$.
   b) $\mathrm{non}(\mathcal{I}) \le \mathrm{cof}(\mathcal{I})$, $\mathrm{add}(\mathcal{I}) \le \mathrm{cf}(\mathrm{cof}(\mathcal{I}))$, $\mathrm{cov}(\mathcal{I}) \le \mathrm{cof}(\mathcal{I})$.
   c) $\mathcal{I}$ *is an ideal if and only if* $\mathrm{add}(\mathcal{I}) \ge \aleph_0$.
   d) $\mathcal{I}$ *is a κ-additive ideal if and only if* $\mathrm{add}(\mathcal{I}) \ge \kappa$.
   e) $\mathrm{add}(\mathcal{I})$ *is a regular cardinal.*
   f) *If* $\mathrm{cov}(\mathcal{I}) < \mathrm{cf}(|X|)$, *then* $\mathrm{size}(\mathcal{I}) > |X|$.

The proof is easy and follows immediately from definition. For example, we show that $\mathrm{add}(\mathcal{I}) \le \mathrm{cf}(\mathrm{non}(\mathcal{I}))$. Assume not, i.e., $\kappa = \mathrm{cf}(\mathrm{non}(\mathcal{I})) < \mathrm{add}(\mathcal{I})$. Let $A \notin \mathcal{I}$ be of cardinality $\mathrm{non}(\mathcal{I})$. Then $A = \bigcup_{\xi < \kappa} A_\xi$ with $|A_\xi| < \mathrm{non}(\mathcal{I})$ for each $\xi < \kappa$. By definition of $\mathrm{non}(\mathcal{I})$ every set A_ξ belongs to $\mathcal{I}$ and therefore $A = \bigcup_{\xi < \kappa} A_\xi \in \mathcal{I}$, a contradiction.

The second inequality of b) follows by Theorem 5.41 and (7.1). $\qquad\square$

Theorem [AC] 7.2. *Let $\mathcal{I}_i$ be an ideal on an infinite set X_i, $C_i \in \mathcal{I}_i$, $i = 1, 2$. If there exists a mapping $f : X_1 \setminus C_1 \xrightarrow[\mathrm{onto}]{1\text{-}1} X_2 \setminus C_2$ such that*

$$B \in \mathcal{I}_1 \equiv f(B) \in \mathcal{I}_2 \text{ for any } B \subseteq X_1 \setminus C_1, \tag{7.2}$$

then $\mathrm{ch}(\mathcal{I}_1) = \mathrm{ch}(\mathcal{I}_2)$ *for* $\mathrm{ch} = \mathrm{add}, \mathrm{cof}, \mathrm{cov}, \mathrm{non}$.

Proof. Let us remark that the condition (7.2) is equivalent to the condition

$$f^{-1}(B) \in \mathcal{I}_1 \equiv B \in \mathcal{I}_2 \text{ for any } B \subseteq X_2 \setminus C_2.$$

Assume that $\mathcal{A} \subseteq \mathcal{I}_2$, $|\mathcal{A}| < \mathrm{add}(\mathcal{I}_1)$. Then $\{f^{-1}(A) : A \in \mathcal{A}\} \subseteq \mathcal{I}_1$ and therefore $\bigcup\{f^{-1}(A) : A \in \mathcal{A}\} \in \mathcal{I}_1$. Since

$$\bigcup \mathcal{A} \subseteq f(\bigcup\{f^{-1}(A) : A \in \mathcal{A}\}) \cup C_2 \in \mathcal{I}_2,$$

we obtain $\mathrm{add}(\mathcal{I}_1) \le \mathrm{add}(\mathcal{I}_2)$.

If $\mathcal{A}$ is a base of the ideal $\mathcal{I}_2$, then $\{f^{-1}(A) \cup C_1 : A \in \mathcal{A}\}$ is a base of $\mathcal{I}_1$. Thus $\mathrm{cof}(\mathcal{I}_1) \le \mathrm{cof}(\mathcal{I}_2)$.

If $\mathcal{A} \subseteq \mathcal{I}_2$ is such that $\bigcup \mathcal{A} = X_2$ and $|\mathcal{A}| = \mathrm{cov}(\mathcal{I}_2)$ then

$$\bigcup\{f^{-1}(A) : A \in \mathcal{A}\} \cup C_1 = X_1.$$

Hence $\mathrm{cov}(\mathcal{I}_1) \le \mathrm{cov}(\mathcal{I}_2)$.

Finally, let $A \notin \mathcal{I}_2$, $|A| = \mathrm{non}(\mathcal{I}_2)$. Then $A \setminus C_2 \notin \mathcal{I}_2$ and $|A \setminus C_2| \le |A|$. Since $f^{-1}(A \setminus C_2) \notin \mathcal{I}_1$ we obtain $\mathrm{non}(\mathcal{I}_1) \le \mathrm{non}(\mathcal{I}_2)$. $\qquad\square$

In Section 3.2 we denoted the set of all meager subsets of a topological space $\langle X, \mathcal{O} \rangle$ by $\mathcal{M}(X, \mathcal{O})$ or simply $\mathcal{M}(X)$. By Theorem 3.31 $\mathcal{M}(X)$ is a σ-ideal of subsets of the corresponding space X, provided that X is not meager. Similarly, for a given diffused Borel measure on X, the family $\mathcal{N}(X)$ of sets of outer measure zero introduced in Section 4.1 by (4.18) is a σ-ideal of subsets of X. We shall be interested in the cardinal invariants

$$\mathrm{add}(\mathcal{M}(X)), \quad \mathrm{cov}(\mathcal{M}(X)), \quad \mathrm{cof}(\mathcal{M}(X)), \quad \mathrm{non}(\mathcal{M}(X))$$

and

$$\mathrm{add}(\mathcal{N}(X)), \quad \mathrm{cov}(\mathcal{N}(X)), \quad \mathrm{cof}(\mathcal{N}(X)), \quad \mathrm{non}(\mathcal{N}(X))$$

of those σ-ideals. Evidently for any Polish space X and any diffused Borel measure on X all those cardinal invariants are uncountable cardinals not greater than $\mathfrak{c}$.

From Theorem 7.2 we easily obtain a folklore result.

Theorem [AC] 7.3.

$$\mathrm{ch}(\mathcal{M}([0,1])) = \mathrm{ch}(\mathcal{M}([0,1]^k)) = \mathrm{ch}(\mathcal{M}(\mathbb{R})) = \mathrm{ch}(\mathcal{M}(\mathbb{R}_k))$$
$$= \mathrm{ch}(\mathcal{M}(^\omega 2)) = \mathrm{ch}(\mathcal{M}((^\omega 2)^k)) = \mathrm{ch}(\mathcal{M}(^\omega \omega))$$
$$= \mathrm{ch}(\mathcal{M}((^\omega \omega)^k)) = \mathrm{ch}(\mathcal{M}(\mathbb{T}))$$

and

$$\mathrm{ch}(\mathcal{N}([0,1])) = \mathrm{ch}(\mathcal{N}([0,1]^k)) = \mathrm{ch}(\mathcal{N}(\mathbb{R})) = \mathrm{ch}(\mathcal{N}(\mathbb{R}_k))$$
$$= \mathrm{ch}(\mathcal{N}(^\omega 2)) = \mathrm{ch}(\mathcal{N}((^\omega 2)^k)) = \mathrm{ch}(\mathcal{N}(\mathbb{T}))$$

for ch = add, cov, cof, non.

Proof. One can easily construct a homeomorphism $g : (0,1) \xrightarrow[\mathrm{onto}]{1-1} \mathbb{R}$ such that

$$\lambda(A) = 0 \equiv \lambda(g^{-1}(A)) = 0 \text{ for any measurable } A \subseteq \mathbb{R}.$$

Similarly, the mapping g_k defined as

$$g_k(x_1, \ldots, x_k) = \langle g(x_1), \ldots, g(x_k) \rangle \text{ for } \langle x_1, \ldots, x_k \rangle \in (0,1)^k \qquad (7.3)$$

is a homeomorphism of $(0,1)^k$ onto $\mathbb{R}^k$ such that

$$\lambda_k(A) = 0 \equiv \lambda_k(g_k^{-1}(A)) = 0 \text{ for any measurable } A \subseteq \mathbb{R}^k.$$

Let h be the function defined in the proof of Theorem 3.25. Then the mapping

$$h_k(\alpha_1, \ldots, \alpha_k) = \langle h(\alpha_1), \ldots, h(\alpha_k) \rangle$$

is a homeomorphism of $(^\omega \omega)^k$ onto $(0,1)^k \setminus E_k$, where

$$E_k = \{ \langle x_1, \ldots, x_k \rangle \in (0,1)^k : x_i \in \mathbb{Q} \text{ for some } i = 1, \ldots, k \}$$

is a meager subset of $(0,1)^k$.

If for $\alpha \in {}^{\omega}2$ we set $f(\alpha) = \sum_{n=0}^{\infty} \alpha(n)2^{-n-1}$, then $f : {}^{\omega}2 \xrightarrow{\text{onto}} [0,1]$ and there exists a countable set $A \subseteq {}^{\omega}2$ such that $f : {}^{\omega}2 \setminus A \xrightarrow{1-1} [0,1]$. Moreover, by Theorem 3.9 the space $({}^{\omega}\omega)^n$ is homeomorphic to ${}^{\omega}\omega$ and $({}^{\omega}2)^n$ is homeomorphic via a measure-preserving homeomorphism to ${}^{\omega}2$ for any $n > 0$. Since g, g_k, h, h_k are homeomorphisms, the theorem follows by Theorems 3.9, 4.30 and 7.2. $\square$

Thus, dealing with ideals $\mathcal{M}(\mathbb{R}_k)$, $\mathcal{M}([0,1]^k)$, $\mathcal{M}({}^{\omega}\omega)$, $\mathcal{M}({}^{\omega}2)$ and $\mathcal{M}(\mathbb{T})$, we shall write simply $\mathcal{M}$. We make a similar convention for the measure case. Thus $\mathcal{N}$ denotes any of the ideals $\mathcal{N}(\mathbb{R}_k)$, $\mathcal{N}([0,1]^k)$, $\mathcal{N}({}^{\omega}2)$ and $\mathcal{N}(\mathbb{T})$ with Lebesgue measure. The cardinal invariants are the common values for those ideals and will be denoted simply as

$$\text{add}(\mathcal{M}), \text{cof}(\mathcal{M}), \text{cov}(\mathcal{M}), \text{non}(\mathcal{M}), \text{add}(\mathcal{N}), \text{cof}(\mathcal{N}), \text{cov}(\mathcal{N}), \text{non}(\mathcal{N}).$$

In a similar way, BAIRE and $\mathcal{L}$ denote one of σ-algebras $\text{BAIRE}(\mathbb{R}_k)$, $\text{BAIRE}([0,1]^k)$, $\text{BAIRE}({}^{\omega}\omega)$, $\text{BAIRE}({}^{\omega}2)$, $\text{BAIRE}(\mathbb{T})$, and $\mathcal{L}(\mathbb{R}_k)$, $\mathcal{L}([0,1]^k)$, $\mathcal{L}({}^{\omega}2)$, $\mathcal{L}(\mathbb{T})$, respectively. Usually it will be clear from the context which of the mentioned spaces we actually deal with. If not, we say that explicitly.

We introduce a terminology, which allows us to study some problems concerning a measure as well as related problems concerning the Baire category. A couple $\langle X, \mathcal{I} \rangle$ is called a **Polish ideal space** if $\mathcal{I}$ is a σ-additive ideal of subsets of a perfect Polish space X such that $\bigcup \mathcal{I} = X$ and $\mathcal{I}$ has a Borel base, i.e., every $A \in \mathcal{I}$ is a subset of some $B \in \mathcal{I} \cap \text{BOREL}(X)$. If X is a Polish group, then we speak about a **Polish ideal group**. We write

$$\text{BOREL}^*(\mathcal{I}) = \{(A \setminus P) \cup Q : A \in \text{BOREL}(X) \wedge P, Q \in \mathcal{I}\}.$$

Evidently $\text{BOREL}^*(\mathcal{I})$ is a σ-algebra and $\mathcal{I} \subseteq \text{BOREL}^*(\mathcal{I})$. Since the ideal $\mathcal{I}$ has a Borel base, for any $A \in \text{BOREL}^*(\mathcal{I})$ there exist Borel sets B, C such that $A \subseteq B$, $C \subseteq A$ and $B \setminus A, A \setminus C \in \mathcal{I}$. Moreover, if $A \notin \mathcal{I}$, then neither is $C \in \mathcal{I}$. A Polish ideal space $\langle X, \mathcal{I} \rangle$ is said to be **homogeneous** if any uncountable Borel set $A \subseteq X$ contains an uncountable Borel subset $B \in \mathcal{I}$.

If X is a perfect Polish space then the couple $\langle X, \mathcal{M}(X) \rangle$ is a Polish ideal space. By definition of the Baire Property we have $\text{BAIRE}(X) = \text{BOREL}^*(\mathcal{M})$. If μ is a diffused Borel measure on X, then $\langle X, \mathcal{N}(\mu) \rangle$ is a Polish ideal space. If $\mathcal{S}$ is the algebra of all μ^*-measurable subsets of X, then by Theorem 4.9 we obtain $\mathcal{S} = \text{BOREL}^*(\mathcal{N}(\mu))$. Moreover, by Corollary 6.8 both Polish ideal spaces are homogeneous.

Theorem [AC] 7.4. *If $\langle X, \mathcal{I} \rangle$ is a Polish ideal space with $\aleph_1$-saturated ideal $\mathcal{I}$, then the algebra $\text{BOREL}^*(\mathcal{I})$ is add$(\mathcal{I})$-additive.*

Proof. Denote by κ the additivity of $\text{BOREL}^*(\mathcal{I})$, i.e., the least cardinal for which there exists a sequence $\langle A_\xi \in \text{BOREL}^*(\mathcal{I}) : \xi \in \kappa \rangle$ with $\bigcup_{\xi < \kappa} A_\xi \notin \text{BOREL}^*(\mathcal{I})$. Then $\bigcup_{\xi < \eta} A_\xi \in \text{BOREL}^*(\mathcal{I})$ for any $\eta < \kappa$ and we can assume that the sets $\langle A_\xi : \xi < \kappa \rangle$ are pairwise disjoint. Since the ideal $\mathcal{I}$ is $\aleph_1$-saturated, there exists a countable set $T \subseteq \kappa$ such that $A_\xi \in \mathcal{I}$ for $\xi < \kappa$, $\xi \notin T$. Since T is countable we have $\bigcup_{\xi \in T} A_\xi \in \mathrm{BOREL}^*(\mathcal{I})$. If $\kappa < \mathrm{add}(\mathcal{I})$, then $\bigcup_{\xi \in \kappa \setminus T} A_\xi \in \mathcal{I}$ and therefore $\bigcup_{\xi \in \kappa} A_\xi \in \mathrm{BOREL}^*(\mathcal{I})$, which is a contradiction. □

Theorem [AC] 7.5. *Both ideals $\mathcal{M}$ and $\mathcal{N}$ are $\aleph_1$-saturated.*

Proof. Let $\{A_s : s \in S\} \subseteq \mathrm{BAIRE} \setminus \mathcal{M}$ be such that $A_s \cap A_t \in \mathcal{M}$ for any $s, t \in S$, $s \neq t$. By the definition of the Baire Property there are open sets G_s such that $A_s \setminus G_s \in \mathcal{M}$ and $G_s \setminus A_s \in \mathcal{M}$. Since $A_s \notin \mathcal{M}$ we obtain $G_s \neq \emptyset$. Since $A_s \cap A_t \in \mathcal{M}$ for $s \neq t$ we obtain that $G_s \cap G_t \in \mathcal{M}$ as well. Then by Baire Theorem 3.27 we have $G_s \cap G_t = \emptyset$ for $s, t \in S$, $s \neq t$. Thus S must be countable.

By Lemma 4.6 we can assume that $\mu(X) = 1$. Let $\{A_s : s \in S\} \subseteq \mathcal{L} \setminus \mathcal{N}$ be such that $A_s \cap A_t \in \mathcal{N}$ for any $s, t \in S$, $s \neq t$. Set $S_n = \{s \in S : \mu(A_s) > 1/n\}$ for each $n > 0$. Since $\mu(A_s \cap A_t) = 0$ for $s \neq t$, one can easily see that $|S_n| < n$. Evidently $S = \bigcup_{n=1}^\infty S_n$ and therefore S is countable. □

Corollary [AC] 7.6. *The algebra BAIRE is $\mathrm{add}(\mathcal{M})$-additive and the algebra $\mathcal{L}$ is $\mathrm{add}(\mathcal{N})$-additive.*

Immediately from the definitions one can easily see the following result.

Theorem [AC] 7.7. *If $\langle X, \mathcal{I} \rangle$ is a Polish ideal space and $\mathcal{I}$ is $\aleph_1$-saturated, then the complete Boolean algebras $\mathrm{BOREL}^*(\mathcal{I})/\mathcal{I}$ and $\mathrm{BOREL}(X)/\mathcal{I}$ are isomorphic.*

Similarly as in the proof of Theorem 7.3 we obtain that the quotient Boolean algebras $\mathrm{BAIRE}(\mathbb{R}_k)/\mathcal{M}(\mathbb{R}_k)$, $\mathrm{BAIRE}([0,1]^k)/\mathcal{M}([0,1]^k)$, $\mathrm{BAIRE}(^\omega 2)/\mathcal{M}(^\omega 2)$, and $\mathrm{BAIRE}(^\omega \omega)/\mathcal{M}(^\omega \omega)$ are isomorphic for any $k > 0$. Also the quotient Boolean algebras $\mathcal{L}(\mathbb{R}_k)/\mathcal{N}(\mathbb{R}_k)$, $\mathcal{L}([0,1]^k)/\mathcal{N}([0,1]^k)$ and $\mathcal{L}(^\omega 2)/\mathcal{N}(^\omega 2)$ are isomorphic for any $k > 0$.

Corollary [AC] 7.8. *The Boolean algebras $\mathrm{BAIRE}/\mathcal{M}$ and $\mathcal{L}/\mathcal{N}$ are $\mathbf{CCC}$ and therefore complete. Moreover, $\mathrm{BAIRE}/\mathcal{M}$ is isomorphic to $\mathrm{BOREL}/\mathcal{M}$ and $\mathcal{L}/\mathcal{N}$ is isomorphic to $\mathrm{BOREL}/\mathcal{N}$.*

We say that a Polish ideal space $\langle X, \mathcal{I} \rangle$ possesses the **hull property** if for any set $A \subseteq X$ there exists a Borel set B such that

$$A \subseteq B \wedge (\forall C \in \mathrm{BOREL}^*(\mathcal{I})) (A \subseteq C \to B \setminus C \in \mathcal{I}).$$

Thus, any subset of X can be approximated from above by a Borel set unique up to $\mathcal{I}$.

By Theorems 3.35 and 4.9 we obtain

Theorem [wAC] 7.9. *Let X be a Polish space.*

a) *The Polish ideal space $\langle X, \mathcal{M}(X) \rangle$ possesses the hull property.*

b) *If μ is a diffused Borel measure on X, then the Polish ideal space $\langle X, \mathcal{N}(\mu) \rangle$ possesses the hull property.*

Theorem [wAC] 7.10. *Assume that the Polish ideal space $\langle X, \mathcal{I} \rangle$ possesses the hull property. Then $L(T, \varphi) \in \mathrm{BOREL}^*(\mathcal{I})$ for any Souslin scheme $\langle T, \varphi \rangle$ with Borel values.*

Proof. We can assume that $T = {}^{<\omega}\omega$ and φ is a monotone Souslin scheme. Let us recall that for a node $s \in T$ we denote by T^s the tree $\{t \in T : s \leq t\}$.

By the hull property for every $s \in T$ there exists a Borel set $B_s \supseteq L(T^s, \varphi)$ such that for any $A \in \mathrm{BOREL}^*(\mathcal{I})$, the inclusion $L(T^s, \varphi) \subseteq A$ implies $B_s \setminus A \in \mathcal{I}$. We can assume that $B_s \subseteq \varphi(s)$ and $B_{s \frown n} \subseteq B_s$ (if not, take $B_{s \frown n} \cap \varphi(s \frown n) \cap B_s$). We write $P_s = B_s \setminus \bigcup_n B_{s \frown n}$. Since $L(T^s, \varphi) \subseteq \bigcup_n B_{s \frown n}$, we obtain $P_s \in \mathcal{I}$. We set $P = \bigcup_{s \in T} P_s$. Then $P \in \mathcal{I}$.

Let

$$B^n = \bigcup_{s \in {}^n\omega} B_s, \quad B = \bigcap_n B^n.$$

Evidently $L(T, \varphi) \subseteq B$. We show that $B \setminus L(T, \varphi) \subseteq P$. So, let $x \in B \setminus P$. We show that $x \in L(T, \varphi)$. Since $x \notin P_\emptyset$, there exists an n such that $x \in B_{\langle n \rangle}$. We write $f(0) = n$. Assume that $f(k)$ is defined and $x \in B_{\langle f(0),...,f(k) \rangle}$. Since $x \notin P_{\langle f(0),...,f(k) \rangle}$, there exists an m such that $x \in B_{\langle f(0),...,f(k),m \rangle}$. By assumption $B_{f|n} \subseteq \varphi(f|n)$ for every n, therefore $x \in \bigcap_n \varphi(f|n) \subseteq L(T, \varphi)$.

Since $L(T, \varphi) \subseteq B$ we obtain $L(T, \varphi) \in \mathrm{BOREL}^*(\mathcal{I})$. □

As an immediate consequence of those two theorems we obtain

Theorem [wAC] 7.11 (N.N. Luzin). *If X is a Polish space, and μ a Borel measure on X, then every analytic and every co-analytic subset of X possesses the Baire Property and is μ-measurable.*

A set $\mathcal{B} \subseteq \mathrm{BOREL}^*(\mathcal{I})$ can be viewed as a subset of the quotient Boolean algebra $\mathrm{BOREL}^*(\mathcal{I})/\mathcal{I}$. Thus $\mathcal{B}$ is a predense subset of $\mathrm{BOREL}^*(\mathcal{I})/\mathcal{I}$ if for any $C \in \mathrm{BOREL}^*(\mathcal{I}) \setminus \mathcal{I}$ there exists a $B \in \mathcal{B}$ such that $B \cap C \notin \mathcal{I}$. Similarly, a predense set $\mathcal{C} \subseteq \mathrm{BOREL}^*(\mathcal{I})$ is a refinement of a predense set $\mathcal{B}$ if for every $C \in \mathcal{C}$ there exists an element $B \in \mathcal{B}$ such that $C \setminus B \in \mathcal{I}$. Let us recall that Boolean algebra $\mathrm{BOREL}^*(\mathcal{I})/\mathcal{I}$ is $\aleph_0$-distributive if every countable family of predense sets has a common refinement.

Theorem 7.12. *Assume $X = {}^\omega 2$, $\mathcal{I}$ contains all one-point sets and none of the sets $\{f \in {}^\omega 2 : f(n) = i\}$, $n \in \omega$, $i = 0, 1$ is in $\mathcal{I}$. Then the quotient Boolean algebra $\mathrm{BOREL}^*(\mathcal{I})/\mathcal{I}$ is not $\aleph_0$-distributive.*

Proof. Let $\mathcal{C}_n = \{C_n^0, C_n^1\}$, where $C_n^i = \{f \in {}^\omega 2 : f(n) = i\}$, $i = 0, 1$. We show that the family of predense sets $\langle \mathcal{C}_n : n \in \omega \rangle$ does not have a common refinement.

Assume that $\mathcal{C}$ is a common refinement of all $\langle \mathcal{C}_n : n \in \omega \rangle$. Take any $A \in \mathcal{C}$. Then $A \notin \mathcal{I}$. For every n there exists an $i_n \in \{0, 1\}$ such that $A \setminus C_n^{i_n} \in \mathcal{I}$. Then also $\bigcup_n (A \setminus C_n^{i_n}) = A \setminus \bigcap_n C_n^{i_n} \in \mathcal{I}$. Since $\bigcap_n C_n^{i_n}$ is a one-point set we have a contradiction. □

Corollary 7.13. *Neither $\mathcal{L}/\mathcal{N}$ nor $\mathrm{BAIRE}/\mathcal{M}$ is $\aleph_0$-distributive.*

We close this section with two simple results.

Theorem 7.14. *If X is a Polish space, then Boolean algebra $\mathrm{BAIRE}(X)/\mathcal{M}(X)$ is isomorphic to Boolean algebra $\mathrm{RO}(X)$ of regular open subsets of X.*

The proof is easy. By Theorem 3.34 any set A with the Baire Property may be expressed as $A = (G \setminus P) \cup Q$, where G is regular open and P, Q are meager. Moreover the set G with this property is uniquely determined. One can easily show that assigning such a G to A is the desired isomorphism. $\square$

If **B** is a complete Boolean algebra, then a function $\mu : B \longrightarrow [0, 1]$ is called a **positive measure** on **B** if

a) $\mu(0_B) = 0,$ $\quad$ $\mu(1_B) = 1,$

b) $\mu(a) > 0$ for any $a \in B$, $a \neq 0_B$,

c) $\mu(\bigvee_n a_n) = \sum_{n=0}^{\infty} \mu(a_n)$ for pairwise disjoint system $\langle a_n : n \in \omega \rangle$.

Since any two measurable sets that are equal modulo ideal $\mathcal{N}$ have equal Lebesgue measure, we immediately obtain that

Theorem 7.15. *There is a positive measure on Boolean algebra $\mathcal{L}/\mathcal{N}$.*

Exercises

7.1 [AC] Sacks' Algebra

a) Show that Boolean algebras

$$\begin{array}{ll}
\text{Borel}([0, 1])/[[0, 1]]^{\leq \aleph_0}, & \text{Borel}(\mathbb{R})/[\mathbb{R}]^{\leq \aleph_0}, \\
\text{Borel}(^\omega 2)/[^\omega 2]^{\leq \aleph_0}, & \text{Borel}(^\omega \omega)/[^\omega \omega]^{\leq \aleph_0}, \\
\text{Borel}([0, 1]^k)/[[0, 1]^k]^{\leq \aleph_0}, & \text{Borel}(\mathbb{R}^k)/[\mathbb{R}^k]^{\leq \aleph_0}, \\
\text{Borel}(^\omega 2^k)/[^\omega 2^k]^{\leq \aleph_0}, & \text{Borel}(^\omega \omega^k)/[^\omega \omega^k]^{\leq \aleph_0},
\end{array}$$

$k > 1$, are isomorphic. The completion of any of them is called a **Sacks' algebra**.

b) Show that the set of all perfect sets is a dense subset of Sacks' algebra (modulo countable sets).

c) Sacks' algebra is $\mathfrak{c}^+$-CC.

d) There exists a set of pairwise disjoint elements of Sacks' algebra of cardinality $\mathfrak{c}$.

7.2 [AC] Homogeneous Boolean Algebras

a) Show that Boolean algebra $\text{Baire}/\mathcal{M}$ and $\mathcal{L}/\mathcal{N}$ are homogeneous.
   Hint: Any open and G_δ set are Borel isomorphic to $[0, 1]$. Borel isomorphism induces an isomorphism of corresponding Boolean algebras.

b) Show that Sacks' algebra is homogeneous.
   Hint: Use the Alexandroff-Hausdorff Theorem, Corollary 6.7.

7.3 [AC] Hull Property

Let $\langle X, \mathcal{I} \rangle$ be a Polish ideal space possessing the hull property.

a) For any $A \in \text{Borel}^*(\mathcal{I})$ there are a Borel set G and sets $P, Q \in \mathcal{I}$ such that $A = (G \setminus Q) \cup P$.

b) If $A \in \text{Borel}^*(\mathcal{I}) \setminus \mathcal{I}$ is uncountable then A contains a perfect subset, i.e., $|A| = \mathfrak{c}$.

c) If $A \notin \mathcal{I}$, then there exists a set $C \subseteq A$ such that $C \notin \text{Borel}^*(\mathcal{I})$.
   Hint: Let B be a Bernstein set – see Section 7.2. Then $A \cap B$ or $A \setminus B$ is not in $\mathcal{I}$.

d) If A has not measure zero, then there exists a non-measurable subset of A.

e) If A is not meager, then there exists a subset of A that does not possess the Baire Property.

7.4 [AC] Szpilrajn-Marczewski Theorem
Let S be a σ-algebra of subsets of X, $S \neq \mathcal{P}(X)$. Set

$$\mathcal{I} = \{A \in S : (\forall B)(B \subseteq A \to B \in S)\}. \tag{7.4}$$

S has the **weak hull property** if

$$(\forall A \subseteq X)(\exists B \in S, A \subseteq B)(\forall C \in S)(C \supseteq A \to B \setminus C \in \mathcal{I}). \tag{7.5}$$

a) Show that $\mathcal{I}$ is a σ-ideal in S.

b) Prove the **Szpilrajn-Marczewski Theorem**: If $\langle T, \varphi \rangle$ is a Souslin scheme with values of φ in S having the weak hull property, then $L(T, \varphi) \in S$.

7.5 [AC] Decided Sets
For a given infinite set X we consider a family $\mathcal{R} \subseteq \mathcal{P}(X)$. We assume that $\emptyset \notin \mathcal{R}$ and define

$$\text{Dec}(\mathcal{R}) = \{A \subseteq X : (\forall B \in \mathcal{R})(\exists C \in \mathcal{R})(C \subseteq B \wedge (C \subseteq A \vee C \cap A = \emptyset))\},$$
$$\text{Id}(\mathcal{R}) = \{A \subseteq X : (\forall B \in \mathcal{R})(\exists C \in \mathcal{R})(C \subseteq B \wedge C \cap A = \emptyset)\}.$$

a) $\text{Dec}(\mathcal{R})$ is an algebra of subsets of X and $\text{Id}(\mathcal{R}) \subsetneq \text{Dec}(\mathcal{R})$ is an ideal.

b) $\text{Dec}(\mathcal{R})$ is an add($\text{Id}(\mathcal{R})$)-additive algebra.
   Hint: Consider $\langle A_\xi \in \text{Dec}(\mathcal{R}) : \xi < \kappa \rangle$, $\kappa < \text{add}(\text{Id}(\mathcal{R}))$ and a $B \in \mathcal{R}$. If there exists a $\xi < \kappa$ such that $C \subseteq A_\xi$ for some set $C \subseteq B$, then we are done. Otherwise $B \cap A_\xi \in \text{Id}(\mathcal{R})$ for every ξ and therefore also $B \cap \bigcup_{\xi < \kappa} A_\xi \in \text{Id}(\mathcal{R})$.

Assume that S is an algebra of subsets of X and $\mathcal{I} \subseteq S$ is an ideal. Let $\mathcal{R} \subseteq S \setminus \mathcal{I}$ be such that for any $A \in S \setminus \mathcal{I}$ there exists a $B \in \mathcal{R}$, $B \subseteq A$.

c) Show that $S \subseteq \text{Dec}(\mathcal{R})$ and $\mathcal{I} \subseteq \text{Id}(\mathcal{R})$.

d) If moreover $\langle X, \mathcal{I} \rangle$ is a Polish ideal space possessing the hull property, and $\mathcal{I}$ is $\aleph_1$-saturated, then $\mathcal{I} = \text{Id}(\mathcal{R})$.
   Hint: If $A \notin \mathcal{I}$, then by the hull property there exists a Borel set $B \supseteq A$ such that $B \setminus C \in \mathcal{I}$ for any Borel $C \supseteq A$ and $B \notin \mathcal{I}$. If $A \in \text{Id}(\mathcal{R})$, then there exists a set $E \subseteq B$, $E \cap A = \emptyset$. Taking $C = B \setminus E$ we obtain a contradiction.

e) Show that $\mathcal{N} = \text{Id}(F_\sigma \setminus \mathcal{N})$ and $\mathcal{L} = \text{Dec}(F_\sigma \setminus \mathcal{N})$.
   Hint: If $A \in \mathcal{L} \setminus \mathcal{N}$, then a maximal family $\mathcal{A}$ of pairwise disjoint subsets of A consisting of elements of $F_\sigma \setminus \mathcal{N}$ is countable. Show that $\bigcup \mathcal{A} \in \text{Dec}(F_\sigma \setminus \mathcal{N})$ and $A \setminus \bigcup \mathcal{A} \in \mathcal{N}$.

f) Show that $\mathcal{M} = \text{Id}(G_\delta \setminus \mathcal{M})$ and $\text{Baire} = \text{Dec}(G_\delta \setminus \mathcal{M})$.
   Hint: Every G_δ non-meager set contains a subset of the form $U \setminus P$, where U is non-empty open and P is an F_σ meager set. Follow the proof of e), however use sets $U \setminus P$ in your proof.

g) Deduce Corollary 7.6 from the results of this exercise.

7.2   Duality Continued

In this section we shall continue with results on measure and category which follow
from general results on a σ-additive ideal of a Polish ideal space. Again, our proofs
are common both for measure and category and therefore dual: any such a proof
for the measure case can be easily translated to a proof for the dual category result
and vice versa. The only exception is the Steinhaus Theorem. However, also in this
case the proofs for measure and category are very close.

At this time, we know examples of sets that are Lebesgue measurable, possess the Baire Property, and are either countable or contain a perfect subset: at
least Borel sets have those nice properties. By Luzin-Sierpiński Theorem 7.11 and
Corollary 6.19 also analytic sets possess those regularity properties. However we do
not know whether there exists a Lebesgue non-measurable set, a set not possessing
the Baire Property or an uncountable set without a perfect subset. It is not an
accident since, as we shall see later, the theory **ZFC+wAC** is not enough to decide
such questions. Assuming that the real line can be well ordered, we show that
there exist a Lebesgue non-measurable set, a set possessing Baire Property
and an uncountable set without perfect subset.

Assume that X is an Abelian topological group, e.g., $\mathbb{R}^k$, $^\omega 2$ or $\mathbb{T}$. We say
that a family $\mathcal{D}$ of subsets of X is **shift invariant** if for any $A \in \mathcal{D}$ also $x + A \in \mathcal{D}$
for every $x \in X$. Evidently BAIRE, $\mathcal{M}$, $\mathcal{L}$ and $\mathcal{N}$, when considered on $\mathbb{R}^k$, $^\omega 2$ or
$\mathbb{T}$, are shift invariant.

Assume now that $\mathcal{I}$ is shift invariant. The Polish ideal group $\langle X, \mathcal{I}\rangle$ possesses
the **Steinhaus Property** if the interior of the set

$$A - A = \{x \in X : (x + A) \cap A \neq \emptyset\}$$

is non-empty for any $A \in \text{BOREL}^*(\mathcal{I}) \setminus \mathcal{I}$.

Theorem 7.16 (H. Steinhaus). *The Polish ideal groups* $\langle \mathbb{T}, \mathcal{M}(\mathbb{T})\rangle$, $\langle \mathbb{T}, \mathcal{N}(\mathbb{T})\rangle$,
$\langle \mathbb{R}, \mathcal{M}(\mathbb{R})\rangle$, $\langle \mathbb{R}, \mathcal{N}(\mathbb{R})\rangle$, $\langle ^\omega 2, \mathcal{M}(^\omega 2)\rangle$, *and* $\langle ^\omega 2, \mathcal{N}(^\omega 2)\rangle$ *possess the Steinhaus Property.*

Proof. a) Assume that $A \subseteq \mathbb{T}$ is Lebesgue measurable and $\lambda(A) > 0$. By Theorem 4.17 there exists an open interval I such that $\lambda(A \cap I)) \geq 3/4\lambda(I)$. Let
$a = -1/2\lambda(I)$, $b = 1/2\lambda(I)$.

If $x \in (a, b)$ then $(x+I) \cup I$ is an interval of length smaller than $3/2\lambda(I)$. Since
$\lambda(A \cap I) = \lambda(x + (A \cap I)) \geq 3/4\lambda(I)$ we obtain $(A \cap I) \cap (x + (A \cap I)) \neq \emptyset$. Thus
there are $y, z \in (A \cap I)$ such that $y = x + z$. Then $x \in A - A$. Hence $(a, b) \subseteq A - A$.

b) Now assume that A possesses the Baire Property and A is not meager.
Then there are an open set U and a meager set P such that $U \setminus P \subseteq A$. Let $I \subseteq U$
be a non-trivial open interval, $\delta = \lambda(I)$. Then for any x we have

$$(I \cap (x + I)) \setminus (P \cup (x + P)) \subseteq A \cap (x + A).$$

If $|x| < \delta$ then $(x + I) \cap I$ is a non-trivial interval and therefore the set

$$(I \cap (x + I)) \setminus (P \cup (x + P))$$

is non-empty. Then also $(x + A) \cap A \neq \emptyset$ and therefore $x \in A - A$.

The other assertions can be easily reduced to the proved ones. $\qquad \square$

Let $\langle X, +, 0 \rangle$ be an Abelian Polish group. A set $V \subseteq X$ is called a **Vitali set** if there exists a countable dense subset $S \subseteq X$ such that

$$(\forall x, y)\, ((x, y \in V \wedge x \neq y) \rightarrow x - y \notin S), \qquad (7.6)$$
$$(\forall x \in X)(\exists y \in V)\, x - y \in S. \qquad (7.7)$$

Thus, for every element $x \in X$ there exists exactly one $y \in V$ such that $x - y \in S$.

Theorem 7.17 (G. Vitali). *If the real line can be well ordered, then there exists a Vitali subset of* $\mathbb{T}$. *Similarly for* $\mathbb{R}$ *and* $^\omega 2$.

Proof. We can take $S = \mathbb{Q} \cap \mathbb{T}$ or $S = \mathbb{D}$. For $x \in \mathbb{T}$ we set

$$v(x) = \{y \in \mathbb{T} : x - y \in S\}. \qquad (7.8)$$

Then for any x, $y \in [0, 1]$ either $v(x) = v(y)$ or $v(x) \cap v(y) = \emptyset$. Thus the family

$$\{v(x) : x \in \mathbb{T}\} \qquad (7.9)$$

is a decomposition of the set $\mathbb{T}$. We call it the **Vitali decomposition**. If $\preceq$ is a well-ordering of the real line we can define

$$V = \{x \in \mathbb{T} : (\forall y)\, (y \in v(x) \rightarrow x \preceq y)\}. \qquad (7.10)$$

The set V is a selector for the Vitali decomposition $\{v(x) : x \in \mathbb{T}\}$, i.e., for any $x \in \mathbb{T}$ we have $|V \cap v(x)| = 1$. One can easily see that the set V is a Vitali set. $\quad \square$

Theorem 7.18. *If a Polish ideal group* $\langle X, \mathcal{I} \rangle$ *possesses the Steinhaus Property, then* $\mathrm{BOREL}^*(\mathcal{I})$ *does not contain a Vitali set.*

Proof. Let V be a Vitali set, S being a countable dense subset of X satisfying (7.6) and (7.7). By (7.7) we have $\bigcup_{r \in S}(r + V) = X$. Since S is countable we obtain that $V \notin \mathcal{I}$.

Assume now that $V \in \mathrm{BOREL}^*(\mathcal{I})$. By the Steinhaus Property the interior of the set $\{x \in X : (x + V) \cap V \neq 0\}$ is non-empty and therefore contains a non-zero element of S, a contradiction with (7.6). $\qquad \square$

Corollary 7.19 (G. Vitali). *A Vitali set is neither measurable for any Borel shift invariant measure nor possesses the Baire Property.*

A subset $B \subseteq X$ of a Polish space X is called a **Bernstein set** if

$$|B| = |X \setminus B| = \mathfrak{c}, \tag{7.11}$$

$$\text{neither } B \text{ nor } X \setminus B \text{ contain a perfect subset.} \tag{7.12}$$

Theorem 7.20 (F. Bernstein). *If a Polish space X can be well ordered, then there exists a Bernstein set $B \subseteq X$.*

Proof. We can assume that X is perfect. By the assumption X can be enumerated as $X = \{a_\xi : \xi < \mathfrak{c}\}$. We know that the set of all perfect subsets of X has cardinality $\mathfrak{c}$. Then again we can enumerate all perfect subsets of X as $\{P_\xi : \xi < \mathfrak{c}\}$. Now we construct two sequences $\{x_\xi : \xi < \mathfrak{c}\}$ and $\{y_\xi : \xi < \mathfrak{c}\}$ of mutually distinct elements of X and such that $x_\xi, y_\xi \in P_\xi$ for any $\xi < \mathfrak{c}$.

We set $x_0 = a_\eta$, where η is the first ordinal such that $a_\eta \in P_0$. Similarly we set $y_0 = a_\zeta$, where ζ is the first ordinal such that $a_\zeta \neq x_0$ and $a_\zeta \in P_0$.

If x_η, y_η are defined for every $\eta < \xi$, we set $x_\xi = a_\zeta$, where ζ is the first ordinal such that $a_\zeta \neq x_\eta, y_\eta$ for any $\eta < \xi$ and $a_\zeta \in P_\xi$. Similarly, $y_\xi = a_\rho$ where ρ is the first ordinal such that $a_\rho \neq x_\eta, y_\eta$ for any $\eta < \xi$, $a_\rho \neq x_\xi$ and $a_\rho \in P_\xi$.

Let $B = \{x_\xi : \xi < \mathfrak{c}\}$. Evidently the condition (7.11) holds true. Let P be a perfect set. Then $P = P_\xi$ for some $\xi < \mathfrak{c}$. Then $x_\xi \in P$, $y_\xi \in P$, $x_\xi \in B$ and $y_\xi \in X \setminus B$. Thus, neither $P \subseteq B$ nor $P \subseteq X \setminus B$. $\qquad\square$

Theorem [wAC] 7.21. *If a Polish ideal space $\langle X, \mathcal{I} \rangle$ possesses the hull property, then a Bernstein set does not belong to $\mathrm{BOREL}^*(\mathcal{I})$.*

Proof. To get a contradiction, assume that A is a Bernstein subset of a Polish space X and $A \in \mathrm{BOREL}^*(\mathcal{I})$. Then there exist Borel sets $B \subseteq A \subseteq C$ such that $C \setminus B \in \mathcal{I}$. Since A does not contain a perfect subset, neither does B contain a perfect subset and therefore, by the Alexandroff-Hausdorff Theorem, Corollary 6.7, B is countable. Then $C = (C \setminus B) \cup B \in \mathcal{I}$. Thus also $A \in \mathcal{I}$.

By the same argument we obtain $X \setminus A \in \mathcal{I}$, a contradiction. $\qquad\square$

Corollary [wAC] 7.22. *Let X be a Polish space. A Bernstein subset of X does not possess the Baire Property. If μ is a diffused Borel measure on X, then a Bernstein subset of X is not μ-measurable.*

We show a stronger result.

Theorem [AC] 7.23. *Let $\langle X, \mathcal{I} \rangle$ be a Polish ideal space. If $X = \bigcup_{s \in S} A_s$, every $A_s \in \mathcal{I}$, and the set $\{s \in S : x \in A_s\}$ is finite for every $x \in X$, then there exists a set $S_0 \subseteq S$ such that $\bigcup_{s \in S_0} A_s \notin \mathrm{BOREL}^*(\mathcal{I})$.*

Proof. Since $|X| = \mathfrak{c}$, also $|S| \leq \mathfrak{c}$. So we can assume that $S \subseteq \mathbb{R}$ and that any analytic subset of S is countable (take S as a subset of a Bernstein set).

To get a contradiction, assume that $\bigcup_{s \in S_0} A_s \in \text{BOREL}^*(\mathcal{I})$ for every $S_0 \subseteq S$. We show that there exists a Borel set $P \in \mathcal{I}$ such that

$$(\forall U \subseteq \mathbb{R} \text{ open}) \quad \bigcup_{s \in (U \cap S)} (A_s \setminus P) \in \text{BOREL}. \tag{7.13}$$

Let $\{U_n : n \in \omega\}$ be an open base of $\mathbb{R}$. By assumption, for every n there exists a Borel set B_n such that $\bigcup_{s \in (U_n \cap S)} A_s \subseteq B_n$ and $B_n \setminus (\bigcup_{s \in (U_n \cap S)} A_s) \in \mathcal{I}$. Hence, there exists a Borel $P_n \in \mathcal{I}$ such that $B_n \setminus (\bigcup_{s \in (U_n \cap S)} A_s) \subseteq P_n$. Set $P = \bigcup_n P_n$. One can easily check that

$$\bigcup_{s \in (U_n \cap S)} (A_s \setminus P) = B_n \setminus P.$$

Since each $B_n \setminus P$ is Borel, the assertion (7.13) follows.

Write

$$F = \{\langle s, x \rangle \in \mathbb{R} \times X : s \in S \wedge x \in (A_s \setminus P)\}.$$

Let $\{r_n : n \in \omega\}$ be an enumeration of all rationals. We denote as E the set

$$\bigcap_n \bigcup_m \left((r_m - 2^{-n}, r_m + 2^{-n}) \times \bigcup \{A_s \setminus P : s \in S \cap (r_m - 2^{-n}, r_m + 2^{-n})\} \right).$$

By (7.13) the set E is Borel. Evidently $F \subseteq E$. On the other hand, if $\langle s, x \rangle \in E$ then for every n there exist an $m_n \in \omega$ and an $s_n \in S \cap (r_{m_n} - 2^{-n}, r_{m_n} + 2^{-n})$ such that $s \in (r_{m_n} - 2^{-n}, r_{m_n} + 2^{-n})$ and $x \in (A_{s_n} \setminus P)$. Since $\lim_{n \to \infty} s_n = s$ and $x \in A_t$ for finitely many t, we obtain that $s = s_n$ for all but finitely many n. Consequently $\langle s, x \rangle \in F$. Thus $E = F$.

Hence the set $S_0 = \{s \in \mathbb{R} : (\exists x) \langle s, x \rangle \in F\}$ is an analytic subset of S and therefore countable. Moreover,

$$\bigcup_{s \in S} (A_s \setminus P) = \bigcup_{s \in S_0} (A_s \setminus P),$$

which is a contradiction with $X \notin \mathcal{I}$. $\qquad\square$

Corollary [AC] 7.24. *Let X be a metric separable space. Let μ be a Borel measure on X. If $X = \bigcup_{s \in S} A_s$ and $\langle A_s : s \in S \rangle$ are μ-measure zero sets such that for every $x \in X$ the set $\{s \in S : x \in A_s\}$ is finite, then there exists a set $S_0 \subseteq S$ such that the set $\bigcup_{s \in S_0} A_s$ is not μ-measurable.*

Corollary [AC] 7.25. *Let X be a Polish space. If $X = \bigcup_{s \in S} A_s$ and $\langle A_s : s \in S \rangle$ are meager sets such that for every $x \in X$ the set $\{s \in S : x \in A_s\}$ is finite, then there exists a set $S_0 \subseteq S$ such that the set $\bigcup_{s \in S_0} A_s$ does not possess the Baire Property.*

Assume that $\langle X, \mathcal{I} \rangle$ and $\langle X, \mathcal{J} \rangle$ are Polish ideal spaces with the same underlying space X. We say that $\mathcal{I}$ and $\mathcal{J}$ are **orthogonal** if there exist sets $A \in \mathcal{I}$ and $B \in \mathcal{J}$ such that $A \cap B = \emptyset$ and $A \cup B = X$.

By Theorem 4.13 the ideals $\mathcal{N}(X)$ and $\mathcal{M}(X)$ are orthogonal for any Polish space with a diffused Borel measure.

Theorem 7.26. *If $\langle X, \mathcal{I} \rangle$ and $\langle X, \mathcal{J} \rangle$ are Polish ideal groups with shift invariant orthogonal σ-ideals, then*[1]

$$\mathrm{cov}(\mathcal{J}) \leq \mathrm{non}(\mathcal{I}).$$

Proof. Let X be an Abelian Polish group. Since the ideals $\mathcal{I}$ and $\mathcal{J}$ are orthogonal, there exist sets $A \in \mathcal{I}$ and $B \in \mathcal{J}$ such that $A \cap B = \emptyset$ and $A \cup B = X$.

If $C \subseteq X$ is such that $C + B = \bigcup_{x \in C}(x + B) \neq X$, then $C \in \mathcal{I}$. Actually, there exists a $y \in X$ such that $y \notin C + B$. Then $(y - C) \cap B = \emptyset$ and hence $(y - C) \subseteq A$. Thus $C \in \mathcal{I}$.

Let $C \subseteq X$ be such that $|C| = \mathrm{non}(\mathcal{I})$ and $C \notin \mathcal{I}$. Then $\bigcup_{x \in C}(x + B) = X$ and therefore $\mathrm{cov}(\mathcal{J}) \leq \mathrm{non}(\mathcal{I})$. $\square$

Corollary 7.27 (Rothberger Theorem).

$$\mathrm{cov}(\mathcal{M}) \leq \mathrm{non}(\mathcal{N}), \quad \mathrm{cov}(\mathcal{N}) \leq \mathrm{non}(\mathcal{M}).$$

The inequalities of Theorem 7.1 together with those of the Rothberger Theorem can be summarized in a picture.

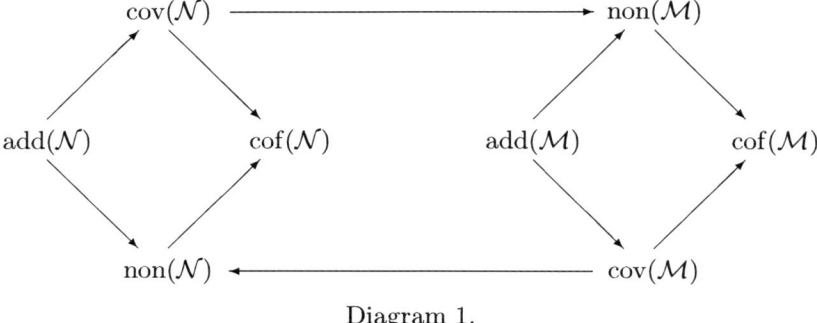

Diagram 1.

As usual, an arrow $\mathfrak{a} \longrightarrow \mathfrak{b}$ means that $\mathfrak{a} \leq \mathfrak{b}$.

We can rearrange the picture in the following more symmetric form:

---

[1]Note that without **AC** we must carefully interpret the following inequality.

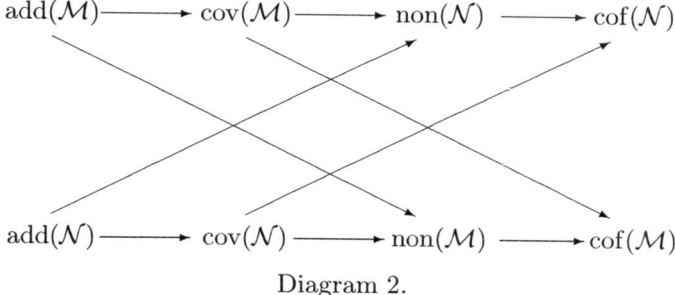

Diagram 2.

Now, we present a result that is "equal" for measure and category. The main tools in a common proof are special sets of reals which are both meager and have measure zero.

Theorem [AC] 7.28.

$$\mathfrak{s} \leq \text{non}(\mathcal{N}), \quad \mathfrak{s} \leq \text{non}(\mathcal{M}), \quad \text{cov}(\mathcal{N}) \leq \mathfrak{r}, \quad \text{cov}(\mathcal{M}) \leq \mathfrak{r}.$$

Proof. Let us recall that for an element $\alpha \in {}^\omega 2$ of Cantor space we denoted the zero-set of α as $Z(\alpha) = \{n \in \omega : \alpha(n) = 0\}$.

For a set $A \in [\omega]^\omega$ we write

$$S(A) = \{\alpha \in {}^\omega 2 : Z(\alpha) \text{ does not split } A\}.$$

For a finite set $s \subseteq A$ we set

$$N(A, s) = \{\alpha \in {}^\omega 2 : Z(\alpha) \cap A = s\}.$$

Since

$$\alpha \in N(A, s) \equiv (\forall n \in A)\, (n \in s \equiv \alpha(n) = 0),$$

one can see that the set $N(A, s)$ is closed nowhere dense and has measure zero. Evidently

$$S(A) = \bigcup\{N(A, s) : s \in [A]^{<\omega}\} \cup \bigcup\{N(\omega \setminus A, s) : s \in [\omega \setminus A]^{<\omega}\}$$

and therefore $S(A)$ is meager and has measure zero.

Let $D \subseteq {}^\omega 2$ be a non-meager set with $|D| = \text{non}(\mathcal{M})$. Then D cannot be a subset of any $S(A)$. Thus

$$(\forall A \in [\omega]^\omega)\, (D \cap ({}^\omega 2 \setminus S(A))) \neq \emptyset.$$

Therefore $\{Z(\alpha) : \alpha \in D\}$ is a splitting family and $\mathfrak{s} \leq \text{non}(\mathcal{M})$ holds true.

By the same argument we obtain $\mathfrak{s} \leq \text{non}(\mathcal{N})$.

Evidently

$$\mathfrak{r} = \min\{|\mathcal{D}| : (\forall \alpha \in {}^\omega 2,\, Z(\alpha) \text{ infinite})(\exists B \in \mathcal{D})\, \alpha \in S(B)\}.$$

Since the set $\{\alpha \in {}^\omega 2 : Z(\alpha) \text{ finite}\}$ is countable we immediately obtain both inequalities $\text{cov}(\mathcal{N}) \leq \mathfrak{r}$ and $\text{cov}(\mathcal{M}) \leq \mathfrak{r}$. $\qquad \square$

All results of this section have been obtained in a common framework for measure and category – a Polish ideal space. Actually, we can give a result with a proof for, say, the category case and then translate it by simple changing of corresponding words in a result and a proof for measure. That is the content of the duality phenomena. Therefore it seems obvious to look for some general transfer principle. Under some circumstances (assuming the continuum hypothesis) such a transform principle – a duality theorem – has been proved by W. Sierpiński [1934a] and then strengthened by P. Erdős [1943]. We present a slight strengthening of both results.

Theorem [AC] 7.29 (P. Erdős – W. Sierpiński). *If*

$$\mathrm{add}(\mathcal{N}) = \mathrm{add}(\mathcal{M}) = \mathrm{cof}(\mathcal{N}) = \mathrm{cof}(\mathcal{M}),$$

then there exists a bijection $f : \mathbb{R} \xrightarrow[\mathrm{onto}]{1-1} \mathbb{R}$ such that for any subset $E \subseteq \mathbb{R}$,

$$f(E) \text{ has Lebesgue measure zero if and only if } E \text{ is meager}$$

and

$$E \text{ has Lebesgue measure zero if and only if } f(E) \text{ is meager}.$$

Proof. Let $\kappa = \mathrm{add}(\mathcal{N}) = \mathrm{add}(\mathcal{M}) = \mathrm{cof}(\mathcal{N}) = \mathrm{cof}(\mathcal{M})$. By Theorem 7.1, e) the cardinal κ is regular.

Let $\{N_\xi : \xi < \kappa\}$ and $\{M_\xi : \xi < \kappa\}$ be a base of $\mathcal{N}$ and a base of $\mathcal{M}$, respectively. By Theorem 4.13 we can assume $N_0 \cup M_0 = \mathbb{R}$ and $N_0 \cap M_0 = \emptyset$. Moreover, we can assume that $N_\eta \subseteq N_\xi$ and $M_\eta \subseteq M_\xi$ for any $\eta < \xi < \kappa$. If not simply take $\bigcup_{\eta \le \xi} N_\eta$ and $\bigcup_{\eta \le \xi} M_\eta$ instead of N_ξ and M_ξ, respectively. Moreover we have

$$\bigcup_{\xi < \kappa} N_\xi = \bigcup_{\xi < \kappa} M_\xi = \mathbb{R}.$$

We show that for any $\eta < \kappa$ there exists a $\xi > \eta$ such that $|N_\xi \setminus \bigcup_{\zeta < \eta} N_\zeta| = \mathfrak{c}$. Actually, since $\bigcup_{\zeta < \eta} N_\zeta \in \mathcal{N}$, by Corollary 6.8 there exists a set $E \in \mathcal{N}$ such that $|E| = \mathfrak{c}$ and $E \cap \bigcup_{\zeta < \eta} N_\zeta = \emptyset$. Then there exists a $\xi > \eta$ such that $E \subseteq N_\xi$. A similar assertion holds true for the sets $\{M_\xi : \xi < \kappa\}$. Hence, we can assume that the sets

$$X_\xi = N_\xi \setminus \bigcup_{\eta < \xi} N_\eta, \quad Y_\xi = M_\xi \setminus \bigcup_{\eta < \xi} M_\eta$$

have cardinality $\mathfrak{c}$ for every $\xi < \kappa$. Note that $X_\xi \cap X_\eta = \emptyset$ and $Y_\xi \cap Y_\eta = \emptyset$ for any $\xi \ne \eta$.

Evidently

$$\bigcup_{\eta < \kappa} X_\eta = \bigcup_{\eta < \kappa} N_\eta = \mathbb{R}, \quad \bigcup_{\eta < \kappa} Y_\eta = \bigcup_{\eta < \kappa} M_\eta = \mathbb{R}.$$

By the choice of N_0 and M_0 we obtain $X_0 = N_0$ and $Y_0 = M_0$. Hence

$$M_0 = \bigcup_{0 < \eta < \kappa} X_\eta, \qquad N_0 = \bigcup_{0 < \eta < \kappa} Y_\eta.$$

Since each X_ξ and Y_ξ has cardinality $\mathfrak{c}$, for each $0 < \xi < \kappa$ there exists a bijection $f_\xi : X_\xi \xrightarrow[\text{onto}]{1-1} Y_\xi$. Set

$$f(x) = \begin{cases} f_\xi(x) & \text{if } x \in X_\xi, \quad \xi > 0, \\ f_\xi^{-1}(x) & \text{if } x \in Y_\xi, \quad \xi > 0. \end{cases}$$

If $A \subseteq \mathbb{R}$ has measure zero, then $A \subseteq N_\xi \subseteq \bigcup_{\zeta \leq \xi} X_\zeta$ for some ξ. Then $f(A)$ is meager since $f(A) \subseteq \bigcup_{\zeta \leq \xi} Y_\zeta$. Vice versa, if A is meager, then $A \subseteq M_\xi \subseteq \bigcup_{\zeta \leq \xi} Y_\zeta$ for some ξ and therefore $f(A) \subseteq \bigcup_{\zeta \leq \xi} X_\zeta$. Thus $f(A)$ has measure zero. $\qquad \square$

Note that the conclusion of the theorem implies that in Diagram 2. any cardinal invariant of measure equals the corresponding cardinal invariant of category. Thus the two lines of the Diagram coincide. The assumption of the theorem asks more: all cardinal invariants of Diagram 2. have to be equal.

Exercises

7.6 [AC] Hamel Base

We assume that $\mathbb{R}$ can be well ordered. We can consider $\mathbb{R}$ as a vector space over $\mathbb{Q}$. A base of $\mathbb{R}$ is called a **Hamel base**. Note that we can always assume that if there exists a Hamel base then it is a subset of $[0, 1]$.

a) There exists a Hamel base.

b) If $A \subseteq \mathbb{R}$ is linearly independent over $\mathbb{Q}$, then there exists a Hamel base $X \supseteq A$.

c) Construct a Hamel base subset of $[0, 1]$ of positive outer Lebesgue measure.
   Hint: Enumerate all G_δ sets of measure zero $\{G_\xi : \xi < \mathfrak{c}\}$ and pick an element of Hamel base $x_\xi \notin G_\xi$.

d) There exists a non-measurable Hamel base.
   Hint: If H is a Hamel base, then the sets $\{rH : r \in \mathbb{Q}\}$ are pairwise disjoint.

e) Show that there exists a Hamel base not possessing the Baire Property.

f) Construct meager sets $A, B \subseteq \mathbb{T}$ of Lebesgue measure zero such that $A + B = \mathbb{T}$.
   Hint: Set

$$A = \left\{ x = \sum_{i=1}^{\infty} x_i 2^{-2i} \in \mathbb{T} : x_i = 0, 1 \right\}, \quad B = \left\{ \sum_{i=1}^{\infty} x_i 2^{-2i+1} \in \mathbb{T} : x_i = 0, 1 \right\}.$$

g) There exists a meager Hamel base of measure zero.
   Hint: Construct a Hamel base that is a subset of $A \cup B$.

7.7 Cauchy Equation

The equation $(\forall x, y)\, f(x + y) = f(x) + f(y)$ is called a **Cauchy equation**. Every function $f(x) = ax$, a real, is a solution of a Cauchy equation. A solution different from those is called non-trivial.

a) Every non-trivial solution of a Cauchy equation is discontinuous at every point.

   Hint: From a Cauchy equation you obtain that $f(r) = rf(1)$ for r rational. So if f is continuous then f is a trivial solution.

b) If the real line can be well ordered, then there exists a non-trivial solution of a Cauchy equation.

   Hint: Use a Hamel base.

c) If f is a non-trivial solution of a Cauchy equation, then $f^{-1}((-n, n)) \notin \mathcal{N}$ for some n.

   Hint: Note that $\mathbb{R} = \bigcup_n f^{-1}((-n, n))$ and $f^{-1}((-n, n)) = nf^{-1}((-1, 1))$.

d) If a solution f of a Cauchy equation is bounded on a neighborhood of 0, then f is continuous.

e) If f is a non-trivial solution of a Cauchy equation, then $f^{-1}((-n, n)) \notin \mathcal{M}$ for some n.

f) Any non-trivial solution of a Cauchy equation is neither Lebesgue measurable nor Baire measurable.

   Hint: Use the Steinhaus Theorem 7.16 and the fact that $U \subseteq f^{-1}((-n, n)) - f^{-1}((-n, n))$ implies the inclusion $f(U) \subseteq (-2n, 2n)$.

g) Find all solutions of a Cauchy equation.

   Hint: Use a Hamel base.

h) A graph of a non-trivial solution of a Cauchy equation is a dense subset of $\mathbb{R}^2$.

   Hint: A non-trivial solution is discontinuous at 0.

7.3   Similar not Dual

Theorem [wAC] 7.30. *Let X, Y be topological spaces with countable bases. A function $f : X \longrightarrow Y$ is Baire measurable if and only if there exists a meager set $D \subseteq X$ such that $f|(X \setminus D)$ is continuous. Especially, for any Borel measurable, i.e., for analytically representable function f, there exists a meager set $D \subseteq X$ such that $f|(X \setminus D)$ is continuous.*

Proof. Assume that f is a Baire measurable function. Let $\{U_n : n \in \omega\}$ be an open base of the topology on Y. Since for any $n \in \omega$ the set $f^{-1}(U_n)$ possesses the Baire Property, there exist an open set V_n and meager F_σ sets $P_n, Q_n \subseteq X$ such that $V_n \subseteq X \setminus Q_n \subseteq f^{-1}(U_n) \subseteq V_n \cup P_n$. Let $D = \bigcup_n P_n \cup \bigcup_n Q_n$. Then D is meager and since $f^{-1}(U_n) \cap (X \setminus D) = V_n \cap (X \setminus D)$ for any $n \in \omega$, the function f is continuous on $X \setminus D$.

Assume that there exists a meager set D such that $f|(X \setminus D) : X \setminus D \longrightarrow Y$ is continuous. Then for any open $U \subseteq Y$ there exists an open set $V \subseteq X$ such that

$$f^{-1}(U) \cap (X \setminus D) = V \cap (X \setminus D).$$

Hence $V \setminus D \subseteq f^{-1}(U) \subseteq V \cup D$, i.e., the set $f^{-1}(U)$ possesses the Baire Property.

$\square$

For a Lebesgue measurable function we have a similar, however weaker result.

Theorem [wAC] 7.31 (N.N. Luzin). *Let X, Y be Polish spaces, μ being a Borel measure on X. A function $f : X \longrightarrow Y$ is μ-measurable if and only if for any positive ε there exists a μ-measurable set $A \subseteq X$ such that $\mu(A) < \varepsilon$ and $f|(X \setminus A)$ is continuous.*

Proof. Assume that f is μ-measurable. Let $\{U_n : n \in \omega\}$ be an open base of the topology on Y. Since $f^{-1}(U_n)$ is measurable, for any $n \in \omega$ there exists an open set $V_n \supseteq f^{-1}(U_n)$ such that $\mu(V_n \setminus f^{-1}(U_n)) < \varepsilon \cdot 2^{-n-1}$. Let $A = \bigcup_n (V_n \setminus f^{-1}(U_n))$. Then $\mu(A) < \varepsilon$ and since

$$f^{-1}(U_n) \cap (X \setminus A) = (X \setminus A) \cap V_n$$

for any $n \in \omega$, the function $f|(X \setminus A)$ is continuous.

Assume now that for any $n \in \omega$ there exists a set A_n such that $\mu(A_n) < 2^{-n}$ and $f|(X \setminus A_n)$ is continuous. Let $U \subseteq Y$ be open. Then there exist open sets V_n such that

$$f^{-1}(U) \cap (X \setminus A_n) = V_n \cap (X \setminus A_n).$$

If we set $A = \bigcap_n A_n$ we obtain $f^{-1}(U) = (f^{-1}(U) \cap A) \cup \bigcup_n (V_n \setminus A_n)$. Since $\mu(A) = 0$ the set $f^{-1}(U)$ is μ-measurable. $\qquad \square$

Luzin's Theorem 7.31 is not dual to Theorem 7.30. Moreover, the dual assertion to Theorem 7.30 is false.

Theorem [wAC] 7.32. *Let X be a Polish space, μ being a diffused probabilistic Borel measure on X with $\operatorname{supp}(\mu) = X$. Then there exists a Borel measurable function f from X into $[0,1]$ such that for any measurable set $A \subseteq X$ the following holds true: if $f|A$ is continuous, then $\mu(A) < 1$.*

Proof. Let $\{U_n : n \in \omega\}$ be an open base. Since $\operatorname{supp}(\mu) = X$, every non-empty open set has positive measure and is not nowhere dense. So by using Lemma 4.12 one can easily construct by induction a sequence of pairwise disjoint closed nowhere dense sets $\langle F_n : n \in \omega \rangle$ of positive measure such that $F_{2n}, F_{2n+1} \subseteq U_n$ for every n. The characteristic function f of the set $F = \bigcup_{n \in \omega} F_{2n}$ is Borel measurable.

Assume that $A \subseteq X$ is a measurable set, $f|A$ is continuous and $\mu(A) = 1$. Since $\mu(\bigcup_n F_{2n+1}) > 0$ and $\bigcup_n F_{2n+1} \subseteq Z(f) \subseteq f^{-1}((-1/2, 1/2))$ we obtain $f^{-1}((-1/2, 1/2)) \cap A \neq \emptyset$. Therefore $U_n \cap A \subseteq f^{-1}((-1/2, 1/2))$ for some n. Since $f(x) = 1$ for each $x \in F_{2n} \subseteq U_n$ we obtain $A \cap F_{2n} = \emptyset$, a contradiction. $\qquad \square$

In Section 4.4 we proved Fubini Theorem 4.28. A similar, actually dual, result holds true for category. However the proof is essentially different.

Theorem [wAC] 7.33 (K. Kuratowski – S. Ulam). *Let X, Y be separable metric spaces. Assume that $A \subseteq X \times Y$ possesses the Baire Property. Then A is meager if and only if the set $\{x \in X : A_x$ is not meager$\}$ is meager.*

Proof. Since $(\bigcup_n A_n)_x = \bigcup_n (A_n)_x$, to prove the implication from left to right we have to show that the set $B = \{x \in X : \mathrm{Int}(A_x) \neq \emptyset\}$ is meager provided that $A \subseteq X \times Y$ is nowhere dense and closed.

Let $\{U_n : n \in \omega\}$ be an open base on Y. Write

$$G_n = \{x \in X : (\exists y \in U_n)\, y \notin A_x\}.$$

If $\langle x, y \rangle \notin A$, $y \in U_n$ then there are open sets G, H such that $(G \times H) \cap A = \emptyset$ and $x \in G$, $y \in H \subseteq U_n$. Thus $G \subseteq G_n$. Consequently G_n is open. Moreover G_n is dense. Actually, for any non-empty open set V we have $(V \times U_n) \setminus A \neq \emptyset$ and therefore $V \cap G_n \neq \emptyset$. Now, if $\mathrm{Int}(A_x) \neq \emptyset$, then there exists an n such that $U_n \subseteq A_x$ and therefore $x \notin G_n$. Thus B is meager.

Assume now that $A \subseteq X \times Y$ possesses the Baire Property and is not meager. Then there exists a non-empty open set $G \subseteq X \times Y$ and a meager set Q such that $G \setminus Q \subseteq A$. Set

$$C = \{x \in X : Q_x \text{ is not meager}\}.$$

We already know that C is meager. Let $U \subseteq X$, $V \subseteq Y$ be non-empty open sets such that $U \times V \subseteq G$. Then

$$\{x \in X : A_x \text{ is not meager}\} \supseteq \{x \in U : V \setminus Q_x \text{ is not meager}\} \supseteq U \setminus C$$

is not meager. $\square$

Corollary [wAC] 7.34. *Let X, Y be metric separable spaces, $A \subseteq X \times Y$ possessing the Baire Property. Then the set $\{x \in X : A_x \text{ is not meager}\}$ is meager if and only if $\{y \in Y : A^y \text{ is not meager}\}$ is meager.*

Theorem 7.35. *Let X be a perfect Polish space. If $W \subseteq X \times X$ is a well-ordering of X of the ordinal type $\mathfrak{c}$, then W is not $\mu \times \mu$-measurable for any diffused Borel measure μ on X and W does not possess the Baire Property.*

Proof. Assume that $W \subseteq X \times X$ is a well-ordering of X of the ordinal type $\mathfrak{c}$.

Let μ be a diffused Borel measure on X. Assume that W is $\mu \times \mu$-measurable. The horizontal section $W_y = \{x \in X : xWy\}$ has cardinality strictly smaller than $\mathfrak{c}$ for any $y \in X$. Since W_y is measurable, by Corollary 6.8 we obtain that $\mu(W_y) = 0$. If $x \in X$, then $\mu(W^x) = \mu(X) - \mu(W_x) = \mu(X) > 0$. Thus, we have obtained a contradiction with the Fubini Theorem 4.29.

Replacing corresponding words, one will obtain a proof for the Baire Property case. $\square$

Let $\langle X, +, 0 \rangle$ be an Abelian Polish group. The arithmetic sum of two subsets can be defined as in Section 3.1. A non-empty set $A \subseteq X$ is called a **tail-set** if there exists a countable dense set $S \subseteq X$ such that $S \subseteq \{x \in X : x + A = A\}$.

We consider $^{<\omega}2$ as a subset of $^{\omega}2$ identifying a finite sequence $s \in {}^{<\omega}2$ with $\alpha \in {}^{\omega}2$ such that $\alpha(n) = s(n)$ for $n \in \mathrm{dom}(s)$ and $\alpha(n) = 0$ otherwise. The set of dyadic numbers $\mathbb{D}$ and $^{<\omega}2$ are dense subsets of the Abelian Polish groups $\mathbb{T}$ and

$^\omega 2$, respectively. The set $\mathcal{P}(\omega)$ can be viewed as an Abelian Polish group with the group operation the symmetric difference

$$A \triangle B = (A \setminus B) \cup (B \setminus A)$$

isomorphic to $^\omega 2$ with a countable dense subset Fin.

Note the following. Let $\langle X, +, 0 \rangle$ be an Abelian Polish group, S being a countable dense subgroup. If $A \subseteq X$ is any non-empty set, then

$$S + A = \bigcup_{x \in S} x + A \tag{7.14}$$

is a tail-set containing A as a subset. If A is meager, then $S + A$ is meager as well. Similarly, if μ is a shift invariant Borel measure on X and $\mu(A) = 0$, then also $\mu(S + A) = 0$. Consequently, any meager or measure zero set is contained in a meager or measure zero tail-set, respectively. We shall use this fact without any commentary.

We begin with folklore results, the so-called "Zero-One Law", which we prove for the Polish group $\mathbb{T}$. One can easily extend them for Polish groups $\mathbb{R}$ and $^\omega 2$. The proofs for the measure case and the category case can be considered as similar, however, one can hardly transform one proof into another one.

Theorem 7.36 (Zero-One Law). *If the set $A \subseteq \mathbb{T}$ is a tail-set, then $\lambda^*(A)$ is either 0 or 1. A similar assertion holds true for $^\omega 2$.*

Proof. Let $C \subseteq \{x \in \mathbb{T} : x + A = A\}$ be countable dense. Assume that $\lambda^*(A) > 0$ and $0 < \varepsilon < 1$ is a real. By Theorem 4.17 there exists an open interval I such that $\lambda(I) < \varepsilon$ and $\lambda(A \cap I) \geq (1 - \varepsilon)\lambda(I)$.

Then there are reals $x_1, \ldots, x_k \in C$ such that $\lambda(\bigcup_{i=1}^{k}(x_i + I)) > 1 - \varepsilon$ and the intervals $x_i + I$ are mutually disjoint. By (4.10) we obtain

$$\lambda^*(A) \geq \lambda^*\left(A \cap \bigcup_{i=1}^{k}(x_i + I)\right) = \sum_{i=1}^{k} \lambda^*(A \cap (x_i + I))$$

$$\geq (1 - \varepsilon) \sum_{i=1}^{k} \lambda(x_i + I) \geq (1 - \varepsilon)^2.$$

Since ε was arbitrary we obtain $\lambda^*(A) = 1$. □

Corollary [AC] 7.37. *There exists a tail-set $A \subseteq \mathbb{T}$ of outer measure 1 of cardinality* $\mathrm{non}(\mathcal{N})$.

Proof. Let $|B| = \mathrm{non}(\mathcal{N})$ and $\lambda^*(B) > 0$. Then the set $A = B + (\mathbb{D} \cup \{0\})$ is a tail-set containing B. □

We have a similar result for category.

Theorem 7.38 (Zero-One Law). *If a tail-set A possesses the Baire Property, then A is either meager or comeager.*

Proof. Let $A \subseteq \mathbb{T}$ possess the Baire Property. Then there are an open set G and meager sets Q, P such that $A = (G \setminus Q) \cup P$. Assume that A is not meager. Then G is non-empty. Since A is a tail-set, there exists a countable dense subset C of $\{x \in \mathbb{T} : x + A = A\}$.

The set $R = \mathbb{T} \setminus \bigcup_{x \in C}(x + G)$ is nowhere dense. We show that

$$\mathbb{T} \setminus A \subseteq \bigcup_{x \in C}(x + Q) \cup R.$$

Actually, if $y \in \mathbb{T} \setminus A$ and $y \notin R$, then there exists an $x \in C$ such that $y \in x + G$. Since $y \notin A = x + A$ we obtain that $y \in x + Q$.

Since Q was meager we have shown that $\mathbb{T} \setminus A$ is meager.

For a subset of ${}^{\omega}2$ the proof goes analogously. Or, by suitable continuous surjection from ${}^{\omega}2$ onto $\mathbb{T}$ you can reduce the problem from a subset of ${}^{\omega}2$ to a subset of $\mathbb{T}$. □

Corollary 7.39. *A free ultrafilter on ω is a Lebesgue non-measurable set and does not possess the Baire Property.*

Proof. If a free ultrafilter $\mathcal{J}$ is considered as a subset of ${}^{\omega}2$, then $\mathcal{J}$ is a tail-set, since $x + \mathcal{J} = \mathcal{J}$ for any finite $x \subseteq \omega$. Evidently the complement ${}^{\omega}2 \setminus \mathcal{J}$ is also a tail-set. $\mathcal{J}$ and ${}^{\omega}2 \setminus \mathcal{J}$ have equal outer measure and are homeomorphic. □

Exercises

7.8 [AC] Innerly Meager Sets
A set $A \subseteq X$ is called **innerly meager** if for any meager set $P \subseteq X$ one has $\text{Int}(A \cup P) = \emptyset$.

   a) If $A \subseteq X$ has the Baire Property and is innerly meager, then A is meager.

   b) If a perfect Polish space X can be well ordered, then there exist innerly meager sets $A, B \subseteq X$ such that $X = A \cup B$.
   Hint: A Bernstein set is innerly meager.

   c) If $A \subseteq {}^{\omega}2$ is a tail-set, then either A is innerly meager or ${}^{\omega}2 \setminus A$ is innerly meager.
   Hint: Follow the proof of Theorem 7.38.

   d) There exists a set $A \subseteq \mathbb{R}$ such that $|A| = \text{non}(\mathcal{M})$ and $\mathbb{R} \setminus A$ is innerly meager.
   Hint: Let $B \notin \mathcal{M}$, $|B| = \text{non}(\mathcal{M})$. Set $A = B + \mathbb{Q}$.

7.9 Other Proofs of the Zero-One Law for Category
   a) If $A \subseteq \mathbb{T}$ possesses the Baire Property and is not meager, then there exist an open interval I and a meager set Q such that $I \setminus Q \subseteq A$.

   b) If $C \subseteq \mathbb{T}$ is dense and $I \subseteq \mathbb{T}$ is an open interval, then there exist finitely many elements $x_1, \ldots, x_n \in C$ such that $\bigcup_{i=1}^{n}(x_i + I) = \mathbb{T}$.

c) If $A \subseteq \mathbb{T}$ is a non-meager tail-set possessing the Baire Property, then $\mathbb{T} \setminus A$ is meager.

Hint: If $C = \{x \in \mathbb{T} : x + A = A\}$, I, Q are those of Exercise a) and $x_1, \ldots, x_n$ are those of Exercise b), then

$$\mathbb{T} \setminus A \subseteq \bigcup_{i=0}^{n} (x_i + Q).$$

We identify sets of the form $A \times B$, where $A \subseteq {}^n 2$ and $B \subseteq {}^{(\omega \setminus n)} 2$ with subsets of ${}^\omega 2$ in the natural way.

d) If $A \subseteq {}^\omega 2$ is a tail-set, then for any $n \in \omega$ there exists a tail-set B such that $A = {}^n 2 \times B$.

Hint: Take $B = \{\alpha | (\omega \setminus n) : \alpha \in A\}$.

e) If a non-meager set $A \subseteq {}^\omega 2$ possesses the Baire Property, then there are an open set G and a meager set Q such that $(G \setminus Q) \subseteq A$.

f) If $A \subseteq {}^\omega 2$ is a non-meager tail-set possessing the Baire Property, then ${}^\omega 2 \setminus A$ is meager.

Hint: If $s \in {}^n 2$ is such that $[s] \subseteq G$, and ${}^n 2 \times B = A$, then

$$ {}^\omega 2 \setminus A \subseteq \bigcup_{t \in {}^n 2} (t + Q).$$

7.4 The Fall of Duality – Bartoszyński Theorem

We begin with a simple application of the results of Theorems 7.36 and 7.38 with different impact to the properties of measure and category. Consider the set

$$A = \left\{ \alpha \in {}^\omega 2 : \lim_{n \to \infty} \frac{\sum_{i=0}^{n} \alpha(i)}{n + 1} = \frac{1}{2} \right\}.$$

The set A is a $\mathbf{\Pi}_3^0$ set. Evidently A is a tail-set. Therefore, by Theorem 7.36 either the set A has measure zero or the set ${}^\omega 2 \setminus A$ has measure zero and by Theorem 7.38 either the set A is meager or the set ${}^\omega 2 \setminus A$ is meager. We show that A is small in the sense of category and large in the sense of measure.

Theorem 7.40. *A is meager and $\lambda(A) = 1$.*

Proof. One can easily check that the set

$$B_k = \left\{ \alpha \in {}^\omega 2 : (\exists n > k) \frac{\sum_{i=0}^{n} \alpha(i)}{n + 1} > \frac{3}{4} \right\}$$

is open and dense in ${}^\omega 2$. Thus the set

$$B = \left\{ \alpha \in {}^\omega 2 : (\forall k)(\exists n > k) \frac{\sum_{i=0}^{n} \alpha(i)}{n + 1} > \frac{3}{4} \right\} = \bigcap_{k} B_k$$

is a G_δ dense set. Since $A \cap B = \emptyset$, the set A is meager.

Set $g(\alpha) = \alpha(0)$. Then g is a continuous and therefore integrable function on $^\omega 2$. The shift T of $^\omega 2$ was defined in Exercise 4.22 by (4.36). Evidently $g(T^k(\alpha)) = \alpha(k)$. By Exercise 4.22, c) the shift T is ergodic and therefore by Ergodic Theorem 4.31 we obtain

$$\lim_{n \to \infty} \frac{\sum_{i=0}^{n} \alpha(i)}{n+1} = \int_{\omega 2} g(x) \, dx = \frac{1}{2}$$

for almost all α, i.e., $\lambda(A) = 1$.   □

We can conclude more: there is no hope to prove an ergodic theorem for category, otherwise A is comeager.

In Exercise 7.10 we present an elementary proof of $\lambda(A) = 1$.

There is an important result on measure that does not have a category analogue.

Theorem [wAC] 7.41 (Egoroff Theorem). *Let $\langle X, \mathcal{S}, \mu \rangle$ be a measure space, μ being finite measure and $|\mathcal{S}| \ll \mathfrak{c}$. If a sequence $\{f_n\}_{n=0}^\infty$ of μ-measurable real-valued functions converges to 0 on a subset $A \subseteq X$, then for every $\varepsilon > 0$ there exists a μ-measurable set $C \subseteq X$, such that $\mu^*(A \setminus C) < \varepsilon$ and $f_n \rightrightarrows 0$ on C.*

Proof. We set

$$B = \{x \in X : (\forall m)(\exists n)(\forall i)\,(i \geq n \to |f_i(x)| < 1/(m+1))\},$$
$$B_{m,n} = \{x \in B : (\forall i)\,(i \geq n \to |f_i(x)| < 1/(m+1))\}.$$

Then B is a measurable set, $A \subseteq B$, and for every m we have $\bigcup_n B_{m,n} = B$.

Since the sets $B_{m,n} \subseteq B_{m,n+1}$ are measurable, for any m there exists an integer n_m such that $\mu(B \setminus B_{m,n_m}) < \varepsilon/2^{m+1}$. Set $C = \bigcap_m B_{m,n_m}$.   □

Corollary [wAC] 7.42. *If a sequence $\{f_n\}_{n=0}^\infty$ of measurable real-valued functions converges to 0 on a subset $A \subseteq X$, then there exists a measurable set C, such that $\mu^*(A \setminus C) = 0$ and $f_n \xrightarrow{\mathrm{QN}} 0$ on C.*

The proof is easy. Let $C_n \subset A$ be such that $\mu^*(A \setminus C_n) < 2^{-n}$ and $f_n \rightrightarrows 0$ on C_n. Set $C = \bigcup_n C_n$.   □

An analogue of Corollary 7.42 for meager sets does not hold true.

Theorem [wAC] 7.43. *Let X be a perfect separable metric space. Then there exists a sequence $\{f_n\}_{n=0}^\infty$ of continuous real-valued functions defined on X such that*

i) *$f_n \to 0$ on X;*

ii) *if $A \subseteq X$, $\{n_k\}_{k=0}^\infty$ is an increasing sequence such that $f_{n_k} \rightrightarrows 0$ on A, then A is nowhere dense.*

Proof. Let $Q = \{r_i : i \in \omega\}$ be a countable dense subset of X. Since no point r_i is isolated there exists a sequence $x_{i,n} \longrightarrow r_i$, such that $x_{i,n} \notin Q$ for each $n \in \omega$. Let

$h_{i,n} : X \longrightarrow [0, 2^{-i}]$ be continuous and such that $h_{i,n}(x_{i,n}) = 2^{-i}$ and $h_{i,n}(x) = 0$ if $\rho(x, x_{i,n}) \geq 1/2\rho(r_i, x_{i,n})$. Set

$$f_n(x) = \sum_{i=0}^{\infty} h_{i,n}(x) \text{ for } x \in X, \, n \in \omega.$$

Each f_n is a continuous function from X into $[0, 2]$. Evidently $h_{i,n} \to 0$ on X for every fixed i. If $x \in X$ and $\varepsilon > 0$, one can find an i_0 such that $\sum_{i > i_0} 2^{-i} < \varepsilon/2$ and such n_0 that $\sum_{i \leq i_0} h_{i,n}(x) < \varepsilon/2$ for every $n \geq n_0$. Thus $f_n \to 0$ on X.

Assume that $\{n_k\}_{k=0}^{\infty}$ is increasing and $f_{n_k} \rightrightarrows 0$ on a set A. We can assume that A is closed. If $\mathrm{Int}(A) \neq \emptyset$ then there exists an $r_i \in \mathrm{Int}(A)$. Hence there exists an m such that $x_{i,n} \in \mathrm{Int}(A)$ for $n \geq m$. Since for $n \geq m$ we have

$$\sup\{f_n(x) : x \in A\} \geq f(x_{i,n}) \geq h_{i,n}(x_{i,n}) = 2^{-i},$$

we get a contradiction. $\qquad \square$

Corollary [wAC] 7.44. *Let $\{f_n\}_{n=0}^{\infty}$ be the sequence of the theorem. If $f_{n_k} \xrightarrow{\mathrm{QN}} 0$ on A, then A is meager.*

Proof. If $f_{n_k} \xrightarrow{\mathrm{QN}} 0$ on A then there are closed sets A_n such that $f_{n_k} \rightrightarrows 0$ on every A_n and $A \subseteq \bigcup_n A_n$. If A is not meager, then there exists an n such that $\mathrm{Int}(A_n) \neq \emptyset$, a contradiction. $\qquad \square$

We show some properties in which the algebras $\mathcal{L}/\mathcal{N}$ and $\mathrm{BAIRE}/\mathcal{M}$ differ.

Theorem [wAC] 7.45. *Boolean algebra $\mathrm{BAIRE}/\mathcal{M}$ contains a countable dense subset. No countable set is dense in Boolean algebra $\mathcal{L}/\mathcal{N}$.*

Proof. Any countable base of the topology is also a dense set in the quotient algebra $\mathrm{BAIRE}/\mathcal{M}$.

Assume that $\langle V_n : n \in \omega \rangle$ is a countable subset of $\mathcal{L}([0, 1]) \setminus \mathcal{N}([0, 1])$. For each n take a set A_n such that $A_n \subseteq V_n$ and $0 < \lambda(A_n) < 2^{-n-2}$. Take $A = \bigcup_n A_n$. Then $\lambda(V_n \cap A) > 0$ for every n. Since $\lambda(A) < 1$ we have $\lambda([0, 1] \setminus A) > 0$ and no V_n is smaller than $[0, 1] \setminus A$ modulo $\mathcal{N}$. Thus $\langle V_n : n \in \omega \rangle$ is not dense in $\mathcal{L}/\mathcal{N}$. $\qquad \square$

In Section 7.2 we have shown that the Boolean algebras $\mathcal{L}/\mathcal{N}$ and $\mathrm{BAIRE}/\mathcal{M}$ are not $\aleph_0$-distributive. We show that the algebras differ in a finer property, weak distributivity. A Boolean algebra $\mathbf{B}$ is **weakly $\aleph_0$-distributive** if for any sequence $\langle A_n : n \in \omega \rangle$ of predense sets there exists a predense set A such that for any $a \in A$ and any n, there exists a finite subset $C \subseteq A_n$ with $a \leq \bigvee C$.

Lemma [wAC] 7.46. *If a complete Boolean algebra $\mathbf{B}$ carries a positive measure, then $\mathbf{B}$ is weakly $\aleph_0$-distributive.*

Proof. Let μ be a positive measure on B. Let $\langle A_n : n \in \omega \rangle$ be predense subsets of B. We can assume that each A_n is countable and contains pairwise disjoint elements. We show that the set

$$A = \{b \in B : (\forall n)(\exists C \subseteq A_n)\,(C \text{ finite and } b \leq \bigvee C)\}$$

is a dense set.

Let $a \in B$, $a \neq 0_B$. Since $\sum_{x \in A_n} \mu(a \wedge x) = \mu(a) > 0$, there exists a finite set $C_n \subseteq A_n$ such that $\mu(a - \bigvee C_n) < 2^{-n-2}\mu(a)$. Then $b = a \wedge \bigwedge_n \bigvee C_n$ is such that $\mu(b) > 0$, $b \leq a$ and $b \in A$. $\qquad\square$

Theorem [wAC] 7.47. *The Boolean algebra $\mathcal{L}/\mathcal{N}$ is weakly $\aleph_0$-distributive and the Boolean algebra $\mathrm{BAIRE}/\mathcal{M}$ is not weakly $\aleph_0$-distributive.*

Proof. The first assertion follows by Theorem 7.15 and Lemma 7.46.

To obtain a contradiction, suppose that $\mathrm{BAIRE}(^{\omega}\omega)/\mathcal{M}$ is weakly distributive. For $i, n \in \omega$ we write $C_i^n = \{\alpha \in {}^{\omega}\omega : \alpha(n) = i\}$. Since $\bigcup_i C_i^n = {}^{\omega}\omega$, the family $\mathcal{A}_n = \{C_i^n : i \in \omega\}$ is predense for every n. Let $A \in \mathrm{BAIRE}(^{\omega}\omega) \setminus \mathcal{M}$. Assume that $\mathcal{C}_n$ is a finite subset of $\mathcal{A}_n$, such that $A \setminus \bigcup \mathcal{C}_n$ is meager, $n \in \omega$. Then also $A \setminus \bigcap_n \bigcup \mathcal{C}_n$ is meager and there exists a function $\alpha \in {}^{\omega}\omega$ such that $\mathcal{C}_n \subseteq \{C_i^n : i < \alpha(n)\}$. Since the intersection $\bigcap_n \bigcup_{i < \alpha(n)} C_i^n$ is a closed nowhere dense set, we get a contradiction. $\qquad\square$

Corollary [wAC] 7.48. *There is no positive measure on $\mathrm{BAIRE}/\mathcal{M}$.*

No assertion on the measure dual to the following fine result on meager sets is known.

Theorem [AC] 7.49 (Z. Piotrowski – A. Szymański).

$$\mathfrak{t} \leq \mathrm{add}(\mathcal{M}).$$

We shall use the terminology and notation introduced in Section 5.4.

Lemma [AC] 7.50. *If $\langle Q_\xi \subseteq \mathbb{Q} : \xi < \kappa \rangle$ is an almost decreasing sequence of dense subsets of $\mathbb{Q}$ of length $\kappa < \mathfrak{t}$, then there exists a dense set $Q \subseteq \mathbb{Q}$ such that $Q \subseteq^* Q_\xi$ for every $\xi < \kappa$.*

Proof. Let $\{r_n : n \in \omega\}$ be an enumeration of $\mathbb{Q}$. We denote by $\mathcal{I}$ the set of all open intervals with rational endpoints. For any $I \in \mathcal{I}$, the sequence $\langle I \cap Q_\xi : \xi < \kappa \rangle$ of infinite subsets of I is almost decreasing. Since $\kappa < \mathfrak{t}$ there exists an infinite set $P_I \subseteq \mathbb{Q} \cap I$ such that $P_I \subseteq^* I \cap Q_\xi$ for every $\xi < \kappa$. Denote by $\alpha_\xi(I)$ the first n such that

$$(\forall m \geq n)\,(r_m \in P_I \to r_m \in I \cap Q_\xi).$$

Since $\mathcal{I}$ is countable and $\kappa < \mathfrak{t} \leq \mathfrak{b}$, the set of functions $\langle \alpha_\xi : \xi < \kappa \rangle$ is bounded, i.e., there exists a function $\beta : \mathcal{I} \longrightarrow \omega$ such that $\alpha_\xi <^* \beta$ for every $\xi < \kappa$. We set

$$Q = \{r_n : (\exists I)\,(r_n \in P_I \wedge n > \beta(I))\}.$$

It is easy to see that Q is the desired set. $\qquad\square$

Proof of the theorem. Let $\langle G_\xi \subseteq \mathbb{R} : \xi < \kappa \rangle$ be open dense sets, $\kappa < \mathfrak{t}$. We show that the intersection $\bigcap_{\xi < \kappa} G_\xi$ is dense in $\mathbb{R}$.

By induction we construct an almost decreasing sequence $\langle Q_\xi : \xi \leq \kappa \rangle$ of dense subsets of $\mathbb{Q}$. Actually, we set $Q_0 = \mathbb{Q}$, $Q_{\xi+1} = Q_\xi \cap G_\xi$ and for limit ξ we apply the lemma. Note that for any $\xi < \kappa$ the set $Q_\kappa \setminus G_\xi$ is finite.

We define functions $\alpha_\xi : Q_\kappa \longrightarrow \omega$ as follows: $\alpha_\xi(r)$ is the first natural number n such that $(r - 1/n, r + 1/n) \subseteq G_\xi$, if $r \in G_\xi$ and $\alpha_\xi(r) = 0$ otherwise. Since $\kappa < \mathfrak{b}$ there exists a function $\beta : Q_\kappa \longrightarrow \omega$ such that $\alpha_\xi <^* \beta$ for every $\xi < \kappa$.

If $s \subseteq Q_\kappa$ is finite then

$$U_s = \bigcup_{r \in Q_\kappa \setminus s} (r - 1/\beta(r), r + 1/\beta(r))$$

is an open dense subset of $\mathbb{R}$. Set $U = \bigcap\{U_s : s \in [Q_\kappa]^{<\omega}\}$. The set U is dense in $\mathbb{R}$ as well.

Consider a $\xi < \kappa$. Let $s = Q_\kappa \setminus G_\xi \cup \{r \in \mathbb{Q} : \beta(r) < \alpha_\xi(r)\}$. The set s is finite and one can easily see that $U \subseteq U_s \subseteq G_\xi$. Thus, $U \subseteq \bigcap_{\xi < \kappa} G_\xi$. $\square$

The obtained results show that the duality measure-category is limited. However for a long time all those anti-dual results were considered rather as the exception that proves the rule. In 1983 Tomek Bartoszyński and independently a bit later Jean Raisonnier and Jaques Stern have announced results, which finally turned out to be equivalent, and refute the measure-category duality. Raisonnier-Stern's result was proved by using inner models of set theory. Bartoszyński's result was formulated in a rather elementary way and the main aim of this section is to prove it. We follow the technology developed in the late 1980s.

Theorem [AC] 7.51 (T. Bartoszyński).

$$\mathrm{add}(\mathcal{N}) \leq \mathrm{add}(\mathcal{M}) \ \ and \ \ \mathrm{cof}(\mathcal{M}) \leq \mathrm{cof}(\mathcal{N}).$$

The theorem follows from Pawlikowski's Theorem 7.56 below by Theorem 5.43 and equalities (7.1). So our main aim reduces to proving Theorem 7.56.

We start with a comment. By Bartoszyński's Theorem we have new arrows in Diagram 2. of Section 7.2:

$$\mathrm{add}(\mathcal{N}) \longrightarrow \mathrm{add}(\mathcal{M}), \ \mathrm{cof}(\mathcal{M}) \longrightarrow \mathrm{cof}(\mathcal{N}).$$

Moreover very soon it turned out that a reverse inequality $\mathrm{add}(\mathcal{M}) \leq \mathrm{add}(\mathcal{N})$ cannot be proved in **ZFC** and the duality fell. Hence, there is a natural question as to whether we have to change Diagrams 1. and 2. that look very dual. A result of such a change is presented in the next section.

We state without a proof the above-mentioned result of Jean Raisonnier and Jaques Stern.

Theorem [AC] 7.52 (J. Raisonnier – J. Stern). *If every Σ_2^1 set of reals is Lebesgue measurable, then every Σ_2^1 set of reals possesses the Baire Property.*

Using some well-known results of the theory of models of set theory, one can deduce the Raisonnier-Stern Theorem from Bartoszyński's Theorem, more accurately, from the existence of Tukey connections used in the proof. Again, using some well-known results of the theory of models of set theory, one can deduce Bartoszyński's Theorem from Raisonnier-Stern's Theorem. Or, one can change Raisonnier-Stern's proof into a proof of Bartoszyński's Theorem.

For a proof of Bartoszyński's Theorem, we need some technical tools.

Lemma 7.53 (D. Fremlin). *If a topological space $\langle X, \mathcal{O} \rangle$ has a countable base of topology, then for any $n > 0$ there exists a countable set $\mathcal{V}$ of open sets such that*

   a) *if $G \subseteq X$ is open dense, then there exists a $V \in \mathcal{V}$ such that $V \subseteq G$;*
   b) *if $V_0, \ldots, V_n \in \mathcal{V}$, then*

$$\bigcap_{i=0}^{n} V_i \neq \emptyset.$$

Proof. Let $\mathcal{U} = \{U_k : k \in \omega\}$ be a countable base of the topology. We can assume that $\mathcal{U}$ is closed under finite unions and $\emptyset \notin \mathcal{U}$. Set

$$A_m = \left\{ k \in \omega : k > m \wedge (\forall Y \subseteq m + 1) \left(\bigcap_{i \in Y} U_i \neq \emptyset \rightarrow U_k \cap \bigcap_{i \in Y} U_i \neq \emptyset \right) \right\},$$

$$\mathcal{V} = \left\{ \bigcup_{i \leq n} U_{s(i)} : s \in {}^{n+1}\omega \wedge (\forall i < n)\, s(i+1) \in A_{s(i)} \right\}.$$

Let G be open dense. Since the base $\mathcal{U}$ is closed under finite unions, for any m there exists a $k \in A_m$ such that $U_k \subseteq G$. Let k_0 be such that $U_{k_0} \subseteq G$ and let $k_{i+1} \in A_{k_i} \cap \{k : U_k \subseteq G\}$ for $i < n$. Then $\bigcup_{i \leq n} U_{k_i} \in \mathcal{V}$ and $\bigcup_{i \leq n} U_{k_i} \subseteq G$. Thus a) holds true.

Let $V_0, \ldots, V_n \in \mathcal{V}$, $V_i = \bigcup_{j \leq n} U_{k_{i,j}}$ with suitable indexes $k_{i,j}$. We can assume that $k_{i,j} \in A_{k_{i,j+1}}$ for any $i \leq n$ and $j < n$. We can re-order $\langle V_j : 0 \leq j \leq n \rangle$ in such a way that $k_{0,0} \leq k_{j,0}$ for any $j \leq m$. By induction, re-order $\langle V_j : i \leq j \leq n \rangle$ in such a way that $k_{i,i} \leq k_{j,i}$ for any $i \leq j \leq n$. Then $k_{i,i} \leq k_{i+1,i} < k_{i+1,i+1}$ and consequently $k_{i+1,i+1} \in A_{k_{i+1,i}} \subseteq A_{k_{i,i}}$. By induction we obtain $\bigcap_{i \leq n} U_{k_{i,i}} \neq \emptyset$. Since $U_{k_{i,i}} \subseteq V_i$ we obtain $\bigcap_{i \leq n} V_i \neq \emptyset$.   □

Let $\langle X, \mathcal{S}, \mu \rangle$ be a measure space with a probabilistic measure. A family $\mathcal{G} \subseteq \mathcal{S}$ is said to be **measure independent** if $0 < \mu(A) < 1$ for any $A \in \mathcal{G}$ and

$$\mu \left(\bigcap_{i=0}^{n} A_i \right) = \prod_{i=0}^{n} \mu(A_i)$$

for any sets $A_0, \ldots, A_n \in \mathcal{G}$. One can easily see that, replacing some element of a measure independent family by its complement, we obtain a measure independent family.

We can easily construct an infinite measure independent family of clopen subsets of the measure space $\langle {}^\omega 2, \mathcal{L}, \lambda \rangle$. Actually, if $\langle s_n \in {}^{[\omega]^{<\omega}} 2 : n \in \omega \rangle$ is a sequence of finite functions with non-empty domains such that $\mathrm{dom}(s_n) \cap \mathrm{dom}(s_m) = \emptyset$ for $n \neq m$, then $\{[s_n] : n \in \omega\}$ is a measure independent family. Note that $\lambda([s]) = 2^{-|\mathrm{dom}(s)|}$.

Let $\{G_{n,m} : n, m \in \omega\}$ be a measure independent family of clopen subsets of ${}^\omega 2$ such that $\lambda(G_{n,m}) = 2^{-n}$. Fix an open base $\{U_n : n \in \omega\}$ on ${}^\omega 2$. For every $A \in \mathcal{N}$ choose (using **AC**) a compact set K_A disjoint with A and of positive measure. Moreover we can assume that if $U_n \cap K_A \neq \emptyset$ then $\lambda(U_n \cap K_A) > 0$. If not, replace K_A by

$$K_A \setminus \bigcup \{U_n : \lambda(U_n \cap K_A) = 0\}.$$

We write

$$\mathrm{Loc} = \{h \in {}^\omega([\omega]^{<\omega}) : (\forall n)\, |h(n)| \leq 2^n\}. \tag{7.15}$$

As we shall see later, the condition $|h(n)| \leq 2^n$ may be replaced by $|h(n)| \leq f(n)$ with a fast growing $f \in {}^\omega\omega$.

If $g, h : \omega \longrightarrow \mathcal{P}(\omega)$ then $g \subseteq^* h$ means that $(\exists k)(\forall n \geq k)\, g(n) \subseteq h(n)$. Similarly, if $\alpha \in {}^\omega\omega$ then $\alpha \in^* h$ means that $(\exists k)(\forall n \geq k)\, \alpha(n) \in h(n)$.

Lemma [AC] 7.54. *There exist functions*

$$\varphi_1 : {}^\omega\omega \longrightarrow \mathcal{N}({}^\omega 2), \quad \varphi_1^* : \mathcal{N}({}^\omega 2) \longrightarrow \mathrm{Loc}$$

such that

$$\varphi_1(\alpha) \subseteq A \to \alpha \in^* \varphi_1^*(A)$$

for any $\alpha \in {}^\omega\omega$ and any $A \in \mathcal{N}({}^\omega 2)$.

Proof. For $A \in \mathcal{N}({}^\omega 2)$, $k, n \in \omega$ we set

$$\Phi(A, k, n) = \{m \in \omega : U_k \cap K_A \neq \emptyset \wedge U_k \cap K_A \cap G_{n,m} = \emptyset\}.$$

If $m \in \Phi(A, k, n)$, then $U_k \cap K_A \subseteq ({}^\omega 2 \setminus G_{n,m})$ and $\lambda(U_k \cap K_A) > 0$. Since the family $\{{}^\omega 2 \setminus G_{n,m} : n, m \in \omega\}$ is measure independent and $\lambda({}^\omega 2 \setminus G_{n,m}) = 1 - 2^{-n}$, the set $\Phi(A, k, n)$ is finite. Evidently

$$(U_k \cap K_A) \cap \bigcup \{G_{n,m} : n \in \omega \wedge m \in \Phi(A, k, n)\} = \emptyset$$

and therefore

$$\lambda(U_k \cap K_A) \leq \lambda \left(\bigcap \{{}^\omega 2 \setminus G_{n,m} : n \in \omega \wedge m \in \Phi(A, k, n)\} \right)$$

$$= \prod_{n \in \omega} (1 - 2^{-n})^{|\Phi(A,k,n)|}. \tag{7.16}$$

If $U_k \cap K_A = \emptyset$, then also $\Phi(A, k, n) = \emptyset$ for any n. If $U_k \cap K_A \neq \emptyset$, then $\lambda(U_k \cap K_A) > 0$ and therefore the infinite product (7.16) converges. Thus by

Theorem 2.24, b) we obtain

$$\sum_{n=0}^{\infty} 2^{-n}|\Phi(A,k,n)| < \infty.$$

Let $N(A,k)$ be the least natural number such that $2^{-n}|\Phi(A,k,n)| \le 2^{-k-1}$ for every $n \ge N(A,k)$. We set $\varphi_1^*(A) = h$ where

$$h(n) = \bigcup_k \{\Phi(A,k,n) : n \ge N(A,k)\}.$$

Since

$$|h(n)| \le \sum_{k=0}^{\infty} 2^n \cdot 2^{-k-1} = 2^n,$$

we have $h \in \mathrm{Loc}$.

If $\alpha \in {}^{\omega}\omega$ we set

$$\varphi_1(\alpha) = \bigcap_k \bigcup_{n \ge k} G_{n,\alpha(n)}.$$

Since for every k,

$$\lambda(\varphi_1(\alpha)) \le \lambda\Big(\bigcup_{n \ge k} G_{n,\alpha(n)}\Big) \le 2^{-k+1},$$

we obtain $\varphi_1(\alpha) \in \mathcal{N}$.

Assume now that $\varphi_1(\alpha) \subseteq A$. Then $\bigcap_k \bigcup_{n \ge k} G_{n,\alpha(n)} \cap K_A = \emptyset$. Since K_A being compact satisfies the Baire Category Theorem, there exists a k_0 such that $\bigcup_{n \ge k_0} G_{n,\alpha(n)}$ is not dense in K_A. Thus there exists a k_1 such that $U_{k_1} \cap K_A \ne \emptyset$ and

$$U_{k_1} \cap K_A \cap \bigcup_{n \ge k_0} G_{n,\alpha(n)} = \emptyset. \tag{7.17}$$

Let $n \ge k_0$, $n \ge N(A,k_1)$. By (7.17) we have $G_{n,\alpha(n)} \cap U_{k_1} \cap K_A = \emptyset$ and therefore $\alpha(n) \in \Phi(A,k_1,n) \subseteq h(n)$. $\qquad\square$

Lemma [AC] 7.55. *There exist functions*

$$\varphi_2 : \mathcal{M}({}^{\omega}2) \longrightarrow {}^{\omega}\omega, \quad \varphi_2^* : \mathrm{Loc} \longrightarrow \mathcal{M}({}^{\omega}2)$$

such that

$$\varphi_2(B) \in^* h \to B \subseteq \varphi_2^*(h)$$

for any $B \in \mathcal{M}({}^{\omega}2)$ and any $h \in \mathrm{Loc}$.

Proof. By Lemma 7.53 for every n there exists a system $\langle V_{n,k} : k \in \omega \rangle$ of open subsets of U_n such that every dense open set contains as a subset some $V_{n,k}$ and $\bigcap_{k \in X} V_{n,k} \ne \emptyset$ whenever $X \in [\omega]^{\le 2^n}$.

If $B \in \mathcal{M}(^\omega 2)$, then there exists a sequence of closed nowhere dense sets $\langle H_n : n \in \omega \rangle$ such that $B \subseteq \bigcup_n H_n$. We can assume that $H_n \subseteq H_{n+1}$. We set $\varphi_2(B) = \alpha$ where $\alpha(n)$ is the least natural number such that $H_n \cap V_{n,\alpha(n)} = \emptyset$. For $h \in \text{Loc}$ we set

$$\varphi_2^*(h) = {}^\omega 2 \setminus \bigcap_n \bigcup_{m \geq n} \bigcap_{k \in h(m)} V_{m,k}.$$

Since $\emptyset \neq \bigcap_{k \in h(m)} V_{m,k} \subseteq U_m$, one can easily see that $\varphi_2^*(h)$ is meager.

Assume that $\alpha = \varphi_2(B) \in^* h$. Then there exists an n_0 such that $\alpha(n) \in h(n)$ for $n \geq n_0$. Hence, for $n \geq n_0$ we have

$$\bigcup_{m \geq n} \bigcap_{k \in h(m)} V_{m,k} \subseteq \bigcup_{m \geq n} V_{m,\alpha(m)}.$$

Because

$$\bigcup_{m \geq n} V_{m,\alpha(m)} \cap H_n = \emptyset,$$

we obtain

$$\varphi_2^*(h) \supseteq \bigcup_{n \geq n_0} H_n \supseteq B$$

and we are done. $\square$

Theorem [AC] 7.56 (J. Pawlikowski). *The pair of mappings $\langle \varphi_2 \circ \varphi_1, \varphi_1^* \circ \varphi_2^* \rangle$ is a Tukey connection from $\langle \mathcal{M}, \subseteq \rangle$ into $\langle \mathcal{N}, \subseteq \rangle$.*

Proof. The theorem follows from Lemmas 7.54 and 7.55. Actually, if $A \in \mathcal{N}$ and $B \in \mathcal{M}$ are such that $\varphi_2 \circ \varphi_1(B) = \varphi_1(\varphi_2(B)) \subseteq A$ then by Lemma 7.54 we obtain $\varphi_2(B) \in^* \varphi_1^*(A)$ and therefore $B \subseteq \varphi_2^*(\varphi_1^*(A))$ by Lemma 7.55. $\square$

Exercises

7.10 An Elementary Proof of Theorem 7.40
For $n, k \in \omega$ and $s \in {}^n 2$ we write

$$N(s) = |\{i < n : s(i) = 1\}|, \quad S(n, k) = \left\{ s \in {}^n 2 : \left| \frac{N(s)}{n} - \frac{1}{2} \right| > 2^{-k} \right\},$$
$$B_k = \{\alpha \in {}^\omega 2 : (\forall m)(\exists n \geq m)\, \alpha | n \in S(n, k)\}.$$

a) Show that ${}^\omega 2 \setminus A = \bigcup_n B_n$.

b) Show that

$$B_k \subseteq \bigcup_{n \geq m} \bigcup_{s \in S(n,k)} [s]$$

for any m.

c) $\binom{2n}{n+k} = \binom{2n}{n-k} < 2^{2n} \exp(-k^2/4n)$ for any $k \geq 2$.

Hint: Note that $\exp(-x) > 1 - x$ for $0 < x < 1$ and $\binom{2n}{n} < \sum_{k=0}^{2n} \binom{2n}{k} = 2^{2n}$. Then

$$\frac{\binom{2n}{n+k}}{\binom{2n}{n}} < \left(1 - \frac{1}{n}\right) \cdots \left(1 - \frac{k-1}{n}\right) < \exp\left(-\frac{1}{n} - \cdots - \frac{k-1}{n}\right) \leq \exp(-k^2/4n).$$

d) Show that $|S(2n,k)| < 2n2^{2n} \exp(-n2^{-2k})$ for every k and for every $n > 2^{k-1}$.

Hint: Note that $|S(2n,k)|$ is

$$\sum_{i>n+n2^{-k}+1} \binom{2n}{i} + \sum_{i<n-n2^{-k}+1} \binom{2n}{i} = \sum_{|i|>n2^{-k}+1} \binom{2n}{n+i}.$$

If $n > 2^{k-1}$, then the sum contains not more than $2n$ non-zero summands. For each of them we have an estimate

$$\binom{2n}{n+i} < 2^{2n} \exp(-i^2/4n) < 2^{2n} \exp(-n^2 2^{-2k+2}/4n) = 2^{2n} \exp(-n2^{-2k}).$$

e) Show that for any given k and any sufficiently large n

$$|S(n,k)| < n2^n \exp(-n2^{-2k-4}).$$

Hint: For n even the result follows by d). Let $n = 2m + 1$ be odd. If $s \in S(n,k)$, then

$$n2^{-k} < \left|N(s) - \frac{n}{2}\right| \leq |N(s) - N(s|2m)| + |N(s|2m) - m| + \left|m - \frac{n}{2}\right|$$
$$< |N(s|2m) - m| + 2,$$

therefore for sufficiently large n

$$|N(s|2m) - m| > n2^{-k} - 2 > 2m2^{-k-1}.$$

Thus $s|2m \in S(2m, k+1)$. Hence $|S(n,k)| \leq 2|S(2m, k+1)|$.

f) For any $\varepsilon > 0$ and any $k > 0$ there exists an m such that

$$\lambda\left(\bigcup_{n \geq m} \bigcup_{s \in S(n,k)} [s]\right) < \varepsilon.$$

Hint: If $s \in {}^n2$, then $\lambda([s]) = 2^{-n}$. Hence by e) we have

$$\lambda\left(\bigcup_{n=0}^{\infty} \bigcup_{s \in S(n,k)} [s]\right) \leq \sum_{n=0}^{\infty} 2^{-n} \cdot n2^n \exp(-n2^{-2k-4}) = \sum_{n=0}^{\infty} n \cdot \exp(-n2^{-2k-4}) < \infty.$$

Thus, there exists an m such that $\sum_{n=m}^{\infty} n \cdot \exp(-n2^{-2k-4}) < \varepsilon$.

g) $\lambda(B_k) = 0$ for every $k \in \omega$. Therefore Theorem 7.40 holds true.

Hint: Use b), f) and then a).

7.11 [AC] Product of Boolean Algebras

We assume that all considered Boolean algebras are complete. A Boolean algebra $\mathbf{B}$ with complete injections $j_i : \mathbf{B}_i \xrightarrow{1-1} \mathbf{B}$, $i = 1, 2$ is said to be a **product of Boolean algebras** $\mathbf{B}_1, \mathbf{B}_2$ if j_1, j_2 are complete homomorphisms, $\mathbf{B}$ is completely generated by the set $j_1(B_1) \cup j_2(B_2)$ and $j_1(x_1) \wedge j_2(x_2) \neq 0$ for any $x_1 \neq 0$, $x_2 \neq 0$. If the set $\{j_1(x_1) \wedge j_2(x_2) : x_1 \in B_1 \wedge x_2 \in B_2\}$ is dense in $\mathbf{B}$, then the product is called **minimal**.

a) The projections of $\mathbb{T} \times \mathbb{T}$ onto $\mathbb{T}$ induce complete injections of $\mathrm{BAIRE}(\mathbb{T})/\mathcal{M}(\mathbb{T})$ into $\mathrm{BAIRE}(\mathbb{T} \times \mathbb{T})/\mathcal{M}(\mathbb{T} \times \mathbb{T})$ such that $\mathrm{BAIRE}(\mathbb{T} \times \mathbb{T})/\mathcal{M}(\mathbb{T} \times \mathbb{T})$ is a product of two copies of $\mathrm{BAIRE}(\mathbb{T})/\mathcal{M}(\mathbb{T})$.

b) Show that the product of a) is minimal.

c) Similarly, the projections proj_i, $i = 1, 2$ induce complete injections of $\mathcal{L}(\mathbb{T})/\mathcal{N}(\mathbb{T})$ into $\mathcal{L}(\mathbb{T} \times \mathbb{T})/\mathcal{N}(\mathbb{T} \times \mathbb{T})$ such that $\mathcal{L}(\mathbb{T} \times \mathbb{T})/\mathcal{N}(\mathbb{T} \times \mathbb{T})$ is a product of two copies of $\mathcal{L}(\mathbb{T})/\mathcal{N}(\mathbb{T})$.

d) If $F \subseteq \mathbb{T}$ has positive measure, then $X = \{\langle x, y \rangle : x - y \in F\}$ is a non-zero element of Boolean algebra $\mathcal{L}(\mathbb{T} \times \mathbb{T})/\mathcal{N}(\mathbb{T} \times \mathbb{T})$.

e) Let $F \subseteq \mathbb{T}$ be closed nowhere dense of positive measure, X being as in d). Then $\lambda_2(A_1 \times A_2 \setminus X) > 0$ for any $A_1, A_2 \subseteq \mathbb{T}$ of positive measure.
   Hint: Let $\langle U_n : n \in \omega \rangle$ be an open base of $\mathbb{T}$. Set

$$B_i = A_i \setminus \bigcup \{U_n : \lambda(U_n \cap A_i) = 0\}.$$

   By the Steinhaus Theorem 7.16 there exists an open set $C \subseteq B_1 - B_2$. Since C cannot be a subset of F we obtain $B_1 \times B_2 \not\subseteq X$. Thus, there exist n, m such that $\lambda(B_1 \cap U_n) > 0$ and $\lambda(B_2 \cap U_m) > 0$. Note that $\lambda_2(A_1 \times A_2 \setminus X) = \lambda_2(B_1 \times B_2 \setminus X)$ and $(B_1 \cap U_n) \times (B_2 \cap U_m) \subseteq B_1 \times B_2 \setminus X$.

f) Conclude that the product of part c) is not minimal.

7.12 [wAC] Tukey connection from $\langle \mathcal{M}, \subseteq \rangle$ into $\langle \mathcal{N}, \subseteq \rangle$

In proofs of Lemmas 7.54 and 7.55 we have essentially exploited the Axiom of Choice. Some weaker version of obtained results can be shown by using **wAC** only.

a) There exists

$$\psi_2 : {}^{\omega}\omega \longrightarrow \mathcal{N}({}^{\omega}2) \cap G_{\delta}, \quad \psi_2^* : \mathcal{N}({}^{\omega}2) \cap G_{\delta} \longrightarrow \mathrm{Loc}$$

   such that $\psi_2(f) \subseteq A \to f \in^* \psi_2^*(A)$ for any $f \in {}^{\omega}\omega$ and any $A \in \mathcal{N}({}^{\omega}2) \cap G_{\delta}$.

b) There exist

$$\psi_3 : \mathcal{M}({}^{\omega}2) \cap F_{\sigma} \longrightarrow {}^{\omega}\omega, \quad \psi_3^* : \mathrm{Loc} \longrightarrow \mathcal{M}({}^{\omega}2) \cap F_{\sigma}$$

   such that $\psi_3(B) \in^* h \to B \subseteq \psi_3^*(h)$ for any $B \in \mathcal{M}({}^{\omega}2) \cap F_{\sigma}$ and any $h \in \mathrm{Loc}$.

c) The pair of mappings $\langle \psi_3 \circ \psi_2, \psi_2^* \circ \psi_3^* \rangle$ is a Tukey connection from $\langle \mathcal{M} \cap F_{\sigma}, \subseteq \rangle$ into $\langle \mathcal{N} \cap G_{\delta}, \subseteq \rangle$.

7.5   Cichoń Diagram

If we change Diagram 2. of Section 7.2 taking into account Bartoszyński's Theorem 7.51 and inequalities between cardinal invariants of the ideal of meager sets and cardinals $\mathfrak{b}$ and $\mathfrak{d}$, we obtain the **Cichoń Diagram**:

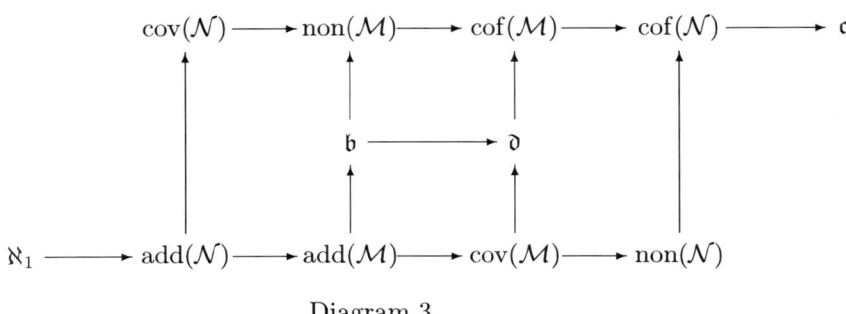

Diagram 3.

The diagram is not more symmetric. As before, an arrow means the inequality between corresponding invariants which can be proved in **ZFC**. All inequalities but those between cardinal invariants of the ideal of meager sets and the cardinals $\mathfrak{b}$ and $\mathfrak{d}$ have been already proved. The remaining inequalities will be proved in this section. Moreover we show

Theorem [AC] 7.57 (A.W. Miller – J. Truss).

$$\mathrm{add}(\mathcal{M}) = \min\{\mathrm{cov}(\mathcal{M}), \mathfrak{b}\}.$$

and

Theorem [AC] 7.58 (D. Fremlin).

$$\mathrm{cof}(\mathcal{M}) = \max\{\mathrm{non}(\mathcal{M}), \mathfrak{d}\}.$$

Let $\mathcal{K}_\sigma(X, \mathcal{O})$, or simply $\mathcal{K}_\sigma(X)$, denote the smallest σ-ideal of subsets of a topological space $\langle X, \mathcal{O} \rangle$ containing the family $\mathcal{K}(X, \mathcal{O})$ of all compact subsets of X. For sake of brevity we shall write $\mathcal{K}_\sigma$ instead of $\mathcal{K}_\sigma(^\omega\omega)$.

For $\alpha \in {}^\omega\omega$ we denote

$$C_\alpha = \{\beta \in {}^\omega\omega : (\forall n)\, \beta(n) \le \alpha(n)\}, \quad C_\alpha^* = \{\beta \in {}^\omega\omega : \beta \le^* \alpha\}.$$

One can easily see that a set C_α^* is a countable union of sets of the form C_γ. Conversely, for any sequence $\langle \beta_n \in {}^\omega\omega : n \in \omega \rangle$ there exists an $\alpha \in {}^\omega\omega$ such that $\bigcup_n C_{\beta_n} \subseteq C_\alpha^*$. By Theorem 3.11 a closed set $A \subset {}^\omega\omega$ is compact if and only if there exists an $\alpha \in {}^\omega\omega$ such that $A \subseteq C_\alpha$. Thus a set A belongs to $\mathcal{K}_\sigma$ if and only if there exists an $\alpha \in {}^\omega\omega$ such that $A \subseteq C_\alpha^*$. Hence, the family $\{C_\alpha^* : \alpha \in {}^\omega\omega\}$ is a base of the σ-ideal $\mathcal{K}_\sigma$.

Theorem [AC] 7.59 (F. Rothberger).

$$\mathrm{add}(\mathcal{K}_\sigma) = \mathrm{non}(\mathcal{K}_\sigma) = \mathfrak{b} \ \text{and} \ \mathrm{cov}(\mathcal{K}_\sigma) = \mathrm{cof}(\mathcal{K}_\sigma) = \mathfrak{d}.$$

Proof. If $B \subseteq {}^\omega\omega$ is unbounded then $B \notin \mathcal{K}_\sigma$ and therefore $\mathrm{non}(\mathcal{K}_\sigma) \leq \mathfrak{b}$. If $\bigcup_{\alpha \in B} C_\alpha^* \notin \mathcal{K}_\sigma$ then B is unbounded and therefore $\mathfrak{b} \leq \mathrm{add}(\mathcal{K}_\sigma)$. The equalities follow by Theorem 7.1, a).

Similarly we can prove the second equalities. If $B \subseteq {}^\omega\omega$ is a dominating family then $\{C_\alpha^* : \alpha \in B\}$ is a base of $\mathcal{K}_\sigma$ and therefore $\mathrm{cof}(\mathcal{K}_\sigma) \leq \mathfrak{d}$. If $\bigcup_{\alpha \in B} C_\alpha^* = {}^\omega\omega$ then B is a dominating family. $\qquad\square$

Since every set $\{\beta \in {}^\omega\omega : \beta \leq \alpha\}$ is nowhere dense, every set C_α^* is meager and hence $\mathcal{K}_\sigma \subseteq \mathcal{M}({}^\omega\omega)$. Thus

Corollary [AC] 7.60.

$$\mathfrak{b} \leq \mathrm{non}(\mathcal{M}) \ \text{and} \ \mathrm{cov}(\mathcal{M}) \leq \mathfrak{d}.$$

Note that by suitable formulation of Theorem 7.59 and Corollary 7.60, the Weak Axiom of Choice **wAC** is enough for a proof of both.

Let ${}^\omega\omega{\uparrow}$ denote the set of all strictly increasing sequences from ${}^\omega\omega$. One can easily see that $\mathfrak{b} = \mathfrak{b}({}^\omega\omega{\uparrow}, \leq^*)$ and $\mathfrak{d} = \mathfrak{d}({}^\omega\omega{\uparrow}, \leq^*)$.

Theorem [AC] 7.61 (T. Bartoszyński). *There is a Tukey connection from $\langle {}^\omega\omega{\uparrow}, \leq^* \rangle$ into $\langle \mathcal{M}({}^\omega 2), \subseteq \rangle$. Hence*

$$\mathfrak{d} \leq \mathrm{cof}(\mathcal{M}), \qquad \mathrm{add}(\mathcal{M}) \leq \mathfrak{b}.$$

Proof. We begin with an observation: if $G \subseteq {}^\omega 2$ is open dense, then for any n there exist $m > n$ and $t \in {}^{m \setminus n}2$ such that $[t] \subseteq G$. Actually, if $s_1, \ldots, s_{2^n}$ is an enumeration of all elements of ${}^n 2$, then one can construct by induction finite functions $t_1, \ldots, t_{2^n}$ such that $[s_i \frown t_1 \frown \cdots \frown t_i] \subseteq G$. Set $t = t_1 \frown \cdots \frown t_{2^n}$.

For $\alpha \in {}^\omega\omega{\uparrow}$ we set

$$\varphi_3(\alpha) = \{\beta \in {}^\omega 2 : (\forall n) \, \beta(\alpha(2n)) = 1\}.$$

Evidently $\varphi_3(\alpha)$ is a closed nowhere dense set.

For a set $A \in \mathcal{M}({}^\omega 2)$ fix a non-decreasing sequence of closed nowhere dense sets $\{H_n\}_{n=0}^\infty$ such that $A \subseteq \bigcup_n H_n$. Since ${}^\omega 2 \setminus H_k$ is open dense, we can define

$$\gamma(0) = 0$$
$$\gamma(k+1) = \min\{n > \gamma(k) : (\exists t)\,(t \in {}^{n \setminus \gamma(k)}2 \wedge [t] \cap H_k = \emptyset\}.$$

We set $\varphi_3^*(A) = \gamma$.

Assume that $\varphi_3(\alpha) \subseteq A$. We want to show that $\alpha \leq^* \gamma = \varphi_3^*(A)$. For every k we fix $t_k \in {}^{\gamma(k+1) \setminus \gamma(k)}2$ such that $[t_k] \cap H_k = \emptyset$ and define

$$\beta(k) = \begin{cases} 1 & \text{if } k = \alpha(2n) \text{ for some } n \text{ or } t_m(k) = 1 \text{ for some } m, \\ 0 & \text{otherwise.} \end{cases}$$

Then $\beta \in \varphi_3(\alpha)$. Therefore there exists a k_0 such that $\beta \in H_k$ for any $k \geq k_0$. Since $[t_k] \cap H_k = \emptyset$, for every $k \geq k_0$ there exists an $i_k \in \gamma(k+1) \setminus \gamma(k)$ such that $\beta(i_k) \neq t_k(i_k)$. If $t_k(i_k) = 1$, then by definition also $\beta(i_k) = 1$, which is impossible. Thus $t_k(i_k) = 0$ and $\beta(i_k) = 1$. By definition of β, for any $k \geq k_0$ there is n_k such that $i_k = \alpha(2n_k)$. Since $i_k < i_{k+1}$ for every k, the sequence $\{n_k\}_{k=k_0}^{\infty}$ is strictly increasing for $k \geq k_0$ (we did not define n_k for $k < k_0$). Thus there exists a k_1 such that $k + 1 \leq 2n_k$ for $k \geq k_1$. Then also $\alpha(k+1) \leq \alpha(2n_k) = i_k < \gamma(k+1)$. Thus $\alpha \leq^* \gamma$. $\qquad\square$

Assume that $\langle V_n : n \in \omega \rangle$ is a decreasing open base of neighborhoods of $0 \in \mathbb{T}$ and $\mathbb{Q} \cap \mathbb{T} = \{r_n : n \in \omega\}$. For a function $\alpha \in {}^{\omega}\omega$, $n \in \omega$ and $\beta \in \mathbb{T}$ we set

$$D_{\alpha}^n = \bigcup_{m \geq n} (r_m + V_{\alpha(m)}),$$

$$W_{\alpha,\beta} = \bigcup_m (\mathbb{T} \setminus (\beta + D_{\alpha}^m)).$$

Then D_{α}^n is an open dense set and $W_{\alpha,\beta}$ is a meager F_{σ} set.

Lemma [wAC] 7.62. *If $H \subseteq \mathbb{T}$ is a meager F_{σ} set, $\beta \notin H - \mathbb{Q}$, then there is a function $\gamma \in {}^{\omega}\omega$ such that for any $\alpha \in {}^{\omega}\omega$ we have*

$$\gamma \leq^* \alpha \rightarrow H \subseteq W_{\alpha,\beta}. \tag{7.18}$$

Proof. Assume that $H = \bigcup_n H_n$, where H_n are closed nowhere dense such that $H_n \subseteq H_{n+1}$. Since $\beta + r_n \notin H$ there exists an m such that $(\beta + r_n + V_m) \cap H_n = \emptyset$. Let $\gamma(n)$ denote the smallest m with this property.

We show that (7.18) holds true. Let $\gamma \leq^* \alpha \in {}^{\omega}\omega$. Consider arbitrary $\delta \in H$. Then there exists an n_0 such that $\gamma(n) \leq \alpha(n)$ and $\delta \in H_n$ for every $n \geq n_0$. Hence $\delta \notin \beta + r_n + V_{\gamma(n)}$ for $n \geq n_0$.

To obtain a contradiction we assume that $\delta \notin W_{\alpha,\beta}$. Then $\delta - \beta \in D_{\alpha}^m$ for any m. By definition of the set $D_{\alpha}^{n_0}$ there exists an integer $n \geq n_0$ such that $\delta - \beta \in r_n + V_{\alpha(n)} \subseteq r_n + V_{\gamma(n)}$, a contradiction $\qquad\square$

Proof of Theorem 7.57. By Theorem 7.1 we have $\mathrm{add}(\mathcal{M}) \leq \mathrm{cov}(\mathcal{M})$ and by Theorem 7.61 we have $\mathrm{add}(\mathcal{M}) \leq \mathfrak{b}$.

Let $\mathcal{A} \subseteq \mathcal{M}$, $|\mathcal{A}| < \mathfrak{b}$, $|\mathcal{A}| < \mathrm{cov}(\mathcal{M})$. We want to show that $|\mathcal{A}| < \mathrm{add}(\mathcal{M})$, i.e., that $\bigcup \mathcal{A}$ is meager.

We can assume that every element of $\mathcal{A}$ is an F_{σ} set. Since $|\mathcal{A}| < \mathrm{cov}(\mathcal{M})$, there exists a real

$$\beta \in \mathbb{T} \setminus \bigcup_{H \in \mathcal{A}} (H - \mathbb{Q}).$$

By Lemma 7.62 for every $H \in \mathcal{A}$ there exists a function $\gamma_H \in {}^{\omega}\omega$ satisfying (7.18). Since cardinality of $\mathcal{A}$ is smaller than $\mathfrak{b}$ there exists a function α such that $\gamma_H \leq^* \alpha$ for every $H \in \mathcal{A}$. Thus $\bigcup \mathcal{A} \subseteq W_{\alpha,\beta}$. $\qquad\square$

Proof of Theorem 7.58. By Theorem 7.1 we have $\mathrm{non}(\mathcal{M}) \leq \mathrm{cof}(\mathcal{M})$ and by Theorem 7.61 we have $\mathfrak{d} \leq \mathrm{cof}(\mathcal{M})$.

Let $A \subseteq \mathbb{T}$ be a non-meager set of cardinality $\mathrm{non}(\mathcal{M})$ and $F \subseteq {}^\omega\omega$ be a dominating family of cardinality $\mathfrak{d}$. If $H \subseteq \mathbb{T}$ is a meager F_σ set, then $A \not\subseteq H - \mathbb{Q}$ and therefore there exists a $\beta \in A \setminus (H - \mathbb{Q})$. Then by Lemma 7.62 there exists a function $\alpha \in F$ such that $H \subseteq W_{\alpha,\beta}$. Thus $\{W_{\alpha,\beta} : \beta \in A \wedge \alpha \in F\}$ is a base of $\mathcal{M}$ of cardinality $\max\{\mathrm{non}(\mathcal{M}), \mathfrak{d}\}$. □

Additivity and cofinality of measure can be characterized in a combinatorial way. We use the obtained result to present such a characterization. Similarly, covering of category and cardinality of the smallest non-meager set can be characterized in a combinatorial way. Theorems 7.63 and 7.66 below can be considered at least as a contribution to measure-category similarity.

We write (compare Exercise 3.12)

$$\ell^1 = \{f \in {}^\omega\mathbb{R} : \sum_{n=0}^{\infty} |f(n)| < \infty\}.$$

The set $\ell^1 \subseteq {}^\omega\mathbb{R}$ can be partially preordered by the eventual domination

$$f \leq^* g \equiv (\exists m)(\forall n)\,(n \geq m \to f(n) \leq g(n)).$$

Our first result is

Theorem [AC] 7.63 (T. Bartoszyński).

$$\mathrm{add}(\mathcal{N}) = \mathfrak{b}(\ell^1, \leq^*) \text{ and } \mathrm{cof}(\mathcal{N}) = \mathfrak{d}(\ell^1, \leq^*).$$

We construct suitable Tukey connections. Instead of ℓ^1 we can consider the set

$$\ell^c = \{\alpha \in {}^\omega\omega : 0 \notin \mathrm{rng}(\alpha) \wedge \sum_{n=0}^{\infty} \frac{1}{\alpha(n)} < \infty\}.$$

One can easily prove that the set $\{\{1/\alpha(n)\}_{n=0}^{\infty} : \alpha \in \ell^c\} \subseteq \ell^1$ is cofinal in ℓ^1. Indeed, if $f \in \ell^1$ take $\alpha \in {}^\omega\omega$ such that $\alpha(n) \leq \max\{1, |1/f(n)|\} < \alpha(n) + 1$. Then $\alpha \in \ell^c$ and $f \leq^* \{1/\alpha(n)\}_{n=0}^{\infty}$. The partial ordering $\leq^*$ on $\{\{1/\alpha(n)\}_{n=0}^{\infty} : \alpha \in \ell^c\}$ as a subset of ℓ^1 corresponds to the inverse ordering $\geq^*$ on $\ell^c \subseteq {}^\omega\omega$. Thus

$$\mathfrak{b}(\ell^1, \leq^*) = \mathfrak{b}(\ell^c, \geq^*) \text{ and } \mathfrak{d}(\ell^1, \leq^*) = \mathfrak{d}(\ell^c, \geq^*).$$

Lemma [AC] 7.64. *There are mappings* $\varphi_4 : \ell^c \longrightarrow {}^\omega\omega$ *and* $\varphi_4^* : \mathrm{Loc} \longrightarrow \ell^c$ *such that*

$$\varphi_4(\alpha) \in^* h \to \alpha \geq^* \varphi_4^*(h) \tag{7.19}$$

for any $\alpha \in \ell^c$ *and* $h \in \mathrm{Loc}$.

Proof. Let $\{s_n : n \in \omega\}$ be a one-to-one enumeration of all functions from a finite subset of ω into $\omega \setminus \{0\}$, i.e.,

$$\{s_n : n \in \omega\} = \{s \in {}^a(\omega \setminus \{0\}) : a \in [\omega]^{<\omega}\}.$$

Let $\alpha \in \ell^c$. We define an increasing function $l \in {}^\omega\omega$ as follows: $l(n)$ is the smallest natural number k such that $\sum_{i=k}^{\infty} 1/\alpha(i) < 2^{-2n-1}$. Then

$$\sum_{i=l(n)}^{l(n+1)-1} \frac{1}{\alpha(i)} < 2^{-2n}.$$

We set $\varphi_4(\alpha) = \gamma$ where $\gamma(n)$ is that m for which $s_m = \alpha | \langle l(n), l(n+1) \rangle$.

Now, let $h \in \text{Loc}$. We write

$$B_{n,k} = \left\{ m \in h(k) : n \in \text{dom}(s_m) \wedge \sum_{i \in \text{dom } s_m} 1/s_m(i) < 2^{-2k} \right\},$$

$$A_n = \left\{ s_m(n) : m \in \bigcup_k B_{n,k} \right\}.$$

We set $\varphi_4^*(h) = \beta$ where

$$\beta(n) = \begin{cases} \min A_n & \text{if } A_n \text{ is non-empty,} \\ 2^n & \text{otherwise.} \end{cases}$$

If $A_n \neq \emptyset$, then there exists an $m_n \in \bigcup_k B_{n,k}$ such that $\beta(n) = s_{m_n}(n)$. We set

$$C_k = \{n : m_n \in B_{n,k}\}, \qquad C_{k,m} = \{n \in C_k : m_n = m\}.$$

Then

$$C_k = \bigcup_m C_{k,m} = \bigcup_{m \in h(k)} C_{k,m}.$$

If $n \in C_{k,m}$, then $\beta(n) = s_m(n)$ and $m \in \bigcup_i B_{i,k}$. Hence

$$\sum_{n \in C_{k,m}} \frac{1}{\beta(n)} \leq \sum_{n \in \text{dom } s_m} \frac{1}{s_m(n)} < 2^{-2k}.$$

Thus

$$\sum_{n \in C_k} \frac{1}{\beta(n)} < \frac{|h(k)|}{2^{2k}} \leq 2^{-k}.$$

If $n \notin \bigcup_k C_k$ then by definition $\beta(n) = 2^n$. Therefore

$$\sum_{n=0}^{\infty} \frac{1}{\beta(n)} < \sum_{n=0}^{\infty} 2^{-n} + \sum_{n=0}^{\infty} \frac{|h(n)|}{2^{2n}} \leq \sum_{n=0}^{\infty} 2^{-n} + \sum_{n=0}^{\infty} 2^{-n} < \infty.$$

Thus $\beta \in \ell^c$.

We want to show that (7.19) holds true. Assume that $\gamma = \varphi_4(\alpha) \in^* h$. Then there exists a k_0 such that $d(k) \in h(k)$ for every $k \geq k_0$. Let $d(k) = m$. Then by the definition $s_m = f|[l(k), l(k+1))$. Assume that $i \in [l(k), l(k+1))$. Then $m \in B_{i,k}$ and therefore $s_m(i) \in A_i$. Thus $d(i) = \min A_i \leq s_m(i) = f(i)$. Hence $d(i) \leq f(i)$ for any $i \geq l(k_0)$.   □

It is easy to show that a set $A \subseteq {}^\omega 2$ has measure zero if and only if there exists a sequence $\langle t_n \in {}^{<\omega}2 : n \in \omega \rangle$ such that

$$A \subseteq \bigcap_m \bigcup_{n \geq m} [t_n] \quad \text{and} \quad \sum_{n=0}^{\infty} \lambda([t_n]) < \infty. \tag{7.20}$$

A similar assertion is true for subsets of $\mathbb{R}$ with open intervals with rational endpoints instead of $[t_n]$.

Lemma [AC] 7.65. *There exist mappings $\varphi_5 : \mathcal{N}({}^\omega 2) \longrightarrow \ell^1$ and $\varphi_5^* : \ell^1 \longrightarrow \mathcal{N}({}^\omega 2)$ such that*

$$\varphi_4(A) \leq^* \alpha \rightarrow A \subseteq \varphi_4^*(\alpha)$$

for any $A \in \mathcal{N}$ and any $\alpha \in \ell^1$.

Proof. Let $\{s_n : n \in \omega\}$ be a one-to-one enumeration of the set ${}^{<\omega}2$. By (7.20), if $A \in \mathcal{N}({}^\omega 2)$, then there exists a sequence $\{t_n\}_{n=0}^{\infty}$ of elements of ${}^{<\omega}2$ such that $A \subseteq \bigcap_m \bigcup_{n \geq m}[t_n]$ and $\sum_{n=0}^{\infty} \lambda([t_n]) < \infty$. We can assume that $t_n \neq t_m$ for any $n \neq m$. We set $\varphi_4(A) = \alpha$, where

$$\alpha(n) = \begin{cases} \lambda([s_n]) & \text{if } s_n = t_m \text{ for some } m, \\ 0 & \text{otherwise.} \end{cases}$$

Since $\sum_{n=0}^{\infty} \alpha(n) = \sum_{n=0}^{\infty} \lambda([t_n])$ we have $\alpha \in \ell^1$.
   For $\alpha \in \ell^1$ we set

$$\varphi_4^*(\alpha) = \bigcap_m \bigcup_{n \geq m} \{[s_n] : \lambda([s_n]) \leq \alpha(n)\}.$$

Evidently $\varphi_4^*(f) \in \mathcal{N}$.
   Assume that $\varphi_4(A) = y$ and $y(n) \leq \alpha(n)$ for any $n \geq n_0$. If $\beta \in A$, then for any n there exists a $k_n \geq n$ and an l_n such that $\beta \in [t_{k_n}] = [s_{l_n}]$. Since $\sum_{n=0}^{\infty} \lambda([t_n]) < \infty$, any $s \in {}^{<\omega}2$ can be repeated in the sequence $\{t_n\}_{n=0}^{\infty}$ only finitely many times, therefore the sequence $\{l_n\}_{n=0}^{\infty}$ is unbounded. Thus for any m there exists a sufficiently large n such that $\beta \in [t_{k_n}] = [s_{l_n}]$ and $l_n \geq \max\{n_0, m\}$. Since $\lambda([s_{l_n}]) = y(l_n) \leq \alpha(l_n)$ we obtain $\beta \in \varphi_4^*(\alpha)$.   □

Proof of Theorem 7.63. According to Lemmas 7.54 and 7.64 the pair of mappings $\langle \varphi_4 \circ \varphi_3, \varphi_3^* \circ \varphi_4^* \rangle$ is a Tukey connection from $\langle \ell^c, \geq^* \rangle$ into $\langle \mathcal{N}, \subseteq \rangle$. Thus by Theorem 5.43 we obtain

$$\text{add}(\mathcal{N}) = \mathfrak{b}(\mathcal{N}) \leq \mathfrak{b}(\ell^c, \geq^*) = \mathfrak{b}(\ell^1, \leq^*)$$

and
$$\mathfrak{d}(\ell^1, \leq^*) = \mathfrak{d}(\ell^c, \geq^*) \leq \mathfrak{d}(\mathcal{N}) = \mathrm{cof}(\mathcal{N}).$$

By Lemma 7.65 the pair of mappings φ_4 and φ_4^* is a Tukey connection from $\mathcal{N}$ into ℓ^1. Hence in both cases we have the equalities. $\qquad\qquad\square$

A careful reader can realize that the constructions, say of mappings φ_2 and φ_3^*, are very close. Even the mapping φ_2^* may be considered as a restriction of the adequately modified mapping φ_3. However for practical reasons, i.e., to obtain a proof as simple as possible, we defined them independently. A reader can easily find a common language for dealing with them.

Two reals $\alpha, \beta \in {}^\omega\omega$ are called **infinitely equal** if $\alpha(n) = \beta(n)$ for infinitely many n. If α, β are not infinitely equal, i.e., if $\alpha(n) \neq \beta(n)$ for all but finitely many n, then α, β are called **eventually different**. If $A \subseteq \omega$ is an infinite set, then we define similarly that α, β are **infinitely equal on** A if $\alpha(n) = \beta(n)$ for infinitely many $n \in A$. Similarly we can define that α, β are **eventually different on** A.

The next result is a combinatorial characterization of $\mathrm{cov}(\mathcal{M})$ and $\mathrm{non}(\mathcal{M})$ in a similar way as those of $\mathrm{cov}(\mathcal{N})$ and $\mathrm{non}(\mathcal{N})$ in Theorem 7.63. One implication of the theorem will be used and also proved in Section 8.4 as Lemma 8.112. A complete proof is presented in the exercises.

Theorem [AC] 7.66.
$$\mathrm{cov}(\mathcal{M}) = \tag{7.21}$$
$$\min\{|X| :\ X \subseteq {}^\omega\omega \wedge (\forall \alpha \in {}^\omega\omega)(\exists \beta \in X)\,(\alpha, \beta\ \textit{are eventually different})\},$$
$$\mathrm{non}(\mathcal{M}) = \tag{7.22}$$
$$\min\{|X| :\ X \subseteq {}^\omega\omega \wedge (\forall \alpha \in {}^\omega\omega)(\exists \beta \in X)\,(\alpha, \beta\ \textit{are infinitely equal})\}.$$

Exercises

7.13 [AC] A Combinatorial Characterization of $\mathrm{add}(\mathcal{N})$ **and** $\mathrm{cof}(\mathcal{N})$

   a) If $\kappa < \mathrm{add}(\mathcal{N})$, $\langle f_\xi \in {}^\omega\omega : \xi < \kappa \rangle$, then there exists a function $h \in \mathrm{Loc}$ such that $f_\xi \in^* h$ for each $\xi < \kappa$.
     Hint: Use Lemma 7.54.

   b) Assume that for any family $\langle f_\xi \in {}^\omega\omega : \xi < \kappa \rangle$ there exists a function $h \in \mathrm{Loc}$ such that $f_\xi \in^* h$ for each $\xi < \kappa$. Then $\kappa < \mathrm{add}(\mathcal{N})$.
     Hint: Use Lemmas 7.64 and 7.65.

   c) $\kappa < \mathrm{add}(\mathcal{N})$ if and only if for any family $\langle f_\xi \in {}^\omega\omega : \xi < \kappa \rangle$ there exists a function $h \in \mathrm{Loc}$ such that $f_\xi \in^* h$ for each $\xi < \kappa$.

   d) Prove the dual assertion: $\kappa < \mathrm{cof}(\mathcal{N})$ if and only if for any family $\langle h_\xi \in \mathrm{Loc} : \xi < \kappa \rangle$ there exists an $f \in {}^\omega\omega$ such that $f \notin^* h_\xi$ for any $\xi < \kappa$.

7.14 [AC] Chopped Reals

A couple $\langle \alpha, \mathbf{I} \rangle$, where $\alpha \in {}^\omega 2$ and $\mathbf{I} \in \mathcal{IP}$ is an interval partition (see Exercise 5.24), is called a **chopped real**. A real $\beta \in {}^\omega 2$ **matches** a chopped real $\langle \alpha, \mathbf{I} \rangle$ if $\alpha|I = \beta|I$ for infinitely many intervals $I \in \mathbf{I}$. We write

$$M(\alpha, \mathbf{I}) = \{\beta \in {}^{\omega}2 : \beta \text{ matches } \langle \alpha, \mathbf{I} \rangle \}.$$

a) $M(\alpha, \mathbf{I})$ is a G_{δ} dense set.

b) If $G \subseteq [0,1]$ is G_{δ} dense, then there exists a chopped real $\langle \alpha, \mathbf{I} \rangle$ such that $M(\alpha, \mathbf{I}) \subseteq G$.
   Hint: Assume that $G = \bigcap_n G_n$, $G_{n+1} \subseteq G_n$ are open. Construct an interval partition $\langle I_n : n \in \omega \rangle$ and a real $\alpha \in {}^{\omega}2$ as follows. If I_i are defined for $i < n$ and values $\alpha(j)$ for $j < k$, where k is the smallest integer not in $\bigcup_{i<n} I_i$, then there exists an $m > k$ and an $s \in {}^{m \setminus k}2$ such that $[s] \subseteq G_n$ – see the proof of Theorem 7.61. Set $I_n = [k, m)$ and $\alpha(j) = s(j)$ for $k \le j < m$.

c) $M(\alpha, \mathbf{I}) \subseteq M(\beta, \mathbf{J})$ if and only if

$$\mathbf{I} \text{ dominates } \mathbf{J} \wedge \alpha | J = \beta | J \text{ for all but finitely many } J \in \mathbf{J} \qquad (7.23)$$

holds true.

d) Show that

$$\mathrm{cov}(\mathcal{M}) = \min\{|X| : X \subseteq {}^{\omega}2 \times \mathcal{IP} \wedge \bigcap_{\langle \alpha, \mathbf{I} \rangle \in X} M(\alpha, \mathbf{I}) = \emptyset \}.$$

e) Show that

$$\mathrm{cof}(\mathcal{M}) = \min\{|X| : X \subseteq {}^{\omega}2 \times \mathcal{IP} \wedge (\forall \langle \beta, \mathbf{J} \rangle)(\exists \langle \alpha, \mathbf{I} \rangle \in X)\,((7.23) \text{ holds true})\}.$$

7.15 [AC] A Proof of (7.21)
Let

$$\kappa = \min\{|X| : X \subseteq {}^{\omega}\omega \wedge (\forall \alpha \in {}^{\omega}\omega)(\exists \beta \in X)\,(\alpha, \beta \text{ are eventually different})\}.$$

Fix an interval partition $\mathbf{J} = \langle J_n : n \in \omega \rangle$ and consider the set

$$\mathcal{H} = \{\{\alpha | J_i \cup J_{i+1} : i \in A\} : A \in [\omega]^{<\omega} \wedge (\forall k)\,(i \in A \rightarrow k + 1 \notin A) \wedge \alpha \in {}^{\omega}2\}.$$

If $\mathbf{I} = \langle I_n : n \in \omega \rangle$ is an interval partition such that $\mathbf{J}$ is not dominated by $\mathbf{I}$, $\alpha \in {}^{\omega}2$, we define a function $f_{\mathbf{I}, \alpha} : \omega \longrightarrow \mathcal{H}$ as follows. Since $\mathbf{J}$ is not dominated by $\mathbf{I}$, by Exercise 5.24, a) there exist infinitely many integers m such that there exists k with $I_k \subseteq J_m \cup J_{m+1}$ and I_k is not a subinterval of any of J_m, J_{m+1}. For any n take a set A_n of such integers of cardinality $2n + 1$ and such that $k \in A_n \rightarrow k + 1 \notin A_n$. Set

$$f_{\mathbf{I}, \alpha}(n) = \{\alpha | J_m \cup J_{m+1} : m \in A_n\}.$$

If $g : \omega \longrightarrow \mathcal{H}$ is such that $g(n)$ has $2n + 1$ elements for each n, we define a real $\tilde{g} \in {}^{\omega}2$ by induction as follows. If $\tilde{g}(i)$, $i < n$ is already defined on at most $2n$ intervals from $\mathcal{J}$, then there exists $s \in g(n)$ such that the interval $\mathrm{dom}(s)$ is disjoint from them. Extend α to agree with s on $\mathrm{dom}(s)$. The other values of α can be chosen arbitrarily.

a) If $g : \omega \longrightarrow \mathcal{H}$ is infinitely equal to $f_{\mathbf{I}, \alpha}$ and such that $g(n)$ has $2n + 1$ elements for each n, then $\tilde{g}$ matches $\langle \alpha, \mathbf{I} \rangle$.

b) Note that $\kappa \le \mathfrak{d}$.

c) Show that $\kappa \leq \mathrm{cov}(\mathcal{M})$.

   Hint: Let $X \subseteq {}^{\omega}2 \times \mathcal{IP}$ be of cardinality $< \kappa$. By b) and Exercise 5.24, c) there exists an interval partition $\mathbf{J}$ which is not dominated by any $\mathbf{I}$ in X. This partition will play the rôle of $\mathbf{J}$ used in definitions. Then there exists a function $g : \omega \longrightarrow \mathcal{H}$ infinitely equal to every $f_{\mathbf{I},\alpha}$, $\langle \alpha, \mathbf{I} \rangle \in X$. We can assume that $|g(n)| = 2n + 1$ for each n. Thus $\bigcap_{\langle \alpha, \mathbf{I} \rangle \in X} M(\alpha, \mathbf{I}) \neq \emptyset$. Consequently $|X| < \mathrm{cov}(\mathcal{M})$.

d) Show that $\kappa \geq \mathrm{cov}(\mathcal{M})$.

   Hint: See Lemma 8.112.

7.16 [AC] **Proof of** (7.22)

a) If α, β are infinitely equal, then $\beta \in M(\alpha, \mathbf{I})$.

b) If $X \subseteq {}^{\omega}\omega$ is such that for every $\alpha \in {}^{\omega}\omega$ there exists a $\beta \in X$ infinitely equal to α, then $X \notin \mathcal{M}$.

   Hint: If X were meager, then $X \cap M(\alpha, \mathbf{I}) = \emptyset$ for some chopped real $\langle \alpha, \mathbf{I} \rangle$. Since α is infinitely equal to some $\beta \in X$, we have a contradiction with $\beta \in M(\alpha, \mathbf{I})$.

c) If $X \subseteq {}^{\omega}\omega$ is not meager, then for every $\alpha \in {}^{\omega}\omega$ there exists a $\beta \in X$ infinitely equal to α.

   Hint: For every chopped real $\langle \alpha, \mathbf{I} \rangle$ there exists a $\beta \in X \cap M(\alpha, \mathbf{I})$. Then β is infinitely equal to α.

Historical and Bibliographical Notes

Polish mathematicians in 1920s have realized that there exists some kind of duality between measure and category. Theorem 7.33 was proved by K. Kuratowski and S. Ulam [1932]. On page 254, the authors have noted that most results of the paper can be proved for measure replacing the corresponding words and that this fact has been already announced by E Szpilrajn-Marczewski [1929]. W. Sierpiński [1934a], p.276 says that "we know several theorems on the meager sets which remain true if one replaces in their statement the meager sets by the sets of measure zero (or inversely)" and assuming **CH** proves that there exists a bijection with the property of Theorem 7.29. Then P. Erdős [1943] has shown that one can assume that the bijection is equal to its inverse.

The investigation of cardinal invariants began in two different ways. K. Kunen and A.W. Miller [1981] studied so-called "Kunen – Miller's chart". In our terminology, they have considered which possibilities of distribution of values of cardinal invariants of measure and category are possible in **ZFC** and $\aleph_1 < \mathfrak{c}$. L. Bukovský [1977] introduced the cardinal invariants of measure and category (maybe in an unhappy notation) and started to investigate the relationships between them. The study was inspired by R. Solovay's talk [1970a]. The notion of a thin set has been implicitly used by J. Arbault [1952] and explicitly was introduced by L. Bukovský [1998]. Most results of Section 7.1 are rather folklore. For Lebesgue measure, Theorem 7.11 was announced already by N.N. Luzin in [1917] and for the Baire Property in [1927]. E. Szpilrajn-Marczewski [1933] proved the results of Exercise 7.4. Exercise 7.5 contains results of W. Kulaga [2006].

Theorems 7.17–7.18, Corollary 7.19, and Theorem 7.20 contain a contemporary treating of classical results by G. Vitali [1905], H. Steinhaus [1920] and F. Bernstein [1908]. Nobody was insightful enough to ask whether the statements of Corollaries 7.24 and 7.25 hold true. K. Kuratowski [1976] (devoted to P.S. Alexandroff on his 80th birthday) noted the simple fact that the continuum hypothesis implies both of them. Then L. Bukovský [1979a] proved both Corollaries for disjoint families by metamathematical methods using fine properties of models of set theory, especially by using a so-called precipitous ultrafilter. R. Solovay proved Corollary 7.24 independently by using some sophisticated facts concerning real-valued measurable cardinals (unpublished). Another independent proof of Corollary 7.24 was presented by D. Fremlin [1987]. K. Prikry independently proved similar results (unpublished). The presented proof follows the idea of J. Brzuchowski, J. Cichoń, E. Grzegorek and C. Ryll-Nardzewski [1979].

Theorem 7.27, proved by F. Rothberger [1938a], should be considered as one of the first results concerning cardinal invariants of the real line. The part a) of Theorem 7.28 is attributed by J. Cichoń [1983] to J. Brzuchowski and part b) was proved by P. Vojtáš [1989].

Results of Exercises 7.6 and 7.7 are mostly due to G. Hamel [1905].

R. Baire [1899] proved the latter part of Theorem 7.30. The former part was proved for real functions by O. Nikodym [1929] and the general form by K. Kuratowski [1930]. N.N. Luzin proved Theorem 7.31. Theorems 7.36 and 7.38 are folklore, however the former one is sometimes attributed to A.N. Kolmogoroff. Corollary 7.39 was proved by W. Sierpiński.

Theorem 7.41 was proved by D.T. Egoroff [1911]. Theorem 7.43 is contained in J. Dieudonné [1969]. For the results concerning Boolean algebra consult R. Sikorski [1964]. Theorem 7.49 was proved by Z. Piotrowski and A. Szymański [1987].

T. Bartoszyński [1984] proved Theorem 7.51 and a bit later J. Raisonnier and J. Stern [1985] proved Theorem 7.52. Then J. Cichoń, investigating the ideas presented by L. Bukovský [1977], obtained the complete relationship picture called by D.H. Fremlin [1984a] "Cichoń's Diagram". J. Pawlikowski [1985] showed Theorem 7.56. Our presentation essentially follows D.H. Fremlin [1984a]. Exercise 7.11 follows results of L. Bukovský [1977].

The inequality ≥ of Theorem 7.57 was implicitly proved by J. Truss [1977] and A.W. Miller [1981] made it explicit including the equality. Theorem 7.58 as a dual result to the preceding result was proved by D.H. Fremlin [1984a]. The statements of Theorem 7.59 are essentially contained in F. Rothberger [1941]. T. Bartoszyński [1983] showed Theorem 7.61. Theorem 7.63 is one of the main results by T. Bartoszyński [1984]. The results contained in Theorem 7.66 are due to T. Bartoszyński [1987] and A.W. Miller [1982]. The proofs presented in Exercises 7.14, 7.15 and 7.16 follow essentially those given by A. Blass [2010].

Chapter 8

Special Sets of Reals

special *adj.* **1.** of a distinct or particular kind or character; ... **3.** having a specific or particular function, purpose, etc.; **4.** distinguished from what is ordinary or usual; **5.** extraordinary; exceptional; ...

Webster's College Dictionary [1990], p. 1284.

In this chapter we shall work in **ZFC**, i.e., we assume that the Axiom of Choice holds true. We shall investigate subsets of a Polish space, especially subsets of the real line with some exceptional, or maybe extremal, properties: small, with nice subsets, with good properties of sequences of real continuous functions on it, good covering properties and also those related to harmonic analysis. According to the extremeness of their properties, one would prefer that some of them do exist (since they are nice) or one would prefer that some of them do not exist (since they are at least strange). As we shall see later, **ZFC** is not strong enough to answer several questions concerning properties of such sets. However, it turned out that there are close relationships among them, which we shall investigate.

8.1 Small Sets

We have introduced the notions of an ideal and a σ-ideal as a tool for measuring smallness of sets. Till now, the main examples of families of small sets were the σ-ideals of measure zero sets, of meager sets, and that of $\mathcal{K}_\sigma$ sets. However, the property "to be meager" is not an internal topological property in any extension **T** of **ZF**. From a similar point of view neither is the property "to have measure zero" an internal property. If we consider the Cantor middle-third set $\mathbb{C}$ as a subset of the "true" real line $\mathbb{R}$ then $\mathbb{C}$ is small in both senses: $\mathbb{C}$ has Lebesgue measure zero and is nowhere dense in $\mathbb{R}$. However when we identify the real line with the set $^\omega 2$, then the set $\mathbb{C}$ cannot be small in any of the two senses. Therefore we shall look for internal topological properties of smallness. Anyway, since not all of the properties which we will introduce will be internal, in the affirmative case we shall state it explicitly (if it is not introduced as a property of a topological space).

Let X be a topological space. A set $A \subseteq X$ is called **perfectly meager** if for any perfect set $P \subseteq X$ the intersection $A \cap P$ is meager in the subspace P. Thus A is perfectly meager if for any perfect set P there are closed sets F_n such that $A \cap P \subseteq \bigcup_n F_n$ and the inclusion $P \cap U \subseteq F_n$ implies $P \cap U = \emptyset$ for any $n \in \omega$ and any open set U. By definition, if X is a perfect Polish space, then any perfectly meager subset of X is meager in X. Moreover, the family $\mathcal{PM}(X)$ of perfectly meager subsets of X is a σ-ideal. If X has an isolated point, then a perfectly meager set need not be meager in X.

We show that the property "to be perfectly meager" is internal for the class of topological spaces with a countable base in the theory **ZFW**.

Theorem [wAC] 8.1. *Let $\langle X, \mathcal{O} \rangle$ be a topological space with a countable base. A set $A \subseteq X$ is perfectly meager in $\langle X, \mathcal{O} \rangle$ if and only if the set A is perfectly meager in $\langle A, \mathcal{O}|A \rangle$.*

Proof. Let $A \subseteq X$ be perfectly meager in X. Assume that $Q \subseteq A$ is a perfect set in A, i.e., Q is closed in A and does not have an isolated point. Then there exists a perfect set $P \subseteq X$ such that $Q = A \cap P$ (take $P = \overline{Q}$ closure in X). By definition, there are closed (in X) sets $\langle F_n : n \in \omega \rangle$ such that $A \cap P \subseteq \bigcup_n F_n$ and F_n are nowhere dense in P, i.e., if U is open in X and $U \cap P \subseteq F_n$ for some n, then also $U \cap P = \emptyset$. Assume that U is open and $Q \cap U \subseteq A \cap F_n$ for some n. Then $Q \cap U \subseteq \overline{Q \cap U} \subseteq F_n$ and therefore $U \cap Q \subseteq U \cap P = \emptyset$. Hence $F_n \cap Q$ is nowhere dense in Q.

Let A be perfectly meager in A. Assume that P is a perfect subset of X. By Theorem 1.24 we have $A \cap P = Q \cup R$, where Q is perfect in A and R is countable. Evidently $R \subseteq P$ is meager in P. Thus we have to show that Q is meager in P. Since Q is meager in A, there are closed sets $\langle F_n : n \in \omega \rangle$ such that $Q \subseteq \bigcup_n F_n$ and for any open subset U of X and any natural number n, the inclusion $U \cap A \subseteq F_n$ implies $U \cap A = \emptyset$. Let $V \supseteq R$ be open and such that $V \cap A = R$. Set $C_n = F_n \setminus V$. Then $Q \subseteq \bigcup_n C_n$. Suppose, to get a contradiction, that U is an open set such that $\emptyset \neq U \cap P \subseteq C_n$ for some n. Then $U \cap P \cap V = \emptyset$ and therefore $U \cap Q \neq \emptyset$. Since $U \cap Q \subseteq U \cap A \subseteq F_n$, we have a contradiction. $\square$

In a certain sense a dual notion is the notion of a universal measure zero set. Let $\langle X, \mathcal{O} \rangle$ be a perfectly normal topological space. A set $A \subseteq X$ has **universal measure zero** if for any finite diffused Borel measure μ on X we have $\mu^*(A) = 0$, i.e., $\mu(B) = 0$ for some Borel $B \supseteq A$. Again, the family $\mathcal{UN}(X)$ of universal measure zero subsets of X is a σ-ideal. Similarly to Theorem 8.1 by Theorem 4.11 we have an equivalent definition which enables us to consider the property "to have universal measure zero" as internal for the class of separable metric spaces in the theory **ZFW**.

Theorem [wAC] 8.2. *Assume that $\langle X, \mathcal{O} \rangle$ is a separable metric space. A subset $A \subseteq X$ has universal measure zero if and only if there exists no finite diffused Borel measure on A.*

Proof. Assume that $A \subseteq X$ has universal measure zero and μ is a diffused finite Borel measure on A. We can extend the measure μ to a diffused finite Borel measure $\bar{\mu}$ on X by setting $\bar{\mu}(B) = \mu(A \cap B)$ for any Borel subset $B \subseteq X$. Then $\mu(A) = \bar{\mu}^*(A) = 0$.

Now assume that μ is a diffused finite Borel measure on X. We can assume that the measure μ is complete. If $\mu^*(A) = \nu(A) \neq 0$, then by Theorem 4.11 the restriction $\nu = \mu^* | \mathrm{BOREL}(A)$ is a diffused finite Borel measure on A. Therefore $\mu^*(A) = \nu(A) = 0$. $\square$

We note the following. If $A \subseteq X$ has universal measure zero then by Lemma 4.6 we have $\mu^*(A) = 0$ for any diffused σ-finite Borel measure as well. Also every diffused σ-finite Borel measure μ on A is trivial.

Of course, every countable set is perfectly meager and has universal measure zero. However, we show that there exists an uncountable set that is perfectly meager and has universal measure zero.

Theorem [AC] 8.3. *In any perfect Polish space there exists a perfectly meager universal measure zero subset of cardinality $\aleph_1$.*

Proof. Let X be a perfect Polish space. By Corollary 6.51 there exists a co-analytic non-analytic subset $A \subseteq X$. Then $A = \bigcup_{\xi < \omega_1} A_\xi$, where $\langle A_\xi : \xi < \omega_1 \rangle$ are Borel sets, the constituents of A. Since A is not analytic, there exists arbitrarily large ξ such that $A_\xi \neq \emptyset$. Let E be a choice set for the system $\{A_\xi : \xi < \omega_1\}$, i.e., $|E \cap A_\xi| = 1$ provided that $A_\xi \neq \emptyset$. Evidently $|E| = \aleph_1$. We show that E is perfectly meager and has universal measure zero.

Let $P \subseteq X$ be a perfect set. Then by Theorem 7.11 the set $A \cap P$ has the Baire Property in P, i.e., there exists a G_δ set G such that $G \cap P \subseteq A \cap P$ and $(A \cap P) \setminus G$ is meager in P. By Boundedness Theorem 6.54 there exists a $\xi_0 < \omega_1$ such that $G \cap P \subseteq \bigcup_{\xi < \xi_0} A_\xi$. Thus $E \cap G \cap P$ is countable. Since

$$E \cap P \subseteq (E \cap G \cap P) \cup ((A \cap P) \setminus G),$$

$E \cap P$ is meager in P.

Now let μ be a diffused finite Borel measure on X. By Lemma 4.4 the measure μ is regular. By Theorem 7.11 the set A is measurable, therefore there exists an F_σ set $F \subseteq A$ such that $\mu(A \setminus F) = 0$. As above, by Boundedness Theorem 6.54 there exists a $\xi_0 < \omega_1$ such that $F \subseteq \bigcup_{\xi < \xi_0} A_\xi$. Then $F \cap E$ is countable. Since

$$E \subseteq (F \cap E) \cup (A \setminus F),$$

the theorem follows. $\square$

Using the idea of the proof of Theorem 8.3 we show

Theorem [AC] 8.4. *Assume that $\mathfrak{c} > \aleph_1$. Then for any $\mathbf{\Pi}_1^1$-subset $A \subseteq X$ of a Polish space X the following are equivalent:*

  a) *$|A| \leq \aleph_1$.*
  b) *A is perfectly meager.*
  c) *A has universal measure zero.*

Proof. We follow the proof of Theorem 8.3. If $|A| = \aleph_1$ then every constituent, being Borel, is countable and therefore there exists arbitrarily large ξ such that $A_\xi \neq \emptyset$. Instead of taking a choice set we let $E = A$ and we obtain that A is perfectly meager and has universal measure zero.

Vice versa, if A is perfectly meager or has universal measure zero, then A cannot contain a perfect subset. Thus every constituent is countable. So $|A| \leq \aleph_1$. $\qquad\square$

We can improve Theorem 8.3 in another direction. We need to know more on both introduced notions, so we start with auxiliary results.

Lemma 8.5. *Let μ be a diffused Borel measure on X, X being a perfect compact Polish space. Then*

$$(\forall \varepsilon > 0)(\exists \delta > 0)(\forall U \subseteq X)\left((U \ open \wedge \operatorname{diam}(U) < \delta) \to \mu(U) < \varepsilon\right). \qquad (8.1)$$

Proof. Let $\varepsilon > 0$. Since the measure of a one-point set is zero, there exists an n_x such that $\mu(\operatorname{Ball}(x, 2^{-n_x})) < \varepsilon$ (to avoid any use of choice we take the minimal integer n_x with this property). Since X is compact there exist finitely many points $x_0, \ldots, x_k \in X$ such that X is covered by $\operatorname{Ball}(x_0, 2^{-n_{x_0}-1}), \ldots, \operatorname{Ball}(x_k, 2^{-n_{x_k}-1})$. Take $\delta = \min\{2^{-n_{x_i}-1} : i = 0, \ldots, k\}$.

Let $\operatorname{diam}(U) < \delta$, $U \neq \emptyset$ being an open set. Then $U \cap \operatorname{Ball}(x_i, 2^{-n_{x_i}-1}) \neq \emptyset$ for some $i = 0, \ldots, k$ and therefore $U \subseteq \operatorname{Ball}(x_i, 2^{-n_{x_i}})$. So we obtain $\mu(U) < \varepsilon$. $\qquad\square$

Theorem 8.6. *A set $A \subseteq [0, 1]$ has universal measure zero if and only if $f(A)$ has Lebesgue measure zero for any homeomorphism $f : [0, 1] \xrightarrow[\text{onto}]{1-1} [0, 1]$.*

Proof. If $f : [0, 1] \longrightarrow [0, 1]$ is a homeomorphism and the image $f(A)$ does not have Lebesgue measure zero, one easily constructs a diffused Borel measure μ on $[0, 1]$ such that $\mu^*(A) > 0$ by setting $\mu(B) = \lambda(f(B))$ for any Borel set B (λ is the Lebesgue measure).

Assume now that μ is a diffused finite Borel measure on $[0, 1]$ such that $\mu^*(A) > 0$. We want to find a homeomorphism f such that $\lambda^*(f(A)) > 0$. Evidently we can assume that $\mu([0, 1]) = 1$. Moreover, we can assume that $\mu((a, b)) > 0$ for any $a < b$, since we could eventually replace μ by $1/2(\mu + \lambda)$. By Lemma 8.5 the increasing function $f : [0, 1] \longrightarrow [0, 1]$ defined as $f(x) = \mu([0, x])$ is continuous and onto $[0, 1]$, thus f is a homeomorphism. Since for any interval $(a, b) \subseteq [0, 1]$ we have $\mu(f^{-1}((a, b))) = \lambda((a, b))$, by Theorem 4.10 we obtain $\mu(f^{-1}(B)) = \lambda(B)$ for any Borel set $B \subseteq [0, 1]$. Thus $\lambda^*(f(A)) = \mu^*(A) > 0$. $\qquad\square$

Lemma [wAC] 8.7. *Assume that $S \subseteq P(X)$ is a σ-algebra with a countable set of σ-generators such that S separates points of X, i.e., for any $x, y \in X$, $x \neq y$ there exists a set $A \in S$ such that $x \in A$ and $y \notin A$.*

a) *There exists an injection $c : X \xrightarrow{1-1} {}^\omega 2$ such that $c^{-1}(A) \in S$ for any Borel set $A \subseteq {}^\omega 2$.*

b) *Set $M = \mathrm{rng}(c) = c(X)$. The set $c(B)$ is Borel in M for any $B \in S$.*

c) *If μ is a finite diffused measure on S then $\mu^*(A) = 0$ for any $A \subseteq X$, $|A| < \mathrm{non}(\mathcal{N})$.*

d) *If there is no finite diffused measure on S, then $M = c(X) \subseteq {}^\omega 2$ has universal measure zero.*

Proof. a) Let $\{E_n : n \in \omega\}$ be the set of σ-generators of S. We define the **characteristic function of the sequence** $\langle E_n : n \in \omega \rangle$ as

$$c(x) = \{\chi_{E_k}(x)\}_{k=0}^\infty \in {}^\omega 2 \text{ for } x \in X.$$

Since the σ-algebra separates points of X one can easily see that for any $x, y \in X$, $x \neq y$ there exists an n such that either $x \in E_n$ and $y \notin E_n$ or $x \notin E_n$ and $y \in E_n$. Consequently, $c : X \longrightarrow {}^\omega 2$ is an injection. Set $M = \mathrm{rng}(c)$. Since

$$c^{-1}(\{\alpha \in {}^\omega 2 : \alpha(n) = 1\}) = E_n \in S,$$

we obtain by induction that $c^{-1}(A) \in S$ for any Borel set $A \subseteq {}^\omega 2$.

b) Since $c(E_n) = \{\alpha \in {}^\omega 2 : \alpha(n) = 1\} \cap M$, the assertion follows.

c) Assume that μ is a diffused finite measure on S. Let $A \subseteq X$ and $\mu^*(A) > 0$. We show that $|A| \geq \mathrm{non}(\mathcal{N})$.

We define a Borel measure ν on ${}^\omega 2$ by setting $\nu(B) = \mu(c^{-1}(B))$ for any Borel set $B \subseteq {}^\omega 2$. Since $\nu^*(c(A)) = \mu^*(A) > 0$, the set $c(A)$ is not of universal measure zero. By Theorem 8.6 there exists a homeomorphism $f : [0, 1] \longrightarrow [0, 1]$ such that $\lambda^*(f(c(A))) > 0$. Because $|A| = |f(c(A))|$, we obtain $|A| \geq \mathrm{non}(\mathcal{N})$.

d) If the set M does not have universal measure zero, then there exists a finite diffused complete Borel measure μ on ${}^\omega 2$ with $\mu^*(M) > 0$. By Theorem 4.11, a) the outer measure μ^* restricted to Borel subsets of M would be a Borel measure on M and ν defined as $\nu(A) = \mu^*(c(A))$ would be a diffused measure on S. $\square$

Theorem [AC] 8.8 (E. Grzegorek). *There exists a universal measure zero set of reals of cardinality $\mathrm{non}(\mathcal{N})$.*

Proof. Assume that $X \subseteq [0, 1]$ is such that $|X| = \mathrm{non}(\mathcal{N}) = \kappa$ and $\lambda^*(X) > 0$. Let $X = \{x_\xi : \xi < \kappa\}$ be a one-to-one enumeration of X. Moreover, we fix an open base $\{U_n : n \in \omega\}$ of the topology on $[0, 1]$.

By assumption we have $\lambda(\{x_\eta : \eta < \xi\}) = 0$ for every $\xi < \kappa$ and therefore there exist open sets $G_{n,\xi}$, $n \in \omega$, $\xi < \kappa$ such that

$$\{x_\eta : \eta < \xi\} \subseteq G_{n,\xi}, \quad \lambda\left(\bigcap_m G_{m,\xi}\right) = 0$$

for any $\xi < \kappa$ and any $n \in \omega$. We can assume that $x_\xi \notin \bigcap_m G_{m,\xi}$ for any $\xi < \kappa$.

Let $\mathcal{S} \subseteq \mathcal{P}(\kappa)$ be the σ-algebra with σ-generators

$$E_{n,m} = \{\xi < \kappa : U_m \subseteq G_{n,\xi}\}, \quad n, m \in \omega.$$

For any $\eta < \xi < \kappa$ we have $x_\eta \notin \bigcap_k G_{k,\eta}$ and $x_\eta \in \bigcap_k G_{k,\xi}$. Hence there exists an n such that $x_\eta \notin G_{n,\eta}$ and there exists an m such that $x_\eta \in U_m \subseteq G_{n,\xi}$. Consequently $\xi \in E_{n,m}$ and $\eta \notin E_{n,m}$. Thus $\mathcal{S}$ separates points.

To get a contradiction assume that there exists a finite diffused measure ν on the σ-algebra $\mathcal{S}$. For simplicity, the restriction of the outer Lebesgue measure λ^* to Borel subsets of X will be denoted by μ. Consider the product measure $\nu \times \mu$ on $\kappa \times X$. We set

$$Z = \left\{\langle\xi, x\rangle : \xi < \kappa \wedge x \in X \cap \bigcap_m G_{m,\xi}\right\}, \quad Z_n = \{\langle\xi, x\rangle : \xi < \kappa \wedge x \in X \cap G_{n,\xi}\}.$$

Evidently $Z = \bigcap_n Z_n$ and it is easy to see that $Z_n = \bigcup_m \left(E_{n,m} \times (X \cap U_m)\right)$. Thus Z is $\nu \times \mu$-measurable. For every $\xi < \kappa$, the vertical section $Z_\xi = X \cap \bigcap_m G_{m,\xi}$ has μ-measure 0. If $x \in X$ then there exists a $\xi < \kappa$ such that $x = x_\xi$. Then $x \in \bigcap_m G_{m,\eta}$ for every $\eta > \xi$ and therefore $Z^x \supseteq \{\eta < \kappa : \eta > \xi\}$. By Lemma 8.7, c) we have $\nu(\{\eta < \kappa : \eta > \xi\}) = 1$ and therefore the set Z^x has positive ν-measure. Since $\mu(X) > 0$ we have got a contradiction with Fubini's Theorem, Corollary 4.28.

Hence there exists no finite diffused measure on $\mathcal{S}$. Then by Lemma 8.7, d) we obtain that there exists a universal measure zero subset of $\mathbb{C}$ of cardinality $|X|$. $\square$

We can prove a dual result for category. The proof is similar.

Theorem [AC] 8.9 (E. Grzegorek). *There exists a perfectly meager set of reals of cardinality* non$(\mathcal{M})$.

We begin with a result similar to that of Theorem 8.6.

Lemma [wAC] 8.10. *Let $X \subseteq {}^\omega 2$. If the inverse image $f^{-1}(X)$ is meager for any Borel measurable bijection $f : {}^\omega 2 \xrightarrow[\text{onto}]{1-1} {}^\omega 2$, then X is perfectly meager.*

The proof is easy. Assume that X is not perfectly meager. Then there exists a perfect set $P \subseteq {}^\omega 2$ such that $X \cap P$ is not meager in P. We can assume that ${}^\omega 2 \setminus P$ is uncountable. Note that every perfect subset of ${}^\omega 2$ is homeomorphic to ${}^\omega 2$. Then one can easily construct a Borel measurable bijection $f : {}^\omega 2 \xrightarrow[\text{onto}]{1-1} {}^\omega 2$ such that

$$f(\{\alpha \in {}^\omega 2 : \alpha(0) = 0\}) = P, \; f(\{\alpha \in {}^\omega 2 : \alpha(0) = 1\}) = {}^\omega 2 \setminus P$$

and $f|\{\alpha \in {}^\omega 2 : \alpha(0) = 0\}$ is a homeomorphism. Then $f^{-1}(X \cap P)$ is a non-meager subset of $\{\alpha \in {}^\omega 2 : \alpha(0) = 0\}$ and therefore $f^{-1}(X)$ is not meager. □

Proof of Theorem 8.9. Assume that $Y \subseteq {}^\omega 2$ is a non-meager subset of cardinality $\kappa = \mathrm{non}(\mathcal{M})$. We can assume that Y is dense in ${}^\omega 2$ – if not add a countable dense set to Y. Fix an open base $\{U_n : n \in \omega\}$ of topology on ${}^\omega 2$.

Note the following: if $A \subseteq Y$ is meager, then A is meager in Y. Actually, there are nowhere dense closed sets $\langle F_n \subseteq {}^\omega 2 : n \in \omega \rangle$ such that $A \subseteq \bigcup_n F_n$. Let U be an open set. Since Y is dense we obtain $U \subseteq \overline{U \cap Y}$. If $U \cap Y \subseteq F_n \cap Y$ then also $\overline{U \cap Y} \subseteq F_n$ and therefore $U = \emptyset$.

Let $Y = \{y_\xi : \xi < \kappa\}$ be a one-to-one enumeration of Y. Then there are closed sets $\langle F_{n,\xi} \subseteq {}^\omega 2 : n \in \omega, \xi < \kappa \rangle$ such that

$$\{y_\eta : \eta < \xi\} \subseteq \bigcup_m F_{m,\xi}, \quad Y \cap F_{n,\xi} \text{ is nowhere dense in } Y$$

for any $\xi < \kappa$ and any $n \in \omega$. We can assume that $y_\xi \notin \bigcup_m F_{m,\xi}$. Let

$$Z = \left\{ \langle \xi, y \rangle : \xi \in \kappa \wedge y \in \bigcup_n F_{n,\xi} \right\} \subseteq \kappa \times {}^\omega 2.$$

We set $E_{n,m} = \{\xi < \kappa : F_{n,\xi} \cap U_m = \emptyset\}$. One can easily see that

$$Z = \bigcup_n \bigcap_m \left(\left(\kappa \times {}^\omega 2 \right) \setminus \left(E_{n,m} \times U_m \right) \right). \tag{8.2}$$

Let $\mathcal{S}$ be the σ-algebra of subsets of κ σ-generated by sets $\langle E_{n,m} : n, m \in \omega \rangle$. One can easily check that $\mathcal{S}$ separates points of κ. By Lemma 8.7, a) there exists a function $c : \kappa \xrightarrow{1-1} {}^\omega 2$ such that $c^{-1}(A) \in \mathcal{S}$ for any Borel set $A \subseteq {}^\omega 2$. Set $M = c(\kappa)$. We show that M is perfectly meager.

Assume that M is not perfectly meager. Then there exists a Borel measurable bijection $f : {}^\omega 2 \xrightarrow[\text{onto}]{1-1} {}^\omega 2$ such that $f^{-1}(M) \subseteq {}^\omega 2$ is not meager. We denote by $\mathcal{V}$ the smallest σ-algebra of subsets of $\kappa \times {}^\omega 2$ containing the sets $\langle E_{n,m} \times U_m : n, m \in \omega \rangle$. By (8.2) we obtain that $Z \in \mathcal{V}$. We define

$$F : f^{-1}(M) \times {}^\omega 2 \longrightarrow \kappa \times {}^\omega 2$$

by setting $F(x, y) = \langle c^{-1}(f(x)), y \rangle$. One can easily see that $F^{-1}(B)$ is Borel for any $B \in \mathcal{V}$. Set

$$X = F^{-1}(Z) \cap (f^{-1}(M) \times Y) = \{\langle x, y \rangle \in f^{-1}(M) \times Y : \langle c^{-1}(f(x)), y \rangle \in Z\}.$$

Then X is a Borel subset of the separable metric space $f^{-1}(M) \times Y \subseteq {}^\omega 2 \times {}^\omega 2$. The vertical section

$$X_x = Y \cap \bigcup_n F_{n, c^{-1}(f(x))}$$

is meager in Y for any $x \in f^{-1}(M)$. On the other hand, for every $y = y_\xi \in Y$ we have
$$X^{y_\xi} \supseteq f^{-1}(M) \setminus \{x : c^{-1}(f(x)) \leq \xi\}.$$
Since $|\{x : c^{-1}(f(x)) \leq \xi\}| < \text{non}(\mathcal{M})$, the set $\{x : c^{-1}(f(x)) \leq \xi\}$ is meager and therefore X^{y_ξ} is not meager in $f^{-1}(M)$. We have reached a contradiction with the Kuratowski-Ulam Theorem 7.33. $\square$

For subsets of a Polish space X there is another notion of smallness related to the measure. A set $A \subseteq X$ has **strong measure zero** if for any sequence $\langle \varepsilon_n : n \in \omega \rangle$ of positive reals there exist open sets $\langle A_n : n \in \omega \rangle$ such that $A \subseteq \bigcup_n A_n$ and $\text{diam}(A_n) < \varepsilon_n$ for each n. Replacing the open sets A_n by open balls $\text{Ball}(a_n, \varepsilon_n)$ with suitably chosen centers $a_n \in X$, we obtain an equivalent definition. We denote by $\mathcal{SN}(X)$ the family of all subsets of X with strong measure zero.

Note the following: if A has strong measure zero and $\{\varepsilon_n\}_{n=0}^{\infty}$ is a sequence of positive reals, then there exists a sequence $\{A_n\}_{n=0}^{\infty}$ of open sets such that $\text{diam}(A_n) < \varepsilon_n$ and every $x \in A$ is contained in infinitely many sets $\{A_n\}_{n=0}^{\infty}$, i.e.,

$$A \subseteq \bigcap_m \bigcup_{n > m} A_n. \tag{8.3}$$

Actually, using a pairing function π satisfying $\pi(n, m) \geq m$, we split the sequence $\{\varepsilon_n\}_{n=0}^{\infty}$ into infinitely many sequences and for each of them we find a corresponding sequence of sets. Then we glue the sequences of sets in one sequence. More precisely, for every fixed m, for the sequence $\{\varepsilon_{\pi(k,m)}\}_{k=0}^{\infty}$ of reals there is a corresponding sequence of open sets $\{B_{k,m}\}_{k=0}^{\infty}$ covering the set A. Set $A_n = B_{\lambda(n), \rho(n)}$ (λ, ρ are inverse to π).

The opposite is trivially true. If for any sequence $\{\varepsilon_n\}_{n=0}^{\infty}$ of positive reals there exists a sequence $\{A_n\}_{n=0}^{\infty}$ of open sets such that $\text{diam}(A_n) < \varepsilon_n$ and (8.3) holds true, then A has strong measure zero.

Theorem [wAC] 8.11. *Any subset of a perfect σ-compact Polish space of strong measure zero has universal measure zero.*

Proof. The theorem follows immediately from Lemma 8.5. $\square$

Theorem [wAC] 8.12. *Let $\langle X, \rho \rangle$, $\langle Y, \sigma \rangle$ be Polish spaces, $\mathcal{O}$ being the topology on X induced by the metric ρ.*

a) *$\mathcal{SN}(X)$ is a σ-ideal.*
b) *If $f : X \longrightarrow Y$ is uniformly continuous and $A \subseteq X$ has strong measure zero, then $f(A)$ has strong measure zero as well.*
c) *If $A \subseteq X$ has strong measure zero, then there exists a clopen countable base of the topology $\mathcal{O}|A$.*

Proof. We show that the union of strong measure zero sets $\langle A_n : n \in \omega \rangle$ is a strong measure zero set. Let π be a pairing function. Assume that a sequence of positive reals $\{\varepsilon_n\}_{n=0}^{\infty}$ is given. For given $n \in \omega$ cover each set A_n with open sets

$\langle U_{n,k} : k \in \omega \rangle$ of diameters less than $\varepsilon_{\pi(n,k)}$. Then $\langle U_{n,k} : n, k \in \omega \rangle$ is a cover of $\bigcup_n A_n$ with diameters less than $\{\varepsilon_{\pi(n,k)}\}_{n,k=0}^{\infty} = \{\varepsilon_n\}_{n=0}^{\infty}$.

Assume that $A \subseteq X$ is a strong measure zero set and $f : X \longrightarrow Y$ is uniformly continuous. We want to cover the set $f(A)$ with open sets of diameters $\{\varepsilon_n\}_{n=0}^{\infty}$. Since f is uniformly continuous, there are positive reals $\langle \delta_n : n \in \omega \rangle$ such that $\rho(f(x), f(y)) < \varepsilon_n/2$ whenever $\rho(x, y) < \delta_n$. The set A has strong measure zero, therefore there exist points $\langle a_n \in X : n \in \omega \rangle$ such that $A \subseteq \bigcup_n \mathrm{Ball}(a_n, \delta_n)$. Then $f(A) \subseteq \bigcup_n \mathrm{Ball}(f(a_n), \varepsilon_n/2)$.

Let S be a countable dense subset of a strong measure zero set A. For any $a \in S$ and any natural number n there exists a real $\delta_{a,n}$ such that $\mathrm{Ball}(a, \delta_{a,n}) \cap A$ is clopen in A and $2^{-n-1} < \delta_{a,n} < 2^{-n}$. Actually, the function $f(x) = \rho(a, x)$ is uniformly continuous and since the interval $(2^{-n-1}, 2^{-n})$ does not have strong measure zero there exists a real $\delta_{a,n} \in (2^{-n-1}, 2^{-n})$ which is not in $f(A)$. The family $\{\mathrm{Ball}(a, \delta_{a,n}) \cap A : a \in S \land n \in \omega\}$ is a countable clopen base of $\mathcal{O}|A$. $\square$

The famous **Borel Conjecture** says that every set of reals of strong measure zero is countable. The Borel Conjecture is neither provable nor refutable in **ZFC**. For further information see Section 9.2. The Borel Conjecture is equivalent to the equality

$$\mathrm{size}(\mathcal{SN}) = \aleph_1.$$

We do not have a category analogue of the notion of a strong measure zero set yet. The following result is a starting point for such a definition.

Theorem [wAC] 8.13 (F. Galvin – J. Mycielski – R.M. Solovay). *A set $A \subseteq \mathbb{T}$ has strong measure zero if and only if for every meager set $F \subseteq \mathbb{T}$ there exists a real a such that $(a + A) \cap F = \emptyset$.*

Let us remark that the statement of the theorem is equivalent to the statement "for every meager set $F \subseteq \mathbb{T}$ we have $A + F \neq \mathbb{T}$". One can replace $\mathbb{T}$ by $\mathbb{R}$ or by $\mathbb{R}^n$.

Corollary [AC] 8.14.

  a) *Let $A, B \subseteq \mathbb{T}$. If $|A| < \mathrm{add}(\mathcal{M})$ and B has strong measure zero, then $A \cup B$ has strong measure zero as well.*

  b) *Any set $A \subseteq \mathbb{T}$ of cardinality $< \mathrm{cov}(\mathcal{M})$ has strong measure zero.*

Proof. Assume that $B \subseteq \mathbb{T}$ has strong measure zero and $|A| < \mathrm{add}(\mathcal{M})$. We can assume that $0 \in A$ and $0 \in B$. Then for every meager set F we have

$$(A \cup B) + F \subseteq (A + B) + F = B + (A + F) \neq \mathbb{T},$$

since $A + F$ is meager.

Assume now, that $|A| < \mathrm{cov}(\mathcal{M})$ and F is meager. The family $\{x + F : x \in A\}$ cannot be a cover of $\mathbb{T}$, therefore $A + F \neq \mathbb{T}$. $\square$

Hence, if $\mathrm{cov}(\mathcal{M}) > \aleph_1$, then there exists an uncountable set of strong measure zero.

We begin a proof of Theorem 8.13 with

Lemma [wAC] 8.15. *Suppose that $0 \leq a < b \leq 1$ and $F \subseteq \mathbb{T}$ is a closed nowhere dense set. Then there exists a finite family $\mathcal{F}$ of closed subintervals of $[a, b]$ and $\varepsilon > 0$ such that for every open interval $I \subseteq [0, 1]$ of length $< \varepsilon$, there exists an interval $J \in \mathcal{F}$ such that $(I + J) \cap F = \emptyset$.*

Proof. For every $x \in [0, 1]$ there are an open interval I_x, $x \in I_x$ and a closed interval J_x such that $(I_x + J_x) \cap F = \emptyset$. By compactness there exists finitely many points $x_0, \ldots, x_n$ such that $[0, 1] \subseteq \bigcup_{i=0}^{n} I_{x_i}$. Let ε be such a positive real that any set of diameter $< \varepsilon$ is contained in one of intervals I_{x_i}, $i = 0, \ldots, n$ ($\varepsilon/2$ is the Lebesgue number of the cover). Set $\mathcal{F} = \{J_{x_i} : i = 0, \ldots, n\}$.

If I is an open interval of length $< \varepsilon$ then there is an $i \leq n$ such that $I \subseteq I_{x_i}$. Then $I + J_{x_i} \subseteq I_{x_i} + J_{x_i} \subseteq [0, 1] \setminus F$. $\square$

Proof of the theorem. We assume that the set $A \subseteq \mathbb{T}$ has strong measure zero and $F = \bigcup_n F_n$, where $\langle F_n : n \in \omega \rangle$ are nowhere dense closed, $F_n \subseteq F_{n+1}$ for every n. We want to show that $A + F \neq \mathbb{T}$.

Using the lemma we construct a Souslin scheme $\langle T, \varphi \rangle$ with vanishing diameter, with a finitely branching subtree $T \subseteq {}^{<\omega}\omega$ and values of φ will be closed intervals. Moreover, for every $s \in T \cap {}^n\omega$ we define a positive real ε_s such that for any open interval I of length $< \varepsilon_s$ there exists an immediate successor $t \in T$ of s such that $(I + \varphi(t)) \cap F_n = \emptyset$.

Set $\varphi(\emptyset) = [0, 1]$. Assume that $T \cap {}^n\omega$ is already constructed and $\varphi(s)$ is defined for $s \in T \cap {}^n\omega$. Fix an $s \in T \cap {}^n\omega$. By the lemma there exists an $\varepsilon_s > 0$ and a finite family $\mathcal{F}_s$ of closed subintervals of $\varphi(s)$ such that for every open interval $I \subseteq [0, 1]$ of length $< \varepsilon_s$ the intersection $(I + J) \cap F$ is empty for some $J \in \mathcal{F}_s$. Take the values $\varphi(s^\frown 0), \ldots, \varphi(s^\frown m)$ exhausting $\mathcal{F}_s$. Of course, we can assume that diameters of $\varphi(s^\frown i)$, $i = 0, \ldots, m$ are less than 2^{-n}.

Set $\delta_n = \min\{\varepsilon_s : s \in T \cap {}^n\omega\}$. Since T is finitely branching we have $\delta_n > 0$ for every n. By assumption A is a strong measure set. Hence there are open intervals $\langle I_n : n \in \omega \rangle$ such that diameter of I_n is $< \delta_n$ and (8.3) holds true, i.e., $A \subseteq \bigcap_m \bigcup_{n>m} I_n$. By induction one can define a branch $f \in [T]$ such that $(I_n + \varphi(f|n)) \cap F_n = \emptyset$. Then the (unique) real $x \in \bigcap_n \varphi(f|n)$ is such that

$$(x + A) \cap F \subseteq \left(x + \bigcap_m \bigcup_{n>m} I_n\right) \cap \bigcup_n F_n = \emptyset.$$

Actually, assume that $x + y \in F_k$ and $y \in A$. Then there exists an $n > k$ such that $y \in I_n$. Since $y + x \in I_n + \varphi(f|n)$, we obtain $x + y \notin F_n$, a contradiction.

Assume now, that for any meager set $F \subseteq \mathbb{T}$ there exists a real a such that $(a + A) \cap F = \emptyset$. Let $\{\varepsilon_n\}_{n=0}^{\infty}$ be a sequence of positive reals. Take open intervals $\langle I_n : n \in \omega \rangle$ such that $\mathrm{diam}(I_n) < \varepsilon_n$ for each n and the set $\bigcup_n I_n$ is dense. Then there exists a real a such that $(a + A) \cap (\mathbb{T} \setminus \bigcup_n I_n) = \emptyset$, i.e., $A \subseteq \bigcup_n (I_n - a)$. $\square$

A set $A \subseteq \mathbb{T}$ of reals is **strongly meager** if for any set $B \subseteq \mathbb{T}$ of Lebesgue measure zero there exists a real a such that $A \cap (a + B) = \emptyset$. We denote by $\mathcal{SM}(\mathbb{T})$ the family of all strongly meager subsets of $\mathbb{T}$. The equivalent definition is as follows: if $A + B \neq \mathbb{T}$ for any Lebesgue measure zero set $B \subseteq \mathbb{T}$. Again, one can replace $\mathbb{T}$ by $\mathbb{R}$ or even by $\mathbb{R}^n$.

Theorem [AC] 8.16.

  a) *A strongly meager set is meager.*

  b) *The family of all strongly meager sets is a family of thin sets.*

  c) *If B is strongly meager and $|A| < \mathrm{add}(\mathcal{N})$, then $A \cup B$ is strongly meager.*

  d) *If $A \subseteq \mathbb{T}$, $|A| < \mathrm{cov}(\mathcal{N})$, then A is strongly meager.*

Proof. a) Let $\mathbb{T} = A \cup B$, A meager, $\lambda(B) = 0$ and $A \cap B = \emptyset$. If E is strongly meager then there exists a real a such that $E \cap (a + B) = \emptyset$. Then $E \subseteq (a + A)$.

    The assertion b) follows from a).

    Proofs of c) and d) are dual to those of Corollary 8.14.   □

A subset A of a Polish space X is said to be κ-**concentrated** on set D if $|A \setminus U| < \kappa$ for every open set $U \supseteq D$. Note that if A is κ-concentrated on D then also $A \cup D$ is κ-concentrated on D. So we can always suppose that $D \subseteq A$. If κ is a regular cardinal, then the family of all subsets of a Polish space X that are κ-concentrated on a set D is a κ-additive ideal. We say simply that a set A is **concentrated**, if A is $\aleph_1$-concentrated on a countable set.

Theorem [AC] 8.17. *There exists a set $A \subseteq [0, 1] \setminus \mathbb{Q}$ of cardinality $\mathfrak{b}$ that is $\mathfrak{b}$-concentrated on the set $\mathbb{Q} \cap [0, 1]$.*

Proof. By Theorem 5.44, there exists an eventually increasing well-ordered unbounded sequence $\{\alpha_\xi : \xi < \mathfrak{b}\} \subseteq {}^\omega\omega$. Let $h : {}^\omega\omega \xrightarrow[\mathrm{onto}]{1-1} [0, 1] \setminus \mathbb{Q}$ be the natural homeomorphism. We set $A = \{h(\alpha_\xi) : \xi < \mathfrak{b}\}$. If $U \supseteq [0, 1] \cap \mathbb{Q}$ is open, then $[0, 1] \setminus U \subseteq [0, 1] \setminus \mathbb{Q}$ is compact. Therefore the set $K = {}^\omega\omega \setminus h^{-1}(U)$ being homeomorphic to $[0, 1] \setminus U$ is compact as well. By Theorem 3.11, the set K is bounded and therefore the set $K \cap h^{-1}(A)$ has cardinality smaller than $\mathfrak{b}$.   □

Theorem [AC] 8.18. *Any set $\mathrm{cov}(\mathcal{M})$-concentrated on a strong measure zero set has strong measure zero.*

Proof. Let $A \subseteq X$ be $\mathrm{cov}(\mathcal{M})$-concentrated on a strong measure zero set B. Assume that $\{\varepsilon_n\}_{n=0}^\infty$ is a sequence of positive reals. Then there are $x_n \in X$ such that $B \subseteq G$ where $G = \bigcup_n \mathrm{Ball}(x_n, \varepsilon_{2n}/2)$. Since G is an open set containing B we obtain that $|A \setminus G| < \mathrm{cov}(\mathcal{M})$. By Corollary 8.14 the set $A \setminus G$ has strong measure zero and therefore it may be covered by balls with diameters $\langle \varepsilon_{2n+1} : n \in \omega \rangle$.   □

Corollary [wAC] 8.19. *Any concentrated set has strong measure zero.*

Thus according to the consistency of the Borel Conjecture one cannot prove that there exists an uncountable set of reals concentrated on a countable set. However, under additional assumptions we can find such sets.

Corollary [AC] 8.20. *If* $\mathrm{add}(\mathcal{M}) = \mathfrak{c}$, *then there exists a strong measure zero set of cardinality* $\mathfrak{c}$.

Proof. If $\mathrm{add}(\mathcal{M}) = \mathfrak{c}$, then both $\mathfrak{b} = \mathfrak{c}$ and $\mathrm{cov}(\mathcal{M}) = \mathfrak{c}$. Let A be the set of Theorem 8.17. By Theorem 8.21 the set A has strong measure zero. $\square$

The proof of the following theorem is based on the so-called Rothberger's trick.

Theorem [AC] 8.21 (F. Rothberger). *If* $\mathfrak{b} = \mathfrak{c}$, *then there exists a set A of reals of cardinality* $\mathfrak{c}$ *such that A is $\mathfrak{c}$-concentrated on the set of rationals and can be mapped continuously onto* $^{\omega}2$.

Proof. Assume $\mathfrak{b} = \mathfrak{c}$. By Theorem 5.44 there exists an eventually increasing unbounded subset $\{\alpha_\xi : \xi < \mathfrak{c}\}$ of $^{\omega}\omega$. Let $\{\beta_\xi : \xi < \mathfrak{c}\}$ be an enumeration of $^{\omega}2$. We use the Rothberger trick to define

$$A = \{2\alpha_\xi + \beta_\xi : \xi < \mathfrak{c}\}.$$

The set A is unbounded as well.

Let $h : {^{\omega}\omega} \xrightarrow{1-1} [0,1]$ be the natural embedding. By the same arguments as in the proof of Theorem 8.17 we obtain that $h(A)$ is $\mathfrak{c}$-concentrated on rationals. For $\alpha \in {^{\omega}\omega}$ we set $f(\alpha) = \beta$, where $\beta \in {^{\omega}2}$ and $\beta(n) = \alpha(n) \mod 2$ for every n. Then $h^{-1} \circ f$ is a continuous surjection of $h(A)$ onto $^{\omega}2$. $\square$

There is another important property of smallness. A topological space X is an **nCM-space** if no continuous mapping from X into $[0,1]$ is a surjection, i.e., if X cannot be continuously mapped onto $[0,1]$.

Theorem 8.22. *Any completely regular* nCM-*space X has a clopen base. Assuming* **AC**, *if X is a normal* nCM-*space, then* $\mathrm{Ind}(X) = 0$, *i.e., for any disjoint closed sets $A, B \subseteq X$ there exists a clopen set U such that $A \subseteq U$ and $U \cap B = \emptyset$.*

Proof. Let $a \in U \subseteq X$, U being open. By complete regularity, there exists a continuous mapping $f : X \longrightarrow [0,1]$ such that $f(a) = 0$ and $f(x) = 1$ for $x \notin U$. Since f is not a surjection there exists an $\varepsilon > 0$ such that $\varepsilon \notin \mathrm{rng}(f)$. Then $f^{-1}(\langle 0, \varepsilon \rangle)$ is a clopen subset of U containing a.

If X is normal, then by the Urysohn Theorem 3.15 there exists a continuous mapping $f : X \longrightarrow [0,1]$ such that $f(x) = 0$ for $x \in A$ and $f(x) = 1$ for $x \in B$. As above, if $\varepsilon \notin \mathrm{rng}(f)$, then $U = f^{-1}(\langle 0, \varepsilon \rangle)$ is the desired clopen set. $\square$

Using Theorem 3.26 we obtain

Corollary 8.23. *Any* nCM-*subset of a metric separable space is homeomorphic to a set of reals, even to a subset of* $^{\omega}2$ *or* $^{\omega}\omega$.

Exercises

8.1 [AC] Carlson Theorem

If there exists an uncountable separable metric space of strong measure zero, then there exists an uncountable set of reals of strong measure zero.

Hint: If $\mathfrak{c} = \aleph_1$, the result follows easily by Theorem 8.26 and Corollary 8.31. If $\mathfrak{c} > \aleph_1$, consider an uncountable separable metric space X of strong measure zero. By Exercise 3.2 there exists a uniformly continuous injection of X into $\mathbb{R}$.

8.2 [AC] Additivity of Strong Measure Zero Sets

Assume that $\{\varepsilon_n\}_{n=0}^\infty$ is a decreasing sequence of positive reals converging to 0. For any n, let $\{I_m^n : m \in \omega\}$ be an enumeration of rational open intervals of length $\leq \varepsilon_{2n+1}$. Let $\kappa < \mathrm{add}(\mathcal{N})$.

a) If $A \subseteq \mathbb{R}$ has strong measure zero, then there exists a function $f \in {}^\omega\omega$ such that

$$A \subseteq \bigcap_m \bigcup_{n>m} I_{f(n)}^n.$$

b) If $\langle A_\xi \subseteq \mathbb{R} : \xi < \kappa \rangle$ are strong measure zero sets, then there exists an $h \in \mathrm{Loc}$ such that

$$\bigcup_{\xi<\kappa} A_\xi \subseteq \bigcap_m \bigcup_{n>m} \bigcup_{k\in h(n)} I_k^n.$$

Hint: Use results of part a) and Exercise 7.13.

c) If $\langle A_\xi \subseteq \mathbb{R} : \xi < \kappa \rangle$ are strong measure zero sets, then there exists a sequence $\{I_n\}_{n=0}^\infty$ of open intervals such that $\bigcup_{\xi<\kappa} A_\xi \subseteq \bigcup_n I_n$ and length of I_n is smaller than ε_n.

Hint: Since $|h(n)| \leq 2^{n+1} - 2^n$ we can order the set $\{I_m^n : m \in h(n) \wedge n \in \omega\}$ into a sequence $\{I_n\}_{n=0}^\infty$ such that the length of I_n is smaller than ε_n.

d) $\mathrm{add}(\mathcal{SN}(\mathbb{R})) \geq \mathrm{add}(\mathcal{N})$.

e) Show that $\mathrm{add}(\mathcal{SN}(X)) \geq \mathrm{add}(\mathcal{N})$ for any metric separable space X.

8.3 [AC] Porous Sets

If A is a subset of a metric separable space X, $x \in X$, r is a positive real, we write

$$\gamma(x, r, A) = \sup\{h \geq 0 : (\exists y \in X)\,\mathrm{Ball}(y, h) \subseteq \mathrm{Ball}(x, r) \setminus A\}.$$

The number

$$\mathrm{por}(x, A) = \limsup_{r\to 0^+} \frac{\gamma(x, r, A)}{r}$$

is called the **porosity** of A at x. A set $A \subseteq X$ is called **porous** if $\mathrm{por}(x, A) > 0$ for every $x \in A$. A set A is σ-**porous** if $A \subseteq \bigcup_n A_n$ with A_n porous.

a) The property "to be porous" is external for any class of topological spaces containing Euclidean spaces in **ZF**.

Hint: Consider $\mathbb{R}$ as a subset of $\mathbb{R}^2$.

b) A porous subset of any Polish space X is nowhere dense.

Hint: Assume that $a \in \overline{A}$, $a \in U$, U open. Then there are $x \in A$ and $r > 0$ such that $\mathrm{Ball}(x, r) \subseteq U$. We can assume that $\gamma(x, r, A) > 0$ and we find a $y \in X$ and an $h > 0$ such that $\mathrm{Ball}(y, h) \subseteq U \cap (X \setminus \overline{A})$.

c) The lower density $\mathrm{dens}_*(a, A)$ was defined in Exercise 4.9. Show that $\mathrm{dens}_*(x, A) \le 1 - \mathrm{por}(x, A)$ for any $x \in \mathbb{R}_k$.

d) Find an example when $\mathrm{dens}_*(x, A) < 1 - \mathrm{por}(x, A)$ for every $x \in \mathbb{R}_k$.

e) A porous subset of $\mathbb{R}_k$ has Lebesgue measure zero.

f) The family of σ-porous subsets of $\mathbb{T}$ is a σ-ideal contained in $\mathcal{N} \cap \mathcal{M}$.

8.4 [AC] Continuous Images of Small Sets

Assume that X is a perfect Polish space.

a) A homeomorphic image of a universal measure zero set of reals is a universal measure zero set.

   Hint: Use Theorem 8.6.

b) Let $A \subseteq X$ have the following property: for any open set U such that $U \cap A \ne \emptyset$ there exists an open set $V \subseteq U$ such that $0 < |A \cap V| \le \aleph_0$. If $h : X \xrightarrow[\text{onto}]{1-1} X$ is a homeomorphism then $h(A)$ possesses also this property.

c) A set A with the property of b) is meager.

   Hint: Let $\mathcal{B}$ be a countable base. For a regular open set U take

   $$\mathcal{V} = \{V \in \mathcal{B} : V \subseteq U \wedge |A \cap V| \le \aleph_0\}.$$

   Show that $\overline{U \setminus \bigcup \mathcal{V}}$ is nowhere dense and therefore $A \cap U$ is meager.

d) Let $B \subseteq {}^\omega 2$ be a Bernstein set. The set $([0, 1] \cap \mathbb{Q}) \cup B$ has the property of b) and is not perfectly meager.

e) An analogue of Theorem 8.6 for perfectly meager sets is false, i.e., there exists a set $A \subseteq [0, 1]$ such that every homeomorphic image of A is meager and A is not perfectly meager.

   Hint: Let $B \subseteq \mathbb{C} \subseteq [0, 1]$ be a Bernstein set. The set $B \cup \mathbb{Q}$ is not meager in $\mathbb{C}$. For every open $U \subseteq [0, 1]$ there exists an open set $V \subseteq U$ such that $V \cap (B \cup \mathbb{Q})$ is countable. A homeomorphism preserves this property. Use c).

8.5 [AC] (S)$_0$-sets

A subset A of a perfect Polish space X is called an (S)$_0$-**set** if for every perfect set $P \subseteq X$ there is a perfect set $Q \subseteq P$ such that $A \cap Q = \emptyset$.

a) The property "to be an (S)$_0$-set" is external for any class of topological spaces containing Euclidean spaces in **ZF**.

b) The set of all (S)$_0$-subsets of a perfect Polish space is a σ-ideal. This ideal is usually called a **Marczewski ideal**.

   Hint: Let A_n be (S)$_0$-sets. By induction construct a regular perfect closed Souslin scheme $\langle {}^\omega 2, \varphi \rangle$ such that $\varphi(s) \cap A_n = \emptyset$ for $s \in {}^n 2$.

c) Any universal measure zero subset A of a perfect Polish space is an (S)$_0$-set.

   Hint: If $P \subseteq X$ is perfect, then we can assume ${}^\omega 2 \subseteq P$. Let μ be Borel measure on X defined as $\mu(E) = \lambda_C({}^\omega 2 \cap E)$. Since $\mu(A) = 0$ there exists a Borel set B such that $A \subseteq B$ and $\lambda_C({}^\omega 2 \setminus B) = 1$.

d) Any perfectly meager subset of a perfect Polish space is an (S)$_0$-set.

   Hint: Proof is dual.

e) There exists an $(S)_0$-subset of $^\omega 2$ of cardinality continuum.

Hint: Assume $^\omega 2 = \{\alpha_\xi : \xi < \mathfrak{c}\}$ and $\{P_\xi : \xi < \mathfrak{c}\}$ is an enumeration of all perfect subsets of $^\omega 2$. The function $\Pi : \,^\omega 2 \times \,^\omega 2 \longrightarrow \,^\omega 2$ is defined by (1.11). If $A \subseteq \,^\omega 2$, $\alpha \in \,^\omega 2$, then $A_\alpha = \{\beta \in \,^\omega 2 : \Pi(\alpha, \beta) \in A\}$ is the vertical section of A. For any $\xi < \mathfrak{c}$ take

$$\beta_\xi \in \,^\omega 2 \setminus \bigcup\{(P_\eta)_{\alpha_\xi} : \eta < \xi \wedge (P_\eta)_{\alpha_\xi} \text{ countable}\}, \beta_\xi \neq \beta_\eta \text{ for } \eta < \xi.$$

Let $X = \{\Pi(\alpha_\xi, \beta_\xi) : \xi < \mathfrak{c}\}$. Remark that $|X_\alpha| = 1$ for any $\alpha \in \,^\omega 2$. Assume $P = P_\xi$ is perfect. If P_α is uncountable for some $\alpha \in \,^\cup 2$, then there exists $Q \subset P_\alpha$ disjoint with X. Assume that every P_α is countable. If $\Pi(\alpha_\eta, \beta_\eta) \in P_\xi \cap X$ then $\eta \leq \xi$. Thus $|P \cap X| < \mathfrak{c}$. Split P into a continuum of many disjoint perfect sets.

8.2 Sets with Nice Subsets

We shall consider sets which have nice subsets from at least three different points of view. We start with the sets in which a "small" set has even a small cardinality. We shall make this approach precise by an example. Consider a subset $A \subseteq X$ of a perfect topological space X. Any countable subset of A is meager in A. It would be very "nice" if the opposite held true: every subset of A, that is meager in A, is countable. By a generalization and an adequate modification we can define a very classical notion of a Luzin set or a Sierpiński set. Another approach demands that a "nice" subset is also "nice" in different ways. We obtain the notion of a σ-set. Finally we can ask that a subset of "small" cardinality is nice. We obtain, e.g., the notion of a λ-set.

There are many such definitions. We shall present and study just a few classical ones.

The first kind of sets will be introduced in a general way as subsets of a Polish ideal space and then we shall specify the ideal. Let κ be an uncountable cardinal not greater than $2^{\aleph_0}$. Let $\langle X, \mathcal{I}\rangle$ be a Polish ideal space. A subset $L \subseteq X$ is called a κ-$\mathcal{I}$-set if $|L| \geq \kappa$ and for any $A \in \mathcal{I}$ one has $|L \cap A| < \kappa$. Note that a κ-$\mathcal{I}$-set does not belong to the ideal $\mathcal{I}$. If L is a κ-$\mathcal{I}$-set, $|L| \geq \lambda > \kappa$, then L is also a λ-$\mathcal{I}$-set. Moreover, every subset of κ-$\mathcal{I}$-set of cardinality $\geq \kappa$ is a κ-$\mathcal{I}$-set. Thus, if L is a κ-$\mathcal{I}$-set, then immediately from the definition we obtain

$$\text{non}(\mathcal{I}) \leq \kappa \ \text{cf}(\kappa) \leq \text{cov}(\mathcal{I}). \tag{8.4}$$

If X is a Polish space, then a κ-$\mathcal{M}(X)$-set is called a κ-**Luzin set**. If μ is a diffused Borel measure on X, then a κ-$\mathcal{N}(\mu)$-set is called a κ-**Sierpiński set**. Note that the notion of a κ-Luzin set is not internal (a κ-Luzin subset of $\mathbb{R}$ is not a κ-Luzin subset of $\mathbb{R}_2$) for any class containing Euclidean spaces in the theory **ZF**. The notion of a κ-Sierpiński set depends on the considered measure. Thus this notion is a priori non-internal. Usually we deal with the real line and we consider the Lebesgue measure. Anyway we can consider any diffused Borel measure on a Polish space.

An $\aleph_1$-Luzin set of cardinality $\mathfrak{c}$ is called a **Luzin set** and an $\aleph_1$-Sierpiński set of cardinality $\mathfrak{c}$ is called a **Sierpiński set**. Thus a Luzin set is a κ-Luzin set and a Sierpiński set is a κ-Sierpiński set for any uncountable $\kappa \leq \mathfrak{c}$.

Theorem [AC] 8.24. *Let $\langle X, \mathcal{I} \rangle$ be a homogeneous Polish ideal space. Then no subset of a κ-$\mathcal{I}$-set of cardinality $\geq \kappa$ belongs to $\mathrm{BOREL}^*(\mathcal{I})$.*

Proof. Let A be a subset of a κ-$\mathcal{I}$-set L, $|A| \geq \kappa$. Assume that $A \in \mathrm{BOREL}^*(\mathcal{I})$. Then there exists a Borel set $B \subseteq A$ such that $A \setminus B \in \mathcal{I}$. Since $A \notin \mathcal{I}$, neither is $B \in \mathcal{I}$. Thus B is uncountable and being Borel, B contains a Borel uncountable subset $D \in \mathcal{I}$. Then $|D \cap L| < \kappa$. Since D is Borel, we obtain $|D| = |D \cap L| = \mathfrak{c} \geq \kappa$, which is a contradiction. $\square$

Corollary [AC] 8.25. *No subset of a κ-Luzin set of cardinality $\geq \kappa$ has the Baire Property. No subset of a κ-Sierpiński set of cardinality $\geq \kappa$ is measurable.*

We begin with the problem of existence of a κ-$\mathcal{I}$-set.

Theorem [AC] 8.26. *If $\kappa = \mathrm{cov}(\mathcal{I}) = \mathrm{cof}(\mathcal{I})$, then there exists a κ-$\mathcal{I}$-set.*

Proof. Let $\{B_\xi : \xi < \kappa\}$ be a base of the ideal $\mathcal{I}$. Since $\kappa = \mathrm{cov}(\mathcal{I})$, for any $\xi < \kappa$ we have $|X \setminus \bigcup_{\eta < \xi} B_\eta| \geq \kappa$. Thus, one can choose an element $x_\xi \in X \setminus \bigcup_{\eta < \xi} B_\eta$, $x_\xi \neq x_\eta$ for $\eta < \xi$. Let $L = \{x_\xi : \xi < \kappa\}$. Then $|L| = \kappa$.

If $A \in \mathcal{I}$, then $A \subseteq B_\xi$ for some $\xi < \kappa$ and therefore $L \cap A \subseteq \{x_\eta : \eta \leq \xi\}$. Thus $|L \cap A| < \kappa$. $\square$

Corollary [AC] 8.27. *If $\kappa = \mathrm{cov}(\mathcal{M}) = \mathrm{cof}(\mathcal{M})$, then there exists a κ-Luzin set. If $\kappa = \mathrm{cov}(\mathcal{N}) = \mathrm{cof}(\mathcal{N})$, then there exists a κ-Sierpiński set.*

Theorem [AC] 8.28. *Assume that $\langle X, \mathcal{I} \rangle$ and $\langle X, \mathcal{J} \rangle$ are Polish ideal groups with shift invariant orthogonal σ-ideals. If κ is an uncountable regular cardinal, then any κ-$\mathcal{I}$-set belongs to $\mathcal{J}$.*

Proof. Let L be a κ-$\mathcal{I}$-set. Since $\mathcal{I}$ and $\mathcal{J}$ are orthogonal, there exist disjoint $A \in \mathcal{I}$ and $B \in \mathcal{J}$ such that $A \cup B = X$. Since $|L \cap A| < \kappa$, by (8.4) and Theorem 7.26 we have

$$|L \cap A| < \kappa \leq \mathrm{cov}(\mathcal{I}) \leq \mathrm{non}(\mathcal{J}).$$

Thus $L \cap A \in \mathcal{J}$. Since $L \subseteq ((L \cap A) \cup B)$, we obtain $L \in \mathcal{J}$. $\square$

Corollary [AC] 8.29. *Any κ-Sierpiński set of reals is meager and any κ-Luzin set of reals has measure zero, provided that κ is an uncountable regular cardinal.*

The result of Corollary 8.29 can be slightly improved.

Theorem [AC] 8.30. *Assume that $\mathrm{cf}(\kappa) > \omega$. Then a set $L \subseteq X$ of cardinality $\geq \kappa$ is a κ-Luzin set if and only if L is κ-concentrated on a countable dense subset of X.*

By Corollary 8.19 we obtain that an internal property could follow from an external one (the exact formulation of the result is left to a reader):

Corollary [AC] 8.31. *Every $\aleph_1$-Luzin set has strong measure zero.*

We have a similar result for a Sierpiński set. However, we omit a quite complicated proof, which can be found, e.g., in T. Bartszynski and H. Judah [1995].

Theorem [AC] 8.32 (J. Pawlikowski). *Every $\aleph_1$-Sierpiński set is strongly meager.*

Corollary 8.31 can be still improved.

Theorem [AC] 8.33. *If κ is regular, then any κ-Luzin set of reals has strong measure zero.*

We begin with a lemma which can be interesting in its own.

Lemma [AC] 8.34. *If there exists a κ-Sierpiński set of reals, then every set of reals of cardinality smaller than $\mathrm{cf}(\kappa)$ is strongly meager. Similarly, if there exists a κ-Luzin set of reals, then every set of reals of cardinality smaller than $\mathrm{cf}(\kappa)$ has strong measure zero.*

Proof. Let S be a κ-Sierpiński set. Assume that $|A| < \mathrm{cf}(\kappa)$ and B has Lebesgue measure zero. For every $x \in A$ we have $|(x + B) \cap S| < \kappa$. Thus $|(A + B) \cap S| < \kappa$ and therefore $A + B \neq \mathbb{R}$.

By Theorem 8.13 we obtain the assertion for the strong measure zero case. $\square$

Proof of Theorem 8.33. Let $L \subseteq \mathbb{R}$ be a κ-Luzin set, $\{\varepsilon_n\}_{n=0}^{\infty}$ being a sequence of positive reals. If $\{r_n : n \in \omega\}$ is dense in $\mathbb{R}$, then $U = \bigcup_n (r_n - \varepsilon_{2n}/2, r_n + \varepsilon_{2n}/2)$ is an open dense set and therefore $|L \backslash U| < \kappa$. By Lemma 8.34 the set $L \backslash U$ has strong measure zero and therefore can be covered by intervals of diameter $\langle \varepsilon_{2n+1} : n \in \omega \rangle$. Then $U \cup (L \backslash U)$ can be covered by intervals of diameter $\langle \varepsilon_n : n \in \omega \rangle$. $\square$

By Lemma 8.34, the existence of a κ-Luzin set implies $\mathrm{cf}(\kappa) \leq \mathrm{non}(\mathcal{N})$. Similarly, the existence of a κ-Sierpiński set implies that $\mathrm{cf}(\kappa) \leq \mathrm{non}(\mathcal{M})$. We have a stronger estimate of the size of κ at least for κ regular.

Theorem [AC] 8.35. *Let $\langle X, \mathcal{I} \rangle$ be a Polish ideal space, κ being an uncountable regular cardinal. If L is a κ-$\mathcal{I}$-set, then $\kappa \leq |L| \leq \mathrm{cov}(\mathcal{I})$.*

Proof. Let L be a κ-$\mathcal{I}$-set. By definition $|L| \geq \kappa$. Let $\{A_\xi : \xi < \mathrm{cov}(\mathcal{I})\} \subseteq \mathcal{I}$ be such that $X = \bigcup_{\xi < \mathrm{cov}(\mathcal{I})} A_\xi$. Then

$$L = \bigcup_{\xi < \mathrm{cov}(\mathcal{I})} (L \cap A_\xi),$$

where $|L \cap A_\xi| < \kappa$ for every $\xi < \mathrm{cov}(\mathcal{I})$. Therefore $|L| \leq \mathrm{cov}(\mathcal{I})$. $\square$

Corollary [AC] 8.36. *Assume that κ is an uncountable regular cardinal. If there exists a κ-$\mathcal{I}$-set L, then $\mathrm{non}(\mathcal{I}) \leq \kappa \leq |L| \leq \mathrm{cov}(\mathcal{I})$.*

Theorem [AC] 8.37. *Assume that $\langle X, \mathcal{I} \rangle$ and $\langle X, \mathcal{J} \rangle$ are Polish ideal groups with shift invariant orthogonal σ-ideals, and κ, λ are regular uncountable cardinals. If there exist both a κ-$\mathcal{I}$-Luzin set L and a λ-$\mathcal{J}$-Luzin set S, then*

$$\kappa = \lambda = \mathrm{non}(\mathcal{I}) = \mathrm{non}(\mathcal{J}) = \mathrm{cov}(\mathcal{I}) = \mathrm{cov}(\mathcal{J}) = |L| = |S|.$$

Proof. By Corollary 8.36 and Rothberger's Theorem, Corollary 7.27 we have

$$\text{non}(\mathcal{I}) \leq \kappa \leq |L| \leq \text{cov}(\mathcal{I}) \leq \text{non}(\mathcal{J}),$$
$$\text{non}(\mathcal{J}) \leq \lambda \leq |S| \leq \text{cov}(\mathcal{J}) \leq \text{non}(\mathcal{I}). \qquad \square$$

Corollary [AC] 8.38. *Assume that κ, λ are regular uncountable cardinals. If there exist both a κ-Luzin set L and a λ-Sierpiński set S, then*

$$\kappa = \lambda = \text{non}(\mathcal{M}) = \text{non}(\mathcal{N}) = \text{cov}(\mathcal{M}) = \text{cov}(\mathcal{N}) = |L| = |S|.$$

Corollary [AC] 8.39 (F. Rothberger). *If there are both a Luzin set and a Sierpiński set, then* **CH** *holds true.*

The σ-ideal $\mathcal{K}_\sigma$ and related notions were introduced in Section 7.5. One can easily see that $\langle {}^\omega\omega, \mathcal{K}_\sigma \rangle$ is a Polish ideal space. By Rothberger's Theorem 7.59 we have

$$\text{cov}(\mathcal{K}_\sigma) = \text{cof}(\mathcal{K}_\sigma) = \mathfrak{d}.$$

Therefore by Theorem 8.26 we immediately obtain the following result.

Theorem [AC] 8.40. *There exists a $\mathfrak{d}$-$\mathcal{K}_\sigma$-set.*

On the other hand, we have another result.

Theorem [AC] 8.41. *There exists a $\mathfrak{b}$-$\mathcal{K}_\sigma$-set.*

Proof. By Theorem 5.44 there exists an eventually increasing well-ordered unbounded sequence $\langle \alpha_\xi \in {}^\omega\omega : \xi < \mathfrak{b} \rangle$. Set $L = \{\alpha_\xi : \xi < \mathfrak{b}\}$. If $A \subseteq {}^\omega\omega$ is a σ-compact set, then there exists an $\alpha \in {}^\omega\omega$ such that $A \subseteq C_\alpha^*$. Then $C_\alpha^* \cap L$ is bounded and therefore of cardinality smaller than $\mathfrak{b}$. $\qquad \square$

Since **ZFC** $+ \, \mathfrak{b} < \mathfrak{d}$ is consistent, see (11.25), we have obtained κ-$\mathcal{K}_\sigma$-sets for eventually different κ's.

Theorem [AC] 8.42. *Assume that κ is an uncountable regular cardinal.*

a) *If there exists a κ-Luzin set L, then there exists a universal measure zero set of cardinality $|L|$.*

b) *If there exists a κ-Sierpiński set S, then there exists a perfectly meager set of cardinality $|S|$.*

Proof. The assertion a) follows by Grzegorek's Theorem 8.8, since by Rothberger's Theorem, Corollary 7.27 we obtain $|L| \leq \text{cov}(\mathcal{M}) \leq \text{non}(\mathcal{N})$.

The dual assertion b) follows by Rothberger's Theorem, Corollary 7.27 and Grzegorek's Theorem 8.9. $\qquad \square$

Now we shall study the second type of sets. A set $A \subseteq X$ is called a σ-**set** if for every $\mathbf{F}_\sigma$ set $F \subseteq X$ there exists a $\mathbf{G}_\delta$ set G such that $F \cap A = G \cap A$. The obvious relativization argument shows that the property is internal for the class of all topological spaces in the theory **ZF**.

Theorem [wAC] 8.43. *Let $A \subseteq X$ be a σ-set, X being a perfect Polish space. Then*

a) *A is perfectly meager,*

b) *for every Borel set B there exists a G_δ set G such that $A \cap B = A \cap G$.*

Proof. Since A is a subset of a Polish space, there exists a countable set $Y \subseteq A$ dense in A. The set Y is F_σ and therefore there exists a G_δ set G such that $Y = A \cap G$. Since $\overline{Y} = \overline{A}$, $Y \subseteq G$, we obtain that the set $\overline{A} \setminus G$ is meager. Thus $A \subseteq (Y \cup (\overline{A} \setminus G))$ is meager.

If $P \subseteq X$ is perfect, then one can easily see that $A \cap P$ is a σ-subset of P. Therefore $A \cap P$ is meager in P.

Part b) follows easily by transfinite induction. $\qquad\square$

Corollary [wAC] 8.44. *No κ-Luzin set is a σ-set.*

On the other hand we have

Theorem [wAC] 8.45. *Every $\aleph_1$-Sierpiński set is a σ-set.*

Proof. Assume that S is an $\aleph_1$-Sierpiński set, B is a G_δ set. Let $F \subseteq B$ be an F_σ set with $\mu(B \setminus F) = 0$. We have

$$S \cap B = S \cap (F \cup (S \cap (B \setminus F))).$$

Since $S \cap (B \setminus F)$ is countable, the set $F \cup (S \cap (B \setminus F))$ is an F_σ set. $\qquad\square$

A subset A of a Polish space X is called a λ-**set** if for every countable $B \subseteq A$ there exists a G_δ set $G \subseteq X$ such that $B = G \cap A$. By definition, every σ-set is a λ-set. Hence, every $\aleph_1$-Sierpiński set is a λ-set. Evidently every countable set is a λ-set and every subset of a λ-set is again a λ-set.

The property "to be a λ-set" is internal for the class of all topological spaces in the theory **ZF**.

Theorem [AC] 8.46. *Any subset of a Polish space X of cardinality less than $\mathfrak{b}$ is a λ-set.*

Proof. Let $A \subseteq X$, $|A| < \mathfrak{b}$, $B \subseteq A$ being countable. We enumerate the set B as $\langle b_n : n \in \omega \rangle$ in such a way that every $b \in B$ occurs infinitely many times in the enumeration. For each n take a decreasing sequence $\langle U_{n,m} : m \in \omega \rangle$ of open sets with $\bigcap_m U_{n,m} = \{b_n\}$. For any $a \in A \setminus B$ we set

$$\gamma_a(n) = \text{the least } m \text{ such that } a \notin U_{n,m}.$$

Since $|A| < \mathfrak{b}$, there exists a function $\alpha \in {}^\omega\omega$ such that $\gamma_a \leq^* \alpha$ for every $a \in A \setminus B$. Now, we set

$$G = \bigcap_k \bigcup_{n \geq k} \bigcap_{m \leq \alpha(n)} U_{n,m}.$$

Evidently G is a G_δ set and we claim that $A \cap G = B$. Since for every $b \in B$ the set $\{n : b = b_n\}$ is infinite, then

$$b \in \bigcup_{n \geq k} \bigcap_{m \leq \alpha(n)} U_{n,m}$$

for every k and therefore $b \in G$.

On the other side, if $a \in A \setminus B$, then there exists a k such that $\gamma_a(n) \leq \alpha(n)$ for every $n \geq k$. Thus a does not belong to $\bigcup_{n \geq k} \bigcap_{m \leq \alpha(n)} U_{n,m}$. Therefore neither a belongs to G.   □

Similarly as above for a σ-set, one can show

Theorem [wAC] 8.47. *Any λ-subset of a perfect Polish space is perfectly meager.*

Proof. Let A be a λ-set, P being a perfect set, $P \cap A \neq \emptyset$. Since A is a subset of a Polish space there exists a countable set $B \subseteq P \cap A$ dense in $P \cap A$. Then there exists a G_δ set G such that $B = G \cap A$. Then $\overline{P \cap A} \setminus G = \bigcup_n F_n$, F_n closed and $P \cap A \subseteq (\overline{P \cap A} \setminus G) \cup B$. Since B is countable it suffices to show that every $F_n \cap P$ is nowhere dense in P.

So let U be open, $U \cap P \subseteq F_n$ and $U \cap P \neq \emptyset$. Then $U \cap P \subseteq \overline{P \cap A} \setminus G$ and therefore $U \cap P \cap A \neq \emptyset$. However, then also $U \cap P \cap B \neq \emptyset$, which contradicts $F_n \subseteq \overline{P \cap A} \setminus B$.   □

Theorem [AC] 8.48. *Every subset of $^\omega\omega$ well-ordered by eventual domination is a λ-set.*

Proof. Let $A = \{\alpha_\xi : \xi < \zeta\}$ be a well-ordered subset of $^\omega\omega$ with an increasing enumeration. We can assume that $\mathrm{cf}(\zeta) \neq \omega$. Indeed, if $\mathrm{cf}(\zeta) = \omega$, then by results of Section 5.4 the set A is bounded by some $\gamma \in {}^\omega\omega$ and we shall consider the set $A \cup \{\gamma\}$.

Assume that $B \subseteq \zeta$ is countable. We can assume that B is closed. If not the closure of B is obtained from B by adding a countable set C and $\{\alpha_\xi : \xi \in C\}$ is an F_σ set. Let η_0 be the least element and η_1 be the greatest element of B.

For every $\alpha \in {}^\omega\omega$ the sets

$$G_\alpha = \{\beta \in {}^\omega\omega : \beta >^* \alpha\}, \ S_\alpha = \{\beta \in {}^\omega\omega : \beta <^* \alpha\}$$

are F_σ sets. E.g., we have

$$G_\alpha = \bigcup_n \bigcap_{m \geq n} \{\beta \in {}^\omega\omega : \beta(m) \geq \alpha(m) + 1\}.$$

Since B is countable, the set

$$F = S_{\alpha_{\eta_0}} \cup G_{\alpha_{\eta_1}} \cup \bigcup_{\eta \in B, \eta \neq \eta_1} (G_{\alpha_\eta} \cap S_{\alpha_{\eta+1}})$$

is an F_σ set. One can easily show that $\{\alpha_\xi : \xi \in B\} = A \setminus F$. Thus we are done.   □

Corollary [AC] 8.49. *In any perfect Polish space there exists a λ-subset of cardinality $\mathfrak{b}$.*

Proof. Since every perfect Polish space topologically contains the Baire space ${}^{\omega}\omega$, it suffices to find a λ-subset of ${}^{\omega}\omega$ of cardinality $\mathfrak{b}$. However, by Theorem 5.44 there exists a well-ordered subset of ${}^{\omega}\omega$ of order type $\mathfrak{d}$. □

Theorem [AC] 8.50. *There exists a λ-set $B \subseteq [0,1]$, $|B| = \mathfrak{b}$ such that $B \cup ([0,1] \cap \mathbb{Q})$ is not a λ-set.*

Proof. Let A be the λ-set constructed in the proof of Corollary 8.49. If h is the homeomorphism constructed in the proof of Theorem 3.25 then $B = h(A)$ is a λ-subset of $[0,1]$.

We show that $\mathbb{Q} \cap [0,1]$ is not a G_δ subset of $B \cup (\mathbb{Q} \cap [0,1])$. Assume that it is. Then there are closed sets $F_n \subseteq [0,1]$, $n \in \omega$ such that $F_n \cap \mathbb{Q} = \emptyset$ and $B \subseteq \bigcup_n F_n$. Then every F_n is a compact subset of $[0,1] \setminus \mathbb{Q}$ and therefore every $h^{-1}(F_n)$ is a compact subset of ${}^{\omega}\omega$. Then by Theorem 3.11 every set $h^{-1}(F_n)$ is a bounded subset of ${}^{\omega}\omega$. Thus, $A \subseteq \bigcup_n h^{-1}(F_n)$ is a bounded subset of ${}^{\omega}\omega$ as well, a contradiction. □

Corollary [AC] 8.51. $\operatorname{non}(\lambda\text{-set}) = \mathfrak{b}$.

The family of λ-sets is a family of thin sets. Since any countable set is a λ-set, according to Theorem 8.50 the family of λ-sets is not an ideal. A slightly modified notion leads even to a σ-ideal of subsets of a Polish space. A subset A of a Polish space X is called a λ'-**set** if for any countable set $B \subseteq X$ there exists a G_δ set G such that $B = G \cap (A \cup B)$. Evidently, any λ'-set is a λ-set. The λ-set of Theorem 8.50 is not a λ'-set. One can easily see that a set A is a λ'-set if and only if $A \cup B$ is a λ-set for each countable $B \subseteq X$. Therefore by Corollary 8.51 we obtain

$$\operatorname{non}(\lambda'\text{-set}) = \mathfrak{b}.$$

The property "to be a λ'-set" is external for the class of all topological spaces in **ZFC**.

Theorem [AC] 8.52. *There exists an uncountable λ'-subset of every perfect Polish space.*

Proof. By Theorem 5.57 there exists a strictly increasing sequence of G_δ sets such that $X = \bigcup_{\xi < \omega_1} X_\xi$. Let $A \subseteq X$ be such that $|A \cap (X_\xi \setminus \bigcup_{\eta < \xi} X_\eta)| = 1$ for each $\xi < \omega_1$.

We show that A is a λ'-set. Actually, if $B \subseteq X$ is countable, then there exists a $\xi < \omega_1$ such that $B \subseteq X_\xi$. Then

$$B = (A \cup B) \cap (X_\xi \setminus ((A \setminus B) \cap X_\xi)).$$

Since X_ξ is a G_δ set and $(A \setminus B) \cap X_\xi$ is countable, we are done. □

Theorem [wAC] 8.53. *The family of all λ'-subsets of a Polish space is a σ-ideal.*

Proof. Evidently a subset of a λ'-set is a λ'-set. If $\langle A_n : n \in \omega \rangle$ are λ'-subsets of X and $B \subseteq X$ is countable, then there are G_δ sets $\langle G_n : n \in \omega \rangle$ such that $B = G_n \cap (A_n \cup B)$. One can easily check that $B = (\bigcap_n G_n) \cap (\bigcup_n A_n \cup B)$. $\qquad\square$

Theorem 8.54. *Any $\aleph_1$-Sierpiński set is a λ'-set.*

The proof is easy. If S is an $\aleph_1$-Sierpiński set and B is countable, then $S \cup B$ is also an $\aleph_1$-Sierpiński set and therefore a λ-set. $\qquad\square$

A subset A of a Polish space X is called a **Q-set** if every subset of the set A is an F_σ set in A. The property is internal for the class of all topological spaces in the theory **ZF**. Evidently every Q-set is a σ-set. Since every subset of a Q-set X is Borel and there are only continuum many Borel sets we have $2^{|X|} \leq \mathfrak{c}$ and therefore $|X| < \mathfrak{c}$. We shall see (in Section 9.1) that it is consistent with **ZFC** that there exist Q-sets of any cardinality $< \mathfrak{c}$. As an easy consequence of Corollaries 7.24 and 7.25 we obtain

Theorem [AC] 8.55. *Every Q-set is perfectly meager and has universal measure zero.*

Proof. Actually, if $\{x_\xi : \xi < \kappa\}$ is an enumeration of a Q-set $A \subseteq X$, then for every $S \subseteq \kappa$ the set $\{x_\xi : \xi \in S\}$, being an F_σ set, has the Baire Property and is μ-measurable for any Borel measure μ. By Corollaries 7.24 and 7.25 the set A is meager and $\mu(A) = 0$, provided that μ is diffused Borel measure.

Let $P \subseteq X$ be a perfect set. Then it is easy to see that $A \cap P$ is a Q-subset of the Polish space P. Therefore $A \cap P$ is meager in P. $\qquad\square$

Exercises

8.6 [wAC] Luzin and Sierpiński Sets

  a) Assume that $L \subseteq X$ is an $\aleph_1$-Luzin set. Then $\mathrm{BAIRE}(X)|L = \mathbf{\Delta}_3^0(X)|L$. Especially

$$\mathbf{\Sigma}_1^1(X)|L = \mathbf{\Pi}_1^1(X)|L = \mathbf{\Delta}_3^0(X)|L.$$

  b) Assume that $S \subseteq \mathbb{R}$ is an $\aleph_1$-Sierpiński set. Then $\mathcal{L}|S = \mathbf{\Delta}_2^0(\mathbb{R})|S$. Especially

$$\mathbf{\Sigma}_1^1(\mathbb{R})|S = \mathbf{\Pi}_1^1(\mathbb{R})|S = \mathbf{\Delta}_2^0(\mathbb{R})|S.$$

8.7 [AC] κ-Luzin and κ-Sierpiński Sets

κ is an uncountable regular cardinal and $\langle \mathbb{R}, \mathcal{I} \rangle$ is a Polish ideal space. Assume that $\mathcal{I}$ is invariant under multiplication by a rational and shift by a real and $\kappa = \mathrm{cov}(\mathcal{I}) = \mathrm{cof}(\mathcal{I})$.

  a) There exists a κ-$\mathcal{I}$-set $A \subseteq \mathbb{R}$ such that A is linearly independent over $\mathbb{Q}$.
  Hint: Let $\{A_\xi : \xi < \kappa\}$ be a base of the ideal $\mathcal{I}$. Construct by transfinite induction an increasing sequence of reals $\langle a_\xi \notin A_\xi : \xi < \kappa \rangle$ such that a_ξ does not belong to the vector space over $\mathbb{Q}$ with base $\{a_\eta : \eta < \xi\}$.

  b) There exists a κ-$\mathcal{I}$-set $L \subseteq \mathbb{R}$ such that L is a vector space over $\mathbb{Q}$.
  Hint: In the construction of a) *take a_ξ not belonging to the vector space generated by $\{a_\eta : \eta < \xi\} \cup A_\xi$. The vector space L with the base $\langle a_\xi \notin A_\xi : \xi < \kappa \rangle$ is a κ-$\mathcal{I}$-set.*

c) Formulate the related results for κ-Luzin sets and κ-Sierpiński sets of reals.

d) If $\kappa = \mathrm{cov}(\mathcal{M}) = \mathrm{cof}(\mathcal{M}) = \mathrm{cov}(\mathcal{N}) = \mathrm{cof}(\mathcal{N})$, then there exist an isomorphism f of the additive group $\langle \mathbb{R}, + \rangle$ onto itself and a κ-Luzin set L such that $f(L)$ is a κ-Sierpiński set.

Hint: Take a κ-Luzin set L which is a vector space over $\mathbb{Q}$. Extend the base of L to a Hamel base – see Exercise 7.6, b). Do the same with a κ-Sierpiński set S which is a vector space.

8.8 [AC] κ-$\mathcal{K}_\sigma$-Sets

Let κ be an uncountable regular cardinal.

a) A κ-Luzin subset of ${}^\omega\omega$ is a κ-$\mathcal{K}_\sigma$-set.

b) If there exists a κ-$\mathcal{K}_\sigma$-set L, then $\mathfrak{b} \leq \kappa \leq |L| \leq \mathfrak{d}$.

Hint: See Rothberger's Theorem 7.59 and Corollary 8.36.

c) If there exists a κ-Luzin subsets of $\mathbb{R}$, then

$$\mathrm{non}(\mathcal{M}) \leq \mathfrak{d} = \mathrm{cof}(\mathcal{M}), \quad \mathrm{add}(\mathcal{M}) = \mathfrak{t} \leq \mathrm{cov}(\mathcal{M}).$$

Hint: Use Theorems 7.57 and 7.58, and Corollary 8.36.

8.9 [AC] Strong $\mathcal{I}$-Sets

Assume that $\langle X, \mathcal{I} \rangle$ and $\langle X, \mathcal{J} \rangle$ are Polish ideal spaces with mutually orthogonal shift invariant ideals. A set $A \subseteq X$ is a **strong $\mathcal{I}$-set** if $A + B \neq X$ for every $B \in \mathcal{J}$.

a) Any strong $\mathcal{I}$-set belongs to $\mathcal{I}$.

b) If A is a strong $\mathcal{I}$-set, $|B| < \mathrm{add}(\mathcal{J})$, then $A \cup B$ is a strong $\mathcal{I}$-set as well.

Hint: See the proof of Corollary 8.14.

c) Any set of cardinality $< \mathrm{cov}(\mathcal{J})$ is a strong $\mathcal{I}$-set.

Hint: See the proof of Corollary 8.14.

d) If there exists a κ-$\mathcal{I}$-set, then every set of cardinality $< \kappa$ is a strong $\mathcal{I}$-set.

Hint: See the proof of Lemma 8.34.

8.10 [AC] Rothberger Family

Let $\langle X, \mathcal{I} \rangle$ be a Polish ideal space, κ being an uncountable regular cardinal. A set $\mathcal{R} \subseteq \mathcal{I}$ is a κ-**Rothberger family for $\mathcal{I}$**, if $|\mathcal{R}| \geq \kappa$ and for any $\mathcal{R}_0 \subseteq \mathcal{R}$, $\bigcup \mathcal{R}_0 \neq X$ implies $|\mathcal{R}_0| < \kappa$.

a) If $\kappa = \mathrm{add}(\mathcal{I}) = \mathrm{cov}(\mathcal{I})$, then there exists a κ-Rothberger family for $\mathcal{I}$.

Hint: Find increasing sequence $\{R_\xi : \xi < \kappa\}$ with $\bigcup_{\xi < \kappa} R_\xi = X$ and $\bigcup_{\eta < \xi} R_\eta \in \mathcal{I}$.

b) If there exists a κ-Rothberger family $\mathcal{R}$ for $\mathcal{I}$, then $\mathrm{cov}(\mathcal{I}) \leq \kappa \leq |\mathcal{R}| \leq \mathrm{non}(\mathcal{I})$.

Hint: Set $\mathcal{R}_x = \{R \in \mathcal{R} : x \notin R\}$. Then $|\mathcal{R}_x| < \kappa$. If $|A| < |\mathcal{R}|$, then $A \subseteq R$ for any $R \in \mathcal{R} \setminus \bigcup_{x \in A} \mathcal{R}_x$.

c) Assume that $\langle X, \mathcal{I} \rangle$ and $\langle X, \mathcal{J} \rangle$ are Polish ideal groups, $\mathcal{I}$ and $\mathcal{J}$ are invariant and orthogonal, and there exists a κ-$\mathcal{I}$-set L. Then there exists a κ-Rothberger family for $\mathcal{J}$ of size $|L|$.

Hint: If $A \in \mathcal{I}$ and $B \in \mathcal{J}$, $A \cup B = X$, consider the family $\mathcal{R} = \{B - x : x \in L\}$. If $E \subseteq L$ and $z \notin \bigcup \{B - x : x \in E\}$, then $E \subseteq A - z$. Hence $|E| < \kappa$.

8.11 [AC] λ-set of Maximal Size

a) If $f : X \xrightarrow{\text{onto}} Y$ is continuous, Y is a λ-set, $f^{-1}(y)$ is a λ-set for every $y \in Y$, then X is a λ-set.

   Hint: If $B \subseteq X$ is countable, there exists a countable $C = \langle y_n : n \in \omega \rangle \subseteq Y$ such that $B \subseteq f^{-1}(C)$. Note that $f^{-1}(C)$ is a G_δ set. There exist G_δ sets $\langle G_n : n \in \omega \rangle$ such that $B \cap f^{-1}(y_n) = G_n \cap f^{-1}(y_n)$. Then $B = f^{-1}(C) \setminus \bigcup_n (f^{-1}(y_n) \setminus G_n)$.

b) Show that a statement similar to a) holds true for λ'-sets as well.

c) If $\kappa = \lim_{\xi < \mu} \kappa_\xi$ and there exist λ-subsets of $^\omega \mathbb{R}$ of cardinalities $\langle \kappa_\xi : \xi < \mu \rangle$ and μ, then there exists a λ-subset of $^\omega \mathbb{R}$ of cardinality κ.

   Hint: Assume that $\langle X_\xi : \xi < \mu \rangle, X$ are λ-sets of cardinalities $\langle \kappa_\xi : \xi < \mu \rangle, \mu$, $X = \{x_\eta : \eta < \mu\}$. Denote $A = \{[x_\xi, y] \in {}^\omega \mathbb{R} \times {}^\omega \mathbb{R} : y \in X_\xi\}$ and $f(x, y) = x$. Note that $|A| = \kappa$ and $A = f^{-1}(X)$ with λ-sets $f^{-1}(x)$ for any $x \in X$.

d) If there exists no inaccessible cardinal $\leq \mathfrak{c}$, then there exists a λ-set of maximal size, i.e., there exists a maximal element of the set $\{|A| : A$ is a λ-set$\}$.

e) Prove a similar result for λ'-sets.

8.12 Hausdorff Gap

Let $\langle \{\alpha_\xi\}_{\xi < \omega_1}, \{\beta_\xi\}_{\xi < \omega_1} \rangle$ be a Hausdorff gap. Let $H = \{\alpha_\xi : \xi < \omega_1\} \cup \{\beta_\xi : \xi < \omega_1\} \subseteq {}^\omega 2$.

a) H is a λ'-set.

   Hint: We know that $A_\xi = \{\gamma \in {}^\omega 2 : \alpha_\xi \leq^* \gamma \leq^* \beta_\xi\}$ is an F_σ set and $\bigcap_{\xi < \omega_1} A_\xi = \emptyset$. If $X \subseteq {}^\omega 2$ is countable, then $X \cap A_\xi = \emptyset$ for some ξ and $X = (H \cup X) \cap (({}^\omega 2 \setminus F_\xi) \setminus Y)$, where $Y = ((H \cup X) \setminus F_\xi) \setminus X)$ is countable.

b) H is perfectly meager.

c) If μ is a diffused Borel measure on $^\omega 2$, then there exists a ξ_0 such that $\mu(A_\xi) = 0$ for every $\xi \geq \xi_0$.

   Hint: The sequence $\mu(A_\xi)$ must stabilize. Assume that $\mu(A_\xi) = \varepsilon > 0$ for $\xi \geq \xi_0$. Let $\beta(n)$ be such that $\mu(\{\alpha \in A_{\xi_0} : \alpha(n) = \beta(n)\}) \geq \varepsilon/2$. Let $\zeta \geq \xi_0$ be such that $\beta \notin A_\zeta$. Set

   $$L = \{n \in \omega : \alpha_{\zeta_0}(n) = \beta_\zeta(n)\}, \quad B_m = \{\alpha : (\forall n \in L, n \geq m)\, \alpha(n) = \alpha_\zeta(n)\}.$$

   We can assume that the set L is infinite and $A_\zeta = \bigcup_m B_m$. There exists an m such that $\mu(B_m) > \varepsilon/2$. Since L is infinite there exists a $k > m$ such that $\beta(k) \neq \beta_\zeta(k)$. The sets $\{\alpha \in A_\zeta : \alpha(k) \neq \beta_\zeta(k)\}, B_m$ are disjoint subsets of A_{ξ_0}, the former has measure $\geq \varepsilon/2$ and the later has measure $> \varepsilon/2$, a contradiction.

d) H has universal measure zero.

   Hint: $H = A_{\xi_0} \cup (H \setminus A_{\xi_0})$, $H \setminus A_{\xi_0}$ is countable.

8.13 Products

a) If A, B have universal measure zero then $A \times B$ has also universal measure zero.

   Hint: If μ is a diffused finite Borel measure on $A \times B$ set $\nu(E) = \mu(E \times B)$. Since B has universal measure zero, ν is diffused.

b) If A, B are λ-sets, then $A \times B$ is a λ-set. The same holds true for λ'-sets.

   Hint: If $E \subseteq A \times B$ is countable take countable $C \subseteq A$ and $D \subseteq B$ such that $E \subseteq C \times D$. $C \times D$ is a G_δ set and $(C \times D) \setminus E$ being countable is an F_σ set.

8.3 Sequence Convergence Properties

By Theorem 3.63 and Lemma 3.62 the space $C_p(X)$ does not possess the sequence selection property and therefore neither is a Fréchet space, provided that the Polish space X contains a perfect subset. However, if X is countable then $C_p(X)$ is Fréchet. Hence a natural question arises whether is there some uncountable X such that $C_p(X)$ is a Fréchet space?

Investigating thin sets of trigonometric series (compare Section 8.6), for a given sequence of continuous functions $\langle f_n : X \longrightarrow [0, +\infty) : n \in \omega \rangle$ converging pointwise to 0 on X, one needs to choose a subsequence $\langle f_{n_k} : k \in \omega \rangle$ such that

$$\sum_{k=0}^{\infty} f_{n_k}(x) < \infty \text{ for every } x \in X. \tag{8.5}$$

A topological space for which it is possible is called a Σ-**space**. Again, every countable set is a Σ-space.

If we replace the pointwise convergence by the uniform one, then one can easily choose a subsequence such that the series $\sum_{k=0}^{\infty} f_{n_k}$ converges even normally on X. However, the uniform convergence is usually a quite strong assumption. Therefore we look for a weaker type of convergence which eventually gives us uncountable sets with the desired property. The quasi-normal convergence is the natural candidate. Actually, if $f_n \xrightarrow{\text{QN}} 0$ on X, then there exists an increasing sequence $\{n_k\}_{k=0}^{\infty}$ such that (8.5) holds true.

Let X be a topological space. X is called a **QN-space** if every sequence of continuous real functions converging pointwise to 0 on X converges quasi-normally to 0 on X as well. This notion turns out to be quite strong. The following notion seems to be much more appropriate. X is called a **weak QN-space** or shortly a **wQN-space** if for every sequence $\{f_n\}_{n=0}^{\infty}$ of continuous real functions converging pointwise to 0 on X there exists an increasing sequence of integers $\{n_k\}_{k=0}^{\infty}$ such that $f_{n_k} \xrightarrow{\text{QN}} 0$ on X. Thus a wQN-space is a Σ-space.

Finally, X is called an **mQN-space** if for every sequence $\{f_n\}_{n=0}^{\infty}$ of continuous functions such that $f_n \searrow 0$ on X also $f_n \xrightarrow{\text{QN}} 0$ on X. By Theorem 3.50 every compact space is an mQN-space. Note that if $f_n \geq f_{r+1}$ for every n and $f_{n_k} \xrightarrow{\text{QN}} 0$ for some increasing sequence $\{n_k\}_{k=0}^{\infty}$, then also $f_n \xrightarrow{\text{QN}} 0$.

Note the following. If $f_n \searrow 0$ on X and $f_{n_k} \xrightarrow{\text{QN}} 0$ on X for some sequence $\{n_k\}_{k=0}^{\infty}$, then $f_n \xrightarrow{\text{QN}} 0$ on X as well. Thus a weak mQN-space is an mQN-space.

Theorem 8.56. *Every Σ-space is an mQN-space.*

Proof. Assume that X is a Σ-space and $f_n \searrow 0$ are continuous. Then the series $\sum_{k=0}^{\infty} f_{n_k}$ converges for some increasing sequence $\{n_k\}_{k=0}^{\infty}$. We show that $f_{n_k} \xrightarrow{\text{QN}} 0$ on X with the control $\{n_k^{-1/2}\}_{k=0}^{\infty}$.

Assume not. Then for some $x \in X$ there exists an arbitrarily large integer k such that $f_{n_k}(x) \geq n_k^{-1/2}$. Then

$$\sum_{i=0}^{n_k} f_{n_i}(x) \geq (n_k + 1) \cdot n_k^{-1/2} \geq n_k^{1/2},$$

which is a contradiction. $\square$

So, we have the implications

$$\text{QN-space} \to \text{wQN-space} \to \Sigma\text{-space} \to \text{mQN-space}.$$

Let us recall that if $A \subseteq \omega$ is an infinite set, then $\{e_A(n)\}_{n=0}^{\infty}$ is the increasing sequence enumerating A.

Lemma 8.57. *Let X be a wQN-space, $f_n \to 0$ on X. If $\{\varepsilon_n\}_{n=0}^{\infty}$ is a given sequence of positive reals converging to 0, then for any infinite set $A \subseteq \omega$ there exists an infinite set $B \subseteq A$ such that $f_{e_B(n)} \xrightarrow{\text{QN}} 0$ with control $\{\varepsilon_n\}_{n=0}^{\infty}$.*

The proof is easy. By assumption we have $f_{e_A(n)} \to 0$ on X. Since X is a wQN-space, there exists an infinite subset $C \subseteq A$ such that $f_{e_C(n)} \xrightarrow{\text{QN}} 0$. By Theorem 3.51 there exists an infinite subset $B \subseteq C$ such that $f_{e_B(n)} \xrightarrow{\text{QN}} 0$ on X with control $\{\varepsilon_n\}_{n=0}^{\infty}$. $\square$

Theorem [AC] 8.58. *Consider the following properties of a topological space X: "$C_p(X)$ is a Fréchet space", "X is a QN-space", "X is a wQN-space", "X is a Σ-space", "X is an mQN-space". If X is perfectly normal, then any of them is preserved under continuous image and under passing to a closed subspace.*

Proof. Assume that $F : X \xrightarrow{\text{onto}} Y$, F continuous and $C_p(X)$ is a Fréchet space. Let $A \subseteq C_p(Y)$, $f \in \overline{A}$. Set $B = \{F \circ g : g \in A\} \subseteq C_p(X)$. If $x_1, \ldots, x_k \in X$, $\varepsilon > 0$ and

$$U = \{h \in C_p(X) : |h(x_1) - f(F(x_1))| < \varepsilon \wedge \cdots \wedge |h(x_k) - f(F(x_k))| < \varepsilon\}$$

is a neighborhood of $F \circ f$ in $C_p(X)$, then

$$V = \{g \in C_p(Y) : |g(F(x_1)) - f(F(x_1))| < \varepsilon \wedge \cdots \wedge |g(F(x_k)) - f(F(x_k))| < \varepsilon\}$$

is a neighborhood of f in $C_p(Y)$. Since $f \in \overline{A}$ there exists a $g \in A \cap V$. Then also $F \circ g \in B \cap U$. Thus $F \circ f \in \overline{B}$. Since $C_p(X)$ is a Fréchet space there exists a sequence $\langle g_n \in B : n \in \omega \rangle$ such that $g_n \to F \circ f$ on X. Every g_n has the form $g_n = F \circ f_n$ with $f_n \in A$. Evidently $f_n \to f$ on Y. So $C_p(Y)$ is a Fréchet space.

For QN-, wQN-, Σ- and mQN-space the proof is even simpler.

Lemma 3.62, a) actually says that the property "$C_p(x)$ is a Fréchet space" is preserved by passing to a closed subspace. In the other cases the assertion easily follows by Corollary 3.21. $\square$

Since the interval $[0, 1]$ does not possess any of these properties, by Theorem 8.22 and Theorem 3.26 we obtain

Corollary [AC] 8.59. *Every completely regular topological space X with any of the properties "$C_p(X)$ is a Fréchet space", "X is a QN-space", "X is a wQN-space", "X is a Σ-space", is an nCM-space and therefore has a clopen base of topology. Thus, a separable metric space with any of those properties is homeomorphic to a set of reals.*

Note that this need not be true for an mQN-space.

Theorem [AC] 8.60. *Let $X = \bigcup_{s \in S} X_s$, $|S| < \mathfrak{b}$. If the sequence $\{f_n\}_{n=0}^{\infty}$ converges quasi-normally to a function f on each $\langle X_s : s \in S \rangle$, then it does so on the union X.*

Proof. Let $\{\varepsilon_n^s\}_{n=0}^{\infty}$ witness the quasi-normal convergence $f_n \xrightarrow{\text{QN}} f$ on X_s for every $s \in S$. We can assume that each $\{\varepsilon_n^s\}_{n=0}^{\infty}$ is non-decreasing. We define

$$\alpha_s(k) = \min\{n : \varepsilon_n^s \leq 2^{-k} \wedge n > \alpha_s(k-1)\}.$$

Since the cardinality of S is smaller than $\mathfrak{b}$, there exists a function $\alpha \in {}^{\omega}\omega$ eventually bounding the set $\{\alpha_s : s \in S\}$. We write

$$\varepsilon_n = \begin{cases} 1 & \text{for } n < \alpha(1), \\ 2^{-k} & \text{for } \alpha(k) \leq n < \alpha(k+1). \end{cases}$$

One can easily see that $f_n \xrightarrow{\text{QN}} f$ on X with the control $\{\varepsilon_n\}_{n=0}^{\infty}$. $\square$

The following result expresses the main additivity properties of considered spaces and their consequences.

Theorem [AC] 8.61.

a) *If $X = \bigcup_{s \in S} X_s$, $|S| < \mathfrak{b}$, each X_s is a QN-space, then X is a QN-space.*

b) *If $X = \bigcup_{s \in S} X_s$, $|S| < \mathfrak{h}$, each X_s is a wQN-space, then X is a wQN-space.*

c) *If $X = \bigcup_{s \in S} X_s$, $|S| < \mathfrak{b}$, each X_s is an mQN-space, then X is an mQN-space.*

d) *Any F_{σ}-subset of a QN-space, a wQN-space or an mQN-space is a QN-space, a wQN-space or an mQN-space, respectively.*

e) *Assume that Y is a metric space and a mapping $f : X \longrightarrow Y$ is a quasi-normal limit of continuous functions. If X is a QN-space, a wQN-space or an mQN-space, then $f(X)$ is a QN-space, a wQN-space or an mQN-space, respectively, as well.*

Proof. The assertions a) and c) follow immediately from definitions by Theorem 8.60. The assertions of d) follow from a), b) and c) by Theorem 8.58, respectively. Thus, we have to prove the assertion b).

Assume that $X = \bigcup_{s \in S} X_s$, $|S| < \mathfrak{h}$ and each X_s is a wQN-space. Let $f_n \to 0$ on X. Set

$$\mathcal{H}_s = \{A \in [\omega]^\omega : f_{e_A(k)} \xrightarrow{\text{QN}} 0 \text{ on } X_s \text{ with control } \{2^{-k}\}_{k=0}^\infty\}.$$

Every space X_s is a wQN-space, so by Lemma 8.57 the family $\mathcal{H}_s$ is a dense subset of $\langle [\omega]^\omega, \subseteq^* \rangle$. Since we assume that $|S| < \mathfrak{h}$ there exists a common refinement $\mathcal{H}$ of all $\mathcal{H}_s$ modulo finite. One can easily see that $f_{e_A(k)} \xrightarrow{\text{QN}} 0$ on X with control $\{2^{-k}\}_{k=0}^\infty$ for any $A \in \mathcal{H}$.

Assume that $f_n \xrightarrow{\text{QN}} f$ on X with control $\{\varepsilon_n\}_{n=0}^\infty$, and f_n are continuous. Then $X = \bigcup_k X_k$, where $X_k = \{x \in X : (\forall m, n \geq k), |f_n(x) - f_m(x)| \leq \varepsilon_n + \varepsilon_m\}$ are closed and $f_n \rightrightarrows f$ on each X_k. The assertion e) follows by Theorem 8.58 and a), b) and c). $\qquad\square$

A discrete space of cardinality $\mathfrak{b}$ is neither a wQN-space nor an mQN-space, thus:

Corollary [AC] 8.62.

$$\text{add(QN-space)} = \text{non(QN-space)} = \text{add(mQN-space)} = \text{non(mQN-space)} = \mathfrak{b},$$
$$\mathfrak{h} \leq \text{add(wQN-space)} \leq \text{non(wQN-space)} = \mathfrak{b}.$$

Theorem [AC] 8.63. *If $\kappa \leq \mathfrak{b}$ and S is a κ-Sierpiński subset of a Polish space X with a finite Borel measure μ, then S is a QN-set.*

Proof. Assume that S is a κ-Sierpiński set, $\langle f_n : n \in \omega \rangle$ are continuous. Let $f_n \to 0$ on S. By the Egoroff Theorem 7.41, for every m there exists a Borel set $C_m \subseteq X$ such that $f_n \rightrightarrows 0$ on $S \setminus C_m$ and $\mu(C_m) < 1/(m+1)$. We can assume that $C_{m+1} \subseteq C_m$. Then $f_n \xrightarrow{\text{QN}} 0$ on

$$\bigcup_m (S \setminus C_m) = S \setminus \bigcap_m C_m.$$

Since $\mu(\bigcap_m C_m) = 0$, we have $|S \cap \bigcap_m C_m| < \kappa \leq \mathfrak{b}$ and therefore $S \cap \bigcap_m C_m$ is a QN-set. So $f_n \xrightarrow{\text{QN}} 0$ on S. $\qquad\square$

An $\mathcal{L}^*$-space X has the **property** (α_i), $i = 1, 2, 3, 4$, if for any $x \in X$ and for any sequence $\{\{x_{n,m}\}_{m=0}^\infty\}_{n=0}^\infty$ of sequences converging to x, there exists a sequence $\{y_m\}_{m=0}^\infty$ such that $\lim_{m \to \infty} y_m = x$ and

(α_1) $\{x_{n,m} : m \in \omega\} \subseteq^* \{y_m : m \in \omega\}$ for each n,

(α_2) $\{x_{n,m} : m \in \omega\} \cap \{y_m : m \in \omega\}$ is infinite for each n,

(α_3) $\{x_{n,m} : m \in \omega\} \cap \{y_m : m \in \omega\}$ is infinite for infinitely many n,

(α_4) $\{x_{n,m} : m \in \omega\} \cap \{y_m : m \in \omega\} \neq \emptyset$ for infinitely many n.

It is easy to see that

$$(\alpha_1) \to (\alpha_2) \to (\alpha_3) \to (\alpha_4). \tag{8.6}$$

There is an important (maybe surprising) relationship between the introduced notions in the case of $C_p(X)$.

Theorem 8.64 (M. Scheepers – D.H. Fremlin). *For a topological space X the following are equivalent:*

a) X *is a* wQN-*space.*

b) *The space* $C_p(X)$ *possesses the sequence selection property.*

c) *The space* $C_p(X)$ *possesses the property* (α_2).

d) *The space* $C_p(X)$ *possesses the property* (α_3).

e) *The space* $C_p(X)$ *possesses the property* (α_4).

Proof. e) $\to$ a). Assume that $C_p(X)$ possesses the property (α_4), $\langle f_n : n \in \omega \rangle$ are continuous, and $f_n \to 0$ on X. We can assume that $f_n > 0$ for every n. Set $f_{n,m} = 2^n \cdot f_{n+m}$. Then $f_{n,m} \to 0$ on X for every n. By (α_4), there exist a sequence $\{m_k\}_{k=0}^{\infty}$ and an increasing sequence $\{n_k\}_{k=0}^{\infty}$ such that $f_{n_k,m_k} \to 0$ on X. We can assume that the sequence $\{m_k + n_k\}_{k=0}^{\infty}$ is increasing. We claim that $f_{m_k+n_k} \xrightarrow{\text{QN}} 0$ on X with control $\{2^{-n_k}\}_{k=0}^{\infty}$. Actually, for any $x \in X$ there exists a k_0 such that $f_{n_k,m_k}(x) < 1$ for $k \geq k_0$. Thus $f_{m_k+n_k}(x) < 2^{-n_k}$ for every $k \geq k_0$.

a) $\to$ b). Assume that X is a wQN-space. Let $f_{n,m} \to 0$ on X for every n. We set

$$g_m(x) = \sum_{n=0}^{\infty} \min\{2^{-n}, f_{n,m}(x)\}, \quad x \in X. \tag{8.7}$$

Let $x \in X$, $\varepsilon > 0$ being a real. Then there exists an n_0 such that $2^{-n_0+2} < \varepsilon$. For every $n < n_0$ there exists an m_n such that $f_{n,m}(x) < \varepsilon/(2n_0)$ for every $m \geq m_n$. We set $k = \max\{m_n : n < n_0\}$. For $m \geq k$ we have

$$g_m(x) \leq \sum_{n<n_0} \frac{\varepsilon}{2n_0} + \sum_{n \geq n_0} 2^{-n} < \varepsilon.$$

Thus $\lim_{m \to \infty} g_m(x) = 0$. Since X is a wQN-space there exists an increasing sequence $\{m_n\}_{n=0}^{\infty}$ such that $g_{m_n} \xrightarrow{\text{QN}} 0$ on X with the control $\{2^{-n}\}_{n=0}^{\infty}$. However by (8.7), if $g_{m_n}(x) < 2^{-n}$, then also $f_{n,m_n}(x) < 2^{-n}$. Thus $f_{n,m_n} \xrightarrow{\text{QN}} 0$ on X.

b) $\to$ c). Assume that $C_p(X)$ possesses the sequence selection property. Let $f_{n,m} \to 0$ on X for every n. We set $h_{n,m} = f_{\lambda(n),m}$, where λ is the left inverse to the pairing function. Therefore the sequence $\{h_{n,m}\}_{n=0}^{\infty}$ contains every $f_{i,m}$ infinitely many times. Let $g_{n,m} = \max\{h_{i,m} : i \leq n\}$. Evidently $g_{n,m} \to 0$ for every n. By Theorem 1.40 there exist increasing sequences $\{n_k\}_{k=0}^{\infty}$ and $\{m_k\}_{k=0}^{\infty}$ such that $\lim_{k \to \infty} g_{n_k,m_k} = 0$. We set $f_i = h_{i,m_k}$ for $n_{k-1} < i \leq n_k$. Then $f_i \leq g_{n_k,m_k}$ for $n_{k-1} < i \leq n_k$ and therefore $f_i \to 0$ on X. Every sequence $\{f_{n,m}\}_{m=0}^{\infty}$ contains infinitely many members of $\{f_i\}_{i=0}^{\infty}$.

Now the theorem follows by (8.6). $\qquad\square$

Since a Fréchet space possesses the sequence selection property we obtain:

Corollary 8.65. *If* $C_p(X)$ *is a Fréchet space, then* X *is a* wQN*-space.*

There is a similar result for QN-space.

Theorem 8.66 (M. Scheepers). *If* $C_p(X)$ *has the property* (α_1), *then the topological space* X *is a* QN*-space.*

Proof. Let $\{f_m\}_{m=0}^\infty$ be a sequence of continuous functions converging pointwise to 0 on X. For each m and n we set $f_{n,m}(x) = 2^n \cdot |f_m(x)|$. Then the sequence $\{f_{n,m}\}_{m=0}^\infty$ converges pointwise to 0 on X for each n. By (α_1) there exists a sequence $\{h_k\}_{k=0}^\infty$ converging to 0 on X and such that the sequence $\{h_k\}_{k=0}^\infty$ contains all but finitely many members of the sequence $\{f_{n,m}\}_{m=0}^\infty$ for every n. Thus, there exists a sequence $\{m_n\}_{n=0}^\infty$ such that $\{f_{n,m}\}_{m=m_n}^\infty$ is a subsequence of $\{h_k\}_{k=0}^\infty$ for each n. We can assume that $\{m_n\}_{n=0}^\infty$ is increasing. Moreover, we can assume that

$$(\forall m \geq m_n)(\forall i)\,(h_i = f_{n,m} \to i \geq n)$$

for every n. Now we set

$$\varepsilon_m = 2^{-k} \text{ for } m_k \leq m < m_{k+1}, \quad \varepsilon_m = 1 \text{ for } m < m_0.$$

Let $x \in X$ be given. Then there exists a k_0 such that $|h_k(x)| < 1$ for $k \geq k_0$. For any $m \geq m_{k_0}$ there exists a $k \geq k_0$ such that $m_k \leq m < m_{k+1}$. Then $f_{k,m} = h_i$ for some $i \geq k \geq k_0$. Hence $f_{k,m}(x) < 1$. Since $\varepsilon_m = 2^{-k}$ we obtain $|f_m(x)| < \varepsilon_m$. $\qquad\qquad\square$

Theorem 8.67 (L. Bukovský – J. Haleš – M. Sakai). *If* X *is a* QN*-space, then* $C_p(X)$ *possesses the property* (α_1).

Proof. Let $\{\{f_{n,m}\}_{m=0}^\infty\}_{n=0}^\infty$ be a sequence of sequences converging to 0 on X. We can assume that values of each $f_{n,m}$ are in $[0,1]$. We define the functions g_m by (8.7). Then g_m are continuous and $g_m \to 0$ on X.

Since X is a QN-space, there exist positive reals $\{\varepsilon_n\}_{n=0}^\infty$, $\varepsilon_n \to 0$ such that

$$(\forall x)(\exists l_x)(\forall m \geq l_x)\, g_m(x) < \varepsilon_m. \tag{8.8}$$

There are also natural numbers m_k such that

$$(\forall k)(\forall m \geq m_k)\, \varepsilon_m < 2^{-k}.$$

We can assume that $m_k < m_{k+1}$ for any k. We claim that the sequence (in any order)

$$\{f_{n,m} : n \in \omega \wedge m \geq m_n\} \tag{8.9}$$

converges to 0 on X.

Let $x \in X$ and $\varepsilon > 0$. Take such a k_0 that $2^{-k_0} < \varepsilon$ and $m_{k_0} > l_x$ (l_x is that of (8.8)). Moreover, take such a natural number p that $f_{n,m}(x) < \varepsilon$ for $m \geq p$ and any $n < k_0$. If $n \geq k_0$ and $m \geq m_n \geq m_{k_0} > l_x$ then

$$g_m(x) < \varepsilon_m < 2^{-n}$$

and therefore

$$\min\{2^{-n}, f_{n,m}(x)\} < 2^{-n} \leq 2^{-k_0} < \varepsilon.$$

Thus $f_{n,m}(x) < \varepsilon$ for any $n \geq k_0$, any $m \geq m_n, p$ and for any $n < k_0$. Therefore for all members of the sequence (8.9) but for finitely many those with $n < k_0$ and $m < p$ we have $f_{n,m}(x) < \varepsilon$. $\qquad\square$

Corollary 8.68. *A topological space X is a QN-space if and only if $C_p(X)$ possesses the property (α_1).*

Theorem 8.69. *The Cantor middle-third set $\mathbb{C}$ is neither a Σ-set nor a wQN-space.*

Proof. The sequence of functions $\langle h_n : n \in \omega \rangle$ constructed in the proof of Theorem 3.63 witnesses that $\mathbb{C}$ is not a Σ-space. $\qquad\square$

By Theorem 3.20 we obtain

Corollary [wAC] 8.70. *If X is a metric Σ-space, then X does not contain a subset homeomorphic to $\mathbb{C}$. Moreover, if we assume **AC**, then no perfectly normal topological Σ-space contains a subset homeomorphic to $\mathbb{C}$.*

If we set

$$g_n(\alpha) = \frac{1}{\min\{k : \alpha(k) + k > n\} + 1} \quad \text{for } \alpha \in {}^{\omega}\omega, \tag{8.10}$$

then we have

Lemma 8.71. *If $X \subseteq {}^{\omega}\omega$, then X is eventually bounded if and only if there exists a sequence $\{n_k\}_{k=0}^{\infty}$ such that $g_{n_k} \xrightarrow{\text{QN}} 0$ on X.*

Theorem 8.72. *A continuous image of a wQN-space into ${}^{\omega}\omega$ is eventually bounded.*

The converse assertion is not true since $\mathbb{C} = {}^{\omega}2$ is a bounded subset of ${}^{\omega}\omega$.

Corollary 8.73. *Let X be a wQN-space. If $f_n \xrightarrow{\text{QN}} \vec{\jmath}$, where $f_n : X \longrightarrow [0,1]$ are continuous and $f : X \longrightarrow {}^{\omega}\omega$, then $f(X) \subseteq {}^{\omega}\omega$ is eventually bounded.*

Proof. By Theorem 8.61, e), $f(X) \subseteq {}^{\omega}\omega$ is a wQN-space. $\qquad\square$

Theorem [wAC] 8.74. *Any wQN-subset A of a separable metric space X is perfectly meager.*

Proof. Let $P \subseteq X$ be a perfect set, $f_n : P \longrightarrow [0,1]$ being functions of Theorem 7.43. Since P is closed, by Corollary 3.21 we can assume that each f_n is defined on the whole space X and $f_n \to 0$ on X. Since A is a wQN-set, then there exists an increasing sequence of natural numbers $\{n_k\}_{k=0}^{\infty}$ such that $f_{n_k} \xrightarrow{\text{QN}} 0$ on A. Then there are sets A_m closed in A such that $A = \bigcup_m A_m$ and $f_{n_k} \rightrightarrows 0$ on each A_m. By Theorem 7.43 every set $P \cap A_m$ is nowhere dense in P and therefore $A \cap P$ is meager in P. $\qquad\square$

Corollary [wAC] 8.75. *Any* wQN-*subset of a perfect separable metric space is meager.*

A topological space X has the **quasi-normal sequence selection property**, shortly **QSSP**, if for any functions $f, f_n, f_m^n : X \longrightarrow \mathbb{R}$, $n, m \in \omega$, such that

(1) $f_n \xrightarrow{\text{QN}} f$ on X,

(2) $f_m^n \xrightarrow{\text{QN}} f_n$ on X for every $n \in \omega$,

(3) every f_m^n is continuous,

there exists an increasing $\beta \in {}^\omega\omega$ such that $f_{\beta(n)}^n \xrightarrow{\text{QN}} f$ on X.

Theorem 8.76 (L. Bukovský – J. Šupina). *Any* QN-*space has the* QSSP *property.*

Proof. Assume that $f, f_n, f_m^n : X \longrightarrow \mathbb{R}$, $n, m \in \omega$ are such that (1)–(3) hold true. We can assume that the control of the quasi-normal convergence in (2) is $\{2^{-2m-1}\}_{m=0}^\infty$ and the control of the quasi-normal convergence in (1) is $\{\varepsilon_n\}_{n=0}^\infty$. Set

$$g_m^n(x) = \min\{|f_m^n(x) - f_{m+1}^n(x)| \cdot 2^m, 1\}.$$

Evidently for a fixed $n \in \omega$ we have $g_m^n \to 0$ on X. Since the space $C_p(X)$ satisfies the condition (α_1), there exists an increasing function $\beta \in {}^\omega\omega$ such that the set $\{g_m^n : m \geq \beta(n) \wedge n \in \omega\}$ converges to 0.

We claim that $f_{\beta(n)}^n \xrightarrow{\text{QN}} f$ with the control $\{2^{-\beta(n)+1} + \varepsilon_n\}_{n=0}^\infty$. Actually, let $x \in X$. Then there exists an n_0 such that $g_m^n(x) < 1$ for any $n \geq n_0$ and any $m \geq \beta(n)$. Moreover, we can assume that $|f_n(x) - f(x)| < \varepsilon_n$ for $n \geq n_0$. Hence $|f_m^n(x) - f_{m+1}^n(x)| < 2^{-m}$ and $|f_{\beta(n)}^n(x) - f_m^n(x)| < 2^{-\beta(n)+1}$ for any $n \geq n_0$ and any $m \geq \beta(n)$. Thus $|f_{\beta(n)}^n(x) - f_n(x)| \leq 2^{-\beta(n)+1}$ for $n \geq n_0$. Therefore, for $n \geq n_0$ we obtain $|f_{\beta(n)}^n(x) - f(x)| < 2^{-\beta(n)+1} + \varepsilon_n$. $\qquad\square$

Theorem [AC] 8.77 (I. Recław). *If a perfectly normal topological space X has the property* QSSP, *then X is a σ-space. Therefore every perfectly normal topological* QN-*space is a σ-space.*

Proof. Assume that $F = \bigcup_n F_n$ is an F_σ set, F_n is closed and $F_n \subseteq F_{n+1}$ for any $n \in \omega$. We have to show that the characteristic function χ_F is G_δ-measurable.

Since X is perfectly normal, there exist closed sets $\langle F_{n,m} : n, m \in \omega \rangle$ such that $F_{n,m} \subseteq F_{n,m+1}$ and $X \setminus F_n = \bigcup_k F_{n,k}$ for any n and m. For any n and m, there exists a continuous function $f_{n,m} : X \longrightarrow [0,1]$ such that $f_{n,m}(x) = 1$ for $x \in F_n$ and $f_{n,m}(x) = 0$ for $x \in F_{n,m}$. Evidently $f_{n,m} \xrightarrow{\text{D}} \chi_{F_n}$ on X. Moreover, $\chi_{F_n} \xrightarrow{\text{D}} \chi_F$ on X.

By QSSP there exists a β such that $f_{n,\beta(n)} \xrightarrow{\text{QN}} \chi_F$. Since $f_{n,\beta(n)}$ are continuous, by Theorem 3.57 the function χ_F is G_δ-measurable. □

Corollary [AC] 8.78. *Every subset $A \subseteq X$ of a metric separable QN-space X is a QN-space.*

Proof. If $\langle f_n : A \longrightarrow [0,1] : n \in \omega \rangle$ are continuous, $f_n \to 0$ on A, then by Kuratowski's Theorem 3.29 there exists a G_δ set $B \supseteq A$ and a sequence of continuous functions $\langle g_n : B \longrightarrow [0,1] : n \in \omega \rangle$ such that $g_n|A = f_n$ for every n. Set $C = \{x \in B : g_n(x) \to 0\}$. Then $C \supseteq A$ is Borel and therefore also an F_σ set. By Theorem 8.61 we obtain $g_n \xrightarrow{\text{QN}} 0$ on C. □

It is easy to see that neither the corollary and hence nor the theorem holds true for any topological space. If $X \subseteq {}^\omega\omega$ is unbounded, then X endowed with the discrete topology is not a QN-space. However, one can easily see that a one-point compactification $X^* = X \cup \{\infty\}$ of X is a QN-space, since for any continuous function $f : X^* \longrightarrow [0,1]$ and any $\varepsilon > 0$ we have $|f(x) - f(\infty)| < \varepsilon$ for all but finitely many $x \in X$.

By simple refining of Lemma 3.74 we obtain

Lemma [AC] 8.79. *If $\operatorname{Ind}(X) = 0$, then every simple $\mathbf{\Delta}_2^0$-measurable function $g : X \longrightarrow [0,1]$ is a discrete limit of a sequence $\{g_n\}_{n=0}^\infty$ of simple continuous functions.*

Proof. Let $g = \sum_{i=0}^k a_i \chi_{A_i}$, $A_i \in \mathbf{\Delta}_2^0$ be pairwise disjoint, $\bigcup_{i=0}^k A_i = X$ and $0 \le a_0 < a_1 < \cdots < a_k \le 1$. Then there exist non-decreasing and non-increasing sequences $\{F_n^i\}_{n=0}^\infty$ and $\{G_n^i\}_{n=0}^\infty$ of F_σ and G_δ sets, respectively, such that $A_i = \bigcup_n F_n^i = \bigcap_n G_n^i$. Since $\operatorname{Ind}(X) = 0$, there exists clopen sets C_n^i such that $F_n^i \subseteq C_n^i \subseteq G_n^i$ for every $i \le k$ and every $n \in \omega$. Replacing eventually C_n^0 by $C_n^0 \cup (X \setminus \bigcup_{0 < i \le k} C_n^i)$, we can assume that $\bigcup_{i \le k} C_n^i = X$. Let $D_n^i = C_n^i \setminus \bigcup_{j < i} C_n^j$. Then D_n^i are pairwise disjoint and $\bigcup_{i \le k} D_n^i = X$. Set $g_n = \sum_{i=0}^k a_i \chi_{D_n^i}$. Since each D_n^i is clopen, g_n is continuous.

Let $x \in X$ and $g(x) = a_i$. Then there exists an n_0 such that $x \in F_n^i$ and $x \notin G_n^j$ for $j < i$, for every $n \ge n_0$. Then $x \in D_n^i$ and therefore $g_n(x) = a_i$. □

Theorem [AC] 8.80. *If X is a normal topological space possessing property QSSP, then any Borel measurable function $f : X \longrightarrow [0,1]$ is a quasi-normal limit of a sequence of continuous functions.*

Proof. If $f : X \longrightarrow {}^\omega\omega \subseteq [0,1]$ is Borel measurable, then by Recław's Theorem 8.77 the function f is $\mathbf{\Delta}_2^0$-measurable.

For any n and any $i < 2^n - 1$, we write

$$A_n^i = \left\{ x \in X : \frac{i}{2^n} \le f(x) < \frac{i+1}{2^n} \right\}, \quad A_n^{2^n-1} = \left\{ x \in X : \frac{2^n-1}{2^n} \le f(x) \right\}.$$

Then the sequence of simple $\mathbf{\Delta}_2^0$-measurable functions

$$f_n = \sum_{i=0}^{2^n-1} \frac{i}{2^n} \chi_{A_n^i}$$

converges uniformly to f with control 2^{-n}. By the lemma, for every n there exists a sequence $\{g_m^n\}_{m=0}^{\infty}$ of simple continuous functions such that $g_m^n \xrightarrow{\mathrm{D}} f_n$ on X. Thus, by Theorem 8.76 there exists an increasing $\alpha \in {}^\omega\omega$ such that $g_{\alpha(n)}^n \xrightarrow{\mathrm{QN}} f$. $\qquad\square$

Theorem [AC] 8.81 (B. Tsaban – L. Zdomskyy). *The image of a perfectly normal topological* QN-*space X by a Borel measurable function into ${}^\omega\omega$ is eventually bounded.*

Proof. Assume that $f : X \longrightarrow {}^\omega\omega \subseteq [0,1]$ is Borel measurable. By Theorem 8.80 there exists a sequence of continuous functions $\langle f_n : X \longrightarrow [0,1] : n \in \omega \rangle$ such that $f_n \xrightarrow{\mathrm{QN}} f$ on X. By Corollary 8.73 the set $f(X) \subseteq {}^\omega\omega$ is eventually bounded. $\square$

Corollary [AC] 8.82. *For a perfectly normal topological space X the following are equivalent:*

a) *X is a QN-space.*

b) *If $\langle f_n : n \in \omega \rangle$ are Borel measurable function from X into $[0,1]$ and $f_n \to f$ on X, then $f_n \xrightarrow{\mathrm{QN}} f$ on X.*

c) *Any Borel measurable image of X into ${}^\omega\omega$ is eventually bounded.*

Proof. a) $\to$ c) follows by the Tsaban-Zdomskyy Theorem, the implication b) $\to$ a) is trivial.

We show that c) $\to$ b). Let $\langle f_n : n \in \omega \rangle$ be Borel measurable functions from X into $[0,1]$ and $f_n \to f$ on X. Set $g_n(x) = \sup\{|f_m(x) - f(x)| : m \geq n\}$. Then g_n is Borel measurable and $g_n \searrow 0$. The function $\psi : X \to {}^\omega\omega$ defined as $\psi(x)(m) = \min\{n : g_n(x) < 2^{-m}\}$ is Borel measurable. By c), the set $\psi(X)$ is eventually bounded by a $\beta \in {}^\omega\omega$. Then $g_{\beta(n)} \xrightarrow{\mathrm{QN}} 0$ with the control $\{2^{-n}\}_{n=0}^{\infty}$. Since $\{g_n\}_{n=0}^{\infty}$ is non-increasing we obtain $g_n \xrightarrow{\mathrm{QN}} 0$ and also $f_n \xrightarrow{\mathrm{QN}} f$. $\square$

In Section 5.4 we have defined a partial preordering of eventual domination on the Baire space ${}^\omega\omega$. We can easily extend this preordering (with the same name) to the set ${}^\omega\mathbb{R}$ of all sequences of reals as

$$f \leq^* g \equiv (\exists n_0)(\forall n \geq n_0)\, f(n) \leq g(n)$$

for any $f, g \in {}^\omega\mathbb{R}$. Evidently ${}^\omega\omega$ considered as a subset of the preordered set $\langle {}^\omega\mathbb{R}, \leq^* \rangle$ is a cofinal subset and therefore

$$\mathfrak{b}({}^\omega\mathbb{R}, \leq^*) = \mathfrak{b}, \quad \mathfrak{d}({}^\omega\mathbb{R}, \leq^*) = \mathfrak{d}.$$

A topological space X is said to have the **property** H^* if for any sequence $\{f_n\}_{n=0}^{\infty}$ of continuous real functions defined on X the family of sequences of reals $\{\{f_n(x)\}_{n=0}^{\infty} : x \in X\} \subseteq {}^{\omega}\mathbb{R}$ is not dominating in the preordering of eventual domination. Similarly, a topological space X has the **property** H^{**} if for any sequence $\{f_n\}_{n=0}^{\infty}$ of continuous real functions defined on X the family of sequences of reals $\{\{f_n(x)\}_{n=0}^{\infty} : x \in X\} \subseteq {}^{\omega}\mathbb{R}$ is eventually bounded. Evidently

$$H^{**} \to H^*.$$

Immediately from the definition one obtains that a continuous image of a topological space with the **property** H^{**} in ${}^{\omega}\omega$ is eventually bounded.

Using the Tietze-Urysohn Theorem 3.20 one can easily see that:

Theorem [AC] 8.83. *The properties* H^{**} *or* H^* *are preserved under continuous image and under passing to a closed subset of a normal topological space.*

If $|X| < \mathfrak{b}$, then the topological space X possesses property H^{**}. If X is an unbounded subset of ${}^{\omega}\omega$ and we set $f_n(\alpha) = \alpha(n)$ for $\alpha \in X$ and $n \in \omega$, then one can easily see that X does not have the property H^{**}. Hence

$$\mathrm{non}(H^{**}) = \mathfrak{b}. \tag{8.11}$$

Similarly we obtain

$$\mathrm{non}(H^*) = \mathfrak{d}. \tag{8.12}$$

Thus, if $\mathfrak{b} < \mathfrak{d}$, then the unbounded subset of ${}^{\omega}\omega$ of cardinality $\mathfrak{b}$ possesses the property H^* and does not possess the property H^{**}. A compact space trivially possesses the property H^{**}. Thus one can easily show that a σ-compact space possesses the property H^{**}. Actually, any set that is a union of less than $\mathfrak{b}$ compact sets possesses the property H^{**}.

Similarly as in the case of a QN-space and a wQN-space, it turned out that the notion of an mQN-space is equal to a classical notion.

Lemma 8.84 (J. Haleš). *If* $f_n \searrow 0$ *on* X *are continuous then there exists a continuous function* $h : X \longrightarrow \mathbb{R}$ *such that*

$$(\forall x \in X)(\forall n > h(x)) \, f_n(x) < 1.$$

Proof. Let

$$g(x) = \sum_{n=0}^{\infty} \min\{1, f_n(x)\} \cdot 2^{-n}.$$

The function $g : X \longrightarrow [0, 2)$ is continuous. Set $h(x) = -\log_2(2 - g(x))$.

If $x \in X$ and $f_n(x) \geq 1$, then using the monotonicity of the sequence $\{f_n\}_{n=0}^{\infty}$ we obtain that $\min\{1, f_0(x)\} = \cdots = \min\{1, f_n(x)\} = 1$. Thus $g(x) \geq 2 - 2^{-n}$ and therefore $n \leq h(x)$. $\qquad \square$

Theorem 8.85. *For a topological space X the following conditions are equivalent:*

a) X *is an mQN-space.*

b) X *has the property* H^{**}.

c) *If for every $n \in \omega$, $\{f_{n,m}\}_{m=0}^{\infty}$ is a sequence of continuous functions defined on X and such that $f_{n,m} \searrow 0$, then there exists an increasing sequence $\{m_n\}_{n=0}^{\infty}$ such that $f_{n,m_n} \to 0$ on X.*

Proof. We begin with proving the implication a) $\to$ b). So assume that a) holds true. Let $\langle f_n : X \longrightarrow \mathbb{R}^+ : n \in \omega \rangle$ be continuous. For $x \in X$ and $n, m \in \omega$ we set

$$h_n^m(x) = \min\{1, f_n(x)/(m+1)\}, \quad g_m(x) = \sum_{n=0}^{\infty} 2^{-n} h_n^m(x). \qquad (8.13)$$

Then $h_n^{m+1}(x) \le h_n^m(x) \le 1$ and $g_{m+1}(x) \le g_m(x)$ for any $x \in X$ and any $n, m \in \omega$. We show that $g_m \searrow 0$ on X.

Let $x \in X$, $\varepsilon > 0$. Let $m_0 \ge \max\{2^k, 2^k \cdot f_0(x), \dots, 2^k \cdot f_k(x)\}$, where k is such that $3 \cdot 2^{-k} < \varepsilon$. For $m \ge m_0$ and $i \le k$ we obtain

$$h_i^m(x) \le f_i(x)/(m+1) \le m_0 2^{-k}/(m+1) \le 2^{-k}.$$

Hence, for $m \ge m_0$ we have

$$g_m(x) = \sum_{i=0}^{k} \frac{h_i^m(x)}{2^i} + \sum_{i>k} \frac{h_i^m(x)}{2^i} \le \frac{2}{2^k} + \frac{1}{2^k} < \varepsilon.$$

By a) there exists a decreasing control sequence of positive reals $\{\varepsilon_n\}_{n=0}^{\infty}$ for $g_n \xrightarrow{\text{QN}} 0$ on X. We set $d_n = \min\{m : \varepsilon_m < 2^{-n}\}$. The sequence $\{d_n\}_{n=0}^{\infty}$ is non-decreasing and unbounded.

Let $x \in X$. Let n_0 be such that $g_n(x) < \varepsilon_n$ for every $n \ge n_0$. Let $n_1 \ge n_0$ be such that $d_n \ge n_0$ for $n \ge n_1$. If $n \ge n_1$ then $g_{d_n}(x) < \varepsilon_{d_n} < 2^{-n}$ and therefore $h_n^{d_n}(x) < 1$. By (8.13) we obtain $f_n(x) < d_n$ for every $n \ge n_1$.

Now we show b) $\to$ c). Let $\{\{f_{n,m}\}_{m=0}^{\infty} : n \in \omega\}$ be a sequence of sequences of continuous functions from X into $\mathbb{R}$ such that $f_{n,m} \searrow 0$ for each $n \in \omega$. By Lemma 8.84 there exists a sequence $\langle h_n : X \longrightarrow \mathbb{R} : n \in \omega \rangle$ of continuous functions such that

$$(\forall n)(\forall x \in X)(\forall m)\, (m > h_n(x) \to f_{n,m}(x) < 2^{-n}).$$

Since the topological space possesses property H^{**}, there exists a sequence of natural numbers $\{m_n\}_{n=0}^{\infty}$ such that

$$(\forall x \in X)(\exists n_0)(\forall n \ge n_0)\, h_n(x) < m_n.$$

Then $f_{n,m_n} \to 0$ on X.

A proof of implication c) $\to$ a) is easy. If $f_n \searrow 0$ on X, we set $f_{n,m} = 2^n \cdot f_m$. Then $f_{n,m} \searrow 0$ on X for each $n \in \omega$ and by c) there exists an increasing sequence

$\{m_n\}_{n=0}^{\infty}$ such that $f_{n,m_n} \to 0$ on X. Then $f_{m_n} \xrightarrow{\text{QN}} 0$ on X with the control 2^{-n}. Since the sequence $\{f_n\}_{n=0}^{\infty}$ is non-increasing, we obtain $f_n \xrightarrow{\text{QN}} 0$ on X. $\qquad\square$

Corollary 8.86. *Every* wQN-*space has the property* H**.

Corollary [AC] 8.87. *If* $\kappa \leq \mathfrak{b}$ *and* A *is a* κ-*Sierpiński subset of a Polish space with a finite Borel measure, then every subset of* A *has the property* H**.

The proof follows immediately from Theorem 8.63 and Corollary 8.78. $\qquad\square$

Theorem [wAC] 8.88. *If* A *is a subset of a Polish space* X *with property* H**, *then there exists a* σ-*compact set* $B \subseteq X$ *such that* $A \subseteq B$.

Proof. Let $\{r_n : n \in \omega\}$ be a countable dense subset of X. Set

$$f_n(x) = \min\{\rho(x, r_i) : i = 0, \dots, n\}.$$

Then $f_n \searrow 0$ on X and therefore $f_n \xrightarrow{\text{QN}} 0$ on A. If $\{\varepsilon_n\}_{n=0}^{\infty}$ is a control sequence for the quasi-normal convergence of $\{f_n\}_{n=0}^{\infty}$ on A, then the set

$$B_n = \{x \in X : (\forall m \geq n)\, f_m(x) \leq \varepsilon_m\}$$

is closed. If $m \geq n$ is such that $\varepsilon_m \leq \varepsilon$, then the set $\{r_0, \dots, r_m\}$ is an ε net on B_n. Thus, B_n is totally bounded and by Theorem 3.36, B_n is compact. Evidently $A \subseteq \bigcup_n B_n$. $\qquad\square$

The following classical result says that a well-definable space with the property H** is always a σ-compact set.

Theorem [wAC] 8.89 (W. Hurewicz). *Any analytic subset* A *of a Polish space* X *possessing the property* H** *is* σ-*compact. However, by Corollary 8.87 this need not be true for any subset of a Polish space.*

Proof. By Theorem 8.88 we can assume that X is σ-compact. Assume that $A \subseteq X$ is an analytic subset of X and A is not σ-compact. Then neither is A an F_σ set. Then by Corollary 6.34 there exists a closed subset of A homeomorphic with ${}^\omega\omega$. Since ${}^\omega\omega$ does not possess the property H**, by Theorem 8.83 neither does the set A possess it. $\qquad\square$

Exercises

8.14 [AC] Σ-spaces

 a) If X is a Σ-space, $A \in [\omega]^\omega$, $f_n \to 0$ on X, then there exists a $B \in [A]^\omega$ such that $\sum_{n=0}^{\infty} f_{e_B(n)}(x) < \infty$ for each $x \in X$.

 b) If $X = \bigcup_{s \in S} X_s$, $|S| < \mathfrak{h}$, each X_s is a Σ-space, then X is a Σ-space.
 Hint: Follow the proof of Theorem 8.61, b).

 c) $\mathfrak{h} \leq \mathrm{add}(\Sigma\text{-space}) \leq \mathrm{non}(\Sigma\text{-space}) = \mathfrak{b}$.
 Hint: See (8.11) and Theorem 8.85.

 d) An F_σ-subset of a Σ-set is a Σ-set.

8.15 [wAC] Continuous Image in $[0, 1]$

a) If $f_n : X \longrightarrow [0, 1]$ are continuous and $\mathrm{ind}(f_n(X)) = 0$, then there exist continuous
$H : X \longrightarrow [0, 1]$ and continuous $h_n : \mathrm{rng}(H) \longrightarrow [0, 1]$ such that $f_n = H \circ h_n$ for
every n.
Hint: Let $F(x) = \{f_n(x)\}_{n=0}^\infty \in {}^\omega[0, 1]$. By Exercise 3.3, h) we have $\mathrm{ind}(F(X)) = 0$.
Thus there exists an embedding $G : F(X) \overset{1\text{-}1}{\longrightarrow} [0, 1]$. Set $h_n = G^{-1} \circ \mathrm{proj}^n$ and
$H = F \circ G$.

b) A topological space X is a QN-space if and only if every continuous image of X
into $[0, 1]$ is a QN-space.

c) A topological space X is a wQN-space if and only if every continuous image of X
into $[0, 1]$ is a wQN-space.

d) In a)–c) one can replace the unit interval $[0, 1]$ by ${}^\omega 2$ or ${}^\omega \omega$.

8.16 (α_i) and Continuous Mappings

a) Assume that $f : X \longrightarrow Y$ is continuous and $(\mathcal{N}(x)$ denotes the filter of neighborhoods of a point $x)$ such that

$$(\forall x \in X)(\forall V \in \mathcal{N}(x))(\exists W \in \mathcal{N}(f(x)))\, f^{-1}(W) \subseteq V. \qquad (8.14)$$

Show that if Y possesses the property (α_i), then X does so, $i = 1, \ldots, 4$.

b) If X is Hausdorff, then a continuous mapping satisfying (8.14) is an injection.

c) Find an injection not satisfying (8.14).

d) If $f : X \longrightarrow Y$ is continuous, we define $F : C_p(Y) \longrightarrow C_p(X)$ by $F(\varphi) = f \circ \varphi$ for
$\varphi \in C_p(Y)$. If f is a surjection, then F satisfies (8.14).
Hint: Let $\varphi \in C_p(Y)$, $y_1, \ldots, y_n \in Y$, $\varepsilon > 0$, and

$$V = \{\psi \in C_p(Y) : |\varphi(y_i) - \psi(y_i)| < \varepsilon \text{ for } i = 1, \ldots, n\}.$$

Let x_i be such that $f(x_i) = y_i$ for $i = 1, \ldots, n$. Take

$$W = \{\theta \in C_p(X) : |\theta(x_i) - \varphi(y_i)| < \varepsilon \text{ for } i = 1, \ldots, n\}.$$

e) If $F : X \overset{\mathrm{onto}}{\longrightarrow} Y$ is a continuous surjection and $C_p(X)$ possesses the property (α_i),
then $C_p(Y)$ does so, $i = 1, \ldots, 4$.

8.17 $\mathcal{L}^*$-group

Let $\langle X, \lim \rangle$ be an $\mathcal{L}^*$-space with a group structure $\langle X, \circ, e \rangle$. X is said to be an $\mathcal{L}^*$-group if
the group operations are continuous, i.e., if for any sequences $\{x_n\}_{n=0}^\infty$, $\{y_n\}_{n=0}^\infty$ such that
$\lim_{n\to\infty} x_n = x$, $\lim_{n\to\infty} y_n = y$ we have $\lim_{n\to\infty} x_n \circ y_n = x \circ y$ and $\lim_{n\to\infty} x_n^{-1} = x^{-1}$.
An $\mathcal{L}^*$-group X has the **property** $(\alpha_i)^*$, $i = 1, 2, 3, 4$, if for any sequence $\{\{x_{n,m}\}_{m=0}^\infty\}_{n=0}^\infty$
of sequences such that $\lim_{m\to\infty} x_{n,m} = x_n$ for every $n \in \omega$ and $\lim_{n\to\infty} x_n = x$, there
exists a sequence $\{y_m\}_{m=0}^\infty$ such that $\lim_{m\to\infty} y_m = x$ and

$(\alpha_1)^*$ $\{x_{n,m} : m \in \omega\} \subseteq^* \{y_m : m \in \omega\}$ for each n,

$(\alpha_2)^*$ $\{x_{n,m} : m \in \omega\} \cap \{y_m : m \in \omega\}$ is infinite for each n,

$(\alpha_3)^*$ $\{x_{n,m} : m \in \omega\} \cap \{y_m : m \in \omega\}$ is infinite for infinitely many n,

$(\alpha_4)^*$ $\{x_{n,m} : m \in \omega\} \cap \{y_m : m \in \omega\} \neq \emptyset$ for infinitely many n.

Show that the properties (α_1)–(α_4) of an $\mathcal{L}^*$-group are equivalent to $(\alpha_1)^*$–$(\alpha_4)^*$, respectively.

Hint: Replace $x_{n,m}$ by $x_{n,m} \circ x_n^{-1}$.

8.18 Properties (α_0) and $(\alpha_0)^*$

An $\mathcal{L}^*$-space X has the **property $(\alpha_0)^*$** if for any $x \in X$ and $\lim_{m \to \infty} x_{n,m} = x_n$ for every $n \in \omega$ and $\lim_{n \to \infty} x_n = x$, there exists an unbounded non-decreasing sequence $\{n_m\}_{m=0}^{\infty}$ of natural numbers such that $\lim_{m \to \infty} x_{n_m, m} = c$. Taking $x_n = x$ one obtains the **property (α_0)**.

a) $(\alpha_1) \to (\alpha_0)$ for any $\mathcal{L}^*$-space.

b) $(\alpha_0) \to (\alpha_0)^*$ for any $\mathcal{L}^*$-group.

c) If $C_p(X)$ has property $(\alpha_0)^*$ then X is a QN-space.
   Hint: Consider $f_{n,m} = 2^n \cdot f_m + 2^{-n}$.

d) The following are equivalent:

   (i) X is a QN-space,

   (ii) $C_p(X)$ has property (α_0),

   (iii) $C_p(X)$ has property $(\alpha_0)^*$,

   (iv) $C_p(X)$ has property (α_1).

8.19 [AC] Property H**

a) A subset A of a Polish space has the property H** if and only if for every G_δ set $G \supseteq A$ there exists a σ-compact set F such that $A \subseteq F \subseteq G$.
   Hint: G is a Polish subspace so apply Theorem 8.88. If $f_n \searrow 0$ on A then by using Theorem 3.29 find a G_δ set $G \supseteq A$ and functions $F_n \searrow 0$ on G extending.

b) If a subset A of a perfectly normal topological space X has the property H**, then for every G_δ set $G \supseteq A$ there exists an F_σ set F such that $A \subseteq F \subseteq G$.
   Hint: Let $G = \bigcap_n G_n$, $G_{n+1} \subseteq G_n$ open. Take $f_n : X \longrightarrow [0,1]$ continuous and such that $Z(f_n) = X \setminus G_n$. Set $h_n(x) = \prod_{i=0}^{n}(1 - f_i(x))$. Then $h_n \searrow 0$ on G. Let $\{\varepsilon_n\}_{n=0}^{\infty}$ be the control of quasi-normal convergence of h_n on A. Assume that $\varepsilon < 1$ and set $F = \bigcup_n \{x \in X : (\forall m \geq n)\, h_m(x) \leq \varepsilon_m\}$.

8.4   Covering Properties

Let X be a topological space. We shall assume that X is infinite, since all considered properties are trivial for finite spaces. Let us recall that $\mathcal{U} \subseteq \mathcal{P}(X)$ is a cover of X if $X = \bigcup \mathcal{U}$ and, to avoid a triviality, we assume that $X \notin \mathcal{U}$. A cover $\mathcal{U}$ of X is said to be an ω-**cover** if every finite subset of X lies in some member of $\mathcal{U}$. An infinite cover $\mathcal{U}$ of a topological space X is said to be a γ-**cover** if every point $x \in X$ lies in all but finitely many members of the cover $\mathcal{U}$. Every γ-cover is an ω-cover. A γ-cover $\mathcal{U}$ is **shrinkable**, if there exists a closed γ-cover $\mathcal{V}$ that is a refinement of $\mathcal{U}$. We shall denote by $\mathrm{O}(X)$, $\Omega(X)$, $\Gamma(X)$ and $\Gamma^{sh}(X)$ the set of all open covers, open ω-covers, open γ-covers, and open shrinkable γ-covers of X, respectively.[1]

Generally an enumeration of a countable cover need not be one-to-one. Often we shall need that an enumeration $\langle U_n : n \in \omega \rangle$ of a cover $\mathcal{U}$ of X is **adequate**, i.e., that every element of $\mathcal{U}$ occurs only finitely many times in the enumeration.

Note that an ω-cover is always infinite. If you split an ω-cover in two parts, then at least one of them is an ω-cover. Any infinite subset of a γ-cover is a γ-cover. Especially any infinite countable subset of a γ-cover is a γ-cover.

Assume now that X has a countable base of topology $\mathcal{B}$. We can assume that $\mathcal{B}$ is closed under finite unions. If $\mathcal{U}$ is an open ω-cover of X, then

$$\mathcal{V} = \{V \in \mathcal{B} : (\exists U)\, V \subseteq U \in \mathcal{U}\}$$

is a countable open ω-cover. If we choose a $U_V \in \mathcal{U}$, $V \subseteq U_V$ for each $V \in \mathcal{V}$ then $\{U_V : V \in \mathcal{V}\}$ is a countable ω-subcover of $\mathcal{U}$. If a γ-cover $\mathcal{V}$ is a refinement of a cover $\mathcal{U}$, then for any $U \in \mathcal{U}$ the set $\{V \in \mathcal{V} : V \subseteq U\}$ is finite. Therefore from any shrinkable γ-cover we can choose a countable shrinkable γ-subcover. Moreover, if a γ-cover $\langle V_n : n \in \omega \rangle$ is a refinement of a γ-cover $\langle U_n : n \in \omega \rangle$, we can assume that $V_n \subseteq U_n$ for each n. Thus dealing with a space with a countable base, we can always assume that the corresponding open cover, open ω-cover, open γ-cover or shrinkable γ-cover is countable. However, the axiom of choice **AC** has been essentially used. For convenience, if $\mathcal{A}(X)$ is a family of covers of a set X we denote by $\mathcal{A}_\omega(X)$ the family of all countable covers belonging to $\mathcal{A}(X)$.

If $\mathcal{U}$ and $\mathcal{V}$ are ω-covers, then $\{U \cap V : U \in \mathcal{U} \wedge V \in \mathcal{V}\}$ is an ω-cover, which is a common refinement of $\mathcal{U}$ and $\mathcal{V}$. If $\mathcal{U} = \langle U_n : n \in \omega \rangle$ and $\mathcal{V} = \langle V_n : n \in \omega \rangle$ are countable open (closed, shrinkable) γ-covers, then the open (closed, shrinkable) γ-cover $\langle U_n \cap V_n : n \in \omega \rangle$ is a common refinement of $\mathcal{U}$ and $\mathcal{V}$. Thus, dealing with a sequence of ω-covers, countable γ-covers or countable shrinkable γ-covers, we can assume that each cover is a refinement of the previous one.

---

[1] We have already in Section 5.2 denoted by $\Gamma(X)$ the set of all subsets of X belonging to a pointclass Γ. The distinction between the notations $\Gamma(X)$ and $\Gamma(X)$ is very small, however we do not want to introduce a notation different from that used in the set-theoretic topology. We hope that a reader can always identify it from the context.

A cover $\mathcal{U}$ of X is called **essentially infinite** if no finite subset of $\mathcal{U}$ is a cover of X. Assume that $\mathcal{U} = \{U_n : n \in \omega\}$ is an essentially infinite cover. Then the increasing sequence $\langle \bigcup_{i \leq n} U_i : n \in \omega \rangle$ is an ω-cover. Moreover, if $\mathcal{U}$ is a γ-cover, then the increasing sequence $\langle \bigcup_{i \leq n} U_i : n \in \omega \rangle$ is a γ-cover as well.

A topological space X is called a γ-**space** if from every open ω-cover of X one can choose a γ-subcover. Note the following: if X is finite, then there exists no ω-cover of X and therefore X is trivially a γ-space.

Theorem [AC] 8.90. *If X has a countable base of the topology and $|X| < \mathfrak{p}$, then X is a γ-space.*

Proof. Let $\mathcal{U}$ be an open ω-cover of X. We can assume that $\mathcal{U}$ is countable. For any $x \in X$ we write
$$\mathcal{U}_x = \{U \in \mathcal{U} : x \in U\}.$$
One can easily see that the family $\{\mathcal{U}_x : x \in X\}$ possesses the finite intersection property. Since the cardinality of this family is smaller than $\mathfrak{p}$ there exists an infinite set $\mathcal{V} \subseteq \mathcal{U}$ such that $\mathcal{V} \setminus \mathcal{U}_x$ is finite for every $x \in X$. One can easily check that $\mathcal{V}$ is the desired γ-cover. $\qquad\square$

Theorem [AC] 8.91. *There exists a subset A of $^{\omega}2$ of cardinality $\mathfrak{p}$ that is not a γ-space.*

Proof. Let $\mathcal{F}$ be the family of subsets of ω of cardinality $\mathfrak{p}$ with the property (5.33), i.e., $\mathcal{F}$ has f.i.p. and has no pseudointersection. Let $A \subseteq {}^{\omega}2$ be the set of all characteristic functions of sets from the family $\mathcal{F}$. Thus, for any finite set $\{\alpha_0, \ldots, \alpha_k\} \subseteq A$, the set $\{n \in \omega : \alpha_0(n) = \cdots = \alpha_k(n) = 1\}$ is infinite and, there is no infinite set $E \subseteq \omega$ such that the set $\{n \in E : \alpha(n) = 0\}$ is finite for any $\alpha \in A$. We claim that A is not a γ-space.

We set $U_n = \{\alpha \in {}^{\omega}2 : \alpha(n) = 1\}$. Then $\{U_n : n \in \omega\}$ is an open ω-cover of A. Assume that there exists a sequence $\{n_k\}_{k=0}^{\infty}$ such that $\{U_{n_k} : k \in \omega\}$ is a γ-cover of A. Let $E = \{n_k : k \in \omega\}$. If $\alpha \in A$ then $\alpha \in U_{n_k}$ for all but finitely many k. Thus the set $\{n \in E : \alpha(n) = 0\}$ is finite, which is a contradiction. $\qquad\square$

Theorem [AC] 8.92 (F. Galvin – A.W. Miller). *If $\mathfrak{p} = \mathfrak{c}$, then there exists a γ-space A of cardinality $\mathfrak{c}$. Moreover, we can assume that A is $\mathfrak{c}$-concentrated on a countable subset.*

Recall that in Section 3.1 we equipped the set $\mathcal{P}(\omega)$ with the topology obtained by identification of $^{\omega}2$ and $\mathcal{P}(\omega)$. The family
$$\{\{C \subseteq \omega : C \cap n = A\} : A \subseteq n \wedge n \in \omega\} \qquad (8.15)$$
is a base of this topology.

If $U \subseteq \mathcal{P}(\omega)$ is open and $U \cap \{C \subseteq \omega : C \cap n = A\} \neq \emptyset$, then there exist a finite set $B \supseteq A$ and an integer $k \geq n$ such that for any $m > k$ we have
$$\{C \subseteq \omega : C \cap m = B \cap m\} \subseteq U \cap \{C \subseteq \omega : C \cap n = A\}.$$

Lemma 8.93. *Let $E \in [\omega]^\omega$, $p \in \omega$, $U \subseteq \mathcal{P}(\omega)$ being open. If $\mathcal{P}(p) \subseteq U$, then there exists a $q \in E$, $q > p$ such that*

$$\{C \subseteq \omega : C \cap q = B\} \subseteq U \text{ for every } B \in \mathcal{P}(p).$$

Thus, if $C \cap [p, q) = \emptyset$, then $C \in U$.

The proof is easy. Since every $B \subseteq p$ belongs to U, there exists a $q_B > p$ and a set $A_B \subseteq q_B$ such that

$$B \in \{C \subseteq \omega : C \cap q_B = A_B\} \subseteq U.$$

Evidently $A_B = B$. We can assume that all q_B have a common value $q \in E$. If $C \cap [p, q) = \emptyset$, then $C \cap q = B$ with $B = C \cap p \in \mathcal{P}(p)$ and therefore $C \in U$. $\square$

If $X \in [\omega]^\omega$ we write

$$X^* = \{Y \in [\omega]^\omega : |Y \setminus X| < \aleph_0\}.$$

Lemma 8.94. *If $\mathcal{U}$ is an open ω-cover of $[\omega]^{<\omega} \subseteq \mathcal{P}(\omega)$ and E is an infinite subset of ω, then there exists a set $Z \in [E]^\omega$ and a γ-cover $\mathcal{V} \subseteq \mathcal{U}$ of $Z^* \cup [\omega]^{<\omega}$.*

Proof. For a given ω-cover $\mathcal{U}$ of $[\omega]^{<\omega}$ and given $E \in [\omega]^\omega$ we construct by induction an increasing sequence $\{k_n\}_{n=0}^\infty$ of elements of E and a sequence $\{U_n\}_{n=0}^\infty$ of elements of $\mathcal{U}$ such that $Y \in U_n$ for any $Y \subseteq \omega$ satisfying $Y \cap [k_n + 1, k_{n+1}) = \emptyset$.

Let k_0 be any element of E. Assume that $U_0, \ldots, U_{n-1}$ and $k_0, \ldots, k_n$ are constructed. Since $\mathcal{U}$ is an ω-cover of $[\omega]^{<\omega}$, there exists a $U_n \in \mathcal{U}$ such that $\mathcal{P}(k_n + 1) \subseteq U_n$. By Lemma 8.93 there exists a $k_{n+1} \in E$, $k_{n+1} > k_n + 1$ such that

$$\{X \subseteq \omega : X \cap k_{n+1} = B\} \subseteq U_n$$

for every $B \in \mathcal{P}(k_n + 1)$. Then $Y \in U_n$ for any Y disjoint with $[k_n + 1, k_{n+1})$.

We claim that $\mathcal{V} = \{U_n : n \in \omega\}$ is a γ-cover of Z^*, where $Z = \{k_n : n \in \omega\}$. Actually, if $Y \in Z^*$ then $Y \setminus Z \subseteq k_n$ for some n. But then $Y \cap [k_m + 1, k_{m+1}) = \emptyset$ for all $m \geq n$ and therefore $Y \in U_m$. If $Y \subseteq \omega$ is finite, then $Y \in U_m$ provided that $k_m > \max Y$. $\square$

Proof of Theorem 8.92. We assume that $\mathfrak{p} = \mathfrak{c}$ and $\langle \mathcal{U}_\xi : \xi < \mathfrak{c} \rangle$ is an enumeration of all countable families of open subsets of $\mathcal{P}(\omega)$. We construct an almost decreasing sequence $\langle X_\xi \subseteq \omega : \xi < \mathfrak{c} \rangle$ such that $\{X_\xi : \xi < \mathfrak{c}\} \cup [\omega]^{<\omega}$ will be a γ-set.

At a limit stage ξ using the assumption $\mathfrak{p} = \mathfrak{c}$ we find a set X_ξ such that $X_\xi \setminus X_\eta$ is finite for every $\eta < \xi$.

If X_η are constructed for every $\eta \leq \xi$ we shall distinguish three cases.

a) If $\mathcal{U}_\xi$ is not an ω-cover of $[\omega]^{<\omega}$, we continue as above: take an infinite set $X_{\xi+1}$ almost contained in all $\langle X_\eta : \eta \leq \xi \rangle$.

b) If $\mathcal{U}_\xi$ is an ω-cover of $[\omega]^{<\omega}$, but it is not an ω-cover of $\{X_\eta : \eta \leq \xi\} \cup [\omega]^{<\omega}$, then by Lemma 8.94 there exists a set $X_{\xi+1} \in [X_\xi]^\omega$ and a γ-cover $\mathcal{V} \subseteq \mathcal{U}$ of the set $X_{\xi+1}^* \cup [\omega]^{<\omega}$.

c) If $\mathcal{U}_\xi$ is an ω-cover of $\{X_\eta : \eta \leq \xi\} \cup [\omega]^{<\omega}$, then by Theorem 8.90 there exists a γ-cover $\mathcal{W} \subseteq \mathcal{U}_\xi$ of this set. Since $\mathcal{W}$ is also an ω-cover of $[\omega]^{<\omega}$, by Lemma 8.94 there exists a set $X_{\xi+1} \in [X_\xi]^\omega$ and a γ-cover $\mathcal{V} \subseteq \mathcal{W}$ of $X^*_{\xi+1}$. Since $\mathcal{V}$ is an infinite subcover of a γ-cover of $\{X_\eta : \eta \leq \xi\} \cup [\omega]^{<\omega}$, we obtain that $\mathcal{V}$ is also a γ-cover of $\{X_\eta : \eta \leq \xi\} \cup X^*_{\xi+1} \cup [\omega]^{<\omega}$.

We claim that $A = \{X_\eta : \eta < \mathfrak{c}\} \cup [\omega]^{<\omega}$ is a γ-set. Note first that for any $\xi < \mathfrak{c}$ we have

$$A \subseteq \{X_\eta : \eta \leq \xi\} \cup X^*_{\xi+1} \cup [\omega]^{<\omega}.$$

Now, let $\mathcal{U}$ be a countable open ω-cover of A. Then there exists a $\xi < \omega_1$ such that $\mathcal{U} = \mathcal{U}_\xi$. Since $\mathcal{U}$ is also an ω-cover of the γ-set $[\omega]^{<\omega} \cup \{X_\eta : \eta \leq \xi\}$, then by part c) of the construction there exists a γ-cover $\mathcal{V} \subseteq \mathcal{U}_\xi$ of the set $\{X_\eta : \eta \leq \xi\} \cup X^*_{\xi+1} \cup [\omega]^{<\omega}$.

We show that the set A is $\mathfrak{c}$-concentrated on $[\omega]^{<\omega}$. Let $U \supseteq [\omega]^{<\omega}$ be an open set. It is easy to find an increasing sequence of open sets $\langle U_n : n \in \omega \rangle$ such that $U = \bigcup_n U_n$. Then $\mathcal{U} = \{U_n : n \in \omega\}$ is an ω-cover of $[\omega]^{<\omega}$. Therefore $\mathcal{U} = \mathcal{U}_\xi$ for some $\xi < \mathfrak{c}$ and case b) or c) occurs.

In case b) we have a γ-cover $\mathcal{V} \subseteq \mathcal{U}$ of $X^*_{\xi+1} \cup [\omega]^{<\omega}$. So we obtain that $A \setminus U \subseteq \{X_\eta : \eta \leq \xi\}$ and therefore $|A \setminus U| < \mathfrak{c}$. In case c) we have a γ-cover $\mathcal{V} \subseteq \mathcal{U}$ of $X^*_{\xi+1}$ and $\{X_\eta : \eta \leq \xi\} \cup [\omega]^{<\omega} \subseteq U$. Therefore $A \setminus U = \emptyset$. $\qquad \square$

Now we shall consider so-called covering selection principles. Let $\mathcal{A}(X)$, $\mathcal{B}(X)$ be families of covers of a topological space X. X is said to be an $S_1(\mathcal{A}, \mathcal{B})$-**space** if for every sequence $\langle \mathcal{U}_n : n \in \omega \rangle$ of covers from $\mathcal{A}(X)$ there exist sets $U_n \in \mathcal{U}_n$ such that $\{U_n : n \in \omega\}$ is a cover belonging to $\mathcal{B}(X)$.

Theorem 8.95. *Assume that the families $\mathcal{A}(X)$ and $\mathcal{B}(X)$ of open covers have the following property:*

i) *if $\mathcal{V} \in \mathcal{B}(X)$ is a refinement of an open cover $\mathcal{U}$, then there exists a subcover of $\mathcal{U}$ which belongs to $\mathcal{B}(X)$,*

ii) *every two covers of $\mathcal{A}(X)$ have a common refinement which belongs to $\mathcal{A}(X)$.*

Then X is an $S_1(\mathcal{A}, \mathcal{B})$-space if and only if for every sequence $\{\mathcal{U}_n\}_{n=0}^\infty$ of covers from $\mathcal{A}(X)$ such that $\mathcal{U}_{n+1}$ is a refinement of $\mathcal{U}_n$ for every n, there exist sets $U_n \in \mathcal{U}_n$ such that $\{U_n : n \in \omega\}$ is a cover belonging to $\mathcal{B}(X)$.

The proof is immediate. $\qquad \square$

Note that the families of covers $O(X)$, $\Omega(X)$, $\Gamma(X)$, and $\Gamma^{sh}(X)$ satisfy both conditions of the theorem.

Theorem [AC] 8.96 (J. Gerlits – Z. Nagy). *A topological space X is a γ-space if and only if X is an $S_1(\Omega, \Gamma)$-space.*

Proof. Evidently any $S_1(\Omega, \Gamma)$-space is a γ-space: take for $\mathcal{U}_n$ the same ω-cover for each n.

Since X is infinite, we choose distinct $\langle x_n \in X : n \in \omega \rangle$. Let $\langle \mathcal{U}_n : n \in \omega \rangle$ be a sequence of ω-covers. By Theorem 8.95 we can assume that $\mathcal{U}_{n+1}$ is a refinement of $\mathcal{U}_n$ for every n. Then

$$\mathcal{U} = \{ U \setminus \{x_n\} : U \in \mathcal{U}_n \wedge n \in \omega \}$$

is an ω-cover. Since X is a γ-space, there exists a γ-subcover $\{V_k : k \in \omega\} \subseteq \mathcal{U}$. Let n_k be such that $V_k = U \setminus \{x_{n_k}\}$, where $U \in \mathcal{U}_{n_k}$. If $\{x_0, \dots, x_n\} \subseteq V_k$, then $n_k > n$. Thus the set $\{n_k : k \in \omega\}$ is infinite and therefore we can assume that the sequence $\{n_k\}_{k=0}^{\infty}$ is increasing and $n_0 = 0$. For any $m < n_k$, $m \geq n_{k-1}$, $k > 0$ take $U_m \in \mathcal{U}_m$ such that $V_k \subseteq U_m \setminus \{x_{n_k}\}$. One can easily see that $\{U_m : m \in \omega\}$ is a γ-cover.   □

Since a γ-cover is an ω-cover as well, we obtain

Corollary [AC] 8.97. *A topological γ-space X is an $S_1(\Gamma, \Gamma)$-space.*

A topological space $\langle X, \mathcal{O} \rangle$ has the **Menger Property** if for every sequence $\langle \mathcal{U}_n : n \in \omega \rangle$ of essentially infinite countable open covers of X there exist finite subsets $\langle \mathcal{V}_n \subseteq \mathcal{U}_n : n \in \omega \rangle$ such that $\{ \bigcup \mathcal{V}_n : n \in \omega \}$ is a cover of X. Similarly, we say that a topological space $\langle X, \mathcal{O} \rangle$ has the **Hurewicz Property** if for every sequence $\langle \mathcal{U}_n : n \in \omega \rangle$ of essentially infinite countable open covers of X there exist finite subsets $\langle \mathcal{V}_n \subseteq \mathcal{U}_n : n \in \omega \rangle$ such that $\{ \bigcup \mathcal{V}_n : n \in \omega \}$ is a γ-cover. Evidently a space with the Hurewicz Property has also the Menger Property. A compact space X has trivially both Menger and Hurewicz Properties, since there is no essentially infinite open cover of X. It is easy to see that a σ-compact space has both Menger and Hurewicz Properties as well.

We note that the Menger and Hurewicz Properties are usually defined without the restriction to countable covers. However, we deal mainly with subsets of Polish spaces and in this case the definitions are equivalent.

Theorem [AC] 8.98. *A topological $S_1(\Gamma, \Gamma)$-space has the Hurewicz Property.*

Proof. Assume that $\langle \{U_{n,m} : m \in \omega\} : n \in \omega \rangle$ is a sequence of essentially infinite countable open covers of a topological space X. If $V_{n,m} = \bigcup_{i \leq m} U_{n,i}$, then the family $\{V_{n,m} : m \in \omega\}$ is a γ-cover of X. By definition there exists a sequence $\{m_n\}_{n=0}^{\infty}$ such that $\{V_{n,m_n} : n \in \omega\}$ is a γ-cover. Setting $\mathcal{V}_n = \{U_{n,i} : i \leq m_n\}$ we obtain a γ-cover $\{\bigcup \mathcal{V}_n : n \in \omega\}$ witnessing the Hurewicz Property.   □

Corollary [AC] 8.99. *A γ-space has the Hurewicz Property.*

Lemma [wAC] 8.100. *If a subset A of a Polish space has the Hurewicz Property and $\mathrm{Int}(A) = \emptyset$, then A is meager.*

Proof. Since $\mathrm{Int}(A) = \emptyset$, there exists a dense set $\{r_n : n \in \omega\}$ disjoint with A. Let

$$U_{n,k} = X \setminus \overline{\mathrm{Ball}}(r_n, 1/(k+1)), \quad \mathcal{U}_n = \{U_{n,k} : k \in \omega\}.$$

Evidently $\mathcal{U}_n$ is an increasing open cover of A. By the Hurewicz Property there exist natural numbers $\langle k_m : m \in \omega \rangle$ such that $A \subseteq \bigcup_n \bigcap_{m \geq n} U_{m,k_m}$. Since the sets

$$\bigcap_{m \geq n} U_{m,k_m} \subseteq (X \setminus \bigcup_{m \geq n} \mathrm{Ball}(r_m, 1/(\bar{\kappa}_m + 1)))$$

are nowhere dense, the set A is meager. □

Theorem [AC] 8.101. *An $\aleph_1$-Luzin subset of a perfect Polish space has the Menger property. A κ-Luzin set does not have the Hurewicz Property.*

Proof. Assume that $L \subseteq X$ is an $\aleph_1$-Luzin set. Let $\{r_n : n \in \omega\}$ be a dense subset of L. Let $\{\mathcal{U}_n\}_{n=0}^{\infty}$ be a sequence of countable open covers of L. For every n there exists a $U_n \in \mathcal{U}_n$ such that $r_n \in U_n$. Since $\overline{L} \setminus \bigcup_n U_n$ is closed nowhere dense, $L \setminus \bigcup_n U_n$ is countable, say $\{q_n : n \in \omega\}$. Then there are sets $V_n \in \mathcal{U}_n$ such that $q_n \in V_n$. Set $\mathcal{V}_n = \{U_n, V_n\}$.

The second assertion follows by Lemma 8.100. □

A subset A of a topological space X has the **Rothberger Property** or the **Property C''** if for every sequence $\langle \mathcal{U}_n : n \in \omega \rangle$ of open covers of A there exist sets $U_n \in \mathcal{U}_n$ such that $\{U_n : n \in \omega\}$ is a cover of A, i.e., if A is an $S_1(\mathrm{O}, \mathrm{O})$-space. Note that a set with the Rothberger Property has also the Menger Property.

Theorem 8.102. *A subset of a metric space with the Rothberger Property has strong measure zero.*

Proof. If $\langle \varepsilon_n > 0 : n \in \omega \rangle$ are given, we let the cover $\mathcal{U}_n$ be the set of all open balls of diameter less than ε_n. □

Corollary 8.103. *Neither Cantor middle-third set $\mathbb{C}$ nor $[0,1]$ has the Rothberger property.*

Since a continuous image of a set with the Rothberger Property is evidently a set with the Rothberger Property, by Theorem 8.22 we obtain

Corollary [AC] 8.104. *A completely regular space with the Rothberger Property has a clopen base. Hence, a metric separable space with Rothberger Property is homeomorphic to a set of reals.*

Theorem [AC] 8.105. *Any set concentrated on a countable subset has the Rothberger Property.*

Proof. If $\langle U_{n,m} : m \in \omega \rangle$, $n \in \omega$ are open covers of A and A is concentrated on a countable set $\{x_n : n \in \omega\}$, then one can choose integers m_{2n} such that $x_n \in U_{2n,m_{2n}}$. Then $A \setminus \bigcup_n U_{2n,m_{2n}}$ is countable and one can choose m_{2n+1} such that $A \subseteq \bigcup_n U_{n,m_n}$. □

By Theorem 8.30 we obtain

Corollary [AC] 8.106. *An $\aleph_1$-Luzin set has the Rothberger Property.*

Corollary [AC] 8.107 (A.S. Besicovitch). *If* $\mathrm{add}(\mathcal{M}) = \mathfrak{c}$, *then there exists a strong measure zero set A of reals of cardinality $\mathfrak{c}$ that is not concentrated on any countable set.*

Proof. By Rothberger's Theorem 8.21 there exists a set A of reals of cardinality $\mathfrak{c}$ that is $\mathfrak{c}$-concentrated on rationals and can be continuously mapped onto Cantor space $^{\omega}2$. Since $\mathrm{add}(\mathcal{M}) = \mathfrak{c}$, by Corollary 8.20 the set A has strong measure zero. Since a continuous image of a set with the Rothberger Property has the Rothberger Property as well, the set A does not possess the Rothberger Property. Therefore the set A is not concentrated on any countable set.   □

Theorem [wAC] 8.108 (D.H. Fremlin – A.W. Miller). *Assume that $\langle X, \mathcal{O} \rangle$ is a separable metrizable topological space. Then X has the Rothberger Property if and only if X has strong measure zero with respect to any metric compatible with the topology $\mathcal{O}$.*

Proof. The implication from left to right follows by Theorem 8.102.

Assume that X has strong measure zero with respect to any metric compatible with the topology $\mathcal{O}$. Let $\langle \mathcal{U}_n : n \in \omega \rangle$ be a sequence of open covers of X. By Theorem 8.12, c) there exists a countable clopen base $\mathcal{B}$ of the topology $\mathcal{O}$. Let ρ be a metric on X compatible with the topology $\mathcal{O}$. One can easily construct a sequence of countable clopen partitions $\langle \mathcal{V}_n : n \in \omega \rangle$ of X such that every $\mathcal{V}_{n+1}$ is a refinement of both $\mathcal{U}_{n+1}$ and $\mathcal{V}_n$, and any $V \in \mathcal{V}_n$ has ρ-diameter less than 2^{-n}. We let $\sigma(x, y) = 2^{-n}$, where n is the least integer such that

$$(\forall U, V \in \mathcal{V}_n)\,((x \in U \wedge y \in V \to U \neq V).$$

One can easily see that σ is a metric. Moreover, if $2^{-n} < \varepsilon$ and $x \in V \in \mathcal{V}_m$ we have

$$\mathrm{Ball}_{\sigma}(x, 2^{-n}) \subseteq \mathrm{Ball}_{\rho}(x, \varepsilon), \quad V \subseteq \mathrm{Ball}_{\sigma}(x, 2^{-m}).$$

Thus, the metric σ induces the same topology as the metric ρ, i.e., $\mathcal{O}$.

Since X has strong measure zero with respect to metric σ, there exist sets $V_n \in \mathcal{V}_n$ of σ-diameters less than 2^{-n} such that $\bigcup_n V_n = X$. Choosing $U_n \in \mathcal{U}_n$ with $V_n \subseteq U_n$, we are ready.   □

Theorem [AC] 8.109 (J. Gerlits – Z. Nagy). *A γ-space has the Rothberger Property.*

Proof. Let $\langle \mathcal{U}_n : n \in \omega \rangle$ be a sequence of open covers of a topological space X. We can assume that each $\mathcal{U}_{n+1}$ is a refinement of $\mathcal{U}_n$. Fix a sequence $\{x_n\}_{n=0}^{\infty}$ of distinct elements of X. Set

$$\mathcal{V}_n = \left\{\bigcup\nolimits_{k \leq 2n} U_k \setminus \{x_n\} : U_k \in \mathcal{U}_{n^2+k} \text{ for } k = 0, \ldots, 2n\right\}.$$

We claim that $\mathcal{V} = \bigcup_n \mathcal{V}_n$ is an ω-cover. Actually, if $Y = \{y_0, \ldots, y_m\}$ is a finite subset of X, then take n such that $m \leq 2n + 1$ and $x_n \notin Y$. Then one can easily find a set $V \in \mathcal{V}_n$ with $Y \subseteq V$.

Since X is a γ-space, there exists a countable γ-subcover $\{V_n : n \in \omega\}$ of $\mathcal{V}$. Let k_n be such that $V_n \in \mathcal{V}_{k_n}$. Since $x_{k_n} \notin V_n$ we can assume that $k_n < k_{n+1}$ for any n. By definition, for every n there exist sets $U_{k_n^2+i} \in \mathcal{U}_{k_n^2+i}, i \leq 2k_n$ such that $V_n = \bigcup_{i \leq 2k_n} U_{k_n^2+i} \setminus \{x_{k_n}\}$.

One can easily see that $\{U_{k_n^2+i} : n \in \omega \wedge i \leq 2k_n\}$ is a cover of X.   □

By Theorem 8.102 we obtain

Corollary [AC] 8.110. *A γ-subspace of a metric space has strong measure zero.*

Theorem [AC] 8.111 (F. Rothberger). *Any topological space X with a countable base of size $|X| < \mathrm{cov}(\mathcal{M})$ has the Rothberger Property.*

We begin with a lemma which is actually one implication of Theorem 7.66.

Lemma [AC] 8.112. *If $A \subseteq {}^\omega\omega$ is such that*

$$(\forall \alpha \in {}^\omega\omega)(\exists \beta \in A)\,(\alpha, \beta \text{ are eventually different}), \qquad (8.16)$$

then $|A| \geq \mathrm{cov}(\mathcal{M})$.

Proof. For a $\beta \in {}^\omega\omega$ we set $E_\beta = \{\alpha \in {}^\omega\omega : \alpha, \beta \text{ are eventually different}\}$. By simple computation one can easily see that every E_β is meager. By (8.16) we obtain that ${}^\omega\omega = \bigcup_{\beta \in A} E_\beta$. Hence $|A| \geq \mathrm{cov}(\mathcal{M})$.   □

Proof of the theorem. Let $|X| < \mathrm{cov}(\mathcal{M})$ be a topological space with a countable base, $\langle \mathcal{U}_n = \{U_{n,m} : m \in \omega\} : n \in \omega \rangle$ being a sequence of open covers of X.

For every $x \in X$ choose a $\beta_x \in {}^\omega\omega$ such that $x \in \bigcap_n U_{n,\beta_x(n)}$. Let us consider the set $A = \{\beta_x \in {}^\omega\omega : x \in X\}$. Since $|A| < \mathrm{cov}(\mathcal{M})$, by the lemma there exists an α such that

$$(\forall x \in X)(\forall m)(\exists n > m)\, \alpha(n) = \beta_x(n).$$

Then $\{U_{n,\alpha(n)} : n \in \omega\}$ is a cover of X. Actually, if $x \in X$, then there exists an $m \in \omega$ such that $\alpha(m) = \beta_x(m)$. Thus

$$x \in \bigcap_n U_{n,\beta_x(n)} \subseteq U_{m,\beta_x(m)} = U_{m,\alpha(m)}. \qquad □$$

It is easy to see that a set $A \subseteq {}^\omega\omega$ satisfying (8.16) does not have the Rothberger Property (take $U_{n,m} = \{\alpha : \alpha(n) = m\}$). Thus by Theorem 7.66 we obtain that

$$\mathrm{non}(\mathrm{C}'') = \mathrm{cov}(\mathcal{M}).$$

Exercises

8.20 [AC] ω-**covers**

Let (ε) denote the following property of a topological space X: any open ω-cover of X contains a countable ω-subcover. Recall that X is Lindelöf if any open cover of X contains a countable subcover, see Exercise 1.19.

a) If X has the property (ε), then X is Lindelöf.

   Hint: If $\mathcal{U}$ is an open cover, consider the ω-cover $\{\bigcup \mathcal{V} : \mathcal{V} \in [\mathcal{U}]^{<\omega}\}$.

b) If $\mathcal{U}$ is an open cover of X^k, then

$$\mathcal{G} = \{G \subseteq X : G \text{ open } \wedge (\exists \mathcal{V} \in [\mathcal{U}]^{<\omega}) \, G^k \subseteq \bigcup \mathcal{V}\}$$

   is an ω-cover of X.

   Hint: If $A \subseteq X$ is finite, find finite $\mathcal{V} \subseteq \mathcal{U}$ such that $A^k \subseteq \bigcup \mathcal{V}$. For every $x \in A$ take a sufficiently small open set G_x such that $G_{x_1} \times \cdots \times G_{x_k} \subseteq \bigcup \mathcal{V}$ for any $x_1, \ldots, x_k \in A$. Then $\bigcup_{x \in A} G_x \in \mathcal{G}$.

c) If X has the property (ε), then X^k is Lindelöf for every k.

d) If $\mathcal{U}$ is an ω-cover of X then $\mathcal{U}^k = \{U^k : U \in \mathcal{U}\}$ is an ω-cover of X^k.

e) If X^k is Lindelöf for every k, then X has the property (ε).

   Hint: For every $k > 0$ find a countable $\mathcal{V}_k \subseteq \mathcal{U}$ such that $\mathcal{V}_k^k$ is a countable subcover of $\mathcal{U}^k$. Then $\bigcup_k \mathcal{V}_k$ is an ω-cover.

f) X has the property (ε) if and only if X^k is Lindelöf for every k.

g) Property (ε) is preserved under passing to closed subspaces and under continuous images.

8.21 [AC] γ-**space**

a) Prove directly that every γ-space is a wQN-space.

   Hint: Let $\{x_n : n \in \omega\}$ be infinite. Set $U_{n,m} = \{x \in X : |f_m(x)| < 1/(n+1)\} \setminus \{x_n\}$ and show that $\{U_{n,m} : n \leq m\}$ is an ω-cover of X.

b) An F_σ subset of a γ-space is a γ-space.

   Hint: If $F = \bigcup_n F_n$, $\langle F_n : n \in \omega \rangle$ non-decreasing, $\mathcal{U}$ an ω-cover of F, set

$$\mathcal{U}^* = \{U \cup (X \setminus F_n) : U \in \mathcal{U} \wedge n \in \omega\}.$$

c) A continuous image of a γ-space is a γ-space.

8.22 [AC] $S_1(\Gamma, \Gamma)$ **and** $S_1(\Gamma^{sh}, \Gamma)$

We suppose that the considered space has a countable base.

a) Show that $S_1(\Gamma_\omega, \Gamma) = S_1(\Gamma, \Gamma)$.

b) If X is an $S_1(\Gamma, \Gamma)$-space and $\langle \{U_{n,k} : k \in \omega\} : n \in \omega \rangle$ is a sequence of γ-covers of X, then there exists an increasing sequence $\{m_n\}_{n=0}^{\infty}$ such that $\{U_{n,m_n} : n \in \omega\}$ is a γ-cover.

   Hint: Apply $S_1(\Gamma, \Gamma)$ to covers $\langle \{U_{\lambda(n),n+m} : m \in \omega\} : n \in \omega \rangle$, where λ is the left inverse to a pairing function.

c) If $X = \bigcup_{s \in S} X_s$, $|S| < \mathfrak{h}$, each X_s is an $S_1(\Gamma, \Gamma)$-space then X is an $S_1(\Gamma, \Gamma)$-space.
Hint: Let $\langle \mathcal{U}_n = \{U_{n,k} : k \in \omega\} : n \in \omega \rangle$ be γ-covers of X. Set

$$\mathcal{H}_s = \{E \in [\omega]^\omega : \{U_{n,\mathbf{e}_E(n)} : n \in \omega\} \text{ is a } \gamma\text{-cover of } X_s\}.$$

The set $\mathcal{H}_s$ is dense in $\langle [\omega]^\omega, \subseteq^* \rangle$. Compare the proof of Theorem 8.61.

d) An F_σ subset of an $S_1(\Gamma, \Gamma)$-space is an $S_1(\Gamma, \Gamma)$-space.

e) Every shrinkable γ-cover contains a countable shrinkable γ-subcover.
Hint: If $\mathcal{V}$ is a closed γ-cover refining a γ-cover $\mathcal{U}$, then $\{U \in \mathcal{U} : (\exists V \in \mathcal{V}) V \subseteq U\}$ is a γ-cover. Take a countable subcover of it.

f) $S_1(\Gamma^{sh}, \Gamma) = S_1(\Gamma^{sh}_\omega, \Gamma)$.

8.23 [AC] $\overline{S}_1(\mathcal{A}, \mathcal{B})$

We suppose that the families of covers $\mathcal{A}$ and $\mathcal{B}$ satisfy conditions of Theorem 8.95.

A cover $\mathcal{V}$ is a **regular refinement** of a cover $\mathcal{U}$, if for every $V \in \mathcal{V}$ there exists a $U \in \mathcal{U}$ such that $\overline{V} \subseteq U$. A topological space X has the property $\overline{S}_1(\mathcal{A}, \mathcal{B})$, if for every sequence of covers $\langle \mathcal{U}_n \in \mathcal{A} : n \in \omega \rangle$ such that every $\mathcal{U}_{n+1}$ is a regular refinement of $\mathcal{U}_n$, there exist $\langle U_n \in \mathcal{U}_n : n \in \omega \rangle$ such that $\{U_n : n \in \omega\} \in \mathcal{B}$.

a) If a cover $\mathcal{U}$ has a regular refinement, then $\mathcal{U}$ is shrinkable.

b) $\overline{S}_1(\mathcal{A}, \mathcal{B}) \to S_1(\mathcal{A}, \mathcal{B})$.

c) If $\mathcal{A}$ consists of clopen covers then $S_1(\mathcal{A}, \mathcal{B}) \to \overline{S}_1(\mathcal{A}, \mathcal{B})$.

8.24 [AC] $U_{\text{fin}}(\mathcal{A}, \mathcal{B})$

Assume that X has a countable base. X is a $U_{\text{fin}}(\mathcal{A}, \mathcal{B})$-**space** if for any sequence $\{\mathcal{U}_n\}_{n=0}^\infty$ of essentially infinite $\mathcal{A}$-covers of X there exist finite $\mathcal{V}_n \subseteq \mathcal{U}_n$ such that $\{\bigcup \mathcal{V}_n : n \in \omega\}$ is a $\mathcal{B}$-cover.

a) If every $\mathcal{A}_1$-cover is also an $\mathcal{A}_2$-cover then $U_{\text{fin}}(\mathcal{A}_2, \mathcal{E}) \to U_{\text{fin}}(\mathcal{A}_1, \mathcal{B})$.

b) If every $\mathcal{B}_1$-cover is also a $\mathcal{B}_2$-cover then $U_{\text{fin}}(\mathcal{A}, \mathcal{B}_1) \to U_{\text{fin}}(\mathcal{A}, \mathcal{B}_2)$.

c) Show that an analogue of Theorem 8.95 holds true for the property $U_{\text{fin}}(\mathcal{A}, \mathcal{B})$.

d) An $U_{\text{fin}}(O, O)$-space is Lindelöf.

e) For a Lindelöf topological space X the following are equivalent:
   (i) X has the Menger Property.
   (ii) X is a $U_{\text{fin}}(\Gamma, O)$-space.
   (iii) X is a $U_{\text{fin}}(\Gamma, \Lambda)$-space.

f) For a Lindelöf topological space X the following are equivalent:
   (i) X has the Hurewicz Property.
   (ii) X is a $U_{\text{fin}}(O, \Gamma)$-space.
   (iii) X is a $U_{\text{fin}}(\Gamma, \Gamma)$-space.

8.25 [AC] Menger and Hurewicz Properties

If $\langle \{U_{n,m} : m \in \omega\} : n \in \omega \rangle$ is a sequence of countable open covers of a topological space X, then we define a mapping $f : X \longrightarrow {}^\omega\omega$ as

$$f(x)(n) = \min\{m : x \in U_{n,m}\}. \tag{8.17}$$

  a) No dominating subset of $^\omega\omega$ has the Menger Property.
     Hint: Use the functions (8.10).

  b) Every topological space of size less than $\mathfrak{d}$ has the Menger Property.
     Hint: Use the mapping (8.17).

  c) non(Menger Property) $= \mathfrak{d}$.

  d) non(Hurewicz Property) $= \mathfrak{b}$.
     Hint: Similarly as in a)–b).

8.26 [AC] **Rothberger Property**

  a) A subset A of a Polish space has the Rothberger Property C″ if and only if A is
     zero-dimensional and every continuous image of A in $^\omega\omega$ has the Property C″.
     Hint: If $U_{n,m}$ are clopen, then mappings defined by (8.17) *are continuous.*

  b) non(C″) is the size of the smallest subset of $^\omega\omega$ which does not have the Rothberger
     Property C″.

  c) If X is a Lindelöf topological space, then the following are equivalent:

     i) X has the Rothberger Property.

     ii) X is an $S_1(\Lambda, O)$-space.

     iii) X is an $S_1(\Lambda, \Lambda)$-space.

     iv) X is an $S_1(\Omega, O)$-space.

     v) X is an $S_1(\Omega, \Lambda)$-space.

  d) add($\mathcal{N}$) $\leq$ add(C″) for any separable metric space X.
     Hint: See Exercise 8.2 *and Theorem* 8.108.

8.5   Coverings versus Sequences

Witold Hurewicz [1927] was probably the first who showed that sequence convergence properties of real continuous functions defined on a metric space X, namely
the above-introduced H* and H**, are equivalent to covering properties of X.

Theorem [AC] **8.113 (W. Hurewicz).** *Let X be a perfectly normal space. Then*

  a) *X has the property* H* *if and only if X has the Menger Property,*
  b) *X has the property* H** *if and only if X has the Hurewicz Property.*

Proof. Let $\langle \{U_{n,k} : k \in \omega\} : n \in \omega \rangle$ be a sequence of countable open covers of X.
We can assume that $U_{n,k} \neq \emptyset$ for every n, k. Let $f_{n,k} : X \longrightarrow [0,1]$ be a continuous
function such that $\mathsf{Z}(f_{n,k}) = X \setminus U_{n,k}$. Set

$$f_n = \sum_{k=0}^{\infty} 2^{-k} \cdot f_{n,k}.$$

Then f_n is continuous and $f_n(x) > 0$ for every $x \in X$.

Assume that the space X has the property H*. Then there exists a sequence $\{a_n\}_{n=0}^{\infty}$ of positive reals such that

$$(\forall x \in X)(\forall m)(\exists n > m)\, 1/f_n(x) < a_n.$$

Let k_n be such that

$$\sum_{i > k_n} 2^{-i} < 1/a_n. \tag{8.18}$$

Set $\mathcal{V}_n = \{U_{n,i} : i \leq k_n\}$.

If $x \in X$, then there exists an n such that $f_n(x) > 1/a_n$. For such an n we have

$$\sum_{i \leq k_n} 2^{-i} \cdot f_{n,i}(x) > 0.$$

Thus $x \in U_{n,i}$ for some $i \leq k_n$ and therefore $x \in \bigcup \mathcal{V}_n$. Hence $\{\bigcup \mathcal{V}_n : n \in \omega\}$ is a cover of X.

Assume now that the space X possesses the property H**. Then there exists a sequence $\{a_n\}_{n=0}^{\infty}$ of positive reals such that $\{1/f_n(x)\}_{n=0}^{\infty} \leq^* \{a_n\}_{n=0}^{\infty}$ for any $x \in X$. Let k_n and $\mathcal{V}_n$ be as above. If $f_n(x) \geq 1/a_n$, then $\sum_{i=0}^{k_n} 2^{-i} \cdot f_{n,i}(x) > 0$ and therefore $x \in \bigcup \mathcal{V}_n$. Since for any $x \in X$, $f_n(x) \geq 1/a_n$ for all but finitely many n, also $x \in \bigcup \mathcal{V}_n$ for all but finitely many n. Thus $\{\bigcup \mathcal{V}_n : n \in \omega\}$ is a γ-cover of X.

Now, let $\{f_n\}_{n=0}^{\infty}$ be a sequence of continuous real functions defined on the topological space X. Set $U_{n,m} = \{x \in X : |f_n(x)| < m\}$. Then for every n, $\mathcal{U}_n = \{U_{n,m} : m \in \omega\}$ is an open cover of X. Moreover, $U_{n,m} \subseteq U_{n,m+1}$ for any n, m. Thus the union of any finite subset of $\mathcal{U}_n$ is equal to its largest element. Also $\mathcal{U}_{n,k} = \{U_{n,m} : m \geq k\}$ is a cover.

Let us assume that X has the Menger Property and consider the sequence of covers $\langle \mathcal{W}_n = \mathcal{U}_{\rho(n),\lambda(n)} : n \in \omega \rangle$, where λ, ρ are the inverse functions to the pairing function π. Then there exist finite sets $\mathcal{V}_n \subseteq \mathcal{W}_n$ such that $\{\bigcup \mathcal{V}_n : n \in \omega\}$ is a cover of X. Since $\bigcup \mathcal{V}_n = U_{\rho(n),m_n}$ for some $m_n \geq \lambda(n)$, we obtain for every $x \in X$ that $|f_n(x)| < m_n$ for infinitely many n.

If X has the Hurewicz Property, then there exist finite subsets $\mathcal{V}_n \subseteq \mathcal{U}_n$ such that $\{\bigcup \mathcal{V}_n : n \in \omega\}$ is a γ-cover of X. Then for every $x \in X$ we have $|f_n(x)| < m_n$ for all but finitely many n. $\qquad\square$

Corollary [AC] 8.114. *A γ-space has the property H**.*

Proof. Note that the last part of the proof of the theorem works for any topological space and use Corollary 8.98. $\qquad\square$

Another result of this type is the characterization of topological spaces for which $C_p(X)$ is a Fréchet space. Actually we have

Theorem [AC] 8.115 (J. Gerlits – Z. Nagy). *A completely regular topological space X is a γ-space if and only if the topological space $C_p(X)$ is Fréchet.*

Proof. If X is finite, then the assertion is trivial.

So let X be an infinite completely regular topological space. Fix a sequence of mutually distinct elements $\langle x_n \in X : n \in \omega \rangle$. For $h \in C_p(X)$ and $n \in \omega$ we write

$$U_{h,n} = \{x \in X : |h(x)| < 2^{-n} \wedge x \neq x_n\}.$$

Evidently $U_{h,n}$ is an open set.

Assume that X is a γ-space. Let $A \subseteq C_p(X)$, $f \in \overline{A}$. We can assume that $f \notin A$. We shall find $g_k \in A$, $k \in \omega$ such that $g_k \to f$ on X.

We start with showing that the family

$$\mathcal{U} = \{U_{f-g,n} : n \in \omega, g \in A\}$$

is an open ω-cover of X.

Let $y_0, \ldots, y_k \in X$ be given. Take an n such that $x_n \neq y_0, \ldots, y_k$. We consider the basic neighborhood

$$V = \{h \in C_p(X) : |h(y_i) - f(y_i)| < 2^{-n} \text{ for } i = 0, \ldots, k\} \qquad (8.19)$$

of f in $C_p(X)$. Since $f \in \overline{A}$ there exists a $g \in V \cap A$. Then we have $y_i \in U_{f-g,n}$ for $i = 0, \ldots, k$. Thus, $\mathcal{U}$ is an ω-cover.

Now, let $\mathcal{G} = \{G_k : k \in \omega\}$ be a countable γ-subcover of $\mathcal{U}$. Then there are $g_k \in A$ and $n_k \in \omega$ such that

$$G_k = U_{f-g_k,n_k} \text{ for each } k \in \omega.$$

We claim that $f = \lim_{k \to \infty} g_k$ on X.

We start with showing that $\lim_{k \to \infty} n_k = \infty$. Actually if not, then there exists a natural number m such that $n_k = m$ for infinitely many k. Thus $x_m \notin G_k$ for infinitely many k contradicting the assumption that $\mathcal{G}$ is a γ-cover.

For given $x \in X$ and $\varepsilon > 0$, one can easily find a k_0 such that $2^{-n_k} < \varepsilon$ and $x \in G_k$ for $k \geq k_0$. Then $|f(x) - g_k(x)| < \varepsilon$ for every $k \geq k_0$.

Assume now that $C_p(X)$ is Fréchet space. Let $\mathcal{U}$ be an open ω-cover of X. We set

$$A = \{f \in C_p(X) : (\exists U \in \mathcal{U}) \{x \in X : |f(x)| < 1\} \subseteq U\}.$$

We show that $0 \in \overline{A}$. Actually, let V be a basic neighborhood of $f = 0$ of the form (8.19). Since $\mathcal{U}$ is an ω-cover there exists a $U \in \mathcal{U}$ such that $y_0, \ldots, y_k \in U$. Since X is completely regular, there exists a function $h \in C_p(X)$ satisfying $h(y_i) = 0$ for $i = 0, \ldots, k$ and $h(x) = 1$ for $x \in X \setminus U$. Then $h \in A \cap V$.

Since $C_p(X)$ is a Fréchet space, there exists a sequence $\langle f_n \in A : n \in \omega \rangle$ such that $\lim_{n \to \infty} f_n = 0$. By definition of the set A, for every n there exists a set $U_n \in \mathcal{U}$ such that $\{x \in X : |f_n(x)| < 1\} \subseteq U_n$. One can easily see that $\{U_n : n \in \omega\}$ is an open γ-cover of X: if $x \notin U_n$, then $|f_n(x)| \geq 1$. $\square$

Hence, by Corollary 8.65 we have:

Corollary [AC] 8.116. *A completely regular γ-space is a* wQN-*space.*

By Theorem 8.74 we obtain:

Corollary [AC] 8.117. *A metric separable γ-space is perfectly meager.*

Now we show similar relationships between wQN-spaces and QN-spaces and covering properties. We begin with wQN-spaces.

Lemma 8.118. *Every* $S_1(\Gamma^{sh}, \Gamma)$-*space is a* wQN-*space.*

Proof. Assume that $\langle f_n : n \in \omega \rangle$ are continuous, $f_n \to 0$ on X. We can assume that $f_n(x) > 0$ for every $x \in X$ and every n (if not, take $|f_n| + 2^{-n}$). We define

$$U_{n,m} = \{x \in X : f_m(x) < 2^{-n}\}, \quad \mathcal{U}_n = \{U_{n,m} : m \in \omega\}. \tag{8.20}$$

If $X \notin \mathcal{U}_n$, then $\mathcal{U}_n$ is a γ-cover of X. Consider the set $L = \{n \in \omega : X \notin \mathcal{U}_n\}$. If $n, m \in L$, $m > n$, then the closed γ-cover $\{\overline{U} : U \in \mathcal{U}_m\}$ is a refinement of $\mathcal{U}_n$.

If $\omega \setminus L = \{n_k : k \in \omega\}$ is infinite, then there exists a sequence $\{m_k\}_{k=0}^{\infty}$ such that $U_{n_k, m_k} = X$ for every k. If the sequence $\{m_k\}_{k=0}^{\infty}$ is bounded then $m_k = m$ for infinitely many k and therefore $f_m = 0$ – a contradiction. Thus we can assume that both sequences $\{n_k\}_{k=0}^{\infty}$ and $\{m_k\}_{k=0}^{\infty}$ are increasing. Since $f_{m_k}(x) < 2^{-n_k}$ for every $x \in X$ we obtain that $f_{m_k} \rightrightarrows 0$ on X.

Assume now that $\omega \setminus L$ is finite. Omitting finitely many members we can assume that $L = \omega$. Then $\mathcal{U}_n$ is a shrinkable γ-cover for every n. Since X is a $S_1(\Gamma^{sh}, \Gamma)$-space there exist $V_n \in \mathcal{U}_n$ such that $\{V_n : n \in \omega\}$ is a γ-cover. Let m_n be such that $V_n = U_{n,m_n}$. Assume that the sequence $\{m_n\}_{n=0}^{\infty}$ is bounded. Then $m_n = m$ for infinitely many n. Take any $x \in X$. Since $\{U_{n,m_n} : n \in \omega\}$ is a γ-cover we obtain $x \in U_{n,m}$ for infinitely many n. Then $f_m(x) = 0$, which is a contradiction. Thus, we can assume that $\{m_n\}_{n=0}^{\infty}$ is increasing. Then one can easily see that $f_{m_n} \xrightarrow{\text{QN}} 0$ on X with the control $\{2^{-n}\}_{n=0}^{\infty}$. $\square$

Corollary 8.119 (M. Scheepers). *Every* $S_1(\Gamma, \Gamma)$-*space is a* wQN-*space.*

Theorem [AC] 8.120 (L. Bukovský – J. Haleš). *A normal topological space X is a* wQN-*space if and only if X is an* $S_1(\Gamma^{sh}, \Gamma)$-*space.*

Proof. Assume that a normal space X is wQN. Let $\langle \mathcal{U}_n : n \in \omega \rangle$ be a sequence of shrinkable γ-covers. Without loss of generality we can assume that every $\mathcal{U}_n$ is countable, thus $\mathcal{U}_n = \{U_{n,m} : m \in \omega\}$. Moreover, by Theorem 8.95 we can assume that every $\mathcal{U}_{n+1}$ is a refinement of $\mathcal{U}_n$. Let $\mathcal{V}_n = \{V_{n,m} : m \in \omega\}$ be a closed γ-cover refining $\mathcal{U}_n$. We can assume that $V_{n,m} \subseteq U_{n,m}$ for every $n, m \in \omega$. Since X is normal, by the Urysohn Theorem 3.15, for every n, m, there exists a continuous function $f_{n,m} : X \longrightarrow [0,1]$ such that $f_{n,m}(x) = 0$ for $x \in V_{n,m}$ and $f_{n,m}(x) = 1$ for $x \in X \setminus U_{n,m}$.

Assume $x \in X$, $n \in \omega$. Since $\{V_{n,m} : m \in \omega\}$ is a γ-cover, there exists an m_0 such that $x \in V_{n,m}$ for every $m \geq m_0$. Thus $f_{n,m}(x) = 0$ for every $m \geq m_0$. Hence

$f_{n,m} \to 0$ on X for every n. By Theorem 8.64 the space $C_p(X)$ has the sequence selection property and therefore there exists an increasing sequence $\{m_n\}_{n=0}^{\infty}$ such that $f_{n,m_n} \to 0$ on X.

We claim that $\{U_{n,m_n} : n \in \omega\}$ is a γ-cover of X. Actually, if $x \in X$ then there exists an n_0 such that $f_{n,m_n}(x) < 1$ for every $n \geq n_0$. Then $x \in U_{n,m_n}$ for all but finitely many n. We need to show that $\{U_{n,m_n} : n \in \omega\}$ is infinite. However, if this set were finite, then $U_{n,m_n} = U \neq X$ for infinitely many n and any $x \in X \setminus U$ does not belong to infinitely many U_{n,m_n}, which is a contradiction. $\square$

Theorem [AC] 8.121. $\mathrm{non}(\mathrm{S}_1(\Gamma,\Gamma)) = \mathfrak{b}$.

Proof. By Corollaries 8.62 and 8.119 we have $\mathrm{non}(\mathrm{S}_1(\Gamma,\Gamma)) \leq \mathfrak{b}$.

Assume that $|X| < \mathfrak{b}$. Let $\langle\{U_{n,m} : m \in \omega\} : n \in \omega\rangle$ be a sequence of γ-covers of X. For every $x \in X$ we define a $\beta_x \in {}^\omega\omega$ as

$$\beta_x(n) = \min\{m : (\forall k \geq m)\, x \in U_{n,k}\}. \tag{8.21}$$

Since the family $\{\beta_x : x \in X\} \subseteq {}^\omega\omega$ is eventually bounded, there exists a $\gamma \in {}^\omega\omega$ such that $\beta_x \leq^* \gamma$ for every $x \in X$. One can easily see that $\{U_{n,\gamma(n)} : n \in \omega\}$ is a γ-cover. $\square$

We show that there exists an uncountable $\mathrm{S}_1(\Gamma,\Gamma)$-space even if $\mathfrak{b} = \aleph_1$. We begin with a technical result.

Lemma [AC] 8.122. *Assume that* $\langle \mathcal{U}_n = \{U_{n,m} : m \in \omega\} : n \in \omega\rangle$ *is a sequence of γ-covers of a topological space X such that each $\mathcal{U}_{n+1}$ is a refinement of $\mathcal{U}_n$, $Y \subseteq X$, $\alpha \in {}^\omega\omega$ is increasing and such that $\{V_n = U_{n,\alpha(n)} : n \in \omega\}$ is a γ-cover of $X \setminus Y$. If $|Y| < \mathfrak{b}$, then there exists an increasing $\gamma \in {}^\omega\omega$ such that $\{U_{n,\gamma(n)} : n \in \omega\}$ is a γ-cover of X.*

Proof. For each $x \in Y$ let $\beta_x(n) = \min\{m : (\forall k \geq m)\, x \in U_{n,k}\}$. Since $|Y| < \mathfrak{b}$, there exists an increasing sequence $\beta \in {}^\omega\omega$ dominating each β_x, $x \in Y$.

Since α is increasing, for every n there exist an $m \geq n$ and a $k \geq \beta(n)$ such that $V_m = U_{m,\alpha(m)} \subseteq U_{n,k}$. Set $\delta(n) = m$ and $\gamma(n) = k$. We can do everything in such a way that both sequences δ and γ are increasing.

Since $V_{\delta(n)} \subseteq U_{n,\gamma(n)}$ for every n, each $x \in X \setminus Y$ is contained in all but finitely many $U_{n,\gamma(n)}$. If $x \in Y$, then there exists an n_0 such that $\beta_x(n) \leq \beta(n)$ for any $n \geq n_0$. Then $x \in U_{n,\gamma(n)}$ for every $n \geq n_0$. $\square$

As usual we identify the set $\mathcal{P}(\omega)$ with Cantor space ${}^\omega 2$. The set $[\omega]^{<\omega}$ is topologically dense in $\mathcal{P}(\omega)$ and can be considered as "the set of rationals" of Cantor space ${}^\omega 2$. A base of the topology of $\mathcal{P}(\omega)$ is described by (8.15).

Theorem [AC] 8.123. *If* $\mathfrak{t} = \mathfrak{b}$, *then there exists a set of reals $X \subseteq {}^\omega 2$ of cardinality $\mathfrak{b}$ such that X is an $\mathrm{S}_1(\Gamma,\Gamma)$-space, therefore also a wQN-space, and $X \setminus [\omega]^{<\omega}$ is not a wQN-space. Hence X is not a QN-space nor is it a λ-space.*

Proof. Let $\kappa = \mathfrak{t} = \mathfrak{b}$. By Theorem 5.51 there exists a tower $\langle A_\xi : \xi < \kappa \rangle$ such that the family of functions $\{e_{A_\xi} : \xi < \kappa\}$ is unbounded in $^\omega\omega$. We show that the set $X = [\omega]^{<\omega} \cup \{A_\xi : \xi < \kappa\} \subseteq \mathcal{P}(\omega)$ is an $S_1(\Gamma, \Gamma)$-space.

Let us consider a $K \in [\omega]^\omega$. We show that there exists a $\xi < \kappa$ such that the set $[e_{A_\xi}(n), e_{A_\xi}(n+1)) \cap K$ has at least two elements for infinitely many n.

Assume not. Then for every ξ there exists an $r_\xi \geq e_{A_\xi}(0)$ such that the intersection $K \cap [e_{A_\xi}(n), e_{A_\xi}(n+1))$ has at most one element for every $n \geq n_\xi$. Then $e_{A_\xi} \leq^* e_{K \setminus L}$ for some finite L. Evidently there exists a $g \in {}^\omega\omega$ such that $e_{K \setminus L}$ for any $L \in [\omega]^{<\omega}$. Then g dominates e_{A_ξ} for every $\xi < \kappa$, a contradiction.

Next we show the following: if $\langle \mathcal{U}_n = \{U_{n,m} : m \in \omega\} : n \in \omega \rangle$ is a sequence of γ-covers of X such that each $\mathcal{U}_{n+1}$ is a refinement of $\mathcal{U}_n$, then there exist a set $Y \subseteq X$, $|Y| < \kappa$, and increasing $\alpha \in {}^\omega\omega$ such that $\{V_n = U_{n,\alpha(n)} : n \in \omega\}$ is a γ-cover of $X \setminus Y$. Then by Lemma 8.122 we obtain that X is an $S_1(\Gamma, \Gamma)$-space.

We construct a γ-cover $\{V_n = U_{n,\alpha(n)} : n \in \omega\}$ of $[\omega]^{<\omega}$ and an increasing sequence $\{k_n\}_{n=0}^\infty$. Assume that $\alpha(i)$, $i < n$ and k_i, $i \leq n$ are already defined. Since $\mathcal{U}_n$ is an ω-cover, there exists an $\alpha(n)$ such that $\mathcal{P}(k_n + 1) \subseteq U_{n,\alpha(n)}$. By Lemma 8.93 there exists a $k_{n+1} > k_n$ such that

$$\{A \subseteq \omega : A \cap k_{n+1} = B \cap k_{n+1}\} \subseteq U_{n,\alpha(n)}$$

for any $B \in \mathcal{P}(k_n + 1)$. Then $A \in U_{n,\alpha(n)}$ for any A disjoint with $[k_n + 1, k_{n+1})$. By Lemma 8.94, for $K = \{k_n : n \in \omega\}$ there exists a $\xi < \kappa$, an infinite $L \subseteq \omega$ and an increasing sequence $\{m_n\}_{n=0}^\infty$ such that $e_{A_\xi}(m_n) \leq k_n < k_{n+1} < e_{A_\xi}(m_{n+1})$ for $n \in L$. Then $A_\eta \in V_n$ for all $\eta \geq \xi$ and for all but finitely many $n \in L$. Thus $\{V_n : n \in L\}$ is a γ-cover of $X \setminus \{A_\zeta : \zeta < \xi\}$.

Since the mapping $F : [\omega]^\omega \longrightarrow {}^\omega\omega$ defined as $F(A) = e_A$ is continuous and the image $F(\{A_\xi, \xi < \kappa\})$ is unbounded in $^\omega\omega$, by Theorem 8.72 the set $X \setminus [\omega]^{<\omega} = \{A_\xi, \xi < \kappa\}$ is not a wQN-space. By Theorem 8.61 d), the set $[\omega]^{<\omega}$ is not G_δ. □

Theorem [AC] 8.124 (I. Recław). *There exists an uncountable $S_1(\Gamma, \Gamma)$-space, hence there exists an uncountable wQN-space.*

Proof. If $\mathfrak{b} > \aleph_1$, then the discrete space of cardinality $\aleph_1$ is an $S_1(\Gamma, \Gamma)$-space. If $\mathfrak{b} = \aleph_1$, then also $\mathfrak{t} = \aleph_1$ and by Theorem 8.123 there exists an $S_1(\Gamma, \Gamma)$-space of cardinality $\aleph_1$. □

We continue by showing that the property to be a QN-space is equivalent to certain covering properties. We need some notions and auxiliary results.

We introduce three covering properties of a topological space:

(β_1) For every sequence of countable open γ-covers $\langle \mathcal{U}_n : n \in \omega \rangle$ there exist finite sets $\mathcal{V}_n \subseteq \mathcal{U}_n$ such that $\bigcup_n (\mathcal{U}_n \setminus \mathcal{V}_n)$ is a γ-cover of X.

(β_2) For every sequence of countable open γ-covers $\langle \mathcal{U}_n : n \in \omega \rangle$ with adequate enumerations $\mathcal{U}_n = \langle U_{n,m} : m \in \omega \rangle$, there exists a non-decreasing unbounded sequence $\{n_m\}_{m=0}^{\infty}$ such that $\langle U_{n_m,m} : m \in \omega \rangle$ is an adequate enumeration of a γ-cover of X.

(β_3) For every sequence of countable open γ-covers $\langle \mathcal{U}_n : n \in \omega \rangle$ with adequate enumerations $\mathcal{U}_n = \langle U_{n,m} : m \in \omega \rangle$, there is an open γ-cover $\langle V_m : m \in \omega \rangle$ such that

$$(\forall n)(\exists k)(\forall m > k)\, V_m \subseteq U_{n,m}. \tag{8.22}$$

Theorem [AC] 8.125 (L. Bukovský – J. Haleš). *If X is a perfectly normal topological space, then the following statements are equivalent:*

(a) $C_p(X)$ *has the property* (α_1).
(b) X *has the property* (β_1).
(c) X *has the property* (β_2).
(d) X *has the property* (β_3).
(e) X *is a* QN-*space.*

We begin with auxiliary results.

Lemma [AC] 8.126. *If X is a normal σ-space with $\mathrm{Ind}(X) = 0$, then for every open countable γ-cover of X with an adequate enumeration $\langle U_n : n \in \omega \rangle$, there exist clopen sets $\langle V_n \subseteq U_n : n \in \omega \rangle$ such that $\langle V_n : n \in \omega \rangle$ is an adequate enumeration of a γ-cover of X.*

Proof. The sequence $\langle G_n = \bigcap_{k \geq n} U_k : n \in \omega \rangle$ is a non-decreasing γ-cover by G_δ sets. Since X is a σ-space, there exist closed sets $\langle F_{n,m} : m \in \omega \rangle$ such that $G_n = \bigcup_m F_{n,m}$. We can assume $F_{n,m} \subseteq F_{n,m+1}$ and $F_{n,m} \subseteq F_{n+1,m}$ for every $n, m \in \omega$. Then $\langle F_{n,n} : n \in \omega \rangle$ is a non-decreasing closed γ-cover. Since $\mathrm{Ind}(X) = 0$, there exist clopen sets $\langle V_n : n \in \omega \rangle$ such that $F_{n,n} \subseteq V_n \subseteq U_n$. It is easy to see that $\langle V_n : n \in \omega \rangle$ is an adequate enumeration of a γ-cover of X. □

Lemma [AC] 8.127. *A topological space with the property (β_1) possesses the property (β_2) as well.*

Proof. Let $\langle \mathcal{U}_n : n \in \omega \rangle$ be a sequence of countable open γ-covers of X with adequate enumerations $\{U_{n,m} : m \in \omega\}$. By (β_1) there are finite sets $\mathcal{V}_n \subseteq \mathcal{U}_n$ such that $\bigcup_n (\mathcal{U}_n \setminus \mathcal{V}_n)$ is a γ-cover. Evidently, there exists a non-decreasing sequence $\{k_n\}_{n=0}^{\infty}$ such that $\mathcal{V}_n \subseteq \{U_{n,0}, \ldots, U_{n,k_n}\}$. By induction we can easily find an increasing sequence $\{l_n\}_{n=0}^{\infty}$ such that $l_n \geq k_n$ for each n, and no set $U_{i,j}, j \leq l_i, i \leq n$ occurs in the set $\{U_{n+1,k} : k > l_{n+1}\}$. Actually, set $l_0 = k_0$ and using the fact that every enumeration is adequate, find the smallest integer l_{n+1} greater than $\max\{l_n, k_{n+1}\}$ and such that no set $U_{i,j}, j \leq l_i, i \leq n$ occurs in the set $\{U_{n+1,k} : k > l_{n+1}\}$. Now let $n_i = m$ for $l_m < i \leq l_{m+1}$ for any m. Evidently $\{n_i\}_{i=0}^{\infty}$ is non-decreasing and unbounded.

Now, the infinite subset $\langle U_{n_i,i} : i \in \omega \rangle$ of the γ-cover $\bigcup_n (\mathcal{U}_n \setminus \mathcal{V}_n)$ is a γ-cover. By the choice of l_n, any set $U_{n_i,i}$ is different from $U_{n_j,j}$ provided that $n_i > n_j$. Thus the enumeration $\langle U_{n_i,i} : i \in \omega \rangle$ is adequate. □

Lemma [AC] 8.128. *Any normal topological QN-space has the property (β_1).*

Proof. Assume that $\langle \mathcal{U}_n : n \in \omega \rangle$ is a sequence of countable open γ-covers of a QN-space X. Let $\langle U_{n,m} : m \in \omega \rangle$ be a bijective enumeration of $\mathcal{U}_n$ for every $n \in \omega$. Since X is a normal σ-space with $\mathrm{Ind}(X) = 0$, by Lemma 8.126 there exist clopen sets $\langle V_{n,m} \subseteq U_{n,m} : n, m \in \omega \rangle$ such that $\langle V_{n,m} : m \in \omega \rangle$ is an adequate enumeration of a γ-cover of X for every $n \in \omega$.

Since X is a normal topological space, for every $n, m \in \omega$ there exists a continuous function $f_{n,m} : X \longrightarrow [0,1]$ such that $f_{n,m}(x) = 0$ for $x \in V_{n,m}$ and $f_{n,m}(x) = 1$ for $x \in X \setminus U_{n,m}$.

Equally as in the proof of Theorem 8.120 one can show that $f_{n,m} \to 0$ on X for every n. Since a QN-space has the property (α_1), there exists a sequence $\{m_n\}_{n=0}^\infty$ of natural numbers such that the sequence $\langle f_{n,m} : n \in \omega \wedge m \geq m_n \rangle$ converges to 0 on X. We show that $\langle U_{n,m} : n \in \omega \wedge m \geq m_n \rangle$ is a γ-cover.

Actually, if $x \in X$ then $f_{n,m}(x) < 1$ for all but finitely many couples $\langle n, m \rangle$ such that $n \in \omega$ and $m \geq m_n$. Therefore also $x \in U_{n,m}$ for all but finitely many couples $\langle n, m \rangle$ such that $n \in \omega$ and $m \geq m_n$. $\square$

Proof of Theorem 8.125. By Corollary 8.68 and Lemma 8.128 we have (e) $\to$ (a) and (a) $\to$ (b). By Lemma 8.127 we have (b) $\to$ (c).

We show the implication (c) $\to$ (d). Let $\langle \mathcal{U}_n : n \in \omega \rangle$ be a sequence of countable open γ-covers with an adequate enumeration $\mathcal{U}_n = \langle U_{n,m} : m \in \omega \rangle$. We set

$$W_{n,m} = \bigcap_{i \leq n} U_{i,m}, \quad \mathcal{W}_n = \{W_{n,m} : m \in \omega\}.$$

Then $\mathcal{W}_n$ is a γ-cover and a refinement of $\mathcal{U}_n$. Then by (c) there is a non-decreasing unbounded sequence $\{n_m\}_{m=0}^\infty$ such that $\langle W_{n_m,m} : n \in \omega \rangle$ is an adequate enumeration of a γ-cover. Since $W_{n_m,m} \subseteq U_{n,m}$ for every m such that $n_m \geq n$, we are done.

Finally, we show the implication (d) $\to$ (e). Assume that X has the property (β_3). Let $f_m \to 0$ on X. We can assume that $0 < f_m(x) \leq 1$ for every $x \in X$ and every m. Let $\langle x_n : n \in \omega \rangle$ be mutually different elements of X. For any $n, m \in \omega$ we define

$$U_{n,m} = \{x \in X : f_m(x) < 2^{-n} \wedge x \neq x_m\}, \quad \mathcal{U}_n = \{U_{n,m} : m \in \omega\}.$$

It is easy to see that every $\langle \mathcal{U}_n : n \in \omega \rangle$ is an open γ-cover with an adequate enumeration $\mathcal{U}_n = \{U_{n,m} : m \in \omega\}$. By (β_3) there exists a γ-cover $\{V_m : m \in \omega\}$ such that (8.22) is satisfied.

If $x \in X$ and $n \in \omega$ are given, then there exists an m_n such that for every $m \geq m_n$ we have $x \neq x_m$, $x \in V_m$ and $V_m \subseteq U_{n,m}$. Then also $f_m(x) < 2^{-n}$ and we are done. $\square$

Corollary [AC] 8.129. *Every normal QN-space is an* $S_1(\Gamma, \Gamma)$*-space.*

Proof. If $\langle \mathcal{U}_n : n \in \omega \rangle$ are γ-covers of X, then by (β_1) there exist finite sets $\langle \mathcal{V}_n \subseteq \mathcal{U}_n : n \in \omega \rangle$ such that $\bigcup_n (\mathcal{U}_n \setminus \mathcal{V}_n)$ is a γ-cover.

Choose a γ-subcover $\{U_m : m \in \omega\}$ of $\bigcup_n (\mathcal{U}_n \setminus \mathcal{V}_n)$ such that $U_m \in \mathcal{U}_m$ for each m. $\square$

By Theorem 8.63 we obtain

Corollary [AC] 8.130. *If $\kappa \leq \mathfrak{b}$ and A is a κ-Sierpiński subset of a Polish space with a Borel measure, then every subset of A is an $\mathsf{S}_1(\Gamma, \Gamma)$-space.*

Exercises

8.27 [AC] Weak Distributivity
Let $\mathcal{A}$ be a family of subsets of X. $\mathcal{A}$ is said to be **weakly distributive** on a set $A \subseteq X$ if for any sequence $\langle A_{n,m} \in \mathcal{A} : n, m \in \omega \rangle$ such that $A \subseteq \bigcap_n \bigcup_m A_{n,m}$, there exists a function $\alpha \in {}^\omega \omega$ such that $A \subseteq \bigcup_k \bigcap_{n \geq k} \bigcup_{m \leq \alpha(n)} A_{n,m}$.

a) We denote by $\mathcal{A}$ the set of all covers of X by sets from $\mathcal{A}$ and by $\mathcal{A}_\Gamma$ the set of all γ-covers by sets from $\mathcal{A}$. Then $\mathcal{A}$ is weakly distributive if and only if $\mathsf{U}_{\mathrm{fin}}(\mathcal{A}, \mathcal{A}_\Gamma)$ holds true.

b) The family of open subsets of a σ-compact space X is weakly distributive.
Hint: If $X = \bigcap_n \bigcup_m A_{n,m}$, $X = \bigcup_k X_k$, X_k compact, we set $\alpha(n) = m$, where m is the smallest integer such that $\bigcup_{k<n} X_k \subseteq \bigcup_{i \leq m} A_{n,i}$.

c) The family of open subsets of X is weakly distributive if and only if X has the Hurewicz Property.

d) If a family $\mathcal{A}$ is weakly distributive, then also the family $\mathbf{\Sigma}^0[\mathcal{A}]$ is weakly distributive.

e) If a family $\mathcal{A} \subseteq \mathcal{P}(X)$ is weakly distributive, then every $\mathbf{\Sigma}^0[\mathcal{A}]$-measurable image of X into ${}^\omega \omega$ is eventually bounded.
Hint: If $f : X \longrightarrow {}^\omega \omega$, set $A_{n,m} = \{x \in X : f(x)(n) = m\}$.

f) Assume that a family $\mathcal{A} \subseteq \mathcal{P}(X)$ satisfies the σ-reduction property. Then $\mathcal{A}$ is weakly distributive if and only if every $\mathbf{\Sigma}^0[\mathcal{A}]$-measurable image of X into ${}^\omega \omega$ is eventually bounded.

g) If X is a perfectly normal wQN-space, then the family of open subsets of X is weakly distributive.
Hint: See Corollary 8.86, Theorem 8.113 and c).

h) If the family of closed subsets of a perfectly normal space X is weakly distributive, then X is a σ-space.
Hint: If $A \subseteq X$ is a G_δ set, $A = \bigcap_n G_n$, $G_{n+1} \subseteq G_n$ open, then there exist $F_{n,k}$ closed such that $G_n = \bigcup_k F_{n,k}$. Since $X = \bigcap_n \bigcup_k (F_{n,k} \cup (X \setminus G_n))$, by weak distributivity there is an $\alpha \in {}^\omega \omega$ such that $X = \bigcup_m \bigcap_{n \geq m} \bigcup_{k \leq \alpha(n)} (F_{n,k} \cup (X \setminus G_n))$. Then $A = \bigcup_m \bigcap_{n \geq m} \bigcup_{k \leq \alpha(n)} F_{n,k}$ is an F_σ set.

i) If X is perfectly normal, then the family of closed sets is weakly distributive if and only if $\mathrm{BOREL}(X)$ is weakly distributive.
Hint: Use the result of d) and h).

j) The family of closed subsets of a perfectly normal topological space X is weakly distributive if and only if X is a QN-space.

   Hint: Set $A_{n,m} = \{x \in X : (\forall k \geq m)|f_k(x)| \leq 1/(n+1)\}$ and assume that α is increasing. For the opposite implication see f) and Theorem 8.81.

k) The family of Borel subsets of a perfectly normal topological space X is weakly distributive if and only if X is a QN-space.

   Hint: See j) and i).

8.28 [AC] Hereditary Properties and σ-space

a) Any open subset A of a perfectly normal nCM-space is a union of countably many clopen sets. Moreover, for any closed $F \subseteq A$ there exists a clopen set $F \subseteq C \subseteq A$.

   Hint: If $A = \bigcup_n F_n$, F_n closed and non-decreasing, construct by induction continuous functions f_n such that $f_n(x) = 0$ for $x \in F_n \cup A_n$ and $f_n(x) = 1$ for $x \notin A$, where $A_n = f_{n-1}^{-1}([0,a))$, $a \notin \mathrm{rng}(f_{n-1})$ and $A_0 = \emptyset$. One can assume that $F \subseteq F_0$. Then A_n is clopen and $F_n \subseteq A_{n+1} \subseteq A$.

b) Assume that X is a perfectly normal nCM-space, $A \subseteq X$ is a G_δ set. If the family of clopen sets is weakly distributive on A, then A is an F_σ set.

   Hint: If $A = \bigcap_n A_n$, A_n open, and $A_n = \bigcup_k U_{n,k}$, non-decreasing and $U_{k,n}$ clopen, by weak distributivity there exists an $\alpha \in {}^\omega\omega$ such that $A = \bigcup_k \bigcap_{n \geq k} \bigcup_{m \leq \alpha(n)} U_{n,m}$.

c) If X is a perfectly normal hereditary wQN-space, then X is a σ-space.

   Hint: Use b) and Exercise 8.27, g).

d) If X is a perfectly normal hereditarily $S_1(\Gamma, \Gamma)$-space, then X is a σ-space.

   Hint: A $S_1(\Gamma, \Gamma)$-space is a wQN-space.

e) If a σ-space X is also an $S_1(\Gamma, \Gamma)$-space, then X is a hereditary $S_1(\Gamma, \Gamma)$-space.

   Hint: If $\{U_{n,m} : m \in \omega\}$ are open γ-covers of A, $n \in \omega$, set $B = \bigcap_n \bigcup_k \bigcap_{m \geq k} U_{n,m}$. Since X is a σ-space, B is an F_σ set and every $\{U_{n,m} : m \in \omega\}$ is a γ-cover of $B \supseteq A$.

f) If $\mathcal{P}$ denotes any of the properties wQN, mQN, $S_1(\neg, \Gamma)$, then a perfectly normal space X is hereditarily $\mathcal{P}$ if and only if X is a σ-space and possesses the property $\mathcal{P}$.

8.29 [AC] wQN

The notion of discrete convergence $f_n \xrightarrow{D} f$ has been defined in Exercise 3.22. A space $C_p(X)$ has the **discrete sequence selection property** if for any sequence of sequences $\langle\{f_{n,m}\}_{m=0}^\infty : n \in \omega\rangle$ of functions from $C_p(X)$ such that $f_{n,m} \xrightarrow{D} 0$ on X for every n, there exists an increasing sequence $\{m_n\}_{n=0}^\infty$ such that $f_{n,m_n} \to 0$ on X.

a) If X is a normal topological space and $C_p(X)$ has the discrete sequence selection property, then X is an $S_1(\Gamma^{sh}, \Gamma)$-space.

   Hint: If $\langle\{U_{n,m} : m \in \omega\} : n \in \omega\rangle$ is a sequence of γ-covers, $\langle\{Z_{n,m} : m \in \omega\} : n \in \omega\rangle$ are closed γ-covers such that $Z_{n,m} \subseteq U_{n,m}$, by the Urysohn Theorem 3.15 take $f_{n,m} : X \longrightarrow [0,1]$ such that $f_{n,m}(x) = 0$ for $x \in Z_{n,m}$ and $f_{n,m}(x) = 1$ for $x \in X \setminus U_{n,m}$.

b) An $\overline{S}_1(\Gamma, \Gamma)$-space is a wQN-space.

   Hint: Follow the proof of Lemma 8.118.

c) For a normal topological space X the following conditions are equivalent:

   (1) X is a wQN-space.

   (2) $C_p(X)$ possesses the sequence selection property.

   (3) $C_p(X)$ possesses the discrete sequence selection property.

   (4) X is an $S_1(\Gamma^{sh}, \Gamma)$-space.

   (5) X is an $\overline{S}_1(\Gamma, \Gamma)$-space.

8.30 [AC] Clopen Refinement Property

A topological space X has the **clopen refinement property**, shortly the **property** $\gamma\gamma_{co}$, if every countable open γ-cover has a clopen γ-refinement.

a) A perfectly normal wQN-space with the property $\gamma\gamma_{co}$ is an $S_1(\Gamma, \Gamma)$-space.
   Hint: See the proof of Theorem 8.120.

b) A perfectly normal γ-space has the property $\gamma\gamma_{co}$.
   Hint: If $\{U_n : n \in \omega\}$ is a γ-cover, by Theorem 8.22 there exist $A_{n,m} \subseteq A_{n,m+1}$ clopen, $\bigcup_m A_{n,m} = U_n$. If $\mathcal{V}$ is a γ-subcover of $\{A_{n,m} : n, m \in \omega\}$ then the set $\{m : A_{n,m} \in \mathcal{V}\}$ is finite.

c) A perfectly normal σ-space with the property nCM has the property $\gamma\gamma_{co}$.
   Hint: See the proof of Lemma 8.126.

d) A perfectly normal hereditarily mQN-space is a σ-space.

e) A perfectly normal hereditarily wQN-space is a hereditarily $S_1(\Gamma, \Gamma)$-space.

8.31 [AC] wQN$_*$ and wQN*

Replacing in the definition of a wQN-space the word "continuous" by "lower semicontinuous" or "upper semicontinuous", one obtains the notion of a wQN$_*$-**space** or a wQN*-**space**, respectively. Similarly, you obtain the **property** SSP$_*$ or the **property** SSP*, respectively.

a) Show that wQN$_*$ $\equiv$ SSP$_*$.
   Hint: Follow the proof of Theorem 8.64.

b) Prove that SSP* $\to$ wQN*.

c) Prove that SSP* $\to$ $S_1(\Gamma, \Gamma)$.
   Hint: If $\{\mathcal{U}_n = \{U_{n,m} : m \in \omega\}\}_{n \in \omega}^{\infty}$ are γ-covers of X, set $f_{n,m}(x) = 0$ for $x \in U_{n,m}$ and $f_{n,m}(x) = 1$ otherwise. If $f_{n,m_n} \to 0$, then $\{U_{n,m_n} : n \in \omega\}$ is a γ-cover.

d) Prove that $S_1(\Gamma, \Gamma) \to$ SSP*.
   Hint: If $f_{n,m}$ are upper semicontinuous functions and $f_{n,m} \to 0$ for every n, write $U_{n,m} = \{x \in X : f_{n,m}(x) < 2^{-n}\}$.

e) Prove that wQN$_*$ $\to$ QN.
   Hint: The function $g_n(x) = \sup\{f_m(x) : m \geq n\}$ is lower semicontinuous.

f) Prove that QN $\to$ wQN$_*$ for any perfectly normal space.
   Hint: Note that a lower semicontinuous function is Borel measurable and use Theorem 8.81.

g) Conclude that $SSP_* \to SSP^*$ and $wQN_* \to wQN^*$ for any normal topological space.
   Hint: See Corollary 8.129.

h) Show that $wQN^* \to S_1(\Gamma, \Gamma)$.
   Hint: If $\{\mathcal{U}_n = \{U_{n,m} : m \in \omega\}\}_{n\in\omega}^{\infty}$ are γ-covers of X, set $V_{n,m} = U_{0,m} \cap \cdots \cap U_{n,m}$ and define upper semicontinuous functions as follows:

$$f_m(x) = \begin{cases} 1 & \text{if } x \in X \setminus V_{0.m}, \\ \frac{1}{k+2} & \text{if } x \in V_{k,m} \setminus V_{k+1,m}, \\ 0 & \text{otherwise.} \end{cases}$$

If $\{m_n\}_{n=0}^{\infty}$ is increasing and such that $f_{m_n} \xrightarrow{QN} 0$ with the control $\{1/n + 2\}_{n=0}^{\infty}$, then $\{V_{n,m_n} : n \in \omega\}$ is a γ-cover.

i) Conclude that for any perfectly normal topological space we have

$$wQN_* \equiv SSP_* \equiv QN, \quad wQN^* \equiv SSP^* \equiv S_1(\Gamma, \Gamma), \quad wQN \equiv SSP \equiv S_1(\Gamma^{sh}, \Gamma).$$

Moreover, all implications hold true for any topological space with the following two exceptions: for $QN \to wQN_*$ we need that X is a perfectly normal topological space and for $wQN \to S_1(\Gamma^{sh}, \Gamma)$ we need that X is a normal topological space.

8.6 Thin Sets of Trigonometric Series

Let us recall that a series

$$\frac{a_0}{2} + \sum_{n=1}^{\infty} (a_n \cos 2\pi nx + b_n \sin 2\pi nx), \tag{8.23}$$

where $\langle a_n, b_n : n \in \omega \rangle$ are reals[2], is called a **trigonometric series**. Since all functions $\sin 2\pi nx$ and $\cos 2\pi nx$ are periodic with the period 1, in all our considerations we can identify the interval $[0,1]$ with the circle $\mathbb{T}$. Moreover, $\mathbb{T}$ is a Polish group.

J. Fourier [1822] in 1812 asserted[3] that every function $f : \mathbb{T} \longrightarrow \mathbb{R}$ can be expressed as a sum of a trigonometric series. This is not true, however many partial results in this direction were proved and the problems related to the convergence of trigonometric series were intensively studied.

A hundred years later, in 1912, A. Denjoy [1912] and N.N. Luzin [1912] independently proved

Theorem [AC] 8.131 (A. Denjoy – N.N. Luzin). *If the series (8.23) absolutely converges on a set of positive Lebesgue measure, then*

$$\sum_{n=0}^{\infty} (|a_n| + |b_n|) < \infty,$$

hence the series (8.23) absolutely converges everywhere.

---

[2] For simplicity we assume that $b_0 = 0$.
[3] J. Fourier announced his assertion in 1812, however, he published it only in 1822.

Later on N.N. Luzin proved

Theorem [AC] 8.132 (N.N. Luzin). *If a series* (8.23) *absolutely converges on a non-meager set, then*

$$\sum_{n=0}^{\infty}(|a_n| + |b_n|) < \infty,$$

hence the series (8.23) *absolutely converges everywhere.*

In the next we shall prove Corollary 8.141, from which both Theorems 8.131 and 8.132 easily follow.

There is a natural question: is the conclusion of either theorem true for some meager or measure zero set? We show that the answer is affirmative, namely the Cantor middle-third set in spite of being meager and measure zero, by Corollary 8.142 satisfies the conclusion of both theorems. On the other hand there are meager and measure zero sets for which the conclusion does not hold true. We show that for every countable set $A \subseteq [0,1]$ there is a trigonometric series (8.23) converging on A absolutely, however

$$\sum_{n=0}^{\infty}(|a_n| + |b_n|) = \infty. \tag{8.24}$$

In connection with such a question we define: a set $A \subseteq [0,1]$ is called an **N-set** if there exists a trigonometric series (8.23) absolutely converging on A and such that (8.24) holds true.

We find a simpler and more convenient equivalent definition of an N-set.

Let us begin with some notation and an important result. By $\|x\|$ we denote the distance of a real x to the nearest integer. Thus

$$\|x\| = \min\{\{x\}, 1 - \{x\}\},$$

where $\{x\} = x - \lfloor x \rfloor$ denotes the fractional part of the real x (see Section 2.1). One can easily see that for any reals x, y and $n \in \omega$ we have

$$\|x + y\| \le \|x\| + \|y\|, \quad \|-x\| = \|x\|, \quad \|nx\| \le n \cdot \|x\|.$$

Moreover

$$\|xy\| \le |x| \cdot \|y\| + (|y| + \|y\|) \cdot \|x\|. \tag{8.25}$$

Since

$$2\|x\| \le |\sin \pi x| \le \pi\|x\| \tag{8.26}$$

for any real x, from the point of view of convergence the functions sin and $\| \ \|$ are equivalent. Thus, instead of the sine function we shall deal with the distance function $\| \ \|$.

Theorem 8.133 (P.G. Lejeune Dirichlet). *Let $\{n_i\}_{i=0}^{\infty}$ be an increasing sequence of natural numbers. For any $\varepsilon > 0$ and for any reals $x_1, \ldots, x_k$, there are $i, j \in \omega$ such that $0 \leq i < j \leq (2/\varepsilon)^k$ and*

$$\|(n_j - n_i)x_l\| < \varepsilon \qquad \text{for } l = 1, 2, \ldots, k. \tag{8.27}$$

Proof. We can assume that $\varepsilon < 1$. Let $m \in \omega$ be such that $\varepsilon/2 \leq 1/m < \varepsilon$. We split the k-dimensional interval $[0,1)^k$ into $t = m^k$ equal intervals of side $1/m$. By the Pigeonhole Principle, Theorem 1.9, there exist $0 \leq i < j \leq t = m^k \leq (2/\varepsilon)^k$ such that the k-tuples

$$[\{n_i x_1\}, \ldots, \{n_i x_k\}], \quad [\{n_j x_1\}, \ldots, \{n_j x_k\}],$$

are in the same interval. Note that if $\{x\} \leq \{y\}$, then $\|x-y\| \leq \{y-x\} = \{y\}-\{x\}$. Therefore (8.27) holds true. □

Corollary 8.134. *For any $m \in \omega$, for any $\varepsilon > 0$ and for any reals $x_1, \ldots, x_k$, there exists an $n \geq m$ such that $n \leq (2/\varepsilon)^k + m$ and*

$$\|nx_l\| < \varepsilon \qquad \text{for } l = 1, 2, \ldots, k.$$

Now we are ready to find an equivalent definition of an N-set.

Lemma 8.135. *Assume that $A = \{x \in [0,1] : \sum_{n=0}^{\infty} \rho_n \|c_n x + \varphi_n\| < \infty\}$, where $\rho_n \geq 0, c_n, \varphi_n$ are reals, $\sum_{n=0}^{\infty} \rho_n = \infty$, and $\lim_{n \to \infty} c_n = \infty$. Then there are non-negative reals d_n such that $\sum_{n=0}^{\infty} d_n = \infty$ and*

$$A \subseteq \left\{ x \in [0,1] : \sum_{n=0}^{\infty} d_n \|nx\| < \infty \right\}. \tag{8.28}$$

Proof. Set $s_n = \sum_{k=0}^{n} \rho_k$. We can assume that $s_0 > 0$, $c_n > 0$. Moreover, we can assume that the sequence $\{\rho_n\}_{n=0}^{\infty}$ is bounded. If not, replace ρ_n by $\min\{1, \rho_n\}$. Then by Theorem 2.23 we obtain

$$\sum_{n=0}^{\infty} \rho_n s_n^{-1} = \infty, \qquad \sum_{n=0}^{\infty} \rho_n s_n^{-3/2} < \infty. \tag{8.29}$$

By Corollary 8.134, for any $k \in \omega$ there exists a positive integer $p_k \leq s_k$ such that $\|p_k c_k\| < 2s_k^{-1/2}$ and $\|p_k \varphi_k\| < 2s_k^{-1/2}$. Let n_k be the natural number nearest to $p_k c_k$, i.e., $\|p_k c_k\| = |p_k c_k - n_k| = \|p_k c_k - n_k\|$. Then for $x \in [0,1]$ we have

$$\|n_k x\| \leq \|n_k x - p_k c_k x\| + \|p_k \varphi_k\| + p_k \|c_k x + \varphi_k\|.$$

By (8.25) we obtain

$$\|xn_k - p_k c_k x\| \leq x\|n_k - p_k c_k\| + (|n_k - p_k c_k| + \|n_k - p_k c_k\|)\|x\| = (x + 2\|x\|)\|p_k c_k\|.$$

Thus

$$\frac{\rho_k}{s_k}\|n_k x\| < \frac{2\rho_k}{s_k^{3/2}}(x + 2\|x\| + 1) + \rho_k \|c_k x + \varphi_k\|$$

and the series $\sum_{k=0}^{\infty} \rho_k s_k^{-1} \|n_k x\|$ converges for any $x \in A$. We can assume that for any natural number $i > 0$ there exists an $x \in A$ such that $\|ix\| \neq 0$. Otherwise the set A is finite and one can easily find the reals d_n. Hence the sum $\sum_{k \in \{j:n_j = i\}} \rho_k s_k^{-1}$ is either finite or a convergent series for any natural $i > 0$. We set $d_0 = 0$ and, for $i > 0$,

$$d_i = \sum_{k \in \{j:n_j = i\}} \rho_k s_k^{-1}.$$

Then $\sum_{k=0}^{\infty} d_k \|kx\| = \sum_{k=0}^{\infty} \rho_k s_k^{-1} \|n_k x\| < \infty$ for $x \in A$ and by (8.29) we obtain $\sum_{k=0}^{\infty} d_k = \infty$. □

Theorem 8.136 (R. Salem). *Let $A \subseteq [0, 1]$. The following conditions are equivalent:*

a) *A is an N-set.*
b) *There exist non-negative reals d_n, $n \in \omega$ such that $\sum_{n=0}^{\infty} d_n = \infty$ and*

$$A \subseteq \left\{ x \in [0, 1] : \sum_{n=0}^{\infty} d_n |\sin(\pi n x)| < \infty \right\}.$$

c) *There exist non-negative reals d_n, $n \in \omega$ such that $\sum_{n=0}^{\infty} d_n = \infty$ and*

$$A \subseteq \left\{ x \in [0, 1] : \sum_{n=0}^{\infty} d_n \|nx\| < \infty \right\}. \tag{8.30}$$

Proof. By the inequalities (8.26) the assertions b) and c) are equivalent.

Assume that A is an N-set. Thus there are reals $\langle a_n, b_n : n \in \omega \rangle$ such that the series (8.23) absolutely converges on A and (8.24) holds true. By elementary trigonometry there are reals φ_n such that

$$a_n \cos 2\pi n x + b_n \sin 2\pi n x = \rho_n \sin(2\pi n x + \varphi_n),$$

where $\rho_n = \sqrt{a_n^2 + b_n^2}$. By (8.26) we have

$$A \subseteq \left\{ x \in [0, 1] : \sum_{n=0}^{\infty} \rho_n \|2nx + \varphi_n/\pi\| < \infty \right\}.$$

By Lemma 8.135 there are non-negative reals d_n such that $\sum_{n=0}^{\infty} d_n = \infty$ and (8.30) holds true.

Assume that b) holds true. Since $|\sin 2\alpha| \leq 2|\sin \alpha|$, we obtain

$$A \subseteq \left\{ x \in [0, 1] : \sum_{n=0}^{\infty} d_n |\sin(2\pi n x)| < \infty \right\}$$

and therefore, A is an N-set. □

A set $A \subseteq [0, 1]$, on which some subsequence of the sequence $\{e^{2\pi inx}\}_{n=0}^{\infty}$ converges uniformly to the constant function 1, plays an important role in harmonic analysis[4] and is called a **Dirichlet set**. Using the inequalities (8.26) and the elementary equality

$$|\sin x| = \sqrt{\frac{1 - \cos 2x}{2}},$$

we obtain that a set $A \subseteq [0, 1]$ is a Dirichlet set if and only if there is an increasing sequence $\{n_k\}_{k=0}^{\infty}$ of natural numbers such that

$$\|n_k x\| \rightrightarrows 0 \text{ on } A.$$

Note that A is a Dirichlet set if and only if

$$(\forall \varepsilon > 0)(\forall m)(\exists n > m)(\forall x \in A) \|nx\| < \varepsilon.$$

There are other notions of sets related to convergence of trigonometric series. A set $A \subseteq [0, 1]$ is called a **pseudo Dirichlet set** if there is an increasing sequence $\{n_k\}_{k=0}^{\infty}$ of natural numbers such that

$$\|n_k x\| \xrightarrow{\text{QN}} 0 \text{ on } A.$$

By Theorem 3.52 we immediately obtain that a set $A \subseteq [0, 1]$ is a pseudo Dirichlet set if and only if there exists a non-decreasing sequence $\langle A_n : n \in \omega \rangle$ of Dirichlet sets such that $A = \bigcup_n A_n$. Actually, if $A = \bigcup_n A_n$ and $A_n \subseteq A_{n+1}$ are Dirichlet sets, then one can easily construct an increasing sequence $\{k_n\}_{n=0}^{\infty}$ such that $\|k_n x\| < 1/n+1$ for every $x \in A_n$. As a consequence we obtain that a union $\bigcup_n A_n$ of a non-decreasing sequence of pseudo Dirichlet sets is a pseudo Dirichlet set.

A set $A \subseteq [0, 1]$ is called an **Arbault set**, shortly an **A-set**, if there is an increasing sequence $\{n_k\}_{k=0}^{\infty}$ of natural numbers such that

$$\|n_k x\| \to 0 \text{ on } A.$$

A set $A \subseteq [0, 1]$ is called a **weak Dirichlet set** if for every Borel measure μ on $[0, 1]$ there exist a Borel set $B \supseteq A$ and an increasing sequence $\{n_k\}_{k=0}^{\infty}$ of natural numbers such that

$$\lim_{k \to \infty} \int_B |e^{2\pi in_k x} - 1| \, d\mu = 0.$$

As above the last condition is equivalent to

$$\lim_{k \to \infty} \int_B \|n_k x\| \, d\mu = 0.$$

---

[4]The function $e^{2\pi inx}$ is a character of the topological group $\mathbb{T}$.

There is another useful notion of a special set. A set $A \subseteq [0,1]$ is called an N_0-**set** if there is an increasing sequence $\{n_k\}_{k=0}^{\infty}$ of natural numbers such that

$$\sum_{n=0}^{\infty} \|n_k x\| < \infty \text{ on } A.$$

We denote the families of N-sets, Dirichlet sets, pseudo Dirichlet sets, N_0-sets, A-sets and weakly Dirichlet sets as $\mathcal{N}$, $\mathcal{D}$, $p\mathcal{D}$, $\mathcal{N}_0$, $\mathcal{A}$, and $w\mathcal{D}$, respectively.

Theorem 8.137. *The following inclusions hold true:*

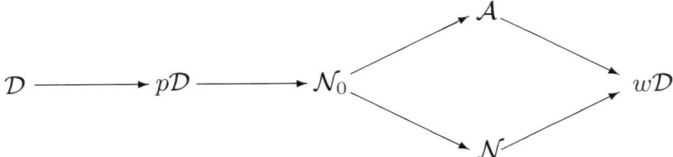

Proof. The inclusion $\mathcal{A} \subseteq w\mathcal{D}$ follows by Lebesgue Dominated Convergence Theorem 4.25.

Now, let A be an N-set. Hence, there exist non-negative reals $\langle a_n : n \in \omega \rangle$ such that

$$A \subseteq B = \left\{ x \in [0,1] : \sum_{n=0}^{\infty} a_n \|nx\| < \infty \right\}$$

and $\sum_{n=0}^{\infty} a_n = \infty$. Thus for every $x \in B$ we have

$$\lim_{n \to \infty} \frac{\sum_{k=0}^{n} a_k \|kx\|}{\sum_{k=0}^{n} a_k} = 0.$$

Since for any n the fraction is bounded by $1/2$, by Lebesgue Dominated Convergence Theorem 4.25 we obtain

$$\lim_{n \to \infty} \frac{\sum_{k=0}^{n} a_k \int_B \|kx\| \, d\mu}{\sum_{k=0}^{n} a_k} = 0$$

for any Borel measure μ. Consequently

$$\liminf_{n \to \infty} \int_B \|nx\| \, d\mu = 0.$$

The other inclusions follow directly from definitions.   □

Theorem 8.138.

a) *Every finite subset of $\mathbb{T}$ is a Dirichlet set. Hence, every countable subset of $\mathbb{T}$ is a pseudo Dirichlet set.*

b) *There exists a perfect Dirichlet set.*

c) *No (non-trivial) interval is a weak Dirichlet set.*

Proof. a) By Corollary 8.134 for every finite set $\{x_0, \ldots, x_m\}$ there exists an increasing sequence $\{n_k\}_{k=0}^\infty$ such that $\|n_k x_i\| < 1/(k+1)$ for $i = 0, \ldots, m$ and each k.

b) Let $\{n_k\}_{k=0}^\infty$, $n_0 = 0$ be an increasing sequence of natural numbers such that the sequence $\{n_{k+1} - n_k\}_{k=0}^\infty$ is also increasing and therefore unbounded. For any $x = \sum_{n=0}^\infty x_n 2^{-n-1}$, $x_n = 0, 1$ we let (we take always infinite expansion)

$$x \in A \equiv x_n = 0 \text{ for any } n_k \leq n < n_{k+1} \text{ and any } k \text{ even.} \tag{8.31}$$

Then for any $x \in A$ and any k even we have $\|n_k x\| \leq 2^{n_k - n_{k+1}}$. Thus A is a Dirichlet set. Evidently A is a perfect set of cardinality $\mathfrak{c}$.

c) Let $[a, b] \subseteq [0, 1]$ be a non-trivial interval. For any sufficiently large positive integer n there exist positive integers k, m such that

$$\frac{k-1}{n} < a \leq \frac{k}{n} < \frac{m}{n} \leq b < \frac{m+1}{n}.$$

Then

$$\frac{m-k}{4n} \leq \int_{[a,b]} \|nx\|\, d\lambda < \frac{m-k+2}{4n}$$

and therefore even for Lebesgue measure we have

$$\lim_{n \to \infty} \int_{[a,b]} \|nx\|\, d\lambda = \frac{b-a}{4} > 0. \qquad \square$$

Consequently every finite set is a Dirichlet set, every countable set is a pseudo Dirichlet set, an N_0-set, an N-set, an A-set, and a weak Dirichlet set. Since the closure of a Dirichlet set is a Dirichlet set as well, not every countable set is a Dirichlet set. On the other hand no non-trivial interval is any of the above-mentioned sets.

Let us recall that a family of thin subsets of a topological space does not contain a non-empty open set, in our case, no open interval. We can summarize.

Theorem 8.139.

a) *The families* $\mathcal{D}$, $p\mathcal{D}$, $\mathcal{N}_0$, $\mathcal{N}$, $\mathcal{A}$, *and* $w\mathcal{D}$ *are families of thin sets.*

b) *The family* $\mathcal{D}$ *has a closed base, the families* $p\mathcal{D}$, $\mathcal{N}_0$, $\mathcal{N}$ *have* F_σ *bases and the family* $\mathcal{A}$ *has a* Σ_3^0 *base.*

Proof. The part a) follows by Theorem 8.138 and by the definition.

If a sequence $\{n_k\}_{k=0}^\infty$ witnesses that $A \subseteq [0, 1]$ is an A-set then

$$A \subseteq \{x \in [0, 1] : (\forall m)(\exists n_0)(\forall n > n_0)\, \|n_k x\| \leq 1/(m+1)\}.$$

The last set is a Σ_3^0 set.

A similar estimate can be done in the remaining cases. $\qquad \square$

We have already said that instead on $[0,1]$ we may work on $\mathbb{T}$. We move to $\mathbb{T}$ since $\mathbb{T}$ is a group.

Theorem 8.140. *If a set A belongs to any of the families $\mathcal{D}$, $p\mathcal{D}$, $\mathcal{N}_0$, $\mathcal{A}$, and $\mathcal{N}$, then $A + A$ and $A - A$ belongs to the corresponding family as well. Moreover, each of the families $p\mathcal{D}$, $\mathcal{N}_0$, $\mathcal{A}$, and $\mathcal{N}$ has a base consisting of Borel subgroups of $\mathbb{T}$.*

Proof. Since $\|x \pm y\| \le \|x\| + \|y\|$ the former assertion holds true for all families.

If $A \in p\mathcal{D}$ we can assume that $0 \in A$. We set $A_0 = A$ and $A_{n+1} = A_n - A_n$. Then $\bigcup_n A_n$ is a subgroup of $\mathbb{T}$ containing the set A and $A_n \subseteq A_{n+1}$ for any n. Since a non-decreasing union of pseudo-Dirichlet sets is a pseudo-Dirichlet set, we obtain $\bigcup_n A_n \in p\mathcal{D}$.

One can easily see that any of the sets

$$\left\{ x \in \mathbb{T} : \sum_{k=0}^{\infty} \|n_k x\| < \infty \right\}, \quad \{ x \in \mathbb{T} : \|n_k x\| \to 0 \}, \quad \left\{ x \in \mathbb{T} : \sum_{k=0}^{\infty} a_k \|kx\| < \infty \right\}$$

is a group. Since sets of those kinds form a base of families $\mathcal{N}_0$, $\mathcal{A}$, and $\mathcal{N}$, respectively, the latter assertion follows. $\qquad\square$

Corollary 8.141. *Every D-set, pD-set, N_0-set, A-set and every N-set is meager and has Lebesgue measure zero.*

Proof. Assume that A is an A-set. We can assume that A is a Borel set. By Theorem 8.140 the arithmetical difference $A - A$ is also an A-set. If A has positive measure then by the Steinhaus Theorem 7.16 the set $A - A$ contains as a subset a non-empty open interval, a contradiction with Theorem 8.138.

If A is an N-set, then again $A - A$ is an N-set and the proof goes as above.

For the families $\mathcal{D}$, $p\mathcal{D}$, and $\mathcal{N}_0$, the assertion follows by Theorem 8.137. $\quad\square$

Corollary 8.142. *The Cantor middle-third set $\mathbb{C}$ is neither an A-set nor an N-set.*

Proof. $\mathbb{C} - \mathbb{C} = \mathbb{T}$ by Theorem 3.6. $\qquad\qquad\qquad\qquad\qquad\qquad\qquad\square$

Note that $\mathbb{C}$ is meager, even closed and nowhere dense, and has Lebesgue measure zero.

Theorem 8.143. *None of the families $\mathcal{D}$, $p\mathcal{D}$, $\mathcal{N}_0$, $\mathcal{N}$, and $\mathcal{A}$ is an ideal.*

Proof. We construct two Dirichlet sets A, B such that $(A \cup B) + (A \cup B) = [0,1]$. Thus $A \cup B$ does not belong to either of those families.

Let $\{n_k\}_{k=0}^{\infty}$, $n_0 = 0$ be an increasing sequence of natural numbers such that the sequence $\{n_{k+1} - n_k\}_{k=0}^{\infty}$ is also increasing and therefore unbounded. The set A is that defined by (8.31) and B is defined similarly just replacing in the definition (8.31) the word "even" by word "odd". Then

$$\|2^{n_k} x\| \le 2^{n_k - n_{k+1}} \text{ for } x \in A \text{ and } k \text{ even,}$$

and

$$\|2^{n_k}x\| \leq 2^{n_k - n_{k+1}} \text{ for } x \in B \text{ and } k \text{ odd.}$$

Thus A and B are Dirichlet sets.

Any real z can be written as $z = x + y$, where $x \in A$ and $y \in B$. Thus $A + B = \mathbb{T}$. Since $A + B \subseteq ((A \cup B) + (A \cup B))$, by Theorem 8.140 the union $A \cup B$ cannot belong to either of the considered families. □

Let $\mathcal{F} \subseteq \mathcal{P}(\mathbb{T})$ be a family of thin subsets of a set X. A set $A \subseteq X$ is said to be $\mathcal{F}$-**permitted** iff $A \cup B \in \mathcal{F}$ for any $B \in \mathcal{F}$. We write

$$\text{Perm}(\mathcal{F}) = \{A \subseteq X : A \text{ is } \mathcal{F}\text{-permitted}\}.$$

Theorem 8.144.

a) $\text{Perm}(\mathcal{F})$ is an ideal,
b) $\text{Perm}(\mathcal{F}) \subseteq \mathcal{F}$,
c) $\text{Perm}(\mathcal{F}) = \mathcal{F}$ if and only if $\mathcal{F}$ is an ideal.

Theorem 8.145. *Any finite set is permitted for any of the families $\mathcal{D}$, $p\mathcal{D}$, $\mathcal{N}_0$, $\mathcal{N}$, $\mathcal{A}$, and $w\mathcal{D}$.*

Proof. Since the family of permitted sets is an ideal it suffices to show that every one-point set is permitted.

Assume that A is a Dirichlet set and $\|n_k x\| \rightrightarrows 0$ on A. Let $y \in \mathbb{T}$, ε being a positive real. By the Dirichlet Theorem 8.133 there are arbitrarily large $k < l$ such that

$$\|n_k x\| < \frac{1}{2}\varepsilon, \quad \|n_l x\| < \frac{1}{2}\varepsilon \text{ for any } x \in A, \quad \|(n_l - n_k)y\| < \varepsilon.$$

Then

$$\|(n_l - n_k)x\| < \varepsilon$$

for any $x \in A \cup \{y\}$. Thus $A \cup \{y\}$ is a Dirichlet set.

For families $p\mathcal{D}$, $\mathcal{A}$, $\mathcal{N}_0$, and $w\mathcal{D}$ the proofs are similar. The case of an N-set is more complicated.

So, let A be an N-set, $\{a_n\}_{k=0}^{\infty}$ being a sequence of non-negative reals such that $\sum_{n=0}^{\infty} a_n = \infty$ and

$$A \subseteq \{x \in [0, 1] : \sum_{n=0}^{\infty} a_n \|nx\| < \infty\}$$

holds true. Let $y \in \mathbb{T}$.

We can assume $a_0 > 0$. Set $s_n = \sum_{k=0}^{n} a_k$. By Theorem 2.23 we have

$$\sum_{n=0}^{\infty} \frac{a_n}{s_n} = \infty, \quad \sum_{n=0}^{\infty} \frac{a_n}{s_n^2} < \infty. \tag{8.32}$$

By the Dirichlet Theorem 8.133 for every k there exists a positive integer $n_k \leq s_k$ such that $\|kn_ky\| < 2/s_k$.

For every n let $M_n = \{k : kn_k = n\}$. Every set M_n is finite and $\bigcup_n M_n = \omega$. Moreover, the sets $\langle M_n : n \in \omega \rangle$ are pairwise disjoint. Set $b_n = \sum_{k \in M_n} a_k/s_k$. If $M_n = \emptyset$ then $b_n = 0$. Then

$$\sum_{n=0}^{\infty} b_n \|ny\| = \sum_{n=0}^{\infty} \sum_{k \in M_n} \frac{a_k}{s_k} \|kn_ky\| \leq \sum_{n=0}^{\infty} \sum_{k \in M_n} \frac{2a_k}{s_k^2} < \infty.$$

Similarly for any $x \in A$ we have

$$\sum_{n=0}^{\infty} b_n \|nx\| = \sum_{n=0}^{\infty} \sum_{k \in M_n} \frac{a_k}{s_k} \|kn_kx\| \leq \sum_{k=0}^{\infty} \frac{a_k}{s_k} n_k \|kx\| \leq \sum_{k=0}^{\infty} a_k \|kx\| < \infty.$$

By (8.32) we have $\sum_{n=0}^{\infty} b_n = \infty$, hence $A \cup \{y\}$ is an N-set.   □

Since the union of an increasing sequence of pseudo Dirichlet sets is a pseudo Dirichlet set, we obtain immediately that every countable set is permitted for the family of pseudo Dirichlet sets. Actually

Theorem 8.146 (J. Arbault – P. Erdős). *Any countable set of reals is permitted for any of the families $\mathcal{N}_0$, $\mathcal{N}$, $\mathcal{A}$, and $w\mathcal{D}$.*

We shall not prove the theorem now, since in Section 10.5 we prove a stronger result.

By Theorem 8.138 there exists a Dirichlet set of cardinality $\mathfrak{c}$. Therefore in any of the introduced families there exists a set of any given cardinality $\leq \mathfrak{c}$. In spite of the Arbault-Erdős Theorem 8.146 the situation with permitted sets is much more complicated. We begin with

Theorem 8.147. *Any universal measure zero set of reals is permitted for $w\mathcal{D}$.*

Proof. Suppose that A is a weak Dirichlet set and B has universal measure zero. Let μ be a Borel measure. If μ is diffused, then $\mu^*(B) = 0$, i.e., there exists a Borel set $C \supseteq B$ such that $\mu(C) = 0$. We can assume that A is Borel and there exists an increasing sequence $\{n_k\}_{k=0}^{\infty}$ such that $\lim_{k \to \infty} \int_A \|n_kx\| \, d\mu = 0$. Since $\int_{A \cup C} \|n_kx\| \, d\mu = \int_A \|n_kx\| \, d\mu$ also

$$\lim_{k \to \infty} \int_{(A \cup C)} \|n_kx\| \, d\mu = 0.$$

Assume now that μ is not diffused. Then the set $D = \{x \in \mathbb{T} : \mu(\{x\}) > 0\}$ is non-empty and countable. The measure $\nu(E) = \mu(E \setminus D)$ is a diffused Borel measure, therefore $\nu(B) = 0$. Since by the Arbault-Erdős Theorem 8.146 the set D is permitted, we obtain that $A \cup B \subseteq (A \cup D) \cup (B \setminus D)$ is weak Dirichlet.

Thus B is permitted.   □

Corollary [AC] 8.148. *There exists an uncountable permitted set for $w\mathcal{D}$.*

Proof. By Theorem 8.3 there exists a universal measure zero set of reals of cardinality $\aleph_1$. $\qquad\square$

Inspired by Theorem 8.146, J. Arbault [1952] tried to construct a perfect set permitted for $\mathcal{N}$. However, N.K. Bari [1961] found an error in his construction. Therefore a natural question remains open:

Problem 8.149. *Does there exist an uncountable permitted set for any of the considered families?*

Or even

Problem 8.150. *Does there exist a permitted set of cardinality continuum?*

In Section 10.5 we show that **ZFC** is not enough to answer the latter question, i.e., the answer is an undecidable statement of **ZFC**.

Exercises

8.32 [AC] $w\mathcal{D}$ **versus** $\mathcal{N}$

We shall need the following form of the **Hahn-Banach Theorem**: If A is a closed convex subset of a Banach space X, $0 \notin A$, then there exists a continuous linear functional $F : X \longrightarrow \mathbb{R}$ and an $\varepsilon > 0$ such that $F(x) \geq \varepsilon$ for each $x \in A$. For a proof see Exercise 3.16, c).

a) A Borel set $A \subseteq \mathbb{T}$ is a weak Dirichlet set if and only if for every Borel measure μ and every $\varepsilon > 0$ there exists a compact Dirichlet set $K \subseteq A$ such that $\mu(A \setminus K) < \varepsilon$.
   Hint: Assume that $\lim_{k \to \infty} \int_A \|n_k x\| \, d\mu = 0$. Then by Exercise 4.13, a) we obtain that $\|n_k x\| \to 0$ almost everywhere on A. Use the Egoroff Theorem 7.41.
   To prove the opposite implication, for a given $\varepsilon > 0$ find a compact Dirichlet set $K \subseteq A$ and an integer n_0 such that $\|nx\| < \varepsilon$ for every $n \geq n_0$ and $\mu(A \setminus K) < \varepsilon$. Then
   $$\int_A \|nx\| \, d\mu \leq \mu(K)\varepsilon + 1/2\varepsilon$$
   for any $n \geq n_0$.

b) If $A_n \subseteq A_{n+1}$ are weak Dirichlet sets then $\bigcup_n A_n$ is a weak Dirichlet set.
   Hint: The union of an increasing sequence of Dirichlet sets is a pseudo Dirichlet set and therefore also a weak Dirichlet set.

c) If $K \subseteq \mathbb{T}$ is a compact weak Dirichlet set then for any m the zero function belongs to the closed convex hull of the set $\{\|nx\| : n \geq m\}$.
   Hint: If not, then by the Hahn-Banach Theorem and Exercise 4.17, f) there exist a Borel measure μ and a positive real ε such that $\int_A \|nx\| \, d\mu > \varepsilon$ for every $n \geq m$.

d) Every F_σ weak Dirichlet set is an N-set.
   Hint: Assume $A = \bigcup_n A_n$, $A_n \subseteq A_{n+1}$, A_n being compact. By induction using c) find an increasing sequence of integers $\{n_k\}_{k=0}^\infty$ and non-negative reals $\{a_k\}_{k=0}^\infty$ such that $\sum_{n_{k-1} \leq i < n_k} a_i = 1$ and $\sum_{n_{k-1} \leq i < n_k} a_i \|ix\| < 2^{-k}$ for $x \in A_k$.

e) A set $A \subseteq \mathbb{T}$ is an N-set if and only if A is a subset of an F_σ weak Dirichlet set.

8.33 [AC] $w\mathcal{D}$ **is Closed Under Arithmetical Difference**

The notion of a capacity was introduced and studied in Exercises 6.7.

a) Assume that $A \subseteq \mathbb{T}$ is a weak Dirichlet set and μ is a Borel measure on $\mathbb{T}$. Then there exists a weak Dirichlet set $K \subseteq A - A$ such that $\mu((A - A) \setminus K) = 0$.
   Hint: Let $B \supseteq A$ be a Borel set such that $\mu(B \setminus A) = 0$. By Exercise 6.8, d) the set B is capacitable, i.e., there exist compact sets $K_n \subseteq K_{n+1} \subseteq B$ such that $\mu(B-B) = \sup_n \mu(K_n - K_n) = \mu(\bigcup_n (K_n - K_n))$. By Exercise 8.32, d) and Theorem 8.140 every $K_n - K_n$ is a compact N-set. Set $K = \bigcup_n (K_n - K_n)$.

b) If A is a weak Dirichlet set, then also $A - A$ is weak Dirichlet.

8.34 H-set

A set $A \subseteq \mathbb{T}$ is called an H-**set** if there exists an increasing sequence $\{n_k\}_{k=0}^{\infty}$ of natural numbers and a non-empty open interval I such that $(n_k \cdot A) \cap I = \emptyset$ for each k. A countable union of H-sets is called an H$_\sigma$-**set**.

a) Show that $A \subseteq \mathbb{T}$ is an H-set if and only if there exists an increasing sequence $\{n_k\}_{k=0}^{\infty}$ of natural numbers and reals $-1/2 \leq a < 1/2, 0 < b < 1/2$ such that $\|n_k x - a\| \leq b$ for each k and each $x \in A$.

b) An H-set is σ-porous and therefore has Lebesgue measure zero and is meager.

c) Any A-set is an H$_\sigma$-set.
   Hint: The set $\{x \in \mathbb{T} : (\forall k \geq n) \|n_k x\| \leq 1/m\}$ is an H-set for any $m > 2$ and any n.

d) Any A-set is σ-porous.

e) There exists an N-set that is not σ-porous.
   Hint: The set $\{x \in \mathbb{T} : \sum_{n=0}^{\infty} \frac{1}{n}\|n!x\| \leq 1\}$ is not σ-porous; see R. Zajíček [1987].

8.35 N-set which is not an A-set

Assume that $\langle a_n \geq a_{n+1} : n \in \omega \rangle$ are positive reals and $\{n_k\}_{k=0}^{\infty}$ is an increasing sequence such that

$$\lim_{n \to \infty} a_n = 0, \quad \sum_{n=0}^{\infty} a_n = \infty, \quad \lim_{k \to \infty} n_k/n_{k+1} = 0, \quad \sum_{n=0}^{\infty} a_n \frac{n_k}{n_{k+1}} < \infty.$$

We assume that $n_k/n_{k+1} < 1/4$ for any k.

a) Let $\{m_j\}_{j=0}^{\infty}$ be an increasing sequence of positive integers. Then there exists an increasing sequence $\{k_i\}_{i=0}^{\infty}$ such that

$$(\forall i)(\exists j) \, n_{k_i} \leq m_j < 4m_j < n_{k_i+1}, \quad \sum_{i=0}^{\infty} a_{k_i} < \infty.$$

b) Construct a sequence of closed intervals $I_0 \supseteq I_1 \supseteq \cdots \supseteq I_k \supseteq \cdots$ of lengths $1/n_k$, respectively, such that

   if $k \notin K = \{k_i : i \in \omega\}$, then $\|n_k x\| \leq \dfrac{n_k}{n_{k+1}}$ for any $x \in I_{k+1}$,

   if $k = k_i \in K$, then $\|m_j x\| \geq 1/4$ for any $x \in I_{k+1}, n_{k_i} \leq m_j < 4m_j < n_{k_i+1}$.

   Hint: If $k \notin K$, take $x_k \in I_k$ such that $\|n_k x_k\| = 0$. Find $I_{k+1} \subseteq I_k$ with $x_k \in I_{k+1}$. If $k = k_i \in K$, $n_{k_i} \leq m_j < 4m_j < n_{k_i+1}$, find $x_k \in I_k$ such that $\|m_j x_k\| = 1/2$, take $I_{k+1} \subseteq I_k$, $x_k \in I_{k+1}$.

c) $\sum_{i=0}^{\infty} a_{k_i} \|n_{k_i} x\| < \infty$ for any $x \in \bigcap_k I_k$.

Hint: Note that $\sum_{i=0}^{\infty} a_{k_i} < \infty$.

d) If $x \in \bigcap_k I_k$ then $\sum_{k \in \omega \setminus K} a_k \|n_k x\| < \infty$.

Hint: If $k \notin K$, then $\|n_k x\| \le \frac{n_k}{n_{k+1}}$.

e) Conclude that $\sum_{k=0}^{\infty} a_k \|n_k x\| < \infty$ and $\lim_{j \to \infty} \|m_j x\| \ne 0$ for $x \in \bigcap_k I_k$.

f) The N-set $\{x \in \mathbb{T} : \sum_{k=0}^{\infty} a_k \|n_k x\| < \infty\}$ is not an A-set.

8.36 A-set which is not an N-set

Assume that $\{n_k\}_{k=0}^{\infty}$ is increasing, $\lim_{k \to \infty} n_k / n_{k+1} = 0$, $a_k \ge 0$ and $\sum_{k=0}^{\infty} a_k = \infty$.

a) Assume that $0 \le b_0 < b_1 < b_2 < b_3 < b_4$, $b_{i+1} - b_i = b_1 - b_0 \le 1/(4n)$ for $i = 1, 2, 3$. Then there exists an $i < 4$ such that $\|nx\| \ge n(b_1 - b_0)$ for each $x \in [b_i, b_{i+1}]$.

Hint: There exists exactly one $x_0 \in [b_0, b_0 + 1/n)$ such that $\|nx_0\| = 0$. If $x_0 \notin [b_0, b_4]$, then take $i = 1$. If $x_0 \in [b_j, b_j + 1]$, take $i \equiv j + 2$ modulo 4.

b) Let $0 < c < d \le 4c$, $0 < \varepsilon \le 1/16$. Assume that $b_0 < b_1 < b_2 < b_3 < b_4$, $b_{i+1} - b_i = b_1 - b_0 \le \varepsilon/c$ for $i = 1, 2, 3$. Then there exists an $i < 4$ such that

$$(\forall x \in [b_i, b_{i+1}]) \sum_{c \le n < d} a_n \|nx\| \ge \varepsilon/4 \sum_{c \le n < d} a_n. \tag{8.33}$$

Hint: If $n < d \le 4c$, then $\varepsilon/c < 1/(4n)$. By a), for each $n \in [c, d]$ there exists an $i_n < 4$ such that $\|nx\| \ge \varepsilon$ for each $x \in [b_{i_n}, b_{i_n+1}]$. There exists an $i < 4$ such that $\sum\{a_n : c \le n < d \wedge i_n = i\} \ge 1/4 \sum\{a_n : c \le n < d\}$.

c) Let $0 < b < c$, $0 < \varepsilon \le 1/16$. Assume that $v - u = 4\varepsilon/b$. Then there exists a closed interval I of length not smaller than ε/c, $I \subseteq [u, v]$ and such that

$$(\forall x \in I) \sum_{b \le n < c} a_n \|nx\| \ge \varepsilon/8 \sum_{b \le n < c} a_n.$$

Hint: Take k such that $b_0 = b$, $b_{i+1} = 4b_i$ for $i < k$ and $b_k = d \le 4b_{k-1}$. Let

$$K = \{n : (\exists i)\,(0 \le i \le k \wedge i \text{ even} \wedge b_i \le n < b_{i+1})\},$$
$$L = \{n : (\exists i)\,(0 \le i \le k \wedge i \text{ odd} \wedge b_i \le n < b_{i+1})\}.$$

Then $\sum_{n \in K} a_n \ge 1/2 \sum_{b \le n < c}$ or $\sum_{n \in L} a_n \ge 1/2 \sum_{b \le n < c}$. Assume that the former inequality holds true. Using the result of b) construct by induction a decreasing sequence of closed intervals I_{2i}, $0 \le 2i < k$, $I_0 \subseteq [u, v]$, of lengths ε/b_{2i+1}, respectively, with the property (8.33). The last member of the sequence is the desired interval I.

d) There exist positive reals $b_k, c_k, \varepsilon_k, \delta_k$, $k \in \omega$, such that

$$\lim_{k \to \infty} \delta_k = 0, \quad \sum_{k=0}^{\infty} \varepsilon_k \cdot \sum_{b_k \le n < c_k} a_n = \infty$$

and for almost all k the following conditions hold true:

$$b_k < c_k < b_{k+1}, \quad 0 < \varepsilon_k \le 1/16, \quad \delta_k \le 1, \quad \delta_k/n_k = 4\varepsilon_k/b_k, \quad 1/n_{k+1} \le \varepsilon_k/c_k.$$

Hint: Set $s(b, c) = \sum_{b \le n < c} a_n$. If there exists a $c > 1/16$ such that

$$\sum_k s(n_k/16, cn_k) = \infty,$$

find $\varepsilon_k \geq cn_k/n_{k+1}$ such that $\sum_k \varepsilon_k s(n_k/16, cn_k) = \infty$ and set $b_k = n_k/16$, $c_k = cn_k$ and $\delta_k = 64\varepsilon_k$.

If for every $c > 1/16$ the series $\sum_k s(n_k/16, cn_k)$ converges, find an increasing unbounded sequence $d_k < n_{k+1}/16n_k$ such that $\sum_k s(d_k n_k, n_{k+1}/16) = \infty$ and set $b_k = d_k n_k$, $c_k = n_{k+1}/16$, $\delta_k = 1/4d_k$ and $\varepsilon_k = 1/16$.

e) The A-set $\{x \in \mathbb{T} : \|n_k x\| \to 0\}$ is not an N-set.

  Hint: Assume that b_k, c_k, ε_k, δ_k, $k \in \omega$ are those of d). *Construct by induction a non-increasing sequence of closed intervals $\{I_k\}_{k=0}^{\infty}$ with the length of I_k equal to $1/n_k$ as follows. If I_k is already defined, find $J_k \subseteq I_k$ of length δ_k/n_k such that $\|n_k x\| \leq \delta_k$ for $x \in J_k$. Since $\delta_k/n_k \geq 4\varepsilon_k/b_k$, by* c) *there exists an interval $I_{k+1} \subseteq J_k$ of length $1/n_{k+1} \leq \varepsilon_k/c_k$ such that*

$$(\forall x \in I_{k+1}) \sum_{b_{k+1} \leq n < c_{k+1}} a_n \|nx\| \geq \varepsilon_k/b \sum_{b_{k+1} \leq n < c_{k+1}} a_n.$$

If $x \in \bigcap_k I_k$ then $\|n_k x\| \to 0$ and $\sum_{k=0}^{\infty} a_k \|kx\| = \infty$.

8.37 Inclusions Between Trigonometric Families

  a) Closure of a Dirichlet set is a Dirichlet set.

  b) A Dirichlet set, which is a group, is finite.

  Hint: Closure of an infinite subgroup of $\mathbb{T}$ is $\mathbb{T}$.

  c) Any countable dense subset of $\mathbb{T}$ is a pseudo Dirichlet set which is not Dirichlet.

  d) There exists an N-set which is not an N_0-set.

  e) Conclude that neither $\mathcal{A} = w\mathcal{D}$ nor $\mathcal{N} = w\mathcal{D}$.

  f) All inclusions provable between families $\mathcal{N}$, $\mathcal{D}$, $p\mathcal{D}$, $\mathcal{N}_0$, $\mathcal{A}$, and $w\mathcal{D}$ are those of the diagram in Theorem 8.137.

  Hint: Use results of Exercises 8.35 *and* 8.36.

8.38 Cardinality of Bases

Let $\{n_k\}_{k=0}^{\infty}$, $n_0 = 0$ be an increasing sequence of natural numbers. If $x \in \mathbb{T}$ we consider its infinite dyadic expansion. For a set $K \subseteq \omega$ we define

$$M_K = \{x \in \mathbb{T} : \text{if } x = \sum_{k=0}^{\infty} \frac{x_k}{2^k} \text{ then } x_n = 0 \text{ for every } n_k \leq n < n_{k+1}, k \in \omega \setminus K\}.$$

  a) If $\omega \setminus K$ is infinite and the sequence $\{n_{k+1} - n_k\}_{k=0}^{\infty}$ is increasing then M_K is a Dirichlet set.

  b) If $K \cap L$ is finite then $M_K + M_L$ contains an open interval.

  c) If $\mathcal{A}$ is an almost disjoint family of subsets of ω and the sequence $\{n_{k+1} - n_k\}_{k=0}^{\infty}$ is increasing then

   1) M_K is a Dirichlet set for any $K \in \mathcal{A}$;

   2) the set $M_K + M_L$ does not belong to any of the families $\mathcal{D}$, $p\mathcal{D}$, $\mathcal{N}_0$, $\mathcal{N}$, $\mathcal{A}$, and $w\mathcal{D}$ for any $K, L \in \mathcal{A}$, $K \neq L$;

   3) the set $M_K \cup M_L$ does not belong to any of the families $\mathcal{D}$, $p\mathcal{D}$, $\mathcal{N}_0$, $\mathcal{N}$, $\mathcal{A}$, and $w\mathcal{D}$ for any $K, L \in \mathcal{A}$, $K \neq L$.

d) Every base of any of the families $\mathcal{D}$, $p\mathcal{D}$, $\mathcal{N}_0$, $\mathcal{N}$, $\mathcal{A}$, and $w\mathcal{D}$ has cardinality at least $\mathfrak{c}$.

Hint: Take an almost disjoint family $\mathcal{A}$ of cardinality $\mathfrak{c}$. If $\mathcal{B}$ is a base then any $B \in \mathcal{B}$ can contain at most one element M_K, $K \in \mathcal{A}$.

8.39 Thin Sets Defined by a Continuous Function

Let $f : [0, 1] \longrightarrow [0, 1]$ be continuous and such that $f(0) = f(1) = 0$. We can consider f as a function from $\mathbb{R}$ into $[0, 1]$ such that $f(x + z) = f(x)$ for any x and any integer z. A set $A \subseteq [0, 1]$ is called an f-**Dirichlet set** if there exists an increasing sequence $\{n_k\}_{k=0}^{\infty}$ of natural numbers such that $f(n_k x) \rightrightarrows 0$ on A. Similarly we can define a **pseudo** f-**Dirichlet set**, an N_f-**set**, an N_{0f}-**set**, an A_f-**set**, and a **weak** f-**Dirichlet set**. E.g., $A \subseteq [0, 1]$ is an N_f-set if there exist non-negative reals $\langle a_n : n \in \omega \rangle$ such that A is a subset of $\{x \in [0, 1] : \sum_{n=0}^{\infty} a_n f(nx) < \infty\}$ and $\sum_{n=0}^{\infty} a_n = \infty$. We denote the corresponding families as $\mathcal{D}_f$, $p\mathcal{D}_f$, $\mathcal{N}_f$, $\mathcal{N}_{0f}$, $\mathcal{A}_f$, and $w\mathcal{D}_f$, respectively.

a) Show that for every interval $\langle a, b \rangle \subseteq [0, 1]$ there exists an n_0 such that

$$\int_{\langle a,b \rangle} f(nx)\, dx \geq 1/2(b - a) \int_{[0,1]} f(x)\, dx \text{ for any } n \geq n_0.$$

Hint: Take $n_0 \geq 4/(b - a)$ and for $n \geq n_0$ take k, l such that $k - 1 \leq na < k < l < nb \leq l + 1$ and estimate the integral $\int_{\langle k/n, l/n \rangle} f(nx)\, dx$.

b) Prove that the following inclusions hold true:

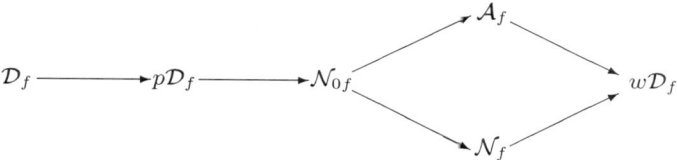

Hint: Follow the proof of Theorem 8.137.

c) Everyone of the families is $\mathcal{D}_f$, $p\mathcal{D}_f$, $\mathcal{N}_f$, $\mathcal{N}_0$, $\mathcal{A}_f$, and $w\mathcal{D}_f$ is a family of thin sets.
Hint: Use a) to show that $w\mathcal{D}_f$ does not contain any open interval.

d) An F_σ weak f-Dirichlet set is an N_f-set.
Hint: See Exercise 8.32.

e) Let $g : [0, 1] \xrightarrow{\text{onto}} [0, 1]$ be continuous, $g(0) = g(1) = 0$, and $f : [0, 1] \xrightarrow{\text{onto}} [0, 1]$. If m is a positive integer such that $m \cdot Z(f) \subseteq Z(g)$, then for every positive $\varepsilon < 1$ there exists a positive δ such that $g(mx) < \varepsilon$ provided that $f(x) < \delta$.
Hint: Set $\delta = \min\{f(x) : g(mx) \geq \varepsilon\}$.

f) If $m \cdot Z(f) \subseteq Z(g)$, then $\mathcal{D}_f \subseteq \mathcal{D}_g$, $p\mathcal{D}_f \subseteq p\mathcal{D}_g$, $\mathcal{N}_f \subseteq \mathcal{N}_g$, $\mathcal{A}_f \subseteq \mathcal{A}_g$, and $w\mathcal{D}_f \subseteq w\mathcal{D}_g$.
Hint: Use e) in the case of $\mathcal{D}_f$, $p\mathcal{D}_f$, $\mathcal{A}_f$, and $w\mathcal{D}_f$. For $\mathcal{N}_f$ use d).

g) If $Z(f)$ is a finite set of rationals, then $\mathcal{D}_f = \mathcal{D}$, $p\mathcal{D}_f = p\mathcal{D}$, $\mathcal{N}_f = \mathcal{N}$, $\mathcal{A}_f = \mathcal{A}$, and $w\mathcal{D}_f = w\mathcal{D}$.

h) If $f(x) = \|x\|$ and $g(x) = \|x\|^2$ then $\mathcal{N}_{0f} \subseteq \mathcal{N}_{0g}$, $\mathcal{N}_{0f} \neq \mathcal{N}_{0g}$.
Hint: The set $\{x \in \mathbb{T} : \sum_{k=0}^{\infty} \|2^{2^k} x\|^2 < \infty\}$ belongs to $\mathcal{N}_{0g}$ and does not belong to $\mathcal{N}_{0f}$. For a proof see J. Arbault [1952].

Historical and Bibliographical Notes

R. Baire [1899] showed that for every Borel measurable function f and for every perfect set P there exists a set $A \subseteq P$, meager in P, such that f is continuous on $P \setminus A$ (that is a consequence of Theorem 7.30). H. Lebesgue [1905] raised a question whether the opposite assertion holds true. N.N. Luzin [1914], assuming **CH**, constructed a perfectly meager subset of $[0, 1]$. The characteristic function of this set is a counterexample to Lebesgue's question. Then N.N. Luzin [1921] constructed an uncountable perfectly meager set by transfinite induction in **ZFC** without the assumption of **CH**. The set of the proof of Theorem 8.3 was constructed by N.N. Luzin and W. Sierpiński [1928].

It seems that the notion of a universal measure zero set appeared for the first time explicitly in W. Sierpiński and E. Szpilrajn [1936] as a property (α), where an uncountable set of universal measure zero is constructed as well. They essentially exploited the main result of F. Hausdorff [1936a]. Lemma 8.5 is proved in E. Szpilrajn [1934], however he attributes the result to V. Saks. Theorem 8.6 was proved very late by E. Szpilrajn [1937]. Therefore the described results were obtained in two lines. Actually, M.M. Lavrentieff [1924], assuming **CH**, found an uncountable set with any homeomorphic image of Lebesgue measure zero. Later on, W. Sierpiński [1925] showed that the set of the proof of Theorem 8.3 possesses that property as well. Moreover, S. Banach [1948] raised a problem that is actually equivalent to the existence of a universal measure zero set of a given cardinality. Actually Theorem 8.8 by E. Grzegorek [1980] was originally formulated as an answer to the problem raised by S. Banach. Lemma 8.7 was essentially proved by E. Szpilrajn [1938]. Then E. Grzegorek [1984] proved Theorem 8.9.

E. Borel [1919] introduced implicitly the notion of a strong measure zero set and conjectured the Borel Conjecture. Theorem 8.11 is proved in [1934], however E. Szpilrajn attributed it to G. Poprougénko. Theorem 8.13 was announced by F. Galvin, J. Mycielski and R.M. Solovay [1973]. The presented proof follows that given by A. Miller [1984a]. Theorems 8.17 and 8.21 are essentially contained in F. Rothberger [1939], see also K. Kuratowski [1958a]. W. Sierpiński [1928] proved Corollary 8.19. The notion of an nCM-space was implicitly used for a long time, compare, e.g., D.H. Fremlin [1994]. The explicit definition was given by J. Haleš [2005]. Anyway, Theorem 8.22 is folklore.

The results of Exercises 8.1 and 8.2 were proved by T.J. Carlson [1993]. For results and a history of porous sets, Exercise 8.3, see, e.g., L. Zajíček [1987]. E. Szpilrajne [1935] introduced the notion of an $(S)_0$-set and essentially showed the results of Exercise 8.5.

P. Mahlo [1913] and N.N. Luzin [1914], assuming **CH**, constructed a Luzin set. Later W. Sierpiński [1924a] constructed a Sierpiński set provided that **CH** holds true. Theorem 8.30 and Corollary 8.31 were essentially proved by E. Szpilrajn [1938]. J. Pawlikowski [1996] proved the dual Theorem 8.32 for a Sierpiński set. The proofs of Theorem 8.37 and Corollary 8.38 are a generalization of the original proof of Corollary 8.39 by F. Rothberger [1938a]. Compare J. Cichoń [1989] and L. Bukovský [∞].

W. Sierpiński proved that every Borel measurable function defined on a Luzin set is of Baire order 2. E. Spilrajn [1930] proved, that a Sierpiński set S is a σ-set (without using that word) and every Borel measurable function defined on S is of Baire order 1. Compare K. Kuratowski [1958a].

K. Kuratowski [1933] introduced the notion of a λ-set. He remarks that an argumentation of N.N. Luzin [1921] proves Theorem 8.48. F. Rothberger [1939] showed Theorem 8.50. N.N. Luzin [1933] essentially showed Corollary 8.49. The notion of a Rothberger family (Exercise 8.10) was introduced and investigated by J. Cichoń [1989]

The notions of a QN-, a wQN- and an mQN-space were introduced and investigated (Theorems 8.56–8.63) by L. Bukovský, I. Recław and M. Repický [1991] and [2001] in connection with some problems of thin sets of trigonometric series. L. Bukovský [1993] introduced and investigated the notion of a Σ-space.

A.V. Arkhangel'skiĭ [1972] defined the properties (α_1)–(α_4). M. Scheepers [1998] proved the equivalence of conditions b)–e) of Theorem 8.64 and then [1999] the implication b) → a). Then D.H. Fremlin [∞] proved the implication a) → b). Theorem 8.66 by M. Scheepers [1999] actually raised a question whether the opposite implication holds true. An affirmative answer was given independently by L. Bukovský and J. Haleš [2007] and M. Sakai [2007] as Theorem 8.67. Theorem 8.74 was already proved by L. Bukovský, I. Recław and M. Repický [1991]. I. Recław [1997] proved Theorem 8.77 for metric spaces. The generalization for perfectly normal spaces is straightforward, see, e.g., L. Bukovský, I. Recław and M. Repický [2001] or J. Haleš [2005]. The presented proof follows L. Bukovský and J. Šupina [∞]. K. Menger [1924] defined and investigated the Menger Property. Since a σ-compact space has trivially the Menger Property, he raised a question whether the opposite implication does hold true. W. Hurewicz [1927] introduced properties H*, H**, the Hurewicz Property of a topological space and proved Theorem 8.113 for metric separable spaces. W. Sierpiński remarked on our Theorem 8.101 (see W. Hurewicz [1927], p. 196) that a Luzin set possesses the Menger Property and does not possess the Hurewicz Property. Theorem 8.85 was proved by L. Bukovský and J. Haleš [2003], for metric separable spaces already in L. Bukovský, I. Recław and M. Repický [2001]. The Key Lemma 8.84 is due to J. Haleš. Theorem 8.89, which is a partial answer to the Menger problem, was proved by W. Hurewicz [1928].

J. Gerlits and Z. Nagy [1982] introduced the notion of a γ-space and proved Theorem 8.96. Then F. Galvin and A.W. Miller [1984] proved Theorems 8.90–8.92. M. Scheepers [1996] started a systematic study of covering properties. Many classically known covering properties of topological spaces are covered by the system of covering properties introduced by M. Scheepers. That is the case of the Menger Property, the Hurewicz Property, and also of the Rothberger Property introduced by F. Rothberger [1938b]. A.W. Miller and D.H. Fremlin [1988] showed Theorem 8.108 and J. Gerlits and Z. Nagy [1982] showed Theorem 8.109. Theorem 8.111 was proved by F. Rothberger [1941].

The results of Exercise 8.20 are due to J. Gerlits and Z. Nagy [1982]. Exercises 8.22–8.26 contain results by M. Scheepers [1996] and W. Just, A.W. Miller, M. Scheepers and P.J. Szeptycki [1996].

Actually Theorem 8.115 is considered as the main result of J. Gerlits and Z. Nagy [1982]. Corollary 8.119 was proved by M. Scheepers [1999]. Then M. Scheepers conjectured that the opposite implication holds true – see Section 9.2. Theorem 8.120 was proved by L. Bukovský and J. Haleš [2007]. However, some weaker versions were already contained in J. Haleš [2005]. Theorem 8.123 was essentially proved by W. Just, A.W. Miller, M. Scheepers and P.J. Szeptycki [1996]. They showed that assuming $\mathfrak{b} = \aleph_1$, the constructed set is the so-called $S_1(\Gamma, \Gamma)^*$-set. Then I. Recław [1997] remarked that this set is actually a wQN-set and showed Theorem 8.124. Then M. Scheepers [1999] proved that $S_1(\Gamma, \Gamma)^* = S_1(\Gamma, \Gamma)$, hence the constructed set is an $S_1(\Gamma, \Gamma)$-set. Note that Recław's result is a typical non-constructive proof: we do not know whether $\mathfrak{b} = \aleph_1$ or not, however in either case we can find a set with desired properties. Theorem 8.125 is one of main results by L. Bukovský and J. Haleš [2007]. Theorem 8.81 was proved by B. Tsaban and L. Zdomskyy [∞]. The presented proof follows L. Bukovský and J. Šupina [∞].

The notion of weak distributivity was introduced by L. Bukovský, I. Recław and M. Repický [1991] and studied also in [2001]. The main results of Exercises 8.28 and 8.30 are due to J Haleš [2005]. Exercise 8.29 contains results of L. Bukovský and J. Haleš [2007]. Exercise 8.31 contains results by L. Bukovský [2008]. The result of Exercise 8.31, h) is due to M. Sakai [2009].

J. Cichoń, A. Karazashvili and B. Węglorz [1995] presented several properties of exceptional sets of reals, however mainly without any proof.

In 1912 in Comptes Rendues d'Académie des Sciences de Paris A. Denjoy [1912] and N.N. Luzin [1912] published independently two papers under equal titles "Sur l'absolue convergence des séries trigonométriques." Both papers contain a proof of Theorem 8.131. Later, N.N. Luzin [1915] showed that Theorem 8.132 is an easy consequence of Theorem 8.131.

The proof of Theorem 8.133 is a typical application of the Dirichlet's Pigeonhole Principle.

The sets of points of absolute convergence of a trigonometric series were studied by several authors and the related set of points of convergence. Probably P. Fatou was the first one in his thesis in 1906. The notion of an N-set explicitly introduced J. Marcinkiewicz [1938] who also essentially proved Theorem 8.143. R. Salem [1941b] proved the equivalence of parts a) and b) of Theorem 8.136. The notions of a Dirichlet set and a weak Dirichlet set were introduced and studied in harmonic analysis. The notion of a pseudo Dirichlet set was independently introduced by Z. Bukovská [1990] (as a D-set) and S. Kahane [1993]. J. Arbault [1952] introduced the notion of an A-set. Theorems 8.137–8.140 are folklore. J. Arbault [1952] introduced the notion of a permitted set and proved Theorem 8.144 for N-sets. P. Erdős (unpublished) and J. Arbault independently proved Theorem 8.146 for N-sets. The other cases are easy. The paper by L. Bukovský, N.N. Kholshchevnikova and M. Repický [1994] surveys the results about trigonometric thin sets obtained before 1995.

Results of Exercise 8.32 were obtained by B. Host, J.-F. Méla and F. Parreau [1991]. See also L.A. Lindhal and F. Poulsen [1971] and a generalization by L. Bukovský [1998]. Exercise 8.33 contains an unpublished result by G. Debs (see S. Kahane [1993]).

It was J. Arbault [1952] who began with showing that some inclusions in the diagram of Theorem 8.137 are proper. S. Kahane [1993] contributed essentially to this problem. The results of Exercises 8.35–8.36 are due to P. Eliaš [1997]. Exercise 8.38 follows L. Bukovský [2003]. Part h) of Exercise 8.39 is due to J. Arbault [1952], the others are due to Z. Bukovská [1999] and [2003].

Chapter 9

Additional Axioms

There are a sort of propositions, which, under the name of maxims and axioms, have passed for principles of science: and because they are self-evident, have been supposed innate, without that anybody (that I know) ever went about to show the reason and foundation of their clearness or cogency. It may, however, be worth while to inquire into the reason of their evidence, and see whether it be peculiar to them alone; and also to examine how far they influence and govern our other knowledge.

John Locke [1690], Book IV, Chapter VII.

As we have already mentioned there are many problems that the **ZFC** set theory is not able to decide. In the first third of the 20th century, mathematicians often used the continuum hypothesis **CH** to solve a problem. The continuum hypothesis **CH** was used as an additional axiom before its consistency was established. Maybe the mathematicians believed they could prove **CH** in **ZFC**. Actually G. Cantor tried to do so. The undecidability of **CH** was established much later.

According to K. Gödel [1931] the **ZFC** theory is non-complete, provided it is consistent. Even any consistent extension of **ZFC** obtained by adding a finite number of axioms is non-complete. Mathematicians looked for possible finite extensions of **ZFC**, i.e., by adding an additional axiom, which would better describe the properties of the abstraction of the notion of infinity and allows solution of an important problem. The consequences of an additional axiom are a measure of its plausibility. We make our explanation precise.

An **additional axiom of ZFC** or **ZF** should be a formula φ with interesting consequences, about which we know (assuming that **ZF** or some stronger theory, e.g., **ZFC + IC**, is consistent) that neither φ nor $\neg\varphi$ is provable in set theory, i.e., an undecidable statement of **ZFC** or **ZF**, respectively. Of course, we do not consider every undecidable statement of **ZFC** as an additional axiom. An additional axiom is usually an undecidable formula solving important problems that remain unanswered in **ZFC**. Often, a mathematician prefers a positive or negative answer to a problem. Then the plausibility of an additional axiom is measured by the result: does it imply what we wanted? Note that if φ is an undecidable statement then $\neg\varphi$ is as well and may become an additional axiom.

We recall some facts. If $\neg\varphi$ is not provable in **ZFC** and we show (in **ZFC**) that $\varphi \to \psi$, then neither is $\neg\psi$ provable in **ZFC**. E.g., (see Section 9.2) if from the Borel Conjecture it follows that every QN-space is countable, then one cannot prove (in **ZFC**) that there exists an uncountable QN-space. Moreover, if φ_1, φ_2 are undecidable in **ZFC** and we show (in **ZFC**) that $\varphi_1 \to \psi$ and $\varphi_2 \to \neg\psi$, then ψ is undecidable in **ZFC** as well.

In the second half of the twentieth century, after invention of the forcing method for construction of models of set theory, mathematicians posited plenty of interesting sentences with interesting consequences that are undecidable in **ZF**, even in **ZFC**, respectively. Many of those sentences became good candidates to be an additional axiom of **ZF** or **ZFC**.

From a large variety of very different possibilities we have chosen four areas. Actually one of the first alternative axioms to the continuum hypothesis was Martin's Axiom. We present some of its consequences in Section 9.1. The simplest relationship between several properties of the real line and related combinatorial structures may be expressed by cardinal invariants, small uncountable cardinals, introduced in Sections 5.4 and 7.1. A possibility or an impossibility to prove a result depends often on inequality or equality of some cardinal invariants. Since the undecidability of such equalities or inequalities are often already known, we propose them as an additional axiom in Section 9.2. In this Section we also present a combinatorial principle related to cardinal invariants. The consistency of several regularity properties of sets of reals with **ZF**, usually contradicting the axiom of choice, were established. We shall investigate their consequences, especially those related to the cardinal arithmetics in Section 9.3. Finally, J. Mycielski and H. Steinhaus [1962] proposed the Axiom of Determinacy **AD** as an alternative to the Axiom of Choice. The set theory **ZF + AD** was extensionally studied and many interesting results were obtained. The basic consequences of **AD** are presented in Section 9.4.

Many other axioms were formulated and were intensively used. Let us mention just the Proper Forcing Axiom introduced by S. Shelah [1982] – for some details see T. Jech [2006], Martin's Maximum introduced by M. Foreman, M. Magidor and S. Shelah [1988] – see also T. Jech [2006], and the Covering Property Axiom introduced by K. Ciesielski and J. Pawlikowski [2004]. Each of them tries to formulate an essence of properties of a particular model of set theory. Unfortunately, in the above-mentioned cases, every such model, in spite of being very interesting and important, satisfies $\mathfrak{c} = \aleph_2$ and each of those axioms actually implies that $\mathfrak{c} = \aleph_2$.

9.1 Continuum Hypothesis and Martin's Axiom

In this section we assume the Axiom of Choice **AC**, i.e., we work in **ZFC**. Moreover we assume that a reader is acquainted with the material of Section 5.3, where the cardinal invariant $\mathfrak{m}$ was introduced and studied.

Martin's Axiom **MA** is the statement $\mathfrak{m} = \mathfrak{c}$. Thus a special case of **MA** is the continuum hypothesis **CH**. We present some consequences of **MA**. Of course each of them is also a consequence of **CH**. Sometimes mathematicians assume **MA** + ¬**CH** to obtain better results. Such a situation will be explicitly indicated. By D.A. Martin and R.M. Solovay's result (11.21) we can assume that $\mathfrak{m} = \mathfrak{c} = \kappa$, where κ is any well-defined uncountable regular cardinal (see Section 11.5).

Martin's Axiom was invented in the time when Bartoszyński's theorem was not known and the mathematicians believed in the duality of measure and category. The main goals of Martin's Axiom were the dual consequences for measure and category. On the other hand, Martin's Axiom is considered as an interesting alternative to the continuum hypothesis **CH**. In his famous book, Wacław Sierpiński [1934b] deduced 82 propositions C_1–C_{82} from **CH**. It turned out that slight modifications of almost any of them (usually replacing the word "countable" by "of cardinality less than continuum" or $\aleph_1$ by $\mathfrak{c}$) follow from Martin's Axiom. E.g., the proposition C_1 says that there exists an $\aleph_1$-Luzin set. We already know (see Theorem 8.26) that **CH** implies C_1. If we modify C_1 as "there exists a $\mathfrak{c}$-Luzin set", then by Theorems 8.26 and 9.1, using the inequalities of the Cichoń Diagram, we obtain that **MA** implies C_1. It is easy to check that, similarly as C_1 implies propositions C_2–C_{14}, the modified C_1 implies the modified propositions C_2–C_{14}.

By Theorems 5.47 and 5.46 we immediately obtain

$$\aleph_0 \leq \kappa < \mathfrak{m} \rightarrow 2^{\kappa} = \mathfrak{c}. \tag{9.1}$$

Theorem [AC] 9.1.
$$\mathfrak{m} \leq \mathrm{add}(\mathcal{N}).$$

Proof. Let $\mathcal{F} \subseteq \mathcal{N}(\mathbb{T})$, $|\mathcal{F}| < \mathfrak{m}$.

We fix a real $\varepsilon > 0$ and consider the **Amoeba forcing**

$$\mathbb{A} = \{A \subseteq \mathbb{T} : A \text{ open} \wedge \lambda(A) < \varepsilon\}.$$

We order $\mathbb{A}$ by inverse inclusion. We show that $\mathbb{A}$ is a **CCC** poset. Assume that $\mathcal{D} \subseteq \mathbb{A}$ is an uncountable antichain. We can assume that for some $0 < \delta < \varepsilon$ we have $\lambda(A) < \delta$ for every $A \in \mathcal{D}$. Actually, if $\mathcal{D}_n = \{A \in \mathcal{D} : \lambda(A) < \varepsilon - 2^{-n}\}$, then there exists an n such that $\mathcal{D}_n$ is uncountable. We replace $\mathcal{D}$ by $\mathcal{D}_n$. Let $\mathcal{U}$ be a countable base of the topology closed under finite unions, e.g., $\mathcal{U}$ is the set of all finite unions of open intervals with rational endpoints. For every $A \in \mathbb{A}$ there exists a set $U_A \in \mathcal{U}$ such that $U_A \subseteq A$ and $\lambda(A \setminus U_A) < \varepsilon/2 - \delta/2$. Since $\mathcal{U}$ is countable, there exist two distinct sets $A_1, A_2 \in \mathcal{D}$ such that $U_{A_1} = U_{A_2}$. Since

A_1 and A_2 are incompatible in $\supseteq$, we have $A_1 \cup A_2 \notin \mathbb{A}$, therefore

$$\varepsilon \leq \lambda(A_1 \cup A_2) \leq \lambda(U_{A_1}) + \lambda(A_1 \setminus U_{A_1}) + \lambda(A_2 \setminus U_{A_2}) < \delta + 2 \cdot (\varepsilon/2 - \delta/2) = \varepsilon,$$

which is a contradiction.

For any $F \in \mathcal{F}$ the set $\mathcal{H}_F = \{A \in \mathbb{A} : F \subseteq A\}$ is dense in the poset $\langle \mathbb{A}, \supseteq \rangle$. Since $|\mathcal{F}| < \mathfrak{m}$, there exists an $\{\mathcal{H}_F : F \in \mathcal{F}\}$-generic filter $\mathcal{G}$ of $\mathbb{A}$. Set $G = \bigcup \mathcal{G}$. G is an open set and let there exist a countable set $\mathcal{G}_0 \subseteq \mathcal{G}$ such that $G = \bigcup \mathcal{G}_0$. Since $\mathcal{G}$ is a filter, each finite union of elements of $\mathcal{G}_0$ is an element of $\mathcal{A}$ and therefore has measure $< \varepsilon$. Then $\lambda(G) = \bigcup \mathcal{G}_0 \leq \varepsilon$. On the other hand since $\mathcal{G}$ meets every $\mathcal{H}_F$, $F \in \mathcal{F}$ we have $\bigcup \mathcal{F} \subseteq G$.

Since ε was arbitrary we obtain $\lambda(\bigcup \mathcal{F}) = 0$. $\qquad \square$

Corollary [AC] 9.2. MA *implies that* $\mathrm{add}(\mathcal{N}) = \mathfrak{c}$. *Consequently all cardinals in the Cichoń Diagram, with the exception of* $\aleph_1$, *are equal to* $\mathfrak{c}$.

Note that the implication **MA** $\rightarrow \mathrm{add}(\mathcal{M}) = \mathfrak{c}$ follows by Theorem 5.34, f) as well.

By Theorems 8.26, 8.8 and 8.9 and Corollary 9.2, we obtain:

Corollary [AC] 9.3. MA *implies that there exist a* $\mathfrak{c}$-*Luzin set, a* $\mathfrak{c}$-*Sierpinski set, a universal measure zero set of cardinality* $\mathfrak{c}$, *and a perfectly meager set of cardinality* $\mathfrak{c}$.

Corollary [AC] 9.4. *If* **MA**, *then*

a) *every set of reals of cardinality less than* $\mathfrak{c}$ *is a* γ-*set, is strongly meager, has strong measure zero, is a* QN-*set, therefore a* σ-*set and also a* λ-*set;*

b) *there exists a* γ-*set of cardinality* $\mathfrak{c}$;

c) *there exist a universal measure zero set of cardinality* $\mathfrak{c}$ *and a perfectly meager set of cardinality* $\mathfrak{c}$.

If moreover $\mathfrak{c} > \aleph_1$, *then also*

d) *every* $\mathbf{\Sigma}_2^1$ *set of reals is Lebesgue measurable and has the Baire Property;*

e) *there is no* $\mathbf{\Sigma}_2^1$ *well-ordering of the real line.*

Proof. Part a) follows by Theorems 5.49, 8.90, 8.16 d), 8.109, 8.102, 8.61, and 8.77. Part b) follows by Theorem 8.92. Part c) follows by Theorems 8.8 and 8.9.

If $\mathfrak{c} > \aleph_1$, then Part d) follows from the theorem using Theorem 6.58. Part e) follows by Theorem 10.4. $\qquad \square$

Lemma [AC] 9.5 (R.M. Solovay). *Let* $\mathcal{A}, \mathcal{B} \subseteq [\omega]^\omega$ *be families of cardinalities smaller than continuum. Assume that for any* $A \in \mathcal{B}$ *and any finite* $\mathcal{C} \subseteq \mathcal{A}$ *the set* $A \setminus \bigcup \mathcal{C}$ *is infinite. If* **MA** *holds true, then there exists a set* $M \subseteq \omega$ *such that* $A \cap M$ *is finite if* $A \in \mathcal{A}$ *and infinite if* $A \in \mathcal{B}$.

Proof. Let us consider the set $\mathbb{P}$ of all ordered pairs $\langle K, \mathcal{C} \rangle$ with K a finite subset of ω and $\mathcal{C}$ a finite subset of $\mathcal{A}$. We partially order $\mathbb{P}$ as follows:

$$\langle K_2, \mathcal{C}_2 \rangle \preceq \langle K_1, \mathcal{C}_1 \rangle \equiv \left(K_1 \subseteq K_2 \wedge \mathcal{C}_1 \subseteq \mathcal{C}_2 \wedge K_2 \cap \left(\bigcup \mathcal{C}_1 \right) \subseteq K_1 \right).$$

Since $\langle K, \mathcal{C}_1 \cup \mathcal{C}_2 \rangle \preceq \langle K, \mathcal{C}_i \rangle$, $i = 1, 2$, the partially ordered set $\langle \mathbb{P}, \preceq \rangle$ is **CCC**.

For any $A \in [\omega]^\omega$ and $n \in \omega$ we write

$$\mathcal{Y}_A = \{ \langle K, \mathcal{C} \rangle \in \mathbb{P} : A \in \mathcal{C} \},$$
$$\mathcal{X}_{A,n} = \{ \langle K, \mathcal{C} \rangle \in \mathbb{P} : |A \cap K| \geq n \}.$$

For any $A \in \mathcal{A}$, $B \in \mathcal{B}$ and any $n \in \omega$, the families $\mathcal{Y}_A$ and $\mathcal{X}_{B,n}$ are dense in $\mathbb{P}$. Hence, since cardinalities of $\mathcal{A}, \mathcal{B}$ are smaller than continuum and we assume **MA**, there exists a $\{ \mathcal{Y}_A : A \in \mathcal{A} \} \cup \{ \mathcal{X}_{A,n} : A \in \mathcal{B} \wedge n \in \omega \}$-generic filter $\mathcal{F}$ of $\mathbb{P}$. We set

$$M = \{ n \in \omega : (\exists \langle K, \mathcal{C} \rangle) (\langle K, \mathcal{C} \rangle \in \mathcal{F} \wedge n \in K) \}.$$

Let $A \in \mathcal{A}$. Then there exists a $\langle K_1, \mathcal{C}_1 \rangle \in \mathcal{F}$ with $A \in \mathcal{C}_1$. Assume now that $n \in A \cap M$. Then there exists $\langle K_2, \mathcal{C}_2 \rangle \in \mathcal{F}$ such that $n \in K_2$. Since $\mathcal{F}$ is a generic filter there is a $\langle K_3, \mathcal{C}_3 \rangle \in \mathcal{F}$ extending both $\langle K_i, \mathcal{C}_i \rangle$, $i = 1, 2$. Hence $K_2 \cap A \subseteq K_3 \cap A \subseteq K_1$. Thus $A \cap M \subseteq K_1$ is finite.

Now assume that $A \in \mathcal{B}$. Let $n \in \omega$. Then there exists a $\langle K, \mathcal{C} \rangle \in \mathcal{F} \cap \mathcal{X}_{A,n}$. Evidently $K \subseteq M$ and $|A \cap M| \geq n$. Since n was arbitrary the set $A \cap M$ is infinite. $\square$

Theorem [AC] 9.6. *If* **MA** *holds true, then every subset of a Polish space of cardinality less than $\mathfrak{c}$ is a Q-set.*

Proof. Let $Y \subseteq E \subseteq X$, $|E| < \mathfrak{c}$, X being a Polish space. We can assume that X has no isolated points. We construct an F_σ set F such that $Y = E \cap F$.

Let $\{ U_n : n \in \omega \}$ be an open base of topology. We set $A_x = \{ n \in \omega : x \in U_n \}$.

If $x \in E \setminus Y$ and $y_0, \ldots, y_k \in Y$, then the set $A_x \setminus \bigcup_{i=0}^k A_{y_i}$ is infinite. Thus by Lemma 9.5 there exists a set $M \subseteq \omega$ such that $M \cap A_x$ is finite for $x \in Y$ and infinite for $x \in E \setminus Y$. Set

$$G_m = \bigcup \{ U_n : n \in M \wedge n \geq m \}, \quad G = \bigcap_m G_m, \quad F = X \setminus G.$$

If $x \in Y$, then $M \cap A_x$ is finite and therefore $x \in F$. If $x \in E \setminus Y$, then $A_x \cap M$ is infinite and therefore $x \in G$. Hence $Y = F \cap E$. $\square$

Theorem [AC] 9.7. *Assume $\mathcal{A} \subseteq \mathcal{C} \subseteq [\omega]^\omega$, $|\mathcal{C}| < \mathfrak{m}$ and elements of $\mathcal{C}$ are pairwise almost disjoint. Then there exists a set $M \subseteq \omega$ such that*

$$|M \cap A| < \aleph_0 \equiv A \in \mathcal{A}$$

for each $A \in \mathcal{C}$.

Proof. In Lemma 9.5 take $\mathcal{B} = \mathcal{C} \setminus \mathcal{A}$.   □

Corollary [AC] 9.8. *If* **MA** *holds true, then every maximal almost disjoint family of subsets of* ω *has cardinality* $\mathfrak{c}$.

We apply the obtained results to the descriptive set theory.

Theorem [AC] 9.9. *If* **MA** *holds true and* $\mathfrak{c} > \aleph_1$, *then the following assertions are equivalent:*

a) *There exists a* $\mathbf{\Pi}_1^1$ *set* $A \subseteq {}^\omega 2$ *of cardinality* $\aleph_1$.
b) *Every subset of* ${}^\omega 2$ *of cardinality* $\aleph_1$ *is a* $\mathbf{\Pi}_1^1$ *set.*
c) *There exists a* $\mathbf{\Sigma}_2^1$ *set* $A \subseteq {}^\omega 2$ *of cardinality* $\aleph_1$.
d) *Every union of* $\aleph_1$ *many Borel sets is a* $\mathbf{\Sigma}_2^1$ *set.*
e) *Every union of* $\aleph_1$ *many* $\mathbf{\Sigma}_2^1$ *sets is a* $\mathbf{\Sigma}_2^1$ *set.*

Proof. Evidently b) $\rightarrow$ a) and e) $\rightarrow$ d) $\rightarrow$ c) $\rightarrow$ a) (the last implication follows from the Uniformization Theorem 6.62).

We prove b) $\rightarrow$ e). By Theorem 6.40 there exists a $\mathbf{\Sigma}_2^1$ set $U \subseteq {}^\omega \omega \times X$ universal for $\mathbf{\Sigma}_2^1(X)$. If $\langle A_\xi : \xi < \omega_1 \rangle$ are $\mathbf{\Sigma}_2^1(X)$ sets, then there exists a set $B \subseteq {}^\omega \omega$ of cardinality $\aleph_1$ such that $\bigcup_{\xi < \omega_1} A_\xi = \bigcup_{x \in B} U^x$. By b) $B \in \mathbf{\Pi}_1^1$ and

$$y \in \bigcup_{x \in B} U^x \equiv (\exists x)\,(x \in B \wedge \langle x, y \rangle \in U).$$

So we have to prove a) $\rightarrow$ b). Assume that a) holds true and we show that any subset of ${}^\omega 2$ of cardinality $\aleph_1$ is a $\mathbf{\Pi}_1^1$ set. So, let $A, B \subseteq {}^\omega 2$ be such that $|A| = |B| = \aleph_1$, $A \in \mathbf{\Pi}_1^1({}^\omega 2)$. Let $g : A \xrightarrow[\text{onto}]{1-1} B$. We take a bijection $H : \omega \xrightarrow[\text{onto}]{1-1} {}^{<\omega} 2$ and define an injection $F : {}^\omega 2 \xrightarrow{1-1} \mathcal{P}(\omega)$ by setting (compare the proof of Theorem 5.35)

$$F(\alpha) = \{n \in \omega : H(n) \subseteq \alpha\}.$$

Then F is a continuous mapping and values of F are pairwise almost disjoint. We set

$$B_{\alpha, n} = \{k \in \omega : \pi(n, k) \in F(\alpha)\}$$

and for a given n,

$$\mathcal{A}_n = \{B_{\alpha, n} : \alpha \in A \wedge g(\alpha)(n) = 1\}.$$

Since $\mathcal{A}_n$ is a family of almost disjoint sets and $|\mathcal{A}_n| \leq \aleph_1$, by Theorem 9.7 there exists a set $M_n \subseteq \omega$ such that

$$|B_{\alpha, n} \cap M_n| < \aleph_0 \equiv B_{\alpha, n} \in \mathcal{A}_n$$

for any $n \in \omega$ and $\alpha \in A$. We define a subset of ${}^\omega 2 \times {}^\omega 2$ by setting

$$\langle \alpha, \beta \rangle \in G \equiv (\forall n)\,(|B_{\alpha, n} \cap M_n| < \aleph_0 \equiv \beta(n) = 1).$$

Since F is continuous and

$$|B_{\alpha,n} \cap M_n| < \aleph_0 \equiv (\exists k)(\forall m \geq k)\,(m \in M \to \pi(n,m) \notin F(\alpha)),$$

the set G is Borel. Moreover if $\alpha \in A$, then $\langle \alpha, \beta \rangle \in G \equiv \beta = g(\alpha)$. Since the projection $\mathrm{proj}^2 : {}^\omega 2 \times {}^\omega 2 \longrightarrow {}^\omega 2$ is injective on G and $B = \mathrm{proj}^2(G \cap (A \times {}^\omega 2))$ we obtain $B \in \mathbf{\Pi}_1^1$. $\qquad\square$

Together with Theorem 6.58 we obtain

Corollary [AC] 9.10. *Assume that* **MA** *holds true,* $\mathfrak{c} > \aleph_1$ *and there exists a* $\mathbf{\Pi}_1^1$ *set of cardinality* $\aleph_1$*. Then a set is* $\mathbf{\Sigma}_2^1$ *if and only if it is a union of* $\aleph_1$ *Borel sets.*

Now we apply Martin's Axiom to essentially different topics. Namely we shall study a topological product of **CCC** topological spaces. We begin with a combinatorial result proved by N.A. Shanin.

Lemma [AC] 9.11 (Δ-Lemma). *Assume that* $\mathcal{F}$ *is an uncountable family of finite sets. Then there exist an uncountable subfamily* $\mathcal{F}_0 \subseteq \mathcal{F}$ *and a finite set* C *such that* $A \cap B = C$ *for any distinct* $A, B \in \mathcal{F}_0$*.*

Proof. We can assume that there exists a positive integer n such that $|A| = n$ for each $A \in \mathcal{F}$. We prove the assertion by induction. If $n = 1$, take $C = \emptyset$. Assume that the assertion holds true for n and every element of an uncountable family $\mathcal{F}$ has cardinality $n + 1$. If the family $\{A \in \mathcal{F} : a \in A\}$ is uncountable for some a, apply the assertion to the family $\{A \setminus \{a\} : A \in \mathcal{F} \wedge a \in A\}$.

Assume that each element belongs to at most countably many elements of the family $\mathcal{F}$. We can construct an uncountable subfamily $\{A_\xi : \xi \in \omega_1\}$ of $\mathcal{F}$ consisting of pairwise disjoint elements. Actually, if $\{A_\eta : \eta \in \xi\}$, $\xi < \omega_1$ is already constructed, then the set $B = \bigcup_{\eta < \xi} A_\eta$ is countable and therefore at most countably many elements of $\mathcal{F}$ meets the set B. Thus there exists an element $A_\xi \in \mathcal{F}$ such that $B \cap A_\xi = \emptyset$. Set $\mathcal{F}_0 = \{A_\xi : \xi < \omega_1\}$ and $C = \emptyset$. $\qquad\square$

Theorem [AC] 9.12. *Assume that* $\langle\langle X_s, \mathcal{O}_s \rangle : s \in S \rangle$ *are topological spaces satisfying* **CCC***. Then the product space* $\langle \Pi_{s\in S} X_s, \Pi_{s \in S} \mathcal{O}_s \rangle$ *is* **CCC** *if and only if the product* $\langle \Pi_{s\in T} X_s, \Pi_{s \in T} \mathcal{O}_s \rangle$ *is* **CCC** *for every finite* $T \subseteq S$*.*

Proof. Assume that $\mathcal{U}$ is an uncountable family of pairwise disjoint open subsets of $X = \Pi_{s\in S} X_s$. We can assume that every $U \in \mathcal{U}$ is of the form $U = \Pi_{s \in S} U_s$, where $U_s \subseteq X_s$ are open and the set $S_U = \{s \in S : U_s \neq X_s\}$ is finite. One can easily see the following. If $U \cap V = \emptyset$, $V = \Pi_{s\in S} V_s$, then also $\Pi_{s\in T} U_s \cap \Pi_{s\in T} V_s = \emptyset$, where $T = S_U \cap S_V$.

By Δ-Lemma there exist an uncountable family $\mathcal{U}_0 \subseteq \mathcal{U}$ and a finite set $T \subseteq S$ such that $S_U \cap S_V = T$ for any distinct $U, V \in \mathcal{U}_0$. Then $\{\Pi_{s\in T} U_s : U \in \mathcal{U}_0\}$ is an uncountable family of pairwise disjoint open subsets of $\Pi_{s\in T} X_s$.

The opposite implication is trivial. $\qquad\square$

Thus the problem, whether a topological product of topological **CCC** spaces is **CCC**, is a problem about the product of two topological **CCC** spaces. Later in Section 10.3 we show that the answer is undecidable in **ZFC**.

A poset $\langle P, \leq \rangle$ has the **Knaster Property**, if every uncountable set $A \subseteq P$ contains an uncountable subset $B \subseteq A$ such that every two elements $x, y \in B$ are compatible, i.e., there exists a $z \in P$, $z \neq 0_P$, such that $z \leq x$ and $z \leq y$. Evidently, a poset with the Knaster Property is **CCC**. It turned out that the Martin's Axiom **MA** and $\mathfrak{c} \neq \aleph_1$ imply also the opposite assertion.

Theorem [AC] 9.13. *If* **MA** *and* $\mathfrak{c} > \aleph_1$ *hold true, then every* **CCC** *poset has the Knaster Property.*

Proof. Let $\langle P, \leq \rangle$ be a poset satisfying **CCC**, $A = \{a_\xi : \xi < \omega_1\}$ being an uncountable subset of P.

Assume that for every $q \in P$ there exists a $p \in P$, $p \leq q$ compatible with only countably many elements of A. Then for every $\eta < \omega_1$ there exists a $\xi > \eta$ and a $p_\eta \leq a_\eta$ which is incompatible with all a_ζ, $\zeta \geq \xi$. One can easily see that the set $\{p_\eta : \eta < \omega_1\}$ contains an uncountable subset of incompatible elements contradicting **CCC**.

Thus, there exists a $q \in P$ such that the set $D_\xi = \{p \leq q : (\exists \zeta \geq \xi)\, p \leq a_\zeta\}$ is dense below q for every $\xi < \omega_1$. By **MA** there exists a $\{D_\xi : \xi < \aleph_1\}$-generic filter F of P such that $q \in F$. By construction, for every ξ there exists $\zeta \geq \xi$ such that $a_\zeta \in F$. Thus $A \cap F$ is an uncountable set of pairwise compatible elements. $\qquad\square$

Theorem [AC] 9.14. *If* **MA** *and* $\mathfrak{c} > \aleph_1$ *hold true, then product of two* **CCC** *topological spaces is a* **CCC** *topological space.*

Proof. Let X and Y be **CCC** topological spaces. Assume that $\mathcal{U}$ is an uncountable family of open subsets of $X \times Y$. We suppose that every $U \in \mathcal{U}$ has the form $U = V_U \times W_U$, where V_U, W_U are open subsets of X, Y, respectively.

By Theorem 9.13 we can assume that $V_{U_1} \cap V_{U_2} \neq \emptyset$ for any $U_1, U_2 \in \mathcal{U}$. Since Y is **CCC**, there exist $U_1, U_2 \in \mathcal{U}$ such that $W_{U_1} \cap W_{U_2} \neq \emptyset$. Then $U_1 \cap U_2 \neq \emptyset$. $\qquad\square$

As a consequence of Theorems 9.12 and 9.14 we obtain that Martin's Axiom implies an affirmative answer to the question about topological product of **CCC** topological spaces.

Corollary [AC] 9.15. *If* **MA** *and* $\mathfrak{c} > \aleph_1$ *hold true, then the topological product of* **CCC** *topological spaces is a* **CCC** *space.*

Exercises

9.1 [AC] Regularity of $\mathfrak{c}$

   a) Show that $\mathrm{cf}(\mathfrak{c}) \geq \mathfrak{m}$.

      Hint: Use Exercise 1.7, b) and (9.1).

   b) **MA** implies that $\mathfrak{c}$ is a regular cardinal.

      Hint: See Exercise 1.7, c).

9.2 [AC] Probabilistic Borel Measures

We assume that μ is a diffused probabilistic Borel measure on $[0,1]$ and **MA** holds true.

a) The "Amoeba forcing" $\mathbb{A}(\mu) = \{U \subseteq [0,1] : U \text{ open} \wedge \mu(U) < \varepsilon\}$ satisfies **CCC**.
   Hint: Follow the proof of Theorem 9.1.

b) $\operatorname{add}(\mu) = \mathfrak{c}$.
   Hint: For $\xi < \kappa < \mathfrak{c}$, let $\mathcal{U}_\xi = \{U \in \mathbb{A}(\mu) : A_\xi \subseteq U\}$. If $\mathcal{G}$ is $\{\mathcal{U}_\xi : \xi < \kappa\}$-generic filter, then $\bigcup_{\xi<\kappa} A_\xi \subseteq \bigcup \mathcal{G}$ and $\mu(\bigcup \mathcal{G}) < 2\varepsilon$.

9.3 [AC] The Martin Number for a Class of Posets

For simplicity we assume that no poset contains the smallest element. Let $\mathcal{C}$ be a class of separative posets. $\mathfrak{m}(\mathcal{C})$ is the smallest cardinal κ such that there exist an $\langle P, \leq \rangle \in \mathcal{C}$ and a family $\mathcal{D}$ of dense subsets of P, $|\mathcal{D}| = \kappa$ such that there exists no $\mathcal{D}$-generic filter F of P.

Let $\langle P, \leq \rangle$ be a poset without the smallest element. A set $R \subseteq P$ is **linked** if every two elements of R are compatible. A set, that is a countable union of linked subsets, is σ-**linked**. A set $R \subseteq P$ is **centered** if every finitely many members of R have a common lower bound in P. A countable union of centered subsets is a σ-**centered** set.

a) Note that every σ-centered set is σ-linked, every σ-linked poset has the Knaster Property, and a poset with the Knaster Property is **CCC**.

b) $\mathfrak{m}(\mathbf{CCC}) = \mathfrak{m} \leq \mathfrak{m}(\text{Knaster}) \leq \mathfrak{m}(\sigma\text{-linked}) \leq \mathfrak{m}(\sigma\text{-centered})$.

c) $\mathfrak{m}(\text{countable}) = \operatorname{cov}(\mathcal{M})$.
   Hint: If $A \subseteq {}^\omega 2$ is open dense, then $\{s \in {}^{<\omega}2 : [s] \subseteq A\}$ is an open dense subset of the poset $\langle {}^{<\omega}2, \supseteq \rangle$. Conclude that $\mathfrak{m}(\text{countable}) \leq \operatorname{cov}(\mathcal{M})$. Now, let $\langle D_\xi : \xi < \kappa \rangle$ be dense subsets of a countable poset $\langle P, \leq \rangle$. Define $F : {}^\omega P \longrightarrow {}^\omega P$ as $F(\alpha)(0) = \alpha(0)$, $F(\alpha)(k+1) = \alpha(k+1)$ if $\alpha(k+1) \leq F(\alpha)(k)$, and $F(\alpha)(k+1) = F(\alpha)(k)$ otherwise. Set $U_\xi = \{\alpha \in {}^\omega P : (\exists k)\, F(\alpha)(k) \in D_\xi\}$. If $\kappa < \operatorname{cov}(\mathcal{M})$, then $\beta \in \bigcap_{\xi<\kappa} U_\xi$ is such that $F(\beta)$ meets every D_ξ.

d) $\mathfrak{m}(\sigma\text{-centered}) \leq \mathfrak{p}$.
   Hint: The poset P in the proof of Theorem 5.49 is σ-centered.

e) $\mathfrak{m}(\sigma\text{-linked}) \leq \operatorname{add}(\mathcal{N})$.
   Hint: Show that the amoeba forcing $\langle \mathbb{A}, \supseteq \rangle$ is σ-linked. Let $\mathcal{R}$ be the family of all finite unions of rational open intervals with measure less than ε. For every $U \in \mathbb{A}$ there exists an $R_U \in \mathcal{R}$ such that $\lambda(U \setminus R_U) < \varepsilon - \lambda(U)$. Show that $\lambda(U \setminus R_U) < 1/2(\varepsilon - \lambda(R_U))$. The family $\{U \in \mathbb{A} : R_U = R\}$ is linked for any $R \in \mathcal{R}$.

9.4 [AC] Bell's Theorem

We assume that $\langle P, \leq \rangle$ is a σ-centered poset, $P = \bigcup_n P_n$, P_n centered, $\mathcal{D}$ is an infinite family of open dense subsets of P closed under finite intersections, and $|\mathcal{D}| < \mathfrak{p}$.

a) There exists a σ-centered subset $Q \subseteq P$ such that $|Q| < \mathfrak{p}$ and $D \cap Q$ is open dense in $\langle Q, \leq \rangle$ for any $D \in \mathcal{D}$.
   Hint: Set $Q = \bigcup_n Q_n$, where Q_0 is any non-empty finite subset of P. By induction, find $Q_{n+1} \supseteq Q_n$ such that if a finite subset of Q_n has a lower bound in P, then it has a lower bound in Q_{n+1}, and for every $p \in Q_n$ and every $D \in \mathcal{D}$, there exists a $q \in Q_{n+1}$ with $p \geq q \in D$, and $|Q_{n+1}| \leq |\mathcal{D}|$.

b) Assume that Q is the set of a) and $\mathcal{F}$ is a $\{D \cap Q : D \in \mathcal{D}\}$-generic filter on Q. Then $\{p \in P : (\exists q \in Q)\, q \leq p\}$ is a $\mathcal{D}$-generic filter on P.

From now on we assume that $\mathcal{D}$ contains all

$$D_{p,q} = \{r \in P : r \leq p, q \vee r \text{ incompatible with } p \text{ or } q\}$$

with $p, q \in P$ and $|P| < \mathfrak{p}$.

c) If $\mathcal{F}$ is a linked subset of P meeting each $D \in \mathcal{D}$, then the upward closure of $\mathcal{F}$ is a $\mathcal{D}$-generic filter on P.

d) Assume $\mathcal{C} \subseteq P$ is centered. If $\mathcal{C}$ meets each $D \in \mathcal{D}$, then $\mathcal{C}$ is a linked set.

So, we can assume that for every n there exists a set $F_n \in \mathcal{D}$ with $F_n \cap P_n = \emptyset$. For each $p \in P$, $D \in \mathcal{D}$ we set $A(p, D) = \{n \in \omega : (\exists q \in P_n \cap D) q \leq p\}$.

e) The family $\{A(p, D) : p \in P_n \wedge D \in \mathcal{D}\}$ has f.i.p., hence it has a pseudointersection A_n.

Hint: If $p_0, \ldots, p_k \in P_n$, $D_0, \ldots, D_k \in \mathcal{D}$, then there exists $p \leq p_i, i = 0, \ldots, k$. For any m we have $D = \bigcap_{i \leq k} D_i \cap \bigcap_{j \leq m} F_j \in \mathcal{D}$. If $q \in D$, $q \leq p$, then $q \in P_n$ for some $n \geq m$ and also $n \in \bigcap_{i \leq k} A(p_i, D_i)$.

We define a function $g : {}^{<\omega}\omega \longrightarrow \omega$: $g(\emptyset) = 0$ and $g(s{}^\frown k) = n_k$, where $\{n_k\}_{k=0}^\infty$ is an increasing enumeration of $A_{g(s)}$. For any $D \in \mathcal{D}$ we define function $h_D : {}^{<\omega}\omega \longrightarrow P$ as follows: $h_D(\emptyset)$ is an arbitrary element of P_0. If $h_D(s) = p$, $s \in {}^{<\omega}\omega$ is already defined and $g(s{}^\frown k) = n$, then we set $h_D(s{}^\frown k)$ equal to any element of $D \cap P_n$ such that, if $n \in A(p, D)$, then $q \leq p$ (and arbitrary in $D \cap P_n$ otherwise).

f) There exists a branch $\alpha \in {}^\omega\omega$ such that for all $D \in \mathcal{D}$ and for all but finitely many n, $g(n) \in A(\alpha|n, D)$.

Hint: For $D \in \mathcal{D}$ and $s \in {}^{<\omega}\omega$, let $f_D(s)$ be a number so large, that $m \in A(h_D(s), D)$ for every $m \geq f_D(s)$. Since $|\mathcal{D}| < \mathfrak{p} \leq \mathfrak{b}$, there exists an α such that $f_D <^* \alpha$ for any $D \in \mathcal{D}$.

g) There exists a linked subset of P meeting each $D \in \mathcal{D}$.

Hint: If α is as in f), construct a branch β setting $\beta(0) = \emptyset$ and $\beta(n) = g(\beta|n)$. For each $D \in \mathcal{D}$ there exists a natural number m_D such that $A(\beta|m_D, D) \neq \emptyset$. Then $\{f_D(\beta|m_D); D \in \mathcal{D}\}$ meets every $D \in \mathcal{D}$ and is linked.

h) Prove **Bell's Theorem**: $\mathfrak{m}(\sigma\text{-centered}) \geq \mathfrak{p}$.

i) Conclude, that $\mathfrak{m}(\sigma\text{-centered}) = \mathfrak{p}$.

9.5 [AC] $\mathfrak{p}$ is regular

a) Assume that $\mathcal{A}, \mathcal{B} \subseteq [\omega]^\omega$, $|\mathcal{A}|, |\mathcal{B}| < \mathfrak{p}$, and $\mathcal{A}$ is closed under finite intersections. If for every $A \in \mathcal{A}$ and $B \in \mathcal{B}$ the intersection $A \cap B$ is infinite, then there exists an infinite set C such that $C \subseteq^* A$ and $|C \cap B| = \aleph_0$ for every $A \in \mathcal{A}$ and every $B \in \mathcal{B}$, respectively.

Hint: The set $P = \{\langle I, A \rangle : I \subseteq \omega \text{ is finite } \wedge A \in \mathcal{A}\}$ ordered as (compare the proof of Theorem 5.49)

$$\langle I_1, A_1 \rangle \preceq \langle I_2, A_2 \rangle \equiv (I_2 \subseteq I_1 \subseteq (I_2 \cup A_2) \wedge A_1 \subseteq A_2)$$

is σ-centered. Since $\mathfrak{m}(\sigma\text{-centered}) = \mathfrak{p}$, there exists a $\mathcal{D}$-generic filter $\mathcal{F}$, where $\mathcal{D} = \{\{\langle I, B \rangle \in P : I \text{ finite}\} : B \in \mathcal{B}\} \cup \{\{\langle I, A \rangle \in P : |I \cap A| > n\} : n \in \omega \wedge A \in \mathcal{A}\}$. Then $C = \{n \in \omega : (\exists \langle I, A \rangle \in \mathcal{F}) n \in I\}$ is the desired set.

b) If $\mathcal{B} \subseteq [\omega]^\omega$ is closed under finite intersections, $\lambda = \mathrm{cf}(\kappa) < \kappa = |\mathcal{B}|$, then there exists an increasing sequence $\langle \mathcal{B}_\xi : \xi < \lambda \rangle$, every $\mathcal{B}_\xi$ closed under finite intersections, $|\mathcal{B}_\xi| < \kappa$, and such that $\mathcal{B} = \bigcup_{\xi < \lambda} \mathcal{B}_\xi$.

c) Assume that $\mathcal{B}$, $\mathcal{B}_\xi$ are as in b) with $\kappa = \mathfrak{p}$ and $\mathcal{A}$ is closed under finite intersections, $|\mathcal{A}| < \mathfrak{p}$. If for every $A \in \mathcal{A}$ and $B \in \mathcal{B}$ the intersection $A \cap B$ is infinite, then there exists an infinite set C such that $C \subseteq^* A$ and $|C \cap B| = \aleph_0$ for every $A \in \mathcal{A}$ and every $B \in \mathcal{B}$, respectively.

   Hint: Since $\mathcal{B}_\xi \cup \mathcal{A}$ has f.i.p. and has size $< \mathfrak{p}$, there exists a pseudointersection C_ξ. Apply a) with $\{ C_\xi : \xi < \lambda \}$ in the role $\mathcal{B}$.

d) Assume that $\mathcal{B}$, $\mathcal{B}_\xi$ are as in b) with $\kappa = \mathfrak{p}$. Then there exists an almost decreasing sequence $\langle \mathcal{B}_\xi : \xi < \lambda \rangle$ of infinite sets such that each B_ξ is a pseudointersection of $\mathcal{B}_\xi$ and meets each $B \in \mathcal{B}$ in an infinite set.

   Hint: If $\langle B_\eta : \eta < \xi \rangle$ are already defined and moreover, each B_η meets each member of $\mathcal{B}$ in an infinite set, then find B_ξ as in d) taking $\mathcal{A} = \mathcal{B}_\xi \cup \{ B_\eta : \eta < \xi \}$.

e) $\mathfrak{p}$ is regular.

   Hint: If not, there exists a family $\mathcal{B}$ with property of b). However, then a pseudointersection of the sequence $\langle B_\xi : \xi < \lambda \rangle$ of d) is a pseudointersection of $\mathcal{B}$ as well.

9.6 [AC] Knaster Property of a Topological Product

We shall say that a topological space $\langle X, \mathcal{O} \rangle$ has the Knaster Property if the poset $\langle \mathcal{O}, \subseteq \rangle$ has the Knaster Property.

a) If $W \subseteq A \times B$ is an uncountable set such that for any $a \in A$ the vertical section W_a is countable and for any $b \in B$ the horizontal section W^b is countable, then there exists an uncountable set $F \subseteq W$ such that F is a one-to-one function from $\mathrm{dom}(F) \subseteq A$.

b) Let $\langle X_i, \mathcal{O}_i \rangle$, $i = 1, 2$ be topological spaces, $\langle X_2, \mathcal{O}_2 \rangle$ having the Knaster Property. Assume that $\mathcal{W}$ is a family of basic open subsets of the product $X_1 \times X_2$ such that the set $\{ V \in \mathcal{O}_1 : U \times V \in \mathcal{O}_1 \times \mathcal{O}_2 \}$ is uncountable for some $U \in \mathcal{O}_1$. Then there exists an uncountable subset of $\mathcal{W}$ of pairwise compatible sets.

c) The product of two topological spaces with the Knaster Property has the Knaster Property as well.

   Hint: Use a) and b).

b) Any product of topological spaces with the Knaster Property has the Knaster Property.

   Hint: Apply Δ-Lemma.

9.2 Equalities, Inequalities and All That

We recommend that the reader consult Sections 11.4 and 11.5. In this section, once and for all we assume that **ZF** is consistent and we shall not repeat that assumption.

In Section 7.5, several results about relationships between properties of measure and category were described by the Cichoń Diagram. It turned out that often such a simple description of a relationship is enough to solve many important problems. In this section we supply the Cichoń Diagram by relationships with other introduced cardinal invariants.

Taking into account the results of Theorems 5.50, 5.49, 5.52, 9.1, 7.49, and 7.28, we can extend the Cichoń Diagram as follows. Note that the arrow $\mathfrak{b} \longrightarrow \mathfrak{d}$ is missing in Diagram 3 for a typographical reason.

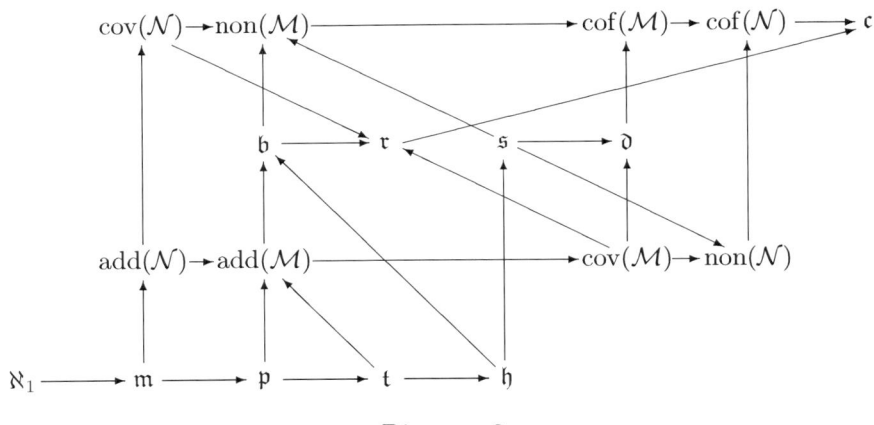

Diagram 3.

Any of the consistent equalities $\mathfrak{f} = \mathfrak{g}$ or the consistent inequalities $\mathfrak{f} < \mathfrak{g}$ of the next Metatheorem can be considered as a candidate for a new axiom of **ZFC**.

Metatheorem 9.1. *Assume that $\mathfrak{f}, \mathfrak{g}$ are cardinal invariants occurring in Diagram 3.*

a) *The theory* **ZFC** $+ \mathfrak{f} = \mathfrak{g}$ *is consistent.*

b) *If the arrow $\mathfrak{f} \to \mathfrak{g}$ occurs in Diagram 3, then the theory* **ZFC** $+ \mathfrak{f} < \mathfrak{g}$ *is consistent, but the case $\mathfrak{f} = \mathfrak{p}$ and $\mathfrak{g} = \mathfrak{t}$. Moreover, we can assume that $\mathfrak{f} = \aleph_1 < \mathfrak{g} = \mathfrak{c}$.*

c) *The following theories are consistent:*

ZFC $+ \mathfrak{h} = \mathfrak{s} < \mathfrak{b}$,	**ZFC** $+ \mathfrak{h} < \mathfrak{s} = \mathfrak{b}$,	**ZFC** $+ \mathfrak{b} < \mathfrak{d}$,
ZFC $+ \operatorname{cov}(\mathcal{M}) < \operatorname{non}(\mathcal{M})$,	**ZFC** $+ \operatorname{cov}(\mathcal{N}) < \operatorname{non}(\mathcal{N})$,	
ZFC $+ \operatorname{non}(\mathcal{M}) < \operatorname{cov}(\mathcal{M})$,	**ZFC** $+ \operatorname{non}(\mathcal{N}) < \operatorname{cov}(\mathcal{N})$.	

The metatheorem actually shows that the obtained results were the best possible and the **ZFC** set theory is not strong enough to answer the delicate (even natural) questions.

The consistency of other theories than those presented in Metatheorem 9.1 are known. Some of them are presented in Metatheorem 11.7. We recommend the monographs T. Bartoszyński and H. Judah [1995], K. Ciesielski and J. Pawlikowski [2004] and recent survey articles T. Bartoszyński [2010], A. Blass [2010], J.E. Vaughan [1990], eventually others, for further information.

Demonstration. Since all cardinals considered in Diagram 3 are not greater than $\mathfrak{c}$, then assuming Martin's Axiom they all are equal to $\mathfrak{c}$

For every inequality $\mathfrak{f} < \mathfrak{g}$ of Parts b) and c), we refer to a corresponding result presented in Appendix 11.5. In some cases the result is not immediate, one must use some inequalities of Diagram 3.

$\aleph_1 < \mathfrak{m}$	(11.21)	$\mathrm{cov}(\mathcal{M}) < \mathfrak{d}$	(11.28), (11.29), (11.36)
$\mathfrak{m} < \mathfrak{p}$	(11.30)	$\mathrm{cov}(\mathcal{M}) < \mathfrak{r}$	(11.23), (11.28)
$\mathfrak{t} < \mathfrak{h}$	(11.29)	$\mathfrak{b} < \mathfrak{r}$	(11.23), (11.26)
$\mathfrak{m} < \mathrm{add}(\mathcal{N})$	(11.30)	$\mathfrak{s} < \mathfrak{d}$	(11.36), (11.27), (11.28)
$\mathfrak{t} < \mathrm{add}(\mathcal{M})$	(11.27)	$\mathfrak{s} < \mathrm{non}(\mathcal{M})$	(11.23), (11.27), (11.28)
$\mathfrak{h} < \mathfrak{s} = \mathfrak{b}$	(11.37)	$\mathfrak{b} < \mathrm{non}(\mathcal{M})$	(11.23)
$\mathfrak{h} = \mathfrak{s} < \mathfrak{b}$	(11.38)	$\mathfrak{b} < \mathfrak{d}$	(11.22), (11.25), (11.36)
$\mathrm{add}(\mathcal{N}) < \mathrm{add}(\mathcal{M})$	(11.34)	$\mathrm{cov}(\mathcal{N}) < \mathfrak{r}$	(11.22)
$\mathrm{add}(\mathcal{M}) < \mathrm{cov}(\mathcal{M})$	(11.26)	$\mathfrak{d} < \mathrm{cof}(\mathcal{M})$	(11.23)
$\mathrm{cov}(\mathcal{M}) < \mathrm{non}(\mathcal{N})$	(11.35)	$\mathrm{cov}(\mathcal{N}) < \mathrm{non}(\mathcal{M})$	(11.27), (11.29)
$\mathrm{add}(\mathcal{M}) < \mathfrak{b}$	(11.29)	$\mathrm{non}(\mathcal{M}) < \mathrm{cof}(\mathcal{M})$	(11.35), (11.36)
$\mathfrak{s} < \mathrm{non}(\mathcal{N})$	(11.27)	$\mathrm{cof}(\mathcal{N}) < \mathfrak{c}$	(11.24)
$\mathfrak{r} < \mathfrak{c}$	(11.36)	$\mathrm{cof}(\mathcal{M}) < \mathrm{cof}(\mathcal{N})$	(11.39)
$\mathrm{add}(\mathcal{N}) < \mathrm{cov}(\mathcal{N})$	(11.23)	$\mathrm{non}(\mathcal{N}) < \mathrm{cof}(\mathcal{N})$	(11.28)
$\mathrm{cov}(\mathcal{M}) < \mathrm{non}(\mathcal{M})$	(11.29)	$\mathrm{cov}(\mathcal{N}) < \mathrm{non}(\mathcal{N})$	(11.22)
$\mathrm{non}(\mathcal{M}) < \mathrm{cov}(\mathcal{M})$	(11.22)	$\mathrm{non}(\mathcal{N}) < \mathrm{cov}(\mathcal{N})$	(11.23) $\qquad\square$

For convenience in the next investigation, we formulate some consistency results as equalities for cardinal invariants. Namely, the results (11.28), (11.31), (11.32), (11.41), and (11.33), can be expressed as

$$\mathbf{ZFC} + \mathrm{size}(\mathcal{SN}) = \aleph_1 < \mathfrak{c} \text{ is consistent,} \qquad (9.2)$$

$$\mathbf{ZFC} + \mathrm{size}(\mathcal{PM}) = \aleph_2 \leq \mathfrak{c} \text{ is consistent,} \qquad (9.3)$$

$$\mathbf{ZFC} + \mathrm{size}(\mathcal{UN}) = \aleph_2 \leq \mathfrak{c} \text{ is consistent,} \qquad (9.4)$$

$$\mathbf{ZFC} + \mathrm{size}(\mathcal{SM}) = \aleph_1 < \mathfrak{c} \text{ is consistent,} \qquad (9.5)$$

$$\mathbf{ZFC} + \mathrm{size}(\sigma\text{-sets}) = \aleph_1 < \mathfrak{c} \text{ is consistent,} \qquad (9.6)$$

respectively.

Metatheorem 9.2. *The following statements are undecidable in* **ZFC***:*

a) *Borel Conjecture.*

b) *There exists a κ-Luzin set, where $\aleph_0 < \kappa \leq \mathfrak{c}$ is a regular cardinal.*

c) *There exists a κ-Sierpiński set, where $\aleph_0 < \kappa \leq \mathfrak{c}$ is a regular cardinal.*

d) *Every universal measure zero set has cardinality $\leq \aleph_1$.*

e) *Every perfectly meager set has cardinality $\leq \aleph_1$.*

f) *Every strongly meager set is countable.*

Demonstration. a) By Corollary 8.14 and Metatheorem 9.1 ($\mathrm{cov}(\mathcal{M}) > \aleph_1$), the negation of the Borel Conjecture is consistent with **ZFC**. Therefore by (9.2) the Borel Conjecture is an undecidable statement of **ZFC**.

b) By Corollary 8.36, the inequalities $\mathrm{non}(\mathcal{M}) \leq \kappa \leq \mathrm{cov}(\mathcal{M})$ follow from the existence of a κ-Luzin set. The inequality $\mathrm{non}(\mathcal{M}) > \mathrm{cov}(\mathcal{M})$ is consistent with **ZFC** by Metatheorem 9.1. On the other hand we know that **CH** or even **MA** implies the existence of a $\mathfrak{c}$-Luzin.

c) Similarly, by Corollary 8.36 the existence of a κ-Sierpiński set implies the inequalities $\mathrm{non}(\mathcal{N}) \leq \kappa \leq \mathrm{cov}(\mathcal{N})$. Again, **MA** implies the existence of a $\mathfrak{c}$-Sierpiński set.

d) By the Grzegorek Theorem 8.8 there exists a universal measure zero set of cardinality $\mathrm{non}(\mathcal{N})$. The assertion follows by Metatheorems 9.1 and (9.4).

e) By the Grzegorek Theorem 8.9 and Metatheorem 9.1 $(\mathrm{non}(\mathcal{M}) > \aleph_1)$, the negation of "every perfectly meager set has cardinality $\leq \aleph_1$" is consistent. Thus, by (9.3) the assertion "every perfectly meager set has cardinality $\leq \aleph_1$" is an undecidable statement of **ZFC**.

f) By Theorem 8.16, d) and Metatheorem 9.1 $(\mathrm{cov}(\mathcal{N}) > \aleph_1)$, the negation of "every strongly meager set is countable" is consistent with **ZFC**. Finally, "every strongly meager set is countable" is consistent with **ZFC** by (9.5). $\square$

Now, we summarize the relationships between special subsets of Polish spaces obtained in Chapter 8 as Diagram 4.

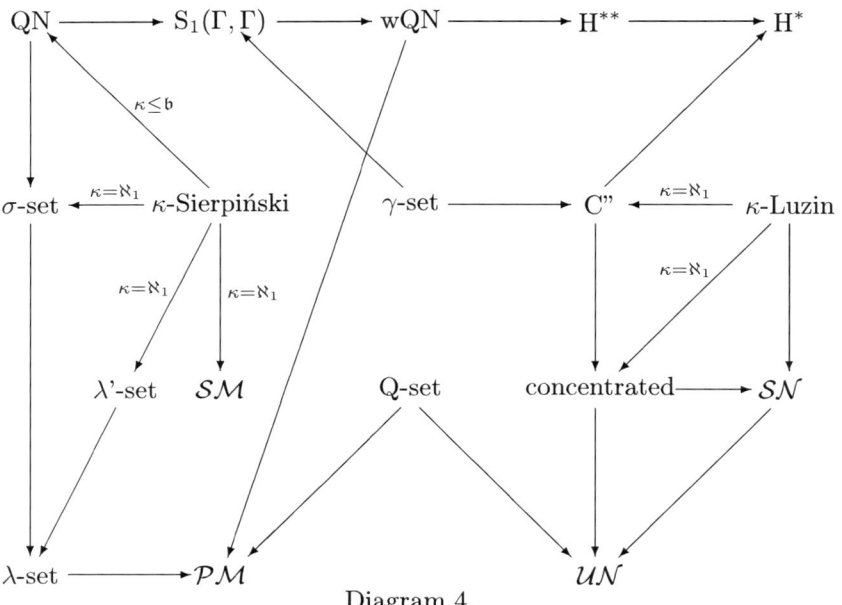

Diagram 4.

We assume that $\kappa \leq \mathfrak{c}$ is an uncountable regular cardinal. An arrow $\mathrm{A} \longrightarrow \mathrm{B}$ of the diagram means that any subset of a Polish space possessing the property A does possess the property B as well. So the negation means that there exists a subset of a Polish space with property A which does not posses the property B.

By Corollaries 8.59 and 8.104, every QN-, wQN-, C"-, and γ-subset of a Polish space being zero-dimensional is homeomorphic to a set of reals, even to a subset of $\mathbb{C}$ or $^\omega\omega$. Any countable set of reals possesses any of the considered properties with the exception of being a κ-Sierpiński or a κ-Luzin set. Hence one cannot prove any implication going into "κ-Sierpiński" or "κ-Luzin". Similarly, since any compact space possesses the property H**, one cannot prove that a set with the property H** possesses any of the considered properties, i.e., no arrow goes from H** except that to H*.

Moreover, we have proved two negative results:

$$\lambda\text{-set} \longrightarrow\!\!\!\!| \quad \lambda\text{'-set} \qquad \kappa\text{-Luzin} \longrightarrow\!\!\!\!| \quad \sigma\text{-set}$$

Using the proposed additional axioms we can show that some relationships between considered sets are not provable in **ZFC**. Namely

Metatheorem 9.3. *None of the implications indicated in Diagram 5 is provable in the theory* **ZFC**.

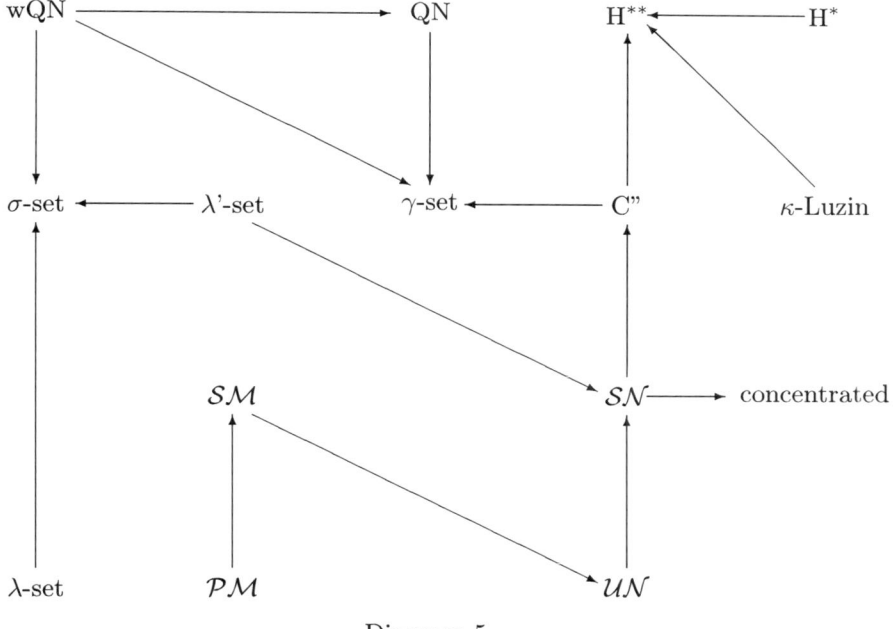

Diagram 5.

Demonstration. By Theorem 8.123, if $\mathfrak{t} = \mathfrak{b}$, then there exists a wQN-set of reals that is not a QN-space. Thus by Metatheorem 9.1 the theory **ZFC** + QN $\neq$ wQN is consistent.

By Corollary 8.27, if $\kappa = \mathrm{cov}(\mathcal{M}) = \mathrm{cof}(\mathcal{M})$, then there exists a κ-Luzin set. However, by Theorem 8.101 a κ-Luzin set does not have the property H^{**}. By Metatheorem 9.1 the theory $\mathbf{ZFC} + \mathrm{cov}(\mathcal{M}) = \mathrm{cof}(\mathcal{M})$ is consistent, consequently the theory $\mathbf{ZFC} + \neg(\kappa\text{-Luzin} \to \mathrm{H}^{**})$ is consistent. Since an $\aleph_1$-Luzin set has the property H^*, neither is $\mathrm{H}^* \to \mathrm{H}^{**}$ provable.[1] Moreover, an $\aleph_1$-Luzin set has the Rothberger property C'' and therefore the theory $\mathbf{ZFC}+\neg(\mathrm{C}'' \to \mathrm{H}^{**})$ is consistent.

By Metatheorem 9.1 the theory $\mathbf{ZFC} + \mathfrak{p} < \mathfrak{b}$ is consistent. By Theorem 8.91 there exists a set $A \subseteq \mathbb{C}$ of cardinality $\mathfrak{p}$ that is not a γ-space. However, since $|A| < \mathfrak{b}$, the set A is a QN-space. Thus by Metatheorem 9.1 the theory $\mathbf{ZFC}+\mathrm{QN} \neq \gamma$-space is consistent. The theory $\mathbf{ZFC}+\mathrm{wQN} \neq \gamma$-space is consistent as well. By Theorem 8.124 and (9.2) the theory $\mathbf{ZFC}+\mathrm{wQN} \neq \sigma$-space is consistent.

By Theorem 8.52 there exists an uncountable λ'-subset of any perfect Polish space and therefore, by (9.6) the theory $\mathbf{ZFC} + \lambda'$-set $\neq \sigma$-set is consistent and by (9.2) the theory $\mathbf{ZFC} + \lambda'$-set $\neq \mathcal{SN}$ is consistent as well.

By the Rothberger Theorem 8.111 and Theorem 8.91, if $\mathfrak{p} < \mathrm{cov}(\mathcal{M})$, then there exists a set of reals of cardinality $\mathfrak{p}$ that is not a γ-space and possesses property C''. Thus, by Metatheorem 9.1 the theory $\mathbf{ZFC} + \mathrm{C}'' \neq \gamma$-space is consistent.

By Corollary 8.49 there exists an uncountable λ-set. Therefore by (9.6) the theory $\mathbf{ZFC} + \lambda$-set $\neq \sigma$-set is consistent.

By the Luzin Theorem 8.3 there exists an uncountable perfectly meager universal measure zero subset of any perfect Polish space. Hence, by (9.5) the theory $\mathbf{ZFC} + \mathcal{PM} \neq \mathcal{SM}$ is consistent. Similarly, by (9.2) the theory $\mathbf{ZFC} +\mathcal{UN} \neq \mathcal{SN}$ is consistent.

By Corollary 8.107, assuming $\mathrm{add}(\mathcal{M}) = \mathfrak{c}$, there exists a strong measure zero set that is not concentrated on any countable set, thus neither possesses the Rothberger property C''.

If $\mathrm{non}(\mathcal{N}) < \mathrm{cov}(\mathcal{N})$, then there exists a set of reals $A \notin \mathcal{N}$ of cardinality $\mathrm{non}(\mathcal{N})$. Then neither is A of universal measure zero, however by Theorem 8.16, A is strongly meager. Thus by Metatheorem 9.1 the theory $\mathbf{ZFC} + \neg(\mathcal{SM} \to \mathcal{UN})$ is consistent.   □

We shall consider the problem of distinguishing Arkhangel'skiĭ's properties (α_1) and (α_2). More precisely, we know that $(\alpha_1) \to (\alpha_2)$ holds true. Can we prove or disprove the opposite implication?

According to (1.18) instead of dealing with sequences we can work with infinite countable subsets of an $\mathcal{L}^*$-space X. Thus, e.g., the property (α_1) reads: if $S_n \subseteq X$ are infinite countable sets, $\lim S_n = x$ for every n, then there exists a countable set S such that $\lim S = x$ and every $S_n \setminus S$ is finite.

We begin with a combinatorial principle related to splitting number. A family $\mathcal{A} \subseteq [\omega]^\omega$ is called an ω-**splitting family**, if for every sequence $\langle A_n : n \in \omega \rangle$ of infinite subsets of ω there exists a set in $\mathcal{A}$ that splits each $A_n, n \in \omega$.

---

[1]Note that this follows also from (8.11), (8.12) and consistency of $\mathfrak{b} < \mathfrak{d}$.

Now we can formulate the **Dow Principle** as follows:

> every ω-splitting family of subsets of ω contains
> a splitting subfamily of cardinality less than $\mathfrak{b}$.

Note that instead of subsets of ω we can deal with subsets of any infinite countable set. A. Dow (11.40) has shown that the Dow Principle is consistent with **ZFC**[2].

We shall need some auxiliary results.

Lemma 9.16. *If $\mathcal{A}$ is a countable family of infinite subsets of a countable set C, then there exists a set $B \in [C]^\omega$ splitting each $A \in \mathcal{A}$.*

The proof is easy and follows the proof of the Bernstein Theorem 7.20. Assume that $\langle A_n : n \in \omega \rangle$ is such an enumeration of $\mathcal{A}$ in which every element of $\mathcal{A}$ occurs infinitely many times. By induction for every $n \in \omega$ find two distinct elements $x_n, y_n \in A_n$ different from x_i, y_i for $i < n$. Set $A = \{x_n : n \in \omega\}$. □

Lemma 9.17. *Let X be an $\mathcal{L}^*$-space with the property (α_2). Assume that $\lim S_n = x$ for every n, where $\langle S_n : n \in \omega \rangle$ are infinite countable and pairwise disjoint subsets of X, $S = \bigcup_n S_n$. Then the family*

$$\mathcal{A} = \{A \in [S]^\omega : (\exists F \in [S]^\omega)(\lim F = x \wedge (\forall n)((A \cap S_n) \setminus F \text{ is finite}))\},$$

is an ω-splitting.

Proof. Note that $A \in \mathcal{A}$ and $B \in [A]^\omega$ imply $B \in \mathcal{A}$.

Assume that $\langle A_n \subseteq S : n \in \omega \rangle$ are infinite. Set

$$J = \{j \in \omega : (\forall n)(A_j \cap S_n \text{ is finite})\}.$$

If J is finite, set $H = \bigcup_{j \in J} A_j$.

If J is infinite, then find an increasing sequence $\langle J_n : n \in \omega \rangle$ of finite sets such that $J = \bigcup_n J_n$. Set

$$H = \bigcup_n \bigcup_{j \in J_n} A_j \cap S_n.$$

If $j \in J$, then there exists an n_0 such that $j \in J_n$ for every $n > n_0$. Therefore the set $A_j \setminus H$ being a subset of $\bigcup_{n \in J_{n_0}} A_j \cap S_n$ is finite.

In both cases we have $|A_j \setminus H| < \aleph_0$ for each $j \in J$ and $|H \cap S_n| < \aleph_0$ for each n. Now, for each $j \in \omega \setminus J$ choose n_j such that $A_j \cap S_{n_j}$ is infinite.

If $\omega \setminus J$ is finite, then the sequence $F = \bigcup_{j \notin J} A_j \cap S_{n_j}$ converges to x and it is easy to see that $F \cup H \in \mathcal{A}$. Moreover, $(F \cup H) \cap A_j$ is infinite for each j.

Assume now that $\omega \setminus J$ is infinite. Since X has the property (α_2), there exists an infinite set $F \subseteq \bigcup_{j \in J}(A_j \cap S_{n_j})$ converging to x and such that $F \cap (A_j \cap S_{n_j})$ is infinite for each $j \in J$. Again $F \cup H \in \mathcal{A}$ and $(F \cup H) \cap A_j$ is infinite for each j.

By Lemma 9.16 there exists a set $B \subseteq F \cup H$ splitting each $\langle A_j : j \in \omega \rangle$. □

---

[2] Actually A. Dow shows that there exists an ω-splitting subfamily of cardinality less than $\mathfrak{b}$.

Now we can show the key result.

Theorem 9.18 (A. Dow). *The Dow Principle implies that every $\mathcal{L}^*$-space possessing the property (α_2) possesses the property (α_1) as well.*

Proof. Let X be an $\mathcal{L}^*$-space possessing the property (α_2). Assume that $S_n \subseteq X$ are countable infinite, $\lim S_n = x$ for every $n \in \omega$. It is easy to see that we can assume that $\langle S_n : n \in \omega \rangle$ are pairwise disjoint. If the family $\mathcal{A}$ is defined as in the lemma, by the Dow Principle there exists an ω-splitting family $\mathcal{A}_0 \subseteq \mathcal{A}$, $|\mathcal{A}_0| < \mathfrak{b}$.

By definition of $\mathcal{A}$, for every $A \in \mathcal{A}_0$ there exists an infinite set $F_A \subseteq S$ converging to x and such that $f_A(n) = (A \cap S_n) \setminus F_A$ is finite for each n. Since $|\mathcal{A}_0| < \mathfrak{b}$, there exists a function $f \in {}^\omega([S]^{<\omega})$ such that $f_A(n) \subseteq f(n)$ for all but finitely many n for each $A \in \mathcal{A}_0$. We show that $F = \bigcup_n (S_n \setminus f(n))$ converges to x. Assume not. Then by the condition (L3) there exists an infinite set $E \subseteq F$ such that no infinite subset of E converges to x. Since $\mathcal{A}_0$ is splitting, there exists an $A \in \mathcal{A}_0$ such that $A \cap E$ is infinite. Since $(A \cap E) \setminus F_A$ is finite, the infinite set $A \cap E \cap F_A$ converges to x, which is a contradiction. $\square$

The equivalence "a topological space X is a QN-space if and only if X is a wQN-space" will be abbreviated as QN = wQN. Similarly for other equivalences.

Metatheorem 9.4. *The equivalences* QN = wQN *and* QN = $S_1(\Gamma, \Gamma)$ *are undecidable in* **ZFC**.

Demonstration. By Theorems 8.64, 8.66, 9.18, and Dow's result (11.40), the theory **ZFC** + QN = $S_1(\Gamma, \Gamma)$ = wQN is consistent.

On the other hand, by Theorem 8.123 there exists an $S_1(\Gamma, \Gamma)$-space that is not a QN-space provided that $\mathfrak{t} = \mathfrak{b}$. By Metatheorem 9.1 the theory **ZFC** + $\mathfrak{t} = \mathfrak{b}$ is consistent too.

Note also the following: by Recław's Theorem 8.124 there exists an uncountable $S_1(\Gamma, \Gamma)$-space. On the other hand, by Miller's result (11.33) the theory **ZFC** + "every σ-set is countable" is consistent. Since every perfectly normal QN-space is a σ-space (Theorem 8.77), the result follows again. $\square$

M. Scheepers [1999] claimed the **Scheepers Conjecture** $S_1(\Gamma, \Gamma)$ = wQN. By Metatheorem 9.4 the Scheepers Conjecture is consistent with **ZFC**. We do not know, whether the negation of the Scheepers Conjecture is consistent with **ZFC**.

Metatheorem 9.5. *The statement* "any $\mathcal{L}^*$-space possessing the property (α_2) possesses the property (α_1) as well" *is undecidable in* **ZFC**.

Exercises

9.7 [AC] Katowice Problem
The **Katowice Problem** is the question

"Are the Boolean algebras $\mathcal{P}(\omega)/\mathrm{Fin}$ and $\mathcal{P}(\omega_1)/[\omega_1]^{<\omega}$ isomorphic?".

a) The affirmative answer to the Katowice Problem implies $2^{\aleph_0} = 2^{\aleph_1}$.

b) The negative answer to the Katowice Problem is consistent with **ZFC**.

c) Prove the **Balcar Theorem**: The affirmative answer to the Katowice Problem implies
$\mathfrak{d} = \aleph_1 < \mathfrak{c}$.

*Hint: Let $h : \mathcal{P}(\omega_1)/[\omega_1]^{<\omega} \xrightarrow[\text{onto}]{1-1} \mathcal{P}(\omega \times \omega)/\text{Fin}$ be an isomorphism. Fix $\langle A_n : n \in \omega \rangle$,
$\langle B_\xi : \xi \in \omega_1 \rangle$ such that $A_n \cap A_m = \emptyset$ for $n \neq m$, $\bigcup_n A_n = \omega_1$, $\bigcup_{\xi < \omega_1} B_\xi = \omega \times \omega$,
$h([A_n]) = [\{n\} \times \omega]$, $[B_\xi] = h([\omega_1 \setminus \xi])$. We can assume that $|A_n| = \aleph_1$ for each n.
Set*

$$\alpha_\xi(n) = \min\{i \in \omega : \langle n, i \rangle \in B_\xi\}.$$

*Since $|A_n \cap (\omega_1 \setminus \xi)| \geq \aleph_0$, we have $B_\xi \cap (\{n\} \times \omega) \neq \emptyset$ for every n and every ξ. Thus
the functions α_ξ are well defined. Evidently $B_\xi \subseteq^* B_\eta$ for $\eta < \xi$. Hence $\alpha_\eta \leq^* \alpha_\xi$.
Let $\beta \in {}^\omega\omega$. Consider $A \subseteq \omega_1$ such that $h([A]) = [\{\langle n, i \rangle : i \leq \beta(n)\}]$. Show that
every $A \cap A_n$ is finite and therefore A is countable. Let ξ be such that $A \subseteq \xi$. Then
$\beta \leq^* \alpha_\xi$.*

9.8 [AC] Ultrafilters on ω

By an ultrafilter we shall understand a free ultrafilter on ω. We introduce necessary
notions. An ultrafilter $\mathcal{F}$ is called **selective**, if for any partition $\{A_n : n \in \omega\} \subseteq \mathcal{P}(\omega) \setminus \mathcal{F}$
of ω, there exists an $A \in \mathcal{F}$ such that $|A \cap A_n| \leq 1$ for every n. An ultrafilter $\mathcal{F}$ is called
a **Q-point**, if for any partition $\{A_n : n \in \omega\}$ of ω into finite sets, there exists an $A \in \mathcal{F}$
such that $|A \cap A_n| \leq 1$ for every n. Finally, an ultrafilter $\mathcal{F}$ is called a **P-point**, if for any
partition $\{A_n : n \in \omega\} \subseteq \mathcal{P}(\omega) \setminus \mathcal{F}$ of ω, there exists an $A \in \mathcal{F}$ such that $|A \cap A_n| < \aleph_0$
for every n. A rapid filter is defined in Section 9.3. $\beta\omega$ is defined in Exercise 5.16.

a) An ultrafilter is selective if and only if it is simultaneously a Q-point and a P-point.

b) An ultrafilter $\mathcal{F}$ is a P-point if and only if for every sequence $\langle A_n \in \mathcal{F} : n \in \omega \rangle$
there exists a set $A \in \mathcal{F}$ such that $A \subseteq^* A_n$ for every n.

c) An ultrafilter $\mathcal{F}$ is a P-point if and only if every G_δ-subset of $\beta\omega \setminus \omega$ containing $\mathcal{F}$
is a neighborhood of $\mathcal{F}$.

d) Every Q-point is a rapid ultrafilter.
*Hint: Let $\alpha \in {}^\omega\omega$, $\alpha(0) = 0$ be increasing. Set $A_n = \{k : \alpha(n) \leq k < \alpha(n+1)\}$. If
A is a selector for this partition, then $\alpha \leq^* e_A$.*

Fix a partition $\langle R_n : n \in \omega \rangle$ of ω such that $|R_n| = n$ for every n. A set $A \subseteq \omega$ is **growing**,
if for every n there exists a k such that $|A \cap R_k| \geq n$. Hence, A is growing if and only if
the set $\{|A \cap R_n| : n \in \omega\}$ is infinite.

e) $\mathcal{G} = \{A \subseteq \omega : (\exists n)(\forall k)\,|R_k \setminus A| < n\}$ is a filter. $A \in \mathcal{G}$ if and only if $\omega \setminus A$ is not
growing.

f) No ultrafilter $\mathcal{F} \supseteq \mathcal{G}$ is a Q-point. Hence, there exists an ultrafilter that is not
a Q-point.

g) If A is growing, $\alpha \in {}^\omega\omega$, then there exists a growing set $B \subseteq A$ such that $e_B >^* \alpha$.

Fix a partition $\langle L_n : n \in \omega \rangle \subseteq [\omega]^\omega$. A set $A \subseteq \omega$ is **large**, if $|A \cap L_n| = \aleph_0$ for infinitely
many n. A family $\mathcal{A} \subseteq \mathcal{P}(\omega)$ is **large** if every element of $\mathcal{A}$ is large.

h) If $\mathcal{A} \subseteq \mathcal{P}(\omega)$ is a large family closed under finite intersections, and $B \subseteq \omega$, then
at least one of the families $\mathcal{A} \cup \{B\}$, $\mathcal{A} \cup \{\omega \setminus B\}$ is large and closed under finite
intersections.

Hint: If $A \cap B$ is large for every $A \in \mathcal{A}$, then $\mathcal{A} \cup \{B\}$ works. If $A \cap B$ is not large for some $A \in \mathcal{A}$, then using the inclusion

$$A \cap C \cap L_n \subseteq (A \cap B \cap L_n) \cup ((C \setminus B) \cap L_n)$$

one can easily show that $C \setminus B$ is large for any $C \in \mathcal{A}$.

i) If $\mathcal{A} \subseteq \mathcal{P}(\omega)$ is a large family closed under finite intersections, then there exists a large ultrafilter $\mathcal{F}$ such that $\mathcal{A} \subseteq \mathcal{F}$.

   Hint: By transfinite induction using the result of h).

j) There exists an ultrafilter that is not a P-point.

   Hint: No large ultrafilter is a P-point.

An infinite set $A \subseteq \omega$ is **thin**, if $\lim_{n \to \infty} e_A(n)/e_A(n+1) = 0$. An ultrafilter $\mathcal{F}$ is **thin**, if for every unbounded $\alpha \in {}^{\omega}\omega$ there exists an $A \in \mathcal{F}$ such that $\alpha(A)$ is thin.

k) Every selective ultrafilter is thin.

   Hint: Assume that $\alpha \in {}^{\omega}\omega$ is unbounded, $A_n = \{k : n! - 1 \leq k < (n+1)! - 1\}$. Then $\langle B_n = \alpha^{-1}(A_n) : n \in \omega \rangle$ is a partition of ω. If there exists an m such that $B_m \in \mathcal{F}$, then, since $\alpha(B_m)$ is finite, you can easily find a set $A \supseteq B_m$ such that $\alpha(A)$ is thin. If no B_n belongs to $\mathcal{F}$, then take a selector $A \in \mathcal{F}$ for this partition. We can assume that $C = A \cap \bigcup_n B_{2n} \in \mathcal{F}$. Then $\alpha(C)$ is thin.

l) Every thin ultrafilter is a Q-point.

   Hint: Let $\langle A_n : n \in \omega \rangle$ be a partition of ω into finite sets. Let $A_n = \{k_n^i : i < m_n\}$. Find an increasing sequence $\{p_n\}_{n=0}^{\infty}$ such that $m_n \leq p_{n+1} - p_n$ and $m_n \leq p_n$ for each n. Define an injection $\alpha : \omega \longrightarrow \omega$ as $\alpha(k_n^i) = p_n + i$ for $i < m_n, n \in \omega$. Let $A \in \mathcal{F}$ be such that $\alpha(A)$ is thin. By the definition of a thin set, there exists an i_0 such that

$$e_{\alpha(A)}(i)/e_{\alpha(A)}(i+1) < 1/2$$

   for every $i \geq i_0$. Set $B = A \setminus \{k : \alpha(k) < e_{\alpha(A)}(i_0)\}$. If $n < m \in A_l \cap B$ for some l, $\alpha(n) = e_{\alpha(A)}(i)$, $\alpha(m) = e_{\alpha(A)}(j)$, then $i < j$ and $\alpha(n)/\alpha(m) \geq 1/2$.

m) We can summarize our results in a picture.

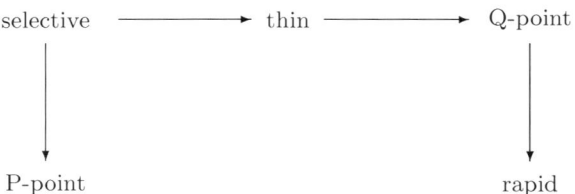

9.9 [AC] Product of Ultrafilters

Let $\pi : \omega \times \omega \xrightarrow[\text{onto}]{1-1} \omega$ be a pairing function. We define the **product of ultrafilters** $\mathcal{F}, \mathcal{G}$ on ω as

$$A \in \mathcal{F} \times \mathcal{G} \equiv \{n \in \omega : \{m \in \omega : \pi(n, m) \in A\} \in \mathcal{G}\} \in \mathcal{F}.$$

a) Show that $\mathcal{F} \times \mathcal{G}$ is an ultrafilter on ω.

b) No product of (free!) ultrafilters is a Q-point.
   Hint: Consider the decomposition $\langle A_n : n \in \omega \rangle$, where

$$A_n = \{\pi(i,j) : i, j \leq n \wedge \max\{i,j\} = n\}.$$

c) If $\mathcal{F}$ is rapid, then $\mathcal{F} \times \mathcal{G}$ is rapid.
   Hint: Assume that π is defined by (1.8), λ, ρ are left and right inverse, respectively. For given increasing $\alpha \in {}^{\omega}\omega$ take an $A \in \mathcal{F}$ such that

$$e_A(k) \geq 2 \max\{\alpha(n) : \lambda(n) + \rho'(n) = k\}$$

   and a set $B \in \mathcal{G}$ such that $0 \in B$. Then $C = \{\pi(n,m) : n \in A \wedge m \in B\} \in \mathcal{F} \times \mathcal{G}$ and for $k = \pi(n,m)$, $n \in A$, $m \in B$ we obtain

$$e_C(k) \geq e_C(\pi(n+m,0)) \geq 1/2 e_A(n+m) \geq \alpha(k).$$

9.10 [AC] Existence of Special Ultrafilters

We shall use the notation and terminology introduced in Exercise 9.8. Especially, we fix a partition $\langle R_n : n \in \omega \rangle$ of ω such that $|R_n| = n$ for every n and a partition $\langle L_n : n \in \omega \rangle \subseteq [\omega]^{\omega}$.

a) If $\mathfrak{t} = \mathfrak{c}$, then there exists a selective ultrafilter.
   Hint: Let $\mathcal{P}(\omega) = \{A_\xi : \xi < \mathfrak{c}\}$. Enumerate the set of all countable partitions of ω as $\{\{B_\eta^n : n \in \omega\} : \eta < \mathfrak{c} \wedge \eta \text{ limit}\}$. Take infinite $C_{\xi+1} \subseteq C_\xi$ such that $C_{\xi+1} \subseteq A_\xi$ or $C_{\xi+1} \cap A_\xi = \emptyset$. For η limit find an infinite set such that $D \subseteq^* C_\xi$ for all $\xi < \eta$. If

$$D \subseteq^* \bigcup_{k<n} B_\eta^k$$

   for some n, set $C_\eta = D$. Otherwise let $C_\eta = \{\min(D \cap B_\eta^n) : D \cap B_\eta^n \neq \emptyset\}$. Define $\mathcal{F} = \{A \subseteq \omega : (\exists \xi < \mathfrak{c}) \, C_\xi \subseteq A\}$.

b) If $\mathrm{cov}(\mathcal{M}) = \mathfrak{d}$, then there exists a rapid ultrafilter.
   Hint: Let $\{\alpha_\xi : \xi < \mathfrak{d}\}$ be a dominating family consisting of increasing functions. If $\langle A_\eta : \eta < \xi \rangle$ are constructed, we set

$$\beta_\eta(n) = \min\{m \in A_\eta : \alpha_\xi(n) \leq m < \alpha_\xi(n+1)\}.$$

   The set

$$G_\eta = \{\alpha \in {}^{\omega}\omega : (\forall m)(\exists n \geq m) \, (n \in \mathrm{dom}(\beta_\eta) \wedge \alpha(n) = \beta_\eta(n))\}$$

   is G_δ dense. There exists an $\alpha \in \bigcap_{\eta < \xi} G_\eta$. Take $A_\xi = \mathrm{rng}(\alpha)$. Take any ultrafilter containing all $A_\xi, \xi < \mathfrak{d}$.

c) If $\mathcal{F}$ is a filter consisting of growing sets with a base $\mathcal{F}_0$ of cardinality less than $\mathfrak{p}$, then there exists a growing set A such that $A \setminus B$ is finite for every $B \in \mathcal{F}$.
   Hint: The poset $\mathbb{P}$ consisting of all ordered triples $\langle p, F, f \rangle$ such that $p \in {}^{n}2$ for some n, F is a finite subset of $\mathcal{F}$, $f \in {}^{F}\omega$ and $p(i) = 0$ for every i such that $E \in F$, $i \notin E$ and $f(E) \leq i < n$, ordered as

$$\langle p, F, f \rangle \leq \langle p', F', f' \rangle \equiv (p' \subseteq p \wedge F' \subseteq F \wedge f' \subseteq f),$$

is σ-centered. For $C \in \mathcal{F}_0$ let $\mathcal{D}_C = \{\langle p, F, f \rangle \in \mathbb{P} : C \in F\}$. Moreover, set

$$\mathcal{E}_n = \{\langle p, F, f \rangle \in \mathbb{P} : (\exists k)\,|\{i \in R_k : p(i) = 1\}| \geq n\}.$$

By Bell's Theorem, see Exercise 9.4, there exists a $\{\mathcal{D}_C : C \in \mathcal{F}_0\} \cup \{\mathcal{E}_n : n \in \omega\}$-generic filter $\mathcal{G}$ of $\mathbb{P}$. One can easily check that

$$A = \{n \in \omega : (\exists \langle p, F, f \rangle \in \mathcal{G})\, p(n) = 1\}$$

is the desired growing set.

d) If $\mathfrak{p} = \mathfrak{c}$, then there exists a rapid ultrafilter that is not a Q-point.
   Hint: Use c) and Exercise 9.8, g). Note that an ultrafilter consisting of growing sets is not a Q-point.

e) If there exists a rapid ultrafilter, then there exists a rapid ultrafilter that is not a Q-point.
   Hint: See Exercise 9.9.

f) Assume that $\mathcal{F} \subseteq \mathcal{P}(\omega)$ is such that $|\mathcal{F}| < \mathfrak{p}$, and for finitely many $B_0, \ldots, B_n \in \mathcal{F}$ the set $\{|R_i \cap B_0 \cap \cdots \cap B_n| : i \in \omega\}$ is infinite. Then there exists a set E such that $E \subseteq^* B$ for every $B \in \mathcal{F}$ and the set $\{|B \cap R_i| : i \in \omega\}$ is infinite.
   Hint: Consider the σ-centered set

$$\mathbb{P} = \{\langle s, F \rangle : s \in [\omega]^{<\omega} \wedge F \in [\mathcal{F}]^{<\omega}\}$$

   ordered as

$$\langle s, F \rangle \leq \langle s', F' \rangle \equiv (s' \subseteq s \wedge F' \subseteq F \wedge s \setminus s' \subseteq \bigcap F').$$

   Find suitable dense sets $\mathcal{D}$ such that the corresponding $\mathcal{D}$-generic filter $\mathcal{G}$ produces a set $\bigcup\{s : \langle s, F \rangle \in \mathcal{G}\} \subseteq^* B$ for any $B \in \mathcal{F}$ and for every n there exists a couple $\langle s, F \rangle \in \mathcal{G}$ such that $|R_i \cap s| \geq n$ for some i.

g) If $\mathfrak{p} = \mathfrak{c}$, then there exists a P-point that is not a Q-point.
   Hint: Similarly as in a) construct by induction an ultrafilter consisting of growing sets: at the limit step apply the result of f).

h) Let $\mathcal{F}$ be the base of a filter of cardinality less than $\mathrm{cov}(\mathcal{M})$, $\alpha \in {}^\omega\omega$ being unbounded. Moreover, assume that $\omega \setminus L_n \in \mathcal{F}$ for every n, the set $\{n : |F \cap L_n| = \aleph_0\}$ is infinite for every $F \in \mathcal{F}$ and for every $K \in [\omega]^{<\omega}$ there exists an $F \in \mathcal{F}$ such that $\alpha^{-1}(K) \cap F \cap L_n$ is infinite for only finitely many n. Then there exists a set A such that $\{n : |A \cap F \cap L_n| = \aleph_0\}$ is infinite for every $F \in \mathcal{F}$ and $\alpha(A)$ is thin or finite.
   Hint: If $\alpha(F)$ is thin or finite for some $F \in \mathcal{F}$, we take $A = F$. Thus we can assume that no $\alpha(F)$ is thin or finite. We must distinguish two cases.
   CASE I. The set $I_F = \{n : \alpha(F) \cap L_n \text{ is infinite}\}$ is infinite for every $F \in \mathcal{F}$. Consider the countable poset

$$\mathbb{P} = \{K \in [\omega]^{<\omega} : (\forall n, m)\,([n, m] \cap \alpha(K) = \{n, m\} \to m > n^2)\}$$

   ordered as

$$K \leq L \equiv (L \subseteq K \wedge \min\{K \setminus L\} > \max L.$$

For any $n \in \omega$, $F \in \mathcal{F}$ and $k \in I_F$ the set $D_{F,n,k} = \{K \in \mathbb{P} : |K \cap F \cap L_n| \geq k\}$ is dense. If $\mathcal{G}$ is the corresponding generic filter, then $A = \bigcup\{K : K \in \mathcal{G}\}$ is the desired set.

CASE II. There exists an $F_0 \in \mathcal{F}$ such that $\{n : \alpha(F) \cap L_n \text{ is infinite}\}$ is finite. Set

$$J_F = \{n : F \cap F_0 \cap L_n \text{ is infinite} \wedge \alpha(F \cap F_0 \cap L_n) \text{ is finite}\}.$$

J_F is infinite and for $n \in J_F$ we can define

$$h(n) = \max\{m \in \alpha(F \cap F_0 \cap L_n) : \alpha^{-1}(\{m\}) \cap F \cap F_0 \cap L_n \text{ is finite}\}.$$

Show that $\{h(n) : n \in J_F\}$ is infinite and choose a sequence

$$H_F = \langle k_i : i \in \omega \rangle \subseteq \{h(n) : n \in J_F\}$$

such that $k_{i+1} > (k_i)^2$. Consider the countable poset

$$\mathbb{P} = \{K \in [\omega]^{<\omega} : (\forall n, m)\, ([n, m] \cap K = \{r, m\} \to m > n^2)\}$$

ordered as above. The set $D_{F,k} = \{K \in \mathbb{P} : |K \cap H_F| \geq k\}$ is dense for every $F \in \mathcal{F}$ and every $k \in \omega$. If $\mathcal{H}$ is the corresponding generic filter and $H = \bigcup\{K : K \in \mathcal{H}\}$, then $A = \alpha^{-1}(H)$ is the desired thin set.

i) If $\mathrm{cov}(\mathcal{M}) = \mathfrak{c}$, then there exists a thin ultrafilter which is not a P-point.

Hint: Enumerate all unbounded elements of $^\omega\omega$ as $\langle \alpha_\xi : \xi < \mathfrak{c} \rangle$. By transfinite induction construct an increasing sequence $\langle \mathcal{F}_\xi : \xi < \mathfrak{c} \rangle$ of filter bases satisfying $|\mathcal{F}_\xi| \leq |\xi| \cdot \aleph_0$ and such that $\{n : |L_n \cap A| = \aleph_0\}$ is infinite for every $\xi < \mathfrak{c}$ and every $A \in \mathcal{F}_\xi$. Moreover, for every $\xi < \mathfrak{c}$, the filter base $\mathcal{F}_{\xi+1}$ contains a set A such that $\alpha_\xi(A)$ is thin. At the non-limit step use the result of 1). Any ultrafilter extending $\bigcup_{\xi < \mathfrak{c}} \mathcal{F}_\xi$ is thin. Since the family $\bigcup_{\xi < \mathfrak{c}} \mathcal{F}_\xi$ is large, by Exercise 9.8, i) there exists a large ultrafilter extending this family, which by Exercise 9.8, j) is not a P-point.

j) If **CH** holds true, then there exists a Q-point that is not a thin ultrafilter.

Hint: Enumerate all partitions of ω into finite sets as $\langle \mathcal{Q}_\xi : \xi < \omega_1 \rangle$. By transfinite induction construct an increasing sequence $\langle \mathcal{F}_\xi : \xi < \omega_1 \rangle$ of countable filter bases such that $|L_n \cap F| = \aleph_0$ for every n, for every $\xi < \omega_1$ and for every $F \in \mathcal{F}_\xi$. Moreover, for every $\xi < \omega_1$, the filter base $\mathcal{F}_{\xi+1} \supseteq \mathcal{F}_\xi$ contains a set A such that $|A \cap B| \leq 1$ for every $B \in \mathcal{Q}_\xi$, and $\mathcal{F}_\xi = \bigcup_{\eta < \xi} \mathcal{F}_\eta$ for ξ limit. $\mathcal{F}_0$ is the Fréchet filter.

Let us assume that $\mathcal{F}_\xi$ is already constructed. Let $\langle M_n : n \in \omega \rangle$ be an enumeration of $\langle L_n : n \in \omega \rangle$ such that each L_n is listed infinitely often, $\langle F_n : n \in \omega \rangle$ being an enumeration of $\mathcal{F}_\xi$, and $\mathcal{Q}_\xi = \langle Q_n : n \in \omega \rangle$.

If for every $F \in \mathcal{F}_\xi$ there exists a set $B \in \mathcal{Q}_\xi$ such that $|F \cap B| > 1$, construct by induction sequences $\{k_i\}_{i=0}^\infty$, $\{n_i\}_{i=0}^\infty$ such that $k_i \in (\bigcap_{j<i} F_j \cap M_i) \setminus \bigcup_{j<i} Q_{n_j}$ and $k_i \in Q_{n_i}$. Set $A = \{k_i : i \in \omega\}$. Then A is compatible with $\mathcal{F}_\xi$, so take for $\mathcal{F}_{\xi+1}$ the base generated by $\mathcal{F}_\xi$ and A. Otherwise set $\mathcal{F}_{\xi+1} = \mathcal{F}_\xi$.

Set $\mathcal{G} = \{\bigcup_{n \in G} L_n : \omega \setminus G \in \bigcup_{\xi < \omega_1} \mathcal{F}_\xi\}$. Any ultrafilter containing $\mathcal{G} \cup \bigcup_{\xi < \omega_1} \mathcal{F}_\xi$ is a Q-point but not a thin ultrafilter.

k) No arrow in the next picture can be proved in **ZFC**.

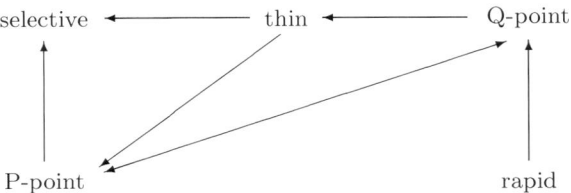

9.3   Assuming Regularity of Sets of Reals

We have proved that many well-defined sets of reals (Borel, analytic) have regularity properties: they are Lebesgue measurable, possess the Baire Property, if
uncountable, they contain a perfect subset. To construct a set that is irregular
in any of the mentioned senses, we needed an additional assumption, e.g., the
existence of a well-ordering of the real line, compare Section 7.2. That is not an
accident, since (see (11.42))

Metatheorem 9.6 (R. Solovay). *If the theory* **ZFC+IC** *is consistent, then the theory*
ZF + *"every set of reals is Lebesgue measurable"*+*" every set of reals possesses*
the Baire Property"+*"every uncountable set of reals contains a perfect subset" is*
consistent as well.

Thus we can assume that all sets of reals are regular. We shall study consequences of such an assumption of regularity.

We begin with a technical result. Let $p : \mathcal{P}(\omega) \xrightarrow{\text{onto}} \mathcal{P}(\omega)/\text{Fin}$ be the quotient
mapping. Thus

$$p(x) = \{y \in \mathcal{P}(\omega) : (x \setminus y) \cup (y \setminus x) \in \text{Fin}\}.$$

For a set $x \subseteq \omega$ we set $m(x) = \{p(x), p(\omega \setminus x)\}$. Let

$$\mathcal{Q} = \{m(x) : x \subseteq \omega\} \subseteq [\mathcal{P}(\omega)/\text{Fin}]^2.$$

Lemma 9.19. *If there exists a selector for the family* $\mathcal{Q}$*, then there exists a subset*
of $^\omega 2$ *that is non-measurable and does not possess the Baire Property.*

Proof. Let $\mathcal{A} \subseteq \mathcal{P}(\omega)/\text{Fin}$ be a selector for $\mathcal{Q}$. One can easily see that

$$\mathcal{B}_0 = \{x \subseteq \omega : p(x) \in \mathcal{A}\}, \quad \mathcal{B}_1 = \{x \subseteq \omega : p(x) \notin \mathcal{A}\}$$

are disjoint tail-sets. Since $\mathcal{B}_0 \cup \mathcal{B}_1 = \mathcal{P}(\omega)$ and $\mathcal{B}_0 = \{x \subseteq \omega : \omega \setminus x \in \mathcal{B}_1\}$, by
Theorems 7.36 and 7.38 we obtain that the sets $\mathcal{B}_0, \mathcal{B}_1$ are neither measurable nor
possess the Baire Property.   ⌐

We can summarize the obtained results as

Theorem 9.20. *If "every set of reals is Lebesgue measurable" or "every set of reals possesses the Baire Property", then:*

a) *there exists no selector for any Vitali decomposition;*
b) *the set of reals $\mathbb{R}$ cannot be well ordered;*
c) *the axiom of choice $\mathbf{AC}$ fails;*
d) *there exists no free ultrafilter on ω;*
e) *a set of cardinality $\mathfrak{k}$ cannot be linearly ordered;*
f) *the axiom of choice for two-elements sets $\mathbf{AC}_2$ introduced in Exercise 1.11 fails;*
g) $\mathfrak{c} < \mathfrak{k} \ll \mathfrak{c}$.

Proof. The assertions a)–d) follow by Theorem 7.17, and Corollaries 7.19 and 7.39.

If the set $\mathcal{P}(\omega)/\mathrm{Fin}$ were linearly ordered, then one could define a selector for the family $\mathcal{Q}$, a contradiction with Lemma 9.19. Thus we have e).

$\mathbf{AC}_2$ implies that there exists a selector for $\mathcal{Q}$, a contradiction. Thus we have f).

By definition we have $\mathfrak{k} \ll \mathfrak{c}$. By Theorem 5.35 there exists an injection of $^{\omega}\omega$ into $\mathcal{P}(\omega)/\mathrm{Fin}$. Thus $\mathfrak{c} \leq \mathfrak{k}$. If $\mathfrak{c} = \mathfrak{k}$, then the set $\mathcal{P}(\omega)/\mathrm{Fin}$ could be linearly ordered. □

A filter $\mathcal{F} \subseteq (\mathcal{P}(\omega) \setminus \mathrm{Fin})$ is called **rapid** if for any function $\alpha \in {}^{\omega}\omega$ there exists a set $F \in \mathcal{F}$ such that for every $k \in \omega$,

$$|\{n \in F : n < \alpha(k)\}| \leq k.$$

If $\tilde{\mathcal{F}} = \{e_A : A \in \mathcal{F}\}$ and $\mathrm{Fin} \cap \mathcal{F} = \emptyset$, then $\mathcal{F}$ is rapid if and only if $\tilde{\mathcal{F}}$ is a dominating family.

If $\mathrm{cov}(\mathcal{M}) = \mathfrak{d}$, then there exists a rapid filter (see Exercise 9.10).

As usual we shall identify subsets of ω with elements of $^{\omega}2$.

Theorem 9.21 (M. Talagrand). *A rapid filter is a non-measurable subset of $^{\omega}2$.*

Actually a rapid filter does not have the Baire Property. However, we shall not use this fact and therefore we do not include a proof.

Lemma 9.22. *Let $\varepsilon > 0$ be given. If $B \subseteq {}^{\omega}2$ is a closed set of positive measure, then there exists arbitrarily large k such that*

$$|\{s \in {}^{k}2 : [s] \cap B \neq \emptyset\}| \cdot 2^{-k} < \lambda(B) + \varepsilon. \tag{9.7}$$

Proof. By definition, there exists a set $T \subseteq {}^{<\omega}2$ such that $B \subseteq \bigcup_{s \in T}[s]$ and $\sum_{s \in T} \lambda([s]) < \lambda(B) + \varepsilon$. Since B is compact we can assume that T is finite. Let k be the largest integer such that $s \in {}^{k}2$ for some $s \in T$. Passing to a refinement we can assume that the length of every finite sequence $s \in T$ is the same, say k. Thus, we have $\lambda([s]) = 2^{-k}$ for every $s \in T$. Since the sets $\langle [t] : t \in {}^{k}2 \rangle$ are pairwise disjoint we obtain (9.7). □

Lemma 9.23. *If $A \subseteq {}^{\omega}2$ has positive measure, then there exists a closed set $B \subseteq A$ and an increasing sequence $\{n_k\}_{k=0}^{\infty}$ such that*

$$(\forall k)(\forall s \in {}^{n_k}2)\,([s] \cap B \neq \emptyset \to \lambda([s] \cap B) > (1 - 2^{-k})\lambda([s])). \tag{9.8}$$

Proof. We construct by induction a decreasing sequence of closed sets $\{B_k\}_{k=0}^{\infty}$ and an increasing sequence $\{n_k\}_{k=0}^{\infty}$ of natural numbers such that, for every k,

$$|\{s \in {}^{n_{k+1}}2 : [s] \cap B_k \neq \emptyset\}| \cdot 2^{-n_{k+1}} < \lambda(B_k) + \lambda(B_0) \cdot 2^{-n_k - 2k - 4}. \tag{9.9}$$

Let $B_0 \subseteq A$ be a compact set of positive measure, $n_0 = 0$. If B_k and n_k are defined, then by Lemma 9.22 there exists a natural number $n_{k+1} > n_k$ such that

$$N_k = |\{s \in {}^{n_{k+1}}2 : [s] \cap B_k \neq \emptyset\}| < (\lambda(B_k) + \lambda(B_0)2^{-n_k - 2k - 4})2^{n_{k+1}}.$$

Set

$$S_k = \{s \in {}^{n_{k+1}}2 : \lambda([s] \cap B_k) \geq (1 - 2^{-k-2})2^{-n_{k+1}}\},$$
$$T_k = \{s \in {}^{n_{k+1}}2 : [s] \cap B_k \neq \emptyset \wedge \lambda([s] \cap B_k) < (1 - 2^{-k-2})2^{-n_{k+1}}\}.$$

Thus $N_k = |S_k| + |T_k|$. We set $B_{k+1} = \bigcup_{s \in S_k}([s] \cap B_k)$. It is easy to see that $2^{-n_{k+1}}|S_k| \geq \lambda(B_{k+1})$ and

$$\lambda(B_k) - \lambda(B_{k+1}) = \lambda(B_k \setminus B_{k+1}) < (1 - 2^{-k-2})|T_k|2^{-n_{k+1}}.$$

By simple computation one obtains that

$$\lambda(B_k) - \lambda(B_{k+1}) < (1 - 2^{-k-2})(N_k - |S_k|)2^{-n_{k+1}}$$
$$\leq (1 - 2^{-k-2})(\lambda(B_k) - \lambda(B_{k+1})) + \lambda(B_0)2^{-n_k - 2k - 4}$$

and therefore
$$\lambda(B_k) - \lambda(B_{k+1}) < \lambda(B_0)2^{-n_k - k - 2}.$$

Set $B = \bigcap_k B_k$. Then

$$\lambda(B) > \lambda(B_k) - \lambda(B_0)\sum_{i=k}^{\infty} 2^{-n_i - i - 2} \geq \lambda(B_k) - \lambda(B_0)2^{-n_k - k - 1}. \tag{9.10}$$

For $k = 0$ we obtain $\lambda(B) > 0$.

If $s \in {}^{n_k}2$, $[s] \cap B \neq \emptyset$, then also $[s] \cap B_k \neq \emptyset$ and therefore

$$\lambda([s] \cap B_k) = \lambda([s] \cap B_{k-1}) \geq (1 - 2^{-k-1})\lambda([s]).$$

Since $\lambda([s]) = 2^{-n_k}$, by (9.10) we obtain

$$\lambda([s] \cap B) > (1 - 2^{-k-1}) \cdot \lambda([s]) - 2^{-n_k - k - 1} = (1 - 2^{-k})\lambda([s]). \qquad \square$$

Proof of Theorem 9.21. Assume that a rapid filter $\mathcal{F}$ is Lebesgue measurable. Then the sets $\mathcal{F}$ and $\{\omega \setminus E : E \in \mathcal{F}\}$ are disjoint and have equal measure. Therefore $\lambda(\mathcal{F}) \leq 1/2$. Thus there exists a closed set $A \subseteq {}^{\omega}2$ of positive measure disjoint with $\mathcal{F}$. Let $B \subseteq A$ and $\{n_k\}_{k=0}^{\infty}$ be those of Corollary 9.8. Since $\mathcal{F}$ is rapid, there exists a $\beta \in \mathcal{F}$ such that

$$|\{i < n_{k+1} : \beta(i) = 1\}| \leq k \qquad (9.11)$$

for each k.

To obtain a contradiction, we construct by inductions $\langle s_k \in {}^{n_k}2 : k \in \omega \rangle$ such that, for each k,

$$s_k \subseteq s_{k+1}, \quad (\forall i < n_k)\,(\beta(i) = 1 \rightarrow s_k(i) = 1), \quad B \cap [s_k] \neq \emptyset. \qquad (9.12)$$

We set $s_0 = \emptyset$. Assume that s_k is already constructed. Since $B \cap [s_k] \neq \emptyset$ we have

$$\lambda([s_k] \cap B) > (1 - 2^{-k})\lambda([s_k]) = (1 - 2^{-k})2^{-n_k}.$$

On the other hand, by (9.11) we obtain

$$\lambda(\{\gamma \in [s_k] : (\forall i < n_{k+1})\,(\beta(i) = 1 \rightarrow \gamma(i) = 1)\}) \geq 2^{-n_k} \cdot 2^{-k}.$$

Thus $B \cap \{\gamma \in [s_k] : (\forall i < n_{k+1})\,(\beta(i) = 1 \rightarrow \gamma(i) = 1)\} \neq \emptyset$. Take an element γ of that set and set $s_{k+1} = \gamma | n_{k+1}$. It is easy to see that conditions (9.12) are satisfied.

Now we set $\alpha = \bigcup_k s_k$. Since B is closed and each $[s_k]$ meets B, we have $\alpha \in B$. By (9.12) we have $\alpha \in \mathcal{F}$. Thus $\alpha \in A \cap \mathcal{F}$, which is a contradiction. $\square$

For distinct $\alpha, \beta \in {}^{\omega}2$ we write $r(\alpha, \beta) = \min\{n : \alpha(n) \neq \beta(n)\}$. Let $X \subseteq {}^{\omega}2$. For an equivalence relation R on ${}^{\omega}2$ we set

$$Z_R = \{r(\alpha, \beta) : \alpha \neq \beta \wedge \alpha, \beta \in X \wedge \langle \alpha, \beta \rangle \in R\}.$$

A Raisonnier filter is the set

$$\mathcal{F}_X = \{A \subseteq \omega : (\exists R)\,(R \text{ is a Borel equivalence relation on}$$
$${}^{\omega}2 \text{ with countably many classes such that } Z_R \subseteq A)\}.$$

Lemma 9.24. *If X is uncountable, then $\mathcal{F}_X$ is a filter containing all cofinite subsets of ω.*

Proof. Note that $Z_{R_1} \cap Z_{R_2} \supseteq Z_{R_1 \cap R_2}$. Since X is uncountable, for every Borel equivalence relation with countable many classes there exist distinct $\alpha, \beta \in X$ such that $\langle \alpha, \beta \rangle \in R$. Then $r(\alpha, \beta) \in Z_R$. Consequently $\emptyset \notin \mathcal{F}_X$.

For any $n \in \omega$ the equivalence relation $R = \{\langle \alpha, \beta \rangle : \alpha | n = \beta | n\}$ is Borel with finitely many classes. Evidently $Z_R \subseteq \omega \setminus n$. $\square$

One can easily show a weaker variant of Lemma 9.23.

Lemma 9.25. *If $A \subseteq {}^{\omega}2$ has measure zero, then there exists a closed set $B \subseteq {}^{\omega}2$ such that $B \cap A = \emptyset$, $\lambda(B) > 1/2$ and*

$$(\forall n)(\forall s \in {}^{n}2)\,([s] \cap B \neq \emptyset \rightarrow \lambda([s] \cap B) > 2^{-3n-3}). \tag{9.13}$$

The proof is simpler than that of Lemma 9.23. Take any closed $B_0 \subseteq {}^{\omega}2 \setminus A$ of measure $\lambda(B_0) \geq 5/6$. By induction, let

$$B_{k+1} = \bigcup\{B_k \cap [s] : s \in {}^{k+1}2 \wedge \lambda(B_k \cap [s]) \geq 2^{-3k-3}\}.$$

We set $B = \bigcap_k B_k$. Then, similarly as above we obtain

$$\lambda([s] \cap (B_k \setminus B_{k+1})) < 2^{-3k-3}$$

for $s \in {}^{k+1}2$ and

$$\lambda(B_k \setminus B_{k+1}) < 2^{-2k-2}.$$

Then

$$\lambda(B) > 5/6 - \sum_{k=0}^{\infty} 2^{-2k-2} = 1/2$$

and (9.13) holds true. $\square$

We shall need a rather technical auxiliary result. Similarly as in Section 7.4, one can construct a measure independent family $\langle G_{s,n,m} : s \in {}^{<\omega}2 \wedge n, m \in \omega\rangle$ of clopen subsets of ${}^{\omega}2$ with $\lambda(G_{s,n,m}) = 2^{-n-m}$. We can assume that each set $G_{s,n,m}$ has the form $[t]$ for suitable $t \in {}^{a}2$ and $a \in [\omega]^{<\omega}$.

Lemma [wAC] 9.26 (J. Raisonnier). *Assume that for every G_{δ} subset H of ${}^{\omega}2 \times {}^{\omega}2$ with measure zero vertical sections the set*

$$H(X) = \{y \in {}^{\omega}2 : (\exists x \in X)\,\langle x, y\rangle \in H\}$$

has measure zero. Then for any increasing $f : \omega \longrightarrow \omega$ there exists a Borel relation $R \subseteq {}^{\omega}2 \times {}^{\omega}2$ such that

$$|Z_R \cap f(n)| < n^2(3n+3)^2 2^{4n} \tag{9.14}$$

for any $n \in \omega$.

Proof. Fix a bijection $h : {}^{<\omega}2 \times \omega \xrightarrow[\text{onto}]{1-1} \omega$ such that $h(s,n) \geq n$ and $h(s,n) \geq k$ for any $s \in {}^{k}2$.

Let $f : \omega \longrightarrow \omega$ be increasing. We set

$$\langle \alpha, \beta\rangle \in H_k \equiv (\exists n, m \geq k)\,\beta \in G_{\alpha|f(n),n,m}$$

and $H = \bigcap_k H_k$. Every H_k is open. It is easy to see that every vertical section H_{α} has measure zero. So by the assumption we have $\lambda(H(X)) = 0$. By Lemma 9.25 there exists a closed set $B \subseteq {}^{\omega}2 \setminus H(X)$ of measure $> 1/2$.

Let $\alpha \in X$. Since every vertical section $(H_k)_\alpha$ is open, the sets $(H_k)_\alpha \cap B$ are open in B. Since $\bigcap_k (H_k)_\alpha = H_\alpha \subseteq H(X)$, we obtain $\bigcap_k (H_k)_\alpha \cap B = \emptyset$. Thus, by the Baire Category Theorem 3.27, there exists an n such that $(H_n)_\alpha \cap B$ is not dense in B, i.e., there exists an $s \in {}^{<\omega}2$ such that $B \cap [s] \neq \emptyset$ and $B \cap [s] \cap (H_n)_\alpha = \emptyset$. Taking into account this fact we define the desired Borel relation.

For any $\alpha \in {}^\omega 2$ we set

$$F(\alpha) = \begin{cases} \text{the least } h(s,n) \text{ such that } B \cap [s] \neq \emptyset \\ \text{and } B \cap [s] \cap (H_n)_\alpha = \emptyset \text{ if there exists any,} \\ \infty \text{ otherwise.} \end{cases}$$

Finally, we set

$$\langle \alpha, \beta \rangle \in R \equiv (F(\alpha) = F(\beta) \wedge (F(\alpha) \neq \infty \to \alpha | f(F'\alpha)) = \beta | f(F(F(\beta))))). \quad (9.15)$$

It is easy to show that R is a Borel equivalence relation with countably many classes. We claim that (9.14) holds true for any $n \in \omega$.

For $s \in {}^l 2, n, m \in \omega$ we let

$$\Delta(s, n, m) = \{t \in {}^{f(n)}2 : B \cap [s] \cap G_{t,n,m} = \emptyset\}.$$

Hence

$$[s] \cap B \subseteq \bigcap_{t \in \Delta(s,n,m)} ({}^\omega 2 \setminus G_{t,n,m}).$$

Using the measure independence we have

$$\lambda(B \cap [s]) \leq (1 - 2^{-n-m})^{|\Delta(s,n,m)|}.$$

Assume now that $B \cap [s] \neq \emptyset$. Then by (9.13) we obtain $\lambda(B \cap [s]) > 2^{-3l-3}$. Since $1 - x \leq 2^{-x}$ for any positive real x, we obtain

$$2^{-3l-3} < 2^{-2^{-n-m} \cdot |\Delta(s,n,m)|}$$

and therefore

$$|\Delta(s, n, m)| < (3l + 3)2^{n+n}. \quad (9.16)$$

We show that $|Z_R \cap f(n)|$ is not greater than the cardinality of the set

$$\{\langle t_1, t_2 \rangle : t_1, t_2 \in \Delta(s, n, m) \wedge s \in {}^l 2 \wedge l < n \wedge m < n \wedge [s] \cap B \neq \emptyset\}. \quad (9.17)$$

By definition

$$Z_R \cap f(n) = \{r(\alpha, \beta) < f(n) : \alpha, \beta \in X \wedge \alpha \neq \beta \wedge \langle \alpha, \beta \rangle \in R\}.$$

Let $r(\alpha, \beta) \in Z_R \cap f(n)$. Then there exist $s \in {}^l 2, m \in \omega$ such that

$$F(\alpha) = F(\beta) = h(s, m) \text{ and } \alpha | f(h(s, m)) = \beta | f(h(s, m)).$$

By definition of r we have $r(\alpha, \beta) \geq f(h(s, m))$. Since f is increasing we obtain $h(s, m) < n$ and therefore $m < n$ and $l < n$. Hence, the couple $\langle \alpha | f(n), \beta | f(n) \rangle$ belongs to the set (9.17). Moreover, for different $r(\alpha, \beta)$ we obtain different couples $\langle \alpha | f(n), \beta | f(n) \rangle$.

The cardinality of the set (9.17) is not greater than

$$\sum \{ |\Delta(s, n, m)|^2 : s \in {}^l 2 \land l, m < n \land B \cap [s] \neq \emptyset \}.$$

Since for $B \cap [s] \neq \emptyset$ we have the estimate (9.16), we obtain

$$|Z_R \cap f(n)| \leq \sum_{l,m<n} (3l + 3)^2 2^{2(n+m)} < n^2 (3n + 3)^2 2^{4n}. \qquad \square$$

Theorem [wAC] 9.27 (J. Raisonnier). *Let $X \subseteq {}^\omega 2$ be uncountable. Assume that for every G_δ subset $H \subseteq {}^\omega 2 \times {}^\omega 2$ with measure zero vertical sections also $H(X)$ is a measure zero set. Then Raisonnier filter $\mathcal{F}_X$ is rapid.*

Proof. Set $g(n) = n^2(3n + 3)^2 2^{4n}$. If $h : \omega \longrightarrow \omega$ is increasing, we set $f(n) = h(g(n + 1))$. By the lemma, there exists a Borel relation R such that (9.14) holds true. If $n \in \omega$ take k such that $g(k) \leq n < g(k + 1)$. Then

$$|Z_R \cap h(n)| \leq |Z_R \cap f(k)| < g(k) \leq n.$$

Thus, $\mathcal{F}_X$ is rapid. $\qquad \square$

Now we are ready to prove important consequences of the assumption that every set of reals is Lebesgue measurable.

Theorem [wAC] 9.28. *The assumption "every set of reals is Lebesgue measurable" implies all of the following assertions:*

  a) $\neg \aleph_1 \leq \mathfrak{c}$, *consequently* $\aleph_1$ *and* $\mathfrak{c}$ *are incomparable,*
  b) $\neg |\mathrm{BOREL}| = \mathfrak{c}$, *i.e.,* $\neg |\mathrm{BOREL}| \leq \mathfrak{c}$,
  c) $\neg \aleph_1^{\aleph_0} \leq \mathfrak{c}$ *and* $\aleph_1^{\aleph_0} \leq |\mathrm{BOREL}|$,
  d) *there is no mapping* $f : [\mathbb{R}]^\omega \longrightarrow \mathbb{R}$ *such that* $f(A) \notin A$ *for every* $A \in [\mathbb{R}]^\omega$.

Proof. Assume that every set of reals is Lebesgue measurable.

Aiming to get a contradiction, assume that $\aleph_1 \leq \mathfrak{c}$. If $\aleph_1 = \mathfrak{c}$, then by Theorem 7.17 and Corollary 7.19 there exists a non-measurable set. Thus $\aleph_1 < \mathfrak{c}$. Then there exists a set $X \subseteq {}^\omega 2$ with $|X| = \aleph_1$. Since X is measurable, X has measure zero. Consider a G_δ relation H with measure zero vertical sections (one can easily find such a relation). By assumption the set $H(X)$ is measurable and therefore by Fubini's Theorem, Corollary 4.28 we obtain $\lambda(H(X)) = 0$. So the assumption of Lemma 9.26 is fulfilled. However, then by Theorem 9.27 there exists a rapid filter, that is by Talagrand's Theorem 9.21 non-measurable, which is a contradiction.

To prove part b) assume that $|\text{BOREL}| \leq \mathfrak{c}$. By Corollary 6.48 there exists an injection $F : \omega_1 \xrightarrow{1-1} \text{BOREL}$ defined as $F(\xi) = \mathbf{WO}_\xi$. Then $\aleph_1 \leq \mathfrak{c}$, a contradiction with a).

Since $\aleph_1 \leq \aleph_1^{\aleph_0}$ the former assertion of c) follows by a). The sets $\mathbf{WO}_\xi$ are pairwise disjoint, the mapping F induces an injection of $[\omega_1]^{\omega_0}$ into BOREL, therefore we obtain the later assertion of c).

Assume that $f : [\mathbb{R}]^\omega \longrightarrow \mathbb{R}$ is such that $f(A) \notin A$ for every $A \in [\mathbb{R}]^\omega$. Then we can define by transfinite induction a mapping $F : \omega_1 \longrightarrow \mathbb{R}$ as follows: for $n \in \omega$ set $F(n) = f(\omega \cup \{F(k) : k < n\})$ and for any $\omega \leq \xi < \omega_1$ set $F(\xi) = f(\{F(\eta) : \eta < \xi\})$. Then F is an injection, a contradiction with a). $\qquad\square$

Raisonnier's Theorem 9.27 has many other important consequences. We have not developed adequate technology for their proofs. So, just for information.

Metatheorem 9.7 (S. Shelah). *If* $\mathbf{ZF} + \mathbf{wAC} +$ *"every* $\mathbf{\Sigma}_3^1$ *set of reals is Lebesgue measurable" is consistent, then also* $\mathbf{ZFC} + \mathbf{IC}$ *is consistent.*

Actually, by Raisonnier's Theorem the assertion "every $\mathbf{\Sigma}_3^1$ set is Lebesgue measurable" implies that $\aleph_1$ is an inaccessible cardinal in the constructible universe. The Metatheorem also shows that in Solovay's Metatheorem 9.6 the assumption of consistency of $\mathbf{ZF}$ is not enough. We need $\mathbf{IC}$.

Exercises

9.11 When is $\aleph_1 \leq \mathfrak{c}$?

a) If $\aleph_1 \leq \mathfrak{c}$, then there exists a mapping $f : [\mathbb{R}]^\omega \longrightarrow \mathbb{R}$ such that $f(A) \notin A$ for every $A \in [\mathbb{R}]^\omega$.
   Hint: If $B \subseteq \mathbb{R}$ is a well-ordered set of the type ω_1, set $f(A) =$ the first $x \in B \setminus A$ for any $A \in [\mathbb{R}]^\omega$.

b) $\aleph_1 \leq \mathfrak{c}$ if and only if there exists a mapping $f : [\mathbb{R}]^\omega \longrightarrow \mathbb{R}$ such that $f(A) \notin A$ for every $A \in [\mathbb{R}]^\omega$.

9.12 Properties of $\mathfrak{k}$

a) Show that the cardinality of the Vitali decomposition $\{v(x) : x \in \mathbb{T}\}$ defined by (7.8) with $S = \mathbb{D}$ is $\mathfrak{k}$.

b) Show that $\mathfrak{k} = |\mathbb{R}/\mathbb{Q}|$.

c) Show that $\mathfrak{k}^{\aleph_0} \ll \mathfrak{k}$.
   Hint: The natural bijection of $\mathcal{P}(\omega \times \omega)$ onto the set $^\omega\mathcal{P}(\omega)$ induces a surjection of $\mathcal{P}(\omega \times \omega)/\text{Fin}$ onto $^\omega(\mathcal{P}(\omega)/\text{Fin})$.

d) Show that $\mathfrak{k} \cdot \mathfrak{k} = \mathfrak{k}$ and $2^{(\mathfrak{k}^{\aleph_0})} = 2^\mathfrak{k}$.
   Hint: Use c) and (1.4).

e) If a Vitali decomposition $\{v(x) : x \in \mathbb{T}\}$ can be linearly ordered, then there exists a non-measurable set and a set without the Baire Property.
   Hint: If $\sqsubseteq$ is a linear ordering of the set $\{v(x) : x \in \mathbb{T}\}$, then consider the tail-set $A = \{\langle x, y \rangle \in \mathbb{T}^2 : v(x) \sqsubseteq v(y)\}$ and apply Theorems 7.36 and 7.38.

f) As a consequence you obtain another proof of the assertion e) of Theorem 9.20.

9.13 Once more $\mathfrak{k}$

$^\omega 2$ is considered as a topological group. Set

$$(^\omega 2)^* = \{\alpha \in {}^\omega 2 : |\{k : \alpha(k) = 1\}| < \aleph_0\}.$$

a) The quotient group $^\omega 2/(^\omega 2)^*$ has cardinality $\mathfrak{k}$.

b) Let $p : {}^\omega 2 \longrightarrow {}^\omega 2/(^\omega 2)^*$ be the quotient mapping. For any $X \subseteq {}^\omega 2/(^\omega 2)^*$ the inverse image $p^{-1}(X)$ is a tail-set.

c) The set $\mathcal{I} = \{X \subseteq {}^\omega 2/(^\omega 2)^* : p^{-1}(X)$ is meager$\}$ is a σ-additive ideal of $^\omega 2/(^\omega 2)^*$.

d) If "every set of reals possesses the Baire Property", then $\mathcal{I}$ is a maximal ideal.
   Hint: Use Theorem 7.38.

e) The set $\mathcal{J} = \{X \subseteq {}^\omega 2/(^\omega 2)^* : p^{-1}(X)$ has measure zero$\}$ is a σ-additive ideal on $^\omega 2/(^\omega 2)^*$.

f) If "every set of reals is Lebesgue measurable", then $\mathcal{J}$ is a maximal ideal.
   Hint: Use Theorem 7.36.

9.14 Choice from Finite Sets

The natural number $n = \{k < n : k \in \omega\}$ is considered as the additive group modulo n. Similarly $^\omega n$. As above, we set

$$(^\omega n)^* = \{\alpha \in {}^\omega n : |\{k : \alpha(k) = 1\}| < \aleph_0\}.$$

a) Show that $|^\omega(2^k)/(^\omega(2^k))^*| \leq \mathfrak{k}$ for any $k > 1$.

b) $|^\omega n/(^\omega n)^*| = \mathfrak{k}$ for any $n > 1$.
   Hint: Use a) and the fact $|^\omega n/(^\omega n)^*| \leq |^\omega m/(^\omega m)^*|$ for any $n \leq m$.

c) Let A_n be the subgroup of $^\omega n/(^\omega n)^*$ generated by $p_n(\alpha)$, where $\alpha(k) = 1$ for every $k \in \omega$ and $p_n : {}^\omega n \longrightarrow {}^\omega n/(^\omega n)^*$ is the quotient map. The quotient group $(^\omega n/(^\omega n)^*)/A_n$ is a family of n-elements sets.
   Hint: Note that $|A_n| = n$.

d) If "every set of reals is Lebesgue measurable", then there exists no selector for the family $(^\omega n/(^\omega n)^*)/A_n$ provided that $n > 1$.
   Hint: Assume that Z is a selector for G_n/A_n. Then $^\omega n = \bigcup_{\alpha \in A_n} p_n^{-1}(\alpha + Z)$. The tail-sets $\langle p_n^{-1}(\alpha + Z) : a \in A_n\rangle$ are pairwise disjoint and have equal measure. Use Theorem 7.36.

e) If "every set of reals is Lebesgue measurable", then the Axiom of Choice from Finite Sets $\mathbf{AC}_n$ fails for any $n > 1$.

f) Show similar conclusions assuming "every set of reals possesses the Baire Property".

9.4 The Axiom of Determinacy

We assume that a reader is familiar with the topic of Section 5.5.

Let X be a set, $|X| > 1$, $n > 0$ and $P \subseteq {}^{2n}X$. We introduce a finite game $\mathrm{FGAME}_{n,X}(P)$ as follows. Player I plays $x_{2i} \in X$, player II plays $x_{2i+1} \in X$ for each $i < n$. Of course each player knows the preceding moves of the other one[3]. Player I wins if $\{x_i\}_{i=0}^{2n-1} \in P$. Otherwise player II wins.

Player I can win if

$$(\exists x_0)(\forall x_1)\dots(\exists x_{2n-2})(\forall x_{2n-1})\,\{x_i\}_{i=0}^{2n-1} \in P. \tag{9.18}$$

Similarly, player II can win if

$$(\forall x_0)(\exists x_1)\dots(\forall x_{2n-2})(\exists x_{2n-1})\,\{x_i\}_{i=0}^{2n-1} \notin P. \tag{9.19}$$

The well-known de Morgan law says that

$$\neg(\forall x)\,\mathcal{V}(x) \equiv (\exists x)\,\neg\mathcal{V}(x).$$

Thus the negation of the assertion "player I can win" is the assertion "player II can win". Therefore by the principle tertium non datur we obtain "player I can win or player II can win".

Similarly as in Section 5.5 we can define a strategy in game $\mathrm{FGAME}_{n,X}(P)$. A strategy is a function $f : {}^{<n}X \longrightarrow X$. Player I follows the strategy f if for each $i < n$, she/he plays $x_{2i} = f([x_1, x_3, \dots, x_{2i-1}])$ (we assume that $f(\emptyset) = x_0$). The strategy f is a winning strategy for player I if she/he wins in each particular game in which she/he follows the strategy f. Similarly we define the notion of a winning strategy for player II. We say that the game $\mathrm{FGAME}_{n,X}(P)$ is determined if at least one of the two players has a winning strategy.

If player I has a winning strategy, then (9.18) holds true. Generally, those two assertions are not equivalent. However if the set X can be well ordered, then (9.18) implies that there exists a winning strategy for player I. Similarly for player II. Thus, if the set X can be well ordered, then the game $\mathrm{FGAME}_{n,X}(P)$ is determined for any $P \subseteq {}^{2n}X$. This result is known as **Zermelo's Theorem**.

We can by analogy formulate similar assertions for infinite games. The fact that "player I can win in $\mathrm{GAME}_X(P)$" can be expressed as

$$(\exists x_0)(\forall x_1)\dots(\exists x_{2n})(\forall x_{2n+1})\dots\,\{x_n\}_{n=0}^{\infty} \in P. \tag{9.20}$$

Extending the de Morgan law for the infinite sequence of quantifiers it seems natural to assume that the negation of assertion (9.20) is

$$(\forall x_0)(\exists x_1)\dots(\forall x_{2n})(\exists x_{2n+1})\dots\,\{x_n\}_{n=0}^{\infty} \notin P. \tag{9.21}$$

---

[3]They play a game with perfect information.

This assertion however expresses that "player II can win in $\mathrm{GAME}_X(P)$". If player I or player II has a winning strategy in $\mathrm{GAME}_X(P)$, then (9.20) or (9.21) holds true, respectively. If X can be well ordered we can proceed as above. Unfortunately, we do not have any background for extending the de Morgan law for an infinite string of quantifiers.

However, in spite of all those unclear formulations, J. Mycielski and H. Steinhaus [1962] proposed an axiom of set theory which is called the **Axiom of Determinacy**, denoted **AD**, saying that "the game $\mathrm{GAME}_\omega(P)$ is determined for any $P \subseteq {}^\omega\omega$". In notation of Section 5.5, **AD** says that $\mathrm{DET}_\omega(\mathcal{P}({}^\omega\omega))$.

Some consequences of **AD** were already implicitly proved in Section 5.5.

Theorem 9.29. *If* **AD** *holds true, then*

  a) *the weak axiom of choice* **wAC** *holds true,*
  b) *every uncountable set of reals contains a perfect subset,*
  c) *every set of reals possesses the Baire Property,*
  d) *the axiom of choice* **AC** *fails.*

Proof. a) We have to show that for any $\langle X_n \subseteq {}^\omega\omega, X_n \neq \emptyset : n \in \omega \rangle$ there exists a function $g : \omega \longrightarrow {}^\omega\omega$ such that $g(n) \in X_n$ for each n. We set

$$P = \{\{x_n\}_{n=0}^\infty \in {}^\omega\omega : \{x_{2n+1}\}_{n=0}^\infty \notin X_{x_0}\}.$$

The set P "depends" only on odd members of sequences. Thus player I has no possibility to influence the game and actually cannot win (player II simply plays an element $\{x_{2n+1}\}_{n=0}^\infty \in X_{x_0}$). Thus there exists a winning strategy f for player II. Let $\{x_k\}_{k=0}^\infty$ be the run in game $\mathrm{GAME}_\omega(P)$ in which player II follows the strategy f and player I plays the sequence $\langle n, n, \ldots, n, \ldots \rangle$. We set

$$g(n) = \{x_{2k+1}\}_{k=0}^\infty = R(\bar{f}_{II}(\{n\}_{k=0}^\infty)).$$

One can easily see that g is the desired selector, i.e., $g(n) \in X_n$ for every n.

b) By Theorem 5.58, b) we have $\mathrm{DET}_\omega^*(\mathcal{P}({}^\omega\omega))$ and by Morton Davis' Theorem 5.61 we obtain the assertion.

c) By Theorem 5.58, b) we have $\mathrm{DET}_\omega^{**}(\mathcal{P}({}^\omega\omega))$. If the set P is meager, we are done. So assume that P is not meager. We set

$$G = \bigcup\{[t] : t \in {}^{<\omega}\omega \wedge [t] \setminus P \text{ is meager}\}.$$

Then

$$G \setminus P \subseteq \bigcup\{[t] \setminus P : [t] \setminus P \text{ is meager}\}$$

and therefore $G \setminus P$ is meager.

If $P \setminus G$ is not meager, then by the Banach-Mazur Theorem 5.62 and the determinacy of $\mathrm{GAME}_\omega^{**}(P \setminus G)$ there exists a non-empty open set U such that $U \setminus (P \setminus G)$ is meager. Then there exists a finite sequence $t \in {}^{<\omega}\omega$ such that

$[t] \setminus (P \setminus G)$ is meager. Then $[t] \setminus P$ is meager as well and therefore $[t] \subseteq G$. However, then $[t] \subseteq [t] \setminus (P \setminus U)$, which is a contradiction. Thus P possesses the Baire Property.

d) The assertion follows by any of b) or c). $\qquad\square$

There is another important consequence of **AD**.

Theorem 9.30 (J. Mycielski – S. Swierczkowski). *If* **AD** *holds true, then every set of reals is Lebesgue measurable.*

We present a proof based on a "covering game" invented by L. Harrington. We need a technical result and some notation. Let $\{S_n : n \in \omega\}$ be an enumeration of the set $[^{<\omega}2]^{<\omega}$ of finite subsets of finite sequences of 0 and 1. If you prefer to work with true $[0,1]$, then $\{S_n : n \in \omega\}$ is an enumeration of all finite sets of open intervals with rational endpoints. Let μ be a Borel measure on $^{\omega}2$. Note the following. If $A \subseteq {}^{\omega}2$ has measure zero, then for any decreasing sequence $\{\varepsilon_n\}_{n=0}^{\infty}$ of positive reals there exists a sequence $\{m_k\}_{k=0}^{\infty}$ such that

$$A \subseteq \bigcup_k \bigcup_{s \in S_{m_k}} [s], \qquad \mu\left(\bigcup_{s \in S_{m_k}} [s]\right) < \varepsilon_k \text{ for every } k. \tag{9.22}$$

Actually, take a sequence $\{t_n\}_{n=0}^{\infty}$ of elements of $^{<\omega}2$ such that $A \subseteq \bigcup_n [t_n]$ and $\sum_{i=0}^{\infty} \mu([t_i]) < \varepsilon_0$. Now, let $n_0 = 0$ and

$$n_{k+1} = \min\left\{n > n_k : \sum_{i \geq n} \mu([t_i]) < \varepsilon_{k+1}\right\}.$$

Let m_k be such that $S_{m_k} = \{t_i : n_k \leq i < n_{k+1}\}$.

Evidently $\bigcup_k S_{m_k} = \{t_i : i \in \omega\}$ and the inequality of (9.22) holds true.

For a given set $A \subseteq {}^{\omega}2$ and $\varepsilon > 0$, the **covering game** is defined as follows: player I plays an $\alpha \in {}^{\omega}2$, player II plays a $\beta \in {}^{\omega}\omega$ such that

$$\mu\left(\bigcup_{t \in S_{\beta(n)}} [t]\right) < \varepsilon \cdot 2^{-2n-1}$$

for each n. Player I wins if $\alpha \in A \setminus \bigcup_n \bigcup_{t \in S_{\beta(n)}} [t]$.

It is easy to see that the covering game is equivalent to the game $\mathrm{GAME}_{\omega}(P)$, where

$$P = \left\{ \Pi(\alpha, \beta) \in {}^{\omega}\omega : \alpha \in A \setminus \bigcup_n \bigcup_{s \in S_{\beta(n)}} [s] \vee \beta \notin C \right\}, \tag{9.23}$$

and where

$$C = \left\{ \beta \in {}^{\omega}\omega : (\forall n)\, \mu\left(\bigcup_{s \in S_{\beta(n)}} [s]\right) < \varepsilon \cdot 2^{-2n-1} \right\}.$$

Lemma [wAC] 9.31 (L. Harrington). *Let Γ be a pointclass containing all open and closed sets. Let μ be a Borel measure on $^\omega 2$. If $A \in \Gamma(^\omega 2)$, $\mu_*(A) = 0$ and $\mathrm{DET}_\omega(\Gamma)$ holds true, then $\mu^*(A) = 0$.*

Proof. The set P defined by (9.23) belongs to Γ. Let us consider the game $\mathrm{GAME}_\omega(P)$.

We show that player I does not have a winning strategy. We show that for any strategy f of player I, there exists a play β of player II such that player II wins.

We may assume that $B = \Lambda(\bar{f}_I(C)) \subseteq A$, otherwise f is not a winning strategy. One can easily see that C is a Borel set. Thus B is an analytic set and therefore $\mu(B) = 0$. If player II plays $\beta \in C$ such that $B \subseteq \bigcup_n \bigcup_{s \in S_{\beta(n)}} [s]$, then she/he wins.

Thus by $\mathrm{DET}_\omega(\Gamma)$, player II has a winning strategy f. To any partial play $t \in {}^n 2$ of player I player II following this strategy answers by $\beta(n) = f(t)$ such that the set $G_t = \bigcup_{s \in S_{\beta(n)}} [s]$ has measure less than $\varepsilon \cdot 2^{-2n-1}$. Then

$$\mu\left(\bigcup_{t \in {}^n 2} G_t\right) < \varepsilon \cdot 2^{-n-1}.$$

Since f is a winning strategy for player II, we obtain $A \subseteq \bigcup_n \bigcup_{t \in {}^n 2} G_t$ and therefore $\mu(A) < \varepsilon$. □

Proof of Theorem 9.30. Let $A \subseteq [0,1]$. By Theorems 4.16 and 4.8 there exists a Borel set $C \subseteq A$ such that $\lambda_*(A) = \lambda(C)$. Then $\lambda_*(A \setminus C) = 0$. By Lemma 9.31 the set $A \setminus C$ is Lebesgue measurable and therefore also A is Lebesgue measurable. □

Thus, the Axiom of Determinacy **AD** implies all assertions of Theorems 9.20 and 9.28, including those concerning the cardinal arithmetics. However, **AD** influences the cardinal arithmetics in a much stronger way. We present such a consequence of **AD**.

Theorem 9.32 (R.M. Solovay). *If* **AD** *holds true, then there exists a surjection of $\mathcal{P}(\omega)$ onto $\mathcal{P}(\omega_1)$, i.e., $2^{\aleph_1} \ll \mathfrak{c} = 2^{\aleph_0}$.*

Proof. We code a set $A \subseteq \omega_1$ by a set $A^* \subseteq {}^\omega 2$ using the Lebesgue decomposition (6.37) of $\mathcal{P}(\omega)$ and a suitable game. The game may be described as follows. Player II wins if either player I plays an $\alpha \notin \mathbf{WO}$ or if player I plays a code of a well-ordering $\alpha \in \mathbf{WO}_\zeta$ for some $\zeta < \omega_1$ and player II can decide for any ordinal $\xi \leq \zeta$ whether $\xi \in A$. Trying to do that, player II plays $II(\beta, \gamma)$ with $\beta \in \mathbf{WO}_\eta$ that is a code of a well-ordering of the length $\eta > \zeta$, and with γ coding the set $A \cap \eta$. Thus we can decide for any $\xi < \eta$ whether $\xi \in A$ or not from the run $\delta = II(\alpha, II(\beta, \gamma))$.

We denote by $A^* \subseteq {}^\omega 2$ the set of all runs in such a game, which lead to the success of player I:

$$A^* = \{ \Pi(\alpha, \Pi(\beta, \gamma)) \in {}^\omega 2 : \alpha \in \mathbf{WO} \wedge \neg(\beta \in \mathbf{WO} \wedge \mathrm{ot}(\alpha) < \mathrm{ot}(\beta)$$
$$\wedge (\exists h)\,(h \text{ is an isomorphism of } \langle \{d_n \in \mathbb{D} : \beta(n) = 1\}, \preceq \rangle$$
$$\text{onto } \mathrm{ot}(\beta) \wedge (\forall m)\,(\gamma(m) = 1 \equiv h(m) \in A \cap \mathrm{ot}(\beta)))) \}.$$

Hence our game is actually the game $\mathrm{GAME}_2(A^*)$.

We claim that player I does not have a winning strategy in $\mathrm{GAME}_2(A^*)$. Assume that f is a strategy for player I. Then the function $\bar{f}_I$ defined by (5.37) is continuous and $\bar{f}_I({}^\omega 2) \subseteq A^*$ is compact. Then $\Lambda(\bar{f}_I({}^\omega 2)) \subseteq \mathbf{WO}$ and by the Boundedness Lemma 6.55 there exists an ordinal $\zeta < \omega_1$ such that

$$\Lambda(\bar{f}_I({}^\omega 2)) \subseteq \bigcup_{\xi < \zeta} \mathbf{WO}_\xi.$$

Thus, if player I follows f, all his plays α are of order type less than ζ. Since player II could play $\Pi(\beta, \gamma)$ such that $\mathrm{ot}(\beta) > \zeta$ and γ codes the set $A \cap \mathrm{ot}(\beta)$, the strategy f is not winning.

By determinacy for any set $A \subseteq \omega_1$ there exists a winning strategy for player II in game $\mathrm{GAME}_2(A^*)$. From such a strategy f we can decode the set A. Namely, for any $f \in {}^{(<\omega 2)}2$ we set

$$\sigma(f) = \{ \xi \in \omega_1 : (\forall \alpha)\,((\alpha \in \mathbf{WO} \wedge \mathrm{ot}(\alpha) > \xi) \rightarrow (\mathrm{ot}(\Lambda(R(\bar{f}_{II}(\alpha)))) > \mathrm{ot}(\alpha)$$
$$\wedge (\exists h)\,(h \text{ is an isomorphism of } \langle \{d_n \in \mathbb{D} : R(R(\bar{f}_{II}(\alpha)))(n) = 1\}, \preceq \rangle$$
$$\text{onto } \mathrm{ot}(\Lambda(R(\bar{f}_{II}(\alpha)))) \wedge (\forall m)\,(h(m) = \xi \equiv \Lambda(R(\bar{f}_{II}(\alpha)))(m) = 1)))) \}$$

If f is a winning strategy for player II in the game $\mathrm{GAME}_2(A^*)$, then $\sigma(f) = A$. Hence $\sigma : {}^{(<\omega 2)}2 \xrightarrow{\text{onto}} \mathcal{P}(\omega_1)$. $\qquad \square$

Theorem 9.33. *If* **AD** *holds true, then*

a) *there exists no cardinality between $\aleph_0$ and $\mathfrak{c}$, i.e. there is no set X such that $\aleph_0 < |X| < \mathfrak{c}$,*

b) *there exists no well-ordering of the real line,*

c) *$\aleph_1$ and $\mathfrak{c}$ are incomparable,*

d) *there is no selector for Lebesgue decomposition,*

e) *$\aleph_1 < \aleph_1 + \mathfrak{c} < \aleph_1 + \mathfrak{k}$,*

f) *$2^{\aleph_1} \ll \mathfrak{c} < \mathfrak{c} + \aleph_1 < 2^{\aleph_1} < 2^{\aleph_1} + \mathfrak{k} < 2^{\mathfrak{k}} = 2^{\mathfrak{c}}$,*

g) *the relation $\ll$ is not antisymmetric.*

Proof. a) By b) of Theorem 9.29 an uncountable set of reals contains a subset of cardinality $\mathfrak{c}$, thus there exists no uncountable set of cardinality smaller than $\mathfrak{c}$.

b) The assertion follows immediately by Bernstein's Theorem 7.20 and by b) of Theorem 9.29.

c) If $\mathfrak{c}$ and $\aleph_1$ were comparable, then by b) we would obtain $\aleph_1 < \mathfrak{c}$ which contradicts a).

d) A selector of Lebesgue decomposition is a set of reals of cardinality $\aleph_1$. By c) such a set does not exist.

e) The inequalities follow by c) and Theorems 9.29, c) and 9.20, e).

f) The first inequality follows by Theorem 9.32. The second inequality follows from c). Evidently $\mathfrak{c} + \aleph_1 \leq 2^{\aleph_1}$. Since $2^{\aleph_1} + \aleph_1 = 2^{\aleph_1}$, if $\mathfrak{c} + \aleph_1 = 2^{\aleph_1}$, then by Tarski's Theorem 1.8 we obtain $2^{\aleph_1} = \mathfrak{c}$, contradicting c). The inequality $\mathfrak{k} \leq 2^{\aleph_1}$ contradicts Theorem 9.20, e), so we obtain the fourth inequality of f). If $2^{\mathfrak{k}} = 2^{\aleph_1} + \mathfrak{k}$, then again by Tarski's Theorem 1.8 we obtain $2^{\aleph_1} = 2^{\mathfrak{k}}$. Since $\mathfrak{k} \leq 2^{\mathfrak{k}}$, we obtain a contradiction with Theorem 9.20, e). The last equality follows from Theorem 9.20, g) by (1.4).

g) By Theorem 9.32 we have $2^{\aleph_1} \ll \mathfrak{c}$ and trivially $\mathfrak{c} \ll 2^{\aleph_1}$. By f) we obtain $\mathfrak{c} \neq 2^{\aleph_1}$. $\square$

Note that by f) we have

$$2^{\aleph_1} \ll \mathfrak{c} < 2^{\aleph_1}.$$

We present without proof another interesting result (the notion of a measurable cardinal is defined in Section 10.2):

Theorem 9.34 (R.M. Solovay). *If* **AD** *holds true, then* $\aleph_1$ *is a measurable cardinal. Actually the filter generated by closed unbounded subsets of* ω_1 *is an ultrafilter.*

This result shows that the consistency strength of **AD** is very high. We can formulate it as

Metatheorem 9.8. *If* **ZF** + **AD** *is consistent, then* **ZFC** + *"there exists a measurable cardinal" is consistent as well.*

Actually the consistency strength of **AD** is much higher – see (11.46). We recommend the reader to T. Jech [2006] and A. Kanamori [2009].

Many important results follow already from some weaker modifications of the Axiom of Determinacy. We present some examples.

By a result by D.A. Martin [1970] we have

Theorem [AC] 9.35. *Assume that there exists a measurable cardinal. Then the game* $\mathrm{GAME}_\omega(A)$ *is determined for any analytic* $A \subseteq {}^\omega\omega$, *i.e.,* $\mathrm{DET}_\omega(\mathbf{\Sigma}_1^1({}^\omega\omega))$ *holds true.*

Using this result we obtain

Theorem [AC] 9.36. *If there exists a measurable cardinal, then every uncountable* $\mathbf{\Sigma}_2^1$ *set of reals contains a perfect subset.*

Proof. Let $A \in \mathbf{\Sigma}_2^1({}^\omega\omega)$. Similarly as in proof of Corollary 6.63 we can show that there exists a set $C \in \mathbf{\Pi}_1^1({}^\omega\omega \times {}^\omega\omega)$ such that $A = \mathrm{proj}_2(C)$ and $|C^\alpha| \leq 1$ for any $\alpha \in {}^\omega\omega$, or in other words, proj_2 is injective. If A is uncountable also C is uncountable. By Theorem 5.58 game $\text{GAME}_\omega(C)$ is determined and by M. Davis' Theorem 5.61 the set C contains a perfect subset. Its injective image is a perfect subset of A. $\qquad\square$

We present one of the key results of descriptive set theory in "a playful universe", i.e., a consequence of **AD**, the so-called Coding Lemma. A proof uses quite a great deal of general recursion theory and therefore is out of the scope of this book. A proof can be found in Y.N. Moschovakis [1970] or [1980]. We need a technical notion.

We suppose that X, Y are Polish spaces, $S \subseteq X$ and $\rho : S \xrightarrow{\text{onto}} \lambda$ is a regular rank. Let $f : \lambda \longrightarrow \mathcal{P}(Y)$. A set $C \subseteq X \times Y$ is called a **choice set** for f if

$$\langle x, y \rangle \in C \to (x \in S \wedge y \in f(\rho(x)), \quad f(\xi) \neq \emptyset \to (\exists x)(\exists y)\, (\rho(x) = \xi \wedge \langle x, y \rangle \in C).$$

Lemma 9.37 (The Coding Lemma). *Assume* **AD***. Let* X*,* Y *be Polish spaces. Let* $S \subseteq X$*,* $\rho : S \xrightarrow{\text{onto}} \lambda$*. If* Γ *is a* $^\omega\omega$*-parametrized pointclass closed under countable unions and intersections, closed under* $\exists^{\,{}^\omega\omega}$ *and such that* $\boldsymbol{\Sigma}_1^1 \subseteq \Gamma$*,* $<_\rho \in \Gamma(X \times X)$*, then every function* $f : \lambda \longrightarrow \mathcal{P}(Y)$ *has a choice set in* $\Gamma(X \times Y)$*.*

One of the most interesting consequences of the Coding Lemma is the next result, which has been proved, under essentially different assumptions, as Corollary 9.10.

Theorem 9.38 (Y.N. Moschovakis). *If* **AD** *holds true, then a subset* A *of a perfect Polish space* X *is a* $\boldsymbol{\Sigma}_2^1$ *set if and only if* A *is a union of* $\aleph_1$ *Borel sets. Thus*

$$|\boldsymbol{\Sigma}_2^1(X)| \ll \mathfrak{c} = 2^{\aleph_0} < 2^{\aleph_1} \leq |\boldsymbol{\Sigma}_2^1(X)|.$$

Proof. If A is a $\boldsymbol{\Sigma}_2^1$ set, then A is a union of $\aleph_1$ Borel sets by Theorem 6.58.

Let $\langle B_\xi : \xi \in \omega_1 \rangle$ be Borel subsets of X. By Theorem 6.40 there exists a $\boldsymbol{\Sigma}_2^1$-universal set $U \subseteq {}^\omega\omega \times X$. We define

$$f(\xi) = \{\alpha \in {}^\omega\omega : U^\alpha = B_\xi\}.$$

By the Coding Lemma 9.37 there exists a choice set $C \in \boldsymbol{\Sigma}_2^1({}^\omega\omega \times X)$ for f. One can easily see that

$$x \in \bigcup_{\xi < \omega_1} B_\xi \equiv (\exists y)(\exists \alpha)\, (\langle y, \alpha \rangle \in C \wedge \langle \alpha, x \rangle \in U).$$

Hence $\bigcup_{\xi < \omega_1} B_\xi \in \boldsymbol{\Sigma}_2^1(X)$.

Let S be a $\boldsymbol{\Pi}_1^1$-subset of X, which is not analytic. Then $S = \bigcup_{\xi < \omega_1} S_\xi$, where $\langle S_\xi : \xi < \omega_1 \rangle$ are the pairwise disjoint Borel constituents of S. Since S is not analytic, the set $T = \{\xi \in \omega_1 : S_\xi \neq \emptyset\}$ has cardinality $\aleph_1$. The function $F : \mathcal{P}(T) \longrightarrow \boldsymbol{\Sigma}_2^1(X)$ defined as $F(E) = \bigcup_{\xi \in E} \mathbf{WO}_\xi$ for $E \subseteq T$ is an injection. Therefore $2^{\aleph_1} \leq |\boldsymbol{\Sigma}_2^1(X)|$. The other inequalities have been already proved. $\qquad\square$

The **Axiom of Projective Determinacy PD** says that $\text{GAME}_\omega(P)$ is determined for any projective set $P \subseteq {}^\omega\omega$. Thus, in notations of Section 5.5, the axiom **PD** is the assertion $D_\omega(\bigcup_n \mathbf{\Sigma}_n^1({}^\omega\omega))$. Evidently $\mathbf{ZF} \vdash \mathbf{AD} \to \mathbf{PD}$. However, the axiom **wAC** does not follow from **PD** (at least, we do not know it). As above, we obtain the main consequences of the axioms **PD** + **wAC**.

Theorem [wAC] 9.39. *If* **PD** *holds true, then*

  a) *every uncountable projective set of reals contains a perfect subset,*

  b) *every projective set of reals possesses the Baire Property,*

  c) *every projective set of reals is Lebesgue measurable,*

  d) *a set $A \subseteq \mathbb{R}$ is a $\mathbf{\Sigma}_2^1$ set if and only if A is a union of $\aleph_1$ Borel sets,*

  e) *for any uncountable Polish space X we have*

$$|\mathbf{\Sigma}_2^1(X)| \ll \mathfrak{c} = 2^{\aleph_0} < 2^{\aleph_1} \leq |\mathbf{\Sigma}_2^1(X)|.$$

Actually checking the proofs of related consequences of the axiom **AD**, one can verify that in case of a projective set, the corresponding game is "projective", and therefore the axiom **PD** is enough for obtaining the result.

Exercises

9.15 Harrington Lemma

Show that $\text{DET}_\omega(\mathbf{\Sigma}_1^1)$ implies that every $\mathbf{\Sigma}_2^1$ subset of ${}^\omega 2$ is measurable.

Hint: If $A \in \mathbf{\Sigma}_2^1({}^\omega 2)$, then $\alpha \in A \equiv (\exists\gamma) \langle \alpha, \gamma \rangle \in C$ for a $\mathbf{\Pi}_1^1$ set $C \subseteq {}^\omega 2 \times {}^\omega\omega$. Modify the covering game by letting player I play $\pi(\alpha(n), \gamma(n))$. Player I wins if

$$\langle \alpha, \gamma \rangle \in C \wedge \alpha \notin \bigcup_n \bigcup_{s \in S_{\beta(n)}} [s].$$

9.16 Solovay Game

For given $S \subseteq \omega_1$ we introduced a game as follows. Player I plays an element $\alpha \in \mathbf{WO}$ and player II plays a code β of a countable subset $\{\text{Proj}_n(\beta) : n \in \omega\}$ of ω_1. Player II wins if either $\alpha \notin \mathbf{WO}$ or

$$\{\xi \in S : \xi \leq \text{ot}(\alpha)\} \subseteq \{\text{Proj}_n(\beta) : n \in \omega\} \subseteq S.$$

  a) If S is unbounded, then player I does not have a winning strategy.

   Hint: If f is a winning strategy for player I, then $\Lambda(\bar{f}_I({}^\omega 2) \subseteq \mathbf{WO}$ and by the Boundedness Lemma 6.55 there exists an ordinal $\zeta < \omega_1$ such that $\Lambda(\bar{f}_I({}^\omega 2)) \subseteq \zeta$. Player II plays an ordinal greater than ζ – a contradiction.

  b) If S is unbounded and player II has a winning strategy, then $\{\alpha \in \mathbf{WO} : \text{ot}(\alpha) \in S\}$ is a $\mathbf{\Pi}_1^1$ set.

   Hint: If f is a winning strategy for player II, then

$$\text{ot}(\alpha) \in S \equiv (\exists n) (\text{ot}(\alpha) = \text{ot}(\text{Proj}_n(\Lambda(\bar{\alpha}_I({}^\omega 2)))).$$

   Note that $\text{ot}(\alpha)$ is a $\mathbf{\Pi}_1^1$-rank on $\mathbf{WO}$.

  c) If **AD** holds true, then the union $\bigcup_{\xi \in S} \mathbf{WO}_\xi$ is a $\mathbf{\Pi}_1^1$ set for any $S \subseteq \omega_1$.

   Hint: If S is bounded, then S is countable and use Corollary 6.18.

Historical and Bibliographical Notes

It was George Cantor who advanced the hypothesis that $2^{\aleph_0} = \aleph_1$. Moreover he tried to prove it. D. Hilbert in his lecture at the International Congress of Mathematicians in Paris 1900 raised his celebrated Hilbert's Problems. The first problem was the question whether **CH** holds true. Later many mathematicians (e.g., N.N. Luzin [1914], W. Sierpiński [1934b], M.M. Lavrentieff [1924]) used to answer an important problem assuming **CH** when they could not find an answer in the framework of set theory. Of course they considered a proof of an answer without **CH** as a better result. I propose just one example. N.N. Luzin [1914] proved, assuming **CH**, that there exists an uncountable perfectly meager set. Then, the aim of N.N. Luzin [1921], actually published by W. Sierpiński, "est de démontrer le même sans l'hypothèse que $2^{\aleph_0} = \aleph_1$". W. Sierpiński and many others obtained a great deal of consequences of **CH**, which they were unable to prove in **ZFC**. It turned out that many of those consequences are actually equivalent to **CH**. The most important results of this kind were systematically presented by W. Sierpński [1935]. Finally K. Gödel [1944] showed that **CH** cannot be refuted by **ZFC**. However, K. Gödel did not believe that **CH** is "true". As we have already mentioned F. Rothberger [1939] and [1941] was enough, at least implicitly, to assume that **CH** is not true. P.J. Cohen [1963] constructed a model of **ZFC** in which **CH** fails.

R.M. Solovay and S. Tennenbaum [1971] have constructed a model of **ZFC** in which the Souslin Hypothesis holds true. D.A. Martin observed that they have proved more: they have constructed a model, in which the Martin Axiom holds true. The Martin Axiom **MA** was formulated and the main consequences of it were presented by D.A. Martin and R.M. Solovay [1970], including (9.1), Theorem 9.1, Lemma 9.5, and Theorems 9.7 and 9.9. The monograph D.H. Fremlin [1984b] is devoted to a systematic study of consequences of the Martin Axiom and of two of its weaker forms.

Shanin's Δ-lemma 9.11 was proved by N.A. Shanin [1946]. Theorem 9.13 (unpublished) was discovered independently by several authors: K. Kunen, F. Rowbottom and R.M. Solovay. R. Engelking and M. Karlowicz [1965] proved that a topological product of topological spaces with the Knaster Property has the Knaster Property as well. So Theorem 9.14 and Corollary 9.15 follow. See also Exercise 9.6.

For more information related to Exercise 9.3 see D.H. Fremlin [1984b] and A. Blass [2010]. Results of Exercise 9.4 are those of M. Bell [1981]. The presentation in Exercises 9.3–9.5 follows essentially A. Blass [2010].

Results of Metatheorems 9.1–9.3 are attributed to their authors in Section 11.5. Lemma 9.17 and Theorem 9.18 are due to A. Dow [1990].

The Katowice Problem was formulated at Katowice University in the 1970s and is still open. The result of Exercise 9.7, c) is due to B. Balcar (unpublished). The results of Exercise 9.8, k) and l), Exercise 9.10, i) and j) are due to

J. Flašková [2005]. Some results of Exercise 9.8 are folklore. A reader can find more information concerning results of Exercise 9.10 in W. Just and M. Weese [1997].

Metatheorem 9.6 was proved by R.M. Solovay [1970b]. Lemma 9.19 is proved in J. Mycielski [1964b], however the author attributes the result to W. Sierpiński. Theorem 9.21 was proved by M. Talagrand [1980]. Our proof follows his ideas. Lemmas 9.24–9.26 and Theorem 9.27 are due to J. Raisonnier [1984]. Results of Exercises 9.13 and 9.14 are partially contained in J. Mycielski [1964b].

In the theory of models of set theory the following is well known: if $\aleph_1$ is not inaccessible in the constructible universe $\mathbf{L}$, then there exists a set $X \subseteq {}^\omega 2$ of cardinality $\aleph_1$ and a well-ordering $R \in \mathbf{\Sigma}_2^1$. Thus the measurability of all $\mathbf{\Sigma}_3^1$ sets implies consistency of the existence of an inaccessible cardinal. Hence the existence of an inaccessible cardinal is necessary in a demonstration of Metatheorem 9.6. Consequently, the theories $\mathbf{ZFC}$ + "there exists an inaccessible cardinal" and $\mathbf{ZF} + \mathbf{wAC} +$ "every set of reals is Lebesgue measurable" are equiconsistent. On the other side, the consistency of the theory $\mathbf{ZF}$ implies the consistency of the theory $\mathbf{ZF} + \mathbf{wAC} +$ "every set of reals possesses the Baire Property". Thus the consistency strength of $\mathbf{wAC} +$ "every set of reals is Lebesgue measurable" is strictly greater than that of $\mathbf{wAC} +$ "every set of reals possesses the Baire Property". We can consider this result as another failure of the measure-category duality (see also Section 11.5).

As we have said the Axiom of Determinacy was proposed by J. Mycielski and H. Steinhaus [1962]. Since a great deal of results related to infinite games (e.g., those by S. Banach and S. Mazur) was not published, we recommend to the reader J. Mycielski [1964b] for historical remarks. See also a survey by R. Telgársky [1987]. Theorem 9.29 appeared in J. Mycielski [1964b]. Theorem 9.30 was proved by J. Mycielski and S. Swierczkowski [1964]. The presented proof follows L. Harrington (unpublished). The Solovay's game of Theorem 9.32 was invented by R.M. Solovay, see, e.g., Y. Moschovakis [1980] or T. Jech [2006]. Theorem 9.35 was essentially refined by L. Harrington [1978]. The Coding Lemma 9.37 and Theorem 9.38 was proved by Y.N. Moschovakis [1970]. See also Y.N. Moschovakis [1980].

Chapter 10

Undecidable Statements

> ... *le domaine des ensembles projectifs est un domain où le tiers exclu ne s'aplique plus* ...
>
> Nikolaj N. Luzin [1930], p. 323.

> ... *область проективных множеств есть область, где принцип исключенного третьего уже неприменим* ...
>
> Russian translation of Nikolaj N. Luzin [1930], p. 321.

> ... *the domain of projective sets is a domain where the principle of excluded third can no longer be applied* ...

Actually, any of the additional axioms of set theory considered in Chapter 9 is an undecidable statement of set theory. It is known that assuming consistency of **ZF** (sometimes one needs stronger assumptions), one can show consistency of both theories **ZF** + "the additional axiom" and **ZF** + "the negation of the additional axiom". However, neither of those axioms is a statement formulated by a working mathematician non-specialist in set theory[1]. In this chapter we present several questions formulated in some mathematical fields related to the structure of the real line which have no answer in **ZFC**.

10.1 Projective Sets

In the proofs of many results of Section 7.2 we have replaced the axiom of choice by a weaker condition: "there exists a well-ordering of the real line", or equivalently, "$\mathfrak{c}$ is an aleph". The natural question arises immediately: how nice can a well-ordering of the real line be? If the axiom of constructibility holds true (see Section 11.5), then there exists a Σ_2^1 well-ordering. Is this result the best possible?

---

[1]Maybe the assertion "every set of reals is Lebesgue measurable" is an exception, since it essentially simplifies the theory of integration.

Consider a strict well-ordering W of $[0,1]$, i.e., the set $[0,1]$ is well ordered by the relation $\langle x, y \rangle \in W$. We denote the diagonal by $D = \{\langle x, x \rangle : x \in [0,1]\}$. Then the sets W, D and W^{-1} form a partition of $[0,1] \times [0,1]$. The descriptive complexity of the set W is the same as that of W^{-1}. Since the set D is closed, we obtain the following simple fact:

$$W \in \mathbf{\Sigma}_n^1 \equiv W \in \mathbf{\Pi}_n^1,$$

i.e.,

$$W \in \mathbf{\Sigma}_n^1 \to W \in \mathbf{\Delta}_n^1.$$

Thus the existence of a $\mathbf{\Sigma}_n^1$ well-ordering of reals is equivalent to the existence of a $\mathbf{\Delta}_n^1$ well-ordering. Also the existence of a $\mathbf{\Pi}_n^1$ well-ordering of reals is equivalent to the existence of a $\mathbf{\Delta}_n^1$ well-ordering.

We shall use the notation of the proof of Theorem 7.17. We repeat a definition (7.10) of a selector V of the Vitali decomposition using the well-ordering W:

$$x \in V \equiv (\forall y)\,((x \neq y \land x - y \in \mathbb{Q}) \to \langle x, y \rangle \in W).$$

By the Kuratowski–Tarski Theorem 6.45, if W is a $\mathbf{\Delta}_n^1$ set, then V is a $\mathbf{\Pi}_n^1$ set. Moreover, then every set $V + r$ is a $\mathbf{\Pi}_n^1$ set. We know that a countable union of $\mathbf{\Pi}_n^1$ sets is a $\mathbf{\Pi}_n^1$ set (assuming **wAC**). Since

$$V = \mathbb{T} \setminus \bigcup_{r \in (0,1) \cap \mathbb{Q}} V + r,$$

we obtain that V is also a $\mathbf{\Sigma}_n^1$ set. So by Corollary 7.19 we obtain

Theorem [wAC] 10.1. *If there exists a $\mathbf{\Delta}_n^1$ well-ordering of $\mathbb{T}$, then there exists a $\mathbf{\Delta}_n^1$ set which is neither Lebesgue measurable nor possesses the Baire Property.*

Corollary [wAC] 10.2. *There is no $\mathbf{\Delta}_1^1$ well-ordering of $\mathbb{T}$.*

Thus Gödel's result (11.15) about the existence of a $\mathbf{\Sigma}_2^1$ well-ordering of the real line (and consequently of a $\mathbf{\Delta}_2^1$ well-ordering) is the best possible.

Lemma [AC] 10.3. *If $\mathrm{add}(\mathcal{N}) > \aleph_1$, then every $\mathbf{\Sigma}_2^1$ set is Lebesgue measurable. If $\mathrm{add}(\mathcal{M}) > \aleph_1$, then every $\mathbf{\Sigma}_2^1$ set possesses the Baire Property.*

Proof. By Theorem 6.58 every $\mathbf{\Sigma}_2^1$ set is a union of $\aleph_1$ Borel sets. $\square$

We obtain immediately

Theorem [AC] 10.4. *If $\mathrm{add}(\mathcal{N}) > \aleph_1$ or $\mathrm{add}(\mathcal{M}) > \aleph_1$, then there is no $\mathbf{\Delta}_2^1$ well-ordering of the real line.*

The Axiom of Constructibility $\mathbf{V} = \mathbf{L}$ has an important consequence (11.14) which we repeat here for convenience of the reader:

there exists a well-ordering $<_L$ of $^\omega\omega$ in order type ω_1 such that
$R = \{\langle \alpha, \beta \rangle \in {^\omega\omega} \times {^\omega\omega} : \{\gamma \in {^\omega\omega} : \gamma <_L \alpha\} = \{\mathrm{Proj}_n(\beta) : n \in \omega\}\}$
is a $\mathbf{\Sigma}_2^1$ set and $(\forall \alpha \in {^\omega\omega})(\exists \beta \in {^\omega\omega})\,\langle \alpha, \beta \rangle \in R$.   (10.1)

We know that (10.1) implies that the well-ordering $<_L$ is a $\mathbf{\Sigma}_2^1$ set.

The assumption (10.1) implies the continuum hypothesis $2^{\aleph_0} = \aleph_1$ and **wAC**.

By the basic result of K. Gödel [1944] the Axiom of Constructibility is consistent with **ZFC**, therefore (10.1) is consistent with **ZFC**. By Metatheorem 11.7 each of the assumptions $\text{add}(\mathcal{M}) > \aleph_1$, $\text{add}(\mathcal{N}) > \aleph_1$ is consistent with **ZFC**, thus we can summarize:

Metatheorem 10.1. *If* **ZF** *is consistent, then sentences* "there exists a $\mathbf{\Delta}_2^1$ well-ordering of the reals", "there exists a non-Lebesgue measurable $\mathbf{\Sigma}_2^1$ set of reals", "there exists a $\mathbf{\Sigma}_2^1$ set of reals not possessing the Baire Property" *are undecidable in* **ZFC**.

By Corollary 6.63 the existence of a $\mathbf{\Sigma}_2^1$ set of cardinality $\aleph_1$ is equivalent to the existence of a $\mathbf{\Pi}_1^1$ set of cardinality $\aleph_1$. Note that $|{}^\omega\omega \cap \mathbf{L}| = |\aleph_1^{\mathbf{L}}|$. Thus, in any model of **ZFC** $+ \neg$**CH** in which $\aleph_1^{\mathbf{L}}$ is not countable, the set of constructible reals ${}^\omega\omega \cap \mathbf{L}$ is a $\mathbf{\Sigma}_2^1$ set of cardinality $\aleph_1$, since

$$\alpha \in {}^\omega\omega \cap \mathbf{L} \equiv (\exists \beta) \langle \alpha, \beta \rangle \in <_L .$$

Thus in every such model there exists a $\mathbf{\Pi}_1^1$ set of reals of cardinality $\aleph_1$. Actually we know many models of this type, see, e.g., T. Jech [2006]. On the other hand A. Levy and R.M. Solovay [1967] have shown that if **ZFC**+"there exists a measurable cardinal" is consistent, then the theory **ZFC**+"there exists a measurable cardinal"$+\neg$**CH** is also consistent. By Theorem 9.36 in such a theory every $\mathbf{\Pi}_1^1$ set has cardinality $\mathfrak{c}$. Therefore we obtain:

Metatheorem 10.2. *If* **ZFC** $+$ "there exists a measurable cardinal" *is consistent, then the sentence* "there exists a $\mathbf{\Pi}_1^1$ set of reals of cardinality $\aleph_1$" *is undecidable in* **ZFC**.

We present some consequences of the axiom of constructibility concerning the properties of projective sets. Actually, we use the assertion (10.1), which enables us to replace a universal $<_L$-bounded quantifier by existential ones

$$(\forall \beta <_L \alpha)\varphi(\beta) \equiv (\exists \gamma)(\langle \alpha, \gamma \rangle \in R \wedge (\forall n)(\exists \delta)(\text{Proj}_n(\gamma) = \delta \wedge \varphi(\delta))) \qquad (10.2)$$

and an existential $<_L$-bounded quantifier by universal ones

$$(\exists \beta <_L \alpha)\varphi(\beta) \equiv (\forall \gamma)(\langle \alpha, \gamma \rangle \in R \to (\exists n)(\forall \delta)(\text{Proj}_n(\gamma) = \delta \to \varphi(\delta)). \qquad (10.3)$$

Theorem 10.5 (J.W. Addison). *If* (10.1) *holds true, then every pointclass* $\mathbf{\Sigma}_n^1$, $n \geq 2$ *is ranked. Hence, for $n \geq 2$, the pointclass* $\mathbf{\Sigma}_n^1$ *has the reduction property and the pointclass* $\mathbf{\Pi}_n^1$ *has the separation property.*

Proof. The later assertions follow from the former one by Theorems 5.22 and 5.20, b).

Let X be a Polish space, $A \in \mathbf{\Sigma}_n^1(X)$, $n > 1$. We define a $\mathbf{\Sigma}_n^1$-rank on A. By Theorem 6.39, c) there exists a set $B \in \mathbf{\Pi}_{n-1}^1(X \times {}^\omega\omega)$ such that

$$x \in A \equiv (\exists \alpha \in {}^\omega\omega) \langle x, \alpha \rangle \in B.$$

Since $^\omega\omega$ is well ordered by $<_L$ in the order type ω_1, there exists an order preserving mapping $d_L : {}^\omega\omega \xrightarrow[\text{onto}]{1-1} \omega_1$. We define a rank ρ on the set A as follows:

$$\rho(x) = \min\{d_L(\alpha) : \langle x, \alpha\rangle \in B\}.$$

We need to show that ρ is a $\boldsymbol{\Sigma}_n^1$-rank.

One can easily check that for any $x \in X$ and any $y \in A$ the following equivalences hold true:

$$\rho(x) \le \rho(y) \equiv (x, y \in A \wedge (\exists\alpha, \beta \in {}^\omega\omega)\, (\langle x, \alpha\rangle \in B \wedge \langle y, \beta\rangle \in B$$
$$\wedge\, \alpha \le_L \beta \wedge (\forall\delta <_L \beta)\, \langle y, \delta\rangle \notin B)), \tag{10.4}$$
$$\rho(x) \le \rho(y) \equiv (\forall\beta \in {}^\omega\omega)\, (\langle y, \beta\rangle \in B \to (\exists\alpha \le_L \beta)\, \langle x, \alpha\rangle \in B).$$

By (10.2) the formula $(\forall\delta <_L \beta)\, \langle y, \delta\rangle \notin B$ is equivalent to the formula

$$(\exists\gamma)\, (\langle\beta, \gamma\rangle \in R \wedge (\forall n)(\exists\alpha)\, (\mathrm{Proj}_n(\gamma) = \alpha \wedge \langle x, \alpha\rangle \notin B))$$

and by (10.3) the formula $(\exists\alpha \le_L \beta)\, \langle x, \alpha\rangle \in B$ is equivalent to the formula

$$(\forall\gamma)\, (\langle\beta, \gamma\rangle \in R \to (\exists n)(\forall\delta)\, (\mathrm{Proj}_n(\gamma) = \delta \to \langle x, \delta\rangle \in B)) \vee \langle x, \beta\rangle \in B.$$

Thus by Theorems 6.43 and 6.45, the formula $(\forall\delta <_L \beta)\, \langle y, \delta\rangle \notin B$ is $\boldsymbol{\Sigma}_n^1$ and the formula $(\exists\alpha \le_L \beta)\, \langle x, \alpha\rangle \in B$ is $\boldsymbol{\Pi}_n^1$. Consequently the formulas on the right sides of (10.4) are $\boldsymbol{\Sigma}_n^1$ and $\boldsymbol{\Pi}_n^1$, respectively. Thus the relation $\rho(x) \le \rho(y)$) satisfies the condition b) of Lemma 5.21 and ρ is a $\boldsymbol{\Sigma}_n^1$-rank. $\qquad\square$

Theorem 10.6. *Assume* (10.1) *holds true. Then for $n \ge 2$, every pointclass $\boldsymbol{\Sigma}_n^1$ has the uniformization property.*

Proof. By Lemma 5.26 it suffices to consider the sets from $\boldsymbol{\Sigma}_n^1(X \times {}^\omega\omega)$, where X is a Polish space. So let $A \in \boldsymbol{\Sigma}_n^1(X \times {}^\omega\omega)$ and

$$\langle x, \alpha\rangle \in A \equiv (\exists\beta)\, \langle x, \alpha, \beta\rangle \in C,$$

where $C \in \boldsymbol{\Pi}_{n-1}^1(X \times {}^\omega\omega \times {}^\omega\omega)$.

Let Π, Λ and R be the continuous mappings between $^\omega\omega$ and $^\omega\omega \times {}^\omega\omega$ defined by (1.11) and (1.12). Set

$$\langle x, \alpha\rangle \in E \equiv (\langle x, \Lambda(\alpha), R(\alpha)\rangle \in C \wedge (\forall\beta <_L \alpha)\, \langle x, \Lambda(\beta), R(\beta)\rangle \notin C).$$

Using (10.2) one can easily show that $E \in \boldsymbol{\Sigma}_n^1(X \times {}^\omega\omega)$. Now we define

$$\langle x, \alpha\rangle \in B \equiv (\exists\beta)\, \langle x, \Pi(\alpha, \beta)\rangle \in E.$$

Evidently $B \in \boldsymbol{\Sigma}_n^1(X \times {}^\omega\omega)$ and B uniformizes A. $\qquad\square$

We show that the Axiom of Projective Determinacy **PD** has essentially different consequences. We begin with an important result often called the **First Periodicity Theorem**.

Theorem [wAC] 10.7 (D.A. Martin – Y.N. Moschovakis). *Let* $\mathrm{DET}_\omega(\Delta[\Gamma])$ *hold true for a pointclass* Γ. *If every set in* $\Gamma(X \times {}^\omega\omega)$ *admits a* Γ*-rank, then every set in* $\forall^{{}^\omega\omega}[\Gamma](X)$ *admits a* $\forall^{{}^\omega\omega}\exists^{{}^\omega\omega}[\Gamma]$*-rank.*

Proof. Assume that $B \in \Gamma(X \times {}^\omega\omega)$, $A = \{x \in X : (\forall\alpha \in {}^\omega\omega)\, \langle x, \alpha\rangle \in B\}$ and ρ is a Γ-rank on B. For given $x, y \in A$ we set

$$C_{x,y} = \{\alpha \in {}^\omega\omega : \langle y, R(\alpha)\rangle <^*_\rho \langle x, \Lambda(\alpha)\rangle\}.$$

By definition $C_{x,y} \in \Gamma({}^\omega\omega)$.

If $y \in A$, then $(\forall\alpha)\, \langle y, \alpha\rangle \in B$ and therefore

$${}^\omega\omega \setminus C_{x,y} = \{\alpha \in {}^\omega\omega : \langle x, \Lambda(\alpha)\rangle \leq^*_\rho \langle y, R(\alpha)\rangle\}$$

and $C_{x,y} \in \neg\Gamma({}^\omega\omega)$. Thus $\mathrm{GAME}_\omega(C_{x,y})$ is determined.

We define

$$x \sqsubseteq y \equiv (x, y \in A \wedge (\text{player II has a winning strategy in } \mathrm{GAME}_\omega(C_{x,y}))). \quad (10.5)$$

We show that $\sqsubseteq$ is a $\forall^{{}^\omega\omega}[\Gamma]$-rank on A.

Let us note the following simple facts. If α is a run in $\mathrm{GAME}_\omega(C_{x,y})$, then

$$\text{player I wins} \equiv \langle y, R(\alpha)\rangle <^*_\rho \langle x, \Lambda(\alpha)\rangle, \quad (10.6)$$

$$\text{player II wins} \equiv \langle x, \Lambda(\alpha)\rangle \leq^*_\rho \langle y, R(\alpha)\rangle. \quad (10.7)$$

If $x \in A$ and if player II plays the same play as player I, then II wins in $\mathrm{GAME}_\omega(C_{x,x})$. Hence $\sqsubseteq$ is reflexive on A.

Assume that $x, y, z \in A$, $x \sqsubseteq y$ and $y \sqsubseteq z$. Then player II has winning strategies in both games $\mathrm{GAME}_\omega(C_{x,y})$ and $\mathrm{GAME}_\omega(C_{y,z})$. We describe a winning strategy for player II in $\mathrm{GAME}_\omega(C_{x,z})$. Assume that α is a play of player I. Let β be the answer of player II in $\mathrm{GAME}_\omega(C_{x,y})$ following the winning strategy. Now, let player I play β in $\mathrm{GAME}_\omega(C_{y,z})$ and let γ be the answer of player II following the winning strategy. In $\mathrm{GAME}_\omega(C_{x,z})$ player II will play the same play[2] γ as an answer to the play α of player I. Since all x, y, z are in A we know that $\langle x, \alpha\rangle, \langle y, \beta\rangle, \langle z, \gamma\rangle$ are in B and therefore

$$\langle x, \alpha\rangle \leq_\rho \langle y, \beta\rangle, \qquad \langle y, \beta\rangle \leq_\rho \langle z, \gamma\rangle.$$

Hence also $\langle x, \alpha\rangle \leq_\rho \langle z, \gamma\rangle$ and player II wins.

We show that $x \sqsubseteq y$ or $y \sqsubseteq x$ for every $x, y \in A$. Assume that $\neg x \sqsubseteq y$. Then by (10.5) player II does not have a winning strategy in $\mathrm{GAME}_\omega(C_{x,y})$. Since the

---

[2]Of course, everything will be done step by step.

game is determined, player I does have a winning strategy. We describe a winning strategy for player II in $\text{GAME}_\omega(C_{y,x})$. Actually, the player's II answer β to a play α by player I is the same as the answer by player I in $\text{GAME}_\omega(C_{x,y})$ to α played by player II. Since player I followed the winning strategy, by (10.6) we obtain $\langle y, \alpha \rangle <^*_\rho \langle x, \beta \rangle$. Then by (10.7) player II wins in $\text{GAME}_\omega(C_{y,x})$. Thus $y \sqsubseteq x$.

We have to show that there is no strictly $\sqsubseteq$-descending infinite sequence of elements of the set A. To get a contradiction assume that $x_n \in A$, and $x_{n+1} \sqsubset x_n$ for every $n \in \omega$. Thus by the definition (10.5) player I has a winning strategy in each $\text{GAME}_\omega(C_{x_n, x_{n+1}})$, $n \in \omega$. Using **wAC**, we choose a sequence $\langle f_n : n \in \omega \rangle$ of winning strategies of player I in corresponding games. Following those strategies in the first step of games $\text{GAME}_\omega(C_{x_n, x_{n+1}})$ player I plays $\alpha_n(0)$ for any n. We let player II answer $\beta_n(0) = \alpha_{n+1}(0)$. By induction, at the step k player I following the strategy f_n plays $\alpha_n(k) = f_n(\{\beta_n(i)\}_{i=0}^{k-1})$. Again, we let player II answer $\beta_n(k) = \alpha_{n+1}(k)$. Thus in $\text{GAME}_\omega(C_{x_n, x_{n+1}})$ the play of player I is α_n and the play of player II is α_{n+1}. Since player I followed her/his winning strategies in each game of the sequence $\langle \text{GAME}_\omega(C_{x_n, x_{n+1}}) : n \in \omega \rangle$, we obtain $\langle x_{n+1}, \alpha_{n+1} \rangle <_\rho \langle x_n, \alpha_n \rangle$ for each $n \in \omega$, a contradiction.

By definition, a strategy $\text{GAME}_\omega(A)$ is a function $f : {}^{<\omega}\omega \longrightarrow \omega$. Since the space ${}^{({}^{<\omega}\omega)}\omega$ is homeomorphic to ${}^\omega\omega$ we can consider a strategy as an element of the Baire space ${}^\omega\omega$. We recall the notation introduced in Section 5.5: if player I follows a strategy α and player II plays β, then the resulting run is $\bar{\alpha}_I(\beta)$. The mapping $F : {}^\omega\omega \times {}^\omega\omega \longrightarrow {}^\omega\omega$ defined as $F(\alpha, \beta) = \bar{\alpha}_I(\beta)$ is continuous. Similarly for player II.

Taking into account the determinacy assumption, by definition of the ordering $\sqsubseteq$ we obtain

$$x \sqsubseteq y \equiv (x, y \in A \land (\forall \alpha \in {}^{({}^{<\omega}\omega)}\omega)(\exists \beta \in {}^\omega\omega)\, \bar{\alpha}_I(\beta) \notin C_{x,y}),$$

and therefore the relation $\sqsubseteq$ belongs to $\forall^{\omega}{}^\omega \exists^{\omega}{}^\omega[\,\Gamma\,]$. Similarly

$$x \sqsubset y \equiv (x, y \in A \land (\forall \alpha \in {}^{({}^{<\omega}\omega)}\omega)(\exists \beta \in {}^\omega\omega)\, \bar{\alpha}_{II}(\beta) \in C_{y,x}).$$

Thus $\sqsubseteq$ is a $\forall^{\omega}{}^\omega \exists^{\omega}{}^\omega[\,\Gamma\,]$-rank. $\square$

As a main consequence of the First Periodicity Theorem we obtain the following result.

Theorem [wAC] 10.8 (D.A. Martin – Y.N. Moschovakis). *The Axiom of Projective Determinacy* **PD** *implies that the pointclasses* $\mathbf{\Sigma}^1_{2n}$ *and* $\mathbf{\Pi}^1_{2n-1}$ *are ranked for any $n > 0$.*

Proof. The theorem follows easily by Theorems 6.53, 5.24 and 10.7. $\square$

Similarly one can prove the **Second Periodicity Theorem**, see Exercise 10.3 and Y.N. Moschovakis [1980] or A.S. Kechris [1995].

Theorem [wAC] 10.9 (Y.N. Moschovakis). *Let* $\mathrm{DET}_\omega(\Delta[\Gamma])$ *hold true for a point-class* Γ. *If every set in* $\Gamma(X \times {}^\omega\omega)$ *admits a* Γ*-scale, then every set in* $\forall^\omega{}^\omega[\Gamma](X)$ *admits a* $\forall^\omega{}^\omega\exists^\omega{}^\omega[\Gamma]$*-scale.*

Again, the main consequence of this Theorem is

Theorem [wAC] 10.10 (Y.N. Moschovakis). *If the Axiom of Projective Determinacy* **PD** *holds true, then the pointclasses* $\mathbf{\Sigma}_{2n}^1$ *and* $\mathbf{\Pi}_{2n-1}^1$ *have the uniformization property for any* $n > 0$.

We summarize the situation with two different assumptions: the Axiom of Constructibility (the universe of the set theory is the smallest possible) and the Axiom of Projective Determinacy (you have enough strategies for winning your games). A boxed pointclass in the next pictures possesses the uniformization property, is ranked and therefore has the reduction property. Recall that by Corollary 5.23 the pointclasses $\mathbf{\Pi}_n^1$ and $\mathbf{\Sigma}_n^1$ cannot be both ranked.

In **ZFC**+Axiom of Constructibility, even in **ZF**+(10.1)+**wAC**, we have

$\mathbf{\Sigma}_1^1$	$\boxed{\mathbf{\Sigma}_2^1}$	$\boxed{\mathbf{\Sigma}_3^1}$	$\cdots$	$\boxed{\mathbf{\Sigma}_{2n-1}^1}$	$\boxed{\mathbf{\Sigma}_{2n}^1}$	$\boxed{\mathbf{\Sigma}_{2n+1}^1}$	$\cdots$
$\boxed{\mathbf{\Pi}_1^1}$	$\mathbf{\Pi}_2^1$	$\mathbf{\Pi}_3^1$	$\cdots$	$\mathbf{\Pi}_{2n-1}^1$	$\mathbf{\Pi}_{2n}^1$	$\mathbf{\Pi}_{2n+1}^1$	$\cdots$

Assuming the Axiom of Projective Determinacy **PD** and **wAC** we obtain an essentially different picture:

$\mathbf{\Sigma}_1^1$	$\boxed{\mathbf{\Sigma}_2^1}$	$\mathbf{\Sigma}_3^1$	$\cdots$	$\mathbf{\Sigma}_{2n-1}^1$	$\boxed{\mathbf{\Sigma}_{2n}^1}$	$\mathbf{\Sigma}_{2n+1}^1$	$\cdots$
$\boxed{\mathbf{\Pi}_1^1}$	$\mathbf{\Pi}_2^1$	$\boxed{\mathbf{\Pi}_3^1}$	$\cdots$	$\boxed{\mathbf{\Pi}_{2n-1}^1}$	$\mathbf{\Pi}_{2n}^1$	$\boxed{\mathbf{\Pi}_{2n+1}^1}$	$\cdots$

Exercises

10.1 Well-ordering of ${}^\omega\omega$

Assume that $\prec$ is a well-ordering of ${}^\omega\omega$ in type ω_1. The relation R is defined as in (10.1).

a) If $R \in \mathbf{\Sigma}_n^1$, then $\prec\, \in \mathbf{\Delta}_n^1$.

b) Show that

$$(\forall\beta \prec \alpha)\, \varphi(\beta, \dots) \equiv (\forall\gamma)\, ((\alpha, \gamma) \in R \to (\forall n)(\exists\delta)\, (\mathrm{Proj}_n(\gamma) = \delta \wedge \varphi(\delta, \dots)))$$
$$\equiv (\exists\gamma)\, ([\alpha\gamma] \in R \wedge (\forall n)(\exists\delta)\, \mathrm{Proj}_n(\gamma) = \delta \wedge \varphi(\delta, \dots))),$$
$$(\exists\beta \prec \alpha)\, \varphi(\beta, \dots) \equiv (\exists\gamma)\, ((\alpha, \gamma) \in R \wedge (\exists n)(\exists\delta)\, (\mathrm{Proj}_n(\gamma) = \delta \wedge \varphi(\delta, \dots)))$$
$$\equiv (\forall\gamma)\, ((\alpha, \gamma) \in R \to (\exists n)(\exists\delta)\, (\mathrm{Proj}_n(\gamma) = \delta \wedge \varphi(\delta, \dots))).$$

c) If $R \in \mathbf{\Sigma}_n^1$, then every pointclass $\mathbf{\Sigma}_m^1$, $m \geq n$ is ranked and has the uniformization property.

10.2 [wAC] Unions of $\mathbf{\Delta}_n^1$ sets
The Moschovakis numbers δ_n^1 were defined in Exercise 6.13.

a) If $\mathbf{\Pi}_n^1$, $n > 0$ is ranked, then every set $A \in \mathbf{\Sigma}_{n+1}^1(X)$ is a union of δ_n^1 sets from $\mathbf{\Delta}_n^1$.
   Hint: See Exercise 5.8, c).

b) If $\mathbf{\Sigma}_n^1$, $n > 1$ is ranked, then any union $\bigcup_{\xi<\delta_{n-1}^1} A_\xi$ with $A_\xi \in \mathbf{\Pi}_{n-1}^1$ belongs to $\mathbf{\Sigma}_n^1$.

c) If **AD** holds true, then for any $n > 0$,

$$A \in \mathbf{\Sigma}_{2n}^1 \equiv A \text{ is a union of } \delta_{2n-1}^1 \text{ sets from } \mathbf{\Delta}_{2n-1}^1.$$

10.3 Proof of the Second Periodicity Theorem
Let X be a Polish space, $n > 0$, Γ being a pointclass, $\Delta = \Gamma \cap \neg\Gamma$. Assume that $\text{GAME}_\omega(D)$ is determined for every $D \in \Delta(^\omega\omega)$ and $B \in \Gamma(X \times {}^\omega\omega)$ admits a Γ-scale. Then by Exercise 5.9 there exists a very good Γ-scale $\langle \rho_n : n \in \omega \rangle$ on B. Fix an enumeration $^{<\omega}\omega = \{s_n : n \in \omega\}$ such that $s_0 = \emptyset$ and $s_i \subseteq s_j \to i \leq j$.

a) Let
$$C_{x,y}^n = \{\alpha \in {}^\omega\omega : \langle y, s_n{}^\frown R(\alpha)\rangle <_{\rho_n}^* \langle x, s_n{}^\frown \Lambda(\alpha)\rangle\}.$$
   Show that $C_{x,y}^n \in \Delta$.
   Hint: See the proof of Theorem 10.7.

b) Show that
$$x \sqsubseteq_n y \equiv (x, y \in A \wedge (\text{II has a winning strategy in } \text{GAME}_\omega(C_{x,y}^n)))$$
   is a $\forall^{\omega}\omega[\Gamma]$-rank on A.
   Hint: Again, follow the proof of Theorem 10.7.

c) Define
$$x \preceq_n y \equiv (x \sqsubset_0 y \vee (x \sqsubseteq_0 y \wedge y \sqsubseteq_0 x \wedge x \sqsubseteq_n y)).$$
   Then $\langle \preceq_n : n \in \omega \rangle$ is a $\forall^{\omega}\omega[\Gamma]$ scale on $\forall^{\omega}\omega B$.
   Hint: Consult A.S. Kechris [1995], pp. 338–339 or Y.N. Moschovakis [1980], pp. 311–317.

d) Prove the Second Periodicity Theorem 10.10.
   Hint: Apply the results of c) and Exercise 5.10.

10.2   Measure Problem

In Section 7.2 we have shown that if the real line can be well ordered, then there exists a Lebesgue non-measurable set. Actually we have proved a stronger result, Corollary 7.19 saying the following: if $\mathcal{S} \subseteq \mathcal{P}(\mathbb{T})$ is a σ-algebra and ν is a probabilistic shift invariant measure defined on $\mathcal{S}$, then a Vitali set V does not belong to $\mathcal{S}$. Thus, the Axiom of Choice implies that there exists no shift invariant probabilistic measure on $\mathcal{P}(\mathbb{T})$ and therefore neither on $\mathcal{P}([0,1])$. The shift invariance has been essentially used. Thus, a natural question arises.

The Measure Problem. *Is there a probabilistic diffused measure on $\mathcal{P}([0,1])$?*

Let us notice that the Measure Problem is not a question about the real line but about any set of cardinality $\mathfrak{c}$. As we shall see **ZFC** does not decide the answer to the Measure Problem.

We shall consider a diffused probabilistic measure ν defined on $\mathcal{P}(X)$. The measure ν is called κ-**additive** if for every cardinal $\gamma < \kappa$ and for every disjoint family $\langle X_\xi : \xi < \gamma \rangle$ of subsets of X, we have

$$\nu \left(\bigcup_{\xi < \gamma} X_\xi \right) = \sum_{\xi < \gamma} \nu(X_\xi).$$

Similarly as in Theorem 7.4 one can show that the measure ν is κ-additive if and only if $\text{add}(\mathcal{N}(\nu)) \geq \kappa$.

An uncountable cardinal number κ is called **real-valued measurable** if there exists a diffused probabilistic κ-additive measure on $\mathcal{P}(\kappa)$.[3]

Theorem [AC] 10.11. *If the answer to the Measure Problem is affirmative, then there exists a real-valued measurable cardinal κ such that $\aleph_0 < \kappa \leq \mathfrak{c}$.*

Proof. Let ν be a diffused probabilistic σ-additive measure defined on $\mathcal{P}([0,1])$. Since $\nu(\{x\}) = 0$ for any $x \in [0,1]$ and $\nu([0,1]) = 1$, the measure ν is not $\mathfrak{c}^+$-additive. Let κ be the smallest cardinal such that ν is not κ^+-additive. Evidently $\aleph_0 < \kappa \leq \mathfrak{c}$. We show that κ is a real-valued measurable cardinal.

By definition of κ there are pairwise disjoint sets $\langle X_\xi : \xi \in \kappa \rangle$ such that

$$\nu(X_\xi) = 0 \text{ for every } \xi < \kappa \text{ and } \nu \left(\bigcup_{\xi < \kappa} X_\xi \right) > 0.$$

Let $\varepsilon = \nu(\bigcup_{\xi \in \kappa} X_\xi) > 0$. It is easy to see that the function λ defined for $A \subseteq \kappa$ by

$$\lambda(A) = \nu \left(\bigcup_{\xi \in A} X_\xi \right) / \varepsilon.$$

is a diffused κ-additive probabilistic measure on $\mathcal{P}(\kappa)$. Thus κ is real-valued measurable. $\square$

There are classical important results concerning real-valued measurable cardinals. The first one reads as follows.

Theorem [AC] 10.12 (S. Ulam). *A real-valued measurable cardinal is weakly inaccessible.*

The theorem immediately follows from two lemmas.

Lemma 10.13. *Every real-valued measurable cardinal is regular.*

---

[3]Let us recall that κ is the set of all smaller ordinals, i.e., a set of cardinality κ.

Proof. The proof is easy. Assume that κ is real-valued measurable with a measure ν and κ is singular. Then

$$\kappa = \bigcup_{\xi \in \mathrm{cf}(\kappa)} A_\xi, \quad |A_\xi| < \kappa \text{ for each } \xi \in \mathrm{cf}(\kappa).$$

Since ν is κ-additive we have $\nu(A_\xi) = 0$ for any $\xi < \mathrm{cf}(\kappa)$. Since $\mathrm{cf}(\kappa) < \kappa$, then also $\nu(\kappa) = 0$, a contradiction. □

Lemma [AC] 10.14. *If $\kappa = \lambda^+$, then κ is not a real-valued measurable cardinal.*

Proof. We begin with constructing an **Ulam matrix**, i.e., a system

$$\{A_{\xi,\eta} : \xi \in \lambda, \eta \in \kappa\}$$

of subsets of κ such that

1) $A_{\xi,\eta_1} \cap A_{\xi,\eta_2} = \emptyset$ for every $\eta_1 \neq \eta_2$;
2) $|\kappa \setminus \bigcup_{\xi<\lambda} A_{\xi,\eta}| < \kappa$ for every $\eta \in \kappa$.

For every $\eta < \kappa$ choose a one-to-one mapping $f_\eta : \eta \xrightarrow{1-1} \lambda$. Now, set

$$A_{\xi,\eta} = \{\zeta \in \kappa : \eta < \zeta \wedge f_\zeta(\eta) = \xi\}.$$

If $\eta_1 \neq \eta_2$, then $f_\zeta(\eta_1) \neq f_\zeta(\eta_2)$ and therefore $A_{\xi,\eta_1} \cap A_{\xi,\eta_2} = \emptyset$. By the definition we have

$$\zeta \in \bigcup_{\xi<\lambda} A_{\xi,\eta} \equiv (\exists \xi < \lambda)\, f_\zeta(\eta) = \xi \equiv \eta < \zeta.$$

Thus

$$\left| \kappa \setminus \bigcup_{\xi<\lambda} A_{\xi,\eta} \right| \leq |\eta| + 1 < \kappa.$$

Now assume that ν is a κ-additive diffused probabilistic measure on $\mathcal{P}(\kappa)$. For every $\xi < \lambda$ the set $S_\xi = \{\eta : \nu(A_{\xi,\eta}) > 0\}$ is countable. Hence the set $S = \bigcup_{\xi<\lambda} S_\xi$ has cardinality smaller than κ. Thus, there exists an $\eta_0 < \kappa$ such that $\eta_0 \notin S$, i.e.,

$$\nu(A_{\xi,\eta_0}) = 0 \text{ for every } \xi < \lambda.$$

By 2) we have $\nu(\bigcup_{\xi<\lambda} A_{\xi,\eta_0}) = 1$ – a contradiction. □

Corollary [AC] 10.15. *If $2^{\aleph_0}$ is smaller than the first weakly inaccessible cardinal, e.g. if $2^{\aleph_0} = \aleph_1$, or $2^{\aleph_0} = \aleph_{\omega_1+1}$, then the answer to the Measure Problem is negative.*

Thus by the Consequence (11.12) of the Kuratowski Metatheorem 11.3 we obtain

Metatheorem 10.3. *If the theory **ZFC** is consistent, then*

$$\mathbf{ZFC} \nvdash \text{``Affirmative answer to the Measure Problem''}.$$

A cardinal κ is called **measurable** if there exists a κ-additive diffused probabilistic measure on $\mathcal{P}(\kappa)$ admitting only values $0, 1$. In this case we say that the measure is **two-valued**. Evidently a measurable cardinal is also a real-valued measurable. A weaker result holds true in the opposite direction.

Theorem [AC] 10.16 (S. Ulam). *Assume that κ is a real-valued measurable cardinal. Then either $\kappa \leq \mathfrak{c}$ or κ is measurable.*

Proof. Let ν be a κ-additive diffused probabilistic measure on $\mathcal{P}(\kappa)$. Assume that κ is not measurable.

For every set $A \subseteq \kappa$ of positive measure there exists a set $B \subseteq A$ such that $|B| = \kappa$ and $0 < \nu(B) < \nu(A)$. Actually, if not, then $\lambda(Y) = \nu(A \cap Y)/\nu(A)$ is a κ-additive two-valued measure on κ, a contradiction.

Now let $A \subseteq \kappa$ be such that $\nu(A) > \varepsilon > 0$. We claim that

$$\text{there exists a set } B \subseteq A \text{ such that } \frac{1}{2}\varepsilon < \nu(B) \leq \varepsilon. \tag{10.8}$$

Suppose that (10.8) does not hold true. We already know that there is a set $B \subseteq A$ such that $0 < \nu(B) < \nu(A)$. Since (10.8) does not hold true we have $\nu(B) > \varepsilon$ or $\nu(B) \leq \varepsilon/2$. Denote by A_0 the set B in the former case and $A \setminus B$ in the later case. If $\nu(B) \leq \varepsilon/2$, then $\nu(A \setminus B) > \varepsilon/2$ and since (10.8) does not hold true, also $\nu(A \setminus B) > \varepsilon$. Thus, in both cases we have $\varepsilon < \nu(A_0) < \nu(A)$. If A_ξ is defined, $\xi < \omega_1$, then again there is a set $B \subseteq A_\xi$ such that $0 < \nu(B) < \nu(A_\xi)$. As above let $A_{\xi+1}$ be that of the sets B, $A_\xi \setminus B$ which has measure greater than ε and smaller than $\nu(A_\xi)$. For $\xi < \omega_1$ limit we set $A_\xi = \bigcap_{\eta < \xi} A_\eta$. Then $\nu(A_\eta) \geq \varepsilon$. Since (10.8) does not hold true it must be $\nu(A_\eta) > \varepsilon$ and we can continue. So, we have constructed a (strictly) decreasing sequence $\{\nu(A_\xi) : \xi < \omega_1\}$ of reals, which is impossible.

Note that for any $\varepsilon > 0$, using (10.8), one can easily construct finitely many pairwise disjoint sets $A_0, \ldots, A_k$ such that $\kappa = \bigcup_{i=0}^{k} A_i$ and $\nu(A_i) \leq \varepsilon$ for any $i = 0, \ldots, k$.

For every m, let $\langle A_{m,n} : n \in \omega \rangle$ be pairwise disjoint subsets of κ such that $\kappa = \bigcup_{n \in \omega} A_{m,n}$ and $\nu(A_{m,n}) \leq 2^{-m}$ (apply the above note with $\varepsilon = 2^{-m}$ and for $n > k$ set $A_{m,n} = \emptyset$). By the distributive law we obtain

$$\kappa = \bigcap_m \bigcup_n A_{m,n} = \bigcup_{\alpha \in {}^\omega \omega} \bigcap_m A_{m,\alpha(m)}.$$

For every $\alpha \in {}^\omega \omega$ we have $\nu(\bigcap_m A_{m,\alpha(m)}) = 0$. If $\mathfrak{c} < \kappa$, then ν is $\mathfrak{c}^+$-additive and therefore

$$\nu(\kappa) = \sum_{\alpha \in {}^\omega \omega} \nu\left(\bigcap_m A_{m,\alpha(m)}\right) = 0,$$

a contradiction. Thus $\kappa \leq \mathfrak{c}$. $\qquad\square$

Actually, we have proved more. A measure μ on $\mathcal{P}(\kappa)$ is **atomless** if for any $A \subseteq \kappa$ with $\mu(A) > 0$, there exists a set $B \subseteq A$ such that $0 < \mu(B) < \mu(A)$. By the proof of the theorem we obtain e.g., that if κ is real-valued measurable cardinal with an atomless measure, then $\kappa \leq \mathfrak{c}$. We shall not study further properties of such measures, we stay without a proof an important result and we recommend a reader to consult T. Jech [2006] or a survey by D.H. Fremlin [1993].

Theorem [AC] 10.17. *The following are equivalent:*

a) *An affirmative answer to the Measure Problem.*
b) *There exists a real-valued measurable cardinal not greater than* $\mathfrak{c}$*.*
c) *There exists a real-valued measurable cardinal with an atomless measure.*
d) *There exists a measure on* $\mathcal{P}(\mathbb{R})$ *extending the Lebesgue measure.*

For a measurable cardinal Theorem 10.12 can be strengthened.

Theorem [AC] 10.18 (S. Ulam). *Every measurable cardinal is strongly inaccessible.*

Proof. Let κ be a measurable cardinal. Assume that there is a cardinal λ such that $\lambda < \kappa \leq 2^\lambda$. Let $X \subseteq \mathcal{P}(\lambda)$ be such that $|X| = \kappa$. Then there exists a κ-additive two-valued measure ν on $\mathcal{P}(X)$. We extend the measure on $\mathcal{P}(\mathcal{P}(\lambda))$ setting $\nu(Y) = \nu(Y \cap X)$ for any $Y \subseteq \mathcal{P}(\lambda)$. So extended measure is κ-additive and diffused.

For $\xi \in \lambda$ we write

$$A_\xi = \{Y \subseteq \lambda : \xi \in Y\}, \qquad B_\xi = \{Y \subseteq \lambda : \xi \notin Y\}.$$

Evidently either $\nu(A_\xi) = 1$ or $\nu(B_\xi) = 1$. Denote by C_ξ that of sets A_ξ, B_ξ which has measure 1. Let $Y_0 = \{\xi \in \lambda : \nu(A_\xi) = 1\}$. Then

$$Y \in C_\xi \equiv Y \cap \{\xi\} = Y_0 \cap \{\xi\}$$

for any $Y \subseteq \lambda$. Therefore $\bigcap_{\xi \in \lambda} C_\xi = \{Y_0\}$. Since $\nu(C_\xi) = 1$ and $\lambda < \kappa$ we obtain $\nu(\{Y_0\}) = 1$, a contradiction.

The theorem follows by Lemma 10.13. $\square$

For more than 30 years there was an open problem: is the first strongly inaccessible cardinal measurable? The negative answer was obtained in the early 1960s. We present a typical result of this kind.

Theorem [AC] 10.19 (W.P. Hanf). *If* κ *is a measurable cardinal, then there exists* κ *many strongly inaccessible cardinals smaller than* κ*.*

Proof. Let κ be a measurable cardinal, ν being a κ-additive diffused two-valued measure on $\mathcal{P}(\kappa)$. On the set $\mathcal{H} = {}^\kappa\kappa$ we define a preordering $\preceq$ by

$$f \preceq g \equiv \nu(\{\xi \in \kappa : f(\xi) \leq g(\xi)\}) = 1. \tag{10.9}$$

We shall identify in a suitable way the set $\mathcal{H}$ with the corresponding quotient set and therefore we deal with $\preceq$ as with an ordering. The corresponding strict ordering will be denoted by $\prec$ and the quotient equivalence by $\simeq$. Thus, e.g.,

$$f \simeq g \equiv \nu(\{\xi \in \kappa : f(\xi) = g(\xi)\}) = 1.$$

It is easy to see that $\langle \mathcal{H}, \preceq \rangle$ is a linearly ordered set. We show that $\preceq$ is a well-ordering. Assume that $\langle \mathcal{H}, \preceq \rangle$ is not well ordered. Then by Theorem 11.1 there exists a sequence $\langle f_n \in \mathcal{H} : n \in \omega \rangle$ such that $f_{n+1} \prec f_n$ for every $n \in \omega$. Let $A_n = \{\xi \in \kappa : f_{n+1}(\xi) < f_n(\xi)\}$. Then $\nu(A_n) = 1$ for every $n \in \omega$ and therefore also $\nu(\bigcap_n A_n) = 1$. Especially, $\bigcap_n A_n \neq \emptyset$. Thus there exists a $\xi \in \bigcap_n A_n$. Then $f_0(\xi) > f_1(\xi) > \cdots > f_n(\xi) > f_{n+1}(\xi) > \cdots$. Since values of the functions f_n are ordinals, we have a contradiction.

For any $\xi < \kappa$ we denote by $\hat{\xi}$ the function from $\mathcal{H}$ defined by $\hat{\xi}(\eta) = \xi$ for every $\eta \in \kappa$. Then $\hat{\eta} \prec \hat{\xi}$ if and only if $\eta < \xi$. Moreover, if $f \prec \hat{\xi}$, then there exists an $\eta < \xi$ such that $f \simeq \hat{\eta}$. Actually

$$\{\zeta \in \kappa : f(\zeta) < \xi\} = \bigcup_{\eta < \xi} \{\zeta \in \kappa : f(\zeta) = \eta\}.$$

Since $\nu(\{\zeta \in \kappa : f(\zeta) < \xi\}) = 1$ and ν is κ-additive, there exists an $\eta < \xi$ such that $\nu(\{\zeta \in \kappa : f(\zeta) = \eta\}) = 1$.

Let $d(\xi) = \xi$. Then $\hat{\xi} \prec d$ for every $\xi \in \kappa$. Since $\mathcal{H}$ is well ordered, there exists the least $h \in \mathcal{H}$ such that

$$\xi < \kappa \to \hat{\xi} \prec h, \tag{10.10}$$
$$f \prec h \to (\exists \xi < \kappa) f \simeq \hat{\xi}. \tag{10.11}$$

We set

$$A_1 = \{\xi \in \kappa : h(\xi) \text{ is singular}\},$$
$$A_2 = \{\xi \in \kappa : \text{there is a cardinal } \alpha \text{ such that } \alpha < h(\xi) \leq 2^\alpha\},$$
$$A_3 = \{\xi \in \kappa : h(\xi) \text{ is strongly inaccessible}\}.$$

Evidently $A_1 \cup A_2 \cup A_3 = \kappa$.

To get a contradiction assume that $\nu(A_1) = 1$. We define a function f as

$$\begin{aligned} f(\xi) &= 0 &&\text{for } \xi \in \kappa \setminus A_1, \\ f(\xi) &= \operatorname{cf}(h(\xi)) &&\text{for } \xi \in A_1. \end{aligned}$$

Then $f \prec h$ and by (10.11) there exists an $0 < \eta < \kappa$ such that $f \simeq \hat{\eta}$. If we set $B = \{\xi \in \kappa : \operatorname{cf}(h(\xi)) = \eta\}$, then $\nu(B) = 1$. By the definition of the set B, for every $\xi \in B$ there exists an increasing sequence $\{\alpha_\zeta^\xi : \zeta < \eta\}$ such that

$$\lim_{\zeta < \eta} \alpha_\zeta^\xi = h(\xi). \tag{10.12}$$

We set $g_\zeta(\xi) = \alpha_\zeta^\xi$ for $\xi \in B$ and $g_\zeta(\xi) = 0$ otherwise. Then $g_\zeta \prec h$ for every $\zeta < \eta$. By (10.11) there exists a $\beta_\zeta < \kappa$ such that $g_\zeta \simeq \hat{\beta}_\zeta$. Set

$$B_\zeta = \{\xi : g_\zeta(\xi) = \beta_\zeta\}.$$

Then $\nu(B_\zeta) = 1$ for every $\zeta < \eta$. Since κ is regular, there exists a $\beta < \kappa$ such that $\beta_\zeta < \beta$ for every $\zeta < \eta$. If we write $C = \{\xi : \beta < h(\xi)\}$, then by (10.10) we obtain $\nu(C) = 1$. Since the measure ν is κ-additive, we obtain

$$\nu\left(\bigcap_{\zeta<\eta} B_\zeta \cap B \cap C\right) = 1.$$

Especially this intersection is non-empty. Let $\xi_0 \in \bigcap_{\zeta<\eta} B_\zeta \cap B \cap C$. Then $\beta < h(\xi_0)$ and $g_\zeta(\xi_0) = \alpha_\zeta^{\xi_0} = \beta_\zeta < \beta$, a contradiction with (10.12). Thus $\nu(A_1) = 0$.

Now suppose that $\nu(A_2) = 1$. By definition, for every $\xi \in A_2$ there exists a cardinal $\alpha_\xi < h(\xi)$ such that $h(\xi) \leq 2^{\alpha_\xi}$. If we set $g(\xi) = \alpha_\xi$ for $\xi \in A_2$ and $g(\xi) = 0$ otherwise, then $g \prec h$. By (10.11) there exists a $\beta < \kappa$ such that $g \simeq \hat\beta$. Hence the set $B = \{\xi : g(\xi) = \beta\}$ has ν-measure 1. Set $\gamma = 2^\beta$. Since κ is strongly inaccessible we obtain $\gamma < \kappa$. On the other hand, by definitions $h(\xi) \leq 2^{\alpha_\xi} = 2^\beta = \gamma$ for any $\xi \in A_2 \cap B$. Thus $h \preceq \hat\gamma$, a contradiction. Hence $\nu(A_2) = 0$ and therefore $\nu(A_3) = 1$.

We claim that $|\{h(\xi) : \xi \in A_3\}| = \kappa$. Actually, if $|\{h(\xi) : \xi \in A_3\}| < \kappa$, then there exists an $\alpha < \kappa$ such that $h(\xi) < \alpha$ for every $\xi \in A_3$. Then $h \prec \hat\alpha$, a contradiction.

Since $\{h(\xi) : \xi \in A_3\}$ is a set of strongly inaccessible cardinals smaller than κ, we are done.   □

By more subtle reasoning one can show

Theorem [AC] 10.20 (R.M. Solovay). *If κ is a real-valued measurable cardinal, then there exist κ many weakly inaccessible cardinals smaller than κ.*

So the affirmative answer to the Measure Problem is a rather strong assumption, if you want, an axiom of set theory. Its consistency strength (consult Sections 11.4 and 11.5) is much stronger than that of **ZFC**.

The proof of Lemma 10.14 uses **AC** in an essential way, since T. Jech [1968] has shown

Metatheorem 10.4 (T. Jech). *If* **ZFC**+ "there exists a measurable cardinal" *is consistent, then also* **ZF** + "$\aleph_1$ is measurable" *is consistent.*

Exercises

10.4 [AC] Martin Axiom and Measure Problem
We assume that **MA** holds true.

a) If $\kappa \leq \mathfrak{c}$ is real-valued measurable, then $\kappa = \mathfrak{c}$.

   Hint: Assume that ν is a κ-additive probabilistic measure defined on $\mathcal{P}(\kappa)$. Let $f : \kappa \xrightarrow{1-1} [0,1]$. Then $\mu(A) = \nu(f^{-1}(A))$ is an extension of a κ-additive Borel measure on $[0,1]$. If $\kappa < \mathfrak{c}$, then by Exercise 9.2 we obtain $\nu(\kappa) = 0$.

b) $\mathfrak{c}$ is not real-valued measurable.

    Hint: Assume that ν is a real-valued $\mathfrak{c}$-additive measure on $\mathcal{P}([0,1])$. By Corollaries 9.2 and 8.27 there exists a $\mathfrak{c}$-Luzin set L. We can assume that $\nu(L) = 1$. A contradiction with Theorem 8.33. You can find easily a direct proof.

c) There exists no real-valued measurable cardinal $\le \mathfrak{c}$

10.5 [AC] Normal Measure

We follow the notation of the proof of the Hanf Theorem 10.19. Let ν be a two-valued κ-additive measure on a measurable cardinal κ. ν is said to be a **normal measure**, if the function $d(\xi) = \xi$) is the smallest element greater than all $\hat{\xi}$, $\xi < \kappa$ in the preordering (10.9) of ${}^{\kappa}\kappa$.

a) If κ is measurable cardinal, then there exists a normal measure on κ.

    Hint: If ν is a measure, $h \in {}^{\kappa}\kappa$ is the smallest element greater than all $\hat{\xi}$, $\xi < \kappa$, set $\mu(A) = \nu(h^{-1}(A))$.

b) Show that a measure ν on κ is normal if and only if for every sequence $\langle A_{\xi} : \xi < \kappa \rangle$ with $\nu(A_{\xi}) = 1$ we have $\nu(\{\xi \in \kappa : \xi \in \bigcap_{\eta < \xi} A_{\eta}\}) = 1$.

c) If ν is a normal measure, then $\nu(\{\{\xi < \kappa : \xi$ is a cardinal$\}\}) = 1$.

    Hint: Follow the proof of Theorem 10.19.

Assume that ν is a normal measure on κ.

d) We can order the class ${}^{\kappa}\mathbf{On}$ by the relation (10.9). Show that there exists an order preserving isomorphism $\mathrm{col} : {}^{\kappa}\mathbf{On} \xrightarrow[\text{onto}]{1-1} \mathbf{On}$.

e) If ν is a normal measure, then $\mathrm{col}(s) = \kappa^{+}$, where $s(\xi) = |\xi|^{+}$ for $\xi \in \kappa$.

f) If $F : \kappa \longrightarrow \mathcal{P}(\mathbf{On})$, we extend the mapping col as follows:

$$\mathrm{col}(F) = \{\mathrm{col}(f) : \nu(\{\xi \in \kappa : f(\xi) \in F(\xi)\}) = 1\}.$$

    Denote $F(\xi) = A \cap \xi$. If $A \subseteq \kappa$, then $\mathrm{col}(F) = A$.

g) Assume that $g : \kappa \longrightarrow \kappa$ and for every $\xi < \kappa$, $h_{\xi} : \mathcal{P}(\xi) \xrightarrow[\text{onto}]{1-1} g(\xi)$. Then

$$\{\langle \mathrm{col}(F), \mathrm{col}(f)\rangle : (\forall \xi)\,(F(\xi) \subseteq \xi \wedge f(\xi)) = h_{\xi}(F(\xi))\}$$

    is a bijection of $\mathcal{P}(\kappa)$ onto $\mathrm{col}(g) \in \mathbf{On}$.

h) If $2^{\lambda} = \lambda^{+}$ for every cardinal $\lambda < \kappa$, then $2^{\kappa} = \kappa^{+}$. More generally, $2^{\kappa} = \kappa^{+}$ whenever $\nu(\{\lambda : 2^{\lambda} = \lambda^{+}\}) = 1$.

i) If $2^{\lambda} = \lambda^{++}$ for every cardinal $\lambda < \kappa$, then $2^{\kappa} = \kappa^{++}$.

10.6 [AC] Weakly Compact Cardinal

An uncountable cardinal κ is called **weakly compact**, if the partition relation $\kappa \to (\kappa)_{2}^{2}$ holds true (for the definition of the partition relation see Exercise 1.14).

a) Order the set ${}^{\lambda}2$ lexicographically

$$f < g \equiv f(\min\{\xi : f(\xi) \neq g(\xi)\}) < g(\min\{\xi : f(\xi) \neq g(\xi)\}).$$

    Show that every increasing or decreasing sequence of this ordering has cardinality at most λ.

b) If $\kappa = \lambda^+$, then $2^\lambda \nrightarrow (\kappa)_2^2$.

Hint: See Exercise 2.5.

c) If κ is singular, then $\kappa \nrightarrow (\kappa)_2^2$.

Hint: If $\kappa = \sup\{\kappa_\xi : \xi < \lambda\}$, set $F(\{\eta, \zeta\}) = 0$ if $\kappa_\xi \le \eta, \zeta < \kappa_{\xi+1}$ for some $\xi < \lambda$ and 1 otherwise.

d) A weakly compact cardinal is weakly inaccessible.

e) A measurable cardinal is weakly compact.

Hint: If $F : [\kappa]^2 \longrightarrow 2$, let $A_\xi^i = \{\eta \in \kappa : F(\{\eta, \xi\}) = i\}$, $i \in 2$. There exists an $i \in 2$ such that $\nu(B) = 1$, where $B = \{\xi : \nu(A_\xi^i) = 1\}$. If ν is a normal measure, then $\nu(B \cap \{\xi : \xi \in \bigcap_{\eta < \xi} A_\eta\}) = 1$.

f) If ν is a normal measure on a measurable cardinal κ, then

$$\nu(\{\lambda < \kappa : \lambda \text{ is weakly compact }\}) = 1.$$

Hint: A proof can be obtained as a more complicated version of the proof of Theorem 10.19 using ideas of Exercise 10.5, f).

10.3   The Linear Ordering of the Real Line

In this section we assume the Axiom of Choice **AC**.

The basic structure of the real line, Euclidean topology, has been defined by its ordering. Actually, the ordering of the real line is in itself an important structure. In this section we shall study some properties of the ordering of reals and we show that they depend on the assumed set theory, more precisely, on additional axioms.

In Section 2.3 we have proved Theorem 2.29 which says that a linearly ordered set with properties (2.15)–(2.17) is order isomorphic to the real line $\langle \mathbb{R}, \le \rangle$. Of course, the set of reals $\langle \mathbb{R}, \le \rangle$ possesses all those properties. For convenience of the reader we repeat three of the mentioned properties of a linearly ordered set $\langle X, \le \rangle$ and one property that is a consequence of the others.

$\langle X, \le \rangle$ has neither a least nor a greatest element,   (10.13)

$\langle X, \le \rangle$ is densely ordered,   (10.14)

there exists a countable dense subset of $\langle X, \le \rangle$,   (10.15)

every set of pairwise disjoint open intervals in $\langle X, \le \rangle$ is countable.   (10.16)

By a slight modification of the proof of Theorem 2.29 one can show that any linearly ordered set with properties (10.13)–(10.15) is order isomorphic to a subset of the real line $\langle \mathbb{R}, \le \rangle$. Moreover, if X satisfies the Bolzano Principle, then it is order isomorphic to $\mathbb{R}$. Evidently

$$(10.15) \rightarrow (10.16)$$

and any subset of the real line satisfies the condition (10.15).

M.J. Souslin in 1917 raised the natural question: can the condition (10.15) be replaced by a seemingly weaker condition (10.16)? Or equivalently: is any linearly ordered set with properties (10.13), (10.14) and (10.16) order isomorphic to a subset of the real line? It turned out that the problem is much more difficult than the author thought. Now we know that the axioms of set theory **ZFC** are not enough to decide it.

A linearly ordered set $\langle X, \leq \rangle$ is said to be a **Souslin line** if $\langle X, \leq \rangle$ possesses properties (10.13), (10.14)), (10.16) and does not possess property (10.15). The **Souslin Hypothesis** says that there is no Souslin line, equivalently, any linearly ordered set with properties (10.13), (10.14), (10.16) is order isomorphic to a subset of the real line $\langle \mathbb{R}, \leq \rangle$.

The standard completion by adding all Dedekind cuts as presented in Section 2.3 allows us to extend any Souslin line to a Souslin line satisfying the Bolzano Principle.

A linear ordering on a set naturally induces the order topology. The notions "topologically dense" and "order dense" are equivalent. In topology a set with countable dense subset is called "separable". Thus, a linearly dense ordered set is a Souslin line if it generates a non-separable **CCC** topological space. The following rather technical result will be useful in our investigations.

Theorem [AC] 10.21. *If there exists a Souslin line X then there exists a Souslin line $\tilde{X}$ such that no open interval in $\tilde{X}$ is separable.*

Proof. Let $\langle X, \leq \rangle$ be a Souslin line. We define an equivalence relation $\sim$ as

$$x \sim y \equiv \text{the open interval with endpoints } x, y \text{ is separable.}$$

Note that a subinterval of a separable interval is separable as well. If $x < y$ and $\neg x \sim y$ then for every element of the equivalence classes $u \in [x]_\sim, v \in [y]_\sim$ we have $u < v$. Thus the set $\tilde{X} = \{[x]_\sim : x \in X\}$ is linearly ordered by the relation

$$[x]_\sim < [y]_\sim \equiv (x < y \wedge \neg x \sim y).$$

We show that for every $x \in X$ there is a countable dense subset $H_x \subseteq [x]_\sim$. Actually either $[x]_\sim = \{x\}$ or $[x]_\sim$ contains an open interval. In the former case set $H_x = \{x\}$. In the latter case take a maximal family $\mathcal{A}$ of pairwise disjoint open subintervals of $[x]_\sim$. By (10.15), $\mathcal{A}$ is countable. Every interval from $\mathcal{A}$ contains a countable dense subset. Let H_x be the union of all them. If an interval (u, v) meets $[x]_\sim$, then it meets some interval from $\mathcal{A}$ and therefore also H_x.

Since X is not separable, $\tilde{X}$ has more than one point.

Assume now that $[x]_\sim < [y]_\sim$. Since $(x, y) \cap (H_x \cup H_y)$ is not dense in (x, y) there exists an open interval $(a, b) \subseteq (x, y)$ disjoint with $(x, y) \cap (H_x \cup H_y)$. If $z \in (a, b)$ then $[x]_\sim < [z]_\sim < [y]_\sim$. Thus (10.14) holds true.

Assume that $[x]_\sim < [y]_\sim$ and the open interval $([x]_\sim, [y]_\sim)$ in $\tilde{X}$ is separable. Thus there is a countable set $B \subseteq X$ such that $\{[z]_\sim : z \in B\}$ is dense in $([x]_\sim, [y]_\sim)$. Take a maximal family $\mathcal{A}$ of pairwise disjoint open separable subintervals of (x, y). Again, this family is countable. For every $I \in \mathcal{A}$, let G_I denote a countable dense subset of I. Set

$$D = \bigcup_{y \in B} H_y \cup \bigcup_{I \in \mathcal{A}} G_I.$$

Let $(u, v) \subseteq (x, y)$. If (u, v) is separable, then (u, v) meets some $I \in \mathcal{A}$ and then also $(u, v) \cap G_I \neq \emptyset$. Otherwise, $([u]_\sim, [v]_\sim)$ is a non-trivial interval and therefore contains some $y \in B$. Then $(u, v) \cap H_y \neq \emptyset$. Thus, D is a countable dense subset of (x, y), a contradiction with $\neg x \sim y$.

If $([x]_\sim, [y]_\sim)$, $([u]_\sim, [v]_\sim)$ are disjoint open intervals in $\tilde{X}$ then (x, y), (u, v) are disjoint open intervals in X. One can easily conclude that $\tilde{X}$ has the property (10.16).

It may happen that $\tilde{X}$ has a least or a greatest element. We can simply omit it.   □

Now we give a reformulation of the Souslin Hypothesis in terms of trees. A pruned tree $\langle T, \leq \rangle$ is called **a Souslin tree** if

a) the height of T is $\aleph_1$,

b) every branch in T is countable,

c) every antichain in T is countable.

Evidently the condition a) is equivalent to $|T| > \aleph_0$. A pruned subtree $S \subseteq T$ of a Souslin tree T is again a Souslin tree provided that $|S| > \aleph_0$.

Let us recall that the ξth level $T(\xi)$ is the set of all $x \in T$ of the height $\mathrm{hg}(x) = \xi$. If T is a Souslin tree, then every $T(\xi)$, $\xi < \omega_1$ is non-empty and countable and

$$T = \bigcup_{\xi < \omega_1} T(\xi), \quad |T| = \aleph_1.$$

Let $\xi \leq \omega_1$. A tree T of height ξ is called a **normal tree**,

d) if each level of T is countable,

e) for every $x \in T$ and every ordinal $\xi > \eta > \mathrm{hg}(x)$ there exists a $y \in T(\eta)$ such that $x < y$,

f) if $x \in T$ is not maximal, then x has infinitely many immediate successors,

g) if $\eta < \xi$ is limit, $x, y \in T(\eta)$ and $\{u \in T : u < x\} = \{u \in T : u < y\}$, then $x = y$.

Lemma [AC] 10.22. *If there exists a Souslin tree, then there exists a normal Souslin tree.*

Proof. Let T be a Souslin tree. Take (note that $T^x = \{y \in T : x \leq y\}$)

$$S = \{x \in T : |T^x| = \aleph_1\}.$$

The set S contains with any x all $y < x$. Thus the height of an element in S is the same as in T. We show that S satisfies the condition e). So let $x \in S$ and $\mathrm{hg}(x) < \eta < \omega_1$. Then

$$T^x \setminus \bigcup_{y \in T(\eta)} T^y$$

is countable. Since T^x is uncountable there exists a $y \in T(\eta)$, $y > x$ with $|T^y| = \aleph_1$, i.e., $y \in S(\eta)$.

For every $\eta < \omega_1$ limit, we eventually add new nodes to S. Consider a node $x \in S(\eta)$ and denote

$$A = \{y \in S : y < x\}, \qquad B = \{z \in S(\eta) : (\forall_y)\,(y \in A \to y < z)\}.$$

If x is the only element of B, we do not do anything. If $|B| > 1$ we add a node a_A and we extend the ordering as follows:

$$(\forall y)\,(y \in A \to y < a_A), \quad (\forall z)\,(z \in B \to a_A < z).$$

So obtained tree W satisfies the conditions d), e) and g).

Now one can easily see that the subtree

$$W_1 = \{x \in W : x \text{ is a branching point}\}$$

satisfies the conditions d), e) and g) as well. Finally, the set $W_2 \subseteq W_1$ consisting of nodes of W_1 at limit levels and the root is a normal Souslin tree. $\qquad \square$

A complete Boolean algebra $\mathbf{B}$ is said to be a **Souslin algebra** if $\mathbf{B}$ is **CCC**, $\aleph_0$-distributive and has no atom.

Lemma 10.23. *A Souslin algebra is not $\aleph_1$-distributive. Actually there exists a sequence $\langle A_\xi, \xi < \omega_1 \rangle$ of refining partitions of $\mathbf{B}$ without a common refinement.*

Proof. Let $\mathbf{B}$ be a Souslin algebra. Then $\mathbf{B}$ is infinite and moreover uncountable. Since $\mathbf{B}$ is not atomic, by Theorem 5.32 there exists an $\xi > 0$ such that $\mathbf{B}$ is not $\aleph_\xi$-distributive. Let ξ be the least ordinal with this property. Then $\xi \geq 1$. By definition, there exists a sequence $\langle A_\eta \subseteq B : \eta < \aleph_\xi \rangle$ of dense subsets of $\mathbf{B}$ without a common refinement. We can assume that elements of every A_η are pairwise disjoint. Since $\aleph_\xi$ is the least cardinal for which $\mathbf{B}$ is not $\aleph_\xi$-distributive, we can assume that A_η is a refinement of A_ζ for any $\zeta < \eta < \omega_1$. Moreover, since $\mathbf{B}$ has no atom, we can assume that A_ζ is a strict refinement of A_η, i.e., if $y \in A_\eta$, $x \in A_\zeta$ and $x \leq y$ then actually $x < y$.

Now, if $\xi > 1$, then $\mathbf{B}$ is $\aleph_1$-distributive and therefore there exists a common refinement C of all $\langle A_\eta : \eta < \omega_1 \rangle$. Let $c \in C$, $c \neq 0$. Then for every $\eta < \omega_1$ there exists $x_\eta \in A_\eta$, $x_\eta \geq c$. The sequence $\langle x_\eta : \eta < \omega_1 \rangle$ is a strictly decreasing chain contradicting the **CCC**. Thus $\xi = 1$. $\qquad \square$

All introduced notions are closely related. Namely

Theorem [AC] 10.24. *The following are equivalent:*

  (i) *There exists a Souslin line.*
 (ii) *There exists a Souslin algebra.*
(iii) *There exists a Souslin tree.*

Proof. Let $\langle X, \leq \rangle$ be a Souslin line. By Theorem 10.21 we can assume that no interval in X is separable. Let $B = \mathrm{RO}(X, \mathcal{O})$, where $\mathcal{O}$ is the topology induced by the order $\leq$. Since open intervals form a base of the topology $\mathcal{O}$ and an open interval is a regular open set, we obtain that the set of open intervals is dense in the complete Boolean algebra $\mathbf{B}$. Thus $\mathbf{B}$ is **CCC** and has no atoms.

We have to show that $\mathbf{B}$ is $\aleph_0$-distributive. Let $\langle A_n : n \in \omega \rangle$ be a family of dense subsets of $\mathbf{B}$. We can assume that elements of each A_n are pairwise disjoint open intervals. Let D be the set of all endpoints of intervals $\langle A_n : n \in \omega \rangle$. Consider an open interval (x, y). Since we suppose that no open interval is separable, there exists an open interval $(u, v) \subseteq (x, y)$ such that $(u, v) \cap D = \emptyset$. However every set A_n is dense. Therefore for every n there exists an interval $(x_n, y_n) \in A_n$ such that $(u, v) \subseteq (x_n, y_n)$. Thus $\mathbf{B}$ is $\aleph_0$-distributive.

Assume now that there exists a Souslin algebra $\mathbf{B}$. Since $\mathbf{B}$ is not $\aleph_1$-distributive, there exists a sequence of partitions $\langle A_\xi : \xi < \omega_1 \rangle$ of $\mathbf{B}$ such that every A_ξ is a strict refinement of A_η, provided that $\eta < \xi < \omega_1$. It is easy to see that the set $\{1_B\} \cup \bigcup_{\xi < \omega_1} A_\xi$ ordered by the inverse ordering of $\mathbf{B}$ is a Souslin tree.

Finally, assume that there exists a normal Souslin tree T. By the condition f) the set $\mathrm{IS}(T, x)$ is countably infinite for each $x \in T$. We order every $\mathrm{IS}(T, x)$ by a relation $\sqsubset$ in the order type of $\mathbb{Q}$.

Let X be the set of all branches of the tree T. If $A, B \in X$ and $A \neq B$, then there exist $a \in A$, $b \in B$, $a \neq b$ are such that $\{x \in A : x < a\} = \{x \in B : x < b\}$. By the condition g) we obtain $a, b \in T(\xi)$ for a ξ non-limit. Thus there exists a node $x \in A \cap B$ such that $a, b \in \mathrm{IS}(T, x)$. We set $x = \mathrm{d}(A, B)$. Define the ordering of X as

$$A < B \equiv a \sqsubset b.$$

It is easy to see that $\langle X, < \rangle$ is a linearly densely ordered set.

If (A, B) is an open interval in $\langle X, < \rangle$ and $x = \mathrm{d}(A, B)$, then there exists an element $y(A, B) \in \mathrm{IS}(T, x)$ such that

$$\{E \in X : y(A, B) \in E\} \subseteq (A, B).$$

If (A, B), (C, D) are disjoint, then $y(A, B)$ and $y(C, D)$ are incomparable elements of T. Thus, any set of pairwise disjoint intervals in X is countable.

It remains to show that X is not separable. Assume that $Y \subseteq X$ is a countable set of branches in T. Then there exists an ordinal $\xi < \omega_1$ such that every branch of Y is shorter than ξ. Take any two branches A, B going through $x, y \in T(\xi + 1)$, $x \neq y$. Then $Y \cap (A, B) = \emptyset$. Thus, Y is not dense in X. $\qquad\square$

We are ready to show one of the main results of this section.

Theorem [AC] 10.25. *If* **MA** *holds true and* $c > \aleph_1$, *then there is no Souslin tree.*

Proof. The proof is simple. To get a contradiction assume that **MA** holds true, $c > \aleph_1$ and there exists a Souslin algebra **B**. By Lemma 10.23 there exists a family $\mathcal{D} = \{A_\xi : \xi < \omega_1\}$ of partitions of **B** such that each A_ξ is a strict refinement of A_η for any $\xi > \eta$. By **MA** there exists a $\mathcal{D}$-generic ultrafilter $G \subseteq B$. If a_ξ is the only element of $G \cap A_\xi$ then $\langle a_\xi : \xi < \omega_1 \rangle$ is a strict decreasing sequence, a contradiction with **CCC**. $\square$

We show that **ZFC** is consistent with the existence of a Souslin line.

Theorem 10.26 (R.B. Jensen). *If* $\mathbf{V} = \mathbf{L}$, *then there exists a Souslin tree.*

For the proof of the theorem we need a simple auxiliary results. First, we introduce a principle $\Diamond$ dealing with stationary sets. So we need the fundamental result about them.

Let κ be an uncountable regular cardinal. A set $A \subseteq \kappa$ is called **closed unbounded** if A is unbounded and $\sup B \in A$ for any non-empty bounded $B \subseteq A$. A set $C \subseteq \kappa$ is called **stationary** if $A \cap C \neq \emptyset$ for every closed unbounded subset $A \subseteq \kappa$.

Theorem [AC] 10.27 (G. Fodor). *Assume that κ is an uncountable regular cardinal. If $f : \kappa \longrightarrow \kappa$ is such that the set $B = \{\xi \in \kappa : f(\xi) < \xi\}$ is stationary, then there exists a stationary set $C \subseteq B$ and an ordinal $\eta < \kappa$ such that $f(\xi) = \eta$ for every $\xi \in C$.*

For a proof see T. Jech [2006], W. Just and M. Weese [1997], or Exercise 10.7.

The **Diamond Principle** $\Diamond$ says that there exists a sequence $\langle S_\xi, \xi < \omega_1 \rangle$ of sets of ordinals such that $S_\xi \subseteq \xi$ for every $\xi < \omega_1$ and for every $X \subseteq \omega_1$ the set $\{\xi < \omega_1 : X \cap \xi = S_\xi\}$ is stationary.

Studying the fine structure of the constructible universe, R.B. Jensen [1968] has proved

Theorem 10.28 (R.B. Jensen). *If* $\mathbf{V} = \mathbf{L}$, *then* $\Diamond$ *holds true.*

For a proof see, e.g., T. Jech [2006].

One can easily show that $\Diamond \rightarrow \mathbf{CH}$. Thus by (11.9) and (11.17), $\Diamond$ is undecidable in **ZFC**.

Lemma 10.29. *Assume that T is a normal tree of height ω_1. If $A \subseteq T$ is a maximal antichain, then the set*

$$C = \{\xi < \omega_1 : A \cap T(< \xi) \text{ is a maximal antichain in } T(< \xi)\}$$

is closed unbounded.

Proof. One can easily see that C is closed. We show that C is unbounded. Let $\eta < \omega_1$. Since the set $A \cap T(< \eta)$ is countable, there exists an $\xi_0 > \eta$ such that every node of $T(< \eta)$ is comparable with some node in $A \cap T(< \xi_0)$. Similarly, if ξ_n is already defined, there exists a $\xi_{n+1} > \xi_n$ such that every node of $T(< \xi_n)$ is comparable with some node of $A \cap T(< \xi_{n+1})$. If $\xi = \lim_{n \in \omega} \xi_n$, then $\xi > \eta$ and $A \cap T(< \xi)$ is maximal in $T(< \xi)$.   □

Now, Jensen's Theorem 10.26 follows immediately from the next result.

Theorem 10.30 (R.B. Jensen). *If $\diamondsuit$ holds true, then there exists a Souslin tree.*

Proof. Assume that the Diamond Principle $\diamondsuit$ holds true. Then there exists a sequence $\langle S_\xi : \xi < \omega_1 \rangle$ of subsets of ω_1 such that the set $\{\xi < \omega_1 : X \cap \xi = S_\xi\}$ is stationary for any $X \subseteq \omega_1$. We define an ordering $\preceq$ on ω_1 such that $\langle \omega_1, \preceq \rangle$ will be a normal Souslin tree. We will stipulate during the construction that the partial trees are normal.

We define by transfinite induction sets $W_\xi \subseteq \omega_1$ and the ordering $\preceq | W_\xi$ for any $\xi < \omega_1$. The set $W_\xi \setminus \bigcup_{\eta < \xi} W_\eta$ will be the ξth level of the constructed tree. Moreover we assume that $\bigcup_{\eta < \xi} W_\eta$ is an initial segment of ω_1, i.e., a countable ordinal, and $\langle \bigcup_{\eta < \xi} W_\eta, \preceq | \bigcup_{\eta < \xi} W_\eta \rangle$ is a normal tree, for every $\xi < \omega_1$.

For $\xi = 0$ we simply set $W_0 = \{0\}$.

$W_{\xi+1}$ is obtained by adding countably infinitely many immediate successors to each node of $W_\xi \setminus \bigcup_{\eta < \xi} W_\eta$.

The main problem consists in constructing W_ξ for ξ limit. We must decide to which ξ-branch of $\bigcup_{\eta < \xi} W_\eta$ do we have to add immediate successors. We use the Diamond Principle to decide the choice.

If S_ξ is not a maximal antichain in $\bigcup_{\eta < \xi} W_\eta$, then to assure normality, we take for every $u \in \bigcup_{\eta < \xi} W_\eta$ a branch B_u of length ξ containing u and add a node to W_ξ greater than any node of B_u. Of course, for different branches we add different nodes to keep the condition f) preserved.

Assume now that S_ξ is a maximal antichain in $\bigcup_{\eta < \xi} W_\eta$. Then for any node $u \in \bigcup_{\eta < \xi} W_\eta$ there exists a node $v \in S_\xi$ compatible with u. Take a branch B_u of length ξ containing both u and v and add a node to W_ξ greater than any node of B_u.

By construction, the tree $\langle \omega_1, \preceq \rangle$ is normal. We have to show that there exists no uncountable antichain in the tree $\langle \omega_1, \preceq \rangle$. So assume that $A \subseteq \omega_1$ is a maximal antichain. If we set $f(\xi) = \bigcup_{\eta < \xi} W_\eta$ then we obtain a continuous function from ω_1 into ω_1. By elementary ordinal arithmetic, $D = \{\xi \in \omega_1 : f(\xi) = \xi\}$ is a closed unbounded set. By Lemma 10.29 the set C is also closed unbounded. Thus, by the Diamond Principle, there is a limit ordinal $\xi \in C \cap D$ such that $\bigcup_{\eta < \xi} W_\eta \cap A = S_\xi$. Then $\bigcup_{\eta < \xi} W_\eta = \xi$ and $\bigcup_{\eta < \xi} W_\eta \cap A = S_\xi$ is a maximal antichain in $\bigcup_{\eta < \xi} W_\eta$. In this situation, for every node $u \in S_\xi$ we added a node in W_ξ comparable with u. So the set A, being an antichain, is contained in $\bigcup_{\eta < \xi} W_\eta$ and therefore countable.   □

We can summarize

Metatheorem 10.5. *The Souslin Hypothesis is undecidable in* **ZFC**.

The result follows by Theorems 10.25 and 10.26, since by (11.21) and (11.9) we know that both **ZFC** $+$ **MA** $+ \mathfrak{c} > \aleph_1$ and **ZFC** $+$ **V** $=$ **L** are consistent.

We finish with an important undecidable statement related to the product of topological spaces.

Theorem [AC] 10.31. *Let $\mathcal{O}$ be the topology induced by the ordering of a Souslin line $\langle X, \leq \rangle$. The topological product $\langle X \times X, \mathcal{O} \times \mathcal{O} \rangle$ is not* **CCC**.

Proof. Let $\langle X, \leq \rangle$ be a Souslin line.

One can construct by transfinite induction a sequence $\langle \mathcal{I}_\xi, \xi < \omega_1 \rangle$ of non-empty sets of left-open intervals in X such that:

(i) every $I \in \mathcal{I}_\xi$ is a union of two disjoint intervals $I_l, I_r \in \mathcal{I}_{\xi+1}$, for every $\xi < \omega_1$,

(ii) if $I \in \mathcal{I}_\xi$ and $J \in \mathcal{I}_\eta$, then $I \cap J = \emptyset$ or $I \subseteq J$.

At the non-limit step, follow the condition (i). At limit step λ the set of all endpoints of intervals from $\bigcup_{\xi < \lambda} \mathcal{I}_\xi$ is not dense in X, therefore there exists an interval I not containing any of those endpoints. Take $\mathcal{I}_\lambda = \{I\}$. One can easily show that (ii) holds true.

Consider the family $\{\text{Int}(I_l) \times \text{Int}(I_r) : I \in \bigcup_{\xi < \omega} \mathcal{I}_\xi\}$ of open sets. Let $I \in \mathcal{I}_\xi$, $J \in \mathcal{I}_\eta$, $I \neq J$. Assume $\eta \leq \xi$. Then either $I \cap J = \emptyset$, or $I \subseteq J$ and $\eta < \xi$. If I, J are disjoint, $I_l \times I_r, J_l \times J_r$ are disjoint as well. If $I \subseteq J$, then $I \subseteq J_l$ or $I \subseteq J_r$. Evidently in both cases the sets $I_l \times I_r, J_l \times J_r$ are disjoint.

Thus the topological product $\langle X \times X, \mathcal{O} \times \mathcal{O} \rangle$ is not **CCC**. $\square$

Metatheorem 10.6. *The assertion* "the topological product of two **CCC** topological spaces is a **CCC** topological space" *is undecidable in* **ZFC**.

Demonstration. By Theorem 9.14, **MA** $+ \mathfrak{c} > \aleph_1$ implies an affirmative answer.

Assuming the existence of a Souslin line, e.g., assuming the Axiom of Constructibility, you obtain a negative answer. $\square$

By Corollary 9.15 we obtain even more.

Metatheorem 10.7. *The assertion* "the topological product of **CCC** topological spaces is a **CCC** topological space" *is undecidable in* **ZFC**.

Exercises

10.7 [AC] A Proof of Fodor's Theorem

κ is an uncountable regular cardinal.

a) The intersection of two closed unbounded subsets of κ is a closed unbounded set.

Hint: If C, D are closed unbounded, $\xi < \kappa$, construct an increasing sequence $\{\eta_n\}_{n=0}^\infty$ of ordinals greater than ξ such that $\eta_{2n} \in C$ and $\eta_{2n+1} \in D$.

b) The intersection of less than κ closed unbounded subsets of κ is a closed unbounded set.

Hint: By transfinite induction.

c) If $\langle A_\xi : \xi < \kappa \rangle$ are closed unbounded, then also $A = \{\eta < \kappa : \eta \in \bigcap_{\xi < \eta} A_\xi\}$ is closed unbounded.

   Hint: One can assume that $A_\xi = \bigcap_{\eta \le \xi} A_\eta$. Let ζ be a limit point of A. If $\xi < \zeta$, let

$$B = \{\eta \in A : \xi < \eta < \zeta\}$$

   and show that $B \subseteq A_\xi$. Then $\zeta = \sup B \in A_\xi$.
   If $\zeta < \kappa$, construct a sequence $\{\xi_n\}_{n=0}^{\infty}$ such that $\xi \in A_0$ and $\xi_n < \xi_{n+1} \in A_{\xi_n}$ for every n. Then $\zeta < \lim_n \xi_n \in A$.

d) Prove Fodor's Theorem.
   Hint: Assume that the set $\{\eta < \kappa : f(\eta) = \xi\}$ is non-stationary for each $\xi < \kappa$. Take a closed unbounded A_ξ such that $\eta \in A_\xi \to f(\eta) \ne \xi$. If $\zeta \in \bigcap_{\xi < \zeta} A_\xi\}$, then $f(\zeta) \ne \xi$ for any $\xi < \zeta$ – a contradiction.

10.8 Aronszajn Tree
A tree T of height ω_1 is called an **Aronszajn tree** if $T(\xi)$ is countable for every $\xi < \omega_1$ and every branch is countable.

a) Note that a Souslin tree is an Aronszajn tree.

b) The set

$$W = \{f \in {}^{<\omega_1}\mathbb{Q} : f \text{ is increasing and } \mathrm{dom}(f) \text{ is non-limit ordinal}\}$$

   is a subtree of the tree $\langle {}^{<\omega_1}\mathbb{Q}, \subseteq \rangle$. For simplicity, for any $f \in W$ we set

$$\max f = f(\xi) \text{ where } \mathrm{dom}(f) = \xi + 1.$$

   Show that every branch of W is countable.

c) Let $\xi < \omega_1$, $T \subseteq W(< \xi + 1)$. Assume that T satisfies the following condition:

$$(\forall f \in T(< \xi))(\forall r > \max f)(\exists g \in T(\xi))\,(f \subseteq g \wedge \max g \le r). \qquad (10.17)$$

   Then there exists a countable set $C \subseteq W(\xi + 1)$ such that $T \cup C$ satisfies condition (10.17) for $\xi + 1$.
   Hint: Take $C = \{f^\frown r : f \in W(\xi) \wedge r \in \mathbb{Q} \wedge r > \max f\}$.

d) Let $\lambda < \omega_1$ be a limit ordinal. Assume that $T \subseteq W(< \lambda)$ is countable and such that (10.17) holds true for every $\xi < \lambda$ and $T(\xi)$. Then there exists a countable set $C \subseteq W(\lambda)$ such that $T \cup C$ satisfies condition (10.17) for λ.
   Hint: Let $\lambda = \lim_{n \in \omega} \lambda_n$, $\lambda_n, n \in \omega$ being increasing. If $f \in T$, $\max f < r$ take an increasing sequence $\lim_{n \to \infty} r_n = r$, $r_0 > \max f$. By (10.17) there exist $g_n \in T(\xi)$ such that

$$f \subseteq g_0 \subseteq \cdots \subseteq g_n \subseteq g_{n+1} \subseteq \cdots$$

   and $\max g_n \le r_n$. For each $f \in T$ and $r > \max f$, $r \in \mathbb{Q}$ choose $g = \bigcup_n g_n$ and put into $C \subseteq W(\lambda)$.

e) There exists an Aronszajn tree.

10.9 Kurepa Theorem
If $\langle X_i, \le_i \rangle$, $i = 1, 2$ are posets, then a strictly increasing mapping $f : X_1 \longrightarrow X_2$ is called a **quasi-embedding** of X_1 into X_2. We say that X_1 is **quasi-embeddable** in X_2 if there exists a quasi-embedding of X_1 into X_2.

a) Every tree T is quasi-embeddable in $hg(T)$.

b) A quasi-embedding need not be an injection.

c) If f is a quasi-embedding of X_1 into X_2, then the set $f^{-1}(\{x\})$ is an antichain, provided that $x \in rng(f)$.

d) If a poset X is a union $X = \bigcup_n X_n$ of antichains, then there exists a quasi-embedding f of X into $^\omega 2$, ordered lexicographically, such that $|rng(f)| \le \aleph_0$.

   Hint: Assume that X_n are pairwise disjoint. For $x \in X_k$ we set $f(x) = h \in {}^\omega 2$, where

$$h(n) = \begin{cases} 1 & \text{if } n \le k \text{ and there exists a } y \le x, y \in X_n, \\ 0 & \text{otherwise.} \end{cases}$$

e) Prove the **Kurepa Theorem**: A partially ordered set X is quasi-embeddable in $\mathbb{Q}$ if and only if X is a union of countably many antichains.

   Hint: Use c), d) and Theorem 11.4.

10.10 [AC] Martin's Axiom and Aronszajn Trees

We assume that T is an Aronszajn tree.

a) Assume that $S \subseteq [T]^n$ is an uncountable family of pairwise disjoint sets. Then there exist $A, B \in S$ such that x, y are incomparable for any $x \in A$ and any $y \in B$.

   Hint: Assume not, i.e., any couple $A, B \in S$ contains comparable elements. Let $\mathcal{F}$ be a uniform ultrafilter of S. Fix an enumeration $\{x_1, \ldots, x_n\}$ of each $A \in S$. For every $x \in T$, $1 \le k \le n$ denote by $S_{x,k}$ the set of all $A \in S$ such that x is comparable with the kth element of A. Then

$$\bigcup_{x \in A} \bigcup_{k=1}^{n} S_{x,k} = S.$$

   For each $A \in S$ pick $x_A \in A$ and k_A such that $S_{x_A, k_A} \in \mathcal{F}$. There exists a k such that $S' = \{A \in S : k_A = k\}$ is uncountable.

   For any $A, B \in S'$, if $C \in S_{x_A,k} \cap S_{x_B,k}$, then the kth element of C is comparable with x_A and x_B. Since $S_{x_A,k} \cap S_{x_B,k} \in \mathcal{F}$, this set is uncountable. Thus there exists a $C \in S$ such that $x_A \le x_C$ and $x_B \le x_C$. Hence x_A and x_B are comparable. Thus $\{x_A : A \in S'\}$ is a chain, a contradiction.

b) Assume that $S \subseteq [T]^{<\omega}$ is an uncountable family of pairwise disjoint sets. Then there exist $A, B \in S$ such that x, y are incomparable for any $x \in A$ and any $y \in B$.

c) Assume that P is the set of all functions s from a finite subset $dom(s) \subseteq T$ into ω and such that $s(x) \ne s(y)$ provided that $x, y \in dom(s)$ are comparable. Show that the poset $\langle P, \supseteq \rangle$ is **CCC**.

   Hint: Follow the proof of Lemma 9.11, c). Apply b) to the family $\{dom(s) \setminus C : s \in S_2\}$ and find two non-disjoint elements of S_2.

d) If $\aleph_1 < \mathfrak{m}$, then there exists a mapping $p : T \longrightarrow \omega$ such that $p(x) \ne p(y)$ for any comparable elements $x, y \in T$.

   Hint: $D_x = \{s \in P : x \in dom(s)\}$ is dense. If $G \subseteq P$ is $\{D_x : x \in T\}$-generic filter, then set $p = \bigcup G$.

10.11 Special Aronszajn Tree
An Aronszajn tree T is called **special** if T is a union of countably many antichains.

a) A Souslin tree is not a special Aronszajn tree.

b) The Aronszajn tree constructed in Exercise 10.8, d) is special.
Hint: The mapping max f *is a quasi-embedding into* $\mathbb{Q}$.

c) If $\mathfrak{m} > \aleph_1$, then every Aronszajn tree is special.
Hint: Use the result of Exercise 10.10, d).

d) The assertion "Every Aronszajn tree is special" is undecidable in **ZFC**.

10.4   Reversing the Order of Integration

We shall deal only with Lebesgue measure on $[0,1]$ and $[0,1]^2$.

Let $f : [0,1] \times [0,1] \longrightarrow \mathbb{R}$ be a function such that the following holds true:

(i) functions $f(\cdot, y)$, $f(x, \cdot)$ are measurable for almost every $x, y \in [0,1]$;

(ii) functions

$$\varphi(y) = \int_{[0,1]} f(x,y)\, d\lambda(x), \quad \psi(x) = \int_{[0,1]} f(x,y)\, d\lambda(y)$$

are defined almost everywhere and are measurable;

(iii) the integrals

$$\int_{[0,1]} \psi(x)\, d\lambda(x), \quad \int_{[0,1]} \varphi(y)\, d\lambda(y)$$

do exist.

Now a natural question arises: does the equality

$$\int_{[0,1]} \psi(x)\, d\lambda(x) = \int_{[0,1]} \varphi(y)\, d\lambda(y), \tag{10.18}$$

hold true?

The main goal of this section is to show that the answer to this question is undecidable in **ZFC**.

By the Fubini-Tonelli Theorem 4.29 we obtain

Theorem [AC] 10.32.

a) *If f is integrable, then φ and ψ are integrable and* (10.18) *holds true.*

b) *If f is a non-negative measurable function, then conditions* (i) *and* (ii) *are fulfilled and* (10.18) *holds true.*

Thus if f is integrable or measurable and bounded from one side, we obtain the affirmative answer by the Fubini-Tonelli Theorem. We present a simple example of an unbounded measurable function for which conditions (i)–(iii) hold true and equality (10.18) fails.

Let $a_n = 2^{-n}$. For any $n > 0$ and $x \in (a_n, a_{n-1})$ we set

$$f(x, y) = \begin{cases} -2^{2n} & \text{for } 0 \leq y < a_n, \\ 2^{2n} & \text{for } a_n \leq y < a_{n-1}, \\ 0 & \text{for } a_{n-1} \leq y \leq 1, \end{cases}$$

and

$$f(0, y) = 0 \quad \text{for any } y \in [0, 1].$$

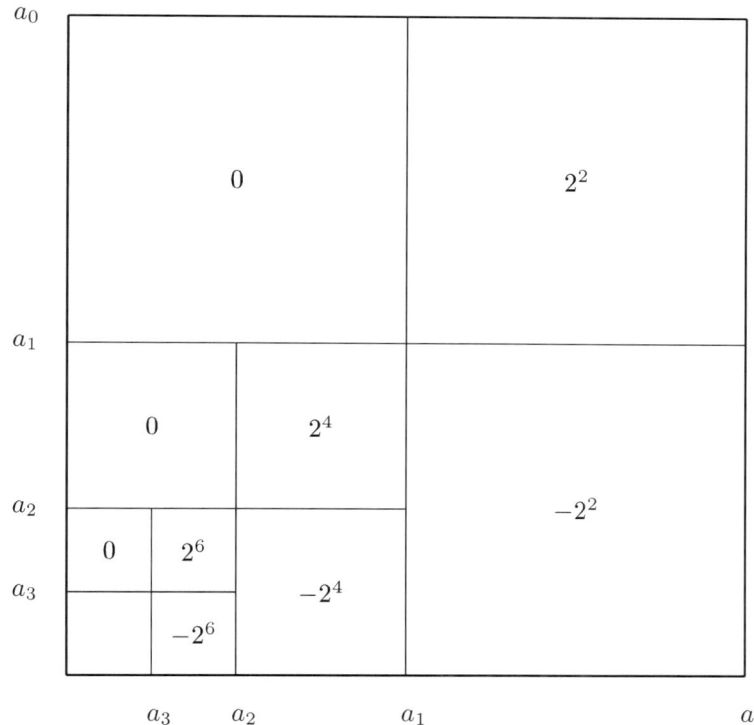

For any $y \in [a_n, a_{n-1})$ we obtain

$$\varphi(y) = \int_{[0,1]} f(x, y) \, d\lambda(x) = 2^{-n} \cdot 2^{2n} - (2^{-(n-1)} \cdot 2^{2(n-1)} + \cdots + 2^{-1} \cdot 2^2) = 2.$$

On the other hand, for any $x \in [0, 1]$,

$$\psi(x) = \int_{[0,1]} f(x, y) \, d\lambda(y) = 0.$$

Hence
$$\int_{[0,1]} \psi(x)\,d\lambda(x) = 0, \quad \int_{[0,1]} \varphi(y)\,d\lambda(y) = 2.$$

So we have the only unanswered case: f non-measurable and bounded. We denote by **NROI** the following sentence:

> *There exists a function $f : [0,1] \times [0,1] \longrightarrow [0,1]$ such that conditions (i), (ii) and (iii) hold true and the equality (10.18) fails.*

Note that condition (iii) is superfluous, since it follows from (i), (ii) and the fact, that f is bounded. Of course the condition $\mathrm{rng}(f) \subseteq [0,1]$ can be replaced by the condition "f is bounded".

The promised main result reads as follows.

Metatheorem 10.8. *If* **ZF** *is consistent, then* **NROI** *is an undecidable statement of* **ZFC**, *i.e.,*
$$\textbf{ZFC} \nvdash \textbf{NROI}, \qquad \textbf{ZFC} \nvdash \neg\textbf{NROI}.$$

We recall some notation. If $A \subseteq X \times Y$, $x \in X$, $y \in Y$, then the vertical and the horizontal section of A were defined as the sets
$$A_x = \{u \in Y : \langle x, u \rangle \in A\}, \qquad A^y = \{v \in X : \langle v, y \rangle \in A\},$$

respectively.

A set $A \subseteq [0,1] \times [0,1]$ is called a **Steinhaus set** if $\lambda(A_x) = 0$ and $\lambda(A^y) = 1$ for almost every $x, y \in [0,1]$. Let us note that if there exists a Steinhaus set, then there exists a set $A \subseteq [0,1] \times [0,1]$ such that $\lambda(A_x) = 0$ and $\lambda(A^y) = 1$ for every $x, y \in [0,1]$. If A is a Steinhaus set, then the characteristic function of A satisfies the conditions of **NROI**, i.e., **NROI** holds true.

Theorem [AC] 10.33. *If* $\mathrm{add}(\mathcal{N}) = \mathrm{cov}(\mathcal{N})$ *or* $\mathrm{non}(\mathcal{N}) = \mathfrak{c}$, *then there exists a Steinhaus set.*

Proof. Assume that $\mathrm{add}(\mathcal{N}) = \mathrm{cov}(\mathcal{N}) = \kappa$. Let $[0,1] = \bigcup_{\xi<\kappa} B_\xi$, $B_\xi \in \mathcal{N}$. We can assume that $\langle B_\xi : \xi \in \kappa \rangle$ are pairwise disjoint. We set

$$A = \bigcup_{\xi<\kappa} \left(B_\xi \times \bigcup_{\eta<\xi} B_\eta \right).$$

If $x \in B_\xi$, then $A_x = \bigcup_{\eta<\xi} B_\eta$. Hence $\lambda(A_x) = 0$. If $y \in B_\xi$ then $A^y = \bigcup_{\eta\geq\xi} B_\eta$. Hence $\lambda(A^y) = 1$.

Now assume that $\mathrm{non}(\mathcal{N}) = \mathfrak{c}$ and $\{x_\xi : \xi < \mathfrak{c}\}$ is a one-to-one enumeration of $[0,1]$. Similarly as above we set

$$A = \{\langle x_\xi, x_\eta \rangle : \eta < \xi < \mathfrak{c}\}.$$

Evidently $\lambda(A_x) = 0$ and $\lambda(A^y) = 1$ for every $x, y \in [0,1]$. $\qquad\square$

Corollary [AC] 10.34. *If* $\mathrm{add}(\mathcal{N}) = \mathrm{cov}(\mathcal{N})$ *or* $\mathrm{non}(\mathcal{N}) = \mathfrak{c}$, *then* **NROI** *holds true.*

Theorem [AC] 10.35. *If* $\mathrm{non}(\mathcal{N}) < \mathrm{cov}(\mathcal{N})$, *then there is no Steinhaus set.*

Proof. Assume that A is a Steinhaus set. Let $B \subseteq [0,1]$ be a non-measure zero set such that $|B| = \mathrm{non}(\mathcal{N})$. We claim that $\bigcup_{x \in B} A_x = [0,1]$. Actually, let $y \in [0,1]$. Since $\lambda(A^y) = 1$ we have $B \cap A^y \neq \emptyset$. Thus there exists an $x \in B$ such that $y \in A_x$. Therefore $\mathrm{cov}(\mathcal{N}) \leq \mathrm{non}(\mathcal{N})$. $\square$

Theorem [AC] 10.36 (C. Freiling – M. Laczkovich). *The assertion* **NROI** *implies that there exists a Steinhaus set.*

Corollary [AC] 10.37. *If* $\mathrm{non}(\mathcal{N}) < \mathrm{cov}(\mathcal{N})$, *then* **NROI** *does not hold true.*

Demonstration of Metatheorem 10.8. Since by Metatheorem 9.1 each of the equalities and inequality $\mathrm{add}(\mathcal{N}) = \mathrm{cov}(\mathcal{N})$, $\mathrm{non}(\mathcal{N}) = \mathfrak{c}$, $\mathrm{non}(\mathcal{N}) < \mathrm{cov}(\mathcal{N})$ is consistent with **ZFC**, the Metatheorem follows by Corollaries 10.34 and 10.37. $\square$

For a proof of Theorem 10.36 we need a special case of Ergodic Theorem 4.31. Let $T : \mathbb{T} \longrightarrow \mathbb{T}$ be a shift by a real a, i.e., $T(x) = x + a$. It is known that T is ergodic if and only if a is irrational, see, e.g., Exercise 4.21, d). Hence if a is irrational, then the function g^* of Theorem 4.31 is constant. Since λ is a probabilistic measure we obtain that $g^*(x) = \int_{[0,1]} g \, d\lambda$ for every $x \in [0,1]$. Thus, as a special case of Ergodic Theorem 4.31 we have

Theorem [AC] 10.38. *Let* $g : \mathbb{T} \longrightarrow \mathbb{R}$ *be an integrable function,* $a \in \mathbb{R}$ *being irrational. Then*

$$\lim_{n \to \infty} \frac{1}{n} \sum_{i=0}^{n-1} g(x + ia) = \int_{[0,1]} g \, d\lambda \text{ for almost every } x.$$

Now we are ready to show the main result of the section.

Proof of Theorem 10.36. Assume that **NROI** holds true. Then there exists a function $f : [0,1] \times [0,1] \longrightarrow [0,1]$ such that (i)–(iii) hold true and the equality (10.18) fails. We can assume that

$$\int_{[0,1]} \psi \, d\lambda < p < \int_{[0,1]} \varphi \, d\lambda \tag{10.19}$$

for suitable real p. The functions φ and ψ being bounded are integrable.

Now, we fix an irrational $a \in [0,1]$ and write

$$X = \{x \in [0,1] : (\forall n) \, f(x + na, \cdot) \text{ is measurable}\}.$$

By **NROI** we have $\lambda(X) = 1$. Set

$$F(x, y) = \liminf_{n \to \infty} \frac{1}{n} \sum_{i=0}^{n-1} f(x + ia, y), \quad G(x, y) = \limsup_{n \to \infty} \frac{1}{n} \sum_{i=0}^{n-1} F(x, y + ia),$$

$$A = \{\langle x, y \rangle \in [0, 1]^2 : G(x, y) < p\},$$

$$B = \left\{ x \in [0, 1] : \lim_{n \to \infty} \frac{1}{n} \sum_{i=0}^{n-1} \psi(x + ia) \neq \int_{[0,1]} \psi(t) \, dt \right\},$$

$$C = \left\{ y \in [0, 1] : \lim_{n \to \infty} \frac{1}{n} \sum_{i=0}^{n-1} \varphi(y + ia) \neq \int_{[0,1]} \varphi(t) \, dt \right\}.$$

By Ergodic Theorem 10.38 we have $\lambda(B) = \lambda(C) = 0$. We show that A is a Steinhaus set.

For any $x \in X$ the function $F(x, \cdot)$ is measurable, being bounded is integrable, therefore by Ergodic Theorem 10.38 for almost every y we obtain

$$G(x, y) = \limsup_{n \to \infty} \frac{1}{n} \sum_{i=0}^{n-1} F(x, y + ia) = \int_{[0,1]} F(x, t) \, dt.$$

If moreover $x \notin B$, then by (ii) and the Fatou Lemma, Corollary 4.24 we obtain

$$\int_{[0,1]} F(x, t) \, dt \leq \liminf_{n \to \infty} \frac{1}{n} \sum_{i=0}^{n-1} \int_{[0,1]} f(x + ia, t) \, dt$$

$$= \lim_{n \to \infty} \frac{1}{n} \sum_{i=0}^{n-1} \psi(x + ia) = \int_{[0,1]} \psi(t) \, dt < p \text{ for almost every } y.$$

Thus, for $x \in X \setminus B$ we have $\lambda(A^x) = 1$.

Again, for any $x \in X$ and any $y \in [0, 1]$, by Ergodic Theorem 10.38 we obtain

$$F(x, y) = \lim_{n \to \infty} \frac{1}{n} \sum_{i=0}^{n-1} f(x + ia, y) = \int_{[0,1]} f(t, y) \, dt = \varphi(y) \text{ for almost every } x \in X.$$

If $y \notin C$, then

$$G(x, y) = \limsup_{n \to \infty} \frac{1}{n} \sum_{i=0}^{n-1} F(x, y + ia)$$

$$= \limsup_{n \to \infty} \frac{1}{n} \sum_{i=0}^{n-1} \varphi(y + ia) = \int_{[0,1]} \varphi(t) \, dt > p \text{ for almost every } x \in X.$$

Thus, for $y \notin C$ we have $\lambda(A_y) = 0$. $\qquad\square$

We have shown that in the set theory **ZFC** one cannot decide a natural question concerning integration.

Exercises

10.12 Nonabsolutely Convergent Series

a) Prove the well-known theorem of analysis: If $\left|\sum_{n=0}^{\infty} a_n\right| < \infty$ and $\sum_{n=0}^{\infty} |a_n| = \infty$, then for any real α there exists a bijection $f : \omega \xrightarrow[\text{onto}]{1-1} \omega$ such that $\sum_{n=0}^{\infty} a_{f(n)} = \alpha$. Moreover, there are bijections g, h such that $\sum_{n=0}^{\infty} c_{g(n)} = +\infty$ and $\sum_{n=0}^{\infty} a_{h(n)} = -\infty$.

b) Show that the example of unbounded function presented in the text is a particular case of this theorem.

10.13 [AC] Steinhaus Set

Let $\mathcal{I}$ be a σ-ideal of a subset of a set X. A set $A \subseteq X \times X$ is called an $\mathcal{I}$-**Steinhaus set**, if $A_x, X \setminus A^x \in \mathcal{I}$ for any $x \in X$. Show that results similar to those of Theorems 10.33 and 10.35 hold true. Especially, they hold true for the ideal of meager sets in a Polish space.

10.14 [AC] Bounded Functions into $^\omega\omega$

Let $\mathcal{I}$ be a σ-ideal of subsets of a set X. We define a cardinal invariant $\mathrm{com}(\mathcal{I})$ of the ideal $\mathcal{I}$ as

$$\mathrm{com}(\mathcal{I}) = \min\{|B| : B \subseteq X \wedge X \setminus B \in \mathcal{I}\}.$$

We denote by $\mathbf{BFB}(X, \mathcal{I})$ the following statement: for every function $f : X \longrightarrow {}^\omega\omega$ there exists a set $A \subseteq X$ such that $X \setminus A \in \mathcal{I}$ and $f(A)$ is eventually bounded, i.e., $f(A) \in \mathcal{K}_\sigma$. The notion of a κ-$\mathcal{K}_\sigma$-set was defined in Section 8.2.

a) If there exists a κ-$\mathcal{K}_\sigma$-subset C of $^\omega\omega$, where $\kappa = \mathrm{com}(\mathcal{I})$, then $\mathbf{BFB}(X, \mathcal{I})$ fails.
   Hint: Let $X \setminus B \in \mathcal{I}$, $|B| = \kappa$. Take any mapping f from X onto C such that $f|B : B \longrightarrow C$ is a bijection.

b) If $\mathrm{com}(\mathcal{I}) = \mathfrak{b}$ or $\mathrm{com}(\mathcal{I}) = \mathfrak{d}$, then $\mathbf{BFB}(X, \mathcal{I})$ fails.
   Hint: See Theorems 8.40 and 8.41.

c) If there exists a κ-Luzin set of reals, where $\kappa = \mathrm{com}(\mathcal{I})$, then $\mathbf{BFB}(X, \mathcal{I})$ fails.
   Hint: Any κ-Luzin subset of $^\omega\omega$ is a κ-$\mathcal{K}_\sigma$-set.

d) If $\mathrm{com}(\mathcal{I}) < \mathfrak{b}$, then $\mathbf{BFB}(X, \mathcal{I})$ holds true.

10.15 [AC] Generalized Egoroff Theorem

Let $\mathcal{I}$ be a σ-ideal of subsets of a set X. We assume that there exists a set $B \subseteq X$ such that $|B| = \mathrm{non}(\mathcal{I})$ and $X \setminus B \in \mathcal{I}$. $\mathbf{GETh}(X, \mathcal{I})$ denotes the following statement: for every sequence of functions $f_n : X \longrightarrow \mathbb{R}$ converging to 0 on X there exists a set $A \subseteq X$ such that $X \setminus A \in \mathcal{I}$ and $f_n \xrightarrow{\mathrm{QN}} 0$ on A.

Let $\mathcal{T}$ denote the set of all sequences $\{f_n\}_{n=0}^\infty$ of functions belonging to $^X\mathbb{R}$. We define a mapping $\Theta : {}^\omega({}^X\mathbb{R}) \longrightarrow {}^X({}^\omega\omega)$ setting $\Theta(\{f_n\}_{n=0}^\infty) = F$, where $F(x) = h \in {}^\omega\omega$, where

$$h(n) = \min\{m : (\forall k \geq m)\, |f_k(x)| < 2^{-n}\}.$$

a) If $Y \subseteq X$, then $f_n \xrightarrow{\mathrm{QN}} 0$ on Y if and only if the function $\Theta(\{f_n\}_{n=0}^\infty)|Y : Y \longrightarrow {}^\omega\omega$ is eventually bounded.

b) Show that $\mathbf{BFB}(X, \mathcal{I})$ is equivalent to $\mathbf{GETh}(X, \mathcal{I})$.
   Hint: Use a) and note that Θ is a surjection.

c) If $\mathrm{com}(\mathcal{I}) = \mathfrak{b}$ or $\mathrm{com}(\mathcal{I}) = \mathfrak{d}$ or there exists a $\mathrm{com}(\mathcal{I})$-Luzin set of reals, then **GETh**$(X, \mathcal{I})$ fails.

d) If $\mathrm{com}(\mathcal{I}) < \mathfrak{b}$, then **GETh**$(X, \mathcal{I})$ holds true.

e) **GETh**$([0,1], \mathcal{N})$ is undecidable in **ZFC**.
   Hint: By Corollary 7.37 we have $\mathrm{non}(\mathcal{N}) = \mathrm{com}(\mathcal{N})$ *and use Metatheorem 9.1.*

f) **GETh**$([0,1], \mathcal{N})$ is equivalent to the statement: for every sequence $\{f_n\}_{n=0}^{\infty}$ of functions from $[0,1]$ into $\mathbb{R}$ and for every $\varepsilon > 0$, there exists a set $A \subseteq [0,1]$ with $\lambda^*(A) > 1 - \varepsilon$ and such that $f_n \rightrightarrows 0$ on A.

10.16 [AC] Weak Generalized Egoroff Theorem

Let us consider a weak generalization of Egoroff Theorem 7.41 referred to as **wGETh**: Let $f_n : [0,1] \longrightarrow \mathbb{R}$, $f_n \to 0$ on $[0,1]$. For every $\varepsilon < 1$ there exist a set A and an increasing sequence $\{n_k\}_{k=0}^{\infty}$ such that $\lambda^*(A) > \varepsilon$ and $f_{n_k} \rightrightarrows 0$ on A. Evidently **GETh**$([0,1], \mathcal{N})$ implies **wGETh**.

a) Assume $\mathcal{G} \subseteq [\omega]^{\omega}$, $|\mathcal{G}| < \mathrm{cov}(\mathcal{M})$. Then for every $\alpha \in {}^{\omega}\omega$ there exists $\beta \in {}^{\omega}\omega$, $\beta \geq^* \alpha$ and such that for any $G \in \mathcal{G}$ the set $G \cap \mathrm{rng}(\beta)$ is infinite.
   Hint: Assume that α is increasing. Set $I_n = \{k \in \mathbb{N} : \alpha(n) \leq k < \alpha(n+1)\}$, $N_G = \{n \in \omega : G \cap I_n \neq \emptyset\}$, $\gamma_G(n) = \min G \cap I_n$ if $n \in N_G$ and $\gamma_G(n) = 0$ otherwise. By Theorem 7.66 there exists a real γ such that

$$(\forall G \in \mathcal{G})(\forall m)(\exists n > m)\,(n \in N_G \wedge \gamma_G(n) = \gamma(n)).$$

   Set $\beta(n) = \gamma(n)$ if $\gamma(n) \in I_n$ and $\beta(n) = \alpha(n)$ otherwise.

b) If $\kappa = \mathfrak{b} = \mathrm{cov}(\mathcal{M})$, then for any $\mathcal{G} \subseteq [\omega]^{\omega}$, $|\mathcal{G}| = \kappa$ there exists a $\leq^*$ increasing sequence $\langle \alpha_\xi \in {}^{\omega}\omega : \xi < \kappa \rangle$ such that

$$(\forall G \in \mathcal{G})(\forall \alpha \in {}^{\omega}\omega)(\exists \eta)(\forall \xi > \eta)\,(\alpha \leq^* \alpha_\xi \wedge |G \cap \mathrm{rng}(\alpha_\xi)| = \aleph_0).$$

   Hint: By transfinite induction using a).

c) If $\mathrm{add}(\mathcal{M}) = \mathfrak{c}$, then **wGETh** does not hold true.
   Hint: Let $\langle \alpha_\xi : \xi < \mathfrak{c} \rangle$ be the sequence of c) with $\mathcal{G} = [\omega]^{\omega}$, $[0,1] = \{x_\xi : \xi < \mathfrak{c}\}$. Define $f_n(x_\xi) = 1/k$ if $k > 0$ and $\alpha_\xi(k) = m$, $f_n(x_\xi) = 0$ otherwise. Evidently $f_n(x_\xi) \to 0$. Suppose $A \subseteq [0,1]$, $\lambda^*(A) > 0$, and $\{n_k\}_{k=0}^{\infty}$ increasing such that

$$(\forall n)(\exists k_0)(\forall x \in A)(\forall k \geq k_0)\,|f_{n_k}(x)| < 1/(n+1).$$

   Since $|A| = \mathfrak{c}$ there exists a ξ such that $x_\xi \in A$, $\{n_k : k \in \omega\} \cap \mathrm{rng}(\alpha_\xi)$ is infinite and $\{n_k\}_{k=0}^{\infty} \leq^* \alpha_\xi$, a contradiction.

d) **wGETh** is undecidable in **ZFC**.

10.5   Permitted Sets of Trigonometric Series

In Section 8.6 we presented some properties of thin sets of trigonometric series and we showed that some sets are permitted for those families. We finished the Section with Problems 8.149 and 8.150 about the existence of a permitted set of large size. Now we show that in **ZFC** there is no answer to the latter problem.

We begin with a simple consequence of the Dirichlet Theorem 8.133.

Lemma 10.39. *Assume that $\{n_i\}_{i=0}^{\infty}$ is an increasing sequence of natural numbers. We write $K = \{n_i - n_j : (\exists k)\, 2^k \leq j < i < 2^{k+1}\}$. Then for any non-empty set $A \subseteq \mathbb{T}$ the function 0 belongs to the closure of the set $\{\|mx\| : m \in K\} \subseteq C_p(A)$ in the product topology.*

Proof. We have to show that for given $x_1, \ldots, x_m \in A$ and $\varepsilon > 0$ there exist natural numbers i, j, k such that $2^k \leq i < j < 2^{k+1}$ and $\|(n_i - n_j)x_l\| < \varepsilon$ for any $l = 1, \ldots, m$.

Let $2^k > (2/\varepsilon)^m$. By the Dirichlet Theorem 8.133 there exist natural numbers i, j such that $2^k \leq i < j \leq 2^k + (2/\varepsilon)^m$ and $\|(n_i - n_j)x_l\| < \varepsilon$ for any $l = 1, \ldots, m$. Since $2^k + (2/\varepsilon)^m < 2^{k+1}$ we are done. $\qquad\square$

Theorem [AC] 10.40. (**L. Bukovský – N.N. Kholshchevnikova – M. Repický**) *Every γ-subset of $\mathbb{T}$ is permitted for any of the families $p\mathcal{D}$, $\mathcal{N}_0$, $\mathcal{N}$, $\mathcal{A}$, and $w\mathcal{D}$.*

Proof. Assume that A is a pD-set and B is a γ-set. By definition there exists an increasing sequence of positive integers $\{n_k\}_{k=0}^{\infty}$ such that $\|n_k x\| \xrightarrow{\text{QN}} 0$ on A. If $K \subseteq \mathbb{N}$ is the set of natural numbers defined in Lemma 10.39, the function 0 belongs to the topological closure of the set $\{\|mx\| : m \in K\} \subseteq {}^B\mathbb{R}$. Since the set B is a γ-set there exists a sequence $\{m_l\}_{l=0}^{\infty}$ of elements of the set K such that $\|m_l x\| \to 0$ on B. Every m_l is of the form $n_{j_l} - n_{i_l}$ with $2^k \leq i_l < j_l < 2^{k+1}$. Thus we may suppose that both sequences $\{i_l\}_{l=0}^{\infty}$ and $\{j_l\}_{l=0}^{\infty}$ are increasing. Since B is also a wQN-set there exists a subsequence converging to 0 quasi-normally. Thus without loss of generality we may suppose that $\|m_l x\| \xrightarrow{\text{QN}} 0$ on B. Since $\|m_l x\| \leq \|n_{j_l} x\| + \|n_{i_l} x\|$ we obtain $\|m_l x\| \xrightarrow{\text{QN}} 0$ on $A \cup B$.

Proofs for $\mathcal{N}_0$-permitted and $\mathcal{A}$-permitted sets are similar. We present a proof for N-set that uses an old idea by J. Arbault.

Assume now that A is an N-set and B is an infinite γ-set. Let $\{n_k\}_{k=0}^{\infty}$, $\{a_n\}_{k=0}^{\infty}$ be such sequences of natural numbers and non-negative reals, respectively, that $\sum_{n=0}^{\infty} a_n = \infty$ and

$$A \subseteq \left\{ x \in [0,1] : \sum_{n=0}^{\infty} a_n \|nx\| < \infty \right\}$$

holds true. Let $\{y_k\}_{k=0}^{\infty}$ be a sequence of distinct elements of B.

We write $s_n = \sum_{k=0}^{n} a_k$ and assume that $a_0 \geq 1$. Hence $s_k \geq 1$ for all k. By Theorem 2.23 we obtain

$$\sum_{n=0}^{\infty} a_n s_n^{-1} = +\infty \quad \text{and} \quad \sum_{n=0}^{\infty} a_n s_n^{-1-\varepsilon} < +\infty$$

for any $\varepsilon > 0$. One can easily construct a non-decreasing unbounded sequence $\{p_n\}_{n=0}^{\infty}$ of positive integers such that

$$\sum_{n=0}^{\infty} \frac{a_n}{s_n^{1+\frac{1}{p_n}}} < +\infty.$$

Set $b_n = a_n/s_n$, $\varepsilon_n = s_n^{-1/p_n}$ and $h(n) = \min\{m : \sum_{k=n}^{m} b_k \geq 1\}$. Let S_n be the set of all sequences $t = \{t(k)\}_{k=n}^{h(n)}$ of positive integers with $t(k) \leq s_k$. For $n \in \omega$ and $t \in S_n$ we set

$$U_{n,t} = \{x \in \mathbb{T} : (\forall k \geq n, k \leq h(n)) \, \|t(k)kx\| < 2\varepsilon_k\} \setminus \{y_n\}.$$

We claim that $\{U_{n,t} : n \in \omega, t \in S_n\}$ is an open ω-cover of $\mathbb{T}$.

Assume $x_1, \ldots, x_m \in \mathbb{T}$. Take n such that $p_n \geq m$ and $x_i \neq y_n$ for any $i = 1, \ldots, m$. For any k, $n \leq k \leq h(n)$, by the Dirichlet Theorem 8.133 there exists a $t(k)$ such that $\|t(k)kx_i\| < 2\varepsilon_k$ for $i = 1, \ldots, m$ and

$$t(k) \leq \left(\frac{2}{2\varepsilon_k}\right)^m = (s_k^{\frac{1}{p_k}})^m \leq s_k.$$

Thus $t \in S_n$ and $x_i \in U_{n,t}$ for any $i = 1, \ldots, m$.

Since B is a γ-set, there exists a sequence $\{[n_k, t_k]\}_{k=0}^{\infty}$ such that

$$B \subseteq \bigcup_{m=0}^{\infty} \bigcap_{k=m}^{\infty} U_{n_k, t_k}.$$

As $y_n \in B$ the equality $n = n_k$ can hold true for finitely many k's. Therefore the sequence $\{n_k\}_{k=0}^{\infty}$ is unbounded and we can assume that $n_{k+1} > h(n_k)$ for every k. By definition of h we obtain $\sum_{k=0}^{\infty} \sum_{n=n_k}^{h(n_k)} b_n = +\infty$. We show that the series

$$\sum_{k=0}^{\infty} \sum_{n=n_k}^{h(n_k)} b_n \|t_k(n)nx\| \tag{10.20}$$

converges on $A \cup B$.

By definition we have $b_n\|t_k(n)nx\| \leq b_n s_n\|nx\| = a_n\|nx\|$. Thus for $x \in A$ the series (10.20) converges. For $x \in B$ there exists an m such that $x \in U_{n_k,t_k}$ for every $k \geq m$. Then

$$\sum_{k=m}^{\infty} \sum_{n=n_k}^{h(n_k)} b_n \|t_k(n)nx\| \leq \sum_{k=m}^{\infty} \sum_{n=n_k}^{h(n_k)} b_n 2\varepsilon_n = 2 \sum_{k=m}^{\infty} \sum_{n=n_k}^{h(n_k)} \frac{a_n}{s_n^{1+\frac{1}{p_n}}} < \infty.$$

For wD-sets the assertion follows by 8.147 since a γ-set has strong measure zero. $\qquad \square$

Since every countable set is a γ-set, the Arbault-Erdős Theorem 8.146 follows.

Since $\mathfrak{p} = \mathfrak{c}$ is consistent with **ZFC** (see e.g. (11.21)), by Theorem 8.92 we obtain

Metatheorem 10.9. ZFC + "there exists a set permitted for every family $\mathcal{N}$, $\mathcal{D}$, $p\mathcal{D}$, $\mathcal{A}$, $\mathcal{N}_0$, $w\mathcal{D}$ of cardinality $\mathfrak{c}$" *is consistent.*

On the other hand one can show that permitted sets are small, actually at least perfectly meager and therefore one cannot prove that there exists a large permitted set. We start with an auxiliary classical result.

We shall consider $\mathbb{R}^n$ as a vector space over the field $\mathbb{Q}$. We define the **scalar product** of $x = \langle x_1, \ldots, x_n \rangle$, $y = \langle y_1, \ldots, y_n \rangle \in \mathbb{R}^n$ as $x \cdot y = \sum_{i=1}^n x_i \cdot y_i$ and the **norm** $|x| = \max\{|x_1|, \ldots, |x_n|\}$. Evidently $|x \cdot y| \le n \cdot |x| \cdot |y|$.

Theorem 10.41 (L. Kronecker). *Assume that the set $\{1, x_1, \ldots, x_n\}$ is linearly independent over $\mathbb{Q}$ and $y_1, \ldots, y_n$ are arbitrary reals. Then for every $\varepsilon > 0$ there exists an arbitrarily large integer k such that $\|kx_i - y_i\| < \varepsilon$ for $i = 1, \ldots, n$.*

We assume that the set $\{1, x_1, \ldots, x_n\}$ is linearly independent over $\mathbb{Q}$. We put

$$L = \{\langle p_1, \ldots, p_n \rangle + q \cdot \langle x_1, \ldots, x_n \rangle : p_1, \ldots, p_n, q \in \mathbb{Z}\}.$$

Lemma 10.42. *Let $\varepsilon > 0$. If $u = \langle u_1, \ldots, u_n \rangle \in \mathbb{R}^n$ is a non-zero element, then there exists a $z \in L$ such that $|z| < \varepsilon$ and $u \cdot z \ne 0$.*

Proof. Assume that for every $z \in L$, $|z| < \varepsilon$ we have $u \cdot z = 0$.

Let $v_1, \ldots, v_m$, $m \le n$ be a base of the smallest vector subspace of $\mathbb{R}$ over $\mathbb{Q}$ containing $u_1, \ldots, u_n$. Then $u = \sum_{j=1}^m v_j \langle a_{1,j}, \ldots, a_{r,j} \rangle$ for some $a_{i,j} \in \mathbb{Q}$, $i \le n$, $j \le m$. We write $a(j) = \langle a_{1,j}, \ldots, a_{n,j} \rangle$. By the Dirichlet Theorem 8.133 there exist a natural number $l > 0$ and integers $p_1, \ldots, p_n$ such that the vector

$$z = \langle z_1, \ldots, z_n \rangle = \langle p_1, \ldots, p_n \rangle + l \cdot \langle x_1, \ldots, x_n \rangle \in L$$

satisfies $|z| < \varepsilon$. So, by the assumption we have $u \cdot z = 0$.

Let us consider arbitrary $\delta < \varepsilon$. Using again the Dirichlet Theorem 8.133 there exist a natural number $k > 0$ and integers $q_1, \ldots, q_n$ such that $|kz_i - q_i| < \delta$. Hence $w = k \cdot \langle z_1, \ldots, z_n \rangle - \langle q_1, \ldots, q_n \rangle \in L$ and $|w| < \delta$. Since $\delta < \varepsilon$ we obtain again $u \cdot w = 0$. As $u \cdot z = 0$ we obtain $u \cdot [q_1, \ldots, q_n] = 0$, i.e.,

$$\sum_{j=1}^m v_j \cdot (a(j) \cdot \langle q_1, \ldots, q_n \rangle) = 0.$$

Each scalar product $a(j) \cdot \langle q_1, \ldots, q_n \rangle$ is a rational number and $v_i, \ldots, v_m$ are independent over $\mathbb{Q}$. Therefore $a(j) \cdot \langle q_1, \ldots, q_n \rangle = 0$ for $j = 1, \ldots, m$. Thus

$$|a(j) \cdot z| \le |ka(j) \cdot z| = |a(j) \cdot (kz - \langle q_1, \ldots, q_n \rangle)| < n|a(j)| \cdot \delta.$$

Since δ was arbitrary small we have $a(j) \cdot z = 0$ for $j = 1, \ldots, m$ and therefore

$$\sum_{i=0}^n a_{i,j}(p_i + lx_i) = 0$$

for $j = 1, \ldots, n$. Since the set $\{1, x_1, \ldots, x_n\}$ is linearly independent over $\mathbb{Q}$ we have $a_{i,j} = 0$ for every $i = 1, \ldots, n$ and every $j = 1, \ldots, m$. Thus $u = 0$, a contradiction. $\square$

Lemma 10.43. *For any given $\varepsilon > 0$ there exists a set $\{z(1), \ldots, z(n)\} \subseteq L$ linearly independent over $\mathbb{R}$ and such that $|z(i)| < \varepsilon$ for $i = 1, \ldots, n$.*

Proof. We construct the elements $z(1), \ldots, z(n) \in L$ by induction. Assume that a linearly independent set $\{z(1), \ldots, z(m)\}$, $m < n$ is already constructed. Then there exists a non-zero $u \in \mathbb{R}^n$ such that $u \cdot z(i) = 0$ for $i = 1, \ldots, m$ (a non-zero solution of m linear equations). By Lemma 10.42 there exists a $z(m+1) \in L$, $|z(m+1)| < \varepsilon$ such that $u \cdot z(m+1) \neq 0$. Evidently, the set $\{z(1), \ldots, z(m+1)\}$ is linearly independent over $\mathbb{R}$. □

Proof of Theorem 10.41. Assume that $\varepsilon > 0$, $y_1, \ldots, y_n \in \mathbb{R}$ are given and the set $\{1, x_1, \ldots, x_n\}$ is linearly independent over $\mathbb{Q}$. By Lemma 10.43 there exists a set $\{z(1), \ldots, z(n)\} \subseteq L$ linearly independent over $\mathbb{R}$ and such that $|z_{j,i}| < \varepsilon/n$ for $j, i = 1, \ldots, n$, where $z(j) = \langle z_{j,1}, \ldots, z_{j,n} \rangle$. Since $\mathbb{R}^n$ has dimension n over $\mathbb{R}$ there exist reals $b_1, \ldots, b_n$ such that $\langle y_1, \ldots, y_n \rangle = \sum_{i=1}^{n} b_i \cdot z(i)$. Let m_i be the nearest integer to b_i. Then $z = \sum_{i=0}^{n} m_i \cdot z(i) \in L$ and therefore there exist integers $l, p_1, \ldots, p_n \in \mathbb{Z}$ such that $z = \langle p_1, \ldots, p_n \rangle + l \cdot \langle x_1, \ldots, x_n \rangle$. Then

$$|(p_i + l x_i) - y_i| = |\sum_{j=1}^{n} m_j z_{j,i} - \sum_{j=1}^{n} b_j z_{j,i}| \leq \sum_{j=1}^{n} |m_j - b_j| |z_{j,i}| < \varepsilon/2,$$

and therefore $\|l x_i - y_i\| < \varepsilon/2$ for any $i = 1, \ldots, n$. By the Dirichlet Theorem 8.133 there exists a natural number $k \geq k_0 + l$ such that $\|(k - l)x_i\| < \varepsilon/2$ for $i = 1, \ldots n$. Then $\|k x_i - y_i\| < \varepsilon$ as well. □

We shall need another two auxiliary results.

Lemma [wAC] 10.44. *Let $P \subseteq \mathbb{T}$ be a perfect set. If a finite set $\{x_0, \ldots, x_n\} \subseteq P$ is linearly independent over $\mathbb{Q}$, then for any $y \in P$ and any $\varepsilon > 0$ there exists an element $x_{n+1} \in P$ such that $\{x_0, \ldots, x_{n+1}\}$ is linearly independent over $\mathbb{Q}$ and $|y - x_{n+1}| < \varepsilon$.*

Proof. The set of all linear combinations of elements of $\{x_0, \ldots, x_n\}$ over $\mathbb{Q}$ is countable. Thus the set $\mathrm{Ball}(y, \varepsilon) \cap P$ contains an element not linearly dependent on $\{x_0, \ldots, x_n\}$. □

Lemma [wAC] 10.45. *If $P \subseteq \mathbb{T}$ is a perfect set, then there exists an increasing sequence of natural numbers $\{n_k\}_{k=0}^{\infty}$ such that the set $P \cap (y - C)$ is dense in P for every $y \in \mathbb{T}$, where $C = \{x \in \mathbb{T} : (\exists k_0)(\forall k \geq k_0) \|n_k x\| \leq 2^{-k}\}$.*

Proof. Let $Q \subseteq P$ be a countable set dense in P. Consider an enumeration $\langle q_n : n \in \omega \rangle$ of Q such that every element occurs in the enumeration infinitely many times. Set $B_k = \{m 2^{-k-1} : m < 2^{k+1}\}$.

By induction we shall construct n_k, $\varepsilon_k > 0$ and finite sets $A_k \subseteq P$ such that $n_k < n_{k+1}$, $\varepsilon_{k+1} \leq \varepsilon_k/2$, $\{q_i : i \leq k\} \subseteq A_k$ and for every $a \in A_k$ and $b \in B_k$ there exists a $c \in A_{k+1}$ satisfying

(1) $|a - c| < \varepsilon_k/2$,
(2) for any x the inequality $|x - c| < \varepsilon_{k+1}$ implies $\|n_{k+1} x - b\| < 2^{-k-2}$,

everything for every k.

Let $A_0 = \{q_0\}$, n_0, $\varepsilon_0 > 0$ being arbitrary. We describe the inductive step. Using Lemma 10.44 for every $a \in A_k$ and $b \in B_k$ pick an $f_k(a,b) \in P$, such that $|f_k(a,b) - a| < \varepsilon_k/2$, f_k is one-to-one, and the set

$$\{1\} \cup \{f_k(a,b) : a \in A_k \wedge b \in B_k\}$$

is linearly independent over $\mathbb{Q}$. By the Kronecker Theorem 10.41 there exists a natural number $n_{k+1} > n_k$ such that $\|n_{k+1}f_k(a,b) - b\| < 2^{-k-2}$ for all $a \in A_k$ and $b \in B_k$. Set

$$A_{k+1} = \{q_0, \ldots, q_{k+1}\} \cup \{f_k(a,b) : a \in A_k \wedge b \in B_k\}.$$

Since A_k, B_k are finite, there exists a positive real $\varepsilon_{k+1} \leq \varepsilon_k/2$ such that the condition (2) is satisfied for $c = f_k(a,b)$.

Assume that $y \in \mathbb{T}$ is given and U is an open set with $U \cap P \neq \emptyset$. We show that $U \cap P \cap (y - C) \neq \emptyset$

For any k take $b_k \in B_k$ such that $\|n_{k+1}y - b_k\| \leq 2^{-k-2}$. Since each element of Q occurs infinitely many times in $\langle q_n : n \in \omega \rangle$, there exists a k_0 such that $\overline{\text{Ball}}(q_{k_0}, \varepsilon_{k_0}) \subseteq U$. Set $p_k = q_{k_0}$ for $k \leq k_0$. By induction find a $p_{k+1} \in A_{k+1}$ such that $|p_{k+1} - p_k| < \varepsilon_k/2$ and $\|n_{k+1}x - b_k\| < 2^{-k-2}$ for all x satisfying $x - p_{k+1}| < \varepsilon_{k+1}$. Then $\{p_k\}_{k=0}^{\infty}$ is a Bolzano-Cauchy sequence. If $p = \lim_{k \to \infty} p_k$, then $p \in P \cap \overline{\text{Ball}}(q_{k_0}, \varepsilon_{k_0}) \subseteq P \cap U$ and $|p - p_k| < \varepsilon_k$ for every k. Moreover, for any $k \geq k_0$ we have

$$\|n_{k+1}(y - p)\| \leq \|n_{k+1}p - b_k\| + \|n_{k+1}y - b_k\| \leq 2^{-k-1}.$$

Hence $y - p \in C$ and consequently $p \in U \cap P \cap (y - C)$. $\qquad \square$

Theorem [wAC] 10.46 (P. Eliaš). *Assume that $\mathcal{F}$ is a family of thin subsets of $\mathbb{T}$ such that $\mathcal{F}$ has an F_σ base, $pD \subseteq \mathcal{F}$ and $A - A \in \mathcal{F}$ for any $A \in \mathcal{F}$. Then every $\mathcal{F}$-permitted set is perfectly meager.*

Proof. Assume that $A \subseteq \mathbb{T}$ is an $\mathcal{F}$-permitted set, $P \subseteq \mathbb{T}$ is perfect. By Lemma 10.45 there exists a pseudo Dirichlet set C such that $P \cap (y - C)$ is dense in P for every $y \in \mathbb{T}$. By assumption about $\mathcal{F}$ we have $C \in \mathcal{F}$ and therefore $A \cup C \in \mathcal{F}$. Then there exists an F_σ set $B \in \mathcal{F}$, $B \supseteq A \cup C$. Since $B - B \in \mathcal{F}$, we obtain $B - B \neq \mathbb{T}$. Thus there exists a $y \in \mathbb{T}$ such that $B \cap (y - B) = \emptyset$. Then also $B \cap (y - C) = \emptyset$ and therefore $P \cap (y - C) \subseteq P \setminus B$. Since $P \cap (y - C)$ is dense in P, the G_δ set $P \setminus B$ is dense in P. Hence $P \cap A \subseteq P \cap B$ is meager in P. $\qquad \square$

Theorem [wAC] 10.47 (P. Eliaš). *Every set permitted for any of the families pD, $\mathcal{N}$, $\mathcal{N}_0$, and $\mathcal{A}$ is perfectly meager.*

Proof. For any of the families pD, $\mathcal{N}_0$ and $\mathcal{N}$ the assertion follows directly from Theorem 10.46. For A-sets we must change the proof a little.

Let A be an $\mathcal{A}$-permitted set, P being a perfect set. By Lemma 10.45 there exists a pseudo Dirichlet set C such that $P \cap (y - C)$ is dense in P for every $y \in \mathbb{T}$.

Then the set $A \cup C$ is an A-set, and therefore there exists an increasing sequence $\{n_k\}_{k=0}^{\infty}$ such that $A \cup C \subseteq \{x \in \mathbb{T} : \|n_k x\| \to 0\}$. Set

$$B_i = \{x \in \mathbb{T} : (\forall k \geq i)\|n_k x\| \leq 1/8\}, \quad B = \bigcup_i B_i.$$

Then B is an F_{σ} set and $A \cup C \subseteq B$. If $x \in B - B$, then there are i_1, i_2 and $x_1 \in B_{i_1}$, $x_2 \in B_{i_2}$ such that $x = x_1 - x_2$. If $i_0 = \max\{i_1, i_2\}$, then $x_1, x_2 \in B_{i_0}$ and therefore $x \in B_{i_0} - B_{i_0}$. Thus $B - B = \bigcup_i (B_i - B_i)$. On the other hand we have $B_i - B_i \subseteq B_{i+1} - B_{i+1}$ and $B_i - B_i \subseteq \{x \in \mathbb{T} : \|n_i x\| \leq 1/4\}$. One can easily see that $\lambda(\{x \in \mathbb{T} : \|nx\| \leq 1/4\}) = 1/2$ and therefore $\lambda(B_i - B_i) \leq 1/2$ for any $i > 0.$[4] Thus $\lambda(B - B) \leq 1/2$. Hence $B - B \neq \mathbb{T}$ and we can continue as in the proof of Theorem 10.46.   □

Using the consistency result (11.31), Theorem 10.46 implies

Metatheorem 10.10. "A permitted set for any of the families $p\mathcal{D}$, $\mathcal{N}$, $\mathcal{N}_0$, and $\mathcal{A}$ has cardinality $\leq \aleph_1$" is consistent with **ZFC** $+ \neg$**CH**.

We obtain two important consequences of this Metatheorem. By (11.21), Theorems 8.90, and 10.40 we have

Metatheorem 10.11. "Every set of cardinality $< \mathfrak{c}$ is permitted for the families $p\mathcal{D}$, $\mathcal{N}$, $\mathcal{N}_0$, and $\mathcal{A}$" is undecidable in **ZFC** $+ \neg$**CH**.

By Metatheorem 10.9 we obtain the promised result concerning an answer to Problem 8.150

Metatheorem 10.12. If $\mathcal{F}$ denotes any of the families $p\mathcal{D}$, $\mathcal{N}$, $\mathcal{N}_0$, or $\mathcal{A}$, then the statement "there exists a permitted set for $\mathcal{F}$ of cardinality $\mathfrak{c}$" is undecidable in **ZFC**.

Note that we have no result about an upper bound of the size of permitted sets for the family of weak Dirichlet sets. Neither have we any answer to Problem 8.149.

Exercises

10.17 [AC] **Permitted Sets**

a) If $|A| < \mathfrak{s}$ is a wQN-space, then A is permitted for $p\mathcal{D}$ and $\mathcal{N}_0$.

   Hint: If $\sum_{k=0}^{\infty} \|n_k x\| < \infty$ on B, then by the Booth Lemma, Exercise 5.28 we can assume that both sequences $\{\sin n_k \pi x\}_{k=0}^{\infty}$, $\{\cos n_k \pi x\}_{k=0}^{\infty}$ converge on A. Then

$$\lim_{k \to \infty} |\sin(n_{k+1} - n_k)\pi x| = 0.$$

   Since A is a wQN-space we obtain $\sum_{k=0}^{\infty} \|(n_{k+1} - n_k)x\| < \infty$. Note that

$$|\sin(n_{k+1} - n_k)\pi x| \leq |\sin n_{k+1}\pi x| + |\sin n_k \pi x|.$$

b) Show that $\mathrm{non}(\mathrm{Perm}(p\mathcal{D})) \geq \min\{\mathfrak{s}, \mathfrak{b}\}$ and $\mathrm{non}(\mathrm{Perm}(\mathcal{N}_0)) \geq \min\{\mathfrak{s}, \mathfrak{b}\}$.

   Hint: Use the result of a) and Corollary 8.62.

---

[4]Actually $\lambda(B_i) = 0$, see Exercise 8.34.

c) Show that $\mathrm{non}(\mathrm{Perm}(\mathcal{A})) \geq \mathfrak{s}$.

Hint: Use the Booth Lemma, Exercise 5.28 as in part a).

d) Show that $\mathrm{non}(\mathrm{Perm}(\mathcal{N})) \geq \mathfrak{t}$.

e) Show that $\mathrm{non}(\mathrm{Perm}(w\mathcal{D})) \geq \mathrm{cov}(\mathcal{M})$.

Hint: See Corollary 8.14.

10.18 [AC] Dirichlet Permitted Sets

For a perfect set $P \subseteq \mathbb{T}$, let $\{n_k\}_{k=0}^{\infty}$ be the sequence of Lemma 10.45. Set

$$D = \{x \in \mathbb{T} : (\forall k)\, \|n_k x\| \leq n^{-k}\}.$$

a) Show that $P + D = \mathbb{T}$.

Hint: Follow the proof of Lemma 10.45.

b) If $A \subseteq \mathbb{T}$ is $\mathcal{D}$-permitted, then there exists no perfect subset of A.

10.19 [AC] Perfectly Meager in the Transitive Sense

A set $A \subseteq \mathbb{T}$ is **perfectly meager in the transitive sense** if for any perfect set $P \subseteq \mathbb{T}$ there exists an F_σ set $F \supseteq A$ such that $(F + y) \cap P$ is meager in P for every $y \in \mathbb{T}$.

a) A set perfectly meager in the transitive sense is perfectly meager.

b) Assume that $\mathcal{F}$ is a family of thin subsets of $\mathbb{T}$ such that $\mathcal{F}$ has an F_σ-base, $p\mathcal{D} \subseteq \mathcal{F}$ and $A - A \in \mathcal{F}$ for any $A \in \mathcal{F}$. Then every $\mathcal{F}$-permitted set is perfectly meager in the transitive sense.

Hint: By Lemma 10.45 there exists a pseudo-Dirichlet set D such that $(D-y)\cap D$ is dense in P for any y. Since $A + D \in \mathcal{F}$, there exists an F_σ set $F \in \mathcal{F}$, $F \supseteq A + D$. Then $D + F \subseteq F + F \neq \mathbb{T}$. Thus $F \cap (D - x) = \emptyset$ for some x. Then $P \setminus (F + y) \supseteq P \cap (D - x + y)$ is dense in P for every y.

c) Every set permitted for any of the families $p\mathcal{D}$, $\mathcal{N}$, $\mathcal{N}_0$, $\mathcal{A}$, and $w\mathcal{D}$ is perfectly meager in the transitive sense.

Historical and Bibliographical Notes

K. Gödel [1938] announced a proof of the consistency of an existence of a $\boldsymbol{\Delta}_2^1$ Lebesgue non-measurable set and a $\boldsymbol{\Delta}_2^1$ set of power of the continuum containing no perfect subset. K. Kuratowski in 1939 realized that for this proof it suffices to show that the Axiom of Constructibility implies the existence of a $\boldsymbol{\Delta}_2^1$ well-ordering of the real line. A. Mostowski carried out a proof of this implication, however during the Second World War the manuscript was destroyed. K. Kuratowski proved similar results for projective set. Again, during the Second World War the manuscript was destroyed and then reconstructed as [1948]. P.S. Novikoff [1951] independently proved that the Axiom of Constructibility implies existence of above mentioned sets. Then J.W. Addison [1958a] formulated (10.1) and proved that it follows from the Axiom of Constructibility. Then he showed the main consequences of (10.1) including Theorems 10.5 and 10.6.

The First Periodicity Theorem 10.7 was proved independently by D.A. Martin [1968] and Y.N. Moschovakis [1968]. The Second Periodicity Theorem 10.10

was proved by Y.N. Moschovakis [1971]. For details concerning both theorems see Y.N. Moschovakis [1980] and A. Kechris [1995].

S. Banach in 1929 formulated the Measure Problem. Immediately assuming **CH**, S. Banach and K. Kuratowski [1929] proved a negative answer. S. Banach [1930], assuming **GCH**, showed that if a measure is κ-additive, then κ is inaccessible. Then S. Ulam [1930] proved Theorems 10.12, 10.16 and 10.18. The natural question whether the first strongly inaccessible cardinal is measurable, was open until 1964, when W.P. Hanf [1964], in his thesis supervised by A. Tarski, proved the negative answer. H.J. Keisler and A. Tarski [1964] contains many related results. Our proof of Theorem 10.19 uses the technology of ultraproducts of models. R.M. Solovay [1971] proved Theorem 10.20. Metatheorem 10.4 was proved by T. Jech [1968].

M.J. Souslin [1920] raised the problem, whether there exists a Souslin line. D. Kurepa [1935] showed the equivalence of conditions (i) and (iii) in Theorem 10.24. The equivalence of (i) and (ii) was essentially proved by H. Gaifman [1964]. T. Jech [1967] and S. Tennenbaum [1968] independently constructed (different) models of **ZFC** in which there exists a Souslin line. R.B. Jensen [1968] proved Theorem 10.26. Theorem 10.27 was proved by G. Fodor [1956]. R.M. Solovay and S. Tennenbaum [1971] constructed a model in which the Souslin Hypothesis holds true. As we have already noted they constructed a model of **ZFC** with **MA** and $2^{\aleph_0} > \aleph_1$. A construction of an Aronszajn tree as in Exercise 10.8 and Kurepa's Theorem of Exercise 10.9 is presented by D. Kurepa [1935]. The main result of Exercise 10.11, part d) was proved by J.E. Baumgartner, J.I. Malitz and W.N. Reinhart [1970].

H. Friedman [1980] constructed a model of **ZFC** in which each function satisfying the conditions (i)–(iii) of Section 10.4 satisfies the equality (10.18) as well. W. Sierpiński [1920] proved that **CH** implies **NROI**. Steinhaus raised the question whether a Steinhaus set does exist. Theorems 10.33 and 10.35 are folklore. Then C. Freiling [1986] and independently M. Laczkovich [1986] showed Theorem 10.36. Our proof follows that of M. Laczkovich. The results of Exercises 10.14, 10.15 and 10.16 were originated by T. Weiss [2004], then generalized and simplified by R. Pinciroli [2006] and M. Repický [2009].

N.N. Kholshchevnikova [1985] extended the Erdős-Arbault Theorem 8.146 replacing "countable" by "less than $\mathfrak{m}$". Then Z. Bukovská [1990] showed that $\mathfrak{m}$ may be replaced by $\mathfrak{p}$. Finally L. Bukovský, N.N. Kholshchevnikova and M. Repický [1994] proved Theorem 10.40. T.W. Körner [1974] proved a stronger variant of Lemma 10.45. Independently P. Erdős, K. Kunen and R.D. Mauldin [1981] proved a similar (weaker) result. Then P. Eliaš [2009], trying to strengthen this result, independently proved Lemma 10.45. As a consequence he obtained Theorem 10.46; see also P. Eliaš [2005]. Results of Exercise 10.17 were proved by L. Bukovský, N.N. Kholshchevnikova and M. Repický [1994]. Results of Exercise 10.19 are due to P. Eliaš [2009].

Chapter 11

Appendix

It may be more interesting to consider a family of problems on the continuum: investigate cardinal invariants. ... There is a myriad of such measures, many of them important in many directions ...; naturally they are uncountable but $\leq 2^{\aleph_0}$.

If the continuum is $\leq \aleph_2$, we will not have (by a trivial pigeonhole principle) many relations, as no three can be simultaneously distinct.

Saharon Shelah [2003], p. 213

Trying to make the book self-contained, in Sections 11.1 and 11.2 we recall some well-known notions and their properties. A reader need not read those sections. If she/he needs to recall something, she/he can use the Index to find it. Section 11.3 summarizes well-known facts from elementary topology showing the close connection between the notions of a compact topological space and the real line. Finally, in Sections 11.4 and 11.5 we summarize basic facts of mathematical logic and main results of the theory of models of set theory. The latter are intensively used in the basic text.

It is worth noting that many consistency results allow that the cardinality of the continuum is anything it ought to be, i.e., any well-defined cardinal not cofinal with $\aleph_0$. On the other hand, several consistency results (e.g., (11.24), (11.27), (11.28), (11.31), (11.34)) were shown just assuming a strong condition: c can be only $\aleph_2$. The question whether we can omit this condition is often an important open problem.

11.1 Sets, Posets, and Trees

An **ordered pair** $\langle x, y \rangle$ is the set $\{\{x\}, \{x, y\}\}$. The basic property of an ordered pair is expressed by the equivalence

$$\langle x, y \rangle = \langle u, v \rangle \equiv (x = u \wedge y = v). \tag{11.1}$$

An **ordered n-tuple** is defined by induction as

$$\langle x_1, \ldots, x_n, x_{n+1} \rangle = \langle \langle x_1, \ldots, x_n \rangle, x_{n+1} \rangle.$$

Actually we do not worry what an ordered n-tuple is, we ask only that a condition similar to (11.1) holds true. The **Cartesian product** or simply the **product** of sets A and B is the set $A \times B$ of all ordered pairs $\langle x, y \rangle$ such that $x \in A$ and $y \in B$. Similarly by induction we define

$$A_1 \times \cdots \times A_n \times A_{n+1} = (A_1 \times \cdots \times A_n) \times A_{n+1}.$$

If $A_1 = \cdots = A_n = A$, then we shall write $A^n = A_1 \times \cdots \times A_n$. If $A \subseteq X \times Y$, $x \in X$, $y \in Y$, then the **vertical** and the **horizontal section** of A are the sets

$$A_x = \{u \in Y : \langle x, u \rangle \in A\}, \qquad A^y = \{v \in X : \langle v, y \rangle \in A\}, \tag{11.2}$$

respectively.

A **binary relation** R is a set of ordered pairs. If $X = \bigcup \bigcup R$, then $R \subseteq X \times X$. The **domain** of R is the set $\mathrm{dom}(R) = \{x \in X : (\exists y)\langle x, y \rangle \in R\}$ and the **range** of R is the set $\mathrm{rng}(R) = \{x \in X : (\exists y)\langle y, x \rangle \in R\}$. The **field** of R is the set $\mathrm{dom}(R) \cup \mathrm{rng}(R)$. Instead of $\langle x, y \rangle \in R$ we shall often write xRy. A **function** f is a binary relation such that

$$(\forall x, y, z)\,((\langle x, y \rangle \in f \wedge \langle x, z \rangle \in f) \rightarrow y = z).$$

If $\langle x, y \rangle \in f$ we write $f(x) = y$. y is the value of the function f at x. Often we shall write $f_x = y$ instead of $f(x) = y$. The set $\mathrm{dom}(f)$ is the **definition domain** of f and $\mathrm{rng}(f)$ is the **range** or **domain of values**. By ${}^X Y$ we denote the set of all functions f with $\mathrm{dom}(f) = X$ and $\mathrm{rng}(f) \subseteq Y$. Evidently ${}^X Y \subseteq \mathcal{P}(X \times Y)$. Instead of the official notion "function" we shall use synonyms **mapping**, **map**, **transformation** etc.

The notation $f : X \longrightarrow Y$ means that $f \in {}^X Y$, i.e., f is a function with $\mathrm{dom}(f) = X$ and $\mathrm{rng}(f) \subseteq Y$. The notations $f : X \xrightarrow{1-1} Y$ and $f : X \xrightarrow{\mathrm{onto}} Y$ mean moreover that f is a **one-to-one function** and f is a **function onto** $Y = \mathrm{rng}(f)$, respectively. A one-to-one function is also called an **injection**, a function onto Y is called a **surjection** and a function which is both injection and surjection is called a **bijection**.

If $f : X \longrightarrow Y$, $A \subseteq X$, then the **image** of A is $f(A) = \{f(x) : x \in A\}$ and if $B \subseteq Y$, then the **inverse image** of B is $f^{-1}(B) = \{x \in X : f(x) \in B\}$. Similarly for a relation $R \subseteq X \times Y$, e.g., $R(A) = \{y \in Y : (\exists x)\,(x \in A \wedge \langle x, y \rangle \in R)\}$. The notation may be sometimes confusing, however from the context it should be clear what it means. The **restriction** of a function f to an $A \subseteq \mathrm{dom}(f)$ is the function $f|A = \{\langle x, y \rangle \in f : x \in A\}$. A function g is an **extension** of the function f if $f \subseteq g$, i.e., if $f = g|\mathrm{dom}(f)$.

Sometimes we shall denote a function $f : X \longrightarrow Y$ as $\langle f(x) : x \in X \rangle$. We shall be free in using such a denotation and sometimes we shall include in it some additional information, e.g., $\langle f(x) = \{y \in A : \varphi(x, y)\} : x \in X \rangle$ or $\langle f(x) \subseteq A : x \in X \rangle$, etc. Moreover, sometimes we shall not distinguish between the function $\langle f(x) : x \in X \rangle$ and its range $\{f(x) : x \in X\}$. Always, everything must be clear from the context.

If $f : X \longrightarrow Y$, $g : Y \longrightarrow Z$, then the **composition** $f \circ g : X \longrightarrow Z$ is defined as $f \circ g(x) = g(f(x))$ for any $x \in X$. The reader should know what an **inverse** of f is. If the inverse of f does exist it will be denoted by f^{-1}. If $X \subseteq Y$, then the **identity mapping** $\mathrm{id}_X : X \longrightarrow Y$ is defined as $\mathrm{id}_X(x) = x$ for any $x \in X$.

If $f : X \times Y \longrightarrow Z$, then instead of $f(\langle x, y \rangle)$ we simply write $f(x, y)$. Moreover, for a fixed $x \in X$ the function $g : Y \longrightarrow Z$ defined as $g(y) = f(x, y)$ will be simply denoted by $f(x, \cdot)$. Similarly $f(\cdot, y)$ for a given $y \in Y$. If $Z = X \times Y$, then the **projections** $\mathrm{proj}_1 : Z \longrightarrow X$ and $\mathrm{proj}_2 : Z \longrightarrow Y$ are defined as

$$\mathrm{proj}_1(x, y) = x, \qquad \mathrm{proj}_2(x, y) = y.$$

If S is a non-empty set and values of the function $\langle X_s : s \in S \rangle$ are non-empty sets, then the **Cartesian product** or simply the **product** $\Pi_{s \in S} X_s$ is the set of all functions $f : S \longrightarrow \bigcup_{s \in S} X_s$ satisfying the condition $(\forall s)(s \in S \to f(s) \in X_s)$. The **projection** $\mathrm{proj}_s : \Pi_{t \in S} X_t \longrightarrow X_s$ is defined as $\mathrm{proj}_s(f) = f(s)$ for any $s \in S$. If $X_s = X$ for any $s \in S$, then $\Pi_{s \in S} X_s = {}^S X$. If the set S is finite, e.g., $S = \{1, \ldots, n\}$, then we often identify the sets $\Pi_{s \in S} X_s$ and $X_1 \times \cdots \times X_n$ in the natural way.

A **sequence** is a function with $\mathrm{dom}(f) = \omega$ or $\mathrm{dom}(f) \in \omega$. In the former case we speak about an **infinite sequence**, in the latter about a **finite sequence**. A value of a sequence f is usually denoted as f_n. An infinite sequence a is also denoted as $\{a_n\}_{n=0}^\infty$ and a finite sequence a with $\mathrm{dom}(a) = n+1$ is denoted as $\{a_i\}_{i=0}^n$. From typographical reasons, we shall often use the denotation as $\langle a_n \in A : n \in \omega \rangle$ or $\langle a_i : i < n \rangle$, etc. We shall also use other related notation, e.g., instead of $\bigcup_{n \in \omega} A_n$ we shall write $\bigcup_{n=0}^\infty A_n$ or simply $\bigcup_n A_n$.

Sometimes we shall identify in the natural way the sets ${}^n A$ and A^n. Thus a finite sequence $\{a_i\}_{i=0}^{n-1}$ will be also denoted as $\langle a_0, \ldots, a_{n-1} \rangle$. Especially we shall use this convention in the case $n = 1$, i.e., $A = {}^1 A = A^1$. Moreover, ${}^0 A = A^0 = \{\emptyset\}$. We use similar notation for an infinite sequence

$$\{a_n\}_{n=0}^\infty = \langle a_0, \ldots, a_n, \ldots \rangle.$$

The set of all finite sequences of elements of A will be denoted by ${}^{<\omega} A = \bigcup_{n \in \omega} {}^n A$. Similarly ${}^{<\xi} A = \bigcup_{\eta < \xi} {}^\eta A$ for any ordinal ξ.

If $\alpha = \{a_i\}_{i=0}^n$ and $\beta = \{b_j\}_{j=0}^m$ are finite sequences, then the **concatenation** $\alpha^\frown \beta$ is the sequence $\{c_i\}_{i=0}^{n+m+1}$, where $c_i = a_i$ for $i \leq n$ and $c_{n+i+1} = b_i$ for $i \leq m$. Similarly in the case when β is an infinite sequence.

Let X be a set. If $\mathcal{F}$ is a subset of $\mathcal{P}(X)$ we say that $\mathcal{F}$ is a **family of sets**. Let $A \subseteq X$ be a non-empty subset of X. A family $\mathcal{F}$ of subsets of X is called a **cover of the set** A if $\bigcup \mathcal{F} \supseteq A$ and $\emptyset, A \notin \mathcal{F}$. If $\mathcal{F}, \mathcal{G}$ are covers of a set A and $\mathcal{G} \subseteq \mathcal{F}$, then we say that $\mathcal{G}$ is a **subcover** of $\mathcal{F}$. A cover $\mathcal{F}$ of the set X is called a **partition** of X if $\mathcal{F}$ is a family of pairwise disjoint subsets of X. Moreover, the one point family $\{X\}$ is a partition of X as well. A cover $\mathcal{G}$ is a **refinement of a cover** $\mathcal{F}$, if every $C \in \mathcal{G}$ is a subset of some $D \in \mathcal{F}$.

If $\mathcal{F}$ is a family of subsets of a set X and $A \subseteq X$ then

$$\mathcal{F}|A = \{B \cap A : B \in \mathcal{F}\} \tag{11.3}$$

denotes the **restriction** of the family $\mathcal{F}$ to the set A. If $\mathcal{F}$ is a cover (a partition) of X and $A \subseteq X$ then $\mathcal{F}|A$ is a cover (a partition) of the set A.

If we deal with subsets of a fixed set X we use the **characteristic function** χ_A of $A \subseteq X$ defined as $\chi_A(x) = 1$ for $x \in A$ and $\chi_A(x) = 0$ for $x \in X \setminus A$.

We shall very often need a standard identification usually made in mathematics. A binary relation E on a set X is called an **equivalence relation** if E is reflexive, symmetric and transitive, i.e.,

$$xEx, \quad xEy \to yEx, \quad (xEy \wedge yEz) \to xEz$$

for any $x, y, z \in X$, respectively. If E is an equivalence relation on X we define the **equivalence class** of an $x \in X$ as

$$\{x\}_E = \{y \in X : xEy\}.$$

The **quotient set** of X by E is the set

$$X/E = \{\{x\}_E : x \in X\}.$$

If E is an equivalence relation on X, then X/E is a partition of X. Vice versa, if $\mathcal{F}$ is a partition of X, then

$$xEy \equiv (\exists A \in \mathcal{F})\, x, y \in A$$

is an equivalence relation on X such that $X/E = \mathcal{F}$. Such a passing to a quotient set will be often used without an explicit notice.

Let $R \subseteq X \times X$ be a reflexive, antisymmetric[1] and transitive relation. The couple $\langle X, R \rangle$ is a **partially ordered set**, shortly a **poset**, and the relation R is called a **partial ordering**. R is a **strict partial ordering** if $xRy \vee x = y$ is a partial ordering and xRx for no x. If R is a partial ordering we often use notations $\leq, \preceq, \sqsubseteq$ and the strict partial ordering $xRy \wedge x \neq y$ is denoted by $<, \prec, \sqsubset$, respectively. If $x < y$, then we say that y is a **successor** of x. If there exists no z such that $x < z < y$ we say that y is an **immediate successor** of x. Two elements $x, y \in X$ are said to be **comparable** if $x \leq y$ or $y \leq x$. Otherwise x, y are **incomparable**. A subset $A \subseteq X$ is called a **chain** if any two $x, y \in A$ are comparable. If the whole set X is a chain we say that $\langle X, \leq \rangle$ is a **linearly ordered set**. If $A \subseteq X$, $a \in A$, then a is called the **greatest element** (the **least element**) of A if $x \leq a$ ($a \leq x$) for every $x \in A$. If $a \in X$, then the set $\{x \in X : x < a\}$ is called an **initial segment**. A poset

---

[1] i.e., xRy and yRx imply $x = y$ for any $x, y \in X$.

$\langle X, \leq \rangle$ is a **well-ordered set** if every non-empty subset $A \subseteq X$ has a least element. A well-ordered set is also linearly ordered.

Theorem 11.1. *Assume* **DC**. *Then a linearly ordered set* $\langle X, \leq \rangle$ *is well ordered if and only if there is no function* $f : \omega \longrightarrow X$ *such that* $f(n+1) < f(n)$ *for every* $n \in \omega$. *If* X *is countable, we do not need any axiom of choice.*

The theorem is a very useful test for showing that a poset is well ordered.

Suppose that $\langle X_1, \leq_1 \rangle$ and $\langle X_2, \leq_2 \rangle$ are posets and $f : X_1 \longrightarrow X_2$ is a mapping. f is said to be **increasing (non-decreasing)** if $f(x) <_2 f(y)$ $(f(x) \leq_2 f(y))$ for any $x <_1 y$. A one-to-one mapping onto X_2 is called an **isomorphism** if both f and f^{-1} are increasing. $\langle X_1, \leq_1 \rangle$ and $\langle X_2, \leq_2 \rangle$ are **isomorphic** if there exists an isomorphism from X_1 onto X_2. If f is an isomorphism from X_1 onto $f(X_1) \subseteq X_2$, then we say that f is an **embedding** of X_1 into X_2.

If X_1 and X_2 are isomorphic we say that X_1 and X_2 have **equal order type**. We shall write $\mathrm{ot}(X_1) = \mathrm{ot}(X_2)$. If there exists an increasing mapping from X_1 into X_2 we shall say that the **order type of** X_1 **is smaller than or equal to the order type of** X_2, write $\mathrm{ot}(X_1) \leq \mathrm{ot}(X_2)$. The meaning of notation $\mathrm{ot}(X_1) < \mathrm{ot}(X_2)$ is clear.

Theorem 11.2. *If* $\langle X_1, \leq_1 \rangle$ *and* $\langle X_2, \leq_2 \rangle$ *are well-ordered sets, then exactly one of the following possibilities occurs:*

a) X_1 *and* X_2 *are isomorphic,*
b) X_1 *is isomorphic with an initial segment of* X_2,
c) X_2 *is isomorphic with an initial segment of* X_1.

Moreover, if there exists an increasing mapping $f : X_1 \longrightarrow X_2$, *then case* a) *or* b) *occurs.*

Let $\langle X, \leq \rangle$ be a poset, $A \subseteq X$. An element $a \in A$ is called **maximal (minimal)** element of A if $a \not< x$ $(x \not< a)$ for every $x \in A$. An element $a \in X$ is an **upper bound (lower bound)** of A if $x \leq a$ $(a \leq x)$ for any $x \in A$. Thus a is the greatest (the least) element of A if $a \in A$ and a is an upper (lower) bound of A. Note that if A is a chain, then a maximal (minimal) element of A is the greatest (the least) element. The greatest (the least) element of poset $\langle X, \leq \rangle$ will be denoted by 1_X (by 0_X) if there exists any. The inequality $x \neq 1_X$ $(x \neq 0_X)$ means that the element x is not the greatest (the least) element of X (independently of whether 1_X or 0_X does exist or does not). Similarly for $x < 1_X$ and $x > 0_X$.

Using Theorem 1.7 one can easily show by induction

Theorem 11.3 (Maximum Principle). *If* A *is a non-empty finite subset of a poset* $\langle X, \leq \rangle$, *then there exists a maximal element of* A.

A set A is **bounded from above(from below)** if there exists an upper (lower) bound of A. An element a is called a **supremum (infimum)** of the set A, written $a = \sup A$ $(a = \inf A)$, if a is the least upper (the greatest lower) bound of the set A. If every non-empty finite subset of a poset $\langle X, \leq \rangle$ has a supremum and an infimum, then $\langle X, \leq \rangle$ is called a **lattice**.

Two elements $x, y \in X$ are called **compatible** if

$$(\exists a \neq 0_X)\,(a \leq x \wedge a \leq y).$$

Otherwise, x, y are **incompatible**. Sometimes we speak about **disjoint elements** instead of incompatible and we write $x \wedge y = 0_X$. A subset $A \subseteq X$ of pairwise incompatible elements is called an **antichain**. A subset $A \subseteq B \subseteq X$ is **cofinal in** B if for any $x \in B$ there exists an $a \in A$ such that $x \leq a$. A cofinal subset of the whole set X is called also a **dominating set**.

The word "dense" will be used in several meanings. We shall introduce the notions "order dense", "linearly dense", and also "topologically dense". When one understand from the context which of those notions we deal with we shall say simply "dense". Note that also the notion "open dense" will have at least two different meanings.

A subset $A \subseteq X$ is **order dense** or simply **dense** in X if $0_X \notin A$ and for any $x \in X$, $x \neq 0_X$ there exists an $a \in A$ such that $a \leq x$. If moreover, for any $y \leq x \in A$, $y \neq 0_X$, also $y \in A$, the set A is called **open dense**. A subset $A \subseteq X$ is **predense** in X if $0_X \notin A$ and for any $x \in X$, $x \neq 0_X$ there exists an $a \in A$ such that a, x are compatible. If A is a predense set, then $\{x \in X : x \neq 0_X \wedge (\exists y \in A)\, x \leq y\}$ is open dense. A predense set A is a **partition** of X if every two distinct elements of A are disjoint. A predense set A is a **refinement** of a predense set C if for every $a \in A$ there exists a $c \in C$ such that $a \leq c$. Note that a dense set A is a refinement of an open dense set C if and only if $A \subseteq C$.

If $\langle X, \leq \rangle$ is a poset, $a, b \in X$, $a < b$ we define the **open** and **closed interval**

$$(a, b) = \{x \in X : a < x < b\}, \qquad [a, b] = \{x \in X : a \leq x \leq b\}.$$

Similarly we define intervals $[a, b)$, $(a, b]$, $(-\infty, b)$, $(a, +\infty)$, $(-\infty, b]$, $[a, +\infty)$, e.g., $(-\infty, b] = \{x \in X : x \leq b\}$. Thus, e.g., the set $(-\infty, a) = \{x \in X : x < a\}$ is an initial segment. If $Y \subseteq X$, then the interval (a, b) in Y need not be equal to the interval (a, b) considered in X. However, it will be always clear from the context which underlying set is considered.

A relation $R \subseteq X \times X$ is a **partial preordering** if R is reflexive and transitive. The **strict part** of a preordering R is the relation

$$\{\langle x, y \rangle \subseteq X \times X : xRy \wedge \neg yRx\}.$$

If $\leq$, $\preceq$, $\sqsubseteq$ are preorderings, then their strict parts are usually denoted as $<$, $\prec$, $\sqsubset$, respectively.

If R is a partial preordering, we can define an equivalence relation E on X by

$$xEy \equiv (xRy \wedge yRx). \tag{11.4}$$

Accordingly defined relation R/E on the quotient set X/E is already a partial ordering. We shall often not distinguish between the partially preordered set $\langle X, R \rangle$ and partially ordered quotient set $\langle X/E, R/E \rangle$.

A preordering R on a set X is called a **prewell-ordering** if every non-empty subset $A \subseteq X$ has a least element[2], i.e., there exists an element $a \in A$ such that for every $x \in A$ we have aRx. One can easily see that R is a prewell-ordering of X if and only if R/E is a well-ordering of X/E, where the equivalence relation E is defined by (11.4). The **order type** of the prewell-ordering R is the ordinal number $\mathrm{ot}(R) = \mathrm{ot}(R/E)$.

A subset $A \subseteq X$ of a linearly ordered set $\langle X, \le \rangle$ is **linearly dense in X**, shortly **dense**, if for any $x, y \in X$, $x < y$ there exists a $z \in A$ such that $x < z < y$. A linearly ordered set $\langle X, \le \rangle$ is said to be **densely ordered** if X is linearly dense in X.

We introduce a technical notion which will be useful in the proof of the next theorem. Let X_1, X_2 be subsets of X. If

$$(\forall x_1, x_2)\,(x_1 \in X_1 \wedge x_2 \in X_2) \to x_1 < x_2,$$

we write $X_1 < X_2$. We say that $x \in X$ **lies between** X_1 and X_2 if

$$(\forall x_1, x_2)\,(x_1 \in X_1 \wedge x_2 \in X_2) \to x_1 < x < x_2.$$

If $\langle X, \le \rangle$ is a densely ordered set, $X_1 < X_2$ are finite subsets, then there exists an $x \in X$ that lies between X_1 and X_2.

Theorem 11.4.

a) *Any countable linearly ordered set can be embedded in any densely ordered countable set.*

b) *Any two densely ordered countable sets without the least and the greatest elements are order isomorphic.*

Proof. Let $\langle X, \le \rangle$ and $\langle Y, \preceq \rangle$ be infinite countable linearly ordered sets. Assume that $X = \{x_n : n \in \omega\}$, $Y = \{y_n : n \in \omega\}$ are one-to-one enumerations of those sets.

Assume that X is a densely ordered set. We construct an increasing mapping $f : Y \longrightarrow X$ by induction. We can assume that X neither has a least nor a greatest element.

Let $f(y_0) = x_0$. Assuming that $f(y_i)$, $i = 0, \ldots, n$ are defined, we define $f(y_{n+1})$. Let $A = \{y_i : i \le n \wedge y_i < y_{n+1}\}$, $B = \{y_i : i \le n \wedge y_i > y_{n+1}\}$. Then $f(A) < f(B)$. Let x_m be the first element lying between $f(A)$ and $f(B)$. Set $f(y_{n+1}) = x_m$. It is easy to check that f is the desired mapping.

Assume now that both $\langle X, \le \rangle$, $\langle Y, \preceq \rangle$ are densely ordered sets, both without the least and the greatest elements. We define an isomorphism $f : X \xrightarrow[\text{onto}]{1-1} Y$ by induction. Actually we define by induction set $X_n \subseteq X$ such that

$$|X_n| = 2n + 2, \qquad x_0, \ldots, x_n \in X_n, \qquad y_0, \ldots, y_n \in f(X_n).$$

---

[2]Note that in a partial preordering a least and a greatest element need not be unique.

Let $f(x_0) = y_0$. If $y_1 > y_0$ take the first x_m such that $x_0 < x_m$ and set $f(x_m) = y_1$ and $X_0 = \{x_0, x_m\}$. Similarly in the case $y_1 < y_0$. Assume that the set X_n is defined and f is defined for $x \in X_n$. Let x_m be the first element of $X \setminus X_n$. Let $A = \{x \in X_n : x < x_m\}$, $B = \{x \in X_n : x_m < x\}$. Then there exists the first $y_k \in Y$ lying between $f(A)$ and $f(B)$. Set $f(x_m) = y_k$. Let y_l be the first element of $Y \setminus (f(X_n) \cup \{y_k\})$. Let $C = \{x \in X_n \cup \{x_m\} : f(x) < y_l\}$, $D = \{x \in X_n \cup \{x_m\} : f(x) > y_l\}$. Then there exists an $x_p \in X \setminus (X_n \cup \{x_m\})$ lying between C and D. Set $f(x_p) = y_l$ and $X_{n+1} = X_n \cup \{x_m, x_p\}$.

One can easily see that $f : \bigcup_n X_n \longrightarrow \bigcup_n f(X_n)$ is an isomorphism and $X = \bigcup_n X_n$, $Y = \bigcup_n f(X_n)$. $\qquad\qquad\qquad\qquad\qquad\qquad\qquad\qquad\qquad\qquad\qquad$ □

A poset $\langle T, \leq \rangle$ with the least element $0_T \in T$ is called a **tree** if every initial segment $\{v \in T : v < u\}$ is a well-ordered set. An element of a tree is called a **node**. The element 0_T is called the **root** of T. The **height** $\mathrm{hg}(u)$ **of a node** $u \in T$ is the order type of the initial segment $\{v \in T : v < u\}$. The ξ**th level** $T(\xi)$ is the set of all nodes $u \in T$ with $\mathrm{hg}(u) = \xi$. We also denote

$$T(< \xi) = \bigcup_{\eta < \xi} T(\eta).$$

The **height** $\mathrm{hg}(T)$ **of a tree** T is the least ordinal ξ such that $\mathrm{hg}(u) < \xi$ for every $u \in T$. Thus the height of a tree T is the least ordinal ξ such that $T = T(< \xi)$. The set of all immediate successors of a node $u \in T$ is denoted as $\mathrm{IS}(u, T)$. The **branching degree of a node** $u \in T$ is the cardinal $\mathrm{BD}(u) = |\mathrm{IS}(u, T)|$. If the branching degree of any $u \in T$ is at most 2 we say that T is a **binary tree**. A node $u \in T$ is called a **leaf** if $\mathrm{IS}(u, T) = \emptyset$. A tree without leafs is **pruned**. A tree T is **perfect** if for any $u \in T$ there exist a $v > u$ with $\mathrm{BD}(v) > 1$. Thus a perfect tree is pruned. A maximal chain in a tree is called a **branch**. The set of all branches of a tree $\langle T, \leq \rangle$ is denoted by $[T]$. If $u \in T$ we shall denote by $[u]$, or by $[u]_T$ in the case of possible misunderstanding, the set of all branches of T containing the node u. If $u \in T$ is not a leaf, we have

$$[0_T] = [T], \qquad [u] = \bigcup\{[v] : v \in \mathrm{IS}(u, T)\}. \tag{11.5}$$

A subset $S \subseteq T$ is called a **subtree** if S has a least element. Often we ask that for a subtree $\mathrm{hg}(S) = \mathrm{hg}(T)$ and $0_T \in S$ hold true. In this case we have

$$[T] = [S] \cup \bigcup_{v \in T \setminus S} [v]. \tag{11.6}$$

For a node $v \in T$ we let $T^v = \{u \in T : v \leq u\}$. Then T^v is a tree. If $\alpha \in [v]_T$ is a branch, then $\alpha \cap T^v$ is a branch of T^v. Moreover, every branch $\beta \in T^v$ has a unique extension $\alpha \in [v]_T$: $\alpha = \beta \cup \{u \in T : u < v\}$.

The set $T = {}^{<\omega}X$ ordered by the inclusion $\subseteq$ is a tree of height ω. The nth level is $T(n) = {}^n X$. The degree of branching of any $s \in T$ is $|X|$. Thus, if $|X| > 1$, then T is a perfect tree. If $|X| = 2$, then $^{<\omega}X$ is a (typical) binary perfect tree with height ω. Actually

Theorem 11.5. *Let T be a countable tree of height ω. Let $E \subseteq T$, $E \neq \emptyset$ be such that for any $u \in E$ there exist two incomparable nodes $v, w \in E$, $u < v, u < w$. Then there exists a subtree S of T, $S \subseteq E$, isomorphic to $^{<\omega}2$.*

Proof. We do not need any form of the axiom of choice for the proof. We can use standard procedure: take fixed enumeration of T and always choose the first element in this enumeration.

We construct the subtree S by induction constructing the levels $S(n)$. Let $u \in E$ be any minimal element. Set $S(0) = \{u\}$. If $S(n) \subseteq E$ is constructed, then by the supposed property of the set E, every node $u \in S(n)$ has at least two incomparable successors $v_u, w_u \in E$. Let

$$S(n + 1) = \{v_u : u \in S(n)\} \cup \{w_u : u \in S(n)\}.$$

Then $S = \bigcup_n S(n) \subseteq E$ is the desired subtree. $\square$

If $\alpha \in {}^{\omega}X$, then $\{\alpha|n : n \in \omega\}$ is a branch of the tree $^{<\omega}X$. Vice versa, each branch of this tree has the form $\{\alpha|n : n \in \omega\}$. Thus we shall identify those two sets $[^{<\omega}X] = {}^{\omega}X$. Similarly, if $s \in {}^{<\omega}X$ then $[s] = \{\alpha \in {}^{\omega}X : s \subseteq \alpha\}$.

One can easily prove

Theorem 11.6. *Let T_1, T_2 be countable trees of height ω. Assume that there exist cofinal subtrees $S_i \subseteq T_i$, $i = 1, 2$ such that S_1 is isomorphic to S_2. Then there exists a bijection $f : [T_1] \xrightarrow[\text{onto}]{1-1} [T_2]$.*

D. König [1926] proved the following simple but important statement.

Theorem 11.7 (König's Lemma). *If T is an infinite tree with finite branching degree of every node and T can be linearly ordered, then T contains an infinite branch.*

Proof. One can easily construct an infinite branch of length ω by induction. Let v_0 be the root of T. Assume that we have already constructed a node v_n such that T^{v_n} is infinite. Since the branching degree of v_n is finite, for at least one node $v \in \text{IS}(v_n, T)$ the set T^v is infinite. Take for v_{n+1} the first (in the linear ordering) such a node in the finite set $\text{IS}(v_n, T)$. $\square$

Note that we have assumed that T can be linearly ordered to avoid any use of the axiom of choice.

Exercises

11.1 Partially Ordered Sets

Let $R \subseteq X \times X$. The inverse R^{-1} of the relation R is defined as

$$R^{-1} = \{\langle x, y \rangle \in X \times X : \langle y, x \rangle \in R\}.$$

If R is a partial ordering, then R^{-1} is too. R^{-1} is called the **dual partial ordering**. Every notion of a poset $\langle X, R \rangle$ has its dual notion.

  a) Find dual notions of the following notions: the least element, maximal element, comparable elements, chain, supremum, disjoint elements.

  b) Prove Theorem 11.3. What is the dual statement?

  c) When are the notions "cofinal" and "dense" mutually dual.

  d) The relation of cofinality is transitive, i.e., if $A \subseteq B \subseteq C$, A is cofinal in B, B is cofinal in C, then A is cofinal in C. Therefore also the notion of density is transitive.

11.2 Directed Set

A poset $\langle E, \leq \rangle$ is said to be **directed**, if for any $x, y \in E$ there is a $z \in E$ such that $x \leq z$ and $y \leq z$.

  a) If $\mathcal{I}$ is an ideal of subsets of a given set X, then $\langle \mathcal{I}, \subseteq \rangle$ is a directed set. Similarly for a filter with the relation $\supseteq$.

  b) If $\langle E, \leq \rangle$ is directed, then the set $\mathcal{F}(E)$ defined as

$$A \in \mathcal{F}(E) \equiv (A \subseteq E \land (\exists a \in E)(\forall x \geq a)\, x \in A)$$

   is a filter. The filter $\mathcal{F}(E)$ is called **Fréchet filter**. Show that $\langle E, \leq \rangle$ is order isomorphic to a cofinal subset of $\langle \mathcal{F}(E), \supseteq \rangle$.

  c) Any directed countable set without the greatest element contains a cofinal subset of the order type ω.

11.3 Ramsey Theorem and Trees

A tree may be used in proving theorems of Ramsey type.

  a) Show the particular case of Ramsey Theorem: $\aleph_0 \to (\aleph_0)_n^2$ for any natural number n.
   Hint: Let $F : [X]^2 \longrightarrow n$ be a coloring, X being infinite countable. Define a mapping $G : {}^{<\omega}n \longrightarrow \mathcal{P}(X)$ as follows: if $G(s)$ is finite, set $G(s^\smallfrown i) = \emptyset$ for any $i < n$. Otherwise choose a $g(s) \in G(s)$ and set $G(s^\smallfrown i) = \{x \in G(s) : F(\langle g(s), x \rangle) = i\}$. By the König Lemma the subtree $\{s \in {}^{<\omega}n : G(s) \text{ is infinite}\}$ has an infinite branch $B \subseteq {}^{<\omega}n$. Note that for $s, s^\smallfrown i \in B$ we have $F(\langle g(s), g(t) \rangle) = i$ for any $t \in B, t > s$. Thus there exists an $i_0 < n$ and infinite $A \subseteq B$ such that $F(g(s), g(t)) = i_0$ for any distinct $s, t \in A$.

  b) Prove **Ramsey Theorem**: $\aleph_0 \to (\aleph_0)_n^k$ for any natural numbers k, n.

11.4 Well-founded Relation

An antireflexive ($x \prec x$ for no $x \in X$) transitive relation $\prec$ on X is **well founded** if every non-empty subset of X has a minimal element.

  a) Assuming **DC**, show that $\prec$ is well founded if and only if there is no infinite decreasing chain $x_0 \succ x_1 \succ \cdots \succ x_n \succ \cdots$. If $|X| \ll \mathfrak{c}$, then **DC** can be replaced by **wAC**.

  b) If $\prec$ is well founded, then there exists a unique ordinal ξ and a mapping $\rho : X \xrightarrow{\text{onto}} \xi$ such that $x \prec y \equiv \rho(x) < \rho(y)$ for any $x, y \in X$. ρ is called a **rank function**, $\rho(x)$ is the **rank of** x, and ξ is the **rank of** $\prec$.
   Hint: Set $\rho(x) = \sup\{\rho(y) + 1 : y \prec x\}$.

c) Show that $\xi = \sup\{\rho(x) + 1 : x \in X\}$.

d) Show that the relation $\sqsubset$ defined on $^{<\omega}\omega$ by

$$x \sqsubset y \equiv (x \neq y \wedge (x \subseteq y \vee (\exists n)(\forall k < n)\,(x(k) = y(k) \wedge x(n) < y(n))))$$

is a linear well-founded ordering.

e) Compute the rank of $\sqsubset$.

f) Do the same for the set $^{<\omega}n$, $n > 1$.

g) If $\langle T, \leq \rangle$ is a tree, then $<$ is a well-founded relation

h) If $\langle T, \leq \rangle$ is a tree, then the height of a node is its rank and the height of the tree is its rank.

11.5 König's Lemma

Prove König's Lemma replacing the assumption that T can be linearly ordered by the assumption $\mathbf{AC}_{<\omega}$ defined in Exercise 1.8.

11.2 Rings and Fields

In algebra a function $f : X^n \longrightarrow X$ is called an n-**ary operation**. In practice we consider mainly **unary** (= 1-ary) and **binary** (= 2-ary) operations. In the case of a binary operation $\diamond \colon X \times X \longrightarrow X$ we usually denote the value of the operation $\diamond$ as $x \diamond y$ or $(x \diamond y)$ instead of $\diamond(x, y)$. A binary operation $\diamond$ is called **associative** if $(x \diamond y) \diamond z = x \diamond (y \diamond z)$ for any $x, y, z \in X$. $\diamond$ is called **commutative** if $x \diamond y = y \diamond x$ for any $x, y \in X$.

One of the simplest algebraic structures is a group. A set G endowed with a binary operation $\circ$ and a particular element e, usually denoted as $\mathcal{G} = \langle G, \circ, e \rangle$ is a **group** if the operation $\circ$ is associative, e is the **identity**, i.e.,

$$(\forall x \in G)\,(x \circ e = e \circ x = x)$$

and every element $x \in G$ has an **inverse element** y, i.e., such that $x \circ y = y \circ x = e$. Since the inverse element is unique we often denote it x^{-1}. If the group operation $\circ$ is commutative we say that $\mathcal{G}$ is a **commutative** or an **Abelian group**.

If $\mathcal{O}$ is a topology on G such that the binary operation $\circ$ is a continuous mapping from $G \times G$ into G and the inverse operation is a continuous mapping from G into G, then we say that $\mathcal{G} = \langle G, \mathcal{O}, \circ, e \rangle$ is a **topological group**. One can easily see that for given $a \in G$ the right multiplication $f(x) = x \circ a$ is a homeomorphism of G. Since $U \subseteq G$ is a neighborhood of a if and only if $f^{-1}(U)$ is a neighborhood of e, the topology of a topological group is determined by neighborhoods of the identity element e. Similarly for the left multiplication $a \circ x$.

A set K endowed with two binary operations $+$ and $\cdot$, with two particular elements $0, 1 \in K$, $0 \neq 1$, denoted as $\mathcal{K} = \langle K, +, \cdot, 0, 1 \rangle$, is called a **ring** if both operations $+$ and $\cdot$ are associative and commutative, the **distributive law** holds true

$$(\forall x, y, z)\,x \cdot (y + z) = (x \cdot y) + (x \cdot z),$$

the elements $0, 1$ are **additive** and **multiplicative identity**, respectively, i.e.,

$$(\forall x)\, x + 0 = x, \qquad (\forall x)\, x \cdot 1 = x,$$

and for any element there exists an **additive inverse**

$$(\forall x)(\exists y)\, x + y = 0.$$

When we deal with several rings (or similar structure), then to avoid a misunderstanding we will write $\mathcal{K} = \langle K, +_K, \cdot_K, 0_K, 1_K \rangle$.

If $\langle K, +, \cdot, 0, 1 \rangle$ is a ring, then the structure $\langle K, +, 0 \rangle$ is a commutative group. A ring $\mathcal{K} = \langle K, +, \cdot, 0, 1 \rangle$ is an **integral domain** if for any $x, y \in K$ the implication

$$x \cdot y = 0 \rightarrow (x = 0 \vee y = 0)$$

holds true. Finally, a ring $\mathcal{K} = \langle K, +, \cdot, 0, 1 \rangle$ is called a **field** if for any non-zero element there exists a **multiplicative inverse**, i.e.,

$$(\forall x \in K, x \neq 0)(\exists y \in K)\, x \cdot y = 1.$$

Every field is an integral domain. Moreover, if $\mathcal{K} = \langle K, +, \cdot, 0, 1 \rangle$ is a field, then $\langle K \setminus \{0\}, \cdot, 1 \rangle$ is a commutative group.

The additive inverse of an $x \in K$ is unique and is usually denoted by $-x$. Similarly the multiplicative inverse is unique and is denoted as x^{-1} or $1/x$. Instead of $x + (-y)$ we write $x - y$ and instead of $x \cdot y^{-1} = x \cdot 1/y$ we write x/y. We suppose that the reader knows elementary properties of those notions from an elementary course in algebra.

If $\langle K_i, +_i, \cdot_i, 0_i, 1_i \rangle$, $i = 1, 2$ are rings, then a mapping $f : K_1 \longrightarrow K_2$ is called a **homomorphism** if

$$f(1_1) = 1_2, \quad f(0_1) = 0_2, \quad f(x +_1 y) = f(x) +_2 f(y), \quad f(x \cdot_1 y) = f(x) \cdot_2 f(y)$$

for any $x, y \in K_1$. A homomorphism, that is a bijection is called an **isomorphism**. Then the inverse mapping f^{-1} is a homomorphism as well. If K_1, K_2 are fields and f is a homomorphism, then also $f(1/x) = 1/f(x)$ for any $x \in K_1, x \neq 0_1$. Two rings (or fields) are **isomorphic** if there exists an isomorphism from one of them onto the another one. Evidently isomorphic rings possess the same properties expressible in terms of the corresponding algebraic structure.

Let $\langle K, +, \cdot, 0, 1 \rangle$ be a ring. A subset $L \subseteq K$ is a **subring** of $\mathcal{K}$ if $0, 1 \in L$ and L is closed under operations $+$, $\cdot$ and $-$, i.e., for any $x, y \in L$ also $x + y, x \cdot y, -x \in L$. One can easily see that $\langle L, +|L \times L, \cdot|L \times L, -|L, 0, 1 \rangle$ is a ring. If L is a field we say that L is a **subfield of** K. Any subring of a field is an integral domain.

If K is a ring and $A \subseteq K$, then there exists the smallest subring L containing A, i.e., such that L is a subring of K, $A \subseteq L$ and if $M \supseteq A$ is a subring of K, then $L \subseteq M$. Actually, it suffices to take the intersection of all subrings of K containing

A. We say that the subring L is **generated by** A. Similarly we define the notion of a field generated by a subset.

Let K_i, $i = 1, 2$ be rings. We say that K_2 is an **extension of** K_1 if there exist a subring $L \subseteq K_2$ of K_2 and an isomorphism $f : K_1 \xrightarrow[\text{onto}]{1-1} L$. In other words, if K_1 can be identified with a subring of K_2.

Theorem 11.8. *For any integral domain L there exists a field K that is an extension of L and every element of K admits the form $f(x)/f(y)$, $x, y \in L$, $y \neq 0$, where f is the isomorphism of L onto a subring of K. Moreover, the field K is unique up to isomorphism.*

Proof. We sketch a proof. Set $M = L \times (L \setminus \{0\})$. An element $\langle x, y \rangle$ of M is intended to be the fraction x/y of the field extension of L. So we define a relation E on M as follows:

$$\langle x_1, y_1 \rangle E \langle x_2, y_2 \rangle \equiv x_1 \cdot y_2 = x_2 \cdot y_1.$$

It is easy to see that E is an equivalence relation on M. Let $K = M/E$, i.e.,

$$K = \{\{a\}_E : a \in M\}.$$

We define operation $+$ and $\cdot$ on K as follows:

$$\{\langle x_1, y_1 \rangle\}_E + \{\langle x_2, y_2 \rangle\}_E = \{\langle x_1 \cdot y_2 + x_2 \cdot y_1, y_1 \cdot y_2 \rangle\}_E,$$
$$\{\langle x_1, y_1 \rangle\}_E \cdot \{\langle x_2, y_2 \rangle\}_E = \{\langle x_1 \cdot x_2, y_1 \cdot y_2 \rangle\}_E.$$

It is easy to see that $\langle K, +, \cdot, \{\langle 0, 1 \rangle\}_E, \{\langle 1, 1 \rangle\}_E \rangle$ is a field. We define an isomorphism f of L onto a subring of K as

$$f(x) = \{\langle x, 1 \rangle\}_E.$$

Then every element of K has the desired form. □

The unique field of Theorem 11.8 is called the **field of fractions** of the integral domain L.

Let K be a ring. Assume that we have a partial ordering $\leq$ on K. We shall say that $\langle K, \leq, +, \cdot, 0, 1 \rangle$ is an **ordered ring** if the partial ordering is compatible with ring structure, more precisely, if the following hold true:

$$0 < 1,$$
$$(\forall x, y, z \in K)\, x \leq y \rightarrow x + z \leq y + z,$$
$$(\forall x, y, z \in K)\, (z > 0 \wedge x \leq y) \rightarrow x \cdot z \leq y \cdot z.$$

If $\leq$ is a linear ordering we speak about a **linearly ordered ring**. If moreover K is a field we speak about a **linearly ordered field**.

Let K be a ring. For an element $n \in \omega$ we define $n \times x \in K$ for an $x \in K$ by induction: $0 \times x = 0_K$, $(n + 1) \times x = n \times x +_K x$. A linearly ordered ring K

has the **Archimedes Property** if for any $x, y \in K$, $x >_K 0_K$ there exists a natural number $n \in \omega$ such that $n \times x >_K y$.

If K is a field we define similarly $1/n \times x$ for $x \in K$ and a natural $n > 0$ as

$$1/n \times x = x/_K (n \times 1_K),$$

provided that $n \times 1_k \neq 0_K$.

Assume now that $\langle K, +_K, \cdot_K, 0_K, 1_K \rangle$ is a field and $\langle V, +_V, 0_V \rangle$ is a commutative group. $\langle V, +_V, 0_V \rangle$ is called a **vector space over** K if an external operation $\cdot : K \times V \longrightarrow V$ is given such that the following equalities hold true:

$$\alpha \cdot (x +_V y) = (\alpha \cdot x) +_V (\alpha \cdot y), \quad \alpha \cdot (\beta \cdot x) = (\alpha \cdot_K \beta) \cdot x,$$
$$(\alpha +_K \beta) \cdot x = (\alpha \cdot x) +_V (\beta \cdot x), \quad 0_K \cdot x = 0_V, \quad 1_K \cdot x = x$$

for any $\alpha, \beta \in K$ and any $x, y \in V$. Since usually from the context it is clear whether we are dealing with $+_K$ or $+_V$, or with $\cdot_K$ or $\cdot$, we shall omit the subscripts and write simply $+$ and $\cdot$.

If $A = \{x_0, \ldots, x_n\} \subseteq V$ is a finite subset, we say that an element $x \in V$ is a **linear combination** of A if there exist elements $\alpha_0, \ldots, \alpha_n \in K$ such that $x = \alpha_0 \cdot x_0 + \cdots + \alpha_n \cdot x_n$. For any such set A the zero element $0 \in V$ is a **trivial linear combination** of A: every α_i is 0_K. If $0 = \alpha_0 \cdot x_0 + \cdots + \alpha_n \cdot x_n$ is a linear combination of A with some $\alpha_i \neq 0_K$, we say that it is a **non-trivial linear combination**. A set $B \subseteq V$ is **linearly independent** if 0 is not a non-trivial linear combination of any non-empty finite subset A of B. A maximal (according to inclusion) linearly independent subset of a vector space V is called a **base**. It is well known that a linearly independent subset $A \subseteq V$ is a base if and only if every element of V is a linear combination of a finite subset of A. Note that generally for proving that a vector space over a field has a base, we need **AC**.

Theorem [AC] 11.9. *Any two bases of a vector space have equal cardinality.*

If there exists a finite base of a vector space, then one can prove Theorem 11.9 in **ZF**. The common cardinality of bases is called the **dimension** of the vector space.

If K is a field, then for a given natural number $n > 0$ one can define in a natural way a structure of a vector space on K^n:

$$\langle x_1, \ldots, x_n \rangle + \langle y_1, \ldots, y_n \rangle = \langle x_1 + y_1, \ldots, x_n + y_n \rangle,$$
$$z \cdot \langle x_1, \ldots, x_n \rangle = \langle z \cdot x_1, \ldots, z \cdot x_n \rangle,$$

where $x_1, \ldots, x_n, y_1, \ldots, y_n, z \in K$.

The set $\{\langle 1_K, 0, \ldots, 0 \rangle, \langle 0, 1_K, 0 \ldots, 0 \rangle, \ldots, \langle 0, \ldots, 0, 1_K \rangle\}$ is a base of K^n. Thus the vector space K^n over K is n-dimensional. Moreover, any n-dimensional vector space over K is isomorphic to K^n with endowed vector space structure.

Exercises

11.6 Ideal

Let $\langle K, +, \cdot, 0, 1 \rangle$ be a ring. A set $J \subseteq K$, $J \neq K$ is an **ideal**[3] of K if for any $x, y \in J$ and any $a \in K$ we have $x + y \in J$ and $a \cdot x \in J$.

a) If $f : K_1 \longrightarrow K_2$ is a homomorphism from the ring K_1 into the ring K_2, then the **kernel** of f is $\mathrm{Ker}\,(f) = \{a \in K_1 : f(a) = 0\}$. Show that $\mathrm{Ker}\,(f)$ is an ideal provided that $\mathrm{rng}(f) \neq \{0\}$.

b) More generally, if $f : K_1 \longrightarrow K_2$ is a homomorphism from K_1 into K_2, $J \subseteq K_2$ is an ideal, then $f^{-1}(J)$ is an ideal, provided that $\mathrm{rng}(f) \not\subseteq J$.

c) Let $a \in K$. The set $a \cdot K = \{a \cdot x : x \in K\}$ is an ideal if and only if there is no multiplicative inverse of a in K.

d) If K is a field, then the only ideal is the set $\{0\}$.

e) **AC** implies that every ideal is contained in a maximal ideal[4].
   Hint: Every chain in $\langle \{J \subseteq K : J \text{ is an ideal}\}, \subseteq \rangle$ is bounded and apply Zorn's Lemma.

11.7 Quotient Ring

If $J \subseteq K$ is an ideal we define an equivalence relation as

$$a \sim_J b \equiv a - b \in J.$$

a) If $x \sim_J y$ and $u \sim_J v$, then also $x + u \sim_J y + v$ and $x \cdot u \sim_J y \cdot v$.

b) The quotient set $K/\sim_J = \{\{x\}_{\sim_J} : x \in K\}$ is a ring with operations

$$\{x\}_{\sim_J} + \{y\}_{\sim_J} = \{x + y\}_{\sim_J}, \quad \{x\}_{\sim_J} \cdot \{y\}_{\sim_J} = \{x \cdot y\}_{\sim_J}.$$

$K/\sim_J = \{\{x\}_{\sim_J} : x \in K\}$ is called the **quotient ring** and denoted simply by K/J.

c) The mapping $\{\ \}_{\equiv_J} : K \longrightarrow K/J$ is a homomorphism with kernel J.

d) An ideal J is maximal if and only if K/J is a field.

11.8 Euclidean Domain

An integral domain K is a **Euclidean domain** if there is a mapping $\nu : K \setminus \{0\} \longrightarrow \omega$ such that

1) $\nu(a) \leq \nu(a \cdot b)$ for any non-zero $a, b \in K$;

2) For any non-zero $a, b \in K$ there are $q, r \in K$ such that $a = b \cdot q + r$ and either $r = 0$ or $\nu(r) < \nu(b)$.

An element $a \in K$ **divides** an element $b \in K$ if there is a $c \in K$ such that $b = a \cdot c$. Two elements $a, b \in K$ are **associated** if both a divides b and b divides a. An element associated with the identity 1 is called a **unit**. A non-unit element $a \in K$ is **prime** if for any $b, c \in K$, if a divides $b \cdot c$, then a divides b or a divides c. An element $a \in K$ is a **common divisor** of $b, c \in K$, if a divides both b and c. A common divisor a of b, c is a **greatest common divisor** if any common divisor d of b, c divides a.

---

[3]Sometimes also $J = K$ is allowed and if $J \neq K$ we speak about a proper ideal.
[4]Maximal in the $\subseteq$ order.

a) Euclidean domain K is a **principal ideal domain**, i.e., any ideal $J \subseteq K$ has the form aK for some $a \in K$.
   Hint: Consider an element $a \in J$ with the minimal value of $\nu(a)$.

b) If K is a principal ideal domain, then any two elements $b, c \in K$ have a greatest common divisor $a \in K$.
   Hint: The set $\{b \cdot x + c \cdot y : x, y \in K\}$ either contains the multiplicative identity or is an ideal.

c) If K is a principal ideal domain and a is a greatest common divisor of $b, c \in K$, then there exist elements $x, y \in K$ such that $a = b \cdot x + c \cdot y$.
   Hint: The set $\{b \cdot x + c \cdot y : x, y \in K\}$ is either an ideal or contains the multiplicative identity.

d) If K is a principal ideal domain, then the ideal aK is maximal if and only if a is **irreducible**, i.e., a is not a unit and every non-unit dividing a is associated with a.

e) Show that in a principal ideal domain any irreducible element is prime.

f) A principal ideal domain is a **unique factorization domain**, i.e., every non-zero element $a \in K$ not associated with 1 can be decomposed as $a = a_0^{k_0} \cdots a_n^{k_n}$, where $a_i \in K$ is irreducible, $k_i > 0$ is an integer, $i = 0, \ldots, n$ and this decomposition is unique up to a permutation and replacing the elements a_i by associated ones.

11.9 Polynomials

Let K be a field, $x \notin K$. In algebra a symbol x is called a **variable** and a finite sequence $\langle a_0, \ldots, a_n \rangle$ of elements of K is written as

$$p(x) = a_0 + a_1 \cdot x + \cdots + a_n \cdot x^n$$

and called a **polynomial over** K. If $a_n = 0$ we identify the polynomial $p(x)$ with the polynomial $a_0 + a_1 \cdot x + \cdots + a_{n-1} \cdot x^{n-1}$. The set of all polynomials over K is denoted by $K[x]$. If $a_n \neq 0$, then $n = \deg(p)$ is the **degree** of the polynomial p. We suppose that the reader knows how to introduce the addition and the multiplication in $K[x]$. Then $K[x]$ is a ring, actually an integral domain.

a) If $p, q \in K[x]$, then $\deg(p + q) \leq \max\{\deg(p), \deg(q)\}$ and $\deg(p \cdot q) = \deg(p) + \deg(q)$.

b) If $p, q \in K[x]$, then there are polynomials $s, r \in K[x]$ such that $p = s \cdot q + r$ and $\deg(r) < \deg(q)$.

c) $K[x]$ is a Euclidean domain.
   Hint: $\nu(p(x))$ is the degree of $p(x)$.

d) Any two polynomials p, q over K have a common divisor of maximal degree.

e) If r is a common divisor of polynomials p, q, and $a \in K$ is such that $p(a) = q(a) = 0$, then also $r(a) = 0$.
   Hint: Use the result of Exercise 11.8, c).

f) If a polynomial p over K of degree > 1 is irreducible in $K[x]$, then there exists no $a \in K$ such that $p(a) = 0$.

g) Show that $K[x]$ is a unique factorization domain.

11.10 Vector Space

a) If K is a subfield of a ring L, then L is a vector space over K.

b) $K[x]$ is an infinite-dimensional vector space over the field K.

Assume that V_1, V_2 are vector spaces over a field K. A mapping $f : V_1 \longrightarrow V_2$ is said to be **linear** if for any $x, y \in V_1$, $\alpha \in K$ we have $f(x+y) = f(x) + f(y)$ and $f(\alpha \cdot x) = \alpha \cdot f(x)$.

c) If $e_1, \ldots, e_n$ is a base of V_1, then a linear mapping $f : V_1 \longrightarrow V_2$ is uniquely determined by the values $f(e_1), \ldots, f(e_n)$.

d) If $f : V_1 \longrightarrow V_2$ is a linear mapping, then the kernel $\{x \in V_1 : f(x) = 0\}$ is a vector subspace of V_1.

e) If $f : V_1 \longrightarrow V_2$ is linear, then $\mathrm{rng}(f)$ is a vector subspace of V_2.

f) If V_1 is a vector subspace of V_2, define the quotient vector space V_2/V_1.

11.11 The Existence of a Base

a) Assume that the vector space V over K has a finite dimension n, $k \leq n$. If $\{e_1, \ldots, e_k\}$ is a linearly independent subset of V, then there are elements e_{k+1}, $\ldots, e_n \in V$ such that $\{e_1, \ldots, e_n\}$ is a base of the vector space V.
Hint: Consider the set $\{B \subseteq V : B$ is linearly independent and $\{e_1, \ldots, e_k\} \subseteq B\}$ ordered by inclusion. Take the set of maximal cardinality.

b) If a vector space V has a finite dimension, then any linearly independent set is a subset of a base of V.

c) If V_1 is a vector subspace of a finite-dimensional V_2, then the dimension of V_2 is the sum of dimensions of V_1 and V_2/V_1.

d) If $V_1 \subseteq V_2$ are vector spaces over a field K, V_2 has a finite dimension, then the dimension of V_1 is not greater than the dimension of V_2.

e) Assuming **AC**, prove the assertion of a) for any vector space: if $D \subseteq V$ is a linearly independent set, then there exists a base B of V such that $D \subseteq B$.
Hint: Consider the poset $\langle \{A \subseteq V : D \subseteq A \wedge A$ is linearly independent$\}, \subseteq \rangle$ and apply Zorn's Lemma.

f) **AC** implies that any vector space possesses a base.

g) **AC** implies that any linearly independent set is a subset of a base.

11.12 Proof of Theorem 11.9

a) Let $\{e_1, \ldots, e_n\}$ be a base of V, $\{b_1, \ldots, b_m\}$ being linearly independent. By induction construct a permutation $\langle k_1, \ldots, k_n \rangle$ of $\langle 1, \ldots, n \rangle$ such that

$$\{b_1, \ldots, b_i, e_{k_{i+1}}, \ldots, e_{k_n}\}$$

is a base of V.
Hint: If $k_1, \ldots, k_i$ are constructed, then $b_{i+1} = \sum_{j=1}^{i} x_j b_j + \sum_{j=i+1}^{n} x_j e_{k_j}$. There exists a $j > i$ such that $x_j \neq 0$. Take $k_{i+1} = j$.

b) Prove Theorem 11.9 for finite-dimensional space: if a vector space V has a finite base $e_1, \ldots, e_n$, then any other base has n elements.

c) In an n-dimensional vector space any $n + 1$ vectors are linearly dependent.

d) If B, C are bases of V, then $|[B]^{<\omega}| = |[C]^{<\omega}|$.
Hint: Any finite subset of B is contained in a subspace of V generated by a (uniquely determined) finite subset of C.

e) Assuming **AC** prove Theorem 11.9 for V with an infinite base.

11.13 Algebraic Extension of a Field

Let $L \subseteq K$ be a subfield of a field K. Let $a \in K$. Similarly as in Exercise 11.9 we denote

$$L[a] = \{p(a) \in K : p \text{ is a polynomial over } L\}.$$

By $L(a)$ we denote the smallest subfield of K containing a as an element and L as a subset. The element $a \in K$ is called **algebraic over** L if there exists a polynomial p over L such that $p(a) = 0$. K is called an **algebraic extension of** L if every element of K is algebraic over L.

a) $L[a]$ is a subring of K.

b) If $a \in K$ is algebraic over L, then there exists an irreducible polynomial p over L such that $p(a) = 0$. Moreover, this polynomial has minimal degree among polynomials q with $q(a) = 0$.
   Hint: Use Exercise 11.9, g).

c) Assume that a is algebraic over L, $p(a) = 0$, where p is an irreducible polynomial over L. Then for any polynomial q over L there exists a polynomial r over L with $\deg(r) < \deg(p)$ such that $q(a) = r(a)$.
   Hint: Use Exercise 11.9, b).

d) If $a \in K$ is algebraic over L, then $L(a) = L[a]$.
   Hint: Use the result of Exercise 11.8, c) for the ring $L[x]$.

e) $a \in K$ is algebraic over L if and only if $L(a)$ as a vector space over L has a finite dimension.
   Hint: By c), if $p(a) = 0$ and n is the degree of p, then the dimension of the vector space $L[a]$ over L has a dimension $\leq n$. Vice versa, if the dimension of $L(a)$ over L is n, then the elements $1, a, a^2, \ldots, a^n$ are linearly dependent over L.

f) If a is algebraic over L, $b \in K$, then a is algebraic over $L(b)$ as well.

g) If $a \in K$ is algebraic over L and $b \in L(a)$, then b is algebraic over L as well.
   Hint: Use the result of e).

h) If a is algebraic over L and b is algebraic over $L(a)$, then b is algebraic over L. Moreover, if dimension of $L(a)$ over L is n, and dimension of $L(a)(b)$ over $L(a)$ is m, then dimension of $L(b)$ over L is not greater than $1 + (n-1)(m-1)$.
   Hint: Consider the dimension of the vector space $L(b)$ over L. If $1, e_1, \ldots, e_{n-1}$ is a base of $L(a)$ over L, $1, f_1, \ldots, f_{m-1}$ is a base of $L(a)(b)$ over $L(a)$, then every element of $L(a)(b)$ is a linear combination of elements $1, e_i \cdot f_j, i < n,, j < m$ over L. Use the result of Exercise 11.12, b).

i) The set of all elements of K algebraic over L is a subfield of K.
   Hint: If $a, b \in K$, $b \neq 0$ are algebraic over L, then the vector space $L(a)(b)$ over L has finite dimension and $a + b, a - b, a \cdot b, a/b \in L(a)(b)$.

11.3   Topology and the Real Line

Maybe the title of the section should be "Adequacy of the topology in investigation of the real line" or something similar. The main aim of the section is to show how in the very abstract theory of topological spaces the real line appears as an important tool. In this section we shall use the axiom of choice in essential ways.

We present three Theorems 11.11, 11.12 and 11.14, which have in common the following. All say that two conditions are equivalent. The first condition is formulated in a purely topological language. The second condition essentially uses the real line or the unit interval.

The notion of a normal topological space was dealt with in Section 1.2. We know that this notion is not hereditary, i.e., there exists a normal topological space with a non-normal subspace. It turned out that there exists a weaker property of a topological space than normality which is hereditary: the notion of a completely regular topological space. However, in the definition we need the real line.

Theorem 11.10. *Any subspace of a completely regular topological space is completely regular.*

Let $\langle X, \mathcal{O} \rangle$ and $\langle Y, \mathcal{Q} \rangle$ be topological spaces, $j : X \xrightarrow{1-1} Y$ being a homeomorphism of $\langle X, \mathcal{O} \rangle$ onto $\langle f(X), \mathcal{Q}|f(X) \rangle$. The triple $\langle Y_i, \mathcal{Q}_i, f_i \rangle$ is called a **compactification** of $\langle X, \mathcal{O} \rangle$ if Y is compact and $\overline{f(X)} = Y$. Compare Exercise 1.20.

The first main result reads as follows.

Theorem [AC] 11.11. *Let X be a topological space. Then the following are equivalent:*

a) *There exists a compactification of X.*

b) *X is completely regular.*

Proof. According to Theorems 1.27, c), 3.15 and 11.10, if $\langle X, \mathcal{O} \rangle$ has a compactification, then $\langle X, \mathcal{O} \rangle$ is completely regular.

We just sketch the idea of a proof of the opposite implication, which can be found in any textbook of topology.

Let $\langle X, \mathcal{O} \rangle$ be a completely regular topological space. We denote by $\mathcal{F}$ the set of all continuous functions from X into $[0,1]$. Let $Z = {}^{\mathcal{F}}[0,1]$ be endowed with the product topology. By the Tychonoff Theorem 1.37 the topological space Z is compact. Let us remark that a basic open subset of Z is a set of the form

$$\{\varphi \in {}^{\mathcal{F}}[0,1]; \varphi(f_0) \in U_0 \wedge \cdots \wedge \varphi(f_n) \in U_n\},$$

where $f_0, \ldots, f_n \in \mathcal{F}$ and $U_0, \ldots, U_n$ are open subsets of $[0,1]$. We define an injection $j : X \xrightarrow{1-1} Z$ setting $j(x) = \varphi$, where $\varphi(f) = f(x)$ for each $f \in \mathcal{F}$. Let $Y = \overline{j(X)} \subseteq Z$. Then Y, being a closed subset of Z, is compact and $\langle Y, j \rangle$ is a compactification of X. $\square$

If $\kappa = |\mathcal{F}|$ then the product spaces ${}^{\kappa}[0,1]$ and ${}^{\mathcal{F}}[0,1]$ are homeomorphic and we can replace the compact space Y by a subspace of ${}^{\kappa}[0,1]$.

Theorem [AC] 11.12. *Let X be a topological space. Then the following are equivalent:*

a) *X is compact.*
b) *There is a cardinal κ and a closed subset $Y \subseteq {}^\kappa[0,1]$ such that X and Y are homeomorphic.*

Thus the real line, actually the unit interval $[0,1]$, plays an important rôle in description of all compact spaces: compact spaces are, up to a homeomorphism, exactly closed subsets of some power ${}^\kappa[0,1]$ of the unit interval $[0,1]$. In other words, the unit closed interval generates all compact spaces by two operations: (infinite) topological product and taking closed subspaces.

One can easily see that usually a completely regular space has several compactifications, e.g., both $\mathbb{T}$ and $[0,1]$ are compactifications of $(0,1)$ with the natural injections. A compactification $\langle Y_1, \mathcal{Q}_1, f_1 \rangle$ is **weaker** (compare Exercise 1.20) than the compactification $\langle Y_2, \mathcal{Q}_2, f_2 \rangle$ if there exists a continuous mapping $g : Y_2 \xrightarrow{\text{onto}} Y_1$ such that $f_1 = f_2 \circ g$. We shall also say that the compactification $\langle Y_2, \mathcal{Q}_2, f_2 \rangle$ is **stronger** than $\langle Y_1, \mathcal{Q}_1, f_1 \rangle$.

Theorem [AC] 11.13 (E. Čech – M.H. Stone).

a) *The strongest compactification of a completely regular topological space is up to homeomorphism unique.*
b) *The compactification of the proof of Theorem 11.11 is the strongest one.*

Proof. If $\langle Y_i, \mathcal{Q}_i, f_i \rangle$, $i = 0, 1$ are two strongest compactifications of a completely regular space X, then there exist continuous mappings $g_i : Y_i \longrightarrow Y_{1-i}$ such that $f_i \circ g_i = f_{1-i}$, $i = 0, 1$. Since $g_1(g_0(f_0(x))) = f_0(x)$ for any $x \in X$ and $f_0(X)$ is dense in Y_0, we obtain that $g_0 \circ g_1 = \mathrm{id}_{Y_0}$. Similarly for $g_1 \circ g_0$. Thus g_0 and g_1 are homeomorphisms.

Assume that $\langle Y, k \rangle$ is a compactification of X, and $\langle Z, j \rangle$ is the compactification constructed in the proof of Theorem 11.11. We have to find a continuous mapping $g : Z \xrightarrow{\text{onto}} Y$ such that $j \circ g = k$. If $x \in X$ then of course we set $g(j(x)) = k(x)$. Now assume that $\varphi \in Z \setminus X \subseteq {}^\mathcal{F}[0,1]$.

For any continuous function $f : Y \longrightarrow [0,1]$ we define $\overline{f} : Z \longrightarrow [0,1]$ by $\overline{f}(\varphi) = \varphi(k \circ f)$ for $\varphi \in Z$.

If $x \in X$ then by definition $j(x) = \psi$, where $\psi(h) = h(x)$ for $h \in \mathcal{F}$. Thus

$$\overline{f}(j(x)) = \psi(k \circ f) = f(k(x)).$$

Consequently

$$k \circ f = j \circ \overline{f}. \tag{11.7}$$

If $f, f_1, f_2 : Y \longrightarrow [0,1]$ are continuous and $f = \min\{1, f_1 + f_2\}$, then we have also $\overline{f} = \min\{1, \overline{f_1} + \overline{f_2}\}$. Actually, for $x \in X$ we obtain the equality by (11.7) and since $j(X)$ is dense in Z the equality holds on Z.

For a $\varphi \in Z$ we write

$$\mathcal{A}_\varphi = \{f : Y \longrightarrow [0,1]; f \text{ continuous } \wedge \overline{f}(\varphi) = 0\}.$$

One can easily show that there exists a unique point $y \in Y$ such that $y \in Z(f)$ for any $f \in \mathcal{A}_\varphi$. We set $g(\varphi) = y$. $\qquad\square$

The unique strongest compactification (up to homeomorphism) of a completely regular topological space $\langle X, \mathcal{O}\rangle$ is denoted by $\langle \beta X, \beta \mathcal{O}, j\rangle$ and is called **Čech-Stone compactification**.

We close with the important characterization of the Čech-Stone compactification, which can be easily proved using the previous constructions.

Theorem [AC] 11.14. *Let us assume that $\langle X, \mathcal{O}\rangle$ is a completely regular topological space, $\langle Y, \mathcal{Q}\rangle$ is compact, $j : X \xrightarrow{1-1} Y$ is a homeomorphism of X onto $j(X)$, and $j(X)$ is dense in Y. Then the following are equivalent:*

a) *$\langle Y, \mathcal{Q}, j\rangle$ is the Čech–Stone compactification of $\langle X, \mathcal{O}\rangle$.*

b) *For every continuous function $f : X \longrightarrow [0,1]$ there exists a continuous function $F : Y \longrightarrow [0,1]$ such that $j \circ F = f$.*

11.4 Some Logic

Similarly as in other scientific disciplines, a mathematician, like a meteorologist, geologist, chemist, physician, physicist, historian, tries to predict what will happen or, to explain what actually has happened (with the hope that in future we can avoid such an unhappy event or repeat a happy one). However, the attempt of a mathematician in his prediction or/and presentation of results is rather specific. A mathematician introduces abstractions of considered items and tries to deduce new knowledge from assumed properties of such abstractions.

A historically developed attempt by mathematicians to make predictions is based on an abstraction leading to a formulation of primitive notions and axioms, which describe the basic properties of primitive notions, completed by necessary definitions as abbreviations for more complicated notions, and usually continues with a series of theorems with their proofs. The short name for this attempt is usually expressed as "Definition, Theorem, Proof".[5] Essentially this attempt was first presented by Euclid about 300 B.C. in his Στοιχεῖα (Stoichea). Till now mathematicians have not essentially changed this attempt in presentations of their results, but have instead made non-trivial improvements and have developed the method presented by Euclid. However, do not forget that mathematics is more than a proof of a theorem. Usually an abstraction is inspired by a real situation or problem.

---

[5]I suppose that it should be "Axiom, Definition, Theorem, Proof".

Logic of the 20th century uses a formal language to describe precisely one side of this attempt. Since we are going to investigate the work of mathematicians, we work in **metamathematics**. We must be very careful in our choice of methods. We shall try to use in metamathematics only methods that are finite. If we use some notions of set theory in our investigations, i.e., in metamathematics, we use them just as a convenient way of speaking and no property of the set we consider, we prefer to say **collection**, can be based on an assumption concerning infinity.

We recall the structure of the most used type, which is called the **first-order predicate calculus**. We essentially follow J.R. Shoenfield [1967].

A first-order language $\mathcal{L}$ has **logical symbols**:

a) the variables $x, y, z, \ldots, x_1, x_2, \ldots$;

b) logical connectives $\neg, \wedge, \vee, \rightarrow, \equiv$;

c) the quantifiers $\exists, \forall$

and may contain non-logical symbols:

d) n-ary predicate symbols $P, Q, R, \ldots$ for some $n \geq 0$;

e) n-ary function symbols $f, g, h, \ldots$ for some $n > 0$;

f) constants $a, b, c, \ldots$.

Usually we demand that a language contain at least one predicate symbol. We assume that the reader is familiar with the definitions of a term, free variable in a term, atomic formula, formula, free and bound variable in a formula. Writing $\varphi(x_1, \ldots, x_k)$ we say that the formula φ does not contain other free variables than those occurring in the list $x_1, \ldots, x_k$ and not necessarily all of them. Let us recall that a formula not containing any free variable is called a **closed formula** . The metamathematical collection of all closed formulas of a language $\mathcal{L}$ is denoted by $\mathcal{CF}(\mathcal{L})$. **The universal closure** $(\forall \ldots) \varphi$ of a formula $\varphi(x_1, \ldots, x_k)$ is the closed formula $(\forall x_1) \ldots (\forall x_k) \varphi(x_1, \ldots, x_k)$.

Given a language $\mathcal{L}$, some formulas expressing true statements are called **logical axioms**. A finite sequence of formulas of the form

$$\frac{\varphi_1, \ldots, \varphi_k}{\varphi},$$

which leads from true statements to a true statement, is called a **rule of inference**. The intended interpretation is such that a rule of inference produces from provable formulas above the line a provable formula written below the line. We leave it to the taste of the reader to choose her/his favorite system of logical axioms and rules of inference.[6] Usually **modus ponens** ,

$$\frac{\varphi, \varphi \rightarrow \psi}{\psi},$$

---

[6]I prefer that of S.C. Kleene [1952]. Another plausible system is that of J.R. Shoenfield [1967] or M. Goldstern and H. Judah [1995].

is a rule of inference for any formulas φ, ψ. In S.C. Kleene [1952] the following sequences are rules of inference:

$$\frac{\psi(x_1, \ldots, x_k) \to \varphi(x, x_1, \ldots, x_k)}{\psi(x_1, \ldots, x_k) \to (\forall x)\, \varphi(x, x_1, \ldots, x_k)}, \qquad \frac{\varphi(x, x_1, \ldots, x_k) \to \psi(x_1, \ldots, x_k)}{(\exists x)\, \varphi(x, x_1, \ldots, x_k) \to \psi(x_1, \ldots, x_k)},$$

provided that ψ does not contain the variable x.

A **mathematical theory** or shortly a **theory T** is a subcollection of the collection $\mathcal{CF}(\mathcal{L})$ of all closed formulas in a language $\mathcal{L}$. The elements of **T** are called **axioms of the theory T**. Since we want to use only finite methods, in the case of an infinite theory **T** we must have an algorithm deciding if a given formula is an axiom or it is not. If **T** is a theory and φ is a closed formula, then $\mathbf{T} + \varphi$ is the theory $\mathbf{T} \cup \{\varphi\}$.

If a mathematician tries to express the symmetry of an equality, he usually says "$x = y \to y = x$" instead of "$(\forall x)(\forall y)\,(x = y \to y = x)$". Generally, if a formula φ does not contain any quantifier, then an axiom of the form $\varphi(x_1, \ldots, x_k)$ means actually the universal closure $(\forall x_1) \ldots (\forall x_k)\, \varphi(x_1, \ldots, x_k)$. Such a convention, called the **interpretation of generality**, is freely used.

Usually a language of a mathematical theory contains the predicate symbol $=$ for an equality. Then the theory must contain corresponding axioms for equality: reflexivity, symmetry, transitivity and substitution rules for every predicate symbol and every function symbol. We illustrate it in the following examples.

Example 11.1. We describe the **Peano arithmetic PA**. The language $\mathcal{L}_{\mathrm{PA}}$ of Peano arithmetic consists of the predicate symbol of equation $=$, two binary function symbols $+, \cdot$ and two constants $0, 1$. Thus (not very correctly written)

$$\mathcal{L}_{\mathrm{PA}} = \{=, +, \cdot, 0, 1\}.$$

We start with the axioms for equality. The reflexivity, symmetry and transitivity of the equation is expressed by the universal closure of the formulas

$$x = x, \quad x = y \to y = x, \quad x = y \to (y = z \to x = z).$$

The substitution rules for function symbols $+, \cdot$ are the universal closures of the formulas

$$(x_1 = x_3 \wedge x_2 = x_4) \to x_1 + x_2 = x_3 + x_4, \quad (x_1 = x_3 \wedge x_2 = x_4) \to x_1 \cdot x_2 = x_3 \cdot x_4.$$

The next axioms of **PA** are the formulas (their universal closure):

$$x + 0 = x,$$
$$x + (y + 1) = (x + y) + 1,$$
$$x \cdot 0 = 0,$$
$$x \cdot (y + 1) = (x \cdot y) + x,$$
$$\neg(\exists x)\, x + 1 = 0,$$
$$x + 1 = y + 1 \to x = y,$$
$$\neg x = 0 \to (\exists y)\, x = y + 1.$$

Finally, **PA** contains infinitely many axioms described by the **scheme of mathematical induction**: if $\varphi(x, x_1, \ldots, x_k)$ is a formula of the language of **PA** then

$$(\varphi(0, x_1, \ldots, x_k) \wedge (\forall x) (\varphi(x, x_1, \ldots, x_k) \to \varphi(x + 1, x_1, \ldots, x_k)))$$
$$\to (\forall x) \varphi(x, x_1, \ldots, x_k)$$

(its universal closure) is an axiom of **PA**.

Example 11.2. The language $\mathcal{L}_{\text{ZF}}$ of Zermelo–Fraenkel set theory **ZF** has two binary predicate symbols $=, \in$. The axioms of equality are those of reflexivity, symmetry, transitivity and the substitution rule for predicate symbol $\in$:

$$(x_1 = x_3 \wedge x_2 = x_4) \to (x_1 \in x_2 \to x_3 \in x_4).$$

The other axioms of **ZF** are those of Section 1.1.

The theory **ZFC** is **ZF** completed by the axiom of choice **AC**.

A sequence of formulas $\varphi_1, \ldots, \varphi_n$ is a **proof in the theory T** if for any member of the sequence, i.e., for any $i = 1, \ldots, n$, at least one of the conditions holds true:

p1) φ_i is a logical axiom,

p2) φ_i is an axiom of the theory **T**,

p3) there are indexes $j_1, \ldots, j_k < i$ such that

$$\frac{\varphi_{j_1}, \ldots, \varphi_{j_k}}{\varphi_i}$$

is an inference rule (of your chosen logical system).

A formula φ is **provable in T**, written $\mathbf{T} \vdash \varphi$, if there exists a proof in the theory **T** with the last member φ. We used to say that φ is a **theorem** of the theory **T**.

A theory **T** is **inconsistent** if there exists a closed formula φ such that both $\mathbf{T} \vdash \varphi$ and $\mathbf{T} \vdash \neg\varphi$. Otherwise, the theory **T** is **consistent**. Note that in an inconsistent theory **T** any closed formula is provable, or equivalently, a theory **T** is consistent if and only if there exists a closed formula non-provable in **T**. When the paradoxes of the naive set theory developed in 1870s appeared, the mathematical theory **ZF** was formulated and mathematicians hoped to show that **ZF** is consistent. As we shall see, it is impossible.

Let $\mathcal{L}_1$, $\mathcal{L}_2$ be two finite languages, $\mathcal{L}_1 = \{P_1, \ldots, P_n, f_1, \ldots, f_k, a_1, \ldots, a_l\}$. Assume that $\psi, \varphi_1, \ldots, \varphi_n$ are formulas of $\mathcal{L}_2$ and $t_1, \ldots, t_k, s_1, \ldots, s_l$ are terms of $\mathcal{L}_2$ such that ψ has one free variable, the number of free variables in φ_i is the arity of the predicate P_i, $i = 1, \ldots, n$, the number of free variables in t_i is the arity of f_i, $i = 1, \ldots, k$, and s_i is a term without a free variable, $i = 1, \ldots, l$. The **interpretation** Θ of $\mathcal{L}_1$ in $\mathcal{L}_2$ given by formulas $\psi, \varphi_1, \ldots, \varphi_n$ and terms $t_1, \ldots, t_k, s_1, \ldots, s_l$ is a metamathematical mapping which assigns to any formula φ of $\mathcal{L}_1$ a formula $\Theta(\varphi)$ of $\mathcal{L}_2$ obtained by replacing each occurrence of $P_1, \ldots, P_n, f_1, \ldots, f_k, a_1, \ldots, a_l$ in φ by corresponding expressions $\varphi_1, \ldots, \varphi_n, t_1, \ldots, t_k, s_1, \ldots, s_l$ of the language $\mathcal{L}_2$

and replacing each quantifier $(\forall x)\,\vartheta$, $(\exists x)\,\vartheta$ by $(\forall x)\,(\psi \to \vartheta)$, $(\exists x)\,(\psi \wedge \vartheta)$, respectively. More precisely $\Theta((\forall x)\,\vartheta)$ is the formula $(\forall x)\,(\psi \to \Theta(\vartheta))$ and $\Theta((\exists x)\,\vartheta)$ is the formula $(\exists x)\,(\psi \wedge \Theta(\vartheta))$. For details we recommend any standard textbook in mathematical logic, e.g., J.R. Shoenfield [1967] or P. Hájek and P. Pudlák [1993].

If $\mathbf{T}_1$, $\mathbf{T}_2$ are theories in the languages $\mathcal{L}_1$, $\mathcal{L}_2$, respectively, then an interpretation Θ of the language $\mathcal{L}_1$ in the language $\mathcal{L}_2$ is a **syntactic model of the theory $\mathbf{T}_1$ in the theory $\mathbf{T}_2$** if the following conditions are satisfied:

a) $\mathbf{T}_2 \vdash (\exists x)\psi(x)$,

b) $\mathbf{T}_2 \vdash (\psi(x_1) \wedge \cdots \wedge \psi(x_{p_i})) \to \psi(t_i(x_1, \ldots, x_{p_i}))$ for $i = 1, \ldots, k$,

c) $\mathbf{T}_2 \vdash \psi(s_i)$ for $i = 1, \ldots, l$,

d) $\mathbf{T}_2 \vdash \Theta(\varphi)$ for any axiom φ of $\mathbf{T}_1$.

Usually we speak about a **model** instead of a syntactic model. J.R. Shoenfield [1967] uses the notion of an "interpretation of the theory $\mathbf{T}_1$ in $\mathbf{T}_2$" instead of a syntactic model. The basic property of a syntactic model is expressed by

Metatheorem 11.1. *If Θ is a syntactic model of a theory $\mathbf{T}_1$ in a theory $\mathbf{T}_2$, and $\mathbf{T}_1 \vdash \varphi$, then $\mathbf{T}_2 \vdash \Theta(\varphi)$.*

Example 11.3. We describe an interpretation of the language of Peano arithmetic **PA** into the language of Zermelo–Fraenkel set theory **ZF**. We use the definitions in the theory **ZF** introduced in Section 1.1.

$$\psi(x) = x \in \omega, \qquad \Theta(x = y) = x = y,$$
$$\Theta(0) = \emptyset, \qquad \Theta(1) = \{\emptyset\},$$
$$\Theta(t_1 + t_2) = \Theta(t_1) + \Theta(t_2), \qquad \Theta(t_1 \cdot t_2) = \Theta(t_1) \cdot \Theta(t_2).$$

Actually we have misused the notation. We have used the sign "=" in at least three different meanings. The sign "+" on the left side of the fifth equality denotes the name of an operation of the language of Peano arithmetic, and the same sign "+" on the right side of this line denotes the "term" of **ZF** introduced in Section 1.1. Similarly for the sign "$\cdot$".

The interpretation Θ is a syntactic model of the theory **PA** in the theory **ZF**.

The theory $\mathbf{T}_2$ is said to be **stronger** than the theory $\mathbf{T}_1$, or equivalently, the theory $\mathbf{T}_1$ is said to be **weaker** than the theory $\mathbf{T}_2$, if there exists a syntactic model of $\mathbf{T}_1$ in $\mathbf{T}_2$. Thus by Example 11.3, the theory **ZF** is stronger than the theory **PA**. If the language of a theory $\mathbf{T}_1$ is a subcollection of the language of a theory $\mathbf{T}_2$ and every axiom of $\mathbf{T}_1$ is provable in $\mathbf{T}_2$, then the identical mapping is a syntactic model of $\mathbf{T}_1$ in $\mathbf{T}_2$ and the theory $\mathbf{T}_2$ is stronger than $\mathbf{T}_1$. In this case we shall say that $\mathbf{T}_2$ is an **extension of the theory** $\mathbf{T}_1$.

Example 11.4. The interpretation of **ZFC** in **ZFC** + "κ is a strongly inaccessible cardinal" given by formulas[7]

$$|\mathbf{TC}(x)| < \kappa, x = y, x \in y$$

---

[7] For the definition and properties of the transitive closure $\mathbf{TC}(x)$ see Exercise 1.1.

is a syntactic model of **ZFC** in

$$\mathbf{ZFC} + \text{"there exists a strongly inaccessible cardinal"}.$$

We can assume that κ is the smallest strongly inaccessible cardinal. Then this interpretation is a syntactic model of **ZFC** + "there is no strongly inaccessible cardinal" in the theory **ZFC** + "there exists a strongly inaccessible cardinal".

The main result connecting consistency of a theory and a syntactic model is the following.

Metatheorem 11.2. *If the theory* $\mathbf{T}_2$ *is consistent and there exists a syntactic model* Θ *of* $\mathbf{T}_1$ *in* $\mathbf{T}_2$, *then* $\mathbf{T}_1$ *is consistent as well.*

Demonstration. Actually, assume that $\mathbf{T}_1$ is inconsistent. Then there is a closed formula φ in the language of $\mathbf{T}_1$ such that $\mathbf{T}_1 \vdash \varphi$ and $\mathbf{T}_1 \vdash \neg\varphi$. Then $\mathbf{T}_2 \vdash \Theta(\varphi)$ and $\mathbf{T}_2 \vdash \Theta(\neg\varphi)$. By the definition of an interpretation we have $\Theta(\neg\varphi) = \neg\Theta(\varphi)$, hence $\mathbf{T}_2 \vdash \neg\Theta(\varphi)$, which is impossible.   □

Thus, as the main consequence of Example 11.4 we obtain

Metatheorem 11.3 (K. Kuratowski). *If* **ZFC** *is consistent, then*

$$\mathbf{ZFC} \nvdash \text{"there is a strongly inaccessible cardinal"}.$$

When we say that "there exists a (syntactic) model Θ of $\mathbf{T}_1$ in $\mathbf{T}_2$" we want actually to say that "if the theory $\mathbf{T}_2$ is consistent, then $\mathbf{T}_1$ is consistent as well".

All our metamathematical reasonings can be encoded by natural numbers like a computer encodes any communication in English by sequences of zeros and ones, which can be considered as binary expansions of natural numbers. Any formula of the language of **PA** is encoded by a natural number and this number has its name as a term without free variables in the language of **PA**. The property "n is the number of a logical axiom" can be expressed by a formula of **PA**. Going on we can construct a formula expressing "m is the number of a proof of a formula with the number n", finishing with the formula $\mathrm{Cons}_{\mathrm{PA}}$ expressing that "the formula $0 = 1$ is not provable in **PA**". This formula is an encoding of the assertion that "**PA** is consistent".

In similar way, in the language of **PA** or **ZF**, we can encoded as a formula $\mathrm{Cons}_{\mathrm{ZF}}$ the assertion "**ZF** is consistent". By Example 11.3 we obtain

$$\mathbf{PA} \vdash \mathrm{Cons}_{\mathrm{ZF}} \rightarrow \mathrm{Cons}_{\mathrm{PA}}.$$

Since **ZF** is stronger than **PA**, we have also

$$\mathbf{ZF} \vdash \mathrm{Cons}_{\mathrm{ZF}} \rightarrow \mathrm{Cons}_{\mathrm{PA}}.$$

On the other hand, by the elementary model theory developed in **ZF**, one can easily show that

$$\mathbf{ZF} \vdash \Theta(\mathrm{Cons}_{\mathrm{PA}}),$$

where Θ is the syntactic model of Example 11.3, and

$$\mathbf{ZF} + \text{"there exists a strongly inaccessible cardinal"} \vdash \mathrm{Cons}_{\mathrm{ZF}}.$$

We can consider any theory $\mathbf{T}$ stronger than $\mathbf{PA}$ satisfying the above-mentioned finiteness condition: there exists an algorithm deciding about a given formula, whether it is an axiom of the theory $\mathbf{T}$ or it is not. Such a theory is called **axiomatizable**. The theory $\mathbf{ZF}$ is an axiomatizable theory stronger than $\mathbf{PA}$. We can also construct a formula $\mathrm{Cons}_{\mathbf{T}}$ expressing the consistency of $\mathbf{T}$.

K. Gödel [1931] proved an important result.

Metatheorem 11.4 (Gödel's Second Theorem). *If the theory $\mathbf{PA}$ is consistent, then $\mathbf{PA} \nvdash \mathrm{Cons}_{\mathrm{PA}}$. More generally, if $\mathbf{T}$ is a consistent axiomatizable theory stronger than $\mathbf{PA}$, then $\mathbf{T} \nvdash \mathrm{Cons}_{\mathbf{T}}$.*

According to our restriction to finite metamathematics, we assume that any metamathematical consideration can be formalized in the theory $\mathbf{PA}$. So, if we prove the consistency of $\mathbf{PA}$ using our metamathematical finite tools, we could translate the consideration in a proof of $\mathrm{Cons}_{\mathrm{PA}}$ in $\mathbf{PA}$, contradicting the Second Gödel's Theorem. Hence, though we believe that $\mathbf{PA}$ is consistent, we cannot prove its consistency.

We can allow metamathematical reasoning based on the notion of an actual infinity. Again, we must impose some restriction. The most natural one is that any metamathematical consideration can be formalized in the theory $\mathbf{ZF}$. Then any metamathematical proof of consistency of $\mathbf{ZF}$ could be translated into a proof of $\mathrm{Cons}_{\mathrm{ZF}}$ in $\mathbf{ZF}$, again contradicting Gödel's Second Theorem.

We briefly describe an impact of the existence of a strongly inaccessible cardinal on the existence of sets of natural numbers. Now we work in a set theory, say $\mathbf{ZFC}$. By the basic results of model theory, actually by Gödel's Completeness Theorem (see again J.R. Shoenfield [1967]), a theory is consistent if and only if it has a model. By the Löwenheim-Skolem Theorem, see, e.g., J.R. Shoenfield [1967], if a theory has a model then it has a countable model. A model of the set theory formalized in $\mathbf{ZFC}$ is a relation on a countable set, that can be encoded as a set $R \subseteq \omega$. Since $\mathrm{Cons}_{\mathrm{ZFC}}$ is not provable in $\mathbf{ZFC}$, it is not provable that there exists such a set R. However, in $\mathbf{ZFC} + \text{"there exists a strongly inaccessible cardinal"}$ one can prove $\mathrm{Cons}_{\mathrm{ZFC}}$ and therefore one can prove that there exists a special set $R \subseteq \omega$. Thus the existence of a large (inaccessible) cardinal allows us to prove the existence of a set of natural numbers, or equivalently, the existence of a real, that cannot be proved in $\mathbf{ZFC}$ alone.

Gödel's Second Theorem can be generalized in the following way.

Metatheorem 11.5 (Gödel's Generalized Second Theorem). *If $\mathbf{T}$ is a consistent axiomatizable theory stronger than $\mathbf{PA}$, Θ is a syntactic model of $\mathbf{T}$ in $\mathbf{T}$, then $\mathbf{T} \nvdash \Theta(\mathrm{Cons}_{\mathbf{T}})$.*

For details we recommend the reader to P. Hájek and P. Pudlák [1993].

It is natural to say that a theory $\mathbf{T}_1$ is consistently stronger than a theory $\mathbf{T}_2$ if the consistency of $\mathbf{T}_2$ follows from that of $\mathbf{T}_1$. However, there are at least two basic problems. By Gödel's Second Theorem we cannot prove the consistency of the majority of theories we shall deal with. Secondly, what do we mean by "the consistency ... follows from ..."? To be precise and to avoid both problems, we introduce a modified notion that is sufficient for our purposes. A theory $\mathbf{T}_1$ is **consistently stronger** than the theory $\mathbf{T}_2$ if there exists a syntactic model of $\mathbf{T}_2$ in $\mathbf{T}_1$. We shall say that the theories $\mathbf{T}_1$ and $\mathbf{T}_2$ are **equiconsistent** if both $\mathbf{T}_1$ is consistently stronger than $\mathbf{T}_2$ and $\mathbf{T}_2$ is consistently stronger than $\mathbf{T}_1$. Finally, $\mathbf{T}_1$ is **strictly consistently stronger** than the theory $\mathbf{T}_2$ if $\mathbf{T}_1$ is consistently stronger than $\mathbf{T}_2$ and $\mathbf{T}_1$ and $\mathbf{T}_2$ are not equiconsistent.

An inconsistent theory is a consistently strongest theory. Actually any two inconsistent theories are equiconsistent. Of course, we shall not consider inconsistent theory. Assuming that the considered theories are consistent, Gödel's Second Theorem gives us a tool for strict inequality. We illustrate this phenomena by examples.

Example 11.5. If the theory **PA** is consistent then by Gödel's Second Theorem one cannot prove $\mathrm{Cons}_{\mathbf{PA}}$ in **PA**. By Example 11.3 there is a natural syntactic model of **PA** in **ZF**. Moreover, in **ZF** one can prove $\Theta(\mathrm{Cons}_{\mathbf{PA}})$, where Θ is the syntactic model of Example 11.3. If $\varXi$ were a syntactic model of **ZF** in **PA**, then $\Theta \circ \varXi$ is a syntactic model of **PA** in **PA** and $\mathbf{PA} \vdash \varXi(\Theta(\mathrm{Cons}_{\mathbf{PA}}))$, contradicting Gödel's Generalized Second Theorem 11.5. Thus, if **PA** is consistent, then the theory **ZF** is strictly consistently stronger than **PA**.

Example 11.6. If the theory **ZFC** is consistent then by Gödel's Second Theorem one cannot prove $\mathrm{Cons}_{\mathbf{ZFC}}$ in **ZFC**. By Example 11.4 there is a syntactic model of **ZFC** in **ZFC** + "there exists a strongly inaccessible cardinal". Similarly as above we obtain that if **ZFC** is consistent, then **ZFC** + "there exists a strongly inaccessible cardinal" is strictly consistently stronger than **ZFC**.

We close the section with another important topic, which was practically the main theme of the book. Let $\mathbf{T}$ be a theory. For simplicity we assume that $\mathbf{T}$ is consistent. A closed formula φ in the language of $\mathbf{T}$ is called an **undecidable statement of $\mathbf{T}$** if neither $\mathbf{T} \vdash \varphi$ nor $\mathbf{T} \vdash \neg\varphi$. Theory $\mathbf{T}$ is **complete** if there is no undecidable statement of $\mathbf{T}$. Otherwise $\mathbf{T}$ is **incomplete**.

At the beginning of 1930s, K. Gödel surprised the mathematicians with an unexpected result.

Metatheorem 11.6 (Gödel's Incompleteness Theorem). *If the theory **PA** is consistent, then **PA** is incomplete, i.e., there exists an undecidable statement in **PA**. More generally, if $\mathbf{T}$ is a consistent axiomatizable theory stronger than **PA**, then $\mathbf{T}$ is incomplete. Especially, if **ZF** or **ZFC** is consistent, then **ZF** or **ZFC** is incomplete, respectively.*

Note the following consequence of the Metatheorem. If $\mathbf{T}$ is a consistent axiomatizable theory stronger than **PA**, then there exists a closed formula φ undecidable in $\mathbf{T}$. Consequently, both $\mathbf{T} + \varphi$ and $\mathbf{T} + \neg\varphi$ are consistent and incomplete. Thus, there exist formulas ψ_1 and ψ_2 undecidable in $\mathbf{T} + \varphi$ and $\mathbf{T} + \neg\varphi$, respectively. We can go on.

As of now we know only quite complicated undecidable statements of $\mathbf{PA}$. The example presented below is a nice exception. In the case of $\mathbf{ZFC}$ or $\mathbf{ZF}$, we know that plenty of very important statements are undecidable, starting with $\mathbf{CH}$ and finishing with the problem of reversing the order of integration. In the next section we present those undecidable statements of $\mathbf{ZF}$ or $\mathbf{ZFC}$ which are closely related to the topics of the book.

We present a combinatorial principle that is undecidable in Peano arithmetic $\mathbf{PA}$. We slightly change the definition of the partition relation $m \to (r)_n^k$ presented in Exercise 1.14. The strong partition relation $m \longrightarrow_* (r)_n^k$ denotes the following: for any mapping $F : [m]^k \longrightarrow n$ there exists a set $A \subseteq m$ and an $i < n$ such that $|A| = r$, $\min A \leq r$ and $F(x) = i$ for any $x \in [A]^k$.

From the infinite Ramsey Theorem proved in Exercise 11.3 we can easily conclude that

$$\mathbf{ZFC} \vdash (\forall k)(\forall n)(\forall r)(\exists m)\, m \longrightarrow_* (r)_n^k.$$

On the other hand, J. Paris and L. Harrington [1977] showed that

if $\mathbf{PA}$ is consistent, then $\mathbf{PA} \nvdash (\forall k)(\forall n)(\forall r)(\exists m)\, m \longrightarrow_* (r)_n^k$.

Consequently,

if $\mathbf{ZF}$ is consistent, then $(\forall k)(\forall n)(\forall r)(\exists m)\, m \longrightarrow_* (r)_n^k$ is undecidable in $\mathbf{PA}$.

For further information we recommend the monograph by P. Hájek and P. Pudlák [1993].

11.5 The Metamathematics of the Set Theory

In spite of the fact that quite satisfactory theories of real numbers were developed in the second half of the 19th century before establishing set theory as a theory of infinity, further investigations of the real line needed a notion of infinity. The set theory developed by G. Cantor turned out to be a very convenient framework for such a job, that was done mainly in the 20th century. The deep results of set theory, after the invention by P. Cohen [1963] of forcing, made an essential contribution to the knowledge of the structure of the real line. Therefore we present some basic results concerning the forcing and models of the set theory, that have an impact on the properties of reals.

Basically, two types of axiomatization of the theory were developed. Zermelo-Fraenkel type of axiomatization is usually formulated as $\mathbf{ZF}$ or $\mathbf{ZFC}$, and was presented in Section 1.1. Von Neumann-Bernays-Gödel axiomatization is based on the notions of a set and a class. For more details we recommend K. Gödel [1938], T. Jech [2006], or P. Vopěnka and P. Hájek [1972]. The theory **ZF** needs infinitely many axioms, the theory **GB** has only finitely many axioms. The basic result (at least under some assumptions) connecting both theories reads as follows:

If φ is a formula in the language of **ZF***, then* **GB** $\vdash \varphi$ *if and only if* **ZF** $\vdash \varphi$.

Thus, **ZF** describes the behavior of sets as well as theory **GB**. Mathematicians usually prefer to work in **ZF** or **ZFC**. Therefore the presented results will be formulated for Zermelo-Fraenkel type of axiomatization.

In the late 1960s a manuscript written by A.R. Mathias with the name "Surrealistic Landscape with Figures" was widespread and intensively used. The paper was a survey of recent results obtained mainly by forcing construction of (syntactic) models of set theory, shortly called **independence results**. Later, the paper was published as A.R. Mathias [1979]. We need a shorter version of such a survey, however, containing more recent results. This section tries to be such a survey written according to our needs. We shall try to attribute to each presented result the source where it is presented, if possible the original one. However, many results were obtained by several authors independently and they are often considered as a folklore result.

To simplify our presentation we shall assume that

$$\textbf{ZF is consistent.} \tag{11.8}$$

Thus a metatheorem of the form "**ZF** $+ \varphi$ is consistent" is an abbreviation for "If **ZF** is consistent, then **ZF** $+ \varphi$ is consistent as well". If in a particular metatheorem we need the consistency of **ZF** $+ \psi$ then, for the sake of brevity, we say "ψ is consistent". Hence a metatheorem of the form "If **ZF** $+ \psi$ is consistent, then also **ZF** $+ \varphi$ is consistent" will be expressed as "If ψ is consistent, then also **ZF** $+ \varphi$ is consistent" or as "If ψ is consistent, then also φ is consistent". According to Metatheorem 11.2 and our assumption, if we say "there is a model in which φ holds true" we mean the main consequence of this statement saying that "if **ZF** is consistent, then **ZF** $+ \varphi$ is consistent as well".

K. Gödel [1938], [1944] defined a class **L** of **constructible sets**. Then he showed that the interpretation of **ZF** in **ZF** (actually, K. Gödel worked with **GB**) defined simply by restricting the variables to **L** is a syntactic model of **ZF**.[8] Moreover, in this model **V** = **L** holds true, i.e., every set in this model is a constructible set. According to our convention we can express this result as

$$\textbf{ZF} + \textbf{V} = \textbf{L} \text{ is consistent.} \tag{11.9}$$

Then K. Gödel showed that

$$\textbf{ZF} + \textbf{V} = \textbf{L} \vdash \textbf{GCH} + \textbf{AC}. \tag{11.10}$$

---

[8]Thus, Gödel's model is the triple of formulas $x \in \textbf{L}, x = y, x \in y$.

Thus

$$\text{ZF} + \text{AC and ZFC} + \text{GCH are consistent.} \tag{11.11}$$

According to our convention the last result reads fully as follows: "If **ZF** is consistent, then **ZF** + **AC** and **ZFC** + **GCH** are consistent as well."

If **GCH** holds true, then any inaccessible cardinal is strongly inaccessible. Thus any inaccessible cardinal is a strongly inaccessible cardinal in Gödel's model. Therefore by the Kuratowski Metatheorem 11.3 we obtain

$$\text{ZFC} \nvdash \text{"there exists an inaccessible cardinal"}, \tag{11.12}$$

i.e., if **ZFC** is consistent, then we cannot prove the existence of an inaccessible cardinal.

If $\leq$ is a well-ordering of $^{\omega}\omega$, we write

$$R(\leq) = \{\langle \alpha, \beta \rangle \in {}^{\omega}\omega \times {}^{\omega}\omega : \{\gamma \in {}^{\omega}\omega : \gamma < \alpha\} = \{\text{Proj}_n(\beta) : n \in \omega\}\}. \tag{11.13}$$

J.W. Addison [1958b] continued in the study of the constructible universe **L** and showed that

$$\begin{aligned}\text{ZF} + \text{V} = \text{L} \vdash &\text{"there exists a well-ordering } \leq_L \text{ of } {}^{\omega}\omega \text{ in order} \\ &\text{type } \omega_1 \text{ such that the set } R(\leq_L) \text{ is a } \mathbf{\Sigma}_2^1 \text{ set} \\ &\text{and } (\forall \alpha \in {}^{\omega}\omega)(\exists \beta \in {}^{\omega}\omega) \langle \alpha, \beta \rangle \in R(\leq_L)." \end{aligned} \tag{11.14}$$

Since

$$\alpha <_L \beta \equiv (\exists \gamma)\,(\langle \beta, \gamma \rangle \in R(\leq_L) \wedge (\exists n)\, \alpha = \text{Proj}_n(\gamma))$$

we obtain Gödel's result

$$<_L \in \mathbf{\Sigma}_2^1. \tag{11.15}$$

Let us remark that it is easy to show that (11.15) implies that the set

$$\{\langle \alpha, \beta \rangle \in {}^{\omega}\omega \times {}^{\omega}\omega : \{\gamma \in {}^{\omega}\omega : \gamma <_L \alpha\} \supseteq \{\text{Proj}_n(\beta) : n \in \omega\}\}$$

is $\mathbf{\Sigma}_2^1$. To show that the set defined by the opposite inclusion is $\mathbf{\Sigma}_2^1$ as well, one needs to know more about the fine structure of the constructible universe **L**.

D. Scott [1961] proved that

$$\text{ZF} + \text{V} = \text{L} \vdash \text{"There is no measurable cardinal."} \tag{11.16}$$

The last assertion follows also from a result of P. Vopěnka [1962].

A. Fraenkel [1922] and A. Mostowski [1939], [1948] developed a method of permutation models allowing one to show that the axiom of choice **AC** is not provable in a weak version of **ZF** (allowing so-called individuals), neither in **ZF** + "Every set can be linearly ordered" nor in **ZF** + **DC**.

The definite answer about provability of **CH** and **AC** was given by P.J. Cohen [1963] and [1966]:

$$\text{The theories ZFC} + \neg\text{CH and ZF} + \neg\text{AC are consistent.} \tag{11.17}$$

Moreover, the technology invented for the construction of corresponding models by P.J. Cohen, called forcing, turned out to be very fruitful and was further independently improved by D. Scott [1971], R.M. Solovay [1970b] and P. Vopěnka [1962] and [1967], and is still successfully used. Moreover the forcing was combined with Fraenkel-Mostowski permutation models. There exists a general method, invented by T. Jech and A. Sochor [1966], for transforming a Fraenkel-Mostowski permutation model in a model of **ZF**. Using forcing, many results of independence were obtained. The forcing construction is investigated in several monographs, of which we recommend T. Jech [2006] or K. Kunen [1980].

We would like to say that "**ZFC**$+2^{\aleph_0} = \kappa$ is consistent, where κ is a cardinal". Such a sentence has no sense, since we mix two levels of communication: the words before the comma are spoken in metamathematics and the sentence "κ is a cardinal" is a formula of **ZF**, roughly speaking, "κ is a cardinal" is said inside mathematical theory **ZF**. We may avoid such a problem, using a new defined constant in **ZF**, by a formula φ, provided that

$$\mathbf{ZF} \vdash (\exists x)\, \varphi(x) \land (\forall x)\, (\varphi(x) \to x \text{ is a cardinal}) \land (\forall x, y)\, (\varphi(x) \land \varphi(y) \to x = y).$$

The same holds true for any extension of **ZF**, e.g., for the theory **ZFC**. Thus we can say that "**ZFC** $+ 2^{\aleph_0} = \aleph_2$ is consistent" or "**ZFC** $+ 2^{\aleph_0} = \aleph_{\omega_1}$ is consistent", since both $\aleph_2$ and $\aleph_{\omega_1}$ are constants defined in **ZF** and expressing a cardinal. Actually $\aleph_2$ is the second uncountable cardinal and $\aleph_{\omega_1}$ is the first cardinal such that there exist uncountably many smaller cardinals. In theory **ZFC** + **IC** one can define a new constant "the first inaccessible cardinal". In the next, if κ is a constant defined in **ZF** (or its extension) expressing a cardinal, we shall briefly say that κ is a **well-defined cardinal**.

Cohen's result (11.17) can be improved as

> *If κ is a well-defined cardinal such that*
> $$\mathbf{ZFC} \vdash \mathrm{cf}(\kappa) > \omega, \text{ then } \mathbf{ZFC} + 2^{\aleph_0} = \kappa \text{ is consistent.} \qquad (11.18)$$

As usual, a subset of ω is identified with a real (e.g., using the dyadic expansion). According to this convention, Cohen's model was constructed "by adding κ Cohen reals".

Immediately after Cohen's result, S. Feferman and A. Levy [1963] constructed a model of the theory

$$\mathbf{ZF} + \text{``}\mathcal{P}(\omega) \text{ is a union of countably many countable sets''}. \qquad (11.19)$$

T. Jech [1967] and S. Tennenbaum [1968] constructed independently different models to show

$$\mathbf{ZFC} + \text{``there exists a Souslin line''} \text{ is consistent.} \qquad (11.20)$$

Then R.M. Solovay and S. Tennenbaum [1971] constructed a model of **ZFC**, in which there exists no Souslin line, i.e., in which the Souslin Hypothesis holds true.

D.A. Martin realized, see D.A. Martin and R.M. Solovay [1970], that they have actually shown that for any well-defined uncountable regular cardinal κ,

$$\mathbf{ZFC} + \mathfrak{c} = \mathfrak{m} = \kappa \text{ is consistent.} \qquad (11.21)$$

Several mathematicians independently have shown that in Cohen's model the following equalities hold true:

$$\mathrm{non}(\mathcal{M}) = \aleph_1 \wedge \mathrm{cov}(\mathcal{M}) = \mathfrak{c}. \qquad (11.22)$$

Thus, using results summarized in Diagram 3 of Section 9.2, in Cohen's model we have (compare T. Bartoszyński and H. Judah [1995], A. Blass [2010])

$$\mathfrak{m} = \mathfrak{p} = \mathfrak{t} = \mathfrak{h} = \mathrm{add}(\mathcal{N}) = \mathrm{add}(\mathcal{M}) = \mathrm{cov}(\mathcal{N}) = \mathrm{non}(\mathcal{M}) = \mathfrak{b} = \mathfrak{s} = \aleph_1$$

and

$$\mathrm{cov}(\mathcal{M}) = \mathfrak{r} = \mathfrak{d} = \mathrm{cof}(\mathcal{M}) = \mathrm{cof}(\mathcal{N}) = \mathrm{non}(\mathcal{N}) = \mathfrak{c} > \aleph_1.$$

R. Solovay [1970b] constructed another model for ¬**CH**. He "added Random reals" to the ground model. He essentially showed (compare T. Bartoszyński and H. Judah [1995], A. Blass [2010]) that for a well-defined regular cardinal κ,

$$\mathfrak{c} = \kappa \wedge \mathrm{cov}(\mathcal{N}) = \mathfrak{c} \wedge \mathfrak{d} = \mathrm{non}(\mathcal{N}) = \aleph_1 \text{ is consistent.} \qquad (11.23)$$

Then G.E. Sacks [1971] constructed a model of **ZFC** ("adding Sacks reals") in which

$$\mathfrak{c} = \aleph_2 \wedge \mathrm{cof}(\mathcal{N}) = \aleph_1. \qquad (11.24)$$

For given well-defined cardinals λ, κ, μ satisfying

$$\aleph_1 \le \lambda \le \mathrm{cf}(\mu) \le \mu \le \kappa \wedge \mathrm{cf}(\kappa) > \aleph_0,$$

S.H. Hechler [1974] by "adding Hechler reals" to a model of $\mathfrak{d} = \mu \wedge \mathfrak{c} = \kappa$ constructed a model of **ZFC**, in which

$$\mathfrak{b} = \lambda \wedge \mathfrak{d} = \mu \wedge \mathfrak{c} = \kappa. \qquad (11.25)$$

Moreover, if, e.g., $\lambda = \aleph_1, \mu = \kappa$, then in Hechler's model we have (for details see T. Bartoszyński and H. Judah [1995] or A. Blass [2010])

$$\mathrm{cov}(\mathcal{N}) = \mathfrak{b} = \mathfrak{s} = \aleph_1 \wedge \mathrm{non}(\mathcal{M}) = \mathrm{cov}(\mathcal{M}) = \mathfrak{c}. \qquad (11.26)$$

By adding Hechler reals to a model of **GCH**, one obtains a model of **ZFC** in which

$$\mathrm{add}(\mathcal{M}) = \mathfrak{c} = \aleph_2 \wedge \mathrm{cov}(\mathcal{N}) = \mathfrak{s} = \aleph_1. \qquad (11.27)$$

Another basic result is a construction of a model of **ZFC** by R. Laver [1976] ("by adding Laver reals") in which the Borel Conjecture holds true:

$$\mathbf{ZFC} + \mathfrak{b} = \mathfrak{c} = \aleph_2 + \mathrm{cov}(\mathcal{N}) = \mathrm{non}(\mathcal{N}) = \aleph_1 + \text{"Borel Conjecture"}$$
$$\text{is consistent.} \qquad (11.28)$$

Adding Mathias reals, A.R.D. Mathias [1977] has constructed a model of **ZFC** in which

$$\mathfrak{h} = \mathfrak{c} = \aleph_2 \wedge \mathrm{cov}(\mathcal{N}) = \mathrm{non}(\mathcal{N}) = \aleph_1. \tag{11.29}$$

Later on, R. Laver showed that the Borel Conjecture holds true in Mathias' model.
K. Kunen and F. Tall [1979] showed that

$$\textbf{ZFC} + \mathfrak{m} < \mathrm{add}(\mathcal{N}) + \mathfrak{m} < \mathfrak{p} \ \textit{is consistent.} \tag{11.30}$$

J. Baumgartner and R. Laver [1979] showed that

ZFC $+ \mathfrak{c} = \aleph_2 +$ "every perfectly meager set has cardinality $\leq \aleph_1$"

is consistent $\tag{11.31}$

and (see A.W. Miller [1983])

ZFC $+ \mathfrak{c} > \aleph_1 +$ "every universal measure zero set has cardinality $\leq \aleph_1$"

is consistent. $\tag{11.32}$

Finally, there is no hope to prove (in **ZFC**) that there exists a large σ-set, since by A.W. Miller [1979] we have

ZFC $+$ "every σ-set of reals is countable" *is consistent.* $\tag{11.33}$

A.W. Miller [1981] (alternatively adding a Cohen and a Laver real) constructed a model of **ZFC**, in which

$$\mathrm{cov}(\mathcal{N}) = \aleph_1 \wedge \mathrm{add}(\mathcal{M}) = \mathfrak{c} = \aleph_2. \tag{11.34}$$

In the same paper A.W. Miller [1981] constructed a model of **ZFC** in which

$$\mathrm{cov}(\mathcal{M}) = \mathrm{non}(\mathcal{M}) < \mathfrak{d} = \mathrm{non}(\mathcal{N}). \tag{11.35}$$

Then A.W. Miller [1984b] constructed a model of **ZFC** (adding Miller reals) in which (see also T. Bartoszyński and H. Judah [1995] and A. Blass and S. Shelah [1989])

$$\mathfrak{c} = \mathfrak{d} = \aleph_2 \wedge \mathfrak{r} = \mathrm{non}(\mathcal{N}) = \mathrm{non}(\mathcal{M}) = \aleph_1. \tag{11.36}$$

S. Shelah [1984b] showed that

$$\mathfrak{h} < \mathfrak{s} = \mathfrak{b} \ \textit{is consistent.} \tag{11.37}$$

On the other hand, J. Baumgartner [1984], p. 128 and independently A. Dow [1989] constructed a model of **ZFC**, in which

$$\aleph_1 = \mathfrak{h} = \mathfrak{s} < \mathfrak{b}. \tag{11.38}$$

A. Kamburelis [1989] (also A. Krawczyk, unpublished) showed that

$$\mathbf{ZFC} + \mathrm{cof}(\mathcal{M}) < \mathrm{cof}(\mathcal{N}) \text{ is consistent.} \tag{11.39}$$

In [1990] A. Dow showed that in Laver's model for (11.28)

$$\text{"Dow Principle" holds true.} \tag{11.40}$$

T. Carlson [1993] showed the consistency of the dual Borel Conjecture:

$$\mathbf{ZFC} + \text{"every strongly meager set is countable"} + \mathfrak{c} > \aleph_1 \text{ is consistent.} \tag{11.41}$$

A Cichoń Diagram contains ten cardinal invariants. Each of them may have the value $\aleph_1$ or $\mathfrak{c} > \aleph_1$ (eventually others). There are 23 possible ways in which to distribute the values $\aleph_1$ and $\mathfrak{c} > \aleph_1$ without contradicting the inequalities of the Cichoń Diagram. T. Bartoszyński and H. Judah [1995] present the construction of 23 syntactic models of **ZFC** in **ZFC** for all those possibilities (several of them were constructed by other authors). We present it as

Metatheorem 11.7. *Any of 23 distributions of values $\aleph_1$ and $\mathfrak{c} > \aleph_1$ to cardinal invariants in a Cichoń Diagram which does not contradict the proved inequalities, is consistent with* **ZFC.**

R.M. Solovay [1970b] proved

> *If* **ZFC + IC** *is consistent, then the theory*
> **ZF + DC +** *"every set of reals is Lebesgue measurable"*
> *+ "every set of reals has the Baire Property"* $\tag{11.42}$
> *+ "every uncountable set of reals contains a perfect set"*
> *is consistent as well;*

and

> *If* **ZFC + IC** *is consistent, then the theory*
> **ZFC +** *"every projective set of reals is Lebesgue measurable"*
> *+ "every projective set of reals has the Baire Property"* $\tag{11.43}$
> *+ "every uncountable projective set of reals contains a perfect set"*
> *is consistent as well.*

Later S. Shelah [1984a] (see also J. Raisonnier [1984]) showed that

If **ZF + wAC +** *"every $\mathbf{\Sigma}_3^1$ set of reals is Lebesgue measurable" is consistent, then* **ZFC + IC** *is consistent.* $\tag{11.44}$

Actually the result easily follows from Raisonnier's Theorem 9.27. If every $\mathbf{\Sigma}_3^1$ set of reals is measurable, then for every $a \subseteq \omega$, the cardinal $\aleph_1^{L(a)}$ must be countable (otherwise by Theorem 9.27 we can construct a non-measurable $\mathbf{\Sigma}_3^1$ set of reals).

Using some elementary reasoning (see Historical and Bibliographical Notes to Chapter 9) it implies that $\aleph_1$ is inaccessible in L. The results can be refined in many directions, see J. Raisonnier [1984].

On the other hand, S. Shelah announced – for a proof see H. Judah and S. Shelah [1993] – that

$$\text{``every set of reals has the Baire Property''} \wedge \mathbf{DC} \ \textit{is consistent.} \qquad (11.45)$$

More information can be found, e.g., in T. Bartoszyński and H. Judah [1995].

The former results can be summarized as follows:

> *The theories* $\mathbf{ZF} + \mathbf{wAC} +$ "every set of reals is Lebesgue measurable" *and* $\mathbf{ZFC} + \mathbf{IC}$ *are equiconsistent*

and

> *The theories* $\mathbf{ZF} + \mathbf{DC} +$ "every set of reals has the Baire Property" *and* $\mathbf{ZFC}$ *are equiconsistent.*

Let us consider the theory $\mathbf{ZFC} +$ "there exists a measurable cardinal". In this theory the constant "the first measurable cardinal κ" can be well defined. Assume that λ is a well-defined cardinal such that

$$\mathbf{ZFC} + \text{``there exists a measurable cardinal''} \vdash \lambda < \kappa.$$

By similar reasoning as in Example 11.6 and using the Hanf-Tarski Theorem 10.19 one obtains

> *The consistency strength of the theory*
> $\mathbf{ZFC} +$ "there exists a measurable cardinal"
> *is strictly greater than the consistency strength of the theory*
> $\mathbf{ZFC} +$ "there exist λ strongly inaccessible cardinals."

If λ_1, λ_2 are well-defined cardinals such that $\mathbf{ZFC} \vdash \lambda_1 < \lambda_2 < \kappa$, then

> *The consistency strength of the theory*
> $\mathbf{ZFC} +$ "there exist λ_2 strongly inaccessible cardinals"
> *is strictly greater than the consistency strength of the theory*
> $\mathbf{ZFC} +$ "there exist λ_1 strongly inaccessible cardinals."

Since one can easily well define potentially infinitely many cardinals less than the first measurable one (e.g., $\aleph_1$, $\aleph_2$, $\aleph_\omega$, $\aleph_{\omega_1}$ etc.), we conclude that there exists potentially infinitely many strengths of consistency between the consistency of $\mathbf{ZFC}$ and consistency of $\mathbf{ZFC} +$ "there exists a measurable cardinal", i.e., the consistency strength of the latter theory is very high.

By Metatheorem 9.8 the consistency strength of **AD** is at least as much as that of the existence of a measurable cardinal, therefore it is very high. Indeed, it is much higher. W.H. Woodin [1999] in a deep investigation of the consistency strength of the Axiom of Determinacy isolated the notion of a Woodin cardinal. We did not develop technology for explaining this notion. Remark just the following: if κ is a Woodin cardinal, then the set $\{\lambda < \kappa : \lambda$ is a measurable cardinal$\}$ has cardinality κ. The main result by W.H. Woodin reads as follows.

> The theory **ZF** + **AD** *is equiconsistent with the theory*
> **ZFC** + "there exists infinitely many Woodin cardinals". (11.46)

More surprising is another result by W.H. Woodin [1999] which connects the consistency of the Axiom of Determinacy with the consistency of a combinatorial property of the power set of ω_1. We denote by NS the ideal of non-stationary subsets of ω_1, i.e., the ideal of those subsets that are disjoint with some closed unbounded subset of ω_1. W.H. Woodin [1999] has shown that

> The theory **ZFC** + **AD** *is equiconsistent with the theory*
> **ZFC** + "Boolean algebra $\mathcal{P}(\omega_1)/$NS has a dense subset of cardinality $\aleph_1$".

For details see A. Kanamori [2009].

Bibliography

Aczel D.A.

[2000] The Mystery of the Aleph, Washington Square Press, New York 2000.

Addison J.W.

[1958a] *Separation principles in the hierarchies of classical and effective descriptive set theory*, Fund. Math. **46** (1958), 123–135.

[1958b] *Some consequences of the axiom of constructibility*, Fund. Math. **46** (1958), 337–357.

Addison J.W. and Moschovakis Y.N.

[1968] *Some consequences of the axiom of definable determinateness*, Proc. Nat. Acad. Sci. U.S.A. **59** (1968), 708–712.

Alexandroff P.S.

[1916] *Sur la puissance des ensembles mesurables* B, C. R. Acad. Sci. Paris **162** (1916), 323–325.

Arbault J.

[1952] *Sur l'ensemble de convergence absolue d'une série trigonométrique*, Bull. Soc. Math. France **80** (1952), 253–317.

Arkhangel'skiĭ A.V. (Архангельский А.В.)

[1972] *Спектр частот топологического пространства и классификация пространств*, ДАН СССР, **206**:2 (1972), 265–268. English translation: *The frequency spectrum of a topological space and the classification of spaces*, Soviet Math. Dokl. **13** (1972), 1185–1189.

Baire R.

[1898a] *Sur les fonctions discontinues développables en séries de fonctions continues*, C. R. Acad. Sci. Paris **126** (1898), 884–887.

[1898b] *Sur les fonctions discontinues qui se rattachent aux fonctions continues*, C. R. Acad. Sci. Paris **126** (1898), 1621–1623.

[1899] *Sur la théorie des fonctions discontinues*, C. R. Acad. Sci. Paris **129** (1899), 1010–1013.

Balcar B., Pelant J. and Simon P.

[1980]   *The space of ultrafilters on* N *covered by nowhere dense sets*, Fund. Math. **110** (1980), 11–24.

Balcar B. and Simon P.

[1989]   *Disjoint refinement,* in: Handbook of Boolean algebras, (Monk J.D. and Bonnet R., editors), North Holland Publishing Co., Amsterdam, 1989.

[1995]   *Baire number of the spaces of uniform ultrafilters*, Israel J. Math. **92** (1995), 263–272.

Balcar B. and Štěpánek P.

[2000]   Teorie Množin, (Czech, Set Theory), second edition, Academia, Prague 2000.

Banach S.

[1930]   *Über additive Massfunktionen in abstrakten Mengen*, Fund. Math. **15** (1930), 97–101.

[1932]   Théorie des opérations linéaires, Monografie Matematyczne **1**, Warszawa 1932.

[1948]   *Sur les suites d'ensembles excluant l'existence d'une mesure*, (Note posthume avec préface et commentaire de E. Marczewski), Coll. Math. **1** (1948), 103–108.

Banach S. and Kuratowski K.

[1929]   *Sur une généralisation du problème de la mesure*, Fund. Math. **14** (1929), 127–131.

Bartoszyński T.

[1983]   *On subideals of the ideal of meager sets*, in: Open days in Model Theory, Jadwisin 1981, (Guzicki W. et al., editors), University of Leeds, 1983, 61–65.

[1984]   *Additivity of measure implies additivity of category*, Trans. Amer. Math. Soc. **281** (1984), 209–213.

[1987]   *Combinatorial aspects of measure and category*, Fund. Math. **127** (1987), 225–239.

[2010]   *Invariants of Measure and Topology*, in: Handbook of Set Theory, (Foreman M. and Kanamori A., editors), Springer Verlag 2010, 491–556.

Bartoszyński T. and Judah H.

[1995]   Set Theory, On the Structure of the Real Line, A.K. Peters, Wellesley, Massachusetts 1995.

Bary N.K. (Бары Н.К.)

[1961] *Тригонометрические ряды,* GIFML, Moskva 1961; English translation: A Treatise on Trigonometric Series, The Macmillan Co., New York, 1964.

Baumgartner J.E. and Laver R.

[1979] *Iterated perfect set forcing,* Ann. Math. Logic, **17** (1979), 271–288.

Baumgartner J.E., Malitz J.I. and Reinhart W.N.

[1970] *Embedding trees in the rationals,* Proc. Nat. Acad. Sci. U.S.A. **67** (1970), 1748–1753.

Bell M.

[1981] *On the combinatorial principle $P(\mathfrak{c})$.* Fund. Math. **114** (1981), 149–157.

Bendixson I.

[1883] *Quelques théorèmes de la théorie des ensembles de points,* Acta Math. **2** (1883), 415–429.

Bernays P.

[1937] *A system of axiomatic set theory, I.* J. Symbolic Logic **2** (1937), 65–67.

[1942] *A system of axiomatic set theory, III.* J. Symbolic Logic **7** (1942), 65–89.

Bernstein F.

[1908] *Zur Theorie der trigonometrischen Reihen,* Sitzungber. Sachs. Akad. Wiss. Leipzig **60** (1908), 325–338.

Billingsley P.

[1965] Ergodic Theory and Information, John Wiley and Sons, New York 1965.

Birkhoff G.D.

[1931] *Proof of the ergodic theorem,* Proc. Nat. Acad. Sci. USA, **17** (1931), 656–660.

Blass A.

[1989] *Applications of superperfect forcing and its relatives,* in: Set Theory and its Applications, (Steprāns J. and Watson W.S., editors), Lecture Notes in Mathematics **1401** (1989), 18–40.

[1993] *Simple cardinal characteristics of the continuum,* in: Set Theory of the Reals (Judah H., editor), Israel Math. Conf. Proc. **6** (1993), 63–90.

[2010] *Combinatorial Cardinal Characteristics of the Continuum,* in: Handbook of Set Theory, (Foreman M. and Kanamori A., editors), Springer Verlag 2010, 395–490.

Blass A. and Shelah S.

[1989] *Near coherence of filters III: A simplified consistency proof,* Notre Dame J. Formal Logic, **30** (1989), 530–538.

Błaszczyk A. and Turek S.

[2007]   Teoria mnogości, (Polish, The Set Theory), Wydawnictwo Naukowe PWN, Warszawa 2007.

Bolzano B.

[1817]   *Rein analytischer Beweis des Lehrsatzes, daß zwischen je zwey Werthen, die ein entgegengesetztes Resultat gewähren, wenigsten eine reelle Wurzel der Gleichung liege*, Abh. König. Böhm. Ges. der Wiss., (3), **5**, (1814–17), 1–60. See also Ostwald's Klassiker der exakten Wissenschaften #153 (1905), 3–43.

[1950]   Paradoxien des Unendlichen, Leipzig 1851, English translation: Paradoxes of the infinite, Routledge and Kegan Paul, 1950.

[1831]   Funktionenlehre, Manuscript of 1831–34, published by Král. Č. Spol. Nauk., JČMF, Praha 1930.

Boole G.

[1847]   The mathematical analysis of logic, 1847.

Booth D.

[1970]   *Ultrafilters on a countable set*, Ann. Math. Logic **2** (1970), 1–24.

Borel É.

[1895]   *Sur quelques points de la théorie des fonctions*, Ann. Sci. École Norm. Sup., ser. 3, **12** (1895), 9–55.

[1898]   Leçons sur la Théorie des Fonctions, Gauthier-Villars, Paris 1898.

[1919]   *Sur la classification des ensembles de mesure nulle*, Bull. Soc. Math. France, **47** (1919), 97–125.

Bourbaki N.

[1940]   Topologie générale, Chapitres 1–2. Structures topologiques, Structures uniformes, Hermann, Paris 1940.

[1952]   Intégration, Chapitres 1–5, Inégalités de convexité, Espaces de Riesz, Mesures sur les espaces localement compacts, Prolongement d'une mesure, Intégration des mesures, Hermann, Paris 1952.

[1961]   Topologie générale, Chapitre 9, Utilisation des nombres réels en topologie générale, (second ed.), Hermann, Paris 1961.

Brzuchowski J., Cichoń J., Grzegorek E. and Ryll-Nardzewski C.

[1979]   *On the existence of measurable unions*, Bull. Acad. Polon. Sci. Sér. Sci. Math. **27** (1979), 447–448

Bukovská Z.

[1990]   *Thin sets in trigonometrical series and quasinormal convergence*, Math. Slovaca **40** (1990), 53–62.

[1991]   *Quasinormal convergence*, Math. Slovaca **41** (1991), 137–146.

[1999]   *Thin sets defined by a sequence of continuous functions*, Math. Slovaca **49** (1999), 323–344.

[2003]   *Relationships between families of thin sets*, Math. Slovaca **53** (2003), 385–391.

Bukovský L.

[1965]   *The continuum problem and the powers of alephs*, Comment. Math. Univ. Carolinae **6** (1965), 181–197.

[1977]   *Random forcing*, in: Set Theory and Hierarchy Theory V (Lachlan A. et al., editors), Lecture Notes in Math. **619**, Springer-Verlag, Berlin 1977, 101–117.

[1979a] *Any Partition into Lebesgue Measure Zero Sets Produces a Non–measurable Set*, Bull. Acad. Polon. Sci. Sér. Sci. Math. **27** (1979), 431–435.

[1979b] Štruktúra Reálnej Osi, (The Structure of the Real Line, Slovak), Veda, Bratislava 1979.

[1993]   *Thin sets related to trigonometrical series*, in: Set Theory of the Reals (Judah H., editor), Israel Math. Conf. Proc. **6** (1993), 107–118.

[1998]   *Thin Sets in a General Setting*, Tatra Mt. Math. Publ. **14** (1998), 241–260.

[2003]   *Cardinality of Bases and Towers of Trigonometric Thin Sets*, Real Anal. Exchange **29** (2003/2004), 147–153.

[2007]   *Convergences of Real Functions and Covering properties*, in: Selection Principles and Covering Properties in Topology (Kočinac L., editor) Quaderni di Matematica, **18** (2007), 107–132.

[2008]   *On* wQN$_*$ *and* wQN* *spaces*, Topology Appl. **156** (2008), 24–27.

[∞]   *Generalized Luzin Sets*, to appear in Acta Univ. Carolinae – Math. Phys.

Bukovský L. and Haleš J.

[2003]   *On Hurewicz Properties*, Topology Appl. **132** (2003), 71–79.

[2007]   QN-*spaces,* wQN-*spaces and covering properties*, Topology Appl. **154** (2007), 848–858.

Bukovský L., Kholshchevnikova N.N. and Repický M.

[1994]   *Thin sets of harmonic analysis and infinite combinatorics*, Real Anal. Exchange **20** (1994–95), 454–509.

Bukovský L., Recław I. and Repický M.

[1991]   *Spaces not distinguishing pointwise and quasinormal convergence of real functions*, Topology Appl. **41** (1991), 25–40.

[2001]   *Spaces not distinguishing convergences of real-valued functions*, Topology Appl. **112** (2001), 13–40.

Bukovský L. and Šupina J.

[∞]   *Sequence Selection Principle for Quasi-normal Convergence*, to appear.

Cantor G.

[1870]   *Beweis, daß eine für jeden reellen Wert von x durch eine trigonometrische Reihe gegebene Funktion f(x) sich nur auf eine einzige Weise in dieser Form darstellen läßt*, J. Reine Angew. Math. **72** (1870), 139–142.

[1872]   *Über die Ausdehnung eines Satzes aus der Theorie der Trigonometrischen Reihen*, Math. Ann. **5** (1872), 123–132.

[1874]   *Über eine Eigenschaft des Inbegriffes aller reellen algebraischen Zahlen*, J. Reine Angew. Math. **77** (1874), 258–262.

[1878]   *Ein Beitrag zur Mannigfaltigkeitslehre*, J. Reine Angew. Math. **84** (1878), 242–258.

[1883]   *Über unendliche, lineare Punktmannigfaltigkeiten, Teil* 4, Math. Ann. **21** (1883), 51–58.

[1892]   *Über eine elementare Frage der Mannigfaltigkeitslehre*, Jahresber. d. Deutsch. Math.-Verein., **1** (1892), 75–78.

[1895]   *Beiträge zur Begründung der transfiniten Mengenlehre I*, Math. Ann. **46** (1895), 481–512.

Carathéodory C.

[1918]   Vorlesungen über reelle Funktionen, Teubner, Leipzig 1918.

Carlson T.J.

[1993]   *Strong measure zero and strongly meager sets*, Proc. Amer. Math. Soc. **118** (1993), 577–586.

Cartan H.

[1937a] *Théorie des filtres*, C. R. Acad. Sci. Paris **205** (1937), 595–598.

[1937b] *Filtres et ultrafiltres*, C. R. Acad. Sci. Paris **205** (1937), 777–779.

Cauchy A.L.

[1821]   Cours d'Analyse de l'École Royale Polytechnique; Ire Partie, Analyse algébrique, Paris, 1821.

Čech E.

[1932]   *Sur la dimension des espaces parfaitement normaux*, Bull. Inter. Acad. Tchéque Sci. **33** (1932), 38–55

[1937]   *On bicompact spaces*, Ann. of Math. **38** (1937), 823–844.

[1966]   Topological Spaces, revised by Z. Frolík and M. Katětov, Academia, Prague 1966.

Choquet G.

[1953]   *Theory of Capacities*, Ann. Inst. Fourier (Grenoble) **V** (1953–1954), 131–295.

Church A.

[1927]   *Alternatives to Zermelo's assumption*, Trans. Amer. Math. Soc. **29** (1927), 178–208.

Cichoń J.

[1983]   *On bases of ideals*, in: Open days in Model Theory, Jadwisin 1981, (Guzicki W. et al., editors), University of Leeds, 1983, 61–65.

[1989]   *On two cardinal properties of ideals*, Trans. Amer. Math. Soc. **314** (1989), 693–708.

Cichoń J., Kharazishvili A. and Węglorz B.

[1995]   Subsets of the Real Line, Wydawnictwo Uniwersytetu Łódzkiego, Łódź 1995.

Ciesielski K. and Pawlikowski J.

[2004]   Covering Property Axiom CPA, Cambridge Univ. Press, Cambridge 2004.

Cohen P.J.

[1963]   *The independence of the continuum hypothesis*, Proc. Nat. Acad. Sci. USA **50** (1963), 1143–1148, and **51** (1964), 105–110.

[1966]   Set theory and the continuum hypothesis, Benjamin, New York 1966.

Császár Á. and Laczkovich M.

[1975]   *Discrete and Equal Convergence*, Studia Sci. Math. Hungar. **10** (1975), 463–472.

[1979]   *Some remarks on discrete Baire classes*, Acta Math. Acad. Sci. Hungar. **33** (1979), 51–70.

[1990]   *Discrete and Equal Baire Classes*, Acta Math. Hungar. **55** (1990), 165–178.

Davis Morton

[1964]   *Infinite games of perfect information*, in: Advances in Game Theory, Ann. Math. Stud. **52** (1964), 85–101.

Dedekind R.

[1888]   Was sind und was sollen die Zahlen, (1888), 6th edition, Braunschweig 1930, English translation: Essays on the Theory of Numbers, Beman W.W., editor, Chicago and London 1901.

Denjoy A.

[1912]   *Sur l'absolue convergence des séries trigonométriques*, C. R. Acad. Sci. Paris **155** (1912), 135–136.

Dieudonné J.

[1969]   Foundations of Modern Analysis, Academic Press, New York 1969.

Dow A.

[1989]   *Tree π-bases for $\beta\mathbb{N} - \mathbb{N}$*, Topology Appl. **33** (1989), 3–19.

[1990]   *Two classes of Fréchet–Urysohn spaces*, Proc. Amer. Math. Soc. **108** (1990), 241–247.

Egoroff, D.T.

[1911]   *Sur les séries des fonctions mesurables*, C. R. Acad. Sci. Paris **152** (1911), 226–227.

Eliaš P.

[1997]   *A classification of trigonometrical thin sets and their interrelations*, Proc. Amer. Math. Soc. **125** (1997), 1111–1121.

[2005]   *Arbault permitted sets are perfectly meager*, Tatra Mount. Math. Publ. **30** (2005), 135–148.

[2009]   *Dirichlet Sets and Erdős-Kunen-Mauldin Theorem*, Proc. Amer. Math. Soc. **128** (2009), 1111–1121.

Engelking R.

[1977]   General Topology, Monografie Matematyczne **60**, Warszawa 1977, revised edition: Heldermann Verlag, Berlin, 1989.

Engelking R. and Karłowicz M.

[1965]   *Some theorems of set theory and their topological consequences*, Fund. Math. **57** (1965), 275–285.

Erdős P.

[1943]   *Some remarks on set theory*, Ann. Math **44** (1943), 643–646.

Erdős P., Hajnal A., Máté A. and Rado R.

[1984]   Combinatorial Set Theory: Partition Relations for Cardinals, Akadémiai Kiadó, Budapest 1984.

Erdős P., Kunen K. and Mauldin R.D.

[1981]   *Some additive properties of sets of real numbers*, Fund. Math. **113** (1981), 187–199.

Euler L.

[1737]   *De Fractionibus Continuis*, Comm. Acad. Sci. Petrop. **9** (1737), 98–137.

[1748]   Introductio in Analysis Infinitorum, Lausanne 1748. English translation: Blanton J.D., Introduction to analysis of the infinite, Springer, 1988.

Feferman S.

[1965]   *Some applications of the notions of forcing and generic sets*, Fund. Math. **56** (1965), 325–345.

Feferman S. and Lévy A.

[1963]   *Independence results in set theory by Cohen's method II (abstract)*, Notices Amer. Math. Soc. **10** (1963), 592.

Fichtenholz G. and Kantorovitch L.

[1934]   *Sur les opérations linéaires dans l'espace des fonctions bornées*, Studia Math. **5** (1934), 69–98.

Flašková J.

[2005]   *Thin Ultrafilters*, Acta Univ. Carolinae – Math. Phys. **46** (2005), 13–19.

Fodor G.

[1956]   *Eine Bemerkung zur Theorie der regressiven Funktionen*, Acta Sci. Math. (Szeged) **17** (1956), 139–142.

Foreman M., Magidor M. and Shelah S.

[1988]   *Martin's maximum, saturated ideals and nonregular ulttrafilters. I*, Ann. of Math. **127** (1988), 1–47.

Fourier J.

[1822]   Théorie analytique de la chaleur, Paris, 1822.

Fraenkel A.A.

[1921]   *Zu den Grundlagen der Cantor-Zermeloschen Mengenlehre*, Math. Ann. **86** (1921/22), 230–237.

[1922]   *Über den Begriff "definit" und die Unabhängigkeit des Auswahlaxioms*, Sitzungber. Akad. Wiss. Berlin, 1922, 253–257.

Fraenkel A.A. and Bar-Hillel Y.

[1958]   Foundations of Set Theory, North-Holland, Amsterdam 1958.

Fréchet M.

[1906]   *Sur quelques points du calcul fonctionel*, Rend. Circ. Mat. Palermo **22** (1906), 1–74.

[1910]   *Les ensembles abstraits et le calcul fonctionnel*, Rend. Circ. Mat. Palermo **30** (1910), 1–26.

[1928]   Les Espaces abstraits, Gauthier–Villars, Paris 1928.

Freiling C.

[1986]   *Axioms of Symmetry: throwing Darts at the Real Number Line*, J. Symbolic Logic **51** 1986, 190–200.

Fremlin D.H.

[1984a] *Cichoń's diagram*, in: Séminaire Initiation à l'Analyse (Choquet G. et al., editors), Publications Mathématiques de l'Université Pierre et Marie Curie, Paris 1984, 5.01–5.23.

[1984b] Consequences of Martin's Axiom, Cambridge Tracts in Math. **84**, Cambridge 1984.

[1987] *Measure-additive coverings and measurable selectors*, Diss. Math. **260** (1987), 1–121.

[1993] *Real-valued-measurable cardinals,* in: Set Theory of the Reals (Judah H., editor), Israel Math. Conf. Proc. **6** (1993), 151–304.

[1994] *Sequential convergence in* $C_p(X)$, Comment. Math. Univ. Carolinae **35** (1994), 371–382.

[∞] SSP *and* wQN, preprint 2003.

Friedman H.

[1971] *Higher set theory and mathematical practice*, Ann. Math. Logic **2** (1971), 325–357.

[1980] *A consistent Fubini–Tonelli Theorem for nonmeasurable functions*, Illinois J. Math. **24** (1980), 390–395.

Furstenberg H.

[1981] Recurrence in Ergodic Theory and Combinatorial Number Theory, Princeton University Press, Princeton 1981.

Gaifman H.

[1964] *Concerning measures on Boolean algebras*, Pacific J. Math. **14** (1964), 61–73.

Gale D. and Stewart F.M.

[1953] *Infinite games with perfect information*, in: Contributions to the theory of games, Ann Math. Studies **28** (1953), 245–266.

Galvin F. and Miller A.W.

[1984] *γ-sets and other singular sets of real numbers*, Topology Appl. **17** (1984), 145–155.

Galvin F., Mycielski J. and Solovay R.M.

[1973] *Strong measure zero sets*, Notices Amer. Math. Soc. **26** (1973), A-280.

Gardner R.J. and Pfeffer W.F.

[1984] *Borel Measures*, in: Handbook of Set-Theoretic Topology (Kunen K. and Vaughan J.E., editors), North-Holland, Amsterdam, 1984, 961–1043.

Gauntt R.J.

[1970] *Axiom of choice for finite sets – A solution to a problem of Mostowski*, Notices Am. Math. Soc. **17** (1970), 454.

Gerlits F. and Nagy Z.

[1982] *Some properties of C(X), part I*, Topology Appl. **14** (1982), 151–161.

Gödel K.

[1931] *Über formal unentscheidbare Sätze der Principia Mathematica und verwandter Systeme I*, Monatsh. Math. und Phys. **38** (1931), 137–198.

[1938] The consistency of the axiom of choice and of the generalized continuum hypothesis, Proc. Nat. Acad. Sci. U.S.A., **24** (1938), 556–557.

[1944] The consistency of the axiom of choice and of the generalized continuum hypothesis with the axioms of set theory, Annals of Math. Studies **3**, Princeton 1944.

Goldstern M. and Judah H.

[1995] The Incompleteness Phenomenon, A New Course in Mathematical Logic, A.K. Peters, Wellesley, Massachusetts 1995.

Grzegorek E.

[1980] *Solution of a problem of Banach on σ-fields without continuous measures*, Bull. Acad. Polon. Sci. Sér. Sci. Math. **28** (1980), 7–10.

[1984] *Always of the first category sets*, Rend. Circ. Mat. Palermo (1984), Supl. No. 6, 139–147.

Hájek P.

[1966] *The consistency of Church's alternatives*, Bull. Acad. Polon. Sci. Sér. Sci. Math. **14** (1966), 424–430.

Hájek P. and Pudlák P.

[1993] Metamathematics of First-Order Arithmetic. Perspectives in Mathematical Logic, Springer Verlag, Berlin 1993.

Haleš J.

[2005] *On Scheepers' Conjecture*, Acta Univ. Carolinae – Math. Phys. **46** (2005), 27–31.

Halmos P.R.

[1950] Measure Theory, University Series in Higher Mathematics, Van Nostrand Co., New York 1950.

[1956] Lectures on Ergodic Theory, The Mathematical Society of Japan, Tokyo 1956.

Halpern J.D. and Lévy A.

[1971] *The Boolean prime ideal theorem does not imply the axiom of choice*, in: Axiomatic Set Theory, Proc. Symp. Pure Math. **13**, (Scott D., editor), Amer. Math. Soc., Providence 1971, 83–134.

Hamel G.

[1905] *Eine Basis aller Zahlen und die unstetigen Lösungen der Funktionalgleichung $f(x + y) = f(x) + f(y)$*, Math. Ann. **60** (1905), 495–462.

Hanf W.P.

[1964] *Incompactness in languages with infinitely long expressions*, Fund. Math. **53** (1964), 309–324.

Harrington L.

[1978] *Analytic games and $0^\sharp$*, J. Symbolic Logic **43** (1978), 685–693.

Hartogs F.

[1915] *Über das Problem der Wohlordnung*, Math. Ann. **76** (1915), 438–443.

Hausdorff F.

[1904] *Der Potenzbegriff in der Mengenlehre*, Jahresber. Deutsch. Math.-Verein. **13** (1904), 569–571.

[1909] *Die Graduierung nach dem Endverlauf*, Abh. König. Sächs. Gesellschaft Wiss. (Math.-Phys.Kl.), **31** (1909), 296–334.

[1914] Grundzüge der Mengenlehre, Leipzig 1914, English edition AMS Chelsea, New York 1949.

[1916] *Die Mächtigkeit der Borelschen Mengen*, Math. Ann. **77** (1916), 430–437.

[1936a] *Summen von $\aleph_1$ Mengen*, Fund. Math. **26** (1936), 241–255.

[1936b] *Über zwei Sätze von G. Fichtenholz and L. Kantorovitch*, Studia Math. **6** (1936), 18–19.

Hechler S.

[1973] *Powers of singular cardinals and a strong form of the negation of the generalized continuum hypothesis*, Z. Math. Logik Grundlagen Math. **19** (1973), 83–84.

[1974] *On the existence of certain cofinal subsets of $^\omega\omega$*, in: Axiomatic Set Theory, Part II, (Jech T., editor), Amer. Math. Soc., Providence 1974, 155–173.

Heine E.

[1870] *Über trigonomische Reihen*, J. reine and angew. Math., **71** (1870), 335–365.

Hessenberg G.

[1906] Grundbegriffe der Mengenlehre, Abhandlungen der Friesschen Schule, Göttingen, 1906.

Host B., Méla J.-F. and Parreau F.

[1991] *Non singular transformations and spectral analysis of measures*, Bull. Soc. Math. France **119** (1991), 33–90.

Hurewicz W.

[1925] *Über die Verallgemeinerung des Borelschen Theorems*, Math. Zeit. **24** (1925), 401–421.

[1927] *Über Folgen stetiger Funktionen*, Fund. Math. **9** (1927), 193–204.

[1928] *Relativ perfekte Teile von Punktmengen und Mengen* (A), Fund. Math. **12** (1928), 78–108.

Jech T.

[1967] *Nonprovability of Souslin's problem*, Comment. Math. Univ. Carolinae, **8** (1967), 291–305.

[1968] ω_1 *can be measurable*, Israel J. Math., **6** (1968), 363–367.

[1973] The Axiom of Choice, North-Holland Publ. Co., Amsterdam 1973.

[2006] Set Theory, Academic Press, New York 1978, Third edition, Springer-Verlag, Berlin 2006.

Jech T. and Sochor A.

[1966] *On Θ-model of the set theory*, Bull. Acad. Polon. Sci. Sér. Sci. Math. **14** (1966), 297–303.

Jensen R.B.

[1968] *Souslin's hypothesis is incompatible with* V=L, (Abstract), Notices Amer. Math. Soc. **15** (1968), 935.

[1972] *The fine structure of the constructible universe*, Ann. Math. Logic **7** (1972), 229–308.

Jourdain P.E.B.

[1908] *On the multiplication of alephs*, Math. Ann. **65** (1908), 506–512.

Judah H. and Shelah S.

[1993] *Every set of reals has the Baire property and the axiom of choice*, Israel J. Math., **84** (1993), 435–450.

Just W., Miller A.W., Scheepers M. and Szeptycki P.J.

[1996] *Combinatorics of open covers* (II), Topology Appl. **73** (1996), 241–266.

Just W. and Weese M.

[1996]   Discovering Modern Set Theory, I., American Mathematical Society, Rhode Island 1996.

[1997]   Discovering Modern Set Theory, II., American Mathematical Society, Rhode Island 1997.

Kahane S.

[1993]   *Antistable classes of thin sets in harmonic analysis*, Illinois J. Math. **37** (1993), 186–223.

Kamburelis A.

[1989]   *Iterations of Boolean algebras with measures*, Arch. Math. Logic **29** (1989), 21–28.

Kanamori A.

[2009]   The Higher Infinite, Large Cardinals in Set Theory from Their Beginnings, second edition, Springer Verlag, Berlin 2009.

Kchintchin A.Ya. (Хинчин А.Я.)

[1949]   Цепные дроби, GITTL, Moskva 1949, Russian, second edition, English translation: Continued Fractions, University of Chicago Press, Chicago 1961.

Kechris A.S.

[1995]   Classical Descriptive Set Theory, Springer Verlag, New York 1995.

Kechris A.S., Louveau A. and Woodin W.H.

[1987]   *The structure of σ-ideals of compact sets*, Trans. Amer. math. Soc. **301** (1987), 263–288.

Keisler H.J. and Tarski A.

[1964]   *From accessible to inaccessible cardinals. Results holding for all accessible cardinal numbers and the problem of their extension to inaccessible ones*, Fund. Math. **53** (1964), 225–308.

Kelley J.L.

[1955]   General Topology, Van Nostrand, New York 1955.

Kholshchevnikova N.N. (Холщевникова Н.Н.)

[1985]   *О несчетных R- и N-множествах*, Mat. Zametki **38** (1985), 270–277, English translation: *Uncountable R- and N-sets*, Math. Notes **38** (1985), 847–851.

Kleene S.C.

[1943]   *Recursive predicates and quantifiers*, Tran. Amer. Math. Soc. **53** (1943), 41–73.

[1952]   Introduction to Metamathematics, Van Nostrand, New York 1952.

Kline M.

[1972]   Mathematical Thought from Ancient to Modern Time, Oxford University Press, New York 1972.

Kondô M.

[1938]   *Sur uniformization des complémentaires analytiques et les ensembles projectifs de la seconde classe*, Jap. J. Math. **15** (1938), 197–230.

König D.

[1926]   *Sur les corrrespondences multivoques des ensembles*, Fund. Math. **8** (1926), 114–134.

König J.

[1905]   *Zum Kontinuumproblem*, Math. Ann. **60** (1905), 177–180.

Körner T.W.

[1974]   *Some results on Kronecker, Dirichlet and Helson sets* II, J. Anal. Math. **27** (1974), 260–388.

Kulaga W.

[2006]   *On fields and ideal connected with notions of forcing*, Coll. Math. **105** (2006), 271–281.

Kunen K.

[1980]   Set theory, An introduction to independence proofs, North-Holland Publishing Co., Amsterdam 1980.

Kunen K. and Tall F.

[1979]   *Between Martin's axiom and Souslin's hypothesis*, Fund. Math. **102** (1979), 173–181.

Kuratowski K.

[1922]   *Une méthode d'élimination des nombres transfinis des raisonnement mathématiques*, Fund. Math. **3** (1922), 76–108.

[1930]   *La propriété de Baire dans les espaces métrique*, Fund. Math. **16** (1930), 391–395.

[1931]   *Valuation de la class borélienne ou projective d'un ensemble de points à l'aide des symboles logiques*, Fund. Math. **17** (1931), 249–272.

[1933]   *Sur une famille d'ensembles singuliers*, Fund. Math. **21** (1933), 127–128.

[1936]   *Sur les théorèmes de séparation dans la théorie des ensembles*, Fund. Math. **26** (1936), 183–191.

[1948]   *Ensembles projectifs et ensembles singuliers*, Fund. Math. **35** (1948), 131–140.

[1958a] Topologie, Volume I, Fourth edition, Monografie Matematyczne **20**, Warszawa 1958, revised and augmented English edition, Academic Press, New York, 1966.

[1958b] Topologie, Volume II, Fourth edition, Monografie Matematyczne **20**, Warszawa 1958.

[1976] Одна теорема об идеалах и ее применение к свойству Бера в польских пространствах, (Russian, A certain theorem on ideals, and its applications to the Baire property of Polish spaces), Uspekhi Mat. Nauk **31** (1976), No. 5, 108–111.

Kuratowski K. and Mostowski A.

[1952] Teoria Mnogości, (Polish), Monografie Matematyczne **28**, Warszawa – Wrocław 1952, English translation: Set Theory, with an introduction to descriptive set theory, North-Holland, Amsterdam 1968, second revised edition 1976.

Kuratowski K. and Tarski A.

[1931] *Les opérations logiques et les ensembles projectifs*, Fund. Math. **17** (1931), 240–248.

Kuratowski K. and Ulam S.

[1932] *Quelques propriétés topologiques du produit combinatoire*, Fund. Math. **19** (1932), 248–254.

Kurepa D.

[1935] *Ensembles ordonnés et ramifiés*, Publ. Math. Univ. Belgrade **4** (1935), 1–138.

Laczkovich M.

[1986] *Fubini's theorem and Sierpinski's set*, preprint 1986.

Laver R.

[1976] *On the consistency of Borel's conjecture*, Acta Math. **137** (1976), 151–169.

Lavrentieff M.M.

[1924] *Contribution à la théorie des ensembles homéomorphes*, Fund. Math. **6** (1924), 149–160.

Lebesgue H.

[1902] *Intégrale, longuer, aire*, Annali di Mat. **VII** (1902), 231–359.

[1905] *Sur les fonctions représentables analytiquement*, J. Math. Pures Appl., sér. 6, **1** (1905), 139 – 216.

[1918] *Remarques sur les théories de la mesure et de l'intégration*, Ann. Sci. de l'École Norm. Sup. **35** (1918), 191–250.

Lévy A.

[1979]   Basic Set Theory, Springer-Verlag, Berlin 1979.

Lévy A. and Solovay R.M.

[1967]   *Measurable cardinals and the continuum hypothesis*, Israel J. Math. **5** (1967), 234–248

Lindenbaum A. and Tarski A.

[1926]   *Communication sur les recherches de la théorie des ensembles*, Ann. Soc. Polon. Math. **5** (1926), 101.

Lindhal L.A. and Poulsen F.

[1971]   Thin Sets in Harmonic Analysis, Marcel Decker, New York, 1971.

Liouville J.

[1844]   *Sur des classes très étendues de quantités dont valeur n'est ni algébrique, ni même réducible à des irrationnneles algébriques*, C. R. Acad. Sci. Paris **18** (1844), 883–885.

Locke J.

[1690]   An Essay Concerning Human Understanding, London 1690.

Luzin N.N. (Лузин Н.Н.)

[1912]   *Sur l'absolue convergence des séries trigonométriques*, C. R. Acad. Sci. Paris **155** (1912), 580–582.

[1914]   *Sur un problème de M. Baire*, C. R. Acad. Sci. Paris **158** (1914), 1258–1261.

[1915]   Интеграл и тригонометрический ряд, (Russian, Integral and trigonometric series), Moskva, 1915.

[1917]   *Sur la classification de M. Baire*, C. R. Acad. Sci. Paris **164** (1917), 91–94.

[1921]   *Sur l'existence d'un ensemble non dénombrable qui est de première catégorie dans tout ensemble parfait*, Fund. Math. **2** (1921), 155–157.

[1925a] *Sur un problème de M. Émil Borel et les ensembles projectifs de M. Henri Lebesgue; les ensembles analytiques*, C. R. Acad. Sci. Paris **180** (1925), 1318–1320.

[1925b] *Sur les ensembles projectifs de M. Henri Lebesgue*, C. R. Acad. Sci. Paris **180** (1925), 1572–1574.

[1925c] *Les propriétés des ensembles projectifs*, C. R. Acad. Sci. Paris **180** (1925), 1817–1819.

[1927]   *Sur les ensembles analytiques*, Fund. Math. **10** (1927), 1–95.

[1930]   Leçons sur les ensembles analytiques, Gauthier–Villars, Paris 1930. Russian translation: Лекции об аналитических множествах и их приложениях, GITTL, Moskva 1955.

[1933] *Sur les ensembles toujours de première catégorie*, Fund. Math. **21** (1933), 114–126.

[1935] *Sur les ensembles analytiques nuls*, Fund. Math. **25** (1935), 109–131.

Lusin N. and Novikoff P.

[1935] *Choix effectif d'un point dans un complémentaire analytique arbitraire donné par un crible*, Fund. Math. **25** (1935), 559–560.

Luzin N.N. and Sierpiński W.

[1918] *Sur quelques propriétés des ensembles* (A), Bull. Intern. Acad. Sci. Cracovie A (1918), 35–48.

[1922] *Sur une décomposition du continue*, C. R. Acad. Sci. Paris **175** (1922), 357–359.

[1923] *Sur un ensemble non mesurable* B, J. Math. Pures Appl., sér. 9, **2** (1923), 53–72.

[1928] *Sur un ensemble non dénombrable qui est de première catégorie sur tout ensemble parfait*, Atti Accad. Naz. Lincei Rend. Cl. Sci. Fis. Mat. Natur., 6ᵉ sér., **7** (1928), 214–215.

[1929] *Sur les classes des constituant d'un complémentaire analytique*, C. R. Acad. Sci. Paris **189** (1929), 794–796.

Mahlo P.

[1913] *Über Teilmengen des Kontinuums von dessen Mächtigkeit*, Sitzungber. Sachs. Akad. Wiss. Leipzig **65** (1913), 283–315.

Marcinkiewicz J.

[1938] *Quelques théorèmes sur les séries et les fonctions*, Bull. Sém. Math. Univ. Wilno **1** (1938), 19–24.

Marczewski E. = Szpilrajn E.

[1929] *O mierzalności i wlasności Baire'a*, (Polish, On measurability and the Baire Property), C.R. du I. Congrès Math. des Pays Slaves, Varsovie 1929, Książnica Atlas, 297–303.

[1930] *Sur un problème de M. Banach*, Fund. Math. **15** (1930), 212–214.

[1933] *Sur certain invariants de l'opération* (A), Fund. Math. **21** (1933), 229–235.

[1934] *Remarques sur les fonctions complètement additives d'ensembles et sur les ensembles jouissant de la propriété de Baire*, Fund. Math. **22** (1934), 303–311.

[1935] *Sur une classe de fonctions de M. Sierpiński et la classe correspondante d'ensembles*, Fund. Math. **24** (1935), 17–34.

[1937]   *O zbiorach i funkcjach bezezglednie mierzalnych*, (Polish, On absolutely measurable sets and functions), III. Compt. F. Soc. Sci. Vaesovie, Classe III, **30** (1937), 39–68.

[1938]   *The characteristic function of a sequence of sets and some of its applications*, Fund. Math. **31** (1938), 207–223.

Martin D.A.

[1968]   *The axiom of determinateness and reduction principles in the analytical hierarchy*, Bull. Amer. Math. Soc. **74** (1968), 687–689.

[1970]   *Measurable cardinals and analytic games*, Fund. Math. **66** (1970), 287–291.

[1975]   *Borel Determinacy*, Ann. of Math. **102** (1975), 363–371.

Martin D.A. and Solovay R.M.

[1970]   *Internal Cohen extensions*, Ann. Math. Logic **2** (1970), 143–178.

Mathias A.R.D.

[1977]   *Happy families*, Ann. Math. Logic **12** (1977), 59–111.

[1979]   *Surrealistic landscape with figures (a survey of recent results in set theory)*, Period. Math. Hungar. **10** (1979), 109–175.

Menger K.

[1923]   *Über die Dimensionalität von Punktmengen I*, Monatsh. Math. Phys. **33** (1923), 148–160.

[1924]   *Einige Überdeckungssätze der Punktmengenlehre*,   Sitzungber. Wiener Akad. **133** (1924), 421–444.

[1928]   Dimensionstheorie, Teubner, Leipzig 1928.

Miller A.W.

[1979]   *On generating the category algebra and the Baire order problem*, Bull. Acad. Polon. Sci. Sér. Sci. Math. 27 (1979), 751–755.

[1981]   *Some properties of measure and category*, Trans. Amer. Math. Soc. **266** (1981), 93–114 *Corrections and additions*, Trans. Amer. Math. Soc. **271** (1982), 347–348

[1982]   *A characterization of the least cardinal for which the Baire category theorem fails*, Proc. Amer. Math. Soc. **86** (1982),498–502.

[1983]   *Mapping a set of reals onto the reals*, J. Symbolic Logic **48** (1983), 575–584.

[1984a] *Special subsets of the real line*, in: Handbook of Set-Theoretic Topology (Kunen K. and Vaughan J.E., editors), North-Holland, Amsterdam, 1984, 201–233.

[1984b] *Rational perfect set forcing*, in: Axiomatic Set Theory, Part II, (Baumgartner J.E., Martin D.A., and Shelah S., editors), Contemporary Mathematics 31, AMS Providence 1984, 143–159.

Miller A.W. and Fremlin D.H.

[1988]   *On some properties of Hurewicz, Menger, and Rothberger*, Fund. Math.
         129 (1988), 17–33.

Monk J.D. and Bonnet R.

[1989]   Handbook of Boolean algebras, I – III, (Monk J.D and Bonnet R., editors),
         North-Holland Publishing Co., Amsterdam, 1989

Moore E.H. and Smith H.L.

[1922]   *A general theory of limits*, Amer. J. Math. **44** (1922), 527–533.

Moschovakis Y.N.

[1970]   *Determinacy and prewellorderings of the continuum*, in: Mathematical
         Logic and Foundations of Set Theory, (Bar-Hillel Y., editor), North-
         Holland Publishing Co., Amsterdam 1970, 24–62.

[1971]   *Uniformization in a playful universe*, Bull. Amer. Math. Soc. **77** (1971),
         731–736.

[1980]   Descriptive Set Theory, North-Holland Publishing Co., Amsterdam, 1980.

Mostowski A.

[1939]   *Über die Unabhängigkeit des Wohlordnungssatzes vom Ordnungsprinzip*,
         Fund. Math. **32** (1939), 201–252.

[1945]   *Axiom of choice for finite sets*, Fund. Math. **33** (1945), 137–168.

[1946]   *On definable sets of positive integers*, Fund. Math. **34** (1946), 81–112.

[1948]   *On the principle of dependent choice*, Fund. Math. **35** (1948), 127–130.

Mycielski J.

[1964a] *Continuous games with perfect information*, in: Advances in game theory,
         Ann. Math. Stud. **52** (1964), 103–112.

[1964b] *On the Axiom of Determinateness*, Fund. Math. **53** (1964), 205–224.

[1966]   *On the Axiom of Determinateness II*, Fund. Math. **59** (1966), 203–212.

Mycielski J. and Steinhaus H.

[1962]   *A mathematical axiom contradicting the axiom of choice*, Bull. Acad.
         Polon. Sci. Sér. Math. Astronom. Phys. **10** (1962), 1–3.

Mycielski J. and Świerczkowski S.

[1964]   *On the Lebesgue measurability and the axiom of determinateness*, Fund.
         Math. **54** (1964), 67–71.

Mycielski J. and Zięba A.

[1955]   *On infinite games*, Bull. Acad. Polon. Sci. **3** (1955), 133–136.

Natanson I.P. (Натансон И.П.)

[1957] Теория функций вещественной переменной. (Russian, Theory of Functions of Real Variable), GITTL, Moskva 1957.

Nikodym O.

[1929] *Sur la condition de Baire*, Bull. Acad. Pol. 1929, 591–595.

Novikoff P.S. (Новиков П.С.)

[1951] *О непротиречивости некоторых положений дескриптивной теории множеств*, (Russian, On consistency of some statements of the descriptive set theory), Trudy Mat. Inst. Steklov **38** (1951), 279–316.

Oxtoby J.C.

[1970] Measure and Category, Springer Verlag, Berlin 1970, new edition in Graduate Texts in Mathematics, Springer Verlag, New York-Berlin 1980.

Paris J.B. and Harrington L.

[1977] *A mathematical incompleteness in Peano airhmetic*, in: Handbook of Mathematical Logic, (Barwise J., editor), North-Holland, Amsterdam 1977, 1133–1142.

Pawlikowski J.

[1985] *Lebesgue measurability implies Baire property*, Bull. Sci. Math., 2e Série, **109** (1985), 321–324.

[1996] *Every Sierpiński set is strongly meager*, Arch. Math. Logic **35** (1996), 281–285.

Pinciroli R.

[2006] *On the independence of a generalized statement of Egoroff's theorem from ZFC after T. Weiss*, Real Anal. Exchange **32** (2006/2007), 225–232.

Piotrowski Z. and Szymański A.

[1987] *Some remarks on category in topological spaces*, Proc. Amer. Math. Soc. **101** (1987), 805–808.

Pospíšil B.

[1937] *Remark on bicompact spaces*, Ann. of Math. **38** (1937), 845–846.

Raisonnier J.

[1984] *A mathematical proof of S. Shelah's theorem on the measure problem and related results*, Israel J. Math. **48** (1984), 48–56.

Raisonnier J. and Stern J.

[1985] *The strength of measurability hypothesis*, Israel J. Math **50** (1985), 337–349.

Ramsey F.P.

[1930]   *On a problem of formal logic*, Proc. London Math. Soc. **30** (1930), 264–286.

Random House

[1990]   Random House Webster's College Dictionary, Random House, New York 1990.

Rasiowa H. and Sikorski R.

[1950]   *A proof of the completeness theorem of Gödel*, Fund. Math. **37** (1950), 193–200.

[1963]   The Mathematics of Metamathematics, Monografie Matematyczne **41**, Warszawa 1963.

Recław I.

[1997]   *Metric spaces not distinguishing pointwise and quasinormal convergence of real functions*, Bull. Acad. Polon. Sci. **45** (1997), 287–289.

Repický M.

[2009]   *Generalized Egoroff's Theorem*, Comment. Math. Univ. Carolinae **44** (2009), 81–96.

Rothberger F.

[1938a] *Eine Äquivalenz zwischen der Kontinuumhypothese und der Existenz der Lusinschen und Sierpińskischen Mengen*, Fund. Math. **30** (1938), 215–217.

[1938b] *Eine Verschärfung der Eigenschaft C*, Fund. Math. **30** (1938), 50–55.

[1939]   *Sur les ensembles de première catégorie qui est dépourvu de la propriété λ*, Fund. Math. **32** (1939), 294–300.

[1941]   *Sur les familles indénombrables de suites de nombres naturels et les problèmes concernant la propriété C*, Proc. Cambridge Phil. Soc. **37** (1941), 109–126.

[1948]   *On some problems of Hausdorff and Sierpiński*, Fund. Math. **35** (1948), 29–46.

Rudin W.

[1966]   Real and Complex Analysis, McGraw–Hill, New York 1966.

[1973]   Functional Analysis, McGraw–Hill, New York 1973.

Sacks G.E.

[1971]   *Forcing with perfect closed sets*, Axiomatic Set Theory, Proc. Symp. Pure Math. **13**, (Scott D., editor), Amer. Math. Soc., Providence 1971, 331–355.

Sakai M.

[2007] *The sequence selection properties of* $C_p(X)$, Topology Appl. **154** (2007), 552–560.

[2009] *Selection principles and upper semicontinuous functions*, Colloq. Math. **117** (2009), 251–256.

Salem R.

[1941a] *On some properties of symmetrical perfect sets*, Bull. Amer. Math. Soc. **47** (1941), 820–828.

[1941b] *The absolute convergence of trigonometric series*, Duke Math. J. **8** (1941), 317–334.

Scheepers M.

[1996] *Combinatorics of open covers I: Ramsey theory*, Topology Appl. **69** (1996), 31–62

[1997a] *A sequential property of* $C_p(X)$ *and a covering property of Hurewicz*, Proc. Amer. Math. Soc. **125** (1997), 2789–2795.

[1997b] *Combinatorics of open covers III: games,* $C_p(X)$, Fund. Math. **152** (1997), 231–262.

[1998] $C_p(X)$ *and Arhangel'skiĭ's* α_i*-spaces*, Topology Appl. **89** (1998), 265–275.

[1999] *Sequential convergences in* $C_p(X)$ *and a covering property*, East-West J. Math. **1** (1999), 207–214.

Scott D.

[1961] *Measurable cardinals and constructible sets*, Bull. Acad. Polon. Sci. Sér. Math. Astronom. Phys. **7** (1961), 145–149.

[1971] *A proof of the independence of the continuum hypothesis*, in: Axiomatic Set Theory, Proc. Symp. Pure Math. **13**, (Scott D., editor), Amer. Math. Soc., Providence 1971, 89–111.

Shanin N.A. (Шанин Н.А.)

[1946] *Одна теорема из общей теории множеств*, ДАН СССР, **53** (1946), 403–404. English translation *A theorem from the general theory of sets*, Soviet Math. Dokl. **53** (1946), 399–400.

Shelah S.

[1982] Proper Forcing, Lecture Notes in Mathematics **940**, Springer Verlag 1982.

[1984a] *Can you take Solovay inaccessible away?*, Israel J. Math. **48** (1984), 1–47.

[1984b] *On cardinal invariants of the continuum*, in: Axiomatic Set Theory, (Baumgartner J.E., Martin D.A., and Shelah S., editors), Contemporary Mathematics 31, AMS Providence 1984, 183–207.

[1994]   Cardinal Arithmetics, Oxford University Press, Oxford 1994.

[2003]   *Logical Dreams*, Bull. Amer. Math. Soc. **40** (2003), 203–228.

Shoenfield J.R.

[1967]   Mathematical Logic, Addison–Wesley Publishing Company, Massachusetts 1967.

Sierpiński W.

[1920]   *Sur les rapports entre l'existence des intégrales $\int_0^1 f(x,y)\,dx$, $\int_0^1 f(x,y)\,dy$ et $\int_0^1 dx \int_0^1 f(x,y)\,dy$*, Fund. Math. **1** (1920), 142–147.

[1924a] *Sur l'hypothèse du continue $(2^{\aleph_0} = \aleph_1)$*, Fund. Math. **5** (1924), 177–187.

[1924b] *Sur une propriété des ensembles ambigus*, Fund. Math. **6** (1924), 544–548.

[1925]   *Sur un ensemble non dénomrable dont tout homéomorphe est de mesure nulle*, Fund. Math. **7** (1925), 188–190.

[1926]   *Sur une propriété des ensembles (A)*, Fund. Math. **8** (1925), 362–369.

[1928]   *Sur un ensemble non dénombrable, donc tout image continue est de mesure nulle*, Fund. Math. **11** (1928), 302–304.

[1929]   *Sur les familles inductives et projectives d'ensembles*, Fund. Math. **13** (1929), 228–239.

[1934a] *Sur la dualité entre la première catégorie et la mesure nulle*, Fund. Math. **22** (1934), 276–280.

[1934b] Hypothèse du Continu, Monografie Matematyczne **6**, Warszawa 1934.

[1935]   *Sur une Hypothèse de M. Luzin*, Fund. Math. **25** (1935), 132–135.

[1947]   *Démonstration de l'égalité $2^{\mathfrak{m}} - \mathfrak{m} = 2^{\mathfrak{m}}$ pour les nombres cardinaux transfinis*, Fund. Math. **34** (1947), 113–118.

[1950]   Les Ensembles projectifs et analytiques, Mémorial des Sciences Mathématiques, CXII, Gauthier–Villars, Paris 1950.

[1965]   Cardinal and ordinal Numbers, Monografie Matematyczne **34**, Warszawa 1965.

Sierpiński W. and Szpilrajn E.

[1936]   *Remarque sur le problème de la mesure*, Fund. Math. **26** (1936), 256–261.

Sikorski R.

[1964]   Boolean Algebras, Second Edition, Springer-Verlag, Berlin 1964.

Solovay R.M.

[1967]   *Measurable Cardinals and the Axiom of Determinateness* Berkley Workshop 1967, mimeographed notes III-H-1–11.

[1970a] *Metamathematics of measure theory*, Invited Lecture to International Congress of Mathematicians, Nice, 1970.

[1970b] *A model of set theory in which every set of reals is Lebesgue measurable*, Ann. of Math. **92** (1970), 1–56.

[1971] *Real-valued measurable cardinals*, in: Axiomatic Set Theory, Proc. Symp. Pure Math. **13**, (Scott D., editor), Amer. Math. Soc., Providence 1971, 397–428.

Solovay R.M. and Tennenbaum S.

[1971] *Iterated Cohen extensions and Souslin's problem*, Ann. Math. **94** (1971), 201–245.

Souslin M.J. (Суслин М. Я.)

[1917] *Sur une définition des ensembles mesurables B sans nombres transfinis*, C. R. Acad. Sci. Paris **164** (1917), 88–91.

[1920] *Problème 3*, Fund. Math. **1** (1920), 223.

Staš M.

[2008] *Hurewicz Scheme*, Acta Univ. Carolinae – Math. Phys. **49** (2008), 75–78.

Steinhaus H.

[1920] *Sur les distances des points dans les ensembles de mesure positive*, Fund. Math. **1** (1920), 93–104.

Stone M.H.

[1936] *The theory of representations for Boolean algebras*, Trans. Amer. Math. Soc. **40** (1936), 37–111.

[1937] *Applications of the theory of Boolean rings to general topology*, Trans. Amer. math. Soc. **41** (1937), 375–481.

[1947] *The generalized Weierstrass approximation theorem*, Math. Mag. **21** (1947–48), 167–183 and 237–254.

Talagrand M.

[1980] *Compacts de fonctions mesurables et filtres non mesurable*, Stud. Math. **67** (1980), 13–43.

Tarski A.

[1924] *Sur les ensembles finis*, Fund. Math. **6** (1924), 45–95.

[1925] *Quelques théorèmes sur les alephs*, Fund. Math. **7** (1925), 1–14.

[1930] *Une contribution à la théorie de la mesure*, Fund. Math. **15** (1930), 42–50.

[1939] *Ideale in vollständigen Mengenkörpern*, Fund. Math. **32** (1939), 45–63 and **33** (1945), 51–65.

Telgársky R.

[1987] *Topological games: On the 50th anniversary of the Banach-Mazur game,* Rocky Mount. J. Math. **17** (1987), 227–276.

Tennenbaum S.

[1968] *Souslin's Problem,* Proc. Nat. Acad. U.S.A. **59** (1968), 60–63.

Tietze H.

[1915] *Über Funktionen, die auf einer abgeschossenen Menge stetig sind,* J. Reine Angew. Math. **145** (1915), 9–14.

Truss J.

[1977] *Sets having calibre* $\aleph_1$, in: Logic Colloquium 76, (Gandy R.O. and Hyland J.M.E., editors), Studies in Logic and Foundations of Mathematics, North-Holland, Amsterdam 1977, 595–612.

Tsaban B. and Zdomskyy L.

[∞] *Hereditary Hurewicz spaces and Arhangel'skii sheaf amalgamations,* J. European Math Soc., to appear.

Tychonoff A.N.

[1930] *Über die topologische Erweiterung von Räumen,* Math. Ann. **102** (1930), 544–561.

Ulam S.

[1930] *Zur Masstheorie in der allgemeinen Mengenlehre,* Fund. Math. **16** (1930), 140–150.

Urysohn P.

[1922] *Les multiplicités Cantoriennes,* C.R. Acad. Sci. **175** (1922), 440–442.

[1925] *Über die Mächtigkeit der kompakten topologischen Räume,* Math. Ann. **92** (1925), 309–315.

van Douwen E.K.

[1984] *The integers and topology,* in: Handbook of Set-Theoretic Topology, (Kunen K. and Vaughan J.E., editors), North-Holland, Amsterdam, 1984, 111–167.

Vaughan J.E.

[1990] *Small Uncountable Cardinals and Topology,* in: Open Problems in Topology, (van Mill J. and Reed G.M., editors), North–Holland, Amsterdam 1990, 195–218.

Vitali G.

[1905] *Sul problema della misura dei gruppi di punti di una retta,* Tipogr. Gamberini e Parmeggiani, Bologna 1905.

Vojtáš P.

[1989]   *Cardinalities of noncentered systems of subsets of N which reflect some qualities of ultrafilters, p-points and rapid filters*, in: International Conference on Topology and its Applications, Baku, 1987, Elm Baku, 1989, 263–268.

[1993]   *Generalized Galois–Tukey connections between explicit relations on classical objects of real analysis*, in: Set Theory of the Reals (Judah H., editor), Israel Math. Conf. Proc. **6** (1993), 619–643.

von Neumann J.

[1923]   *Zur Einführung der transfiniten Zahlen*, Acta Sci. Math. (Szeged) **1** (1923), 199–208.

[1928]   *Über die Definition durch transfinite Induktion und verwandte Fragen der allgemeinen Mengenlehre*, Math. Ann. **99** (1928), 373–391.

Vopěnka P.

[1962]   *Построение модели теории множеств методом ультрапроизведениа*, Russian, (Construction of models of set theory by the method of ultraproducts), Z. Math. Logic **8** (1962), 293–304.

[1967]   *General theory of ∇-models*, Comment. Math. Univ. Carolinae, **8** (1967), 145–170.

[1979]   Mathematics in the Alternative Set Theory, Teubner-Texte zur Mathematik, Leipzig 1979.

Vopěnka P. and Bukovský L.

[1964]   *The existence of a* **PCA**-*set of cardinality* $\aleph_1$, Comment. Math. Univ. Carolinae, **5** (1964), 125–128.

Vopěnka P. and Hájek P.

[1972]   The Theory of Semisets, Academia, Prague 1972.

Weiss T.

[2004]   *A note on generalized Egorov's theorem*, preprint 2004.

Wikipedia

[Wikipedia] http://en.Wikipedia.org; Duality (mathematics), April 25, 2010.

Woodin W.H.

[1999]   The Axiom of Determinacy, Forcing Axioms, and the Nonstationary Ideal, De Gruyter Series in Logic and Applications 1. Berlin, Walter de Gruyter & Co. 1999.

Zajíček L.

[1987]   *Porosity and σ-porosity*, Real Anal. Exchange **13** (1987–88), 314–350.

Zermelo E.

[1904]   *Beweis, dass jede Menge wohlgeordnet werden kann*, Math. Ann. **59** (1904), 514–516.

[1908]   *Untersuchungen über die Grundlagen der Mengenlehre I*, Math. Ann **65** (1908), 261–281.

Zorn M.

[1935]   *A remark on method in transfinite algebra*, Bull. Amer. Math. Soc. **41** (1935), 667–670.

Index of Notation

Index

Monografie Matematyczne

[1] S. Banach, *Théorie des opérations linéaires*, 1932
[2] S. Saks, *Théorie de l'integrale*, 1933
[3] C. Kuratowski, *Topologie I*, 1933
[4] W. Sierpiński, *Hypothèse de continu*, 1934
[5] A. Zygmund, *Trigonometrical Series*, 1935
[6] S. Kaczmarz, H. Steinhaus, *Theorie der Orthogonalreihen*, 1935
[7] S. Saks, *Theory of the integral*, 1937
[8] S. Banach, *Mechanika*, T. I, 1947
[9] S. Banach, *Mechanika*, T. II, 1947
[10] S. Saks, A. Zygmund, *Funkcje analityczne*, 1948
[11] W. Sierpiński, *Zasady algebry wyższej*, 1946
[12] K. Borsuk, *Geometria analityczna w n wymiarach*, 1950
[13] W. Sierpiński, *Działania nieskończone*, 1948
[14] W. Sierpiński, *Rachunek różniczkowy poprzedzony badaniem funkcji elementarnych*, 1947
[15] K. Kuratowski, *Wykłady rachunku różniczkowego i całkowego*, T. I, 1948
[16] E. Otto, *Geometria wykreślna*, 1950
[17] S. Banach, *Wstęp do teorii funkcji rzeczywistych*, 1951
[18] A. Mostowski, *Logika matematyczna*, 1948
[19] W. Sierpiński, *Teoria liczb*, 1950
[20] C. Kuratowski, *Topologie I*, 1948
[21] C. Kuratowski, *Topologie II*, 1950
[22] W. Rubinowicz, *Wektory i tensory*, 1950
[23] W. Sierpiński, *Algèbre des ensembles*, 1951
[24] S. Banach, *Mechanics*, 1951
[25] W. Nikliborc, *Równania różniczkowe*, Cz. I, 1951
[26] M. Stark, *Geometria analityczna*, 1951
[27] K. Kuratowski, A. Mostowski, *Teoria mnogości*, 1952
[28] S. Saks, A. Zygmund, *Analytic functions*, 1952
[29] F. Leja, *Funkcje analityczne i harmoniczne*, Cz. I, 1952
[30] J. Mikusiński, *Rachunek operatorów*, 1953
* [31] W. Ślebodziński, *Formes extérieures et leurs applications*, 1954
[32] S. Mazurkiewicz, *Podstawy rachunku prawdopodobieństwa*, 1956
[33] A. Walfisz, *Gitterpunkte in mehrdimensionalen Kugeln*, 1957
[34] W. Sierpiński, *Cardinal and ordinal numbers*, 1965
[35] R. Sikorski, *Funkcje rzeczywiste*, 1958
[36] K. Maurin, *Metody przestrzeni Hilberta*, 1959
[37] R. Sikorski, *Funkcje rzeczywiste*, T. II, 1959
[38] W. Sierpiński, *Teoria liczb II*, 1959

* [39] J. Aczél, S. Gołąb, *Funktionalgleichungen der Theorie der geometrischen Objekte*, 1960

[40] W. Ślebodziński, *Formes extérieures et leurs applications*, II, 1963

[41] H. Rasiowa, R. Sikorski, *The mathematics of metamathematics*, 1963

[42] W. Sierpiński, *Elementary theory of numbers*, 1964

* [43] J. Szarski, *Differential inequalities*, 1965

[44] K. Borsuk, *Theory of retracts*, 1967

[45] K. Maurin, *Methods of Hilbert spaces*, 1967

[46] M. Kuczma, *Functional equations in a single variable*, 1967

[47] D. Przeworska-Rolewicz, S. Rolewicz, *Equations in linear spaces*, 1968

[48] K. Maurin, *General eigenfunction expansions and unitary representations of topological groups*, 1968

[49] A. Alexiewicz, *Analiza funkcjonalna*, 1969

* [50] K. Borsuk, *Multidimensional analytic geometry*, 1969

* [51] R. Sikorski, *Advanced calculus. Functions of several variables*, 1969

[52] W. Ślebodziński, *Exterior forms and their applications*, 1971

[53] M. Krzyżański, *Partial differential equations of second order*, vol. I, 1971

[54] M. Krzyżański, *Partial differential equations of second order*, vol. II, 1971

[55] Z. Semadeni, *Banach spaces of continuous functions*, 1971

[56] S. Rolewicz, *Metric linear spaces*, 1972

[57] W. Narkiewicz, *Elementary and analytic theory of algebraic numbers*, 1974

[58] Cz. Bessaga, A. Pełczyński, *Selected topics in infinite dimensional topology*, 1975

* [59] K. Borsuk, *Theory of shape*, 1975

[60] R. Engelking, *General topology*, 1977

[61] J. Dugundji, A. Granas, *Fixed point theory*, 1982

* [62] W. Narkiewicz, *Classical problems in number theory*, 1986

The volumes marked with * are available at the exchange department of the library of the Institute of Mathematics, Polish Academy of Sciences.

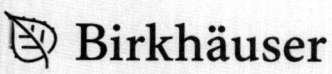 **Birkhäuser** | **www.birkhauser-science.com**

Monografie Matematyczne, New Series (MMNS)

Starting in the 1930s with volumes written by such distinguished mathematicians as Banach, Saks, Kuratowski, and Sierpinski, the original series grew to comprise 62 excellent monographs up to the 1980s. In cooperation with the Institute of Mathematics of the Polish Academy of Sciences (IMPAN), Birkhäuser now resumes this tradition to publish high quality research monographs in all areas of pure and applied mathematics.

Edited by
Przemysław Wojtaszczyk, IMPAN and Warsaw University, Poland

■ **Vol. 70: Positselski, L.**, Homological Algebra of Semimodules and Semicontramodules. Semi-infinite Homological Algebra of Associative Algebraic Structures

2010. 274 pages. Hardcover.
ISBN 978-3-0346-0435-2

This monograph deals with semi-infinite homological algebra. Intended as the definitive treatment of the subject of semi-infinite homology and cohomology of associative algebraic structures, it also contains material on the semi-infinite (co)homology of Lie algebras and topological groups, the derived comodule-contramodule correspondence, its application to the duality between representations of infinite-dimensional Lie algebras with complementary central charges, and relative non-homogeneous Koszul duality.

The book explains with great clarity what the associative version of semi-infinite cohomology is, why it exists, and for what kind of objects it is defined. Semialgebras, contramodules, exotic derived categories, Tate Lie algebras, algebraic Harish-Chandra pairs, and locally compact totally disconnected topological groups all interplay in the theories developed in this monograph. Contramodules, introduced originally by Eilenberg and Moore in the 1960s but almost forgotten for four decades, are featured prominently in this book, with many versions of them introduced and discussed.

Rich in new ideas on homological algebra and the theory of corings and their analogues, this book also makes a contribution to the foundational aspects of representation theory. In

particular, it will be a valuable addition to the algebraic literature available to mathematical physicists.

■ **Vol. 69: Panchapagesan, T.V.**, The Bartle-Dunford-Schwartz Integral.

2008. 313 pages. Hardcover.
ISBN 978-3-7643-8601-6

■ **Vol. 68: Grigoryan, S.A.**, Shift-invariant Uniform Algebras on Groups

2006. 294 pages. Hardcover.
ISBN 978-3-7643-7606-2

■ **Vol. 67: Zoladek, H.**, The Monodromy Group

2006. 592 pages. Hardcover.
ISBN 978-3-7643-7535-5

■ **Vol. 66: Müller, P.F.X.**, Isomorphisms between H^1 Spaces

2005. 472 pages. Hardcover.
ISBN 978-3-7643-2431-5

■ **Vol. 65: Badescu, L.**, Projective Geometry and Formal Geometry

2004. 228 pages. Hardcover.
ISBN 978-3-7643-7123-4

■ **Vol 64: Walczak, P.**, Dynamics of Foliations, Groups and Pseudogroups

2004. 240 pages. Hardcover.
ISBN 978-3-7643-7091-6

■ **Vol. 63: Schürmann, J.**, Topology of Singular Spaces and Constructible Sheaves

2003. 464 pages. Hardcover.
ISBN 978-3-7643-2189-5